A Specialist Periodical Report

Molecular Spectroscopy

Volume 1

A Review of the Recent Literature

The period of coverage extends in all chapters to December 1971 and in some chapters additionally into 1972.

Senior Reporters

R. F. Barrow, *Physical Chemistry Laboratory, University of Oxford*

D. A. Long, *School of Chemistry, University of Bradford*

D. J. Millen, *Department of Chemistry, University College, London*

Reporters

B. M. Chadwick, *University of Bradford*
J. A. Coxon, *Queen Mary College, University of London*
A. J. Downs, *University of Oxford*
H. G. M. Edwards, *University of Bradford*
V. Fawcett, *University of Bradford*
M. J. French, *University of Bradford*
J. M. Hollas, *University of Reading*
A. C. Legon, *University College, London*
R. J. Le Roy, *University of Toronto, Ontario, Canada*
S. C. Peake, *University of Oxford*

The Chemical Society
Burlington House, London, W1V 0BN

ISBN: 0 85186 506 2
Library of Congress Catalog Card No. 72–92545

Printed in Great Britain by Billing & Sons Limited
Guildford and London

Foreword

This is the first of a series of annual volumes relating to molecular spectroscopy. This is a most extensive subject with an enormous literature. It pervades every aspect of chemistry. Not surprisingly, other titles in the Specialist Periodical Reports series necessarily have dealt with applications of molecular spectroscopy to a number of areas of chemistry. However, the information in the literature may with advantage have imposed upon it more than one pattern. While any such pattern has a subjective element, the new perspective of the subject that it affords can be valuable.

In these Reports, the emphasis is on the principles and practice of spectroscopy. Some of the generally accepted subdivisions of molecular spectroscopy, such as, for example, microwave spectroscopy, electronic spectra of large molecules, far-infrared spectroscopy, and rotation and vibration–rotation infrared and Raman spectroscopy, have been used as a basis for classification and correlation of the literature. In addition, to emphasize particular developments, it has proved convenient to use the spectroscopic treatment of a particular class of compounds, *e.g.* the halogens, macromolecules, or the simple cyanides, as a basis for organizing the material.

We make no claim to complete coverage of the literature relating to molecular spectroscopy. That would require many volumes of this size and would inevitably lead to duplication of topics treated in other reports. However, for the topics dealt with here, we have tried to cover the literature as completely as possible, treating the material in depth and subjecting it to critical appraisal. As this is the first volume in this series, the literature in some sections has been surveyed over rather more than the usual twelve month period. Consequently some of our chapters are longer than would normally be the case. This is one reason why some aspects of molecular spectroscopy, particularly of infrared and Raman spectroscopy, could not be covered. However, while the Senior Reporters, who made the choice of topics, recognize certain omissions, some of which will be made good in the next volume, it seems likely that in pursuing a policy of thorough and critical analysis the Senior Reporters will always have to exercise a degree of selectivity in the areas chosen for treatment.

We are pleased to record our thanks to our Reporters and the Editorial Staff of the Chemical Society for their help and collaboration in the preparation of this volume.

January 1973

R. F. B.
D. A. L.
D. J. M.

Contents

Chapter 2 Electronic Spectra of Large Molecules

By J. M. Hollas

Chapter 3 Energy Levels of a Diatomic near Dissociation

By R. J. Le Roy

Chapter 4 The Low-lying Electronic States of Diatomic Halogen Molecules

By J. A. Coxon

Chapter 5 Far-infrared Molecular Spectroscopy

By M. J. French

1

Microwave Spectroscopy

BY A. C. LEGON AND D. J. MILLEN

This review is the first of a series dealing with developments in the field of microwave spectroscopy and covers papers included in *Chemical Titles* for 1971. Although this has the result that some papers published towards the end of 1971 have not been included, continuity will be preserved since such papers will be included in *Chemical Titles* for 1972 and therefore covered in the next review. Because of the detailed treatment which can now be made for diatomic and triatomic molecules, these are discussed as separate topics and are followed by sections on inorganic and organic molecules. Developments in treating vibration–rotation interaction are conveniently taken together, and lastly a number of areas of growing interest are reviewed.

1 Diatomic Molecules

Equilibrium internuclear distances for diatomic molecules, or at least their relative values in the absence of a more accurate value of Planck's constant, are among the most accurately determined physically significant properties of molecules. This year has seen the establishment of the same value of r_e for four isotopic species of one molecule within very close limits.

Equilibrium internuclear distances have been evaluated for carbon monoxide [1] ($r_e = 1.12823 \pm 0.00005$ Å) and for hydrogen chloride [2] ($r_e = 1.27460 \pm 0.00005$ Å). A sub-millimetre-wave investigation has been made of a number of isotopic species of hydrogen halides.[3] Transitions in the region 0.38—1.0 mm wavelength have been measured using a spectrometer which employs a klystron-driven crystal harmonic generator and a 1.6 K InSb photoconducting detector. As a result of this work, r_e values have now been obtained for all of the hydrogen halides using microwave measurements (Table 1). The r_e values are calculated using the relationship $BI_b = 5.05376 \times 10^5$ a.m.u. Å^2 MHz. For hydrogen chloride, four isotopic species have been examined and all the r_e values lie within 10^{-6} Å. Further investigations have been made of silver halides using a specially designed high-

[1] P. R. Bunker, *J. Mol. Spectroscopy*, 1971, **39**, 90.
[2] P. R. Bunker, *J. Mol. Spectroscopy*, 1971, **37**, 197.
[3] F. C. de Lucia, P. Helminger, and W. Gordy, *Phys. Rev. (A)*, 1971, **3**, 1849.

temperature Stark cell.[4] Improved quadrupole coupling constants have been obtained from observations on low transitions for AgCl ($J = 1 \leftarrow 0$) and AgBr ($J = 3 \leftarrow 2$). Transitions in the microwave spectrum of silver iodide have been obtained for the first time. Nuclear quadrupole coupling

Table 1 *B_0, D_0, and r_e values for the hydrogen halides*

Molecule	B_0/MHz	D_0/MHz	r_e/Å
$H^{35}Cl$	312989.297 ±0.20	15.836	1.2745991
$H^{37}Cl$	312519.121 ±0.020	15.788	1.2745990
$D^{35}Cl$	161656.238 ±0.014	4.196 ±0.003	1.2745990
$D^{37}Cl$	161183.122 ±0.016	4.162 ±0.003	1.2745988
$H^{127}I$	192657.577 ±0.019	6.203 ±0.003	1.609018
$D^{127}I$	97537.092 ±0.009	1.578 ±0.001	1.609018
$D^{79}Br$	127357.639 ±0.012	2.6529 ±0.0014	1.4144698
$D^{81}Br$	127279.757 ±0.017	2.6479 ±0.002	1.4144698
$D^{19}F$	325584.98 ±0.300	17.64	0.916914

coefficients reported for the three molecules are given in Table 2. Equilibrium internuclear distances have been obtained for the bromide and iodide: r_e(Ag—Br) = 2.393100 ±0.000029 Å and r_e(Ag—I) = 2.544611 ±0.000031 Å.

Table 2 *Nuclear quadrupole coupling coefficients for* AgX *species*

	$^{107,109}Ag^{35}Cl$	$^{107,109}Ag^{79}Br$	$^{107,109}Ag^{127}I$
$v = 0$	−36.50 ±0.10	297.10 ±0.15	−1062.17 ±0.40
$v = 1$	−36.50 ±0.10	297.65 ±0.15	−1064.81 ±0.40

2 Triatomic Molecules

Linear triatomic molecules for which new information has been obtained include HCN, HCP, OCS, and SCSe. New microwave measurements have been made of rotational transitions within vibrationally excited states of four isotopic species of HCN ($H^{12}C^{14}N$, $D^{12}C^{14}N$, $D^{13}C^{14}N$, and $D^{12}C^{15}N$) and have led to improved values of the relevant B_v constants.[5] Values for r_0 internuclear distances have been obtained for all possible pairs of B_0 values and compared with r_s and r_e values. Conclusions from this work are summarized in the section on vibrational influence on structural determination. Further investigation of the spectrum of the phosphorus analogue of HCN, methylidene phosphine (HCP), has been reported.[6] A millimetre-wave investigation has been made of transitions in the ground state and in the $v_2 = 1$ state. Previous studies have given enough information to determine B_e for HCP, but not for DCP, where α_2 and α_3 remained undetermined.

[4] J. Hoeft, F. J. Lovas, E. Tiemann, and T. Törring, *Z. Naturforsch.*, 1971, **26a**, 240.
[5] G. Winnewisser, A. G. Maki, and D. R. Johnson, *J. Mol. Spectroscopy*, 1971, **39**, 149.
[6] J. W. C. Johns, J. M. R. Stone, and G. Winnewisser, *J. Mol. Spectroscopy*, 1971, **38**, 437.

The new investigation has led to a value for α_2 but α_3 still remains to be determined. A value of q_l has also been obtained for DCP. The microwave spectrum of thiocarbonyl selenide (SCSe) has been observed in the $(0, 1^{\pm}, 0)$ and $(0, 2^{\pm}, 0)$ vibrational states.[7] Ground-state rotational constants have been obtained by extrapolation from the observations on the two vibrationally excited states. The rotational constants which are given in Table 3 differ considerably from those suggested previously. A lack of isotopic information

Table 3 *Rotation constants for various* SCSe *species*

	$^{32}S^{12}C^{80}Se$	$^{32}S^{12}C^{78}Se$	$^{32}S^{12}C^{76}Se$	$^{32}S^{12}C^{82}Se$
B_0/MHz	2043.310	2060.321	2078.190	2027.113
α_2/MHz	−4.133	−4.169	−4.215	−4.116
q_l/MHz	1.005	1.021	1.041	0.996
D_J/kHz	0.163	0.22	—	—

for sulphur species prevents the determination of an r_s-structure, but the following internuclear distances have been obtained from the ground-state rotational constants: $r(C—S) = 1.553$ Å and $r(C—Se) = 1.695$ Å. Of these, the latter is an r_s-value; it is significantly smaller than the corresponding bond distance for OCSe (1.708 Å). By using the value found for q_l and the three vibrational frequencies, the molecular force field has been calculated: $f_{C-Se} = 5.72$, $f_{C-S} = 7.97$, $f_{rr'} = 0.59$, and $f_{\alpha/r_1r_2} = 0.20$ mdyn Å^{-1}. The dipole moment was measured in the $(0, 2^{\pm2}, 0)$ state, transitions for the $(0, 1^{\pm1}, 0)$ state not being fully modulated, and found to be 0.031 D.

For OCS the absolute signs of dipole moment derivatives have been obtained.[8] The determination is based on the use of four pieces of information: the change in dipole moment on excitation to the state with one quantum of the bending mode, the intensities of the two stretching fundamentals, and the newly measured intensity of the bending overtone. These four serve to overdetermine three quantities: the first derivative of the dipole moment with respect to each of the two stretching co-ordinates and the second derivative with respect to bending. For the carbonyl bond, the oxygen becomes more negative as the bond is stretched, and the dipole is estimated to change at a rate of 7 D Å^{-1}. For the C=S bond a similar result is found and the gradient is estimated as 4 D Å^{-1}.

Among bent symmetric XY_2 molecules studied were GeF_2, HDO, HDS, D_2S, and O_3. A detailed study has been made of the GeF_2 molecule by Takeo, Curl, and Wilson [9] in which rotational constants have been obtained for $^{76}GeF_2$, $^{74}GeF_2$, $^{72}GeF_2$, and $^{70}GeF_2$. Changes in each of the rotational constants (α, β, and γ) on vibrational excitation have been obtained for each of the three fundamentals and the equilibrium parameters were found to be:

[7] C. Hirose and R. F. Curl, *J. Chem. Phys.*, 1971, **55**, 5120.
[8] A. Cunnington and D. H. Whiffen, *Chem. Comm.*, 1971, 981.
[9] H. Takeo, R. F. Curl, and P. W. Wilson, *J. Mol. Spectroscopy*, 1971, **38**, 464.

r_e(Ge—F) = 1.7321 Å and α_e = 97°10′; the dipole moment was found to be 2.61 D. Changes in inertial defect on vibrational excitation have been used to evaluate Coriolis constants, and then by combining $\zeta_{13}^{(c)}$ with the three vibrational frequencies, harmonic force constants have been obtained: $f_r = 4.08$, $f_\alpha/r_e^2 = 0.32$, $f_{rr} = 0.26$, and $f_{r\alpha}/r_e = -0.01$ mdyn Å^{-1}. The vibration–rotation interaction constants α, β, and γ for each of the vibrational states have further been used to obtain cubic force constants.

Millimetre and submillimetre studies have been made for HDS and HDO, and in both cases improved centrifugal distortion analyses have been made. For HDS it was found necessary to include both P^4 and P^6 distortion effects [10] and for HDO a partial set of P^8 and P^{10} terms were needed as well.[11] For HDO, 21 new transitions have been measured and a total of 53 were used in the calculation of 15 rotation and distortion constants. In a related study, Bellet and Steenbeckeliers [12] have made distortion treatments for D_2O and HDO using 24 transitions for the former and 30 for the latter to fit to 20 parameters. For H_2O, far-i.r. transitions were included and a 16-parameter model was employed. Two other papers have also appeared on isotopic species of water.[13, 14] Using a millimetre beam maser the nuclear hyperfine structure of the $1_{10} \rightarrow 1_{01}$ transition of D_2S at 91.4 GHz has been resolved.[15] By combining the data with that previously obtained for H_2S and HDS, the deuterium quadrupole coupling tensor has been obtained. The components in its principal axis system are $\chi_{xx} = 149.0$, $\chi_{yy} = -59.8$, and $\chi_{zz} = -89.2$ kHz. The z-axis is perpendicular to the plane of the molecule and the x-axis is rotated 1°35′ from the D—S bond in the direction away from the D—S—D obtuse angle. The rotation is similar to that of 1°20′ found for D_2O.[16] A millimetre-wave study of ozone has led to a recalculation of the rotational constants.[17] From Stark-effect measurements on six transitions, an r.m.s. value of 0.5324 ± 0.0024 D is found for the dipole moment.

Further investigations of the microwave spectrum of hypochlorous acid have extended previous studies, which had obtained $B+C$ values. Through the observation of the a-type transitions, the separate rotational constants B and C have been evaluated for four isotopic species (HO^{35}Cl, HO^{37}Cl, DO^{35}Cl, and DO^{37}Cl).[18] The following r_s-structural parameters were obtained: r(O—H) = 0.959 ± 0.005 Å, r(O—Cl) = 1.689$_5$ Å, and $\widehat{\text{HOCl}}$ = 102°29′ ± 27′. Nuclear quadrupole coupling coefficients have also been evaluated. Stark-effect measurements have been made on the $1_{01} \leftarrow 0_{00}$ transitions for HO^{35}Cl and DO^{35}Cl and values of μ_a for the two species are

[10] P. Helminger, R. L. Cook, and F. C. de Lucia, *J. Mol. Spectroscopy*, 1971, **40,** 125.
[11] F. C. de Lucia, R. L. Cook, P. Helminger, and W. Gordy, *J. Chem. Phys.*, 1971, **55,** 5334.
[12] J. Bellet and G. Steenbeckeliers, *Compt. rend.*, 1971, **271,** *B*, 1208.
[13] G. Steenbeckeliers and J. Bellet, *Compt. rend.*, 1971, **273,** *B*, 471.
[14] W. Lafferty, J. Bellet, and G. Steenbeckeliers, *Compt. rend.*, 1971, **273,** *B*, 388.
[15] F. C. de Lucia and J. W. Cederberg, *J. Mol. Spectroscopy*, 1971, **40,** 52.
[16] H. Bluyssen, J. Verkhoeven, and A. Dymanus, *Phys. Letters* (*A*), 1967, **25,** 214.
[17] M. Lichtenstein, J. J. Gallagher, and S. A. Clough, *J. Mol. Spectroscopy*, 1971, **40,** 10.
[18] A. M. Mirri, F. Scappini, and G. Cazzoli, *J. Mol. Spectroscopy*, 1971, **38,** 218.

found to be 0.367 ±0.008 and 0.412 ±0.15 D.[19] The difference is attributed largely to the different orientation of the principal axes systems in the two molecules. On this basis the dipole moment of the hypochlorous acid molecule is calculated to be 1.3 ±0.3 D and to be inclined at approximately 73° to the O—Cl bond direction. For the structurally related molecule $NHCl_2$, μ_a has been found to have a value of 1.22 ±0.03 D.[19] Rotational constants have been obtained for four isotopic species of NSCl ($^{14}N^{32}S^{35}Cl$, $^{14}N^{32}S^{37}Cl$, $^{14}N^{34}S^{35}Cl$, and $^{15}N^{32}S^{35}Cl$).[20] From these constants an r_s-structure has been obtained: r(N—S) = 1.450 Å, r(S—Cl) = 2.161 Å, and $\widehat{NSCl}$ = 117°42′. The S—Cl bond distance is rather longer than those observed for S_2Cl_2, SCl_2, or $SOCl_2$. Nuclear quadrupole coupling coefficients were also obtained: $\chi_{aa} = -38.51$, $\chi_{bb} = 23.51$, and $\chi_{cc} = 15.00$ MHz. These lead to $\chi_{bond} = -43.36$ MHz. In another investigation of the NSCl molecule, the Stark effect for $1_{11} \leftarrow 0_{00}$, $2_{11} \leftarrow 2_{02}$, and $3_{12} \leftarrow 3_{03}$ transitions has been used to evaluate the dipole moment and leads to a value of μ = 1.87 ±0.02 D.[21]

3 Inorganic Molecules

An earlier study of *trans*-nitrous acid [22] has now been extended and a detailed study has also been made of the *cis*-isomer.[23] For the *trans*-isomer isotopic substitution has been made for all four atoms and an r_s-structure has been evaluated. For the *cis*-isomer the *a* and *b* co-ordinates of the hydrogen and nitrogen atoms have been obtained from Kraitchman's equations, and the remaining co-ordinates from I_b and the usual moment conditions. It is seen from the two structures (1) and (2) shown below that there are significant

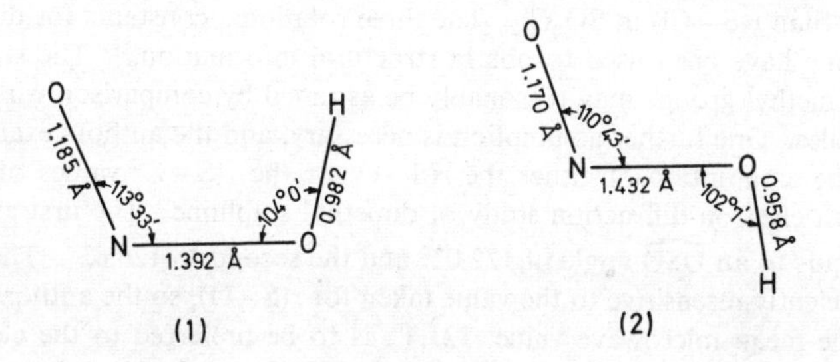

differences in the corresponding nitrogen–oxygen bond lengths. The central N—O bond for the *cis*-isomer is *ca.* 0.04 Å shorter than for the *trans*-isomer

[19] D. G. Lister and D. J. Millen, *Trans. Faraday Soc.*, 1971, **67**, 601.
[20] T. Beppu, E. Hirota, and Y. Morino, *J. Mol. Spectroscopy*, 1970, **36**, 386.
[21] A. Guarnieri, *Z. Naturforsch.*, 1971, **26a**, 1246.
[22] A. P. Cox and R. L. Kuczkowski, *J. Amer. Chem. Soc.*, 1966, **88**, 5071.
[23] A. P. Cox, A. H. Brittain, and D. J. Finnigan, *Trans. Faraday Soc.*, 1971, **67**, 2179.

and this is attributed to increased double-bond character, which is in accord with the higher OH torsional frequency for *cis*-nitrous acid. The N=O bond lengths for both isomers are noticeably longer than those for the nitrosyl halides, and this is regarded as being consequent on some double-bond character in the central bond. The dipole moments are found to be: *cis*-isomer, 1.423 ±0.005 D; *trans*-isomer, 1.855 ±0.016 D. Nitrogen nuclear quadrupole coupling coefficients are also reported.

The C_{4v} symmetry of BrF_5 is confirmed from the statistical weighting of the different K-levels arising from nuclear spin degeneracy.[24] Rotational constants, centrifugal distortion constants, and nuclear quadrupole coupling constants have been obtained for both ^{79}Br and ^{81}Br species. An analogous study has confirmed the C_{4v} symmetry of the IF_5 molecule and again the rotational and quadrupole coupling constants have been obtained.[25] Another molecule of C_{4v} symmetry, $SF_5{}^{35}Cl$, has been further studied.[26] Transitions for four singly excited vibrational states have been assigned, and B_v, D_{JJ}, and α_v evaluated for each of these (ν_4, ν_6, ν_{10}, and ν_{11}).

Structure and bonding in molecules with a central sulphur atom continues to attract interest, and microwave studies have been made on sulphuryl chloride and three derivatives of the sulphuryl halides, namely sulphuryl fluoride chloride (SO_2FCl), dimethyl sulphone [$(CH_3)_2SO_2$] and methanesulphonyl fluoride (CH_3SO_2F). Rotational constants for two isotopic species of sulphuryl chloride fluoride, $^{32}S^{16}O_2{}^{35}Cl^{19}F$ and $^{32}S^{16}O_2{}^{37}Cl^{19}F$, have been obtained and chlorine nuclear quadrupole coupling coefficients have been evaluated.[27] Although a complete structural determination has not been made, some structural information has been obtained. By making assumptions about the SO bond length, from comparison with SO_2Cl_2 and SO_2F_2, two values have been obtained: r(S—F) = 1.55 and r(S—Cl) = 1.985 Å. The former is a little longer than r(S—F) in SO_2F_2 and the latter a little longer than r(S—Cl) in SO_2Cl_2. The three rotational constants for dimethyl sulphone have been used to obtain structural information.[28] The structure of the methyl groups may reasonably be assumed by comparison with other molecules. One further assumption is necessary, and the authors examine in turn the assumption of either the r(S—O) or the r(S—C) values obtained from an electron-diffraction study of dimethyl sulphone. The first assumption leads to an $\widehat{OSO}$ angle of 122.02° and the second to 120.82°. The angle is sufficiently insensitive to the value taken for r(S—O), so the authors argue that the mean microwave value, 121.4°, is to be preferred to the electron-diffraction value. For the CH_3SO_2F molecule,[28] the barrier to methyl rotation is determined from relative-intensity measurements for ground- and excited-state lines to be 2.52 ±0.35 kcal mol^{-1}. The molecular dipole moments are

[24] M. J. Whittle, R. H. Bradley, and P. N. Brier, *Trans. Faraday Soc.*, 1971, **67**, 2505.
[25] R. H. Bradley, P. N. Brier, and M. J. Whittle, *Chem. Phys. Letters*, 1971, **11**, 192.
[26] R. Jurek and J. Chanussot, *Compt. rend.*, 1971, **272**, *B*, 941.
[27] C. W. Holt and M. C. L. Gerry, *Chem. Phys. Letters*, 1971, **9**, 621.
[28] E. J. Jacob and D. R. Lide, *J. Chem. Phys.*, 1971, **54**, 4591.

found to be: $(CH_3)_2SO_2$, $\mu = 4.50 \pm 0.10$ D; CH_3SO_2F, $\mu = 3.88 \pm 0.04$ D. The dramatic difference between these and that for SO_2F_2 ($\mu = 1.11$ D) is rationalized on the basis of comparatively large bond moments for S—O and S—F bonds and a much smaller bond moment for H_3C—S. A further microwave study of sulphuryl chloride has also been made,[29] and centrifugal distortion constants have been obtained for the species $SO_2{}^{35}Cl_2$. First-order Watson theory was first applied to transitions below $J = 40$ and Δ_{JK}, Δ_K, ζ_J, and ζ_K were evaluated. After higher J transitions had been assigned using these values, first- and second-order theory was used for all transitions between $J = 10$ and $J = 70$ to obtain first- and second-order distortion constants.

A value has been obtained for the boron–nitrogen distance in the trimethylamine–boron trifluoride complex.[30] Rotational constants have been determined for the three isotopic species $(CH_3)_3{}^{14}N^{11}BF_3$, $(CH_3)_3{}^{15}N^{11}BF_3$, and $(CH_3)_3{}^{14}N^{10}BF_3$. These lead to an r_s-value for the boron–nitrogen internuclear distance of 1.636 ± 0.004 Å, which is in accord with the trend in boron–nitrogen distances for complexes of trimethylamine with boron trichloride (1.610 Å) and tri-iodide (1.584 Å) found by X-ray studies. A structural study has also been made of a compound containing a phosphorus–boron bond.[31] The structure of difluorophosphineborane (HF_2P—BH_3) has been determined from the rotational constants of six isotopic species. The value of $I_a + I_c - I_b$ establishes a structure with a plane of symmetry. The structure of the molecule has been evaluated using Kraitchman's equations, with first- and second-moment equations being used to determine phosphorus and fluorine co-ordinates. As expected, the molecule has a staggered equilibrium conformation with C_s symmetry. The structural parameters are: r(P—H) $= 1.409 \pm 0.004$, r(P—F) $= 1.552 \pm 0.006$, r(P—B) $= 1.832 \pm 0.009$, r(B—H$_a$) $= 1.226 \pm 0.005$, and r(B—H$_s$) $= 1.200 \pm 0.007$ Å; $\widehat{H_aBH_a} = 112.7 \pm 0.50$, $\widehat{H_aBH_s} = 115.9 \pm 0.4$, $\widehat{FPF} = 100.0 \pm 0.5$, $\widehat{FPH} = 98.62 \pm 0.25$, and $\widehat{BPF} = 117.7 \pm 0.30°$. Here H_s refers to hydrogen atoms in the plane and H_a to those out of the plane of symmetry. The molecular structure shows a very pronounced tilt of the borane group away from the fluorine atoms: the angles $\widehat{PBH_s}$ and $\widehat{PBH_a}$ are $109.9 \pm 0.3°$ and $99.89 \pm 0.26°$, respectively. Comparison with other structures leads to the conclusion that fluorophosphorus molecules, on co-ordination, show bond lengths to phosphorus which decrease and bond angles about phosphorus which increase as more electronegative groups are attached. The barrier to internal rotation is estimated to fall within the range 3.6—4.5 kcal mol^{-1}, which is high by comparison with the isoelectronic molecule HF_2Si—CH_3, for which the value is 1.25 kcal mol^{-1}.

An investigation has been made[32] of the phosphinodifluorophosphine molecule, H_2P—PF_2. I.r. and Raman spectra of the related molecules P_2H_4

[29] J. Burie, J.-L. Destombes, A. Dubrulle, G. Journel, and C. Marliére, *Compt. rend.*, 1970, **271**, *B*, 1197.
[30] P. S. Bryan and R. L. Kuczkowski, *Inorg. Chem.*, 1971, **10**, 200.
[31] J. P. Pasinski and R. L. Kuczkowski, *J. Chem. Phys.*, 1971, **54**, 1903.
[32] R. L. Kuczkowski, H. W. Schiller, and R. W. Rudolph, *Inorg. Chem.*, 1971, **10**, 2505.

and P_2F_4 have been interpreted in terms of a *gauche* configuration for P_2H_4 and a *trans* configuration for P_2F_4. A *trans* configuration for P_2Cl_4 has also been inferred from vibrational spectroscopy and an *X*-ray determination confirms a *trans* configuration for P_2I_4. The microwave study of $H_2P—PF_2$ leads to the conclusion that this molecule has the *trans* configuration. The authors have obtained structural parameters including the dihedral angle and show that dipole-moment considerations confirm the *trans* configuration. For a *gauche* configuration with no symmetry, there will in general be non-zero components of dipole moment, μ_a, μ_b, and μ_c. For a *trans* configuration, C_s symmetry allows two non-zero components of dipole moment. For a species with a C_2 axis, as for the planar C_{2v} conformation, there is only one non-zero component of dipole. Experiment shows that the spectrum contains μ_a- and μ_b-transitions but no μ_c-transitions. From Stark-effect measurements it is concluded that $\mu_c \leqslant 0.01$ D. All of this is consistent with a *trans* structure, whereas a structure with a C_2 axis is excluded. From the rotational constants for H_2PPF_2 and D_2PPF_2, the following structural parameters were evaluated with the assumption $r(\text{P—H}) = 1.42 \pm 0.01$ Å: $r(\text{P—P}) = 2.218 \pm 0.038$ and $r(\text{P—F}) = 1.587 \pm 0.013$ Å; $\widehat{\text{HPH}} = 93.2 \pm 1$, $\widehat{\text{HPP}} = 90.3 \pm 0.4$, $\widehat{\text{FPF}} = 98.2 \pm 1.2$, and $\widehat{\text{FPP}} = 97.2 \pm 1.6°$.

Other phosphorus-containing molecules which have been examined include phosphorus trifluoride and trimethylphosphine. Bryan and Kuczkowski have obtained rotational constants for $(CH_3)_3P$, $(CD_3)(CH_3)_2P$, and $(CH_2D)(CH_3)_2P$, the latter occurring in two rotameric forms, in one of which the deuterium is in the plane of symmetry.[33] The results lead to the conclusion that the methyl group is tilted towards the phosphorus lone pair of electrons. For phosphorus trifluoride l-type doubling transitions in the ν_4 state have been observed and analysed; values for two cubic constants k_{444} and k_{443} have been obtained.[34] In a later paper the transitions are re-examined by taking into account third-order terms.[35]

A detailed study [36] has been made of ethynyldifluoroborane, $F_2BC{\equiv}CH$, for which the microwave spectra of ten isotopic species were measured and rotational constants evaluated using a least-squares Watson-type of analysis of centrifugal distortion on lines involving states up to $J = 40$. The derived centrifugal distortion constants were transferred to other isotopic species where sufficient transitions for a full treatment were not available. Intensity alternation arising from $g = 1$ for even K_{-1} states and $g = 3$ for odd K_{-1} states for the ground vibrational state establishes that the molecule has C_{2v} symmetry. Intensity measurements also lead to the conclusion that for the lowest vibrational state $\nu = 144 \pm 20$ cm^{-1}, and it is assigned as an in-plane skeletal mode. The molecular structure obtained is essentially an r_s-structure, and the fluorine atom co-ordinates are obtained from first-moment conditions:

[33] P. S. Bryan and R. L. Kuczkowski, *J. Chem. Phys.*, 1971, **55**, 3049.
[34] E. Hirota, *J. Mol. Spectroscopy*, 1971, **37**, 20.
[35] E. Hirota, *J. Mol. Spectroscopy*, 1971, **38**, 195.
[36] W. J. Lafferty and J. J. Ritter, *J. Mol. Spectroscopy*, 1971, **38**, 181.

r(C—H) = 1.058 ±0.003, r(C≡C) = 1.206 ±0.003, r(C—B) = 1.513 ±0.005, and r(B—F) = 1.323 ±0.005 Å; $\widehat{FBF}$ = 116.5 ±1.0°. The acetylenic bond distance is closely similar to those of many other such bonds, and the same is true of the C—H bond distance. The C—B bond distance is the shortest measured to date, and this is consistent with the general finding of small radii for carbon in acetylenic compounds. No other such C—B bond is available for comparison, the nearest being that in borane carbonyl, which can be written O≡C—BH_3 and for which the C—B distance is 1.540 Å.

Three halogenostannanes, SnH_3Cl,[37] SnH_3Br,[38] and SnH_3I,[39] have been investigated during the course of the year. The study of the spectra of these very unstable species has made it possible to make a comparison of corresponding compounds of the elements of Group IV: the halogenomethanes (CH_3X), the halogenosilanes (SiH_3X), the halogenogermanes (GeH_3X), and the halogenostannanes (SnH_3X). The nuclear quadrupole coupling coefficients are listed in Table 4. Wolf, Krisher, and Gsell make the usual assumptions

Table 4 *Nuclear quadrupole coupling coefficients of Group IV* MH_3X *compounds*/MHz

X	CH_3X	SiH_3X	GeH_3X	SnH_3X
^{35}Cl	−74.77	−40.0	−46	−41.62 ±0.3
^{37}Cl	−58.93	−30.8	−36	−33.28 ±0.5
^{79}Br	577.15	336	380	350 ±6
^{81}Br	482.16	278	321	291 ±6
^{127}I	−1934	−1240	—	−1273 ±8

that the halogen quadrupole coupling is due solely to contributions of p-orbitals to the molecular bond and that the halogen bond can be characterized by the hybridization of s and p atomic orbitals only. This leads to the conclusion for the halogenostannanes that the importance of hybridization decreases along the series Cl > Br > I. The same is true for the corresponding halogen compounds of the other elements of Group IV, as is pointed out by Krishnaji and Chandos.[40] However, for any given halide, the new results lead to the conclusion that the s-hybridization increases in the order C, Ge, Sn or Si. Wolf, Krisher, and Gsell[39] emphasize that the results make it clear that it is the silicon compounds that are out of line with the order of the elements in Group IV, and that this is the case for the coupling coefficients themselves even if one leaves aside any question of interpretation. The authors point out that the order is in agreement with the ordering of the electronegativities, C > Ge > Sn or Si. It is also of interest to note that a comparison of bond energies of Group IV elements with the halogens again shows silicon to be anomalous. The bond energies for bonds formed by

[37] L. C. Krisher, R. A. Gsell, and J. M. Bellamy, *J. Chem. Phys.*, 1971, **54**, 2287.
[38] S. N. Wolf, L. C. Krisher, and R. A. Gsell, *J. Chem. Phys.*, 1971, **54**, 4605.
[39] S. N. Wolf, L. C. Krisher, and R. A. Gsell, *J. Chem. Phys.*, 1971, **55**, 2106.
[40] Krishnaji and S. Chandos, *J. Sci. Ind. Res., India*, 1968, **27**, 135.

silicon with chlorine and bromine are greater in fact than for bonds formed by any other of the Group IV elements with chlorine or bromine. Bond distances have also been obtained for the tin–halogen bonds: r(Sn—I) = 2.327 ±0.0012, r(Sn—Br) = 2.4691 ±0.0003, and r(Sn—I) = 2.674 ±0.002 Å.

4 Organic Molecules

Acyclic Molecules without Internal Rotation.—The equilibrium geometries of the methyl halides have been evaluated by Duncan.[41] Although very accurate B_0 values and some of the α's are available from microwave spectroscopy, recourse to high-resolution i.r. spectroscopy is essential for the A_0 values and the remaining α's necessary for a complete r_e-structure. Even so, not all six α_i^A and α_i^B values are available for all the methyl halides, and the experimentally undetermined values are estimated by consideration of the steady trends exhibited by α_i^A/A_0 and α_i^B/B_0 among the halides for a given mode of vibration i. Particular attention is paid to the determination of A_0 values from the perpendicular bands of symmetric-top molecules and a check on the A_e values using the ratio $A_e(CH_3X)/A_e(CD_3X)$, which should be just the ratio of the masses of the deuteron and the proton, indicates doubt only in the case of the CD_3Cl value. The equilibrium structures reported for CH_3X are given in Table 5. The results show that within the quoted error, the

Table 5 *Equilibrium structures for* CH_3X *compounds*

	Methyl fluoride	*Methyl chloride*	*Methyl bromide*	*Methyl iodide*
r_e(CH)/Å	1.095	1.086 ±0.004	1.086 ±0.003	1.085 ±0.003
r_e(CX)/Å	1.382	1.778 ±0.002	1.933 ±0.002	2.133 ±0.002
$\widehat{(HCH)}_e$	110°28′	110°40′ ±40′	111°10′ ±25′	111°17′ ±25′

structure of the CH_3 group is identical when X = Cl, Br, or I. In the case X = F, however, the C—H bond appears to be slightly longer than in the other halides. No estimate of uncertainty is made in this case, as the results depend entirely on the validity of the extrapolation for A_e and B_e.

The methyl halides CH_3X (X = F, Cl, Br, or I) together with fluoroform (CF_3H) also feature in a study [42] wherein rotational transitions in the frequency range 120—250 GHz are measured with a view to improving the accuracy of the known rotational and centrifugal distortion constants. For X = Br or I, the fourth-order distortion constant H_{JKK} was required in addition to D_J and D_{JK} to fit the observed spectra within the experimental error. Because of the precision of measurement the nuclear quadrupole

[41] J. L. Duncan, *J. Mol. Structure*, 1970, **6**, 447.
[42] T. E. Sullivan and L. Frenkel, *J. Mol. Spectroscopy*, 1971, **39**, 185.

coupling corrections necessary to calculate unperturbed line centres were made to second order. The rotational and distortion constants are in good agreement with previously known values, but their precision is somewhat improved. In the case of CH_3Cl, the analysis is extended to the $\nu_3 = 1$ state and the doubly degenerate $\nu_6 = 1$ state. The component pattern of a given rotational transition in the latter state is unexpected and its assignment by laser-microwave double-resonance experiments is reported in another paper (see ref. 220).

Some interesting structural conclusions about the methyl group are suggested by a molecular beam maser study [43] of the hyperfine structure of the $J = 1 \rightarrow 0$ transitions of CH_2DCl and CD_3Cl arising from the D and Cl nuclear quadrupole coupling. Differences in the ^{35}Cl quadrupole coupling constant along the C—Cl bond between these two isotopic species are attributed to differences of molecular structure in the methyl group. On the basis of the known decrease of 0.0008 Å between the C—Cl bond length in CH_3Cl and CD_3Cl, it is argued that the bond decreases in length by 0.0005 Å between CH_2DCl and CD_3Cl. Pauling's estimate that a 0.0005 Å decrease of C—Cl could be caused by a 0.006 V change in the difference of electronegativity between CH_3 and Cl is used with the Townes–Dailey model for quadrupole coupling to predict a coupling constant along the C—Cl bond which is 0.4% greater in CH_2DCl than in CD_3Cl, a value consistent with the experimental difference of 0.15%. The theory necessary to treat the problem of three equivalent and one inequivalent coupling nuclei (as in CD_3Cl) is developed according to the general formulation due to Bersohn.[44]

The effects of interactions of rotation with doubly degenerate vibrations in the symmetric-top molecules acetonitrile (CH_3CN) and trifluoroacetonitrile (CF_3CN) have been analysed to test some current theories of such interactions which apply to C_{3v} molecules. Bauer [45] reports an interpretation of the rotational spectrum of acetonitrile ($CH_3C^{14}N$ and $CH_3C^{15}N$) in the $\nu_8 = 2$ state (ν_8 is the —C—C≡N bend) based on an extension of the Amat–Goldsmith–Nielsen theory (see ref. 183) by Tarrago [46] in which vibration–rotation energy up to fourth order in vibration and sixth order in rotation is calculated. The assignment of quantum numbers k and l to the rather scattered components of rotational transitions in this state is a problem of some difficulty which is overcome by graphical methods suggested by the theoretical expressions. Several vibration–rotation interaction constants are precisely determined by fitting the many components of several $J+1 \leftarrow J$ transitions to the theory. However, during the course of the graphical assignment a strong resonance due to the accidental near-degeneracy of the $l = 0, k = \pm 4$ and $l = \pm 2, k = \mp 2$ levels was noted. Such a resonance invalidates the usual perturbation expressions and therefore transitions involving these

[43] S. G. Kukolich, *J. Chem. Phys.*, 1971, **55**, 4488.
[44] R. Bersohn, *J. Chem. Phys.*, 1950, **18**, 1124.
[45] A. Bauer, *J. Mol. Spectroscopy*, 1971, **40**, 183.
[46] G. Tarrago, *J. Mol. Spectroscopy*, 1970, **34**, 23.

levels are omitted from the fitting procedure. In another paper, Bauer and Maes [47] treat this resonance by a method involving diagonalization of the energy matrix. In so doing they are able to obtain the energy separation between the resonant levels and to account accurately for the observed deviations of transitions involving these levels from Tarrago's theory.

For trifluoroacetonitrile, Whittle, Baker, and Corbelli [48] have analysed the central group of components having $(kl-1) \neq 0$ in several rotational transitions for each of the doubly degenerate vibrational states $\nu_8 = 1$ and $\nu_7 = 1$ according to the theoretical treatment of Grenier-Besson and Amat [49] to give the vibration–rotation constants (in the notation of the latter authors) B^*_v, D^v_{JK}, D^v_J, ρ^*_v, $q^2_{vl}/(B-A-\zeta_v A)$, q_v, and ζ_v. Here, ν_8 refers to the —C—C≡N bending mode whose fundamental is at 187.5 cm^{-1}, and ν_7 refers to the CF_3 rocking mode at 462.7 cm^{-1}. Of particular value because of their uses in force-field determinations are the Coriolis constants ζ_v. The value $\zeta_8 = 0.527$ so determined is in poor agreement with the 0.16 ± 0.01 recently obtained from an i.r.-band contour analysis.[50] It is interesting that when the above vibration–rotation parameters derived from lower J transitions of ν_8 are used to predict higher transitions there are considerable differences from observed values, especially for high J, low K transitions. The authors discuss the reasons for the inapplicability of the theory in this region.

Structural information about another substituted methane, dibromomethane, has been derived from the moments of inertia [51] and quadrupole coupling constants.[52] The moments of inertia of four isotopic species, $CH_2{}^{79}Br_2$, $CH_2{}^{81}Br_2$, $CD_2{}^{79}Br_2$, and $CD_2{}^{81}Br_2$, are used in several ways to obtain the internuclear distances and angles. In a first method, a set of internally consistent structures is obtained whichever isotopic molecule is used as a basis when all the atom co-ordinates are determined from the data using a combination of essentially r_0 and r_s methods. The best structure by this method is determined to be: r(C—Br) $= 1.930 \pm 0.003$ and r(C—H) $= 1.079 \pm 0.01$ Å; $\widehat{\mathrm{BrCBr}} = 112.5 \pm 0.3$ and $\widehat{\mathrm{HCH}} = 113.6 \pm 1.5°$. The errors are those suggested by the range of the small r_s b-co-ordinates of the hydrogen atoms, which as usual are poorly determined by isotopic substitution of D for H. The r(C—H) is suspiciously short compared with that of similar molecules thought to have been reliably determined. Because of this, an alternative structure of dibromomethane has been calculated from the experimental information using the reasonable assumption r(C—H) $= 1.097 \pm 0.005$ Å to obviate the use of the unreliable hydrogen b co-ordinate. The mean of four such structures calculated using each of the four isotopic species as basis molecules is then: r(C—Br) $= 1.925 \pm 0.002$ and r(C—H) $= 1.097 \pm 0.005$ Å; $\widehat{\mathrm{BrCBr}} = 112.9 \pm 0.2$ and $\widehat{\mathrm{HCH}} = 110.9 \pm 0.8°$. Thus the

[47] A. Bauer and S. Maes, *J. Mol. Spectroscopy*, 1971, **40**, 207.
[48] M. J. Whittle, J. G. Baker, and G. Corbelli, *J. Mol. Spectroscopy*, 1971, **40**, 388.
[49] M. L. Grenier-Besson and G. Amat, *J. Mol. Spectroscopy*, 1962, **8**, 22.
[50] J. A. Faniran and H. F. Shurvell, *Spectrochim. Acta*, 1971, **27A**, 1945.
[51] D. Chadwick and D. J. Millen, *Trans. Faraday Soc.*, 1971, **67**, 1539.
[52] D. Chadwick and D. J. Millen, *Trans. Faraday Soc.*, 1971, **67**, 1551.

choice of method little affects the CBr_2 parameters but seriously changes the HCH angle. However, with either method both $\widehat{HCH}$ and $\widehat{BrCBr}$ exceed 109°28′, a result also encountered when similar experimental data for CH_2Cl_2 are treated likewise. Although such behaviour is suggestive of bent bonding,[53] interpretation of nuclear quadrupole coupling in both molecules indicates that the explanation is more likely to be found elsewhere, such as in the use of effective moments of inertia to determine the structure. Thus, in dibromomethane,[52] the nuclear quadrupole coupling has been analysed to give not only the diagonal elements of the quadrupole coupling tensor for the bromine nuclei referred to the principal inertial axis system, but also the only allowed off-diagonal element, χ_{ab}, which is evaluated when second-order perturbation theory is used to account for significant second-order shifts of many transitions. Diagonalization of the complete tensor shows in the case of each of the four isotopic species that the electric field gradient in two of the principal directions is almost equal and that the C—Br bond direction and the direction of the unique principal field gradient coincide to within 1°. The C—Br bond is therefore not bent, but rather the direction of maximum electron probability lies along the C—Br internuclear line.

The hyperfine structure of the $J = 1 \rightarrow 0$ transition of both the CF_3H and CF_3D species of fluoroform has been observed by Kukolich *et al.*[54] under very high resolution (linewidth at half height ~6 KHz) with a molecular-beam maser spectrometer. Splittings were analysed using a Hamiltonian which included terms to account for nuclear electric quadrupole coupling (in the case of CF_3D only), magnetic coupling of nuclear spin to molecular rotation, and nuclear magnetic spin–spin interactions. Of the parameters obtained, that of most chemical interest is $q_{zz}Q$, the quadrupole coupling constant of D along the C—D bond. The value of 170.8 ± 2 kHz compares with 191.5 ± 0.8 kHz in CH_3D from molecular electric beam resonance experiments. Similar studies of the hyperfine structure of the $J = 1 \rightarrow 0$ transition of methyl isocyanide (CH_3NC) [55] and the $1_{01} \rightarrow 0_{00}$ and $5_{14} \rightarrow 5_{15}$ transitions of formamide (NH_2CHO) [56] by the same group give, among other things, the nitrogen nuclear quadrupole coupling constants for these molecules.

Microwave investigations of one molecule containing the C=S group and several containing the C=O group have been published. An exhaustive study of thioformaldehyde is reported by Johnson, Powell, and Kirchoff.[57] This new and reactive species (half-life 6 min) can be prepared in several ways, the best being the pyrolysis of $CH_3S—SCH_3$ at low pressure. For the normal isotopic species, $H_2{}^{12}C^{32}S$, the three rotational constants and four of the five constants (τ's) of the Watson first-order centrifugal distortion model were

[53] R. J. Myers and W. D. Gwinn, *J. Chem. Phys.*, 1952, **20**, 1420.
[54] S. G. Kukolich, A. C. Nelson, and D. J. Ruben, *J. Mol. Spectroscopy*, 1971, **40**, 33.
[55] S. G. Kukolich, *Chem. Phys. Letters*, 1971, **10**, 52.
[56] S. G. Kukolich and A. C. Nelson, *Chem. Phys. Letters*, 1971, **11**, 383.
[57] D. R. Johnson, F. X. Powell, and W. H. Kirchoff, *J. Mol. Spectroscopy*, 1971, **39**, 136.

determined. The fifth τ (τ_{aaaa}) and the rotational constant A were highly correlated and the former was constrained to the value indicated by the high-resolution i.r. study of thioformaldehyde by Johns and Olson [58] in order to obtain a well-conditioned solution. For the other three isotopic species investigated, $H^{13}C^{32}S$, $H_2{}^{12}C^{34}S$, and $D_2{}^{12}C^{32}S$, less transitions were available so that corrections for centrifugal distortion had to be made by consideration of the corrections applying to the parent species and only the rotational constants B and C could be determined. The moments of inertia for the isotopic species allow the structure to be much over-determined. In whichever way the structure is derived, the heavy-atom co-ordinates have high internal consistency whereas the hydrogen co-ordinates are not so well behaved. The complete r_s-structure is that ultimately reported: $r(S{=}C) = 1.6108(9)$ Å, $r(C{-}H) = 1.0925(9)$ Å, and $\widehat{HCH} = 116.87(5)°$. The dipole moment of the ground vibrational state of the parent species is measured to be $\mu = \mu_a = 1.6474 \pm 0.0014$ D.

For formaldehyde itself, Kukolich and Ruben [59] have analysed, in very high resolution, the six components of the $1_{10} \rightarrow 1_{11}$ transition which result from magnetic interactions of the hydrogen nuclear spins with molecular rotation and from the hydrogen spin–spin interactions. They describe a two-cavity beam maser spectrometer with an intrinsic line width of 0.35 kHz. By combination of the $1_{10} \rightarrow 1_{11}$ transition data with that measured for the $2_{11} \rightarrow 2_{12}$ transition by Thaddeus, Krisher, and Loubser,[60] the diagonal elements of the magnetic coupling tensor referred to the principal inertial axis system are obtained as: $M_{aa} = -4.0 \pm 0.7$, $M_{bb} = 2.4 \pm 0.7$, and $M_{cc} = -2.1 \pm 0.7$ kHz.

Formyl fluoride has also been subjected to a high-resolution study [61] using a beam maser spectrometer to give the nuclear quadrupole coupling constant of D along the D—C bond direction in DFCO, $\chi_{D-C} = 205 \pm 4$ kHz, with the assumption that the field gradient tensor is symmetrical about the D—C bond. The signs and magnitudes of the magnetic susceptibility anisotropies, $2\chi_{aa}-\chi_{bb}-\chi_{cc}$ and $2\chi_{bb}-\chi_{aa}-\chi_{cc}$, and the magnitudes of the molecular g values have been determined for formyl fluoride by Flygare, Rock, and Hancock.[62] These and several other magnetic quantities are also reported by the same authors [62] for vinyl fluoride, trifluoroethylene, ethyl fluoride and 1,1-difluoroethane.

Phosgene ($COCl_2$) is the latest of a series of planar tetratomic molecules of C_{2v} symmetry for which quadratic force fields have been determined by Mirri *et al.*[63] from combinations of centrifugal distortion constants and vibrational frequencies. The four distortion constants of the Kivelson–Wilson first-order treatment which remain after application of the planarity

[58] J. W. C. Johns and W. B. Olson, *J. Mol. Spectroscopy*, 1971, **39**, 479.
[59] S. G. Kukolich and D. J. Ruben, *J. Mol. Spectroscopy*, 1971, **38**, 130.
[60] P. Thaddeus, L. C. Krisher, and J. H. N. Loubser, *J. Chem. Phys.*, 1964, **40**, 257.
[61] S. G. Kukolich, *J. Chem. Phys.*, 1971, **55**, 610.
[62] S. L. Rock, J. K. Hancock, and W. H. Flygare, *J. Chem. Phys.*, 1971, **54**, 3450.
[63] A. M. Mirri, L. Ferretti, and P. Forti, *Spectrochim. Acta*, 1971, **27A**, 937.

condition (τ_{aaaa}, τ_{bbbb}, τ_{aabb}, and τ_{abab}) are well determined from a large selection of microwave and millimetre transitions by a least-squares analysis. Unfortunately, the six A_1 species symmetry force constants (F_{11}, F_{12}, F_{13}, F_{22}, F_{23}, and F_{33}) cannot be determined uniquely from the three τ values (τ_{aaaa}, τ_{bbbb}, τ_{aabb}) and the three vibrational fundamentals (ν_1, ν_2, and ν_3) which belong to the A_1 species. However, if the three constraints $F_{12} > 0$ with F_{13} and F_{23} of opposite sign to each other, suggested by similar molecules, are assumed, acceptable solutions for the A_1 force constants are contained within narrow limits. Similarly, assumption of a suitable range for F_{45} is necessary if the B_1 species constants F_{44}, F_{45}, and F_{55} are to be determined from ν_4, ν_5, and τ_{abab}.

Lees [64] has established the planarity of carbonyl cyanide, $CO(CN)_2$, from the small positive inertia defect $\Delta = 0.433$ a.m.u. Å^2 of the normal isotopic species. This unusually large quantity reduces to 0.08 a.m.u. Å^2 (within the generally accepted range for effective moments of inertia) when the contribution (0.35 a.m.u. Å^2) of a nearly harmonic, low-frequency in-plane bending vibration is subtracted out. Vibrational excitation increases Δ by 0.7 a.m.u. Å^2 per quantum and therefore its contribution to the zero-point constants is approximately half this quantity. Insufficient moments of inertia are available for a complete structure determination, but with the assumptions $r(C{\equiv}N) = 1.165$ Å, $r(C{=}O) = 1.22$ Å, and a linear CCN chain, the values $r(C{-}C) = 1.45$ Å and $\widehat{CCC} = 115°19'$ best reproduce the observable results. The assumption of CCN chain linearity is open to question in view of the non-linear nature [65] of SCN in $S(CN)_2$; the author therefore presents the results of a similar calculation, but with the CN bonds making an angle with the molecular symmetry axis which is 3° greater than that made by the CC bonds. The $r(C{-}C)$ so calculated is little changed but the $\widehat{CCC}$ angle is reduced by 2°. Some puzzling information about θ_{CN}, the angle between the CN bonds, comes from the first-order nuclear quadrupole coupling constants, χ_{aa}, χ_{bb}, and χ_{cc}, for the two equivalent nitrogen nuclei. If the electric field about the CN bond is assumed to be cylindrically symmetric, it follows that

$$\chi_{bond} = -2\chi_{cc}, \text{ and } \chi_{aa} = \chi_{bond}[3\sin^2(\theta_{CN}/2) - 1]/2$$

from which $\chi_{bond} = -5.6$ MHz and $\theta_{CN} = 110°08'$. The former is higher than expected, while the latter is smaller than the values calculated from inertial data with either assumption about the $\widehat{CCN}$. The discrepancies are taken to invalidate the assumption of cylindrical symmetry of the CN bond. The dipole moment, $\mu_a = 0.704 \pm 0.007$ D, is at serious variance with the benzene-solution value [66] of 1.35 D.

The microwave spectrum of formic acid has been reinvestigated in greater

[64] R. M. Lees, *Canad. J. Phys.*, 1971, **49**, 367.
[65] L. Pierce, R. Nelson, and C. Thomas, *J. Chem. Phys.*, 1965, **43**, 3423.
[66] O. Glemser and V. Hausser, *Z. Naturforsch.*, 1948, **3b**, 159.

detail.[67,68] Precise rotational constants are derived for the vibrational ground states of the isotopic species $H^{12}C^{16}O_2H$, $D^{12}C^{16}O_2D$, $H^{12}C^{16}O_2D$, $D^{12}C^{16}O_2H$, and $H^{13}C^{16}O_2H$. In the first two species,[67,68] the spectrum analysis uses Watson's centrifugal distortion treatment to second order, whereas for the remaining species[68] the data is sufficient to allow the use only of Kivelson's first-order approach. The complete r_s-structure calculated using the data for $H^{12}C^{18}O^{16}OH$ and $H^{12}C^{16}O^{18}OH$ given by Kwei and Curl[69] in combination with that of the above species is shown in (3), which differs

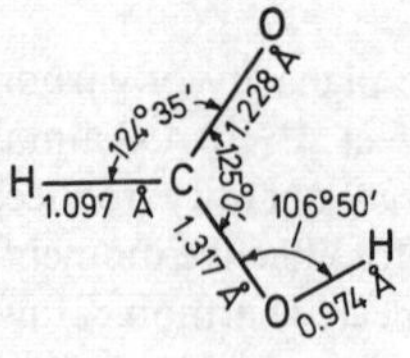

(3)

significantly in the two carbon–oxygen distances from the previously determined structures.[69]

The spectra of three halogenated alkenes and of two systems containing conjugated triple and double carbon–carbon bonds are reported. Gerry[70] has reinvestigated vinyl chloride with the purpose of refining the chlorine nuclear quadrupole coupling constants and determining the first-order centrifugal distortion constants. In respect of the former, the determined value of χ_{aa} agrees well with previous values, whereas the asymmetry parameter η is somewhat different. With knowledge of the geometrical structure and ignoring off-diagonal elements, the inertial axis coupling tensor is transformed to the field gradient principal axis system in which z is assumed parallel to

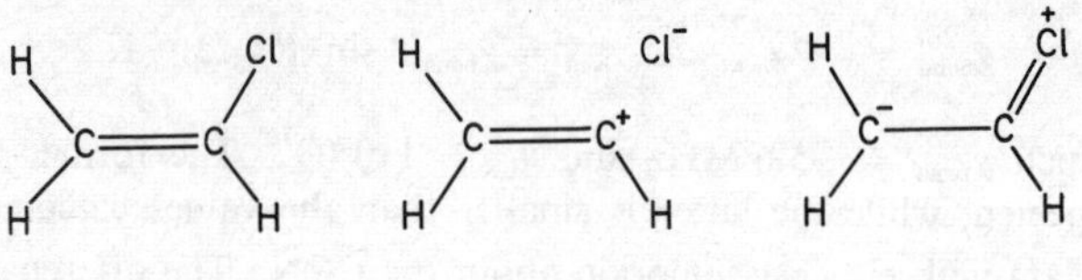

(4) (5) (6)

[67] J. Bellet, C. Samson, G. Steenbeckeliers, and R. Wertheimer, *J. Mol. Structure*, 1971, **9**, 49.
[68] J. Bellet, C. Samson, G. Steenbeckeliers, R. Wertheimer, and A. Deldalle, *J. Mol. Structure*, 1971, **9**, 65.
[69] G. H. Kwei and R. F. Curl, *J. Chem. Phys.*, 1960, **32**, 1592; R. G. Lerner, B. P. Dailey, and J. P. Friend, *J. Chem. Phys.*, 1957, **26**, 680.
[70] M. C. L. Gerry, *Canad. J. Chem.*, 1971, **49**, 255.

the C—Cl bond and *y* perpendicular to the molecular plane. The principal coupling constants are then employed to determine the relative contributions of the main valence bond structures, (4), (5), and (6), according to the methods described by Townes and Schawlow [71] for (4) and (5) and by Goldstein [72] for (6). The four centrifugal distortion constants τ_{aaaa}, τ_{bbbb}, τ_{aabb}, and τ_{abab} determined from a first-order treatment of observed transitions are reasonably well reproduced by a recently determined quadratic force field. Chlorotrifluoroethylene [73] and 3,3,3-trifluoropropene [74] have also been studied.

The *a*-type spectra of two species of vinylacetylene, $HC{\equiv}C{\cdot}CH{=}CH_2$ and $DC{\equiv}C{\cdot}CH{=}CH_2$, in the ground state and in one quantum of each of ν_1 (C≡C—C in-plane deformation) and ν_2 (C≡C—C out-of-plane deformation) are assigned by Hirose.[75] The following vibrational ground-state dipole moments (in D) are also measured:

	$CH_2{=}CHC{\equiv}CH$	$CH_2{=}CHC{\equiv}CD$
μ_a	0.223 ± 0.02	0.206 ± 0.008
μ_b	0.02	0.02

These values differ significantly from $\mu = 0.43$ D obtained from Stark-effect measurement on the assumption of $\mu_b = 0$ by Sobolev, Scherbakov, and Akishin.[76] The change in μ_a on deuteriation is similar in sign and magnitude to the difference in μ between $CH_3C{\equiv}CH$ and $CH_3C{\equiv}CD$.[77] No structural information is recorded in ref. 75, but the B_0 and C_0 values therein derived have been used elsewhere [78] in combination with electron-diffraction data to determine a structure for vinylacetylene.

Flygare *et al.*[79] have extended their studies of a group of closely related molecules with the publication of the microwave spectrum of pent-1-en-3-yne. The main interest centres on the measured dipole moments $\mu_a = 0.571 \pm 0.002$, $\mu_b = 0.334 \pm 0.041$, and $\mu_{total} = 0.663 \pm 0.023$ D, of which μ_{total} compares favourably with the value $\mu = 0.57$ D from previous dielectric measurements. Arguments concerned with the vector sum of bond dipole moments applied to this and related molecules lead to signs for the dipole moments of some of them. Moreover, a comparison is made of measured dipoles with those calculated from an empirically determined set of atomic dipoles, details of which method of calculation are discussed in a later paper [80] wherein the convention for empirically assigning atomic dipole moments and justification

[71] C. H. Townes and A. L. Schawlow, 'Microwave Spectroscopy', McGraw-Hill, New York, 1955.
[72] J. H. Goldstein, *J. Chem. Phys.*, 1956, **24,** 106.
[73] Krishnaji and S. Chandra, *Indian J. Pure and Appl. Phys.* (*B*), 1970, **8,** 634.
[74] I. A. Mukhtarov and V. A. Kudiev, *Izvest. Akad. Nauk Azerb. S.S.R., Ser. Fiz. Mat. i Tech. Nauk*, 1970, **no. 6,** 146.
[75] C. Hirose, *Bull. Chem. Soc. Japan*, 1970, **43,** 3695.
[76] G. A. Sobolev, A. M. Scherbakov, and P. A. Akishin, *Optics and Spectroscopy*, 1962, **12,** 78.
[77] J. S. Muenter and V. W. Laurie, *J. Chem. Phys.*, 1966, **45,** 855.
[78] T. Fukuyama, K. Kuchitsu, and Y. Morino, *Bull. Chem. Soc. Japan*, 1969, **42,** 379.
[79] T. D. Gierke, S. L. Hsu, and W. H. Flygare, *J. Mol. Spectroscopy*, 1971, **40,** 328.
[80] T. D. Gierke, H. L. Tigelaar, and W. H. Flygare, *J. Amer. Chem. Soc.*, 1972, **94,** 330.

of their use are also given. Not only do the observed magnitudes of the molecular dipole moments agree quite well with those so calculated but also the signs determined from bond dipole moments are reproduced.

An investigation of fulminic acid (HNCO),[81] with a newly designed millimetre-wave spectrometer, has led to the evaluation of centrifugal distortion parameters. Rotational constants and centrifugal distortion constants have also been evaluated for two excited vibrational states. Two molecules containing the NO_2 group which have been investigated are nitromethane, CH_3NO_2, and methyl nitrate, CH_3ONO_2. Cox and Waring have made a detailed study of methyl nitrate [82] and have obtained rotational constants for seven isotopic species. The planarity of the heavy-atom skeleton and the conformation of the methyl group as staggered with respect to the *cis* oxygen atom, which had been reported previously,[83] have been confirmed and structural parameters have been evaluated; the structure is shown in (7). The determination of this

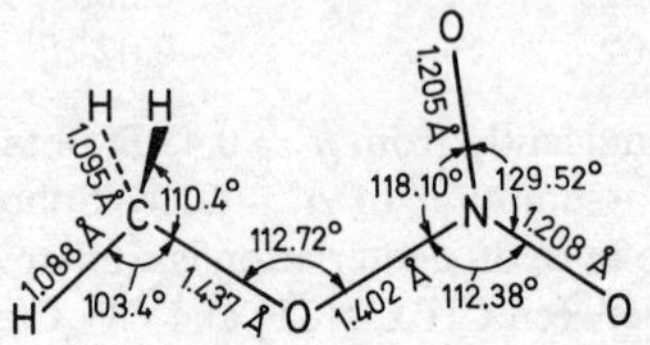

(7)

structure allows comparison with the nitric acid molecule. The NO_2 group angle and bond lengths are in fact closely similar to those of nitric acid [84] (130.3°, 1.211 Å, and 1.199 Å). Similarly, there is a tilt of the NO_2 group as found earlier for nitric acid. For methyl nitrate the axis is tilted by 2.9° as against 1.0° in nitric acid; the methyl group is also found to be tilted away from the *cis* oxygen (4.8°). As in other cases of tilted methyl groups, the tilt is towards an unshared pair of electrons. The molecular dipole moment has been determined as 3.08 D and its orientation in the molecule has also been obtained. For nitromethane,[85] hyperfine structure of some low-J transitions has been studied and nitrogen nuclear quadrupole coupling constants have been evaluated.

Acyclic Molecules with Internal Rotation.—*Ethane Derivatives.* One of the notable achievements of the past year has been the observation by Hirota and Matsumura [86] of the spectrum of the isotopically substituted ethane molecule CH_3CD_3. The $J = 2 \leftarrow 1$, $K = 1$ transition has been identified

[81] M. Winnewisser and B. P. Winnewisser, *Z. Naturforsch.*, 1971, **26a**, 128.
[82] A. P. Cox and S. Waring, *Trans. Faraday Soc.*, 1971, **67**, 3441.
[83] W. B. Dixon and E. B. Wilson, *J. Chem. Phys.*, 1961, **35**, 191.
[84] A. P. Cox and J. M. Riveros, *J. Chem. Phys.*, 1965, **42**, 3106.
[85] A. P. Cox, S. Waring, and K. Morgenstern, *Nature Phys. Sci.*, 1971, **229**, 22.
[86] E. Hirota and C. Matsumura, *J. Chem. Phys.*, 1971, **55**, 981.

by its Stark effect, which consists simply of two components, one moving to higher frequency and the other to lower frequency, as the electric field is increased. The dipole moment is found to be 0.01078 ±0.00009 D, which is about twice as large as that of CH_3D. The result indicates that microwave spectroscopy may be applied to a number of non-polar symmetric rotors, provided isotopic species are available.

An investigation of deuterioethanol (CH_3CHDOH) [87] confirms that the *trans*-form (10) has a plane of symmetry, and shows the existence of two *gauche*-forms, (8) and (9). Rotational constants are obtained for the three

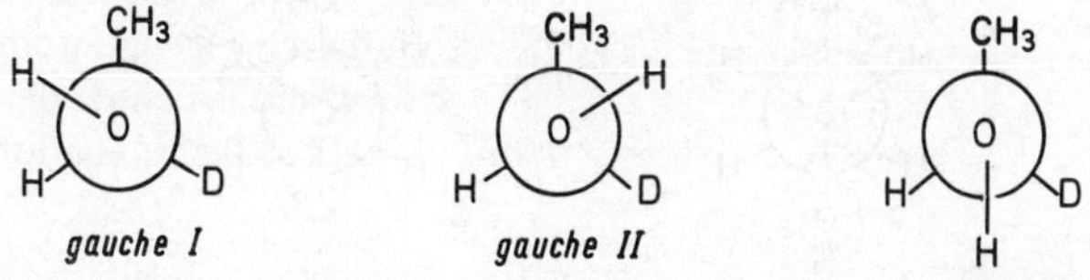

gauche I (8) *gauche II* (9) *trans* (10)

rotamers and structural information is obtained on the basis of transferring certain angles and distances from methanol. The dihedral angle of the *gauche*-forms is *ca.* 126° (from the *trans* position), and $\widehat{CCO}$ is *ca.* 5° larger for the *gauche-* than for the *trans*-rotamer. Dipole moments are found to be as follows: *gauche* I, 1.62 ±0.05 D; *gauche* II, 1.65 ±0.7 D; and *trans*, 1.52 ±0.02 D. Investigations have also been made of 2-amino- and 2-fluoro-ethanol. In a detailed study of the former, Curl and Penn have examined six isotopic species.[88] The molecule exists in a *gauche*-form (12) [in projection in (11)]. The dihedral angle between the CCN and CCO planes is 55.4°. The

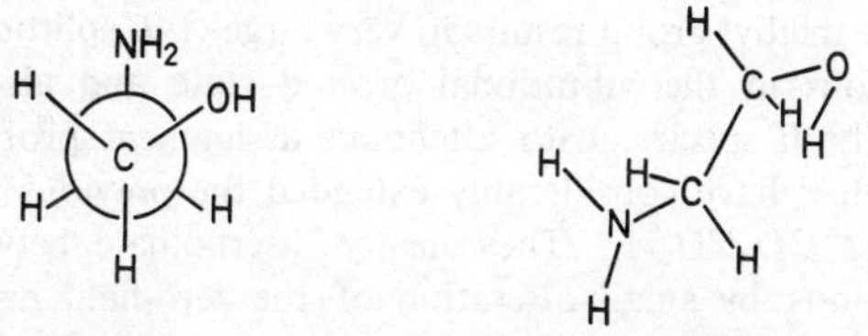

(11) (12)

conformation of the molecule minimizes the distance between the hydroxy-proton and the lone pair of electrons of the nitrogen. An r_s-structure for the groups involved in hydrogen-bonding shows an unusually long O—H bond distance of 1.14 Å. For 2-fluoroethanol it is found that the molecule has

[87] Y. Sasada, M. Takano, and T. Satoh, *J. Mol. Spectroscopy*, 1971, **38**, 33.
[88] R. F. Curl and R. E. Penn, *J. Chem. Phys.*, 1971, **55**, 651.

C—F and C—O bonds in a *gauche* conformation,[89] and a detailed structure has been obtained.[90] Rotational transitions have been assigned for the *gauche*-rotamer of 1,2-difluorethane.[91] With certain assumptions, made by reference to ethyl fluoride, the dihedral angle between the two CCF planes is found to be 73 ±4°. The sensitivity of the value to the assumed parameters is investigated and a significant correlation with the $\widehat{CCF}$ value (assumed) is the main limit on the reliability of the dihedral angle.

For pent-1-yne, *trans*- and *gauche*-rotamers [(13) and (14)] have been

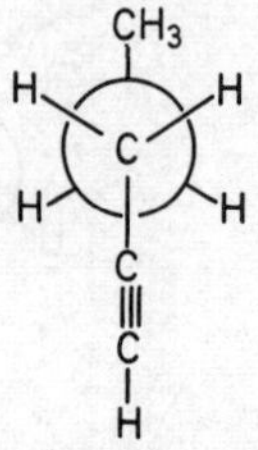

trans (13) *gauche* (14)

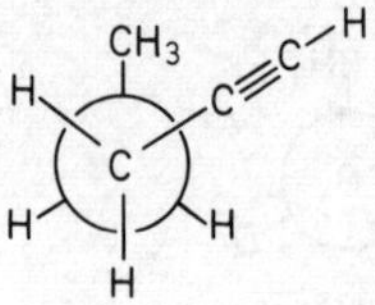

identified.[92] With certain assumptions the dihedral angle between the CH_3—C—C and the C—C—C≡H planes is found to be 115.2°. The dipole moment of each rotamer has been obtained: *trans*, 0.853 ±0.001 D, and *gauche*, 0.760 ±0.006 D. By using the value of the diople moment (0.814 D), previously determined from dielectric constant measurements using a precision heterodyne beat method,[93] the authors calculate ΔU for *gauche*- and *trans*-pent-1-yne to be 170 ±60 cal mol^{-1}.

Acids, Esters, and Ethers. For acetic acid the rather low barrier to internal rotation of the methyl group results in very large *A–E* splittings of the rotational transitions in the vibrational ground state and also in non-rigid behaviour of both species, with attendant assignment problems. Krisher and Saegebarth [94] have considerably extended the previous assignments of CH_3CO_2H and CD_3CO_2H. They neatly discriminate between *A* and *E* species transitions by small alteration of the zero-field half-cycle of the square-wave modulation system. *E* Species have a Stark effect predominantly linear in the applied electric field, whereas that of the *A* species is quadratic, so that a small deviation from zero field causes the *E* species to become

[89] Ch. O. Kadzhar, G. A. Abdullaev, and L. M. Imanov, *Optics and Spectroscopy*, 1971, **30**, 529.
[90] K. S. Buckton and R. G. Azrak, *J. Chem. Phys.*, 1970, **52**, 5652.
[91] S. S. Butcher, R. A. Cohen, and T. C. Rounds, *J. Chem. Phys.*, 1971, **54**, 4123.
[92] D. Damiani and A. M. Mirri, *Chem. Phys. Letters*, 1971, **10**, 351.
[93] H. J. G. Haymann and J. Weiss, *J. Chem. Phys.*, 1965, **42**, 3701
[94] L. C. Krisher and E. Saegebarth, *J. Chem. Phys.*, 1971, **54**, 4553.

diffuse and weaker while the *A* species are sensibly unaffected. The initially assigned *A* species are analysed according to the effective Hamiltonian, $H_{\sigma v}$ for a given vibrational state v, as obtained by Herschbach [95] by applying successive van Vleck transformations to the Hamiltonian for the problem in which coupling of the top and the molecular frame angular momenta is treated as a perturbation:

$$H_{\sigma v} = H_r + F \Sigma W_{v\sigma}(n)\mathscr{P}^n$$

In the case of a reasonably high barrier to rotation, termination in the second power of the relative angular momentum of the top and frame ($\mathscr{P}^2$) is sufficient when σ refers to the *A* species, which then follows rigid-rotor behaviour. For acetic acid, however, terms up to $n = 6$ are necessary to account for the behaviour of both *A* and *E* species because of the low barrier. The *A*-species transitions, being less sensitive to the perturbation, are first assigned and analysed to give initial internal rotation parameters which can then be used to predict and assign the more sensitive *E* species. The internal rotation parameters finally determined from all assigned transitions are of high precision. For example, the barriers to internal rotation of the methyl group are:

	CH_3CO_2H	CD_3CO_2H
V_3	480.8 ± 0.5	468 ± 2 cal mole^{-1}

The lower barrier in the CD_3 case is interpreted on the basis of shorter C—D than C—H bonds. Dipole moments, from the usual quadratic Stark effect of *A*-species transitions, are found to be $\mu_a = 0.86 \pm 0.01$, $\mu_b = 1.47 \pm 0.02$, and $\mu_{total} = 1.70 \pm 0.02$ D.

An investigation [96] has been made of the conformations of a series of molecules derived from methyl formate (HCO_2CH_3) by replacement of the formyl hydrogen by the following groups: F, C≡N, C≡CH, CH_3, or $CH{=}CH_2$. The rotational constants show that each of these molecules has a planar heavy-atom framework. It is established that the ester group O=C—O—C has the *cis*-conformation (15) for the last four of the above five groups and it is assumed in the absence of isotopic data on the fluoroformate that this molecule also has the *cis*-conformation. For methyl acrylate (X = $CH{=}CH_2$) internal rotation about the C—C bond gives rise

(15)

[95] D. R. Herschbach, *J. Chem. Phys.*, 1959, **31**, 91.
[96] G. Williams, N. L. Owen, and J. Sheridan, *Trans. Faraday Soc.*, 1971, **67**, 922.

to two possible conformers, *cis* and *trans*. Although there is i.r. evidence for the presence of both rotamers, only the *cis*-form was identified from the microwave spectrum. The failure to identify the *trans*-form is thought to be due to an unfavourable dipole moment. Barriers to internal rotation of methyl groups have also been determined and are included in the Table 6. A planar heavy-atom structure for the related molecule methyl nitrate [82] has been referred to earlier.

Rotational isomerism in vinyl ethers has also been explored.[97, 98] For divinyl ether,[98] rotational constants have been obtained for one conformer, and the values are shown to be in agreement with expectations for the *cis–trans*-form (16). The inertial defect and the dipole moment show that the

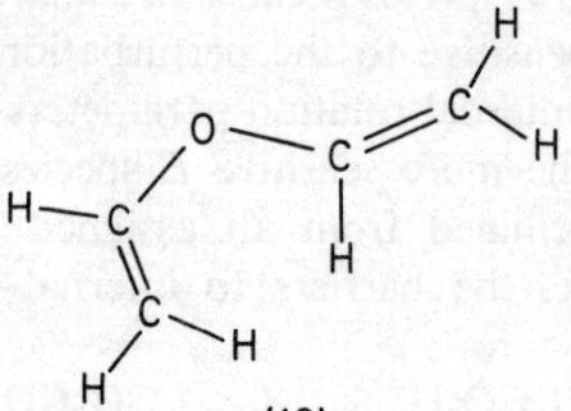

(16)

structure is not planar. The deviation is attributed to strong repulsion between the β-hydrogen of the *cis* vinyl group and the α-hydrogen of the *trans* vinyl group.

Other Molecules. The microwave spectrum of hydroxyacetonitrile [99] establishes a molecular conformation in which the OH and CN groups adopt the *gauche* orientation with respect to each other, there being two equivalent forms, (17) and (18). It is suggested from the direct torsion–rotation transitions in $DOCH_2CN$ that the *cis–gauche* barrier is probably rather higher than that of 90 cm^{-1} for the closely related molecule propy-2-yn-1-ol.

For glycolaldehyde the *cis*-conformer (19) has been identified and the

gauche I
(17)

gauche II
(18)

[97] B. A. Trofimov, N. I. Shergina, A. S. Atavia, A. V. Gushov, and G. K. Gavrilova, *Zhur. priklad. Spectroskopii*, 1971, **14**, 282.
[98] C. Hirose and R. F. Curl, *J. Mol. Spectroscopy*, 1971, **38**, 358.
[99] J. K. Tyler and D. G. Lister, *Chem. Comm.*, 1971, 1350.

majority of stronger lines in the spectrum are assigned, no other conformer being found.[100] Four isotopic species were examined: $CH_2(OD)CHO$, $CH_2(OH)CDO$, CHD(OH)CHO, and CHD(OH)CDO. For the first three species a first-order centrifugal distortion has been made, and for the fourth species rotational constants have been evaluated from low-*J* transitions, and from the rotational constants some structural parameters have been calculated. An investigation has also been made of *trans*-propan-2-ol.[101]

For the very rich spectrum of allylamine, transitions have been assigned by making use of double resonance, and one rotamer has been identified.[102] By using an argument based on the observed rotational constants, the dipole moment, and the nuclear quadrupole hyperfine structure, the conformation of the rotamer is established to be the N-*cis*, N-lone pair *trans*-form (20)

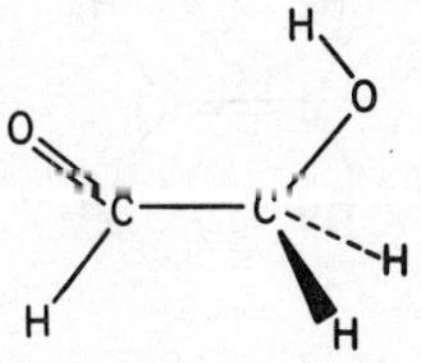

(19) 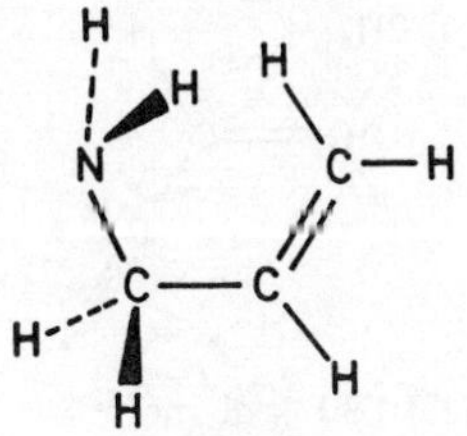 (20)

which is structurally very similar to the *cis*-rotamer of but-1-ene,[103] and some structural comparisons are made. An examination has been made of tackling the problem of assigning transitions, where the spectrum is a rich one, by computer analysis of the observed frequencies.[104] Parameters related to the rotational constants are varied in steps, and at each step the correspondence between observed and calculated spectra is checked. The method has been applied to the identification of a second conformer from the spectrum of fluoroacetic acid.

Barriers. Barriers to internal rotation of methyl groups have been determined for a number of molecules and are collected in Table 6. These include barriers for methyl groups in some derivatives of propene: *cis*-pent-3-ene-1-yne (21),[105] 2-methylacrylonitrile (22),[106] *trans*-crotononitrile (23),[107] and 2-methylacrylaldehyde (24).[108] A considerable number of such

[100] K. M. Marstokk and H. Møllendal, *J. Mol. Structure*, 1971, **7**, 101.
[101] L. M. Imanov, A. A. Abdurakhmanov, and M. M. Elchiev, *Doklady Akad. Nauk Azerb. S.S.R.*, 1971, **27**, 12.
[102] G. Roussy, J. Demaison, I. Botskor, and H. D. Rudolph, *J. Mol. Spectroscopy*, 1971, **38**, 535.
[103] S. Kondo, E. Hirota, and Y. Morino, *J. Mol. Spectroscopy*, 1968, **28**, 471.
[104] B. P. van Eijck, *J. Mol. Spectroscopy*, 1971, **38**, 149.
[105] R. G. Ford and R. A. Beaudet, *J. Chem. Phys.*, 1971, **55**, 3110.
[106] C. L. Norris and W. H. Flygare, *J. Mol. Spectroscopy*, 1971, **40**, 40.
[107] S. L. Hsu and W. H. Flygare, *J. Mol. Spectroscopy*, 1971, **37**, 92.
[108] M. Suzuki and K. Kozima, *J. Mol. Spectroscopy*, 1971, **38**, 314.

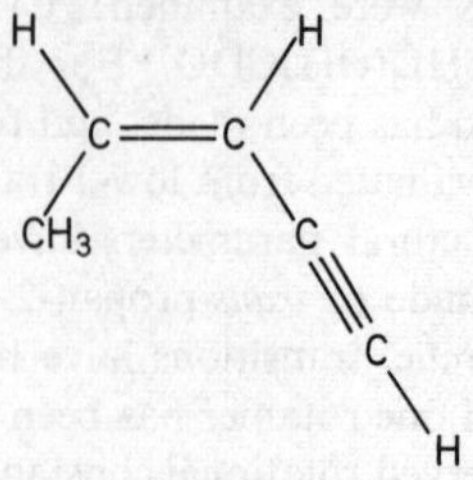

(21) 1.14 kcal mol^{-1}

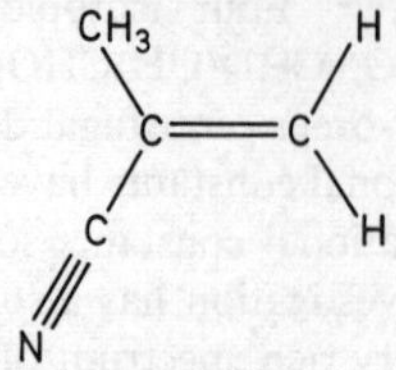

(22) 2.03 kcal mol^{-1}

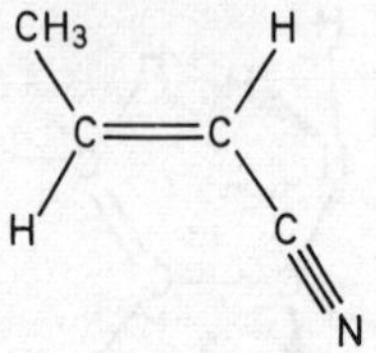

(23) 1.90 kcal mol^{-1}

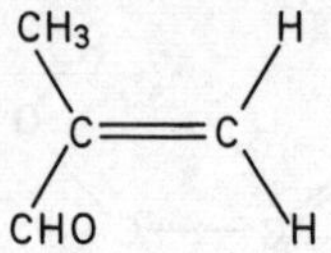

(24) 1.34 kcal mol^{-1}

derivatives of propene have now been studied and Norris and Flygare have discussed [106] the values in terms of the various types of electrostatic interactions contributing to barriers to internal rotation. Other molecules for

Table 6 *Barriers to methyl rotation*

Molecule	*Formula*	*Barrier* /kcal mol^{-1}
Methylamine	CH_3NH_2	1.955
	CH_3ND_2	1.940
	CD_3ND_2	1.920
Acetic acid	CH_3CO_2H	0.481
	CD_3CO_2H	0.468
Dimethylamine	$(CH_3)_2NH$	3.22
Methyl fluoroformate	FCO_2CH_3	1.077
Methyl propiolate	$HC{\equiv}CCO_2CH_3$	1.266
Methyl acetate	$CH_3CO_2CH_3$	1.215
Methyl acrylate	$H_2C{=}CHCO_2CH_3$	1.220
Methyl cyanoformate	$N{\equiv}CCO_2CH_3$	1.172
2-Methylacrylaldehyde	$CH_3C(CHO){=}CH_2$	1.340
2-Methylacrylonitrile	$CH_3C(CN){=}CH_2$	2.030
trans-Crotononitrile	$CH_3CH{=}CHCN$	1.900
cis-Pent-3-ene-1-yne	$CH_3CH{=}CHC{\equiv}CH$	1.140
1,1-Dimethylallene	$(CH_3)_2C{=}C{=}CH_2$	2.025
Methanesulphonyl fluoride	CH_3SO_2F	2.52
3-Methylfuran	CH_3	1.088

which barriers have been reported include 1,1-dimethylallene,[109] methanesulphonyl fluoride,[28] and 3-methylfuran [110] (Table 6).

Barriers to methyl rotation have been obtained by the internal axis method for CH_3NH_2 (684.1 cm^{-1}), CH_3ND_2 (678.7 cm^{-1}), and CD_3ND_2 (671.7 cm^{-1}) by a re-analysis of existing data for CD_3ND_2 and by using some new observation data for the other two isotopic species.[111] Rotational spectra in the first-excited torsional state of a number of isotopic species of dimethylamine have been assigned and the internal rotation splittings lead to a barrier of 3.22 kcal mol^{-1}. Inversion splittings are found to differ significantly from those found for the ground state.[112] The coupling of internal rotation and inversion in methylamine has been examined. It has been shown that the coupling for CH_3NHD is somewhat different from that in CH_3NH_2 or CH_3ND_2 and information on the inversion path in CH_3NHD has been obtained.[113] A hypothetical inversion splitting when coupling is absent is estimated to be 0.34 cm^{-1} for the ground state and 10.0 cm^{-1} for the first-excited state of the amino-wagging vibration. These results lead to a value of 1765 cm^{-1} for the height of the barrier to inversion. The barrier to internal rotation in [$^2H_3$]methyldifluoroborane, CD_3BF_2, has been determined from the $|m| = 3$ spectrum.[114] V_6 is found to have a value of 12.2 ±0.3 cal mol^{-1}, which compares with 13.77 cal mol^{-1} for CH_3BF_2. The decrease on deuteriation parallels the decrease observed for nitromethane on deuteriation.

Origin of Barriers. A number of papers have appeared on the interpretation of barriers and further attempts have been made to calculate barriers theoretically. A Hartree–Fock calculation for hydrogen peroxide carried out for several dihedral angles leads a potential function in agreement with experiment.[115] An *ab initio* calculation of the barrier in ethane,[116] using a minimum basis of Slater-type orbitals, shows that when an attempt to take electron correlation into account is made the barrier is decreased by *ca.* 0.13 kcal mol^{-1}. Allen and co-workers have made *ab initio* LCAO–MO–SFT calculations for a number of single molecules. For ethane and ethyl fluoride [117] the agreement with experiment is sufficiently good that it is concluded that the origin of the barriers is contained within the Hartree–Fock approximation. A physical interpretation is given by separating out two competing and out-of-phase components to the potential, one of which is attractive and the other repulsive. Both ethyl fluoride and ethane are repulsive dominant and

[109] J. Demaison and H. D. Rudolph, *J. Mol. Spectroscopy*, 1971, **40**, 445.
[110] T. Ogata and K. Kozima, *Bull. Chem. Soc. Japan*, 1971, **44**, 2344.
[111] K. Takagi and T. Kojima, *J. Phys. Soc. Japan*, 1971, **30**, 1145.
[112] J. E. Wollrab and V. W. Laurie, *J. Chem. Phys.*, 1971, **54**, 532.
[113] K. Tamagake, M. Tsuboi, K. Takagi, and T. Kojima, *J. Mol. Spectroscopy*, 1971, **39**, 454.
[114] J. Wollrab, E. A. Rinehart, P. B. Rinehart, and P. R. Reed, *J. Chem. Phys.*, 1971, **55**, 1998.
[115] T. H. Dunning and N. W. Winter, *Chem. Phys. Letters*, 1971, **11**, 194.
[116] B. Levy and M. C. Moireau, *J. Chem. Phys.*, 1971, **54**, 3316.
[117] L. C. Allen and H. Basch, *J. Amer. Chem. Soc.*, 1971, **93**, 6373.

this provides a rationalization of the rough equality of the two barriers. Barriers in ethane, propane, and butane have also been discussed by Magnusco and Musso,[118] who use a theoretical model based on the additivity of the intermolecular penetration energy between pairs of different bonds. Calculations have also been made for propene and its *cis* and *trans* fluoroderivatives.[119] The calculated barriers show the correct ordering and are in reasonable agreement with the experimental values. For the *cis*-isomer the barrier is attractive dominant. The nuclear–nuclear repulsions and electron–electron repulsions both quite strongly favour the staggered configuration, but nuclear–electron attractions even more strongly favour the eclipsed configuration. A similar type of calculation has been made for acetaldehyde [120] and it is found that the three-fold barrier is dominated by attractive interactions.

A detailed semi-empirical calculation for hindered rotation in formamide [121] has been made and may be compared with an *ab initio* MO study of formamide.[122] Another paper [123] on *ab initio* calculations adopts the approach of making a very wide ranging set of calculations at a uniform level of approximation. In this way a systematic study is made of many simple molecules containing H, C, N, O, or F atoms and included among the calculations are energies for different conformers.

5 Cyclic Molecules

Three-membered Rings.—Volltrauer and Schwendeman have thoroughly studied the *cis*- and *trans*-conformers of both cyclopropanecarbaldehyde [124] (25 and 26; X = H) and cyclopropanecarbonyl fluoride [125] (25 and 26; X = F). In each case the plane of the XCO group is shown to bisect the cyclopropane ring in the manner indicated. This is interesting because the well-established rule for CH_3COX and RCH_2COX compounds that the CO

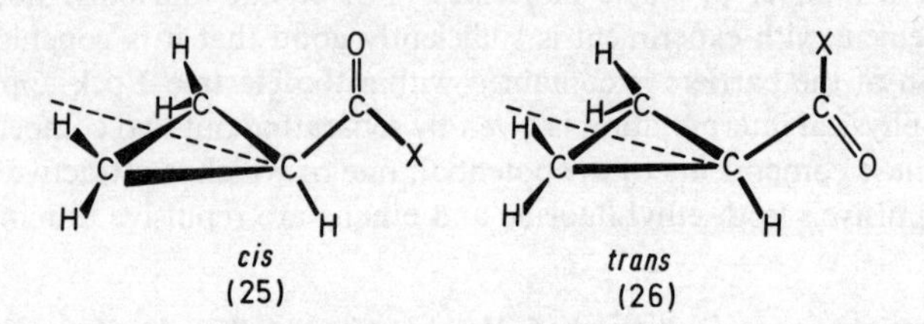

[118] V. Magnusco and G. F. Musso, *Chem. Phys. Letters*, 1971, 4, 433.
[119] E. Scarzafava and L. C. Allen, *J. Amer. Chem. Soc.*, 1971, **93**, 311.
[120] R. B. Davidson and L. C. Allen, *J. Chem. Phys.*, 1971, **54**, 2828.
[121] K. N. Shaw and L. W. Reeves, *Chem. Phys. Letters*, 1971, **10**, 89.
[122] D. A. Christensen, R. N. Kortzeborn, B. Bak, and J. L. Led, *J. Chem. Phys.*, 1970, **53**, 3912.
[123] I. Racom, W. J. Hehre, and J. A. Pople, *J. Amer. Chem. Soc.*, 1971, **93**, 289.
[124] H. N. Volltrauer and R. H. Schwendeman, *J. Chem. Phys.*, 1971, **54**, 260.
[125] H. N. Volltrauer and R. H. Schwendeman, *J. Chem. Phys.*, 1971, **54**, 268.

bond eclipses a C—H or a C—R bond in the equilibrium conformation is apparently contravened in that a considerable fraction of the molecules occur in the *cis* as well as the *trans* configuration.

In both *cis*- and *trans*-species for X = H and F, each rotational transition is accompanied by a strong satellite series assigned to the torsional motion of the CXO group with respect to the cyclopropyl ring. This assignment and the molecular symmetry plane are established from the principal moments of inertia. Thus, for *cis*-cyclopropanecarbaldehyde, and for *cis*- and *trans*-cyclopropanecarbonyl fluoride, the quantity $P_{cc} = (I_a+I_b-I_c)/2 = \sum_i m_i c_i^2$ is 19.8 a.m.u. Å², which establishes that each molecule has the same out-of-plane atoms and therefore a plane of symmetry (*ab*). Moreover, the increase of this quantity with the vibrational quantum number of the satellite in each case indicates that the motion is torsion of the CXO group out-of-plane. For *trans*-cyclopropanecarbaldehyde, the symmetry plane is *ac*, as shown by P_{bb} = 19.9 a.m.u. Å². The smooth, almost linear, variation of the rotational constants with the torsional quantum number is evidence that the motion is nearly harmonic, although *cis*-cyclopropanecarbaldehyde exhibits zig-zag behaviour because of a Fermi resonance involving the $v = 2$ state.

Relative intensity measurements give valuable information about the potential function for the torsional motion of the CXO group. The torsional energy-level separation for each species depends on relative intensity measurements of successive satellites within the given species, whereas the difference in zero-point energy (Δ) between the *cis*- and *trans*-species requires intensity comparison between ground-state transitions of the *cis*- and *trans*-species; the averaged values are given in Table 7. The authors use this information

Table 7 *Energy values for cyclopropanecarbaldehyde and cyclopropanecarbonyl fluoride*

	Cyclopropanecarbaldehyde	*Cyclopropanecarbonyl fluoride*
Excitation energy, *trans*/cm⁻¹	123.6 ± 2	90 ± 5
Excitation energy, *cis*/cm⁻¹	112.6 ± 4	63.3 ± 4
Δ/cm⁻¹	10 ± 20	−200 ± 30

to obtain the potential constants V_1, V_2, and V_3 in the expression

$$V = \tfrac{1}{2}\sum_n V_n(1-\cos n\theta)$$

for the potential energy as a function of θ, the angular co-ordinate describing the rotation of the aldehyde group with respect to the ring; terms V_4 and higher are neglected. Since the torsional motion is essentially harmonic in each well it follows that

$$V-V_0 = \overline{V}_2(\theta-\theta_0)^2$$

where θ_0 is the value of θ at equilibrium in one of the wells and V_0 is the value of the potential energy at θ_0. θ_0 is taken to be 0 for the *trans*-conformer and π for the *cis*-conformer. If $\cos n\theta$ is expressed as a power series in θ, and the coefficients of θ^2 in both equations compared, it is found that

$$\overline{V}_2\,(trans) = \tfrac{1}{4}(V_1+4V_2+9V_3+\cdots)$$
$$\overline{V}_2\,(cis) = \tfrac{1}{4}(-V_1+4V_2-9V_3+\cdots)$$

and the following relationship also exists

$$V_0\,(cis) - V_0(trans) = V_1+V_3+\cdots$$

Then, in the approximation that the torsional oscillation in each well is harmonic, the measured excitation energies and Δ may be related to the V_n with the results given in Table 8. The barriers relative to *trans*-configuration are 4.43 ± 0.41 and 5.17 ± 0.71 kcal mol^{-1} for the aldehyde and acid fluoride, respectively.

Table 8 *Potential constants of cyclopropanecarbaldehyde and cyclopropanecarbonyl fluoride*

V_n	*Cyclopropanecarbaldehyde*	*Cyclopropanecarbonyl fluoride*
V_1/cm^{-1}	-80 ± 40	-530 ± 110
V_2/cm^{-1}	1535 ± 140	1800 ± 240
V_3/cm^{-1}	96 ± 35	345 ± 90

The large value of V_2 in both cases is discussed in terms of both steric and electronic effects. The favoured explanation is that conjugation between the carbonyl group and the cyclopropyl ring through the C—C single bond is more important than steric effects, even though the steric contributions to V_2 are additive. On this basis the authors predict an even larger V_2 term for acrylaldehyde, in which conjugation with the vinyl group should be more important than conjugation with the cyclopropyl ring. The larger V_1 and V_3 for the acid fluoride are attributed to cyclopropyl ring–CH_2–fluorine repulsions, which are greater than the corresponding repulsions involving the hydrogen of the carbaldehyde group. It is interesting that if a similar change in V_1 occurs between acrylaldehyde and acryloyl fluoride, the value of V_1 in the former would be *ca.* +350 cm^{-1}. This would greatly decrease the stability of the *cis*-conformer of acrylaldehyde, in agreement with the fact that only the *trans*-conformer has been observed.[126]

Experimental conclusions about the conformations of two other three-membered-ring derivatives, cyclopropylphosphine and epifluorohydrin [2-(fluoromethyl)oxiran], have also been published. From the study of three isotopic species of cyclopropylphosphine ($C_3H_5PH_2$, C_3H_5PHD, and $C_3H_5PD_2$), Dinsmore, Britt, and Boggs [127] have established that the single observed conformer has the *ac* plane as a plane of symmetry since *b*-type

[126] E. A. Cherniak and C. C. Costain, *J. Chem. Phys.*, 1966, **45**, 104.
[127] L. A. Dinsmore, C. O. Britt, and J. E. Boggs, *J. Chem. Phys.*, 1971, **54**, 915.

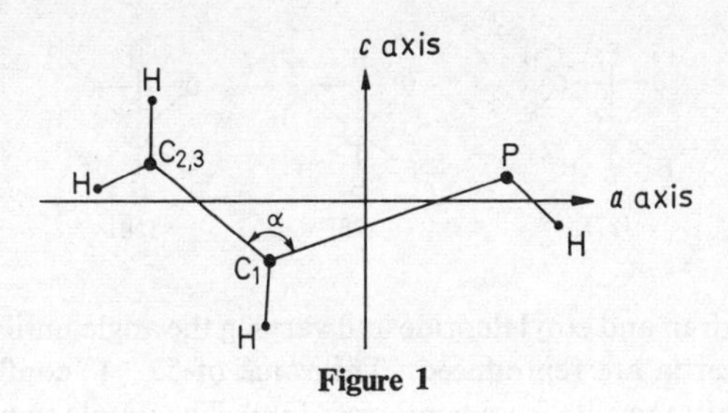

Figure 1

transitions are absent and $\mu_b = 0$ for the parent isotopic species. The r_s-co-ordinates of the phosphine-group hydrogen atoms are shown to be consistent only with the conformation shown projected on the *ac* plane in Figure 1. Using the r_s value for the non-bonded phosphine hydrogen–hydrogen distance and with the assumptions of an equilateral ring in which the plane of the ring bisects the HCP and HCH angles, of *r*(C—H) = 1.080 Å, and of $\widehat{HCH}$ = 116.2° (as in cyclopropyl chloride), the five remaining structural parameters which best reproduce the observed moments of inertia were determined to be: *r*(C—C) = 1.513, *r*(C—P) = 1.834, and *r*(P—H) = 1.413 Å, and $\widehat{CPH}$ = 98.3 and α = 121.1°. The striking feature of this structure is the short *r*(C—P), significantly shorter than in the methylphosphines. The same is true for the nitrogen analogue, cyclopropylamine.[128] The preferred conformation of cyclopropylphosphine and the short *r*(C—P) are rationalized using the Walsh model for cyclopropane in which each carbon atom is considered sp^2 hybridized. Each pure *p*-orbital lies in the plane of the ring and overlaps with a similar *p*-orbital on adjacent carbon atoms to form extra-annular bonds. Two sp^2-orbitals on each carbon form bonds with hydrogen atoms above and below the plane of the ring, whereas the third sp^2-orbital is directed toward the centre of the ring to form an intra-annular, multicentre bond. Conjugation of the phosphorus lone pair with the intra-annular orbitals is thought to stabilize the observed conformation, whereas the C—P bond involving carbon sp^2 hybridized would naturally be short.

Three rotational isomers are possible for epifluorohydrin, (27), (28), and (29). Previous n.m.r. studies [129] have indicated that (28) is the most stable of these. The moments of inertia from the microwave spectrum [130] of the only observed conformer of epifluorohydrin are also compatible only with (28), a somewhat surprising result in view of the proximity of the two electronegative atoms in this situation. Some estimate of the F—C—C—H dihedral angle [0° when fluorine and hydrogen atoms are *cis*, 180° in (27), +60° in (28), and −60° in (29)] is obtained by assuming all other structural para-

[128] D. K. Hendricksen and M. D. Harmony, *J. Chem. Phys.*, 1969, **51**, 700.
[129] W. A. Thomas, *J. Chem. Soc.* (*B*), 1968, 1187.
[130] S. C. Dass, A. Bhaumik, W. V. F. Brooks, and R. M. Lees, *J. Mol. Spectroscopy*, 1971, **38**, 281.

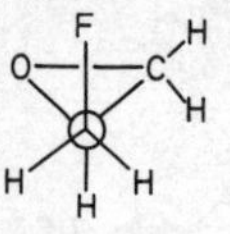

(27) 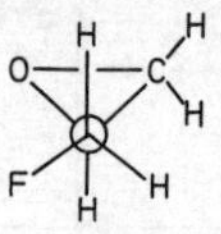 (28) 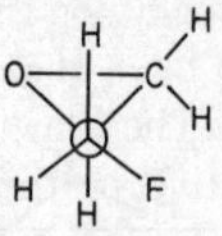 (29)

meters from oxiran and ethyl fluoride and varying the angle until the observed moments of inertia are reproduced. The value of $53 \pm 1°$ confirms that the observed spectrum results from conformer (28). The dipole moment components are: $\mu_a = 1.33 \pm 0.05$, $\mu_b = 2.86 \pm 0.06$, $\mu_c = 0.0$ (assumed), and $\mu_{total} = 3.15 \pm 0.05$ D. The assumed $\mu_c = 0.0$ D is a result of failure to detect any *c*-component of the dipole moment during Stark-effect measurements.

Four-membered Rings.—Interest in the microwave spectra of four-membered rings, particularly with a view to the elucidation of one-dimensional potential-energy functions for the ring-puckering vibration, has continued.

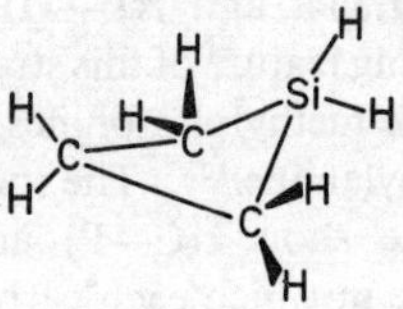

(30)

The investigation of silacyclobutane [siletan (30)] by Pringle [131] may be cited as a good example of the precise and detailed information about ring-puckering vibrations available from microwave spectra. In silacyclobutane, the barrier to ring inversion is 440 cm^{-1}, which is higher than in any four-membered ring previously studied except cyclobutane. The non-planar structure results because the mutual repulsion of the eclipsed methylenic and silylenic hydrogen atoms in the planar structure considerably outweighs the gain from minimizing the ring angle strain. Consequent upon the high barrier, three pairs of vibrational energy levels lie below the maximum, with the nearly degenerate lowest pair predicted to have a separation of only 72 MHz, from far-i.r. work.

The small separation of the lowest pair of inversion levels relative to the rotational constants ensures that rotational transitions in both states obey rigid-rotor theory. However, the rotational constants in each of the $v = 0,1$ states are so similar that the spectra are not resolved by the conventional microwave spectrometer. Nevertheless, the $v = 0,1$ inversion splitting is

[131] W. C. Pringle, *J. Chem. Phys.*, 1971, **54**, 4979.

available from the Stark effect of certain transitions which involve rotational levels connected by μ_c to appropriate nearly degenerate levels in the adjacent inversion state. The Stark effect of the rotational transitions in question is then very sensitive to the inversion splitting and leads to quite accurate values for the $v = 0,1$ separation as well as the transition matrix element $\langle 0|\mu_c|1\rangle$. The intensities and frequencies of suitable direct *c*-type transitions involving $\Delta v = \pm 1$ predicted using these initial quantities are then observed, giving the very accurate values of 75.75 ± 0.03 and 43.06 ± 0.05 MHz for the $v = 0, 1$ separation in silacyclobutane and [1,1-$^2H_2$]silacyclobutane, respectively. Elegant confirmation of these splittings is offered by the observation of two vibration–rotation transitions strictly forbidden in the limit of zero electric field: $5_{15}(0) \leftarrow 4_{04}(1)$ and $5_{05}(1) \leftarrow 4_{14}(0)$. The levels $5_{15}(0)$ and $5_{05}(1)$ occur as a nearly degenerate pair which becomes strongly mixed through the μ_c-component in an applied electric field. The same is true of the $4_{04}(1)$ and $4_{14}(0)$ levels. The mixing gives some *a*-type character to the indicated transitions and thus their Stark components are allowed even though the zero-field lines are not. Extrapolation of the Stark components to zero field gives, for example, 76 ± 1 MHz for the $v = 0,1$ inversion splitting of the parent isotopic species. A similar technique is simultaneously reported [132] for the observation of forbidden transitions in formic acid by mixing the nearly degenerate *K* doublet levels in the field.

The $v = 2,3$ vibrational separation in silacyclobutane is of the same order as the rotational constants and therefore the significant coupling of the vibrational angular momentum owing to the ring-puckering vibration with the overall rotational angular momentum leads to non-rigid behaviour of rotational transitions in these states. Corrections to the rigid rotor are treated by the method used in cyclopentene by Butcher and Costain [133] to second order of perturbation to give the $v = 3 \leftarrow 2$ inversion splittings as 7793 ± 7 and 4480 ± 30 MHz for the parent and deuteriated species, respectively. Assignment difficulties due to the essentially equal intensity of the $v = 2$ and $v = 3$ satellites and their non-rigid behaviour are neatly overcome by microwave–microwave double-resonance techniques.

Silacyclobutane presents an opportunity to investigate how well the low-frequency inversion transitions resulting from a high barrier can alone determine the potential-energy function governing the ring-puckering vibration. The very accurately known $v = 0,1$ and $v = 2,3$ separations are sufficient to allow determination of both potential constants G and η in the type of function $V(X) = G(X^4 - \eta X^2)$ which has hitherto proved very successful for four-membered rings.[134] Here X is a reduced out-of-plane displacement coordinate. However, Pringle finds that using these observed separations alone, the function so derived erroneously predicts both the barrier to inversion and

[132] C. Samson and E. Willemot, *Compt. rend.*, 1971, **271**, *B*, 1204.
[133] S. S. Butcher and C. C. Costain, *J. Mol. Spectroscopy*, 1965, **15**, 40.
[134] See, for example, S. I. Chan, R. T. Borgers, J. W. Russell, H. L. Strauss, and W. D. Gwinn, *J. Chem. Phys.*, 1966, **44**, 1103.

the far-i.r. spectrum and it is concluded that nearly degenerate vibrational separations are insufficient in such cases. On the other hand, a function of the mixed quartic–quadratic type can be determined from combined far-i.r. and microwave data which closely predicts both sets.

The main structural interest in silacyclobutane lies in the degree of non-planarity. By sensible assumption of other parameters, Pringle finds $\widehat{CCC} = 103°$; the distance of the silicon atom from the CCC plane, $z = 0.71$ Å, and r(Si—C) = 1.91 Å. Also, the rapid decrease of the transition moment $|\langle v|\mu_c|v+1\rangle|$ with increasing v is evidence that the molecule becomes increasingly more planar as it approaches the top of the barrier.

The other four-membered ring investigated, 3,3-difluoro-oxetan,[135] is shown to have a planar equilibrium ring structure (31). This should be compared with its precursor, oxetan [trimethylene oxide (32)], which has a slightly but

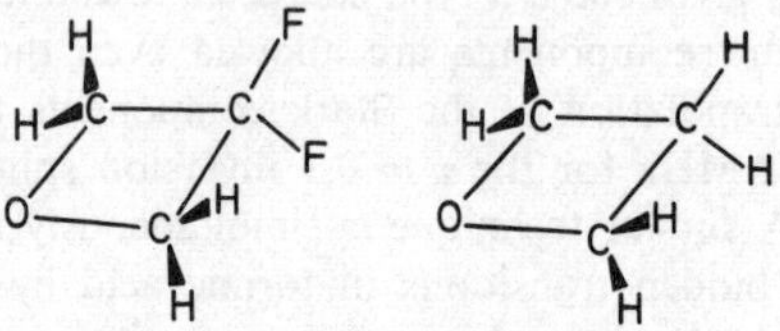

(31) (32)

distinctly non-planar equilibrium structure. The decrease of barrier to the planar structure, noted also between cyclobutane and 1,1-difluorocyclobutane, is attributed to a decreased steric repulsion between the methylene and difluoro-groups in the planar state because the C—F bonds are longer than C—H bonds. The planarity of 3,3-difluoro-oxetan is established from several aspects of its microwave spectrum. Rotational constants are determined for six quanta of the out-of-plane ring bending vibration and the smooth variation of each constant with the vibrational quantum number rules out all possibilities except either a very high barrier to planarity, in which case it must be high enough to make each of the states up to $v = 6$ doubly degenerate, or no barrier to planarity, for otherwise the rotational constants would exhibit a characteristic zig-zag behaviour. The high barrier possibility is ruled out by nuclear spin statistical weight considerations since a planar ring or a very low barrier requires a statistical weight ratio of 9/7 for symmetric to antisymmetric spin states. Only by including the weight ratio in the expression for the relative intensity of adjacent satellites, that is

$$I_{v+1}/I_v = (9/7)^{(-1)^{(v+s)}} \exp(-\Delta E/kT)$$

where v is the vibrational quantum number and $s = 0$ for rotational transitions of the type *ee–eo* and $s = 1$ for *oe–oo* transitions, can sensible and con-

[135] G. L. McKown and R. A. Beaudet, *J. Chem. Phys.*, 1971, **55**, 3105.

sistent vibrational separations be obtained. A grossly non-planar structure is also ruled out by an out-of-plane component of dipole moment less than 0.1 D and by inertial considerations. The curvature of the plot of rotational constants *versus* v leads to the mixed quartic–quadratic potential function for the out-of-plane vibration:

$$V(X) = G(X^4 + 7.0\,X^2)$$

where the scale factor G awaits far-i.r. measurements before it can be evaluated.

Five-membered Rings.—Planar five-membered rings containing one or more nitrogen atoms continue to receive detailed attention.

Nuclear quadrupole coupling in pyrazole (33) has been discussed in three

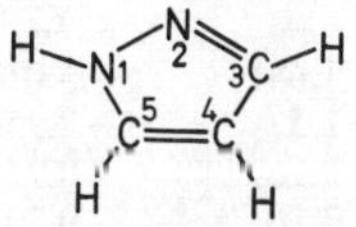

(33)

papers. In the first [136] of two papers from the Monash group, a general expression for the matrix elements of a rotating molecule containing two quadrupolar nuclei is derived using the method described by Curl and Kinsey [137] for matrix elements of Hamiltonians written as spherical tensor products. The theory is then applied to the analysis of the nuclear quadrupole hyperfine structure due to the two ^{14}N nuclei in pyrazole. $\chi_{bb}(1)$ and $\chi_{bb}(2)$ come directly from the $1_{11} \leftarrow 0_{00}$ transition whereas $\chi_{aa}(1)$ and $\chi_{aa}(2)$ are obtained from the $1_{01} \leftarrow 0_{00}$ transition. Of particular help in analysing the quadrupole structure was a computer program which, given χ values, calculates transition frequencies and relative intensities for the hyperfine components and simulates complete rotational transitions by assigning a Lorentzian lineshape function to each component. Illustrations comparing observed and simulated hyperfine patterns of several transitions testify to the usefulness of the program.

The same group discuss the similar determination of the nitrogen nuclear quadrupole coupling constants in the principal inertial axis directions of [1-^{2}H]pyrazole and [4-^{2}H]pyrazole in a second paper.[138] With the assumption that the principal axes, x, y, and z, of the field-gradient tensors are unaffected by isotopic substitution, two sets of values for χ_{ab} (referred to the principal inertial axis system of pyrazole) and the principal quadrupole coupling

[136] G. L. Blackman, R. D. Brown, and F. R. Burden, *J. Mol. Spectroscopy*, 1970, **36**, 528.
[137] R. F. Curl and J. L. Kinsey, *J. Chem. Phys.*, 1961, **35**, 1758.
[138] G. L. Blackman, R. D. Brown, F. R. Burden, and A. Mishra, *J. Mol. Structure*, 1971, **9**, 465.

constants, χ_{xx}, χ_{yy}, and χ_{zz}, are obtained by separate combination of the observed coupling constants of [1-^{2}H]- and [4-^{2}H]-pyrazoles with those of the parent species. Although the agreement between the two sets is good, the experimental results are shown to be in only poor agreement with the results both of *ab initio* and CNDO/2 calculations, particularly for N(1), which has a hydrogen atom attached.

Also published in 1971 is a similar study of quadrupole coupling in pyrazole and [4-^{2}H]pyrazole by Posdeev *et al.*,[139] who likewise derive the complete nuclear quadrupole coupling tensor referred to the principal inertial axis system of the parent species. The agreement between the two groups is quite good (see Table 9).

Table 9 *Nuclear quadrupole coupling tensor of pyrazole*/MHz

	χ_{aa}	χ_{bb}	χ_{cc}	χ_{ab}	*Ref.*
N(1)	1.377	1.641	−3.018	−0.80	*a*
	1.38	1.57	−2.95	−0.634	*b*
N(2)	−3.961	3.167	0.794	−2.00	*a*
	−3.94	3.16	0.79	−2.17	*b*

[a] G. L. Blackman, R. D. Brown, F. R. Burden, and A. Mishra, *J. Mol. Structure*, 1971, **9**, 465; [b] N. M. Posdeev, R. Nasibullin, R. G. Latypova, V. G. Vinokurov, and N. D. Konevskaya, *Optics and Spectroscopy*, 1971, **30**, 32.

Brown *et al.*[140] report also a preliminary investigation of 1,2,4-triazole, for which the possibility of the two tautomeric forms (34) and (35) exists. They conclude that the observed spectrum is due to (35), from the many μ_a and μ_b transitions. Structure (34) is ruled out as its symmetry admits of only one component of dipole moment. That no lines attributable to (34) are observed supports the conclusion of crystallographic work at −155 °C that (35) is the more stable. The measured dipole moment components of (35) are: μ_a = 0.82 and μ_b = 2.59 D.

Two planar five-membered rings containing nitrogen and sulphur atoms, thiazole (36)[141] and 1,3,4-thiadiazole (37),[142] are the subject of further

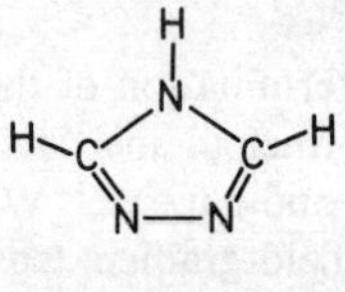

(34) 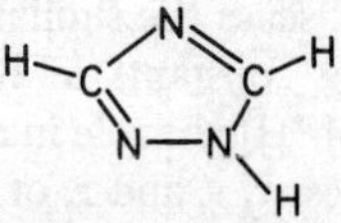 (35)

[139] N. M. Posdeev, R. S. Nasibullin, R. G. Latypova, V. G. Vinokurov, and N. D. Konevskaya, *Optics and Spectroscopy*, 1971, **30**, 32.
[140] K. Bolton, R. D. Brown, F. R. Burden, and A. Mishra, *Chem. Comm.*, 1971, 873.
[141] L. Nygaard, E. Asmussen, J. H. Høg, R. C. Maheshwari, C. H. Nielsen, I. B. Petersen, J. Rastrup-Andersen, and G. O. Sørensen, *J. Mol. Structure*, 1971, **8**, 225.
[142] L. Nygaard, R. L. Hansen, and G. O. Sørensen, *J. Mol. Structure*, 1971, **9**, 163.

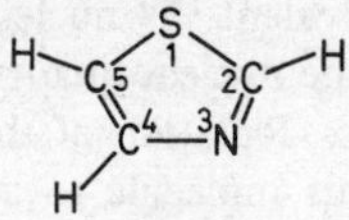

(36) 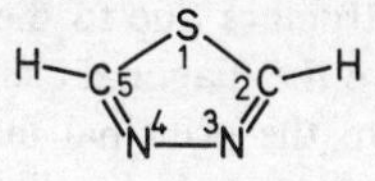 (37)

meticulous studies by the Copenhagen group. The paper on thiazole reports the results of a long and thorough investigation in which the rotational constants of the parent species and each of the eight monosubstituted isotopic species have been derived. In addition, the ^{14}N nuclear quadrupole coupling constants for the [4-^{2}H] and each of the three ^{13}C species, as well as the parent, are obtained. The rotation of the principal inertial axes by 4.46° between the [4-^{2}H] and parent species allows $\chi_{ab} = 2.8 \pm 0.2$ MHz and the elements of the principal quadrupole coupling tensor to be derived. Confirmation of χ_{ab} is yielded by use of [4-^{13}C] species values in combination with those of the parent.

The complete r_s-structure of planar thiazole is considered unreliable because four r_s-co-ordinates are very small and presumably underestimated. Nevertheless, an essentially complete r_s-structure is possible through use of first- and cross-moment conditions. The offending small b-co-ordinate of sulphur is easily settled using $\sum_i m_i b_i = 0$. However, as only two equations, $\sum_i m_i a_i = 0$ and $\sum_i m_i a_i b_i = 0$, are available for the remaining three small a-co-ordinates, a reasonable assumption is necessary for the a-co-ordinate of H(2). The criterion of reasonableness is that when the assumed value is used in the two equations, co-ordinates are obtained which produce distances r[C(2)—H(2)] and r[C(5)—H(5)] and angles $\widehat{\text{NC(2)H(2)}}$ and $\widehat{\text{SC(2)H(2)}}$ comparable with those of similar heterocyclic molecules. The reported structure (Table 10) is closely similar to the hypothetical thiazole structure obtained by adding half a molecule of thiophen to half a molecule of 1,3,4-thiadiazole, the halves being those produced by the vertical symmetry plane in each molecule.

1,3,4-thiadiazole (37) has been reinvestigated [142] with a view to refinement of the spectral constants for the parent isotopic species. The nuclear quadru-

Table 10 *Internuclear distances and bond angles of thiazole*

Distances/Å		*Angles*/°	
S—C(2)	1.7239 ±0.0009	C(5)SC(2)	89.33 ±0.03
C(2)—N	1.3042 ±0.0011	SC(2)N	115.18 ±0.01
N—C(4)	1.3721 ±0.0002	C(2)NC(4)	110.12 ±0.02
C(4)—C(5)	1.3670 ±0.0004	NC(4)C(5)	115.81 ±0.01
C(5)—S	1.7130 ±0.0003	C(4)C(5)S	109.57 ±0.01
C(2)—H(2)	1.0767 ±0.0018	SC(2)H(2)	121.26 ±0.5
C(4)—H(4)	1.0798 ±0.0001	NC(2)H(2)	123.56 ±0.5
C(5)—H(5)	1.0765 ±0.0002	NC(4)H(4)	119.35 ±0.01
		C(5)C(4)H(4)	124.84 ±0.01
		C(4)C(5)H(5)	129.03 ±0.03
		SC(5)H(5)	121.40 ±0.03

pole hyperfine structure due to the two equivalent ^{14}N nuclei is analysed in first order to give the diagonal elements of the nuclear quadrupole coupling tensor referred to the principal inertial axes. Rotation of the inertial axes on isotopic substitution is insufficient in this molecule to allow χ_{ab} to be determined. Analysis of line-centre frequencies for the rotational constants and the four τ values (which in Dowling's approximation of the Kivelson–Wilson first-order treatment for a planar molecule determine the centrifugal distortion corrections) is hampered by ill-conditioning, a problem overcome by use of four linear combinations of the τ values, chosen by the method proposed by Lees [143] so as to be uncorrelated with each other. Comparison of the centrifugal distortion constants so determined with those calculated from a force field based on i.r. frequencies reveals discrepancies serious enough to suggest that the assignment of the fundamentals of 1,3,4-thiadiazole might need revision. An essentially r_s-structure of the molecule recalculated using the refined rotational constants of the parent species with earlier isotopic data is little changed from the previous microwave structure and is in excellent agreement with a recent electron-diffraction structure.

The closely related five-membered cyclic ketones, cyclopent-3-en-1-one (38), cyclopent-2-en-1-one (39), and vinylene carbonate (40) have received some attention. Two groups of workers, Norris, Benson, and Flygare,[144] and Bevan and Legon,[145] report cyclopent-3-en-1-one (38) to have a planar

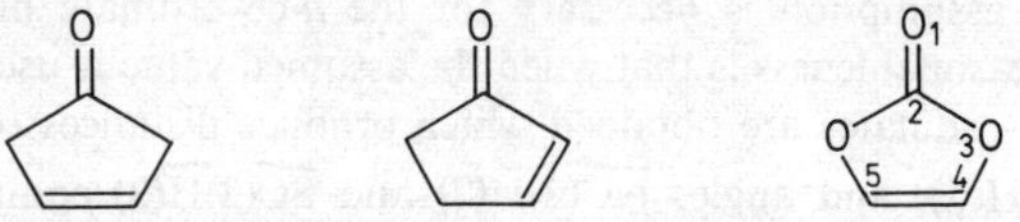

(38) (39) (40)

ring structure. Both use similar arguments. First, the quantity $I_c - I_a - I_b = -6.43$ a.m.u. Å^2 can be accounted for solely by four out-of-plane methylenic hydrogen atoms, and so any structure that does not have a planar, or a very slightly non-planar, ring is excluded. Secondly, analysis of the strong satellite series accompanying each transition shows that the rotational constants vary smoothly with the quantum number of the out-of-plane ring vibration to which the satellite series belongs. The former authors [144] also measure the molecular *g* values, the magnetic susceptibility anisotropies, and the molecular quadrupole moments of cyclopent-3-en-1-one and cyclopent-2-en-1-one. They conclude from their results that neither molecule exhibits a ring current due to delocalized electrons. In the course of the reinvestigation of the spectrum of cyclopent-2-en-1-one for the purpose of making magnetic measurements, these authors comment that experimental moments of

[143] R. M. Lees, *J. Mol. Spectroscopy*, 1970, **33**, 124.
[144] C. L. Norris, R. C. Benson, and W. H. Flygare, *Chem. Phys. Letters*, 1971, **10**, 75.
[145] J. W. Bevan and A. C. Legon, *Chem. Comm.*, 1971, 1136.

inertia previously published by Chadwick, Legon, and Millen [146] do not predict rotational transitions at all accurately. In fact, it appears that the discrepancy arises because Norris, Benson, and Flygare use the conversion factor 505 531 MHz a.m.u. Å^2 based on the ^{16}O = 16.0000 atomic mass scale to obtain rotational constants from the moments of inertia published in ref. 146. The currently accepted value of 505 376 MHz a.m.u. Å^2 based on ^{12}C = 12.0000 gives excellent agreement between the rotational constants obtained by the two groups, as with the case of cyclopent-3-en-1-one.

The r_s-structure for the planar heavy-atom skeleton of vinylene carbonate (40) has been determined by White and Boggs [147] in a study remarkable for the fact that the spectrum of each of the two different singly substituted ^{13}C species, and the two different singly substituted ^{18}O species, is analysed in natural abundance. The *a*-co-ordinate of the ring-oxygen atoms is very small. Hence ambiguity about its sign as well as the usual problem with the uncertain magnitude of small r_s-co-ordinates exists. Appeal to the first-moment condition $\Sigma m_i a_i = 0$ cannot be made, for the hydrogen-atom co-ordinates are not determined in the current study. Instead, the *a* and *b* hydrogen co-ordinates and the ring-oxygen *a*-co-ordinate are simultaneously generated by using all other r_s-co-ordinates with the multiplicity of second-moment conditions. The structure of the heavy-atom skeleton obtained from the O(2) *a*-co-ordinate so generated and the r_s-co-ordinates of the other atoms is given in Table 11. The notable points about this structure are the

Table 11 *Internuclear distances and bond angles of vinylene carbonate*

Distances/Å		*Angles*	
C(4)—C(5)	1.331 ±0.003	C(5)C(4)O(3)	108°40′ ±9′
C(4)—O(3)	1.385 ±0.012	C(4)O(3)C(2)	106°56′ ±24′
C(2)—O(3)	1.364 ±0.006	O(3)C(2)O(1)	108°48′ ±40′
C(2)—O(1)	1.1908 ±0.0006		

rather short C=O length, 1.1908 Å compared with values normally near to 1.21 Å in aldehydes and ketones or 1.23 Å for carboxylic acids and esters, and the nearly pentagonal behaviour.

The rotational spectrum of the symmetric top (C_{5v}) molecule cyclopentadienylthallium in various vibrationally excited states has been closely scrutinized by Whittle, Cox, and Roberts.[148] In particular, they noted an anomalous variation of the centrifugal distortion constant D_{JK} among the vibrational states $\nu_{10} = 1$, $\nu_4 = 1$, $\nu_4+\nu_{10}$, $\nu_4 = 2$, and $\nu_4 = 3$. The vibrations ν_4 (A_1) and ν_{10} (E_1) are respectively the thallium–ring stretch and the thallium–ring tilt. The D_{JK} values, along with other vibration–rotation constants, were

[146] D. Chadwick, A. C. Legon, and D. J. Millen, *Chem. Comm.*, 1969, 1130.
[147] W. F. White and J. E. Boggs, *J. Chem. Phys.*, 1971, **54**, 4714.
[148] C. Roberts, A. P. Cox, and M. J. Whittle, *J. Mol. Spectroscopy*, 1970, **35**, 476.

derived from observed spectra for states involving ν_{10} using a form of the Grenier-Besson and Amat [49] expression modified for C_{5v} molecules. In a theoretical development, Whittle, Cox, and Roberts [149, 150] have now accounted for the erratic behaviour of D_{JK} quantitatively in terms of a second-order Coriolis interaction between the ν_4 (A_1) and ν_{10} (E_1) states. Also, anomalies in the constants B and η_{tJ} are explained when the effect of the accompanying first-order Coriolis interaction is taken into account. It is interesting that their explanation demands that ν_{10} (E_1) should be at a lower frequency than ν_4 (A_1), contrary to the indications from admittedly unreliable microwave intensity measurements. In the later paper,[150] ν_4 (A_1) is reported to be observed at 185 cm^{-1} in the i.r. and Raman spectra, while ν_{10} (E_1) is assigned at 117 cm^{-1}. Confidence in the theory is thus rewarded.

Six-membered Rings.—The large value -0.9 ± 0.5 a.m.u. Å^2 for the inertia defect of nitrobenzene determined some years ago [151] is not unambiguous evidence of molecular planarity. The ambiguity is removed and planarity established by a recent reinvestigation [152] of the microwave spectrum. The assigned spectra in the vibrational ground state and each of the first three excited states of the nitro-group torsional motion exhibit relative intensity variations consistent with statistical weight factors of 10 and 6 for the *ee*,*eo* and *oo*,*oe* rotational levels, respectively. Such weight factors result if two pairs of equivalent hydrogen atoms are exchanged by a two-fold rotation about the symmetry axis of the planar molecule and the barrier to internal rotation of the nitro-group is high. That the latter condition is satisfied is indicated by smooth, almost linear variation of the rotational constants with the torsional quantum number and by the lack of splitting in the torsional satellites. The accurate value of A required for satisfactory determination of the inertia defect necessitates the inclusion of high-J, Q-branch transitions and therefore a centrifugal distortion treatment to determine the rotational constants. The inertia defects, Δ, for the ground and excited torsional states (in a.m.u. Å^2):

	$v = 0$	$v = 1$	$v = 2$	$v = 3$
Δ	-0.4811 ± 0.0003,	-1.8626 ± 0.006,	-3.1858 ± 0.001,	-4.4704 ± 0.0011

are extrapolated to the hypothetical torsionless state to correct for the effects of the large amplitude torsional motion and give the satisfactory value $\Delta_{cor} = +0.249$ a.m.u. Å^2 expected of a planar molecule with only small vibrational amplitudes. From the torsional contribution to Δ, if the motion is assumed harmonic, the torsional frequency of the nitro-group is estimated as 49 cm^{-1}, a value in close agreement with 50 ± 15 cm^{-1} from satellite-intensity measurements. The simple two-fold potential $V = \frac{1}{2}V_0$

[149] M. J. Whittle, A. P. Cox, and C. Roberts, *Chem. Phys. Letters*, 1971, **9**, 42.
[150] M. J. Whittle, A. P. Cox, and C. Roberts, *J. Mol. Spectroscopy*, 1971, **40**, 580.
[151] C. E. Reinert, *Z. Naturforsch.*, 1960, **15a**, 85.
[152] J. H. Høg, L. Nygaard, and G. O. Sørensen, *J. Mol. Structure*, 1970, **7**, 111.

$(1-\cos 2\theta)$ then leads to the barrier to internal rotation of the nitro-group, $V_0 = 1000 \pm 500$ cm^{-1}, which is indeed quite high.

Another monosubstituted benzene, benzonitrile, has also been reinvestigated.[153] The complete r_s-structure previously determined [154] had suggested a ring distortion at the 1-position in which the C(1)—C(2) and C(1)—C(6) bonds were shortened and the angle $\widehat{C(6)C(1)C(2)}$ increased. Unfortunately, the experimental uncertainty was too large to establish definitely this interesting distortion, unlike the case of fluorobenzene.[155] A value of the rotational constant *A*, poorly determined from only *a*-type *R*-branch transitions, was responsible for the uncertainty and in the refined study inclusion of high-*J*, *Q*-branch transitions in a least-squares centrifugal distortion treatment provides the remedy. A two-parameter treatment suffices to determine the small centrifugal distortion corrections and to give the rotational constant *A* with greatly improved accuracy. For the isotopically substituted species, where less transitions are available, one of the two centrifugal distortion parameters is indeterminate and its value is assumed unchanged from the parent species. The r_s structure recalculated from the refined constants still shows the distortion previously noted, although to a slightly smaller extent (Table 12). Some estimates of the order of magnitude of the difference

Table 12 *Bond lengths and angles of benzonitrile*

r_s *Bond lengths*/Å	*Angles*/°
C(1)—C(2) = 1.3876 ± 0.0005	C(6)C(1)C(2) = 121.82 ± 0.05
C(2)—C(3) = 1.3956 ± 0.0004	C(1)C(2)C(3) = 119.00 ± 0.04
C(3)—C(4) = 1.3974 ± 0.0004	C(2)C(3)C(4) = 120.06 ± 0.03
C(1)—C(7) = 1.4509 ± 0.0006	C(3)C(4)C(5) = 120.05 ± 0.03
C(7)—N = 1.1581 ± 0.0002	C(3)C(2)H(2) = 120.36 ± 0.05
C(2)—H(2) = 1.0803 ± 0.0006	C(2)C(3)H(3) = 120.01 ± 0.03
C(3)—H(3) = 1.0822 ± 0.0003	
C(4)—H(4) = 1.0796 ± 0.0004	

between the r_s and r_e co-ordinates lead the authors to the conclusion that the structural distortion is real and not a manifestation of zero-point vibrational effects alone. The errors quoted are those transmitted from the standard deviations of the rotational constants.

Information about the carbon–iodine bond of iodobenzene is reported [156] from an analysis of the ^{127}I nuclear quadrupole coupling. It proved necessary to allow for some rather small second-order effects in order to calculate precise values of the diagonal elements χ_{aa}, χ_{bb}, and χ_{cc} and line-centre fre-

[153] J. Casado, L. Nygaard, and G. O. Sørensen, *J. Mol. Structure*, 1971, **8**, 211.
[154] B. Bak, D. Christensen, W. B. Dixon, L. Hansen-Nygaard, and J. Rastrup-Andersen, *J. Chem. Phys.*, 1962, **37**, 2027.
[155] L. Nygaard, I. Bojesen, T. Pedersen, and J. Rastrup-Andersen, *J. Mol. Structure*, 1968, **2**, 209.
[156] A. M. Mirri and W. Caminati, *Chem. Phys. Letters*, 1971, **8**, 409.

quencies. The asymmetry parameter $\eta = (\chi_{bb} - \chi_{cc})/\chi_{aa} = -0.0319 \pm 0.0025$ indicates that the C—I bond has an essentially cylindrically symmetric charge distribution about the bond direction (*a*-axis) and that the involvement of iodine in double-bonding with the ring is very small. The difference in the number of electrons in the $3p_x$- and $3p_y$-orbitals ($y = b$, $x = c$) given by $\frac{2}{3}(\chi_{cc} - \chi_{bb})/eQq_{n10}$ means 1.75% π character for $3p_x$, the orbital perpendicular to the plane of the ring, assuming that the $3p_y$-orbital in the ring plane cannot be involved in π-bonding.

Internal rotation in the phenol molecule has been subjected to another investigation.[157] This time Gunthard *et al.* apply their method of infinite matrix diagonalization for the rotation–internal rotation problem of a semi-rigid model consisting of a rigid frame of C_s symmetry and a rigid internal rotor of C_{2v} symmetry. The atoms H,O, C(1), C(4), and H(4) [see (41)] are

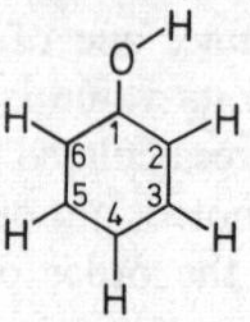

(41)

taken to constitute the rigid C_s frame while C(2), C(3), C(5), C(6), and their accompanying hydrogens define the C_{2v} internal rotor. Many low-*J* doublets associated with the torsional ground state of the molecule as well as the fundamental and hot bands of the O—H torsion are included in the rotation–internal rotation analysis. The essential results of the treatment are a barrier to internal rotation, V_2, of 1207.1 cm^{-1}, confirmation of the assignment of the O—H torsion in the far-i.r. spectrum, and the indication that the principal axis of the frame is parallel to the internal rotation axis within $\pm 0.18°$. The barrier compares with the value 1175 ± 20 cm^{-1} obtained in an analysis in which the OH group is treated as a symmetric internal rotor.[158] The internal rotation splittings in the $v = 1$ torsional state, as yet unmeasured, are predicted.

2-Fluoropyridine has been studied by two groups whose results agree well and are published jointly in a single paper.[159] Structural information is available from the principal moments of inertia of the most abundant isotopic species only with considerable assumption. Combined with the structure of pyridine itself from the microwave study,[160] these moments of inertia give r(C—F) = 1.300 Å and $\widehat{\text{FCN}} = 117.4°$. However, the C—F bond is so short

[157] E. Mathier, D. Welti, A. Bauder, and Hs. H. Gunthard, *J. Mol. Spectroscopy*, 1971, **37**, 63.
[158] T. Pedersen, N. W. Larsen, and L. Nygaard, *J. Mol. Structure*, 1969, **4**, 59.
[159] S. D. Sharma, S. Doraiswamy, H. Legell, H. Mäder, and D. Sutter, *Z. Naturforsch.*, 1971, **26a**, 1342.
[160] B. Bak, L. Nygaard, and J. Rastrup-Andersen, *J. Mol. Spectroscopy*, 1958, **2**, 361.

compared with, for example, 1.350 Å in fluorobenzene [155] as to indicate that the assumption of an undistorted pyridine ring after fluorination is probably invalid. The measured dipole moment components, $\mu_a = 2.78 \pm 0.05$, $\mu_b = 1.84 \pm 0.05$, and $\mu_{total} = 3.33 \pm 0.08$ D are to be compared with the value $\mu_{total} = 3.22$ D from an INDO–MO calculation. The observed ^{14}N nuclear quadrupole coupling constants are transformed from the 2-fluoropyridine principal inertial axis system to that of pyridine (with z as the pyridine symmetry axis) for the purposes of comparison, it being assumed that the off-diagonal term χ_{xy} introduced by fluorine substitution in pyridine is negligible. Then, if the nitrogen atom is assumed to be sp^2 hybridized and one of the sp^2-orbitals is occupied by the lone pair of electrons, the transformed coupling constants may be used to calculate [161] the occupation numbers $a = 1.29$ for the p_x-orbital (perpendicular to the plane of the ring) and $b = 1.36$ for each of the two sp^2 hybrid orbitals involved in σ-bonds with adjacent carbon atoms. These quantities appear to be little changed from pyridine itself, for which the appropriate numbers are $a = 1.23$ and $b = 1.37$. The molecular Zeeman effect in the rotational spectrum of 2-fluoropyridine has been observed by one author,[162] who publishes theoretical expressions for molecular g values and susceptibilities more complete than those hitherto used. The same author is also a participant in some discussion of the theory of the Zeeman effect.[163] Rotational constants for the related 2-cyanopyridine [164] are now available.

2-Pyrone (42) and 4-pyrone (43), although formally heteroaromatic, should be treated as non-aromatic on the basis of magnetic criteria, according to Flygare *et al.*[165] These authors derive the out-of-plane minus the average in-plane molecular magnetic susceptibilities, $\Delta\chi = \chi_{cc} - \frac{1}{2}(\chi_{aa} + \chi_{bb}) = -24.8 \pm 1.3$ and $-22.9 \pm 1.7 \times 10^{-6}$ erg G^{-2} mol^{-1} for (42) and (43), respectively, from magnetic susceptibility anisotropies obtained from the Zeeman effect. They establish for molecules where there is no possibility of aromaticity

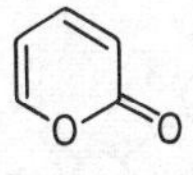

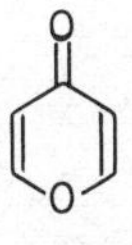

(42) (43)

[161] E. A. C. Lucken, *Trans. Faraday Soc.*, 1961, **57**, 729.
[162] D. Sutter, *Z. Naturforsch.*, 1971, **26a**, 1644.
[163] B. J. Howard and R. E. Moss, *Z. Naturforsch.*, 1970, **25a**, 2004; D. Sutter, A. Guarnieri, and H. Dreizler, *ibid.*, p. 2005.
[164] S. Doraiswamy and S. D. Sharma, *Current Sci.*, 1971, **40**, 398.
[165] R. C. Benson, C. L. Norris, W. H. Flygare, and B. Beak, *J. Amer. Chem. Soc.*, 1971, **93**, 5591.

that $\Delta\chi$ can be reproduced by the sum of $\Delta\chi$ values assigned to the groups which constitute the molecule. The group values are independent of their molecular environment, as evidenced by the good agreement between $\Delta\chi_{calc}$ and $\Delta\chi_{expt}$ for a number of molecules. For furan and benzene, however, the $\Delta\chi_{calc}$ values of -15.6 ± 3.1 and $-26.4 \pm 2.4 \times 10^{-6}$ erg G^{-2} mol^{-1}, calculated from the sum of group $\Delta\chi$ values, are very different from the $\Delta\chi_{expt}$ of -38.7 ± 0.5 and -59.7×10^{-6} erg G^{-2} mol^{-1}, respectively. This exhaltation of the experimental over calculated values is attributed to electron delocalization in the known aromatic molecules. On the other hand, the value $\Delta\chi_{calc} = -26.5 \pm 4.2 \times 10^{-6}$ erg G^{-2} mol^{-1} for each of (42) and (43) is in such good agreement with the experimental values as to dictate that these molecules be classified as non-aromatic. The authors discuss the advantages of this criterion of aromaticity over others currently in use. Of course, this elegant test for aromaticity requires knowledge of the magnetic susceptibilities of a large number of molecules through the measurement of their Zeeman effect. Flygare's group have made the dominant contribution to this field and, fittingly, Flygare and Benson [166] have recently surveyed the theoretical and experimental literature on the subject.

Fused Rings.—The detailed study of the spectra of large cage-like and fused-ring molecules appears to be an area of expanding interest.

The microwave spectra of the symmetric-top molecules 1-fluoro- and 1-chloro-bicyclo[2,2,2]octane (44) have been investigated in some detail by Hirota.[167] In each of the two compounds, rotational transitions are accompanied by a prominent satellite series associated with the torsional motion in which the molecule twists about the symmetry axis. The torsional angle φ is defined as the dihedral angle between the $C_1C_2C_3$ and the $C_2C_3C_4$ planes. When $\varphi = 0$ (C_{3v} structure) the C_2 and C_3 hydrogen atoms eclipse

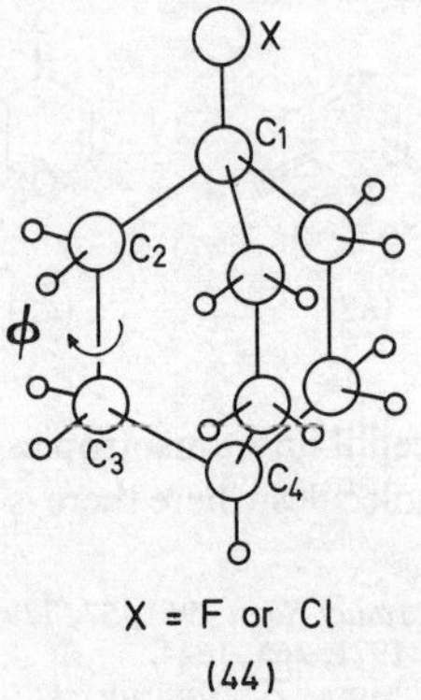

X = F or Cl

(44)

[166] W. H. Flygare and R. C. Benson, *Mol. Phys.*, 1971, **20**, 225.
[167] E. Hirota, *J. Mol. Spectroscopy*, 1971, **38**, 367.

each other, whereas the X and H atoms on C_1 and C_4, respectively, stagger with the C_2 and C_3 hydrogens. The strain inherent in this $\varphi = 0$ situation can be relieved if the molecule is twisted about the symmetry axis, but the relief will eventually be counterbalanced by the extra strain imposed through the necessary eclipsing of the C_2 and C_3 hydrogen atoms by X and H_4, respectively, and the decreased carbon valence angles. A double-minimum potential function with a small maximum at $\varphi = 0$ might therefore be expected for the torsional motion.

Such a double-minimum potential function is betrayed by the unusual behaviour of the satellites assigned to successive quanta of the torsional motion in the symmetric rotor transition (no *K*-splitting evident) of 1-fluoro-bicyclo[2,2,2]octane shown in Figure 2. The $v = 0,1$, and 2 transitions are

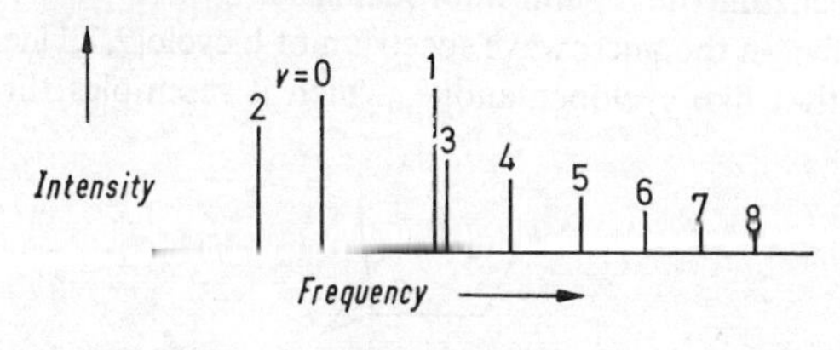

Figure 2

evidently just beneath and just above the top of the barrier and are consequently strongly perturbed. From $v = 3$ onwards, when the vibrational states are increasingly further above the maximum, the perturbing effect rapidly decreases and normal behaviour settles in. The observed variation of the rotational constant B_v with the torsional state v is related to a potential-energy function for the motion by assuming a dynamical model which reproduces the experimental behaviour. In fact, the variation of B_v does not uniquely define the torsional motion and further information, in the form of relative intensities of the vibrational satellites, is necessary. The assumed model is basically one in which C_3 symmetry is maintained during the torsional motion, all bond lengths are independent of φ, all r(C—C) are equal, and the three $\widehat{C_1C_2C_3}$ and the three $\widehat{C_2C_3C_4}$ angles are equal. Then, with this model and a potential function of the form

$$V(\varphi) = -V_2\varphi^2 + V_4\varphi^4 + V_6\varphi^6$$

the potential constants V_2, V_4, and V_6, the barrier height V_0, and the equilibrium value of φ, φ_0, which best reproduce the observables for the 1-fluoro- and 1-chloro-compounds may be obtained; these are given in Table 13. There is a large uncertainty in V_2, V_4, and V_6 due to the inherently inaccurate relative-intensity measurements used to fix uniquely the potential function. V_0 and φ_0 are more accurately determined and show that replacement of F by Cl does not sensibly change the torsional motion of the cage.

It is interesting that the rotational constants for the four different mono-

Table 13 *Potential constants for halogenobicyclo*[2,2,2]*octanes*

	1-*fluorobicyclo*[2,2,2]*octane*	1-*chlorobicyclo*[2,2,2]*octane*
V_2	1477 cm^{-1}	1082 cm^{-1}
V_4	6243 cm^{-1}	5011 cm^{-1}
V_6	22 337 cm^{-1}	14 966 cm^{-1}
V_0	191 ±52 cal mol^{-1}	134 ±67 cal mol^{-1}
ϕ_0	16.4 ±2.6°	16.2 ±4.0°

substituted ^{13}C species suggest that the bicyclo[2,2,2]octane ring is significantly deformed on substitution of F for H at the 1-position. This deformation, consisting of a shortening of the C_1—C_2 bonds and an increase in the bridgehead (C_1) angle, should be compared with the rather similar distortions discussed for benzonitrile [153] and fluorobenzene.[155]

The Stark effect in the microwave spectrum of bicyclo[2,1,1]hexan-2-one [168] (45) indicates that like cyclopentanone, which it resembles through its two

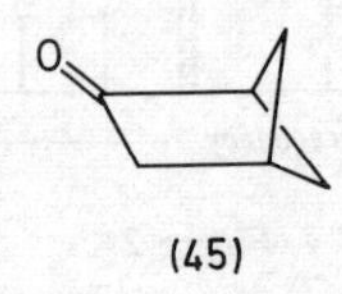

(45)

five-membered rings, it has an anomalously high dipole moment compared with other ketones (see Table 14). The ground-state rotational constants

Table 14 *Dipole moments of various ketones*

Molecule	*Dipole moment*/D
Bicyclo[2,1,1]hexan-2-one	3.35 ±0.02
Cyclopentanone	3.25 ±0.02
Cyclohexanone	2.87 ±0.04
Cyclobutanone	2.89 ±0.03
Cyclopropanone	2.67 ±0.10
Acetone	2.9
trans-Butan-2-one	2.775 ±0.015

of (45) are consistent with a plane of symmetry. The smooth increase of the quantity $(I_a+I_b-I_c)/2$ with the vibrational quantum number for a low-lying vibrational satellite series indicates that the motion of the atoms is perpendicular to the symmetry plane and that the symmetry position represents a potential minimum. Relative-intensity measurements among the satellites establish, from the divergent separation of adjacent vibrational levels, that the walls of the potential function are somewhat steeper than harmonic.

As part of a continuing study of carbaboranes, Beaudet [169, 170] has published

168 D. Coffey, jun. and T. E. Hooker, *J. Mol. Spectroscopy*, 1971, **40**, 158.
169 G. L. McKown and R. A. Beaudet, *Inorg. Chem.*, 1971, **10**, 1350.
170 C. S. Cheung and R. A. Beaudet, *Inorg. Chem.*, 1971, **10**, 1144.

the structures of 2-chloro-1,6-dicarbahexaborane(6) (46) and 2-carbahexaborane(9) (47). These molecules have geometries relatively uncommon in

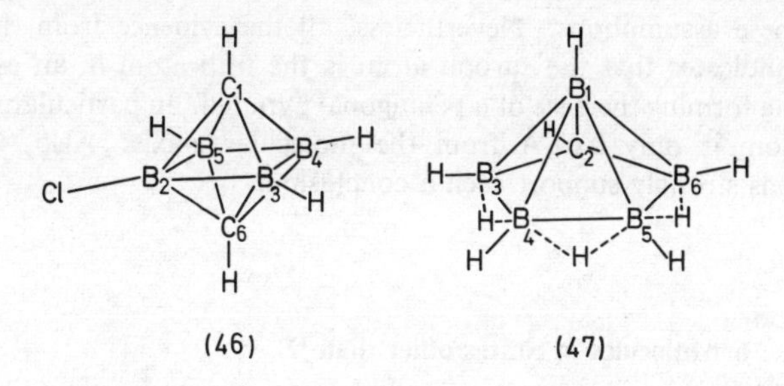

(46) (47)

chemistry. For neither of these molecules is a complete r_s-structure available, as in neither case were ^{13}C or ^{2}H species observed. The skeletal structure for (46) listed below can be derived [169] if C_{2v} symmetry is assumed, if the r_s-co-ordinates of each boron atom and the chlorine atom are used, and if the carbon atom is located according to the conditions

$$I_a + I_b - I_c = 2\sum_i m_i c_i^2$$

and

$$\sum_i m_i a_i = 0$$

where atoms 1, 2, 4, and 6 define the *ac* plane. In order to locate the carbon atom with these equations, some reasonable assumptions about the disposition of the hydrogen atoms in the molecule are made. The derived skeletal structure is given in Table 15. Some minor deviations from other carbaborane

Table 15 *Bond lengths and angles of* 2-*chloro*-1,6-*dicarbahexaborane*(6)

Bond lengths/Å		*Angles*/°	
B(2)—Cl	1.823 ± 0.010	B(2)B(3)B(4)	87.7 ± 0.5
B(2)—B(3)	1.671 ± 0.010	B(3)B(4)B(5)	91.0 ± 0.2
B(3)—B(4)	1.702 ± 0.0005	B(3)B(2)B(5)	93.6 ± 0.5
C(6)—B(2)	1.59 ± 0.04		
C(6)—B(4)	1.61 ± 0.04		
B(5)—C(6)	1.63 ± 0.04		

parameters are discussed. The B—Cl length, essentially equal to the sum of single-bond radii, and the zero value for the asymmetry parameter η, derived from the chlorine nuclear quadrupole coupling constants, are evidence that no π-bonding occurs between the B and Cl atoms.

The B—B bond lengths of (47), calculated from the r_s-co-ordinates of the boron atoms (^{10}B substitution at each position) are (in Å): [170] B(3)—B(4)

= 1.759 ±0.007, B(1)—B(4) = 1.781 ±0.005, B(4)—B(5) = 1.830 ±0.010, and B(1)—B(3) = 1.782 ±0.004. The carbon atom position cannot be calculated from the moments of inertia since assumptions about the hydrogen positions are necessary and it is found that the carbon position is very sensitive to these assumptions. Nevertheless, all the evidence from the boron skeleton indicates that the carbon atom is the fifth atom in an essentially planar ring forming the base of a pentagonal pyramid. In particular, the apex boron atom is only 0.02 Å from the molecular *c*-axis. Also, CNDO/2 calculations strongly support such a conclusion.

6 Molecules in States other than $^1\Sigma$

The spectrum of the formyl free radical, HCO, has been obtained through the use of a continuous flow system in which the radical is produced by the reaction of fluorine atoms with formaldehyde, the reactants being mixed outside the cavity of the e.s.r. spectrometer.[171] The K doubling of the $N = 2$, $K_a = 1$ level leads to a splitting of 3(B—C). Each level is split into two by spin–rotation interaction, and a further doublet splitting due to proton hyperfine interaction leads to a pattern of eight levels characterized by $J = 3/2$, $F = 2,1$ and $J = 5/2$, $F = 3,2$. Each level is split into $(2F+1)$ components in a magnetic field and transitions between the K doublets with selection rules $\Delta F = 0$, $\Delta J = 0$, and $\Delta M_F = \pm 1$ are observed. The value obtained for B—C, 2.84 GHz, is in close agreement with that obtained from an analysis of the electronic spectrum.[172] The spin–rotation parameter, $\varepsilon_{bb} - \varepsilon_{cc}$, is found to be −0.22 GHz and the dipolar hyperfine parameter, B_{bb}–B_{cc}, is 19 MHz, the latter being similar to that obtained from the e.s.r. spectrum of HCO in a carbon monoxide matrix.[173]

Two paramagnetic resonance studies have been made of O_2 ($^1\Delta$). Miller has made measurements for the $J = 3$ level and has also obtained a higher accuracy for the $J = 2$ level than that obtained previously.[174] The combined spectra are analysed to give the rotational constant, the rotation g value ($g_r = -1.234 \pm 0.025 \times 10^{-4}$), and the electronic orbital g value ($g_L = 0.999866 \pm 0.000010$). In another study[175] of O_2 ($^1\Delta$), the ^{17}O hyperfine coupling constant has been determined. A g_L value was also obtained and is in close agreement with that obtained by Miller but, on the other hand, the values found for g_r are not in agreement. A theoretical contribution of interest in this field is that of Brown,[176] who has obtained expressions for rotational energy levels of symmetric rotors in $^2\Sigma$ states.

[171] I. C. Bowater, J. M. Brown, and A. Carrington, *J. Chem. Phys.*, 1971, **54**, 4957.
[172] G. Herzberg and D. A. Ramsay, *Proc. Roy. Soc.*, 1955, **A233**, 34.
[173] F. J. Adrian, E. L. Cochran, and V. A. Bowers, *J. Chem. Phys.*, 1962, **36**, 1661.
[174] T. A. Miller, *J. Chem. Phys.*, 1971, **54**, 330.
[175] C. A. Arrington, A. M. Falick, and R. J. Myers, *J. Chem. Phys.*, 1971, **55**, 909.
[176] J. M. Brown, *Mol. Phys.*, 1971, **20**, 817.

7 Influence of Vibration on Structural Determinations

The problems associated with the effect of vibration on the evaluation of structural parameters continue to receive attention. Kuchitsu [177] has reviewed the position about average bond angles. He emphasizes the need for care in the use of the term and presents a number of definitions with clear physical significance. Simple expressions of the average angles in terms of linear and quadratic mean values of displacements are given. The differences in the various average angles for H_2O, D_2O, CH_4, CD_4, and SO_2 are estimated, and shown to be a small fraction of a degree. Root-mean-square angles for linear molecules (HCN, DCN, CO_2, and CS_2) are also calculated. Kuchitsu and co-workers have made a comparison of microwave and electron-diffraction studies of phosphorus tribromide.[178] Internuclear distances (r_g) and mean vibrational amplitudes have been evaluated from electron-diffraction work. The average structure for the ground vibrational state calculated from the r_g distances, with corrections for vibrational and centrifugal effects, is consistent with the rotational constant B_z calculated from the microwave value for B_0. The experimental values for D_J and mean amplitudes are found to agree with those estimated from quadratic force constants.

Comparison of electron-diffraction and microwave results have also been made for chlorine dioxide [179] and sulphur dioxide.[180] Reduction of the data for chlorine dioxide from both sources, to give zero-point averages, leads to values of r(Cl—O) and r(O—O) distances in excellent agreement from both sources. For larger molecules, rigorous transformations of spectroscopic and diffraction data to an r_z basis are not feasible or, alternatively, would suffer from large uncertainty. In such cases a combination of information from the two sources can be beneficial. Jacob, Thompson, and Bartell [181] have examined this approach for xenon oxide tetrafluoride. Each method has its particular correlations and uncertainties. From the spectroscopic side the main uncertainty is in the position of the xenon atom, which is near the centre of mass and there is correspondingly a correlation between the Xe—O bond distance and $\widehat{OXeF}$ angle obtained in this way. On the other hand, electron diffraction is faced with the problem of resolution of the Xe—O and Xe—F bonds involving similar light atoms attached to a heavy atom, and there is a correlation in this method between these two bond distances. By combining the two sets of data it was found possible to reduce the correlations which occur when either method is used separately.

Winnewisser, Maki, and Johnson [5] have re-examined the effects of vibration on the evaluation of internuclear distances for HCN. New measurements have led to improved values of B_v for a number of vibrational states of $H^{12}C^{14}N$, $D^{12}C^{14}N$, $D^{13}C^{14}N$, and $D^{12}C^{15}N$. Values of r_0 internuclear

[177] K. Kuchitsu, *Bull. Chem. Soc. Japan*, 1971, **44**, 96.
[178] K. Kuchitsu, T. Shibata, A. Yokozeki, and C. Matsumura, *Inorg. Chem.*, 1971, **10**, 2584.
[179] A. H. Clark, *J. Mol. Spectroscopy*, 1971, **7**, 485.
[180] A. H. Clark and B. Beagley, *Trans. Faraday Soc.*, 1971, **67**, 2216.
[181] E. J. Jacob, H. B. Thompson, and L. S. Bartell, *J. Mol. Structure*, 1971, **8**, 383.

distances have been evaluated for all possible pairs of B_0 values. The use of C—D and C—H pairs gives C—H distances consistently lower than the r_e-value by *ca.* 0.003 Å and C—N distances which are consistently higher. The authors go on to use the data to examine the technique of allowing for changes in bond length as isotopic masses are changed. They come to the conclusion that the technique is not yet usable for making structural comparisons because there is no good criterion for assigning bond-shrinkage parameters, their results showing that the resulting internuclear distances are very sensitive to the values of these parameters. Consequently it seems that r_s distances are the most reliable and readily measured quantities for making structural comparisons at the present time.

Helminger, de Lucia, and Gordy [182] have calculated substitution structures for ammonia using four different sets of isotopic data, and find the bond distances obtained to be consistent within 0.0001 Å. They have obtained a more reliable value of B_0 for $^{15}NH_3$, in agreement with a correction of earlier work reported by Krupnov and co-workers, and at the same time have improved the accuracy of B_0 for $^{14}ND_3$. With these improved values, the four substitution structures have been obtained by the various choices of three species from $^{15}NH_3$, $^{15}NH_3$, $^{14}ND_3$, and $^{15}ND_3$.

8 Vibration–Rotation Interaction

Two papers have appeared leading to some simplification of the results of Goldsmith, Amat, and Nielsen [183] on vibration–rotation interactions. Rothman and Clough [184] have made an expansion of the inverse moment of inertia tensor in terms of normal co-ordinates. Using the simplified form of the Darling–Dennison Hamiltonian achieved by Watson,[185] they obtained simplified expressions for the coefficients of vibrational and rotational operators of the expanded Hamiltonian. Chan, Wildardjo, and Parker [186] have moved away from starting with a generalized treatment to consider the particular treatment of the bent XYX molecule itself. They take advantage of the same simplified Hamiltonian and have obtained detailed expressions for all fourth-order coefficients. Strey [187] has discussed the vibration–rotation Hamiltonian for linear molecules. By using commutation relations and sum rules the Hamiltonian is rearranged into a form similar to that given by Watson for non-linear molecules. The Wilson–Howard vibration–rotation Hamiltonian has been obtained in a form in which each term has been related to s-vectors.[188] In the course of the derivation, expressions for Coriolis coupling coefficients are obtained in terms of s-vectors, and the

[182] P. Helminger, F. C. de Lucia, and W. Gordy, *J. Mol. Spectroscopy*, 1971, **39**, 94.
[183] M. Goldsmith, G. Amat, and H. H. Nielsen, *J. Chem. Phys.*, 1956, **24**, 1178.
[184] L. S. Rothman and S. A. Clough, *J. Chem. Phys.*, 1971, **54**, 3246.
[185] J. K. G. Watson, *Mol. Phys.*, 1968, **15**, 479.
[186] M. Y. Chan, L. Wilardjo, and P. M. Parker, *J. Mol. Spectroscopy*, 1971, **40**, 473.
[187] G. Strey, *Z. Naturforsch.*, 1970, **25a**, 1836.
[188] A. M. Walsh-Bakke, *J. Mol. Spectroscopy*, 1971, **40**, 1.

elements of the C matrix of Meal and Polo are obtained in this form as an intermediate step. The coefficients in the expansions of the form

$$I_{\alpha\alpha} = I^e_{\alpha\alpha} + \sum_{\mathbf{s}} a^{\alpha\alpha}_{\mathbf{s}\sigma} Q_{\mathbf{s}\sigma} + \sum_{\mathbf{s}\sigma, \mathbf{s}'\sigma'} A^{\alpha\alpha}_{\mathbf{s}\sigma\, \mathbf{s}'\sigma'} Q_{\mathbf{s}\sigma} Q_{\mathbf{s}'\sigma'}$$

are then obtained in terms of s-vectors and this essentially completes the derivation of the Hamiltonian in terms of s-vectors. The centrifugal distortion coefficients are expressed in terms of the coefficients $a^{\alpha\alpha}_{\mathbf{s}\sigma}$ and $a^{\alpha\beta}_{\mathbf{s}\sigma}$, and so are brought within the s-vector formalism. It thus becomes possible to make the calculation of centrifugal distortion constants a part of the *FG*-matrix calculation.

Further contributions have been made on treating large amplitude vibrational problems. Redding and Hougen [189] have simplified the determination of the Eulerian angles and the large amplitude vibrational co-ordinate associated with an arbitrary configuration of atoms in a non-rigid molecule. The three Eulerian angles and the large amplitude co-ordinate are determined by four equations in four unknowns and a method is presented for reducing the problem to a solution of a single equation in the large amplitude co-ordinate. The vibration–rotation Hamiltonian derived by Hougen, Bunker, and Johns,[190] which allows for large amplitude bending, has been extended to tetratomic quasilinear molecules.[191] A curvilinear quantum-mechanical model for large amplitude bending in triatomic XY_2 molecules of C_{2v} symmetry has been examined.[192] The model has two degrees of freedom: change in the valence angle, and K-type rotation about the z-axis whose moment vanishes in the linear limit case. A comparison is made of two quantum-mechanical treatments of this model, one for rectilinear motion and the other for curvilinear motion. The authors emphasize that it can be a mistake to attribute uneven spacing in vibrational levels entirely to anharmonic terms in the potential function.

Watson [193] has obtained some general quadratic zeta-sum rules. He has developed a method for the evaluation of the sums of squares of Coriolis zeta coupling constants $\sum_k \sum_l [\zeta_{kl}{}^{(\alpha)}]^2$, where each sum extends over vibrations of one species only. These sums are independent of the force field. The procedure employed makes use of the fact that each set of equivalent atoms in a molecule makes an independent additive contribution to the sum. The result is that the procedure parallels a familiar one for obtaining the number of normal vibrations belonging to a given species. The consequence is that the sums themselves are very conveniently obtained in terms of the numbers of vibrations belonging to the two symmetry species and the numbers of equivalent atoms, and expressions in this form have been tabulated.

The dependence of observed rotational constants on vibrational state

[189] R. W. Redding and J. T. Hougen, *J. Mol. Spectroscopy*, 1971, **37**, 366.
[190] J. T. Hougen, P. R. Bunker, and J. W. C. Johns, *J. Mol. Spectroscopy*, 1970, **34**, 136.
[191] K. Sarka, *J. Mol. Spectroscopy*, 1971, **38**, 545.
[192] F. B. Brown and N. G. Charles, *J. Chem. Phys.*, 1971, **55**, 4481.
[193] J. K. G. Watson, *J. Mol. Spectroscopy*, 1971, **39**, 364.

has been further explored. The calculation of the coefficients $\gamma_{r,s}$ in expressions of the form

$$A_v = A - \sum_r \alpha_r^A (v_r + \tfrac{1}{2}) + \sum_{r \leq s}\sum \gamma_{r,s} (v_s + \tfrac{1}{2}) (v_r + \tfrac{1}{2}) + \cdots$$

in terms of molecular force field has been examined.[194] The method used has been to make use of perturbation theory to obtain an effective Hamiltonian, appropriate to rotation in a particular vibrational state. Non-vanishing contributions to the $\gamma_{r,s}$ for each order of perturbation theory are then identified. Understandably, there are a number of such terms but it is an advantage of the perturbation approach that an easy correlation can be made between terms in the final expression and the terms in the Hamiltonian from which they originate. Once the contributions have been identified, the matrix elements are readily obtained in a harmonic oscillator basis set. The method is applied to the torsional dependence of the rotational constant A for glyoxal. It is suggested that one particular interaction is dominant and essentially accounts for the dependence, though it is not possible to assess the magnitude of other contributions to this particular $\gamma_{r,s}$ at the present time. A theoretical examination has also been made of the vibrational dependence of rotational constants for the linear BAB molecule with unequal equilibrium bond lengths.[195] The quantum-mechanical Hamiltonian is obtained in a scheme which allows for large amplitude antisymmetric stretching motion. The dependence of rotational constant on vibrational state is then examined for various heights of the barrier in the double-minimum potential. The plot obtained for B_v again v shows a zig-zag pattern reminiscent of that obtained by Gwinn *et al.*[196] for the case of ring-puckering when there is a double-minimum potential function. A treatment of phosphorus pentafluoride as a non-rigid molecule has been made and the effects of quantum-mechanical tunnelling on the vibronic levels examined.[197] The levels are shown to be split in a characteristic manner and simple results for the expected fine structure and relative intensities of transitions have been obtained.

A new origin of microwave transitions for non-rigid molecules has been found. The selection rules for pure rotational transitions of rigid-rotor molecules are, of course, well known. For example, the rules for the asymmetric rotor have recently been presented [198] in a condensed graphical form. However, a major development is the revelation by Watson [199] and by Fox [200] of a mechanism whereby the non-rigidity of real molecules can lead to transitions forbidden to the rigid rotor. The forbidden spectra in question are allowed by a very small dipole moment, which arises from the small

[194] J. M. Brown, *J. Mol. Spectroscopy*, 1971, **37**, 179.
[195] R. W. Redding, *J. Mol. Spectroscopy*, 1971, **38**, 396.
[196] S. I. Chan, J. Zinn, and W. D. Gwinn, *J. Chem. Phys.*, 1960, **33**, 1643.
[197] B. J. Dalton, *J. Chem. Phys.*, 1971, **54**, 4745.
[198] G. Roussy and J. Demaison, *J. Chim. phys.*, 1970, **67**, 2078.
[199] J. K. G. Watson, *J. Mol. Spectroscopy*, 1971, **40**, 536.
[200] K. Fox, *Phys. Rev. Letters*, 1971, **27**, 233.

displacements in the normal co-ordinates produced by centrifugal distortion. As pointed out by Watson, not only can pure rotational spectra thus be allowed to non-polar molecules but this term can also allow forbidden transitions of polar molecules and can modify intensities of transitions already allowed. Since the effect derives from vibration–rotation interaction the theory applies to any vibrational state, but the main interest of both authors is the forbidden spectra in the ground state of non-polar molecules.

Watson discusses the selection rules, the form of forbidden spectra and also the type of Stark effect allowed to molecules belonging to the point groups D_{3h}, C_{3v}, D_n, D_{2d}, and T_d by the centrifugal distortion dipole moment. Of particular interest is his conclusion that the so-called centrifugal distortion transitions in D_{3h} or C_{3v} molecules would allow determination of the axial rotational constant through the selection rules $\Delta J = 0, \pm 1$ and $\Delta K = 0$. Fox discusses T_d molecules and both he and Watson treat methane in particular. Their calculations show that the strongest rotational transitions would be in the region 110—140 cm^{-1} with integrated absorption coefficients near 10^{-1} cm^{-2} $amagat^{-1}$, which should be detectable. The centrifugal distortion dipole moment for the vibrational ground state of methane is estimated to be 4×10^{-6} and 5.8×10^{-6} D by Fox and Watson, respectively. Excellent agreement therefore exists with the value $(5.38 \pm 0.10) \times 10^{-6}$ D as measured by Ozier [201] from Stark shifts in the ortho–para spectrum of methane in the $J = 2$ rotational state, using a molecular beam–magnetic resonance spectrometer. It should be noted that the Watson–Fox effect differs from that proposed by Mizushima and Venkateswarlu [202] for pure rotational spectra in degenerate vibrational states of non-polar molecules. The latter effect is purely vibrational and generally large, as indicated by the dipole moment of 0.0200 ± 0.0001 D for the degenerate ν_3 state of methane recently observed [203] by non-linear laser absorption spectroscopy.

Some papers dealing with more familiar aspects of centrifugal distortion have appeared. Relationships between centrifugal distortion constants in the I^r ($x = b$, $y = c$, and $z = a$) and the III^r ($x = a$, $y = b$, and $z = c$) representations are given by Typke,[204] who determines the constants in each representation for dimethyl sulphoxide (using Watson's first-order method) and then tests the relationships by transforming from one set to the other. Marstokk and Møllendal [205] apply the Watson first-order centrifugal distortion formula to 25 non-planar molecules and discuss the criteria for good solutions for the constants. In particular, there is high correlation between d_K and d_{WK}, and between d_{JK} and d_{WK}, while for near-prolate rotors, $\kappa < -0.97$, ill-conditioning of the normal equations is a problem.

In connection with the calculation of centrifugal distortion constants from molecular force fields, and *vice versa*, there are three papers. One reports

[201] I. Ozier, *Phys. Rev. Letters*, 1971, **27**, 1329.
[202] M. Mizushima and P. Venkateswarlu, *J. Chem. Phys.*, 1953, **21**, 705.
[203] A. C. Luntz and R. G. Brewer, *J. Chem. Phys.*, 1971, **54**, 3641.
[204] V. Typke, *Z. Naturforsch.*, 1971, **26a**, 175.
[205] K.-M. Marstokk and H. Møllendal, *J. Mol. Structure*, 1971, **8**, 234.

a method [206] whereby the upper limits to the distortion constants τ_{aaaa} and τ_{abab} can be calculated from rotational constants and the lowest fundamental vibrational frequency, whereas another [207] deals with calculation of derivatives of the inertia tensor with respect to normal co-ordinates. In the third, Sørensen [208] gives expressions for derivatives of the distortion constants (τ values) with respect to the elements of the force-constant matrix in the general case of internal co-ordinates with redundancies.

9 Pressure Broadening

Pressure broadening of microwave absorption lines for various molecules arising from collisions with oxygen molecules has been investigated and line-width parameters evaluated.[209] The broadening of ammonia-inversion lines by oxygen was used to obtain the oxygen molecular quadrupole moment. The value (1.6 D Å) is in reasonably good agreement with that found by other methods. This value of the quadrupole moment was used to calculate line broadening in mixtures of oxygen with various substances. Comparison with experiment shows that the calculated values are generally smaller than the experimental values, and it is suggested that higher-order interactions are important in collisions with the oxygen molecule. Two papers have been published on the effect of velocity distribution on pressure broadening of spectral lines.[210, 211] An investigation of the effect of including velocity distribution in the Anderson theory as applied to rotational linewidths of OCS shows that the calculated linewidth is not appreciably affected.[211]

Story, Metchnik, and Parsons have described a microwave spectrometer for observing widths and shifts of absorption lines.[212] Microwave radiation is passed through four waveguide cells in parallel. The cells are arranged in pairs, one member of each pair being a dummy whose function is to balance out the effects of standing waves. A sample of the pure gas under study is introduced into one of the absorption cells while in the other cell the pressure may be varied by the addition of more of the same gas or by a foreign gas. A correction is applied for instrumental shift, due to residual unbalanced standing waves, and values are obtained for the change of centre frequency with increasing pressure; linewidths can also be obtained. Results are reported for the self-broadening of the $(J,K) = (12,12)$ line in the inversion spectrum of NH_3, for the $J = 3 \leftarrow 2$ line for OCS and the $J = 2 \leftarrow 1$ line for CH_3CN. In addition, results are given for the broadening of the NH_3 line by He, CO_2, and CH_3CN, and for the OCS line by He and CH_3CN. No evidence of a shift for either OCS or CH_3CN was found. However, when the ammonia

206 M. R. Aliev, *Optics and Spectroscopy*, 1971, **31**, 20[illegible]
207 V. S. Timoshinin and I. N. Godnev, *Optics and Spectroscopy*, 1971, **31**, 20[illegible]
208 G. O. Sørensen, *J. Mol. Spectroscopy*, 1971, **39**, 533.
209 J. S. Murphy and J. E. Boggs, *J. Chem. Phys.*, 1971, **54**, 2443.
210 M. Mizushima, *J. Quant. Spectroscopy Radiative Transfer*, 1971, **11**, 471.
211 V. Prakash and J. E. Boggs, *J. Chem. Phys.*, 1971, **55**, 1492.
212 I. C. Story, V. I. Metchnik, and R. W. Parsons, *J. Phys. (B)*, 1971, **4**, 593.

line was broadened, a small shift in the centre frequency was observed. In a further paper [213] pressure values of induced shifts of this line are reported and the results compared with theoretically calculated shifts. It is concluded that an important contribution to the shift arises from dipole–induced-dipole dispersion and exchange interactions. It appears that the polarizability of the perturbing molecule plays a dominant role in producing line shifts. The lineshape of the 22 GHz water-vapour absorption line has also been discussed.[214]

10 Double Resonance

Double-resonance experiments are increasingly important, both in the analysis of microwave spectra and in the investigation of molecular collisions. Recently published papers indicate interest in three types of system: those in which microwave radiation is used in combination with radiofrequency or microwave or infrared radiations.

The use of radio frequency–microwave double resonance (RFMDR) as a tool in the analysis of microwave spectra is discussed by Wodarczyk and Wilson.[215] They review quantitatively the expected behaviour of two types of level system. The first is a simple three-level system at and near resonance when a strongly perturbing r.f. source is the pump and a microwave transition is the signal. The second is a four-level system such as is found in near-symmetric rotors where two *K*-type doublets (each with a r.f. doublet splitting) have a microwave separation. When the two *K*-type doublet r.f. splittings are not too different ($\sim$15 MHz or less) there is a frequency range in which the pumping source can be near to resonance with both. For such cases a characteristic RFMDR pattern is observed. In the experimental arrangement described, the strong r.f. pumping source is amplitude modulated by a square-wave, on for one half-cycle and off for the other. If the microwave-signal transition is detected using a narrow-band amplifier tuned to the modulation frequency, only those signal transitions which have a level in common with the pump transition are modulated and therefore observed. The r.f. is imposed on the central septum of a cell purposefully designed to resemble a conventional Stark cell. In order to alleviate some experimental matching problems, the septum is terminated with a pure resistive load equal to the characteristic impedance which the unterminated cell would have. This cell has the advantage that it is easily adapted for conventional Stark spectroscopy.

Several examples of applications of RFMDR are given for symmetric rotor molecules exhibiting either suitable *K*-type doublet splittings or accidental near degeneracies. Among the examples are the determination of resonant

[213] I. C. Story, V. I. Metchnik, and R. W. Parsons, *Phys. Letters* (*A*), 1971, **34**, 59.
[214] T. A. Dillon and H. J. Liebe, *J. Quant. Spectroscopy Radiative Transfer*, 1971, **11**, 1803.
[215] F. J. Wodarczyk and E. B. Wilson, *J. Mol. Spectroscopy*, 1971, **37**, 445.

frequencies for the r.f. transition and the identification of transitions. Particularly elegant in the latter respect is the assignment of vibrationally excited transitions. The r.f. separation of levels involved in a double resonance often changes only a fraction of a MHz on vibrational excitation. Thus, if the pumping source is fixed at a frequency appropriate to the ground state, the microwave frequency can be varied to find a sequence of vibrationally excited transitions. Other lines in the same region are not modulated and therefore not detected because they will not in general have a common level with the r.f. transition.

In the more established field of microwave–microwave double resonance three papers have appeared. Redon and Fourrier [216] have investigated the collisional selection rules which apply among the ground-state inversion doublet levels of ammonia using a method and experimental technique similar to that used previously by Oka.[217] They apply a static electric field to ammonia gas and pump with saturating radiation each of the different M components of a given J,K inversion doublet transition in turn while observing the various M' components of the $J-1,K$ inversion doublet. Oka [217] has already established that the effect of the pumping radiation is transmitted to levels other than those involved in the pump transition by the selection rules $\Delta J = 0, \pm 1, \Delta K = 0, + \leftrightarrow -$. The present authors establish that the additional collisional selection rule $\Delta M = 0, \pm 1$ also holds from observed intensity changes of the M' components of the signal transition while a given M component of the pump transition is irradiated.

A significant development in microwave–microwave double-resonance technique is due to Rudolph *et al.*[218, 219] These authors utilize the fact that the cut-off frequency of the TE_{10} mode of microwave radiation in a rectangular waveguide is determined only by the broad-face dimension, a, whereas the cut-off of the TE_{01} mode, propagated in a plane perpendicular to that of the TE_{10} mode, depends only on the narrow-face dimension, b. Thus, by suitably choosing waveguide component dimensions and polarizing the signal and pump radiations perpendicular to each other, the pump radiation can be very effectively isolated from that of the signal, even when the pump frequency is higher than the signal.[219] The theory of double-resonance lineshapes expected in the case of perpendicularly polarized radiations is given by these authors [218] and the theory compared with experiment.

Microwave–infrared double-resonance experiments present possibilities for studying rates and mechanisms of vibrational-energy transfer in gases. Such possibilities are realized in the case of methyl chloride by Frenkel, Marantz, and Sullivan.[220] They subject the gas to powerful carbon dioxide laser radiation and observe intensity variations among millimetre-wave rotational transitions caused by the laser. Certain selective increases and

[216] M. Redon and M. Fourrier, *Compt. rend.*, 1970, **271**, *B*, 1058.
[217] T. Oka, *J. Chem. Phys.*, 1968, **48**, 4919.
[218] H. D. Rudolph, H. Dreizler, and U. Andresen, *Z. Naturforsch.*, 1971, **26a**, 233.
[219] U. Andresen and H. D. Rudolph, *Z. Naturforsch.*, 1971, **26a**, 320.
[220] L. Frenkel, H. Marantz, and T. Sullivan, *Phys. Rev.* (*A*), 1971, **3**, 1640.

decreases of intensity lead them to the definite conclusion that the $P(26)$ laser line pumps molecules exclusively from the $J = 6, K = 3$ vibrational ground-state level to the $J = 6, K = 4$ level in the $\nu_6 = 1$ vibrationally excited state of $CH_3{}^{35}Cl$. Transmission of the intensity perturbations to rotational transitions of the same K within the same vibrational state is evidence that collisional transfer by the selection rules $\Delta J = 1, \Delta K = 0$ is allowed. Similar evidence of more weakly allowed $\Delta K = 3$ transitions also exists. These conclusions support those of Oka [217] from microwave–microwave double-resonance experiments.

Evidence about vibrational-energy transfer is also available. The authors measure the vibrational relaxation time from the $\nu_6 = 1$ state by observing the time for normalization of the intensity perturbation when the laser is switched off. They find that the relaxation time varies linearly with pressure and with the cell dimensions. In order to explain quantitatively both these results and the way in which rotational transition intensities vary with pressure in the presence of the laser, the authors assume that relaxation from vibrationally excited states occurs only by collision with the cell walls and that relaxation to thermal equilibrium among rotational levels within a vibrational state occurs by intermolecular collision at a rate proportional to the pressure. The observed perturbations of rotational transition intensities in the $\nu_3 = 1$ state requires the additional assumption that vibrational-energy transfer from $\nu_6 = 1$ to $\nu_3 = 1$ occurs. Pressure variation of rotational transition signals in the $\nu_3 = 1$ state of the laser-irradiated gas leads to the conclusion that 1 in 300 collisions results in transfer of vibrational energy from $\nu_6 = 1$ to $\nu_3 = 1$.

Similar, but less extensive, experiments are reported by Lemaire *et al.*[221] They observe rotational transitions in nitrous oxide and methyl bromide while pumping these gases with N_2O and CO_2 laser radiations, respectively. At higher pressures, the observed effects can be explained in both cases by a general heating of the gas. There is some evidence of a selective effect for methyl bromide where the laser pumps $CH_3{}^{79}Br$ molecules from the $J = 7, K = 0$ ground state level to the $J = 6, K = 0$ level of the $\nu_6 = 1$ vibrationally excited state. Although transitions involving these levels were not observed, intensity decreases were noted in the $J = 4 \leftarrow 3, K = 0$ transition for the ground state and increases for the same transition in the $\nu_6 = 1$ and $\nu_3 = 1$ states. Thus the $\Delta J = \pm 1, \Delta K = 0$ selection rules are again at play. Moreover, the authors indicate that there is evidence for a more rapid transfer of population by $\Delta J = \pm 1, \Delta K = 0$ between the levels $J = 7, K = 0$ and $J = 4, K = 0$ than between the former level and $J = 3, K = 0$.

The number of experiments of the type wherein molecules are pumped to vibrationally excited states by i.r. laser radiation is severely limited by the comparative rarity of coincidences between laser transitions and molecular absorptions within the Döppler width of the latter. A neat solution to the

[221] J. Lemaire, J. Houriez, J. Thilbault, and B. Maillard, *J. Phys. (Paris)*, 1971, **32**, 35.

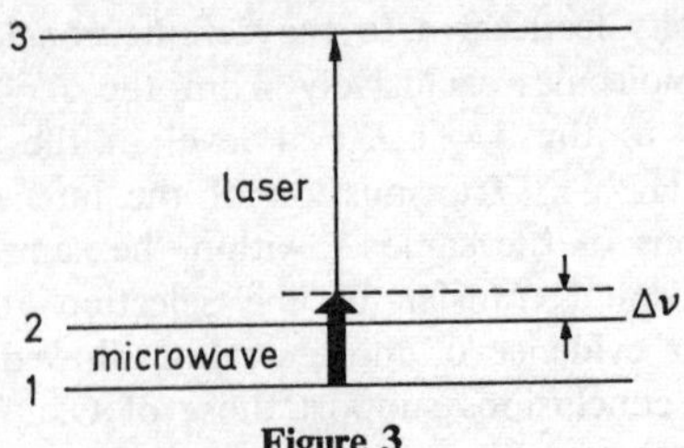

Figure 3

problem is presented by Oka and Shimizu.[222] Even though the i.r. laser transition indicated in Figure 3 is off-resonance by a frequency $\Delta\nu$, the two-photon transition 3 ← 1 can be made to occur if the levels are simultaneously irradiated by the fixed laser frequency and a high-powered microwave radiation at the frequency $\nu_{12}+\Delta\nu$.

The two-photon transition is generally weak, the transition moment varying inversely with $\Delta\nu$, but can become strongly allowed in the presence of a sufficiently high-density radiation at $\nu_{12}+\Delta\nu$, providing $\Delta\nu$ is not too large. Thus in Oka and Shimizu's experiment the i.r. transition $\nu_2[{}^aQ(4,4)]$ in $^{15}NH_3$ gas is some 300 MHz higher in frequency than the $P(15)$ line of the N_2O laser. Nevertheless, when $^{15}NH_3$ gas is simultaneously subjected to the laser radiation and to high-power microwave radiation at *ca.* 300 MHz higher than the frequency of the $J = 4, K = 4$ inversion transition, the two-photon process is strongly allowed and the laser radiation is absorbed. The non-linear process effectively 'tunes' the laser from its fixed frequency into resonance.

Two other i.r.-microwave double-resonance experiments have been described. Takami and Shimoda [223] observe such an effect when a He–Xe gas laser is tuned into resonance with the 6_{06} (ν_5 excited state) ← 5_{15} (ground state) transition at 2850.633 cm^{-1} in low-pressure formaldehyde gas while the $5_{14} \leftarrow 5_{15}$ ground-state transition at 72409.35 MHz is saturated. Tuning of the He–Xe laser is possible over several gigahertz through the Zeeman effect. The laser radiation is set at the centre of the i.r. transition and changes in its intensity are monitored while the high-powered microwave radiation is tuned across the rotational transition. In a second investigation, Fourrier, Van Lerberghe, and Redon [224] saturate the $\nu_2[{}^aQ(8,7)]$ vibration–rotation transition in low-pressure ammonia gas with the $P(13)$ line of an N_2O laser and observe the effect on the $J = 8, K = 7$ ground-state inversion transition in the microwave region.

11 Astrophysics

The detection of small molecules in interstellar space through observa-

[222] T. Oka and T. Shimizu, *Appl. Phys. Letters*, 1971, **19**, 88.
[223] M. Takami and K. Shimoda, *Japan J. Appl. Phys.*, 1971, **10**, 658.
[224] M. Fourrier, A. Van Lerberghe, and M. Redon, *Compt. rend.*, 1971, **273**, *B*, 816.

tion of rotational transitions is a field of rapid growth. Among molecules so detected which have appeared in the literature during 1971 are two isotopic species of hydrogen cyanide ($H^{12}C^{14}N$ and $H^{13}C^{14}N$),[225] cyanoacetylene ($H^{12}C^{12}C^{12}C^{14}N$),[226] formic acid ($H^{12}C^{16}O_2H$),[227] and formamide ($^{14}NH_2{}^{12}CH^{16}O$).[228] Of course, observed transition frequencies are Döppler-shifted in each case to an extent determined by the relative velocity of the source and the observer. The problem of assignment is then either to know the source velocity, in which case the observed frequency can be corrected to rest and compared with the laboratory frequency, or to identify several transitions with a characteristic frequency (and perhaps relative intensity) pattern, in which case all transitions will be shifted by a small constant amount, determined by the source velocity, from the laboratory frequencies. In either case, accurately known laboratory frequencies are crucial.

An example of the first category is the $1_{11} \rightarrow 1_{10}$ transition of formic acid observed [227] in emission in the direction of Sgr B2. The difference between the observed astronomical frequency and the very accurate laboratory frequency specifically determined by the same workers can be accounted for by a Döppler velocity typical of observations in the Sgr B2 direction. However, observations of other lines will be necessary for unambiguous confirmation. Examples in the second category are the characteristic nitrogen nuclear quadrupole components $F = 2 \rightarrow 1$ and $F = 1 \rightarrow 1$ of the $J = 1 \rightarrow 0$ transition of cyanoacetylene ($H^{12}C^{12}C^{12}C^{14}N$),[226] and $F = 1 \rightarrow 1$, $F = 3 \rightarrow 3$, and $F = 2 \rightarrow 2$ of the $2_{11} \rightarrow 2_{12}$ transition of formamide ($^{14}NH_2{}^{12}CH^{16}O$), which have been observed. Naturally, the observation of more transitions of the parent species or of isotopically less-abundant species would further strengthen the assignments. The need for reliable laboratory measurements has specifically stimulated two papers concerned with cyanoacetylene.[229, 230] Also of interest in this respect might be some very accurate molecular-beam work by Kukolich already mentioned (see refs. 55 and 56).

The rotational relaxation of ammonia molecules in interstellar space has previously been discussed in terms of collison-induced $\Delta k = \pm 3$ transitions between metastable rotational levels.[231] However, it has been pointed out recently [232] that *radiative* $\Delta k = \pm 3$ transitions have probabilities not much smaller than those of the collision-induced transitions and that therefore the former may also be important in rotational relaxation of interstellar ammonia.

[225] L. E. Snyder and D. Buhl, *Astrophys. J.*, 1971, **163,** L47.
[226] B. E. Turner, *Astrophys. J.*, 1971, **163,** L35.
[227] B. Zuckerman, J. A. Ball, and C. A. Gottlieb, *Astrophys. J.*, 1971, **163,** L41.
[228] R. H. Rubin, G. W. Swenson, jun., R. C. Benson, H. L. Tigelaar, and W. H. Flygare, *Astrophys. J.*, 1971, **169,** L39.
[229] R. L. de Zafra, *Astrophys. J.*, 1971, **170,** 165.
[230] D. R. Johnson and F. Lovas, *Astrophys. J.*, 1971, **169,** 617.
[231] A. C. Cheung, D. M. Rank, C. H. Townes, S. H. Knowles, and W. T. Sullivan, *Astrophys. J.*, 1969, **157,** L13.
[32] T. Oka, F. O. Shimizu, T. Shimizu, and J. K. G. Watson, *Astrophys. J.*, 1971, **165,** L15.

The mechanism of the $\Delta k = \pm 3$ radiative transitions is that due to centrifugal distortion discussed by Watson [199] (see Section 8).

12 Instrumentation

Several papers describing microwave spectrometers have appeared recently in the literature.

A common dilemma of microwave spectroscopy is how to obtain high sensitivity without the sacrifice of resolution. In the conventional Hughes–Wilson Stark-modulation spectrometer, the high sensitivity achieved through molecular modulation of microwave radiation at rather high frequencies (usually 100 kHz) is at the cost of concomitant modulation broadening and, therefore, resolution. Rudolph and Schwoch [233] are concerned to resolve the dilemma. They discuss and compare the noise figures of three spectrometers: a conventional Stark-modulation spectrometer at various modulation frequencies, a bridge spectrometer with and without Stark modulation, and a bridge spectrometer with Stark modulation but also employing superheterodyne detection with phase-sensitive demodulation at both the Stark-modulation frequency and the intermediate frequency. The last-named spectrometer has the important property that the noise figure is independent of the Stark-modulation frequency so that a frequency low enough to keep modulation broadening small compared with Döppler widths is possible while maintaining optimum sensitivity. The authors conclude that such a bridge spectrometer employing 300 Hz Stark-modulation frequency and superheterodyne detection is superior to a conventional 100 kHz spectrometer at all incident powers, although at the larger powers required for the highest sensitivity the improvement does not warrant the extra complexity. However, at the very low powers and Stark-modulation frequencies necessary for high-resolution performance, the sensitivity of the former is very much superior to the latter. Details of the construction of such a spectrometer are given, together with some impressive recordings which amply verify the authors' conclusions.

The spectrometers used by the group in Berlin for their extensive high-temperature studies are described in two publications.[234, 235] The first [234] is concerned with a detailed description of the construction and performance of heated Stark-absorption cells used in conjunction with essentially conventional spectrometers. This paper is prefaced with a discussion of the sensitivity to be expected when the design parameters of the high-temperature cell are optimized. For such a cell using 100 kHz Stark modulation, the most favourable signal-to-noise ratio predicted for a transition at 10 GHz for a typically involatile diatomic molecule at 1000 °C is *ca.* 200:1. A temperature of 1000 °C is thought to be the upper limit for heated absorption cells

[233] H. D. Rudolph and D. Schwoch, *Z. Angew. Phys.*, 1971, **31**, 197.
[234] J. Hoeft, F. J. Lovas, E. Tiemann, and T. Törring, *Z. Angew. Phys.*, 1971, **31**, 265.
[235] J. Hoeft, F. J. Lovas, E. Tiemann, and T. Törring, *Z. Angew. Phys.*, 1971, **31**, 337.

from consideration of the constructional requirements. Moreover, it is not possible to produce a cell simultaneously suitable for all purposes. Consequently, four different cells are described in detail, each being suitable for a particular purpose. For example, one is designed for use with molecules which would react with a hot metal surface whereas another is intended for dipole moment measurements. The performance of each cell is illustrated with an example.

Although difficulties with heated absorption cells above 1000 °C appear to be insurmountable, a molecular-beam microwave spectrometer described in a second paper [235] by the Berlin group can be used with molecular temperatures of up to 2000 °C. The substance under study is vaporized from an oven into a cooled absorption cell where molecules continuously condense on to the walls, thereby maintaining a small but sufficient vapour density within the cell. Moreover, molecules remain in the vapour for only 10^{-4} s so that the study of unstable species is possible. Details of two cells and two ovens are described and the solution of some awkward experimental problems is reported. For example, above 1300 °C the oven and Stark electrode emit ions and electrons which attenuate the microwave radiation and cause serious background absorption. A superheterodyne detection technique which is insensitive to the background signal neatly overcomes the problem.

A Zeeman-modulated microwave spectrometer designed specifically for investigation of the rotational spectra of free radicals at *K*-band frequencies is described by Hrubesh, Anderson, and Rinehart.[236] A novel feature of the spectrometer is the Fabry–Perot-type resonator used, rather than a waveguide, to provide a compact absorption cell in which there is a high volume-to-surface-area ratio and minimum metal surface. One particular advantage is that the b.w.o. microwave source can be locked to the resonant frequency of the cavity to give high stability. Moreover, the oscillator is made to track the cavity resonance so that very stable, continuous, broad-band sweeps at rates between 5 kHz s^{-1} and 10 MHz s^{-1} are possible by changing the resonator plate separation with a mechanical drive. High sensitivity of the spectrometer is ensured by zero-based Zeeman modulation provided by a pair of Helmholtz coils near the cavity. The utility of the instrument as a stable, sensitive, broad-band search spectrometer is illustrated by recordings of some hitherto-undetected lines of NO_2 and of the hitherto-undetected spectrum of the NF_2 free radical.

Two papers concerned with the ammonia maser have appeared,[237, 238] while a third [239] reports the millimetre- and submillimetre-wave laser radiation produced when parallel bands of molecules such as methyl fluoride and acetonitrile are pumped by carbon dioxide laser radiation.

A scheme to provide frequency markers for microwave spectroscopy is

[236] L. W. Hrubesh, R. E. Anderson, and E. A. Rinehart, *Rev. Sci. Instr.*, 1971, **42**, 789.
[237] D. C. Laine and G. D. S. Smart, *J. Phys. (D)*, 1971, **4**, L23.
[238] W. S. Bardo and D. C. Laine, *J. Phys. (E)*, 1971, **4**, 595.
[239] T. Y. Chang and J. D. McGee, *Appl. Phys. Letters*, 1971, **19**, 103.

described by Smith and Netterfield.[240] The specific advantage is that a good signal-to-noise ratio in the range 30—40 GHz is possible less expensively than previously.

13 Analysis

A study of the isomerization of n-butenes has exemplified the value of microwave spectroscopy as an analytical tool, particularly when combined with gas chromatography and mass spectrometry.[241, 242] In a mechanistic investigation, concentrations of monodeuterio-species were followed during a study of the kinetics of the reactions. Mass spectrometry was used to evaluate overall deuterium content of butene, and microwave spectrocopy to provide quantitive measurements of the deuterium content in particular positions in butene molecules, since each of the five monodeuteriated species of but-1-ene [(48)—(52)] was separately detectable from its microwave spectrum.

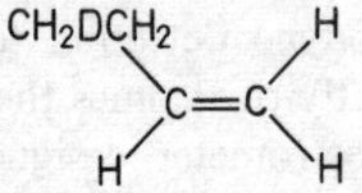

$CH_2DCH_2(H)C{=}C(H)H$ (48)

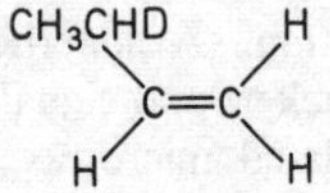

$CH_3CHD(H)C{=}C(H)H$ (49)

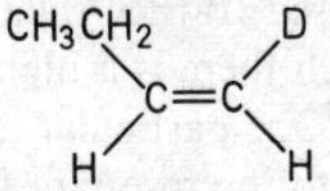

$CH_3CH_2(H)C{=}C(D)H$ (50)

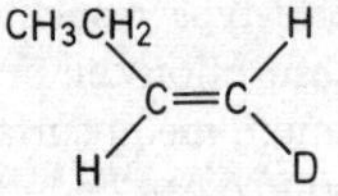

$CH_3CH_2(H)C{=}C(H)D$ (51)

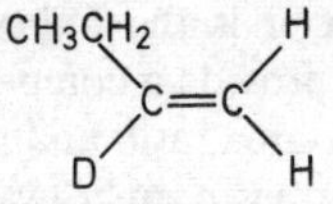

$CH_3CH_2(D)C{=}C(H)H$ (52)

The technique was used to study isomerization of butenes over toluene-*p*-sulphonic acid.[241] The results lead to the conclusion that *cis–trans* isomerization of n-butene under these conditions occurs *via* a carbonium ion, the toluene-*p*-sulphonic acid behaving as a Brönsted acid. The isomerization of butenes over an alumina catalyst has also been studied in the same way.[242] In this case perdeuteriopropene was introduced into the reaction system with the butene under investigation. Microwave spectroscopy combined with gas chromatography and mass spectrometry was used for the analyses of reactants and products. It was found that deuterium on the C_1 atom of the perdeuteriopropene showed extensive change with the hydrogens of the butenes in the reaction mixture. The catalytic isomerization over alumina shows close parallels with that over toluene-*p*-sulphonic acid and this is taken to indicate

[240] A. M. Smith and R. P. Netterfield, *J. Phys. (E)*, 1971, **4**, 618.
[241] Y. Sakurai, Y. Kaneda, S. Kondo, E. Hirota, T. Onishi, and K. Tamaru, *Trans. Faraday Soc.*, 1971, **67**, 3275.
[242] Y. Sakurai, T. Onishi, and K. Tamaru, *Trans. Faraday Soc.*, 1971, **67**, 3094.

that isomerization of butene over alumina occurs on some kind of protonic site formed from the Lewis-acid sites of the alumina surface. The same technique has also been used to study the mercury-photosensitized reactions of *cis*-[1-^{2}H]propene and *cis*-[1-^{2}H]but-1-ene.[243] At pressures above *ca.* 100 mmHg, *cis–trans* isomerization was the main reaction for both propene and butene with a quantum efficiency of nearly unity. At pressures below 30 mmHg, substantial amounts of [2-^{2}H]- and [3-^{2}H]-propene species were detected in the reaction mixture. Decomposition of propene into an allylic radical and a hydrogen atom with subsequent 'recombination' was proposed to account for this observation. Similar results were obtained with *cis*-[1-^{2}H]but-1-ene. Quantitative analysis of alcohols by a microwave spectroscopic method has also been reported.[244] Microwave spectroscopy has been used to follow the pyrolisis of *N*-sulphinylaniline: sulphur monoxide, SO, is identified in the products by the observation of the $J,K = 2,1 \leftarrow 1,0$ transition.[245]

The problem of how many frequency measurements are necessary for the identification by microwave spectroscopy of a substance in a gas mixture has been examined.[246] A computerized analysis has been made for various mixtures on the basis of published data. For a mixture of ten gases, the measurement of a single line within 0.2 MHz resulted in overlap in approximately 15% of the cases. On the other hand, if two lines are measured, then overlap occurs in less than 1% of the cases, providing measurement is within 0.5 MHz. When the number of gases was increased to 33, overlap occurred in 45% of the cases with a precision of measurement of 0.2 MHz. By measuring two lines to the same precision, overlap occurred only in two cases.

[243] Y. Sakurai, T. Onishi, and K. Tamaru, *Bull. Chem. Soc. Japan*, 1971, **44**, 2990.
[244] Ch. O. Kadzhar, A. B. Askerov, and L. M. Imanov, *Azerb. Khim. Zhur.*, 1970, 100.
[245] S. Saito and C. Wentrup, *Helv. Chim. Acta*, 1971, **54**, 273.
[246] G. E. Jones and E. T. Beers, *Analyt. Chem.*, 1971, **43**, 656.

2
Electronic Spectra of Large Molecules

BY J. M. HOLLAS

1 Introduction

This chapter will be concerned mainly with electronic spectra of free molecules, a condition which is most readily attainable in the gas phase at low pressure. Intermolecular interactions are sufficiently strong in the liquid and solid phases as to make information relevant to the free molecule more difficult to obtain and more open to question when it has been obtained. The Report is restricted also to large molecules which, in the context of electronic spectroscopy, are molecules similar in size to benzene and naphthalene, these being model examples both chemically and spectroscopically.

Since this is the first Report on electronic spectra of large molecules in this Specialist Periodical Reports series, it will be necessary first to sketch in the background to the topic. Then it is hoped to cover developments in the subject within the past few years fairly comprehensively, but not to list all published papers irrespective of value.

It was known in the 1920's, in particular from the work of Henri, that some large molecules have absorption spectra which show, in the vapour phase, a considerable number of discrete bands due to molecular vibrations. It was not until after the publication in 1933 of a paper by Herzberg and Teller [1] setting out the vibronic selection rules in an electronic spectrum that detailed interpretation of the spectra could be attempted. The vibrational analysis of the 260 nm* absorption system of benzene attracted a great deal of attention, culminating (in 1948) in a series of papers by Best, Garforth, Ingold, Poole, and Wilson [2] that described the analysis in detail of the absorption and emission spectra of benzene and several deuteriated benzenes. In the 1940's absorption spectra of other large molecules, and in particular of substituted benzenes, were being vibrationally analysed. Much of this work was done under the guidance of Sponer, who had also been associated with the vibrational analysis of the 260 nm system of benzene. A very valuable review entitled 'Electronic Spectra of Polyatomic Molecules' by Sponer and Teller,[3]

* 1 nm = 10 Å.

1 G. Herzberg and E. Teller, *Z. phys. Chem.* (*Leipzig*), 1933, **B21**, 410.
2 A. P. Best, F. M. Garforth, C. K. Ingold, H. G. Poole, and C. L. Wilson, *J. Chem. Soc.*, 1948, 406.
3 H. Sponer and E. Teller, *Rev. Mod. Phys.*, 1941, **13**, 76.

written in 1941, summarizes the work done on spectra of large molecules as well as some smaller molecules. A slightly later review by Sponer [4] is concerned with u.v. spectra of substituted benzenes.

In 1950 Craig [5] was the first to apply the Franck–Condon principle quantitatively to the interpretation of the spectrum of a large molecule. He applied it to the band intensities in a progression of the symmetrical ring-breathing vibration in the 260 nm system of benzene and obtained the result that the carbon–carbon bond length increases in the excited state by 0.0036 nm, a value which has been shown more recently to be a very accurate one. Benzene is a particularly favourable case for this treatment, having only two bond lengths to be determined, but subsequent attempts to use the same or similar methods with other molecules have been disappointing owing to the difficulty of obtaining accurate intensity measurements and to the large number of possible solutions (2^n, where n is the number of totally symmetric vibrations) to the excited-state geometry of molecules having lower symmetry than benzene.

Until the late 1950's little more progress was made in the field. Most of the work was concerned with vibrational analyses of electronic band systems of a wide range of large molecules, but the value of such analyses, once the theories put forward by Herzberg and Teller [1] had been vindicated, was doubtful in many cases. Indeed, the vibrational analysis of the 260 nm system of benzene, although virtually completed by 1948, still remains one of the most difficult and fascinating because the electronic transition is forbidden by symmetry. Only the 500 nm system of *p*-benzoquinone, which was first investigated by Singh [6] in 1959, has provided us with a comparable example of an electronically forbidden system in a large molecule.

In 1959 Mason [7] reported regular rotational structure in the O_0^0 band* of the 552 nm system of *sym*-tetrazine. The structure was analysed in terms of the band approximating to a parallel band of an oblate symmetric rotor: in fact it is a type *C* band† of a slightly asymmetric rotor. Very similar bands

* In this review the notation used for vibrational bands is that suggested by Brand, Callomon, and Watson in an unpublished proposal (1963) and explained in a paper by Callomon *et al.*[8] In this system 12_1^2, for example, implies a transition from a lower electronic state, in which one quantum of ν_{12} is excited, to an upper electronic state, in which two quanta of ν_{12} are excited. The O_0^0 band is what has commonly been called the 0–0 band.

† Throughout this review the recommendation made by Mulliken [9] regarding symbols for rotational constants and axis notation (with regard to symmetry species labels in point groups) will be adhered to. Since Mulliken's recommendations regarding vibrational numbering have very often not been followed, the system in this review will be to follow the numbering used in the publications referred to, and the reader should consult the original papers for an explanation of each numbering system.

[4] H. Sponer, *Rev. Mod. Phys.*, 1942, **16**, 224.
[5] D. P. Craig, *J. Chem. Soc.*, 1950, 2146.
[6] R. S. Singh, *Indian J. Phys.*, 1959, **33**, 376.
[7] S. F. Mason, *J. Chem. Soc.*, 1959, 1269.
[8] J. H. Callomon, T. M. Dunn, and I. M. Mills, *Phil. Trans. Roy. Soc.*, 1966, **A259**, 499.
[9] R. S. Mulliken, *J. Chem. Phys.*, 1955, **23**, 1997.

in spectra of the two diazines pyrazine and pyridazine were observed from 1960 onwards by Innes and co-workers, and the rotational structure was analysed in an analogous way. The first of these analyses to be published was that of pyrazine.[10] However, the treatment in the symmetric-rotor approximation of rotational structure in spectra of asymmetric rotors could not be expected to be satisfactory for more than a few favourable cases. In principle there was no reason to use the symmetric-rotor approximation since the rotational problem of the asymmetric rotor had been solved long before, but in practice the application to spectra of large molecules in which individual rotational lines could not be resolved was quite long delayed. This was because of the very large number of rotational transitions involved; there are of the order of 30000 in a single band of a large asymmetric rotor. The development of large computers in the 1960's meant that the rotational problem in large asymmetric rotors could be tackled, and this was first attempted by Parkin [11] in 1965. It is from this point that the present Report will take up the story of the development of rotational analyses.

Analysis of vibrational structure has been greatly helped by rotational analysis, especially in spectra where non-totally symmetric vibrations are active. The symmetry species of such a vibration can be determined from the rotational selection rules. One of the first attempts to do this was in the 312 nm system of naphthalene by Craig *et al.*,[12] although a full rotational analysis was not achieved. This system was found to be very unusual in that bands involving non-totally symmetric vibrations are considerably more intense than those involving totally symmetric vibrations. Discussion of vibrational analyses in this Report will be confined mainly to work done since rotational analysis became possible.

2 Rotational Band Contour Analysis

Methods.—*Asymmetric Rotors.* Large molecules have large principal moments of inertia I_A, I_B, and I_C and consequently small rotational constants A, B, and C since $A = h/8\pi^2 cI_A$, *etc.* The constants A, B, and C have typical values of the order of 0.1 cm^{-1} in large molecules. This means that the rotational energy levels are very closely spaced, leading to higher population of levels having high values of rotational quantum numbers than occurs in smaller molecules. Appreciable population of levels with J and K_a or, K_{-1}) or K_c (K_1) of the order of 100 is typical. As a result, something like 30000 rotational transitions accompany any electronic or vibronic transition. The rotational transitions usually lie within *ca.* 30 cm^{-1} so, on average, there is one transition for each 0.001 cm^{-1}. Since the Doppler width of transitions is *ca.* 0.02 cm^{-1} there may be about 20 transitions within the Doppler width. So, even with the highest resolution (*ca.* 0.02 cm^{-1}),

[10] J. A. Merritt and K. K. Innes, *Spectrochim. Acta*, 1960, **16A**, 945.
[11] J. E. Parkin, *J. Mol. Spectroscopy*, 1965, **15**, 483.
[12] D. P. Craig, J. M. Hollas, M. F. Redies, and S. C. Wait, jun., *Phil. Trans. Roy. Soc.*, 1961, **A253**, 543.

there is little hope of resolving individual transitions. What is observed is an intensity contour with a characteristic shape. An example is the rotational contour associated with the pure electronic transition in the 294 nm system of aniline [13] shown in Figure 1. The shape of the contour is strongly dependent

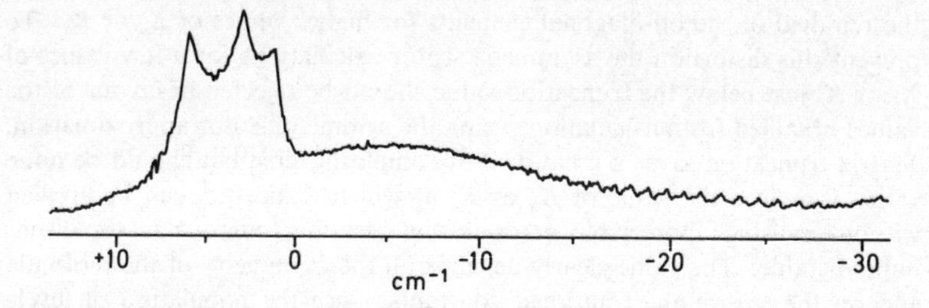

Figure 1 *Observed rotational contour of the* O_0^0 *band of the* 294 nm *system of aniline. Absorption intensity is plotted upwards on the vertical scale*

on the same parameters as one would hope to obtain from a completely resolved spectrum, namely the rotational selection rules (type *A*, *B*, *C*, or hybrid) and the excited-state rotational constants A', B', and C'. However, it is necessary to know the ground-state constants A'', B'', and C'', preferably with microwave accuracy. Crystallographic data or even a geometry extrapolated from those of similar molecules may give adequate accuracy: this is because computed rotational contours are much more sensitive to the changes of rotational constants (ΔA, ΔB, and ΔC) from the ground- to the excited-state than to the absolute values of the constants.

The general theory of rotational transitions in asymmetric rotor molecules is well known, and perhaps best explained by Gordy and Cook.[14] It is because the theory is well known that details of most of the asymmetric-rotor computer programs which have been written have not been published. There are, however, two approximations which have been used in some programs in order to reduce the computing time for a band contour. The first of these involves the calculation of energy levels and intensities for selected values of J only, the remaining levels and intensities being obtained by an interpolation procedure.[11] Comparisons of calculations with and without such interpolation have shown that if one is concerned with fine detail in a contour the results from the interpolation method should be treated with great caution (see, for example, Merer and Innes [15]). The second approximation is that of matrix truncation.[11] Prolate and oblate asymmetric rotors tend to behave more like symmetric rotors as the values of K_a and K_c, respectively, increase,

[13] J. Christoffersen, J. M. Hollas, and G. H. Kirby, *Mol. Phys.*, 1969, **16**, 441.
[14] W. Gordy and R. L. Cook, 'Microwave Molecular Spectra', Interscience, New York, 1970.
[15] A. J. Merer and K. K. Innes, *Proc. Roy. Soc.*, 1968, **A302**, 271.

and matrices can usefully be truncated at an appropriate value of K_a or K_c beyond which the off-diagonal matrix elements are set equal to zero, *i.e.* the symmetric-rotor approximation is used. If simple truncation is used there can be rather severe distortions of the calculations for values of K_a or K_c that are slightly less than that at which truncation has been applied, due to the removal of the off-diagonal elements for higher values of K_a or K_c. To prevent this distortion the asymmetric-rotor calculations for a few values of K_a or K_c just below the truncation value should be rejected in favour of the values obtained from calculations using the symmetric-rotor approximation. Matrix truncation saves a great deal of computing time but should be done with care, since the value of K_a or K_c at which truncation can be applied without causing appreciable error in the resulting contour is somewhat unpredictable. The value clearly depends on the asymmetry of the molecule and on the size of the rotational constants, since the population of levels with K_a or K_c having the truncation value is important. Even if two molecules have the same asymmetry parameters and the same rotational constants, it may be that the ΔA, ΔB, and ΔC in one molecule are such that the structure with K_a or K_c in the region of the truncation value is prominent in the contour whereas in the spectrum of the other molecule the structure may not be prominent. The only certain way of determining the minimum value of K_a or K_c at which matrices can be truncated is by increasing the value until the effect on the contour is negligible. A further improvement is to use, after matrix truncation, the second-order perturbation approximation [16, 17] instead of the symmetric-rotor approximation.

In all band contour analyses of the spectra of large molecules that have been published so far, centrifugal distortion has been neglected. In very few large molecules are centrifugal distortion constants known, even in the ground state, but their expected values, *e.g.* $D_J'' = (2.2 \pm 1.0) \times 10^{-8}$ cm^{-1} in benzene,[18] are so small that they should affect transition energies by at the most only a few hundredths of a cm^{-1} in unfavourable circumstances, *e.g.* the effect of D_{JK}, a symmetric-rotor distortion constant, can be of this order when $\Delta K = \Delta J = \pm 1$ at high J and K.

The approximation, used in the early days of computers, of treating asymmetric rotors with $\kappa \approx \pm 1$ as symmetric rotors and computing complete contours using this approximation [19] is not valid, except in a few favourable cases for distinguishing between band types, for example in *p*-benzoquinone.[20] The fact that we now know that, even for a near-symmetric rotor, the differences between a type *B* and a type *C* band of a prolate asymmetric rotor are very great, whereas in the symmetric-rotor approximation they are identical, illustrates the dangers of using this approximation.

[16] S. R. Polo, *Canad. J. Phys.*, 1957, **35**, 880.
[17] H. C. Allen and P. C. Cross, 'Molecular Vib-Rotors', Wiley, New York, 1963.
[18] B. P. Stoicheff, *Canad. J. Phys.*, 1954, **32**, 339.
[19] J. M. Hollas, *Spectrochim. Acta*, 1966, **22A**, 81.
[20] J. M. Hollas, *Spectrochim. Acta*, 1964, **20A**, 1563.

In an oblate symmetric rotor all energy levels, except those with $K_c = 0$, are doubly degenerate, whereas in a prolate symmetric rotor all except those with $K_a = 0$ are doubly degenerate. In an asymmetric rotor the degeneracies may be split. In an oblate asymmetric rotor the splitting increases as J increases and as K_c decreases: in a prolate asymmetric rotor the splitting increases as J increases and as K_a decreases. At low values of K_c and high values of J an oblate asymmetric rotor is tending to rotate rapidly about the a-axis, and therefore the energy levels resemble those of a *prolate* symmetric rotor. Similarly, at low values of K_a and high values of J the energy levels of a prolate asymmetric rotor resemble those of an *oblate* symmetric rotor. The usefulness of this 'low K, high J approximation' was first recognized by Gora.[21] At the opposite extreme, at low values of J and high values of K_c, in an oblate asymmetric rotor, the energy levels resemble those of an oblate symmetric rotor: similarly, at low values of J and high values of K_a the energy levels of a prolate asymmetric rotor resemble those of a prolate symmetric rotor. Gora [21] has shown that, for a particular value of K, it is only over quite a small range of J that energy levels in any asymmetric rotor are non-degenerate. However, this does not mean to say that, when there is a degeneracy, either symmetric-rotor formulae or the low K, high J formulae derived by Gora will be appropriate, since deviations from either approximation set in before the degeneracies are split.

When the degeneracy of energy levels is split in either the upper or the lower electronic state, the rotational transitions are split. Consequently the region of J, for a particular value of K, in which transitions are split tends to be larger than that in which energy levels are split. When the transitions are not split, a series of them obeying a particular set of rotational selection rules is called a *branch*: when the transitions are split the resulting series are called *sub-branches*. In the system of labelling sub-branches two ante-superscripts indicate the value of ΔK_a, first, and ΔK_c, second (n, o, p, q, r, s, and t indicate -3, -2, -1, 0, 1, 2, and 3, respectively), the main symbol indicates the value of ΔJ (P, Q, and R indicate -1, 0, and 1, respectively) and, where necessary, a post-superscript e or o indicates that $J+K_a+K_c$ in the lower state is even or odd respectively. For example, in a sub-branch labelled ${}^{pr}Q^{e}$, $\Delta K_c = -1$, $\Delta K_a = 1$, $\Delta J = 0$, and $J''+K_a''+K_c''$ is even.

In an asymmetric rotor there is no restriction on ΔK_a (for a prolate rotor) or ΔK_c (for an oblate rotor) except that they are even or odd: in a prolate rotor ΔK_a is even in a type A band and odd in type B and C bands, whereas in an oblate rotor ΔK_c is even in a type C band and odd in type A and B bands. However, in the corresponding symmetric rotors $\Delta K = 0$ or ± 1 only and, even in strongly asymmetric rotors, it is the sub-branches that obey these symmetric-rotor selection rules which contribute most of the intensity. These sub-branches are said to be 'symmetric-rotor allowed', and they are listed for both oblate and prolate asymmetric rotors in Table 1. In this Table, sub-branches which are bracketed become coincident in the type of symmetric

[21] E. K. Gora, *J. Mol. Spectroscopy*, 1965, **16**, 378.

Table 1 *Symmetric-rotor-allowed sub-branches in asymmetric rotors*

	prolate	*oblate*
Type A *bands*	$({}^{qr}Q, {}^{qp}Q)$	$({}^{qr}Q, {}^{or}Q), ({}^{qp}Q, {}^{sp}Q)$
	$({}^{qr}R^e, {}^{qr}R^o)$	$({}^{qr}R^e, {}^{qr}R^o), ({}^{sp}R^e, {}^{sp}R^o)$
	$({}^{qp}P^e, {}^{qp}P^o)$	$({}^{qp}P^e, {}^{qp}P^o), ({}^{or}P^e, {}^{or}P^o)$
Type B *bands*	$({}^{pr}Q^e, {}^{pr}Q^o), ({}^{rp}Q^e, {}^{rp}Q^o)$	$({}^{pr}Q^e, {}^{pr}Q^o), ({}^{rp}Q^e, {}^{rp}Q^o)$
	$({}^{pr}R, {}^{pt}R), ({}^{rp}R, {}^{rr}R)$	$({}^{pr}R, {}^{rr}R), ({}^{rp}R, {}^{tp}R)$
	$({}^{rp}P, {}^{rn}P), ({}^{pr}P, {}^{pp}P)$	$({}^{rp}P, {}^{pp}P), ({}^{pr}P, {}^{nr}P)$
Type C *bands*	$({}^{rq}Q, {}^{ro}Q), ({}^{pq}Q, {}^{ps}Q)$	$({}^{rq}Q, {}^{pq}Q)$
	$({}^{rq}R^e, {}^{rq}R^o), ({}^{ps}R^e, {}^{ps}R^o)$	$({}^{rq}R^e, {}^{rq}R^o)$
	$({}^{pq}P^e, {}^{pq}P^o), ({}^{ro}P^e, {}^{ro}P^o)$	$({}^{pq}P^e, {}^{pq}P^o)$

rotor (oblate or prolate) to which the asymmetric rotor most nearly corresponds. For example, in a prolate asymmetric rotor ${}^{qr}Q$ and ${}^{qp}Q$ sub-branches become coincident as the molecule behaves more like a prolate symmetric rotor: it is often convenient to label the resulting branch ${}^{q}Q$ since the value of ΔK_c is no longer relevant. In general, a single ante-superscript is used to label branches (as opposed to sub-branches) and the label indicates the value of ΔK_a or ΔK_c in a prolate or oblate near-symmetric rotor respectively. It can be seen from Table 1 that the *e* and *o* components of the same branches become coincident in both the prolate and oblate symmetric rotor limits.

The pairs of sub-branches that are connected by an underlining in Table 1 are sub-branches of a prolate asymmetric rotor which become degenerate in the oblate limit. The wavenumbers of the members of these coincident sub-branches are given, for low values of K_a in planar or near-planar molecules only, by simple expressions that have been discussed in some detail by Brown [22] and based on previous work by McHugh *et al.*[23] and, of course, by Gora.[21] The expressions are:

(i) $$\tilde{\nu} = \tilde{\nu}_0 - \tfrac{1}{4}m_2(\Delta A+\Delta B)-(2n+1)C''+n^2\Delta C \quad (1)$$

for coincident ${}^{qp}P^e$ and ${}^{qp}P^o$ sub-branches in a type *A* band and ${}^{rp}P$ and ${}^{pp}P$ sub-branches in a type *B* band, where $\tilde{\nu}_0$ is the wavenumber of the band origin, $m_2 = (m+1)^2+m^2$, $m = J-K_c$, and $n = 2J-K_c$.

(ii) $$\tilde{\nu} = \tilde{\nu}_0 - \tfrac{1}{2}m_2(\Delta A+\Delta B)+(2n+3)C'+(n+1)^2\Delta C \quad (2)$$

for coincident ${}^{qr}R^e$ and ${}^{qr}R^o$ sub-branches in a type *A* band and ${}^{pr}R$ and ${}^{rr}R$ sub-branches in a type *B* band.

(iii) $$\tilde{\nu} = \tilde{\nu}_0 - \tfrac{1}{4}m_2(\Delta A+\Delta B)+(n+1)^2\Delta C \quad (3)$$

for coincident ${}^{rq}Q$ and ${}^{pq}Q$ sub-branches in a type *C* band.

[22] J. M. Brown, *J. Mol. Spectroscopy*, 1969, **31**, 118.
[23] A. J. McHugh, D. A. Ramsay, and I. G. Ross, *Austral. J. Chem.*, 1968, **21**, 2835.

Equations (1)—(3) apply also to oblate asymmetric rotors but then they represent oblate symmetric rotor features.

Transitions given by equations (1)—(3) are intense because they involve only low values of K_a and also because it is common for several transitions with different values of n, arising from different combinations of K_c and J, to be coincident. The separation within such a group of transitions with the same value of n is approximately $(\Delta A+\Delta B)$, which is commonly appreciably smaller than the resolution achieved experimentally.

How do we set about trying to match a computed to an observed band contour of an asymmetric rotor? First we must have a good idea as to what band type (A, B, C, or hybrid) we are dealing with. This information may come from a knowledge of the molecular orbitals concerned, comparison with electronically similar molecules or, at worst, from trial and error with an asymmetric-rotor band-contour computer program. The trial-and-error method may be quite lengthy since we require to know the rotational constants reasonably well before the band type can be ascertained. If we assume that the band type is known, it is possible to conceive of a computer program written in such a way that the computer would be able to 'home in' on the best set of excited-state rotational constants if it were given the ground-state constants, the rotational selection rules, and a digitized observed contour. This would be a very expensive and uninformative way of achieving the result. Over the past few years a considerable amount of information has been built up regarding the structure of asymmetric-rotor band contours. This information can be of great help in getting some idea, and sometimes a very good idea, of the excited-state rotational constants before any contours are computed.

Although complete contours calculated with the symmetric-rotor approximation are of little value, it is still very useful to have some idea of how energy levels and transitions behave at high values of K_a (in a prolate rotor) or K_c (in an oblate rotor), when the symmetric-rotor approximation tends to be at its most acceptable. Sub-band origin positions $\tilde{\nu}$ are given, in a parallel band of a prolate symmetric rotor, by equation (4), where $\Delta(A-\bar{B}) = (A'-\bar{B}')-(A''-\bar{B}'')$ and $\bar{B} = (B+C)/2$: in a symmetric rotor $B = C$ and $\bar{B} = B$, but equation (4) and other symmetric-rotor expressions are also

$$\tilde{\nu} = \tilde{\nu}_0+\Delta(A-\bar{B})K_a^2 \quad (4)$$

useful for near-symmetric rotors for which $B \neq C$ and therefore $\bar{B} \neq B$. The corresponding expression for a parallel band of an oblate symmetric rotor is:

$$\tilde{\nu} = \tilde{\nu}_0+\Delta(C-\bar{B})K_c^2 \quad (5)$$

where $\bar{B} = (A+B)/2$. If structure which is clearly dependent upon K_a or K_c can be observed in a type A band of a prolate asymmetric rotor or a type C band of an oblate asymmetric rotor, and this structure is degraded, say, to low wavenumber, then equations (4) and (5) would indicate that

$\Delta(A-\bar{B})$ or $\Delta(C-\bar{B})$ is negative. Provided the molecule is not too asymmetric the coefficient of K_a^2 or K_c^2 from the observed features might be a quantitative guide to the actual value of $\Delta(A-\bar{B})$ or $\Delta(C-\bar{B})$. Similarly, sub-band origin positions in a perpendicular band are given by equation (6) for a prolate symmetric rotor, and by equation (7) for an oblate symmetric rotor,

$$\tilde{\nu} = \tilde{\nu}_0+(A'-\bar{B}') \pm 2(A'-\bar{B}')K_a+\Delta(A-\bar{B})K_a^2 \qquad (6)$$

$$\tilde{\nu} = \tilde{\nu}_0+(C'-\bar{B}') \pm 2(C'-\bar{B}')K_c+\Delta(C-\bar{B})K_c^2 \qquad (7)$$

where the upper sign refers to ΔK_a or $\Delta K_c = +1$ and the lower to ΔK_a or $\Delta K_c = -1$. Again, an observed degradation can be used to indicate the sign of $\Delta(A-\bar{B})$ or $\Delta(C-\bar{B})$ and perhaps give a guide to the numerical value.

Features dependent upon K_a or K_c, in order to have sufficient intensity to emerge from the contour, must consist of a number of intense transitions that are sufficiently close to be unresolved. In symmetric rotors [19] these may be

(*a*) line-like *Q* branches in which many transitions of a *Q*-branch are coincident. This occurs only when $\bar{B}' \approx \bar{B}''$.

(*b*) *R*-branch heads for which J_{rev}, the value of J at the turning-point or reversal-point of the branches, is given by

$$J_{rev} = (\bar{B}''-3\bar{B}')/2(\bar{B}'-\bar{B}'') \qquad (8)$$

It is clear that, for head formation, $\bar{B}' < \bar{B}''$. For the observation of a long series of *R*-branch heads, J_{rev} must not be too high, otherwise the intensity at the turning-point is low due to the Boltzmann factor, and it must not be too low otherwise, since $J \geqslant K$, not many heads would be observed. In a large molecule $J_{rev} \approx 60$—90 would produce quite a long series of heads. A trial value of J_{rev} in a particular case would give a trial value of $\bar{B}'$.

(*c*) *P*-branch heads for which J_{rev} is given by

$$J_{rev} = (\bar{B}''+\bar{B}')/2(\bar{B}'-\bar{B}'') \qquad (9)$$

For head formation $\bar{B}' > \bar{B}''$, and a trial value of J_{rev} could be used to give a trial value of $\bar{B}'$.

In an asymmetric rotor the features dependent upon K_a or K_c which might emerge in a contour are the same as those discussed for the symmetric rotor in (*a*)—(*c*), with the addition of *Q*-branch heads,[13] possible only in an asymmetric rotor. In an asymmetric rotor, for $\Delta\bar{B} \approx 0$ there can be either a large number of *Q*-branch transitions that are almost coincident but with no actual turning-point, rather similar to the symmetric-rotor case, or there can be a turning-point: either condition can produce a series of intense features in the contour.

In symmetric rotors J_{rev} of equations (8) and (9) is independent of K, but this is not the case in asymmetric rotors except for fairly high K_a (prolate) or K_c (oblate).

Although the symmetric-rotor approach to a series of K_a- or K_c-dependent features is a very useful one in a semi-quantitative way, a much better approach for a quantitative prediction of the positions of such features is through a second-order perturbation treatment.[13, 16, 17] If a series of K_a- or K_c-dependent features is observed in a contour they may be *P*, *Q*, or *R* branches. With

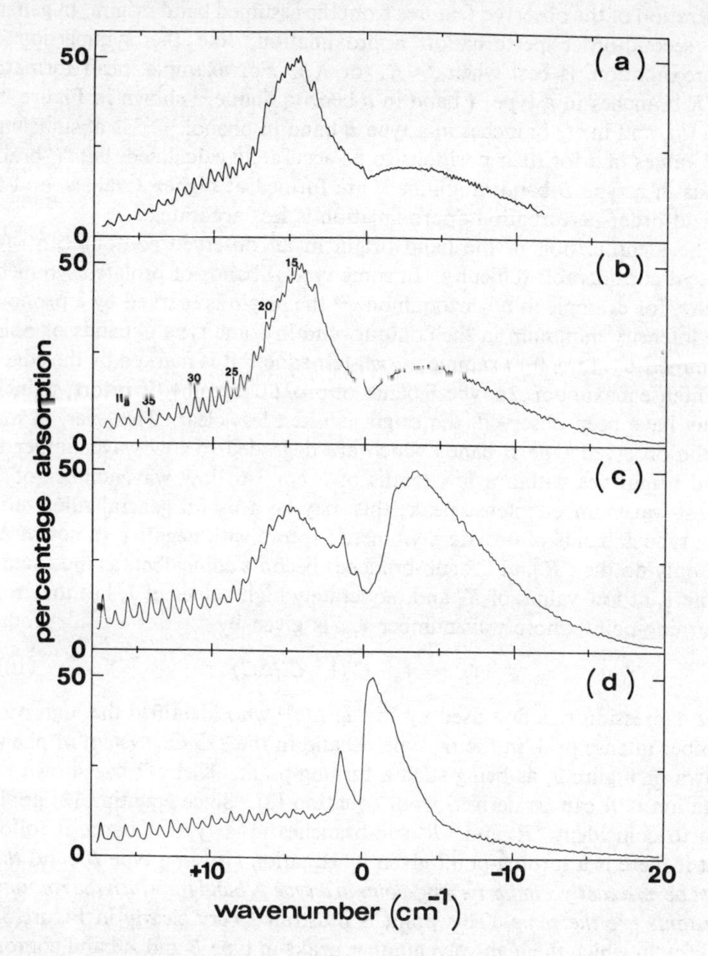

Figure 2 *(a) Observed rotational contour of the type* A 7^1_0 *band in the* 500 nm *system of* p-benzoquinone;
(b) the best computed type A *contour showing the* ^{q}R*-branch heads and their assignments;*
(c) type B, *and*
(d) type C *contours computed with very similar rotational constants to those used for (b).*
(Reproduced by permission from *Mol. Phys.*, 1969, **17**, 655)

any one of these possibilities *and an assumed position of the band origin* $\tilde{\nu}_0$, rotational constants may be varied in a very fast-running second-order perturbation computer program, so that various trial sets of rotational constants may be obtained. Provided the second-order perturbation approximation is a good one, these sets will at least reproduce in computed contours the separation of the observed features from the assumed band origin. In general, the second-order perturbation approximation, like the symmetric-rotor approximation, is best when $J \approx K_a$ (or K_c). For example, head formation in ^qR branches in a type A band in *p*-benzoquinone,[24] shown in Figure 2(*a*) and (*b*), and in ^pQ branches in a type B band in phenol [25–27] is at sufficiently low values of J for their positions to be accurately calculated, but ^pP-branch heads in a type B band in aniline [13] are formed at higher J values, and the second-order perturbation approximation is less accurate.

The identification of the band origin in an observed contour can often present considerable difficulty. In some type A bands of prolate asymmetric rotors, for example in *p*-benzoquinone,[24] the origin is marked by a pronounced intensity minimum in the contour, and in some type C bands of oblate asymmetric rotors, for example in *sym*-tetrazine,[15] it is marked by the edge of an intense maximum. In type B bands of prolate asymmetric rotors, of which many have been observed, the origin is often less clear. However, in most of the observed type B bands which are degraded to low wavenumber the band origin lies within a few tenths of a cm^{-1} to low wavenumber of the lowest-wavenumber intense peak: this may be a useful general rule.

In type B bands of prolate asymmetric rotors with negative values of ΔC, not only do the ^{pr}R and ^{rr}R sub-branches become coincident, as indicated in Table 1, at low values of K_a and moderately high values of J, but they form a turning-point whose wavenumber $\tilde{\nu}_{rev}$ is given by

$$\tilde{\nu}_{rev} = \tilde{\nu}_0 + C'(1 - C'/\Delta C) \qquad (10)$$

This expression was first used by Bist *et al.*,[25] who identified the high-wavenumber intense peak in the O^0_0 type B band in the 275 nm system of phenol, shown in Figure 3, as being such a turning-point. Kirby [27] has shown that equation (10) can be derived from equation (2). Since equation (2) applies also to coincident $^{qr}R^e$ and $^{qr}R^o$ sub-branches in a type A band, it follows that if there is a turning-point given by equation (10) in a type B band *there must be an exactly similar turning-point in a type* A *band for which the rotational constants are the same.* This point is illustrated very clearly in Figure 3(*b*) and (*c*), in which the high-wavenumber peaks in type B and A band contours computed for phenol have the same wavenumber and intensity. This kind of feature in an observed type B or A contour has proved extremely valuable in obtaining a value of ΔC. However, $\tilde{\nu}_0$ must be known for the theory to

[24] J. Christoffersen and J. M. Hollas, *Mol. Phys.*, 1969, **17**, 655.
[25] H. D. Bist, J. C. D. Brand, and D. R. Williams, *J. Mol. Spectroscopy*, 1967, **24**, 413.
[26] J. Christoffersen, J. M. Hollas, and G. H. Kirby, *Proc. Roy. Soc.*, 1968, **A307**, 97.
[27] G. H. Kirby, *Mol. Phys.*, 1970, **18**, 371.

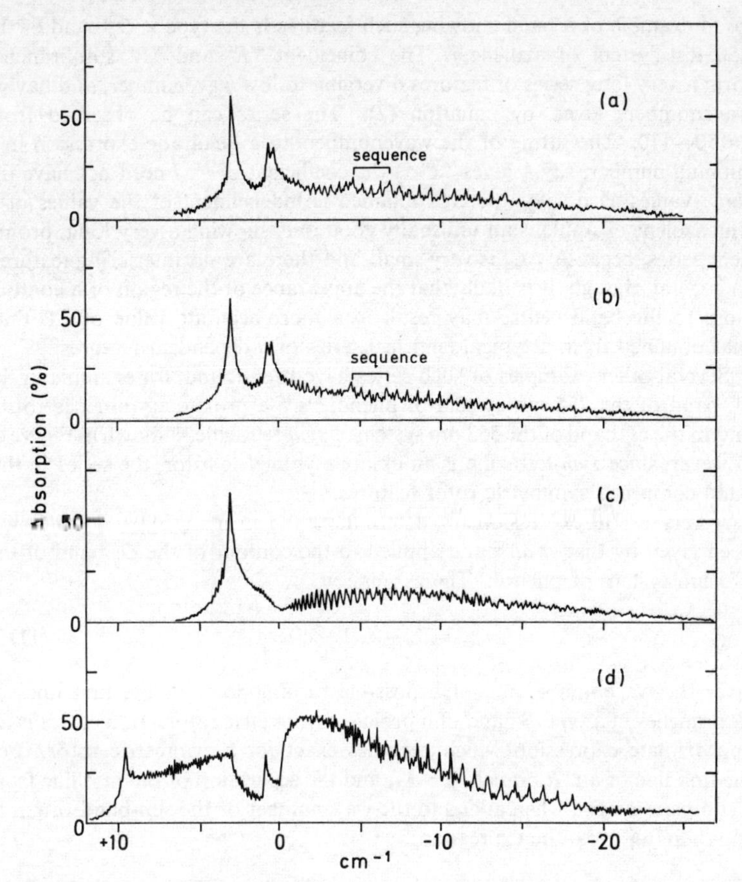

Figure 3 *(a) Observed rotational contour of the type* B O_0^0 *band in the* 275 nm *system of phenol;*
(b) the best computed type B *contour;*
(c) type A, *and*
(d) type C *contours computed with the same rotational constants as those used for* (*b*)
(Reproduced by permission from *Proc. Roy. Soc.*, 1968, **A307**, 97)

be applied, as in the case of the second-order perturbation treatment of features in the spectrum.

If ΔC is positive, rather than negative, a turning-point in P branches with low K_a and high J can be formed: the wavenumber of the turning-point is given by equation (10). No examples of this kind of feature have been observed.

Also important in determining ΔC can be features with low values of K_a and for which the wavenumbers are given by equations (1)—(3). A very good example of a band showing such features is the type A O_0^0 band of the 350 nm system of azulene.[23] The coincident ${}^{qr}R^e$ and ${}^{qr}R^o$ sub-branches form a very long series of features diverging to low wavenumber, and having wavenumbers given by equation (2). The series can be observed from $n \approx 50$—110. The fitting of the wavenumbers to a quadratic expression in a running number, say l, gives ΔC as the coefficient of l^2: l need not have the same values as n since the ΔC obtained is independent of the values of l. The azulene example is an unusually good one, showing a very long, prominent series, because $|\Delta C|$ is very small and there are no interfering features. In general, though, it is likely that the appearance of the region of a contour close to the band centre may result in a more accurate value of ΔC than that obtained from a typical (shorter) series of n-dependent features.

Several other examples of such series have been found, for example in the O_0^0 band of the 275 nm system of phenol,[26,27] a prolate asymmetric rotor, and in the O_0^0 band of the 552 nm system of *sym*-tetrazine,[28] shown in Figure 4: however, since *sym*-tetrazine is an oblate asymmetric rotor, the series in this band comprises symmetric-rotor features.

A very useful expression for determining ΔA in many type B bands has been given by Bist *et al.*[25] and applied to the contour of the O_0^0 band of the 275 nm system of phenol. The expression

$$\tilde{\nu}_{\text{rev}} = \tilde{\nu}_0 + A'(1 - A'/\Delta A) \tag{11}$$

gives the wavenumber $\tilde{\nu}_{\text{rev}}$ of a possible turning-point of the first lines of ${}^rR$ branches in a type B band of a prolate asymmetric rotor. In fact this is an approximate expression which becomes exact for a symmetric rotor. For the first line of an ${}^rR$ branch $J = K_a$, and the separation of the first line from a sub-band origin, when added to the wavenumber of the sub-band origin in equation (6), gives the expression:

$$\begin{aligned}\tilde{\nu}_{\text{R}} = \tilde{\nu}_0 + (A' - \bar{B}') + 2(A' - \bar{B}')K_a + \Delta(A - \bar{B})K_a^2 \\ + \bar{B}'(K_a + 1)(K_a + 2) - \bar{B}''K_a(K_a + 1)\end{aligned} \tag{12}$$

relating the wavenumbers $\tilde{\nu}_{\text{R}}$ of first lines of ${}^rR$ branches to K_a. Differentiating $\tilde{\nu}_{\text{R}}$ with respect to K_a and putting the resulting expression equal to zero gives the value of K_a at the turning-point of first lines. Putting this value into equation (12) gives the wavenumber $\tilde{\nu}_{\text{rev}}$ of the turning-point, as shown in equation (13), which reduces to equation (11) when small quantities are

$$\tilde{\nu}_{\text{rev}} = \tilde{\nu}_0 + A'(1 - A'/\Delta A) + \bar{B}' - A'\Delta\bar{B}/\Delta A - (\Delta\bar{B})^2/4\Delta A \tag{13}$$

neglected. Equation (11) gives quantitatively good agreement with asymmetric-rotor calculations and the reason for this is that when $J = K_a$ the total

[28] J. M. Brown, *Canad. J. Phys.*, 1969, **47**, 233.

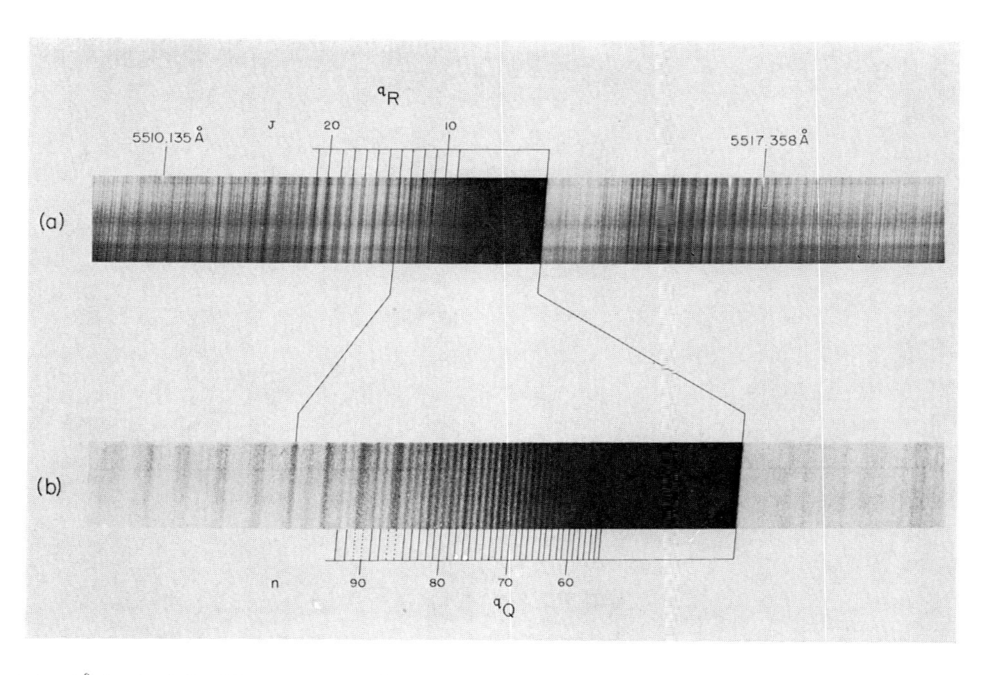

Figure 4 *Part of the* O_0^0 *band of the* 552 nm *system of* sym-*tetrazine. The* n-*dependent features are due to coincident* ^{rq}Q *and* ^{pq}Q *sub-branches in a type* C *band of an oblate asymmetric rotor. Spectrum* (*b*) *is an enlargement of part of* (*a*).
(Reproduced by permission from *Canad. J. Phys.*, 1969, **47**, 233)

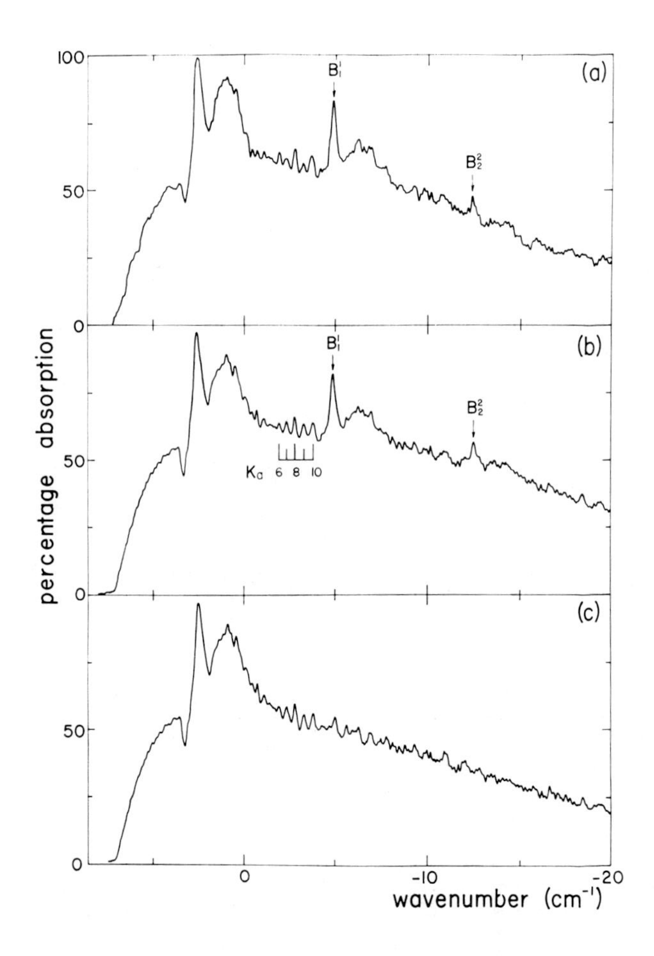

Figure 5 (*a*) *Observed rotational contour of the type* B O_0^0 *band in the* 270 nm *system of chlorobenzene with overlapping sequence bands* B_1^1 *and* B_2^2;
(*b*) *the best computed type* B *contour with similar contours added to simulate sequence bands;*
(*c*) *the best single computed type* B *contour.*
(Reproduced by permission from *Mol. Phys.*, 1970, **18**, 101)

angular momentum vector lies along the *a*-axis, the best condition for an asymmetric rotor to approximate to a symmetric rotor.

It is no accident that equations (10) and (11) are identical in form and differ only in that one equation involves the rotational constant A where the other involves C. The reason for the identical form is that they are both symmetric-rotor expressions giving the wavenumbers of features, the formation of which depends upon whether the molecule concerned, either a prolate or an oblate planar asymmetric rotor, is tending to behave like a symmetric rotor: if it is behaving like a prolate symmetric rotor, equation (11) is appropriate, and equation (10) is appropriate if it is behaving like an oblate symmetric rotor.

The turning-point of the first lines of ^rR branches, given by equation (11), is formed only if ΔA is negative. If ΔA is positive an expression similar to equation (13) results for a turning-point in the first lines of ^pP branches; such an equation again reduces to equation (11) when small quantities are neglected.

A turning-point of either first lines of ^rR or ^pP branches results in the observation of a sharp edge in the contour provided there is no other interfering structure. In the phenol O_0^0 band contour, Bist *et al.*[25] used the non-observation of such an ^rR-branch edge in concluding that it lies under the intense high-wavenumber peak in Figure 3(*a*). Figure 3(*d*), the computed type C contour, shows that this conclusion was correct. Since equation (11) is a symmetric-rotor expression, it must apply equally to type B and type C bands of prolate asymmetric rotors, and Figure 3(*d*) shows a sharp edge at $+3.15$ cm^{-1} which is exactly coincident with the high-wavenumber intense peak of the type B contour. The contour of the O_0^0 band of the 270 nm system of chlorobenzene,[29] shown in Figure 5, is an example of a type B band of a prolate asymmetric rotor which does show a clear separation of the intense high-wavenumber peak [given by equation (10)], at $+2.60$ cm^{-1} from the band origin, from the edge [given by equation (11)], at $+3.53$ cm^{-1} from the band origin.

Type A and type B bands in oblate asymmetric rotors might show turning-points in first lines of ^pP or ^rR branches since a turning-point will be formed having a wavenumber given by equation (11) in which A' and ΔA are replaced by C' and ΔC respectively. However, no examples of such turning-points are known.

A head of R-branch heads may be formed quite commonly for any band type which correlates with a perpendicular band in a symmetric rotor. In a prolate asymmetric rotor this produces a sharp edge to the contour on the high-wavenumber side. Such a feature has been found, consisting of a head of ^rR-branch heads, in the O_0^0 band of the 274 nm system of benzonitrile,[30] in the O_0^0 band of the 279 nm system of phenylacetylene,[31] and in the 44_0^1

[29] T. Cvitaš and J. M. Hollas, *Mol. Phys.*, 1970, **18**, 101.
[30] J. C. D. Brand and P. D. Knight, *J. Mol. Spectroscopy*, 1970, **36**, 328.
[31] G. W. King and S. P. So, *J. Mol. Spectroscopy*, 1971, **37**, 543.

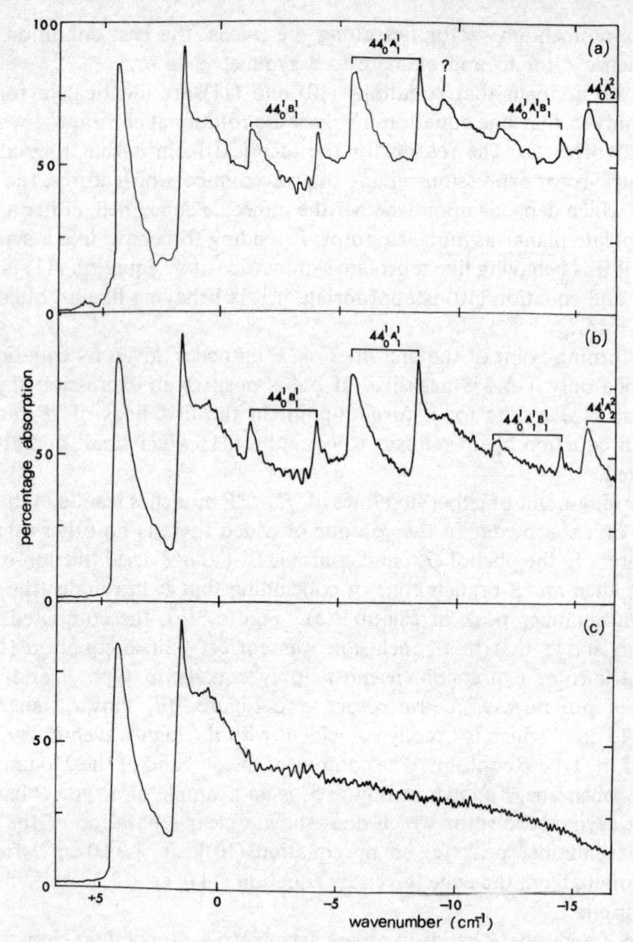

Figure 6 (*a*) *Observed rotational contour of the type* B 44^1_0 *band in the* 312 nm *system of naphthalene with overlapping sequence bands* $44^1_0B^1_1$, $44^1_0A^1_1$, $44^1_0A^1_1B^1_1$, and $44^1_0A^2_2$;
(*b*) *the best computed type* B *contour with similar contours added to simulate sequence bands;*
(*c*) *the best single computed type* B *contour.*
(Reproduced by permission from *Mol. Phys.*, 1971, **22**, 203)

band of the 312 nm system of naphthalene,[32, 33] shown in Figure 6. In naphthalene the feature is at +4.54 cm^{-1} from the band origin. This kind of feature is commonly formed at moderate values of K_a and J and the posi-

[32] K. K. Innes, J. E. Parkin, D. K. Ervin, J. M. Hollas, and I. G. Ross, *J. Mol. Spectroscopy*, 1965, **16**, 406.
[33] J. M. Hollas and S. N. Thakur, *Mol. Phys.*, 1971, **22**, 203.

tion relative to the band origin can be well reproduced by second-order perturbation or symmetric-rotor calculations.[33] It follows that a type *C* contour computed with the same rotational constants should show a high-wavenumber edge in exactly the same position as in the type *B* contour since both band types correlate with a perpendicular band in the prolate-symmetric-rotor approximation. The coincidence of the type *B* and type *C* high-wavenumber edges is evident in Figures 1(*c*) and 2(*d*) of ref. 33.

The expression for the wavenumber of a head of ^rR-branch heads in a perpendicular band of a symmetric rotor can be obtained quite easily. Sub-band origin wavenumbers are given by equation (6) for a prolate symmetric rotor. Differentiating with respect to K_a and putting the resultant expression equal to zero gives the value of K_a at the turning-point of sub-band origins as:

$$(K_a)_{rev} = -(A'-\bar{B}')/\Delta(A-\bar{B}) \tag{14}$$

Using this value of K_a in equation (6) gives the wavenumber of the head of sub-band origins as:

$$\tilde{\nu}_{rev} = \tilde{\nu}_0+(A'-\bar{B}')-(A'-\bar{B}')^2/\Delta(A-\bar{B}) \tag{15}$$

The separation of an ^rR-branch head from a sub-band origin $\Delta\nu_{rev}$ is given by equation (16), where J_{rev} is the value of J at the head given by equation (8). The wavenumber of the head of ^rR-branch heads $(\tilde{\nu}_R)_{rev}$ is then $\tilde{\nu}_{rev}+\Delta\tilde{\nu}_{rev}$, so that $(\tilde{\nu}_R)_{rev}$ is defined by equation (17).

$$\Delta\tilde{\nu}_{rev} = \bar{B}'(J_{rev}+1)(J_{rev}+2)-\bar{B}''J_{rev}(J_{rev}+1) \tag{16}$$

$$(\tilde{\nu}_R)_{rev} = \tilde{\nu}_0+A'+\bar{B}'-(A'-\bar{B}')^2/\Delta(A-\bar{B})-(\bar{B}''-3B')^2/4\Delta\bar{B} \tag{17}$$

The formation of a head of ^rR-branch heads requires that both ΔA and $\Delta\bar{B}$ be negative: if they are both positive a head of ^pP-branch heads may be formed and an expression [equation (18)] for the wavenumber of such a head, analogous to equation (17), can be derived.

$$(\tilde{\nu}_P)_{rev} = \tilde{\nu}_0+A'-\bar{B}'-(A'-\bar{B}')^2/\Delta(A-\bar{B})-(\bar{B}''+\bar{B}')^2/4\Delta\bar{B} \tag{18}$$

In cases where J_{rev} is high, second-order perturbation calculation of the wavenumber of the head of heads might be necessary.

For an oblate asymmetric rotor, *A* in equations (17) and (18) must be replaced by *C*.

If a molecule has an internal torsional motion of one part of the molecule with respect to another which is sufficiently free for it to be treated as an internal rotation rather than a torsional vibration, the rotational problem is considerably more complicated than for a rigid molecule. For a prolate symmetric rotor, for example, in which internal rotation is completely free, the rotational energy levels are given by equation (19), where A_F and A_T

$$F(J,K_a,m) = BJ(J+1)+(A_F-B)K_a^2+(A_F+A_T)m^2-2A_FmK_a \tag{19}$$

are the rotational constants (*A*) of the 'frame' and the 'top' respectively (the 'top' is regarded as rotating, the 'frame' as fixed), and $m = 0, \pm 1, \pm 2\ldots$ is the quantum number associated with the angular momentum due to internal rotation. If there is a non-zero barrier to internal rotation then equation (19) has to be modified and, if the molecule is an asymmetric rotor as well, the rotational problem becomes still more complex. The problem of an asymmetric rotor internally rotating with non-zero barrier has been solved and microwave spectra of such molecules have been interpreted, but no-one has written a computer program to compute rotational band contours for electronic spectra of large, internally rotating, asymmetric rotors. *p*-Fluorotoluene[34] is the only such molecule for which an attempt has been made to obtain rotational information from its band contours. Attempts were made to simulate the O_0^0 band contour at 271 nm by two approximate methods:

(i) treating the molecule as a freely internally rotating symmetric rotor (the barrier is only 4.836 cm^{-1} in the ground state), and
(ii) treating it as a rigid asymmetric rotor.

Method (i) showed that the contour is very sensitive to A_T (the 'top' is the CH_3 group) particularly if, as in this case, ΔA_T is positive and the main part of the band is degraded to low wavenumber. The positive ΔA_T produces a strong degradation to *high* wavenumber of fairly weak, but easily observed, structure and indicates that the hydrogen atoms are closer to the top axis in the excited state than in the ground state. Method (ii) gave values of ΔB, ΔC, and ΔA_F but with greater uncertainties than are usual for a rigid asymmetric rotor. It seems that the barrier to internal rotation, even though it is low in the ground state, and presumably in the excited state as well, may have a serious effect on the contour by strongly affecting those energy levels having very low values of m that are involved in very intense transitions.

Compared with the molecules which are prolate, not many are oblate asymmetric rotors. The only oblate asymmetric rotors whose rotational band contours have been studied are *sym*-tetrazine,[15] pyrazine,[10, 35] and pyridazine.[36, 37] Most of the bands of the long-wavelength singlet systems of all three molecules are type *C* bands, with small changes of rotational constants from the ground to the excited state. As a result, many of the type *C* bands show prominent coincidences of ^qP- and of ^qR-branch transitions. Each resulting group of lines corresponds to a particular value of J and several values of K_c. The symmetric-rotor approximation was used in spectra of these molecules to obtain $\bar{B}'$ (and also $\bar{B}''$, since the ground-state rotational constants were rather uncertain) by the standard method of combination differences. The formation of such line coincidences was shown[15] to be very sensitive to rotational constants. This sensitivity is demonstrated by the fact that the O_0^0 band of the 552 nm system of *sym*-tetrazine does not show

[34] T. Cvitaš and J. M. Hollas, *Mol. Phys.*, 1971, **20**, 645.
[35] K. K. Innes, J. D. Simmons, and S. G. Tilford, *J. Mol. Spectroscopy*, 1963, **11**, 257.
[36] K. K. Innes, J. A. Merritt, W. C. Tincher, and S. G. Tilford, *Nature*, 1960, **187**, 500.
[37] K. K. Innes, W. C. Tincher, and E. F. Pearson, *J. Mol. Spectroscopy*, 1971, **39**, 171.

prominent ${}^qP$- or ${}^qR$-line coincidences whereas some bands involving vibrational excitation do show such coincidences.[15] (The coincidences observed at medium resolution in the O_0^0 band of *s*-tetrazine [7] have been shown [15] at higher resolution to be not as simple as was previously thought.)

In a planar molecule the moment of inertia about the out-of-plane principal axis is equal to the sum of those about the in-plane axes, *i.e.*

$$I_C = I_A + I_B \tag{20}$$

provided that we neglect (*a*) contributions to the inertial defect $\Delta (= I_C - I_A - I_B)$ by out-of-plane vibrations, and (*b*) Coriolis interactions in cases of bands involving vibrational excitation. The effect of out-of-plane vibrations on Δ has been discussed by McHugh and Ross [38] on the basis of theory derived by Oka and Morino.[39] The conclusion is that low-wavenumber out-of-plane vibrations have the greatest effect on Δ, tending to give a non-zero, negative value in any one electronic state. However, what is important in simulation of band contours, which are most sensitive to changes of rotational constants, is the *change* of Δ from the lower to the upper electronic state. This change is sensitive to changes in the wavenumber of low-wavenumber out-of-plane vibrations from the lower to the upper state. The change of Δ in the O_0^0 band of the 350 nm system of azulene is from $\Delta'' = -2.5 \times 10^{-30}$ kg nm^2 (-0.15 u Å^2) in the ground state to $\Delta' = -5 \times 10^{-30}$ kg nm^2 (-0.3 u Å^2) in the excited state due to decreases in the wavenumbers of low-wavenumber out-of-plane vibrations in the excited state.[38] In many cases, however, vibrational effects on the difference between the value of Δ in zero-point levels of the two combining electronic states are too small to be detected. In such cases equation (20) reduces three unknown excited-state rotational constants to two.

In the 7_0^1 band of the 500 nm system of *p*-benzoquinone Δ' was found to be -10×10^{-30} kg nm^2 (-0.8 u Å^2), compared with an assumed Δ'' of zero.[24] This is due to Coriolis interaction between the 7^1 level and an X^1 level, where X is a b_{1u} or b_{2u} vibration. If it were b_{1u}, as seems likely, a weak interaction between this and the a_u vibration ν_1 would affect only the A' rotational constant in giving a non-zero value of Δ'. In the 274 nm system of benzonitrile [30] there is another case of Coriolis interaction, this time between a b_1 and a b_2 vibration, in which either Δ' or Δ'' is non-zero due to the effect of the interaction on the rotational constant A.

During the past ten years much information has been forthcoming on the detailed structure of rotational band contours of asymmetric rotors so that, at the present time, there are many possible ways in which trial excited-state rotational constants may be limited even before any contours are computed. Much of this information has been given in equations (1)—(20).

If the transition moment associated with a band whose rotational contour is to be analysed is not confined by symmetry to any one of the principal

[38] A. P. McHugh and I. G. Ross, *Austral. J. Chem.*, 1969, **22**, 1.
[39] T. Oka and Y. Morino, *J. Mol. Spectroscopy*, 1961, **6**, 472.

axes of the molecule, then a hybrid band may result. The only kind which has been encountered and analysed is the type *A*– type *B* hybrid band in which the transition moment lies in the plane of a planar prolate asymmetric rotor. Such bands take the form of normal type *A* and type *B* bands simply added together in appropriate proportions. The proportions are, as has been pointed out by Ross,[40] related to the direction of polarization of the transition moment. If the transition moment R makes an angle θ with the *a*-axis then the components R_a and R_b, along the *a*- and *b*-axes respectively, are given by

$$R_a = R\cos\theta \tag{21a}$$

$$R_b = R\sin\theta \tag{21b}$$

Since the intensity I is proportional to the square of the transition moment,

$$I_b/I_a = \tan^2\theta \tag{22}$$

where I_a and I_b are intensities of the type *A* and type *B* components of the band respectively. In the computed contour, when the rotational constants have been found, type *A* and type *B* contours are added in various proportions until the resulting contour agrees with that observed. From equation (22) the magnitude *but not the sign* of θ can then be obtained from I_a and I_b.

In principle, if a band is of hybrid character there is a chance of using prominent features in more than one band type to obtain the excited-state rotational constants. In practice, though, the superposition of two band types can lead to a confusion of overlapping fine structure which might have been useful in either band type alone.

The O_0^0 band of the 270 nm system of *m*-chlorofluorobenzene, whose contour was analysed by Wu and Lombardi,[41] provides a good example of a type *A*– type *B* hybrid band. The observed band is shown at the top of Figure 7 and comprises 90% type *A* and 10% type *B* character. Although the proportion of type *B* character is small, it shows clearly in the observed contour as a wing on the high-wavenumber side. The type *B* computed contour shows a prominent head of ^rR-branch heads, producing the high-wavenumber edge given by equation (17). The type *A* computed contour shows a strong peak which is a head of $^{qr}R^e$ and $^{qr}R^o$ coincident sub-branches with low K_a and high J and given by equation (10). In addition, the type *A* contour shows ^qR-branch heads.

Symmetric Rotors. The symmetric rotor is, of course, the exception rather than the rule, but of the few large symmetric-rotor molecules which are known only the spectrum of benzene has been treated by the method of rotational band contour analysis.[8]

Most of the bands in the 260 nm system of benzene are perpendicular bands of an oblate symmetric rotor. Benzene is planar, so the requirement of equation (20) together with the fact that $I_A = I_B$ would apparently make the

[40] I. G. Ross, *Adv. Chem. Phys.*, 1971, **20**, 341.
[41] C-Y. Wu and J. R. Lombardi, *J. Mol. Spectroscopy*, 1971, **39**, 345.

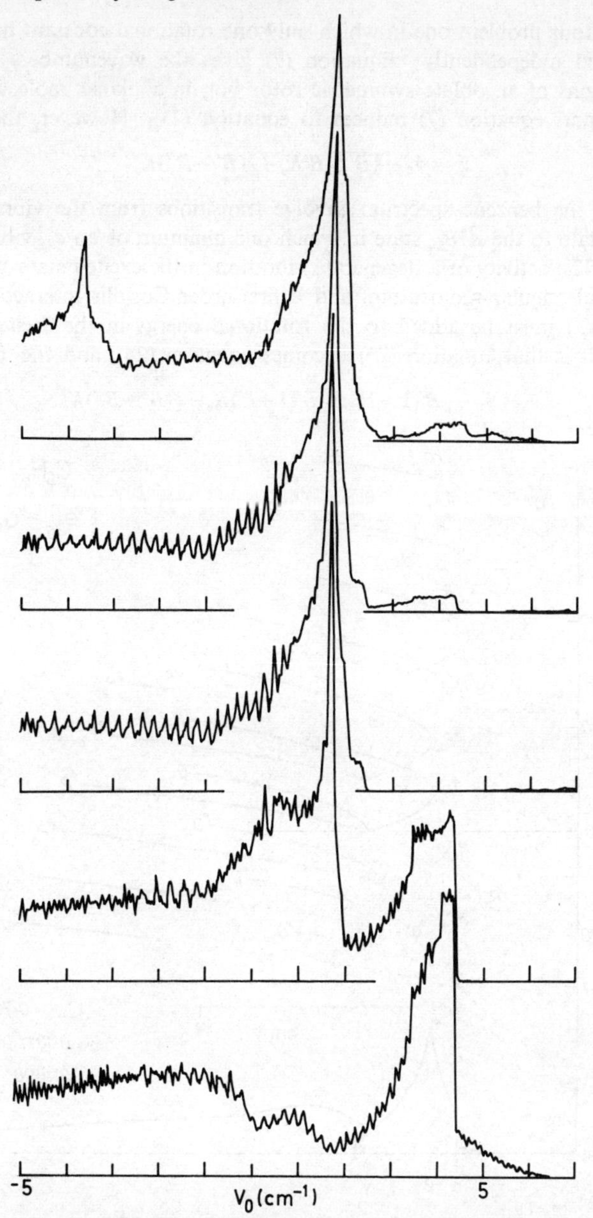

Figure 7 *At the top is the observed rotational contour of the* O_0^0 *band of the* 270 nm *system of* m-*chlorofluorobenzene. Underneath are, in succession, a computed hybrid band contour with* 90% *type* A *and* 10% *type* B *character, and then pure type* A, *type* B, *and type* C *computed contours. Absorption is plotted upwards on the vertical scale.*

(Reproduced by permission from *J. Mol. Spectroscopy*, 1971, **39**, 345)

band contour problem one in which only one rotational constant has to be determined independently. Equation (7) gives the wavenumbers of sub-band origins of an oblate symmetric rotor but, in a planar molecule, $A = B = 2C$ and equation (7) reduces to equation (23). However, the strong

$$\tilde{\nu} = \tilde{\nu}_0 - \tfrac{1}{2}B' \mp B'K_c - \tfrac{1}{2}(B' - B'')K_c^2 \tag{23}$$

bands in the benzene spectrum involve transitions from the vibrationless ground state to the $\tilde{A}^1B_{2u}$ state in which one quantum of an e_{2g} vibration is excited. The activity of a degenerate vibration in the excited state produces vibrational angular momentum, and a first-order Coriolis interaction term ($\mp 2C'\zeta'K_c'$) must be added to the rotational energy in the excited state. The result is that equation (23) becomes equation (24), and the fitting of

$$\tilde{\nu} = \tilde{\nu}_0 - \tfrac{1}{2}B'(1+2\zeta') \mp B'(1+\zeta')K_c - \tfrac{1}{2}(B' - B'')K_c^2 \tag{24}$$

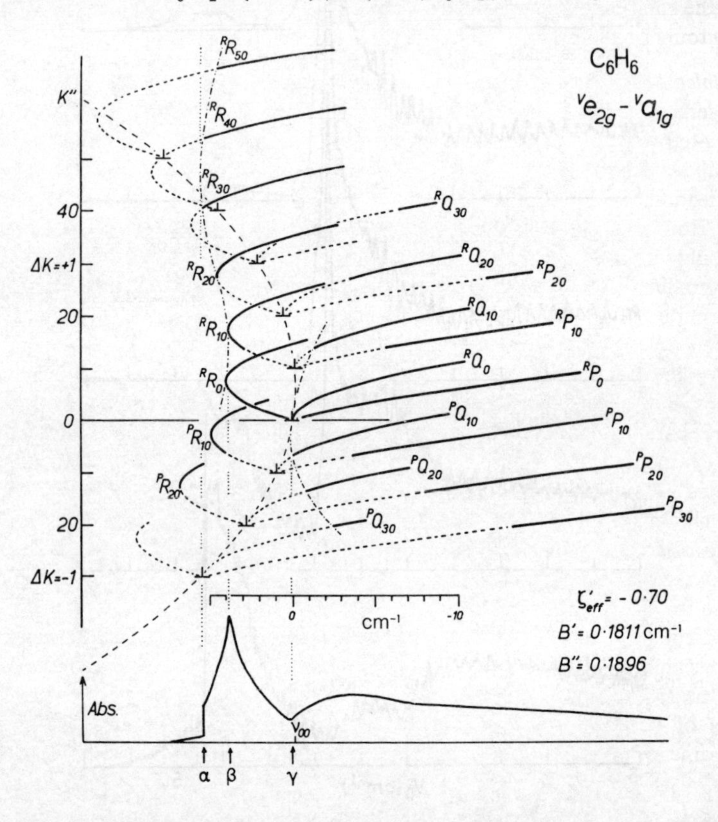

Figure 8 *A contour computed for benzene with rotational constants similar to those in the* 6_0^1 *band of the* 260 nm *system. The Fortrat curves above the contour demonstrate the way in which the prominent features* α, β, *and* γ *are formed.*
(Reproduced by permission from *Phil. Trans. Roy. Soc.*, 1966, **A259,** 499)

rotational contours involves the determination of two parameters, B' and ζ'. Figure 8 shows a computed contour which is very similar to that of the 6_0^1 band, where ν_6 is the e_{2g} vibration most strongly active in the electronically forbidden system. Three prominent features of the contour are labelled α, β, and γ. The feature α is a head of first lines of ^rR branches (called by Callomon *et al.*[8] a 'terminus') and is similar (not completely analogous, because of the terms involving ζ) to such heads formed in some contours of asymmetric rotors discussed in the previous section. β is a head of ^rR-branch heads, also similar to such a feature in asymmetric rotor contours, and γ is an intensity minimum, due to ^rQ branches, which is very close to the band origin. The perpendicular band contours were shown to be very sensitive to the value of ζ' as well as to B'. In bands involving excitation of an e_{2g} vibration in the ground electronic state the term $\mp 2C''\zeta''K_c''$ has to be added to the rotational energy in this state: ζ'' was determined by analysis of the contours of such bands.

Triplet–Singlet Transitions. A few triplet–singlet transitions have been observed in large asymmetric-rotor molecules, for example in pyrazine,[10] *sym*-tetrazine,[43] *p*-benzoquinone,[20] and benzaldehyde,[44] but no band contour analyses have been reported. The rotational structure of triplet–singlet transitions in near-symmetric rotors has been discussed in some detail by Hougen.[45] It is expected that most non-linear polyatomic molecules should, unless they contain a very heavy atom, have triplet states which approximate to Hund's case (*b*) and, for this case, the general rotational selection rules are $\Delta K = 0, \pm 1, \pm 2$ and $\Delta N = 0, \pm 1, \pm 2$ (N replaces J when $S \neq 0$) for a triplet–singlet transition compared to $\Delta K = 0, \pm 1$ and $\Delta J = 0, \pm 1$ for a singlet–singlet transition. Not only rotational selection rules but also intensities are modified: they depend on, at the most, three parameters which are determined by the degree of spin–orbit interaction which may mix any of the three components of the triplet state with singlet states. However, it is likely that one of the mixing mechanisms will be dominant, in which case two of the parameters are approximately zero. The resemblance between triplet–singlet and singlet–singlet band contours in pyrazine,[42] *sym*-tetrazine,[43] and *p*-benzoquinone [20] suggests that this is the case for these three molecules.

Creutzberg and Hougen [46, 47] have given details of rotational energy levels and line intensities for triplet–singlet transitions in molecules belonging to the D_{2h}, C_{2v}, and D_2 point groups, but they point out the possible difficulties of band contour calculations because of the number of adjustable parameters, namely, three triplet-state rotational constants, two triplet-state

[42] K. K. Innes and L. E. Giddings, jun., *Discuss. Faraday Soc.*, 1963, No. 35, p. 192.
[43] D. T. Livak and K. K. Innes, *J. Mol. Spectroscopy*, 1971, **39**, 115.
[44] M. Stockburger, *Z. phys. Chem. (Frankfurt)*, 1962, **31**, 350.
[45] J. T. Hougen, *Canad. J. Phys.*, 1964, **42**, 433.
[46] F. Creutzberg and J. T. Hougen, *Canad. J. Phys.*, 1967, **45**, 1363.
[47] F. Creutzberg and J. T. Hougen, *J. Mol. Spectroscopy*, 1971, **38**, 257.

Table 2 *Summary of rotational band contour analyses*

Molecule[a]	*Bands analysed*[b]	*Band type*	*Major geometry changes in excited state*	*Reference*
(i) *Bicyclic*				
Azulene (P) ([$^1H_8$] and [$^2H_8$])	O_0^0 (350 nm)	*A*	Expansion of all C—C bonds by *ca.* 0.001—0.002 nm, except central bond which contracts by *ca.* 0.003 nm	23, 48
Benzofuran (P)	O_0^0 (278 nm)	90%*A* 10%*B*	General expansion but no detailed interpretation possible	49
2,1,3-Benzo-selenadiazole (P)	O_0^0 (356 nm)	*B*	Similar to benzothiadiazole	50
2,1,3-Benzo-thiadiazole (P)	O_0^0 (328 nm)	*B*	Slight contraction about *z*-axis possibly owing to increase in ∠C(5)C(4)C(9)	51
	12_0^1	*A*		
Indene (P)	O_0^0 (288 nm)	88%*A* 12%*B*	General expansion but no detailed interpretation possible	52
Indole (P)	O_0^0 (284 nm)	80%*A* 20%*B*	General expansion but no detailed interpretation possible	53
Naphthalene (P)	O_0^0 (312 nm)	*A*	Δr[C(1)C(9)] ≈ Δr[C(2)C(3)] ≈ 0, Δr[C(9)C(10)] ≈ 0.005 nm, Δr[C(1)C(2)] ≈ 0.003 nm	32, 33
	44_0^1	*B*		
Thionaphthene (P)	O_0^0 (294 nm)	10%*A* 90%*B*	General expansion but no detailed interpretation possible	49

[a] (P) or (O) implies that the molecule is a prolate or oblate rotor respectively.
[b] (i) The vibrational numbering used here is the same as that used by the authors quoted and is not necessarily systematic.
(ii) When various isotopic substitutions have been made, the wavelength of the O_0^0 band is given for normal isotopes.

[48] A. J. McHugh and I. G. Ross, *Austral. J. Chem.*, 1968, **21**, 3055.
[49] A. Hartford, jun., A. R. Muirhead, and J. R. Lombardi, *J. Mol. Spectroscopy*, 1970, **35**, 199.
[50] J. M. Hollas and R. A. Wright, *Spectrochim. Acta*, 1969, **25A**, 1211.
[51] J. Christoffersen, J. M. Hollas, and R. A. Wright, *Proc. Roy. Soc.*, 1969, **A308**, 537.
[52] A. Hartford, jun. and J. R. Lombardi, *J. Mol. Spectroscopy*, 1970, **34**, 257.
[53] A. Mani and J. R. Lombardi, *J. Mol. Spectroscopy*, 1969, **31**, 308.

Table 2 (*Contd.*)

Molecule[a]	Bands analysed[b]	Band type	Major geometry changes in excited state	Reference
(ii) *Monocyclic*				
2-Aminopyridine (P)	0^0_0 (298 nm)	55%*A* 45%*B*	Similar to aniline	54
Aniline (P)	0^0_0 (294 nm)	*B*	Δr(CN)≈−0.008 nm, all Δr(CC)≈0.004 nm, $\Delta\angle$C(6)C(1)C(2)≈4.1°	13, 55
Benzene (O) ([$^1H_6$] and [$^2H_6$])	6^1_0 (259 nm) and others	⊥	Δr(CC) = 0.0038 nm, Δr(CH) = −0.001 nm	8
Benzonitrile (P) ([$^2H_0$] and [4-$^2H_1$])	0^0_0 (274 nm)	*B*	General expansion	30
p-Benzoquinone (P)	7^1_0 (477 nm)	*A*	Probably C=C and C=O bonds lengthen and all C—C bonds shorten	20, 24
	$7^1_0\,30^1_1$	*A*		
	30^1_0	*B*		
Bromobenzene (P)	0^0_0 (270 nm)	*B*	Δr(CBr)≈−0.004 nm, all Δr(CC)≈0.004 nm, $\Delta\angle$C(6)C(1)C(2)≈1.3°	56, 55
	29^1_0	*A*		
Chlorobenzene (P)	0^0_0 (270 nm)	*B*	Δr(CCl)≈−0.004 nm, all Δr(CC)≈0.004 nm, $\Delta\angle$C(6)C(1)C(2)≈1.2°	29, 55
	29^0_1	*A*		
o-Chlorofluoro-benzene (P)	0^0_0 (270 nm)	65%*A* 35%*B*	General expansion but no detailed interpretation possible	57
m-Chlorofluoro-benzene (P)	0^0_0 (270 nm)	90%*A* 10%*B*	General expansion but no detailed interpretation possible	41
p-Chlorofluoro-benzene (P)	0^0_0 (276 nm)	*B*	Δr(CCl), Δr(CF), Δr(CC), $\Delta\angle$C(6)C(1)C(2), $\Delta\angle$C(3)C(4)C(5) similar to chlorobenzene and fluorobenzene but Δr[C(2)C(3)] may be ≈−0.002 nm	58, 55

[54] J. M. Hollas, G. H. Kirby, and R. A. Wright, *Mol. Phys.*, 1970, **18**, 327.
[55] T. Cvitaš, J. M. Hollas, and G. H. Kirby, *Mol. Phys.*, 1970, **19**, 305.
[56] T. Cvitaš, *Mol. Phys.*, 1970, **19**, 297.
[57] A. Hartford, jun. and J. R. Lombardi, *J. Mol. Spectroscopy*, 1971, **40**, 262.
[58] T. Cvitaš and J. M. Hollas, *Mol. Phys.*, 1970, **18**, 261.

Table 2 (*Contd.*)

Molecule[a]	*Bands analysed*[b]	*Band type*	*Major geometry changes in excited state*	*Reference*
p-Dichlorobenzene (P)	O_0^0 (280 nm)	*B*	Δ*r*(CCl)≈−0.004 nm, all Δ*r*(CC)≈0.004 nm except Δ*r*[C(2)C(3)] ≈0.003 nm, Δ∠C(6)C(1)C(2)≈1.3°	59, 55
	0+538 cm⁻¹	*A*		
o-Difluorobenzene (P)	O_0^0 (264 nm)	*A*	General expansion	57
m-Difluorobenzene (P)	O_0^0 (264 nm)	*A*	General expansion	57
p-Difluorobenzene (P)	O_0^0 (271 nm)	*B*	Δ*r*(CF)≈−0.003 to −0.001 nm, all Δ*r*(CC)≈0.004 nm except Δ*r*[C(2)C(3)] ≈0.003 nm, Δ∠C(6)C(1)C(2)≈2.4°	60, 61, 55
Difluorodiazirine (P)	O_0^0 (352 nm)	*B*	Δ*r*(NN)≈0.006 nm, Δ*r*(FF)≈−0.003 nm	62
p-Fluoroaniline (P)	O_0^0 (306 nm)	*B*	Δ*r*(CN), Δ*r*(CF), Δ∠C(6)C(1)C(2), Δ*r*(CC) similar to aniline and fluorobenzene but Δ*r*[C(2)C(3)] may be as small as 0.001 nm and Δ∠C(3)C(4)C(5)≈0.2°	63, 55
Fluorobenzene (P)	O_0^0 (264 nm)	*B*	Δ*r*(CF)≈−0.003 to −0.001 nm, all Δ*r*(CC)≈0.004 nm, Δ∠C(6)C(1)C(2)≈3.7° to 1.3° when Δ∠C(3)C(4)C(5)≈0° to 2.3°	64, 55
p-Fluorophenol (P)	O_0^0 (285 nm)	*B*	Δ*r*(CO), Δ*r*(CF), Δ*r*(CC), Δ∠C(6)C(1)C(2), Δ∠C(3)C(4)C(5) similar to phenol and fluorobenzene but Δ*r*[C(2)C(3)] may be as small as 0.002 nm	63, 55

[59] T. Cvitaš and J. M. Hollas, *Mol. Phys.*, 1970, **18**, 801.
[60] T. Cvitaš and J. M. Hollas, *Mol. Phys.*, 1970, **18**, 793.
[61] Y. Udagawa, M. Ito, and S. Nagakura, *J. Mol. Spectroscopy*, 1970, **36**, 541.
[62] J. R. Lombardi, W. Klemperer, M. B. Robin, H. Basch, and N. A. Kuebler, *J. Chem. Phys.*, 1969, **51**, 33.
[63] J. Christoffersen, J. M. Hollas, and G. H. Kirby, *Mol. Phys.*, 1970, **18**, 451.
[64] G. H. Kirby, *Mol. Phys.*, 1970, **19**, 289.

Table 2 (*Contd.*)

Molecule[a]	*Bands analysed*[b]	*Band type*	*Major geometry changes in excited state*	*Reference*
p-Fluorotoluene (P)	O^0_0 (271 nm)	*B*	Moment of inertia of CH_3 about top axis decreases. $\Delta r(C—CH_3)$ ≈ -0.003 nm if rest of molecule assumed to behave like fluorobenzene except for $\Delta r[C(2)C(3)] \approx 0.002$ nm	34
Phenol (P)	O^0_0 (275 nm)	*B*	$\Delta r(CO) \approx -0.004$ nm, all $\Delta r(CC) \approx 0.004$ nm, $\Delta\angle C(6)C(1)C(2) \approx 3.7°$	25, 26, 55
	$6b^1_0$	*A*		
Phenylacetylene (P)	O^0_0 (279 nm)	*B*	$\Delta r(\equiv C—C) \approx -0.003$ nm, $\Delta r(C\equiv C) \approx 0.0004$ nm, all $\Delta r(CC) \approx 0.004$ nm	65, 31
	35^1_0	*A*		
Phenyl isocyanide (P)	O^0_0 (272 nm)	*B*	General expansion but detailed interpretation not attempted	66
	35^1_0	*A*		
Pyrazine (O) ([$^{14}N_2$] and [$^{14}N^{15}N$])	O^0_0 (324 nm)	*C*	Moment of inertia about N—N axis decreases from ground state	10, 35, 67, 68
	$6a^1_0$	*C*		
	$16b^1_0$	*C*		
	11^1_0	*C*		
	19^1_0	*C*		
	$6a^0_1$	*C*		
	5^1_0	*A*		
	O^0_0 (370 nm)	Triplet–singlet	No conclusions	42, 67
Pyridazine (O) ([$^2H_0$], [$^2H_4$], and [3,6-$^2H_2$])	O^0_0 (376 nm)	*C*	Molecule elongates along in-plane axis perpendicular to C_2 axis	36, 69, 70, 67
	$6a^1_0$	*C*		
	$16b^2_0$	*C*		
	$6a^0_1$	*C*		
	$6a^1_1$	*C*		
	O^0_0 (440 nm)	Triplet–singlet	No conclusions	64, 61

[65] G. W. King and S. P. So, *J. Mol. Spectroscopy*, 1970, **33**, 376.
[66] A. R. Muirhead, A. Hartford, jun., K.-T. Huang, and J. R. Lombardi, *J. Chem. Phys.*, 1972, **56**, 4385.
[67] K. K. Innes, J. P. Byrne, and I. G. Ross, *J. Mol. Spectroscopy*, 1967, **22**, 125.
[68] K. K. Innes and J. A. Merritt, *J. Mol. Spectroscopy*, 1967, **23**, 280.
[69] K. K. Innes, W. C. Tincher, and E. F. Pearson, *J. Mol. Spectroscopy*, 1970, **36**, 114.
[70] K. K. Innes and R. M. Lucas, jun., *J. Mol. Spectroscopy*, 1967, **24**, 247.

Table 2 (*Contd.*)

Molecule[a]	*Bands analysed*[b]	*Band type*	*Major geometry changes in excited state*	*Reference*
[4-^{2}H]Pyridine	O_0^0 (288 nm)	*C*	General expansion	71
Pyridine *N*-oxide (P)	O_0^0 (342 nm)	*B*	Expansion of ring, contraction of N—O bond	72
Pyrimidine (O) ([^{2}H$_0$] and [^{2}H$_4$])	O_0^0 (322 nm)	*C*	General expansion	73, 67
Styrene (P)	O_0^0 (288 nm)	*A*	General expansion	74
sym-Tetrazine (O) ([^{2}H$_0$], [^{2}H$_1$], [^{2}H$_2$], [1,4-^{15}N$_2$], [1,4-^{15}N$_2$, ^{2}H$_1$], [1,4-^{15}N$_2$, ^{2}H$_2$])	O_0^0 (552 nm)	*C*	Δr(NN) = −0.011 nm, Δr(CC) = 0.010 nm, $\Delta\angle$(NCN) = −8°	7, 15, 28, 67, 75
	A_0^1	*C*		
	O_0^0 (735 nm)	Triplet–singlet	Δr(NN) ≈ −0.006 nm, $\Delta\angle$NCN ≈ 3.4°	43
Tropolone (P)	O_0^0	*B*	Possibly all Δr(CC) ≈ 0.0024 nm, both Δr(CO) ≈ −0.006 nm	76
	H_1^1	*B*		
(iii) *Non-cyclic*				
Acrolein[c] (P) *sym-trans*	O_0^0 (386 nm)	*C*	$\Delta\angle$CCO ≈ $\Delta\angle$CCC ≈ 3.5° Δr(C—C) ≈ −Δr(C=C) but magnitude uncertain	77
sym-cis	O_0^0 (406 nm)	*C*	Similar to *sym-trans*	77
F_2CN (O)	O_0^0 (362 nm)	*A*	Δr(CF) = −0.0024 nm, Δr(CN) = 0.0043 nm, $\Delta\angle$FCF ≈ 1°	78
Thiophosgene (O)	$2_0^1 3_0^1 4_0^1$ (514 nm)	*B*	Planar ground state; out-of-plane angle between C—S bond and bisector of $\angle$ClCCl ≈ 27° in excited state: but symmetric rotor approximation used for computed contour.	79

[c] Acrolein is not a large molecule as defined here, but the rotational half-intensity line-width is so large (2 cm^{-1}) that rotational analysis is possible only by the band contour method.

[71] F. W. Birss, S. D. Colson, J. P. Jesson, H. Kroto, and D. A. Ramsay, to be published.
[72] J. C. D. Brand and K.-T. Tang, *J. Mol. Spectroscopy*, 1971, **39**, 171.

spin-splitting parameters, and two or three spin–orbit coupling intensity parameters.

Results.—Table 2 lists all the molecules whose spectra have been investigated using the rotational band contour simulation method. A few of these molecules are not particularly 'large' but are included because the band contour method had to be used either because the rotational transitions are very crowded or because the rotational linewidth is very large.

The fact that band contour analysis was possible in the cases of difluorodiazirine [62] and *o*-difluorobenzene,[57] both of which have $\kappa'' \approx -0.07$, shows that there is reason to believe that there is no limit to the value of κ for which the band contour method is applicable.

Transition-moment Directions. It comes as no surprise that the electronic transition moment in the first singlet–singlet system of monosubstituted and *p*-disubstituted benzenes (belonging to the C_{2v} or D_{2h} point groups) in the fourteen molecules so far investigated lies in the molecular plane and perpendicular to the carbon–substituent bonds. This result is the one predicted by analogy with the benzene 260 nm system assuming the latter to be $\tilde{A}^1B_{2u}$–$\tilde{X}^1A_{1g}$, an assignment still not confirmed by experimental investigations of the spectrum [8] but unambiguously suggested by theory. It can be argued that the results for the C_{2v} and D_{2h} substituted benzenes suggest that the molecular orbitals are sufficiently similar to unperturbed benzene orbitals that the results provide strong experimental evidence for the benzene $\tilde{A}^1B_{2u}$–$\tilde{X}^1A_{1g}$ assignment.

In some cases transition-moment directions have been determined where either there were no prior indications or where it turned out that such indications were incorrect. The $\tilde{A}^1B_2$–$\tilde{X}^1A_1$ transition in pyridine *N*-oxide [72] was expected to be a π^*–n transition, polarized perpendicular to the molecular plane, but it turned out to be a π^*–π transition polarized in the plane and perpendicular to the N—O bond. The $\tilde{A}^1A'$–$\tilde{X}^1A'$ transition in 2-aminopyridine [54] has been shown to be a π^*–π transition polarized in the molecular plane, whereas a π^*–n transition polarized perpendicular to the plane might have been anticipated. In the first singlet–singlet transitions in 2,1,3-benzothiadiazole,[51] 2,1,3-benzoselenadiazole,[50] and tropolone [76] there were no strong reasons for anticipating any particular polarization direction.

The direction of the transition moment has been determined in the 0^0_0 band of the 265 nm system of *o*-difluorobenzene,[57] the 264 nm system of

[73] K. K. Innes, H. D. McSwiney, jun., J. D. Simmons, and S. G. Tilford, *J. Mol. Spectroscopy*, 1969, **31**, 76.
[74] A. Hartford, jun. and J. R. Lombardi, *J. Mol. Spectroscopy*, 1970, **35**, 413.
[75] K. K. Innes, A. Y. Khan, and D. T. Livak, *J. Mol. Spectroscopy*, 1971, **40**, 177.
[76] A. C. P. Alves and J. M. Hollas, *Mol. Phys.*, 1972, **23**, 927.
[77] A. C. P. Alves, J. Christoffersen, and J. M. Hollas, *Mol Phys.*, 1971, **20**, 625.
[78] R. N. Dixon, G. Duxbury, R. C. Mitchell, and J. P. Simons, *Proc. Roy. Soc.*, 1967 **A300**, 405.
[79] J. R. Lombardi, *J. Chem. Phys.*, 1970, **52**, 6126.

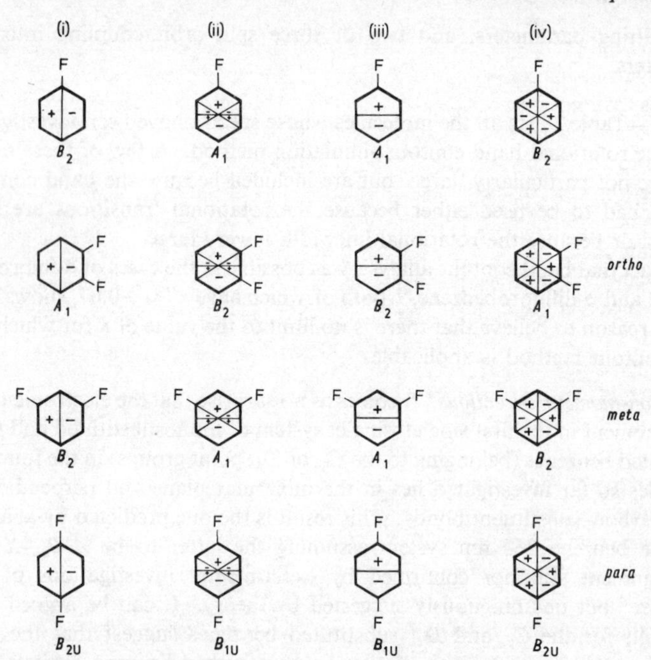

Figure 9 *Transition densities for the first four excited states of fluorobenzene and* o-, m-, *and* p-*difluorobenzene*

m-difluorobenzene,[57] the 264 nm system of fluorobenzene,[64] and the 271 nm system of *p*-difluorobenzene.[60, 61] Figure 9 illustrates the nodal characteristics of the transition density for transitions from the ground state to each of the four excited states resulting from an electron promotion from either of the two highest-energy occupied ground π-orbitals (derived from the degenerate e_{1g} orbitals in benzene) to either of the two lowest excited π-orbitals (derived from the degenerate e_{2u} orbitals in benzene) for mono- and all the di-fluorobenzenes. If the nodal characteristics of the transition density for the lowest-energy electronic transition in all these molecules were those labelled (i) in Figure 9, the first excited state would be $^1B_2$, $^1A_1$, $^1B_2$, $^1B_{2u}$ in fluorobenzene, and *o*-, *m*-, and *p*-difluorobenzene respectively; experimentally this is found to be the case. In the cases of fluorobenzene and *p*-difluorobenzene, simple perturbation theory applied to the interaction of the fluorine $2p_x$ orbitals and the benzene π-orbitals leads to the attribution of the nodal characteristics of (i) rather than (iv) to the transition densities. Although it would be natural to make similar attributions in the cases of *o*- and *m*-difluorobenzene, no simple molecular orbital arguments enable us to distinguish (i) from (iv). However, the fact that the O_0^0 bands in fluorobenzene and *o*-, *m*-, and *p*-difluorobenzene are at very similar wavelengths

(264—271 nm) indicates that the nodal characteristics of the first excited state are likely to be the same [namely, those of (i)] in all these molecules: this also indicates that, because of the small shift of the O_0^0 bands from that of benzene itself, we are probably justified in transferring orbitals, unperturbed, from benzene to the fluorobenzenes.

As well as the curiosity of the lowest electronic transition of *o*-difluorobenzene being $^1A_1$–$^1A_1$ while that in *m*-difluorobenzene is $^1B_2$–$^1A_1$, although the nature of the excited state may well be the same in both cases, there is a further curiosity in that the C_2-axis is the *a*-axis in *o*-difluorobenzene but the *b*-axis in *m*-difluorobenzene. As a result both molecules show a type *A* O_0^0 band.

McHugh and Ross [80] have computed type *A*, *B*, and *C* contours for ground-state constants typical of naphthalene or azulene and various changes of constants in the excited states. These contours could prove useful in determining directions of polarization of transition moments in spectra of the aza-naphthalenes, whose ground-state rotational constants are of similar magnitudes to those of naphthalene or azulene.

When hybrid bands are present in a spectrum there is, in general, no evidence from other sources to indicate the polarization direction of the electronic transition moment, and determination by the band contour method is especially valuable. Styrene [74] is a very interesting example. The direction of the $\tilde{A}^1A'$–$\tilde{X}^1A'$ transition moment is along the *a*-axis, which is very nearly along the ring–substituent bond, so that the direction of polarization has been rotated by almost 90° relative to the direction of polarization along the *b*-axis in monosubstituted benzenes which belong to the C_{2v} point group. It is likely that the polarization of the corresponding transition in indene [52] is very similar to that in styrene, indicating that the CH_2 group in the five-membered ring in indene perturbs the π-electron system only very little.

However, equation (22) shows that, in hybrid bands, only the magnitude and not the sign of the angle θ between the transition-moment direction and an inertial axis can be determined: this results in an ambiguity in the transition-moment direction as determined by analysis of hybrid band contours. Taking the sign of θ to be the same in indene,[52] indole,[53] and benzofuran,[49] the transition-moment direction in the $\tilde{A}^1A'$–$\tilde{X}^1A'$ transition is then very similar in all of them: θ, the angle between the transition moment and the *a*-axis, which is in an almost identical position in all three molecules, is $\pm(20\pm5)°$, $\pm(26\pm7)°$, and $\pm(18\pm5)°$ respectively. In thionaphthene [49] $\theta = \pm(72\pm15)°$ and, even allowing for the rotation of the *a*-axis by 34° compared to the other three molecules, this indicates quite a strong perturbation by the S atom in relation to CH_2, NH, and O.

In 2-aminopyridine [54] the transition moment in the $\tilde{A}^1A'$–$\tilde{X}^1A'$ transition is at an angle of $\pm(42\pm6)°$ to the *a*-axis, this axis being almost along the C—NH_2 bond. It is probable that the correct sign of the angle is the one which results in the direction being nearly perpendicular to the C_2-axis of the

[80] A. J. McHugh and I. G. Ross, *Spectrochim. Acta*, 1970, **26A**, 441.

pyridyl group since the corresponding transition in pyridine is polarized in this direction.

Information on transition-moment directions in hybrid bands should provide a useful test for electronic wavefunctions but little has been done in calculating these for unsymmetrical molecules.

Herzberg–Teller Intensity Stealing. In Table 2 are given many examples of electronic systems in which there are bands of a type different from that of the O_0^0 band, *i.e.* bands which obtain their intensity by Herzberg–Teller intensity stealing.[1]

Identification of type *A* bands in the $\tilde{A}^1B_2$–$\tilde{X}^1A_1$ (C_{2v}) or $\tilde{A}^1B_{2u}$–$\tilde{X}^1A_g$ (D_{2h}) systems of bromobenzene,[56] chlorobenzene,[29] phenol,[25] phenylacetylene,[31, 65] phenyl isocyanide,[66] and *p*-dichlorobenzene [59] confirms the recognition by Sponer [4] that the non-totally-symmetric component (in the C_{2v} and D_{2h} point groups) of the e_{2g} vibration, which is strongly active in making the electronically forbidden $\tilde{A}^1B_{2u}$–$\tilde{X}^1A_{1g}$ system of benzene vibronically allowed, is also active in many substituted benzenes. Sufficient information is now available that a theoretical interpretation of the results should be possible in order to explain why type *A* bands are observed very strongly, in fact more strongly than the type *B* O_0^0 band, in phenylacetylene [31, 65] and phenyl isocyanide,[66] strongly in bromobenzene [49] and several other molecules, but not at all in aniline [13] and others.

The $\tilde{A}^1B_{1g}$–$\tilde{X}^1A_g$ transition in *p*-benzoquinone is electronically forbidden and all of the intensity in the system is due to the activity of non-totally-symmetric vibrations of a_u and b_{3u} symmetry.[20, 24]

Naphthalene,[12, 32, 33] phenylacetylene [31, 65] and phenyl isocyanide [66] are the only examples known in which non-totally-symmetric vibrations produce, in an electronically allowed transition, bands which are stronger than the O_0^0 band.

Geometry of Excited States. Complete geometry determinations for excited states of most large molecules will remain elusive in the foreseeable future. A sufficient number of isotopic substitutions is difficult but not impossible to make, as has been shown by microwave spectroscopists in determining complete ground-state structures of, for example, some substituted benzenes. However, the difficulties in electronic spectroscopy are enhanced by the fact that very high isotopic purity is necessary: this is because band contours of molecules with two different isotopic species of atoms other than hydrogen will not, in general, be resolvable.

So far there are only two large molecules for which a complete excited-state geometry has been determined. The first of these is the $\tilde{A}^1B_{2u}$ state of benzene,[8] for which there are only two independent geometrical parameters. The C—C bonds were found to be 0.0038 nm longer and the C—H bonds 0.001 nm shorter than in the ground state. The second example is the $\tilde{A}^1B_{3u}$ state of *sym*-tetrazine,[15] for which six isotopic species were used to obtain the excited- *and* ground-state geometries. The contraction in the excited state of the N—N bond by 0.011 nm is surprising since a simple molecular orbital picture predicts no change in the bond length. However, the analysis by

Merer and Innes [15] used the symmetric-rotor approximation and, as has been pointed out,[75] an asymmetric-rotor band contour analysis would produce not only more reliable constants but two excited-state constants from each contour instead of the one ($\bar{B}'$) obtained previously.[15]

In difluorodiazirine two geometrical parameters, the F—F (non-bonded) and the N—N distances, can be calculated directly from the $\tilde{A}^1B_2$ excited-state constants.[62] The results show that the N—N bond expands by 0.006 nm, consistent with what is expected in this π^*–n transition, but two rather drastic approximations were used to obtain two of the three excited-state constants: (i) $\Delta(A-\bar{B})$ was obtained from ${}^rR$-branch heads assuming that the molecule behaves like a symmetric rotor (in fact $\kappa'' \approx -0.07$), and therefore that equation (6) holds, and (ii) C' was obtained using equation (1), which holds only for a planar molecule, whereas difluorodiazirine is by no means planar. However, the general agreement between the computed and observed O_0^0 band contours indicates that the estimated geometrical parameters may not be far wrong.

For five monosubstituted and five *p*-disubstituted benzenes, interpretation of the rotational constants in the $\tilde{A}^1B_2$ or $\tilde{A}^1B_{2u}$ states in terms of molecular geometry [55] depends on a close comparison of all the data *i.e.* two independent rotational constants for each of the molecules. In this way it was concluded that a probable consistent picture of the excited states includes the following:

(i) The ring expands in a similar way to that of benzene itself in the $\tilde{A}^1B_{2u}$ state except that, in the *p*-disubstituted benzenes, the expansion along the axis containing the substituents is not as great as in benzene: it may be that the central bonds [C(2)—C(3) and C(5)—C(6)] do not expand in these molecules by as much as they do in benzene. This effect of a lesser expansion of the central bonds may also occur in the monosubstituted benzenes but it would be to a smaller degree, and it does not appear to be as necessary a feature of a reasonable interpretation of the rotational constants as it is in the *p*-disubstituted benzenes. The effect of a lesser expansion of the central bonds than in benzene has been referred to as 'quinoid' or 'quinonoid' character of the excited state, but it should be made clear that this does not imply a contraction of the central bond in the excited state but an expansion which is smaller than in benzene.

(ii) The carbon–substituent bond contracts in the excited state, the contraction being greater the lower the electronegativity of the substituent. For example, the contraction is estimated to be *ca.* 0.002 nm in fluorobenzene and *ca.* 0.008 nm in aniline.

(iii) The internal ring angle adjacent to the substituent opens in the excited state by a few degrees, the opening being greater for less electronegative substituents.

(iv) The excited-state geometry of each end of a *p*-disubstituted benzene is very similar to that of the corresponding monosubstituted benzene.

In aniline three excited-state rotational constants were determined independently and it was found that $\Delta' = -4.5 \times 10^{-30}$ kg nm^2 (-0.27 u Å^2) and that the out-of-plane angle for the bisector of $\angle$HNH is 30°. This compares with 39° or 46° in the ground state.[81, 82]

Nuclear Spin Statistical Weights. If there is, in a rotational contour, a prominent series to each member of which can be attributed a single value of K_a or K_c then, if there are nuclei with non-zero spin and equivalent dispositions with respect to either the *a*- or the *c*-axis respectively, an intensity alternation will be observed in the series. Such an alternation can be observed, for example, in the series of ^{q}R-branch heads in the *p*-benzoquinone contour in Figure 2(*a*). In this molecule there are two pairs of protons symmetrically disposed about the *a*-axis, giving an intensity alternation of 10:6 for K_a even:odd provided the lower state does not involve any vibration which is antisymmetric to the C_2 operation about the *a*-axis: if it does involve one quantum of such a vibration then the alternation is reversed to 6:10 for K_a even:odd. It follows that an observed alternation can be used to help in identifying a ground-state vibration that is active in a particular band. For example, an X_1^1 sequence band in the 500 nm system of *p*-benzoquinone was found to have the opposite intensity alternation to that of the parent band.[24] This shows that the symmetry species of vibration X is antisymmetric to the C_2 operation about the *a*-axis.

Nuclear spin statistical weights have also been used in problems involving double-minimum potentials to give information on the degree of tunnelling through the barrier. For example aniline, in its equilibrium ground-state configuration, belongs to the C_s point group. If there is sufficient tunnelling through the NH_2-inversion barrier to produce resolvable doubling of the $v_I = 0$ level, where ν_I is the —NH_2 inversion vibration, in either the ground or the excited electronic state, or both, then there will be a 7:9 intensity alternation for K_a even:odd in the O_0^0 band. This intensity alternation was observed [13] and supports other evidence [82] that the $v_I = 0$ level is split in *both* electronic states. It was also shown [13] that the $I_1^1(0^-\text{–}0^-)$ band shows the opposite intensity alternation, in accordance with the identification of the band.[82]

In tropolone there is the possibility of a double-minimum potential in the internal hydrogen-bonding vibration ν_H. If the tunnelling through the barrier is sufficient to produce resolvable doubling of the $v_H = 0$ level in either the ground or the excited electronic state, or both, there will be a 10:6 intensity alternation for K_a even:odd in the O_0^0 band. In the H_1^1 ($0^-\text{–}0^-$) band in this vibration there should be an opposite alternation. These alternations have been observed and used to help in identifying the O_0^0 ($0^+\text{–}0^+$) and H_1^1 ($0^-\text{–}0^-$) bands.[76]

[81] D. G. Lister and J. K. Tyler, *Chem. Comm.*, 1966, 152.
[82] J. C. D. Brand, D. R. Williams, and T. J. Cook, *J. Mol. Spectroscopy*, 1966, **20**, 359.

3 Electric and Magnetic Effects

Excited-state Dipole Moments from the Stark Effect.—The Stark effect is the name given to the effect of an electric field on energy levels and transitions. In symmetric rotors the rotational energy F in a particular vibronic state is changed, in the presence of an electric field E, by ΔF [as defined in equation (25)] where $\mu_{a,c}$ is the dipole moment along the *a*- or *c*-axis, M is the quantum

$$\Delta F = -\frac{\mu_{a,c} K_{a,c} M E}{J(J+1)} \tag{25}$$

number associated with the component of the total angular momentum in the direction of the field, and the 'a' and 'c' subscripts of K refer to prolate and oblate symmetric rotors respectively. Since ΔF, in equation (25), depends linearly on the field the effect is called a first-order Stark effect. Equation (25) also applies to an asymmetric rotor and results in a first-order Stark effect provided that the asymmetry splitting of levels in K_a (or K_c), which are degenerate in the symmetric rotor, is small compared with the perturbation of the energy levels by the electric field. The reason for this is that the electric field causes mixing only of the two components of a K_a (or K_c) level but does not mix levels with different values of K_a (or K_c). Thus equation (25) can be used even for highly asymmetric rotors: for example it has been used in the O_0^0 band of the 352 nm system of difluorodiazirine [62] where $\kappa'' = -0.07$ but in which ^rR-branch heads are formed with $J \approx K_a$ and in which the asymmetry splitting of the energy levels is very small. Indeed, in this band the Stark effect was used to distinguish transitions between levels with very small asymmetry splitting from those between levels with asymmetry splitting which is large compared to the electric field effects.

Lombardi [83] has considered the Stark effect in asymmetric rotors in which the asymmetry splitting is large, but the effect has not been investigated in these conditions in large molecules.

In the difluorodiazirine O_0^0 band contour, as in most large molecule contours, the features strongly affected by an electric field are not individual rotational transitions but many coincident, or nearly coincident, transitions.[62] In difluorodiazirine these are ^rR-branch heads in a type B band and the heads are broadened, rather than split, by the field. A computer simulation of a broadened head depends on the value of $\Delta\mu_a$ and on the ground- and excited-state rotational constants, obtained from a rotational band contour analysis. As in many other cases, the sign of $\Delta\mu_a$ could not be determined. Such a determination depends on the Stark effect being measured in either two different types of branch, *e.g.* in an R branch and a Q branch, or in the same type of branch but for widely differing values of J.

Figure 10 illustrates, as a particularly nice example of the Stark effect, the splitting of a single ^pP-branch line with $K_a'' = 19, J'' = 28$ in the contour

[83] J. R. Lombardi, *J. Chem. Phys.*, 1968, **48**, 348.

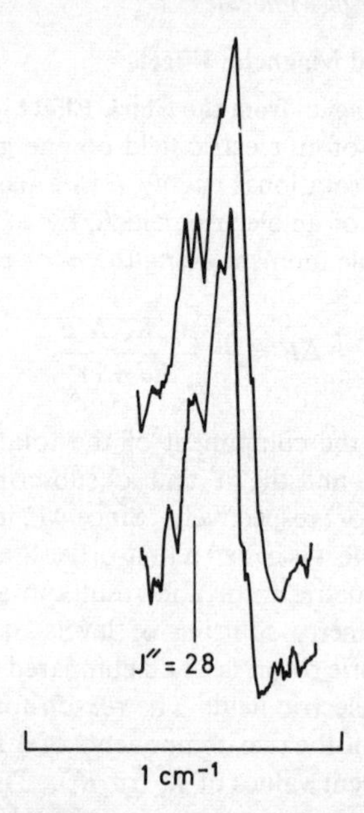

Figure 10 *The lower trace shows a part of the field-free spectrum of the* O_0^0 *band of the* 306 nm *system of* p-*fluoroaniline and, in particular, the* ^PP*-line with* $J'' = 28$, $K_a'' = 19$. *In the upper trace a* 7180 V cm^{-1} *electric field has been applied and the splitting of this line, in polarization perpendicular to the field, is shown. Absorption is plotted upwards on the vertical scale.*
(Reproduced by permission from *J. Chem. Phys.*, 1969, **51,** 1228)

of the O_0^0 band of *p*-fluoroaniline [84] in perpendicular polarization with an applied field of 7180 V cm^{-1}.

In Table 3 are listed all the dipole-moment changes which have been determined from electronic spectra of large molecules.

$\Delta\mu$ is a measure of the flow of charge, along the axis of the dipole moment, due to an electronic transition. In all the cases of substituted benzenes which have been investigated the transitions concerned are $\pi^*-\pi$ and therefore, although μ is a function of σ- and π-electrons, $\Delta\mu$ can be taken to be a function of π-electrons only. All the $\Delta\mu$'s for the substituted benzenes in Table 3 have been taken by the authors to have a positive sign, although this was shown unambiguously only in the case of fluorobenzene. In fluorobenzene, chlorobenzene, benzonitrile, phenylacetylene, and phenyl isocyanide the dipole moment in the ground state has its negative end at the substituent, so a positive $\Delta\mu$ in these cases represents a flow of negative charge *from the ring to the substituent*. In phenol and aniline the positive end is at the substituent, so that a positive $\Delta\mu$ represents a flow of negative charge *from the*

[84] K-T. Huang and J. R. Lombardi, *J. Chem. Phys.*, 1969, **51,** 1228.

Table 3 *Changes of dipole moment from the ground to the first excited electronic state in some large molecules*

Molecule	$\Delta\mu_a/10^{-30}$ C m(*or*/D)[a]	*Rotational structure used*	*Ref.*
Aniline	$2.8\pm0.5(0.85\pm0.15)$[b]	^pP lines	85
Benzonitrile	$1.0\pm0.2(0.31\pm0.07)$	^qQ lines	86
Chlorobenzene	$0.83\pm0.27(0.25\pm0.08)$	^qQ and ^qR lines	87
m-Chlorofluorobenzene	$\pm[0.7\pm0.2(0.20\pm0.05)]$	^qQ lines	88
p-Chlorofluorobenzene	$1.2\pm0.3(0.36\pm0.10)$	^pP lines	89
Difluorodiazirine	$\pm[5.0\pm0.7(1.5\pm0.2)]$	^rR heads	62
p-Fluoroaniline	$2.7\pm0.3(0.82\pm0.09)$	^pP lines	84
Fluorobenzene	$1.0\pm0.2(0.30\pm0.07)$	^rR and ^pP lines	90
p-Fluorophenol	$1.5\pm0.3(0.44\pm0.10)$	^pP lines	84
Phenol	$0.67\pm0.67(0.20\pm0.20)$	^pP lines	85
Phenylacetylene	$0.47\pm0.23(0.14\pm0.07)$	^rR and ^qQ lines	86
Phenyl isocyanide	$0.43\pm0.23(0.13\pm0.07)$	^rR lines	86
Styrene	$\pm[0.43\pm0.10(0.13\pm0.03)]$	^qQ and ^qR lines	91

[a] The sign of $\Delta\mu$ has been determined only for fluorobenzene and phenylacetylene. In other cases the sign was selected, in the references quoted, as the one most likely, except in the cases of styrene, *m*-fluorochlorobenzene, and difluorodiazirine, where it was not possible to make a choice.
[b] See, however, the top of p. 101.

substituent to the ring. In *p*-chlorofluorobenzene, *p*-fluoroaniline, and *p*-fluorophenol the dipole-moment direction is determined by the substituent at the opposite end of the molecule to the fluorine atom: so, in these molecules, the positive $\Delta\mu$ indicates a flow of negative charge from either Cl, NH_2, or OH to the ring that is reinforced by any flow from the ring to the fluorine. *m*-Chlorofluorobenzene is a special case since the carbon–substituent bonds are not on the axis of the dipole moment. In the case of styrene $\Delta\mu_a$ is so small that the sign could not be guessed with any reliability.

Dipole-moment changes have been correlated with structural changes by Lombardi [92] and, including more up-to-date information, by Muirhead *et al.*[86] $\Delta\mu$ has been shown to vary systematically with $\Delta B/B''$ for fluorobenzene, phenol, and aniline, for which ΔB has been obtained by band contour analysis. The reason for this systematic variation is that the carbon–substituent bond is perpendicular to the *b*-axis and any decrease or increase in the bond length strongly affects ΔB: any flow of electrons *in either direction* across the carbon–substituent bond should be reflected in a decrease of the bond length. It would therefore be more meaningful to correlate the change in bond length, rather than $\Delta B/B''$, with $\Delta\mu$ but, whereas the ΔB's are found

85 J. R. Lombardi, *J. Chem. Phys.*, 1969, **50**, 3780.
86 A. R. Muirhead, A. Hartford, jun., K.-T. Huang, and J. R. Lombardi, *J. Chem. Phys.*, 1972, **56**, 4385.
87 C. Y. Wu and J. R. Lombardi, *J. Chem. Phys.*, 1971, **55**, 1997.
88 C. Y. Wu and J. R. Lombardi, *J. Chem. Phys.*, 1971, **54**, 3659.
89 M. J. Janiak, A. Hartford, jun., and J. R. Lombardi, *J. Chem. Phys.*, 1971, **54**, 2449.
90 K.-T. Huang and J. R. Lombardi, *J. Chem. Phys.*, 1970, **52**, 5613.
91 H. Parker and J. R. Lombardi, *J. Chem. Phys.*, 1971, **54**, 5095.
92 J. R. Lombardi, *J. Amer. Chem. Soc.*, 1970, **92**, 1831.

with considerable accuracy, the changes in bond length are obtained only after several assumptions have been made.[55] However, $\Delta B/B''$ can be correlated reliably with carbon–substituent bond-length changes, and therefore with dipole-moment changes, only for molecules having substituents with similar masses, *e.g.* F, OH, NH_2. This is illustrated by the fact that in chlorobenzene ΔB is *negative* [29] whereas in phenol, where the mass of the substituent is considerably different, ΔB is positive [25, 26] and yet the likely structural changes involve a contraction of the carbon–substituent bonds of *ca.* 0.004 nm in *both* molecules.[55] It follows that the choice of the positive sign of $\Delta\mu$ for chlorobenzene,[87] mainly on the basis of a systematic variation between $\Delta\mu$ and $\Delta B/B''$ for molecules having substituents with substantially different masses, should not be considered to be final. A similar doubt should also be expressed regarding the choice of the positive sign for $\Delta\mu$ in *p*-chlorofluorobenzene.

Electric-field-induced Perturbations.—In the O_0^0 band of the 274 nm system of benzonitrile an abnormally strong Stark effect was observed in some regions of the rotational contour.[93] Normally fields of *ca.* 15 kV cm^{-1} have to be applied to observe a measurable Stark effect but in this case only *ca.* 1 kV cm^{-1} was necessary to produce a large effect. The conditions under which such an effect may result require that two rotational levels lie close together at zero field and that there is a non-zero transition dipole moment between them. Under these conditions strong perturbations between the pairs of levels can be induced by even a small electric field and the Stark effect observed will be non-linear. Analysis of the effect results in the determination of (*a*) the dipole moment of the perturbing state, (*b*) the transition moment between the mutually perturbing states, and (*c*) the zero-field separation of the perturbing states. From the effect in the O_0^0 band of benzonitrile these quantities are (*a*) $(4.4 \pm 1.0) \times 10^{-30}$ C m [1.35 ±0.3 D], (*b*) $(3.3 \pm 1.7) \times 10^{-30}$ C m [1.0 ±0.5 D], (*c*) 0.01 ±0.01 cm^{-1}. The symmetry species of the perturbing vibronic state was shown to be B_2 and it was suggested that this is a state involving an a_2 vibration that is active in the $^1B_1(\pi^*n)$ electronic state. If so, this would mean that the $^1B_1$–$^1A_1$ transition is, perhaps surprisingly, the lowest-energy electronic transition, and it should be observable in absorption if one uses a long pathlength of vapour. No similar effect was observed in any other bands of the 274 nm system, so it appears that the accidental near-degeneracy occurs only in the O_0^0 band.

A large non-linear Stark effect has also been observed in the O_0^0 band of the 294 nm system of aniline.[94] From an analysis of the effect it is deduced that there is a perturbing vibronic state of B_2 symmetry only 0.10 cm^{-1} away from the $\tilde{A}^1B_2$ electronic state. The dipole moment of the perturbing state is $(20 \pm 1) \times 10^{-30}$ C m [6.0 ±0.4 D] and the transition moment between the two mutually perturbing states is $(1.7 \pm 1.3) \times 10^{-30}$ C m [0.5 ±0.4 D].

[93] K.-T. Huang and J. R. Lombardi, *J. Chem. Phys.*, 1971, **55**, 4072.
[94] J. R. Lombardi, *J. Chem. Phys.*, 1972, **56**, 2278.

However, the nature of the perturbing state is not clear since there appears to be no possibility of a nearby $^1B_1$ (π^*n) state, as there might be in benzonitrile. The observation of the large non-linear Stark effect requires that the previous results [85] for the dipole moment of the $\tilde{A}^1B_2$ excited state be slightly altered. The new value of $\Delta\mu$ is $(3.1 \pm 0.3) \times 10^{-30}$ C m [0.92 ±0.10 D] compared with $(2.8 \pm 0.5) \times 10^{-30}$ C m [0.85 ±0.15 D], given in Table 3.

Zeeman Effect.—Triplet–singlet transitions can be diagnosed by the observation of the Zeeman effect when the sample is in a magnetic field. Following the observation of the Zeeman effect in the $\tilde{a}^3B_{3u}$–$\tilde{X}^1A_g$ system (370 nm) in pyrazine by Douglas and Milton,[95] little has been done in large molecule spectra except for the observation by Huang and Lombardi [96] of such an effect in the $\tilde{a}^3A_2$–$\tilde{X}^1A_1$ system (550 nm) in thiophosgene.

Magnetic rotation spectroscopy, a variation of the standard Zeeman experiment, in which the spectrum is that of the light transmitted through the sample between crossed polaroids when a magnetic field has been applied, has also been little used recently in diagnosing triplet–singlet transitions. In 1961, Eberhardt and Renner [97] found that the $\tilde{a}^3A_u$–$\tilde{X}^1A_g$ (536 nm) system of *p*-benzoquinone shows a magnetic rotation spectrum, and in 1965 Snowden and Eberhardt [98] found a magnetic rotation spectrum corresponding to the $\tilde{a}^3B_{3u}$–$\tilde{X}^1A_g$ (370 nm) system in pyrazine, but no other work has been reported on large molecules. Detailed interpretation of rotational structure in magnetic rotation spectra of large molecules remains elusive.

4 Vibrational Bands in Electronic Systems

Vibrational Analyses of Absorption Spectra.—No attempt is made in this section to make a comprehensive survey of the very numerous vibrational analyses which have been made in the past few years. Instead a few analyses have been selected which, in one way or another, attempt to go beyond the usual sorting of wavenumber separations from the O_0^0 band into fundamental- and combination-vibration wavenumbers.

The $\tilde{A}^1B_2$–$\tilde{X}^1A_1$ System (275 nm) *of Phenol.* The vibrational analysis of this system by Brand and co-workers [99, 100, 25, 101] represents, after that of the 260 nm system of benzene, the most detailed for any large molecule. The analysis was helped by the use of the isotopically substituted species [$^2H_1$]phenol and [*Ar*-$^2H_5$]phenol, by the observation of type *A* bands as well as the electronically allowed type *B* bands, and by the fact that substituted benzenes are vibrationally quite well understood in the ground state. Results

[95] A. E. Douglas and E. R. V. Milton, *Discuss. Faraday Soc.*, 1963, No. 35, p. 235.
[96] K.-T. Huang and J. R. Lombardi, *J. Chem. Phys.*, 1970, **53**, 460.
[97] W. H. Eberhardt and H. Renner, *J. Mol. Spectroscopy*, 1961, **6**, 483.
[98] B. S. Snowden and W. H. Eberhardt, *J. Mol. Spectroscopy*, 1965, **18**, 372.
[99] H. D. Bist, J. C. D. Brand, and D. R. Williams, *J. Mol. Spectroscopy*, 1966, **21**, 76.
[100] H. D. Bist, J. C. D. Brand, and D. R. Williams, *J. Mol. Spectroscopy*, 1967, **24**, 402.
[101] J. C. D. Brand, S. Califano, and D. R. Williams, *J. Mol. Spectroscopy*, 1968, **26**, 398.

so far produced from a vibrational analysis of the 270 nm system of chlorobenzene by Bist *et al.*[102, 103] suggest that a very detailed analysis of this system also might be achieved: the system, like that of phenol, shows type A as well as type B bands.[29]

In the $\tilde{A}^1B_2$ state of phenol, all vibrational fundamentals (a_1, a_2, b_1, and b_2) with wavenumbers of less than 2000 cm^{-1} were identified by Brand and co-workers, as well as all fundamental vibrations in the $\tilde{X}^1A_1$ state with wavenumbers less than 1000 cm^{-1}, the latter from hot bands.

The barrier to torsion about the C—O bond was shown to be in the range 4300—4700 cm^{-1} in the excited state [25] compared to *ca.* 1200 cm^{-1} in the ground state.[100] The increase in the barrier height in the excited state is consistent with the strong indication from rotational band contour analysis [25, 26] that the C—O bond contracts in the excited state by *ca.* 0.004 nm.[55]

Many bands of the type $X_0^n Y_n^0$, usually called 'cross sequences', have been observed in the 275 nm system: [25] since such bands are usually weak they are most often observed for $n = 1$ only. The extremely important point has been made [25] that normal modes of vibration are not the same in different electronic states, which means that it is an approximation (that may not always be valid) to make a one-to-one correlation of ground- and excited-state normal modes. In principle a normal mode X in the ground electronic state is mixed, in an excited electronic state, with all other normal modes of the same symmetry. It is this mixing which allows cross sequences to have non-zero intensity. In practice the mixing is often small, giving the low intensity characteristic of such bands. However, in the phenol spectrum there are several beautiful examples of families of sequences in which $X_0^1 Y_1^0$ bands are observed for nearly all the vibrations X and Y with the symmetry species b_1. For example, the following family is observed in which all bands have the lower state $16b_1$ (one quantum of vibration $16b$) and the intensities are given in brackets: $16b_1^1(12)$; $4_0^1 16b_1^0(3)$; $17b_0^1 16b_1^0(0.2)$; $10b_0^1 16b_1^0(0.1)$; $\tau_0^1 16b_1^0(0.05)$; $11_0^1 16b_1^0(0.05)$. All the vibrations involved in these transitions are b_1 vibrations: τ represents the C—OH torsional vibration. The fairly high intensity of the $4_0^1 16b_1^0$ band indicates quite strong mixing of these two vibrations in the $\tilde{A}^1B_2$ state.

If the force field were known in both the ground and excited electronic states the intensities of cross sequences could be calculated, and such calculations could possibly help in checking the assignments. Force-constant calculations for the out-of-plane vibrations in the $\tilde{A}^1B_2$ state of phenol [101] have shown that the intensities can be estimated reasonably well even with the usual limitations of such calculations. However, the point is made that although force fields can be transferred usefully from one substituted benzene to another in the ground electronic state this is not expected to be so for excited electronic states, in which interaction between the ring and the substituent is typically much larger.

[102] H. D. Bist, V. N. Sarin, A. Ojha, and Y. S. Jain, *Spectrochim. Acta*, 1970, **26A,** 841.
[103] H. D. Bist, V. N. Sarin, A. Ojha, and Y. S. Jain, *Appl. Spectroscopy*, 1970, **24,** 292.

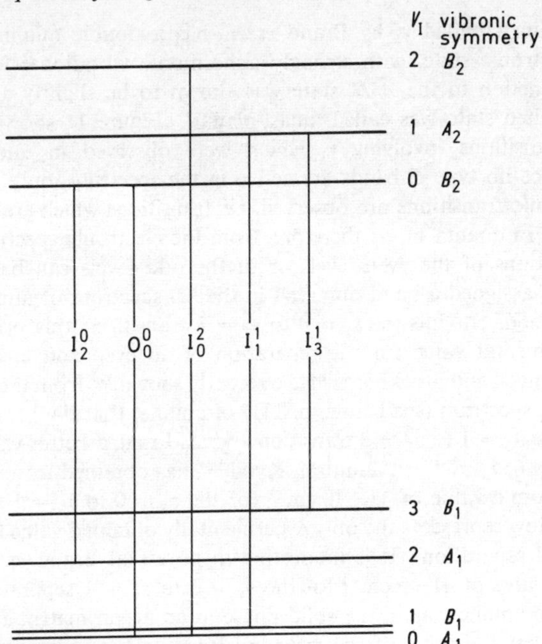

Figure 11 *Transitions involving the out-of-plane* —NH_2 *vibration* I *in undeuteriated aniline in the* $\tilde{A}^1B_2$–$\tilde{X}^1A_1$ (294 nm) *electronic system*

The $\tilde{A}^1B_2$–$\tilde{X}^1A_1$ *System* (294 nm) *of Aniline.* The O^0_0 band of this system has been shown to be a type *B* band [13] but, unlike the corresponding system of phenol, no type *A* bands have been observed. For this reason such a detailed vibrational analysis has not been possible. However, in aniline the electronic spectrum provides much information on the configuration of the amino-group. Before the almost simultaneous work of Lister and Tyler [81] on the microwave spectrum and of Brand *et al.*[104, 82] on the electronic spectrum there was little evidence to support the belief that, in the ground electronic state, the two hydrogen atoms of the amino-group are not coplanar with the rest of the molecule. Lister and Tyler obtained a value of 39°21′ and Brand *et al.* a value of 46° for the angle between the bisector of the HNH angle and the plane of the rest of the molecule. The discrepancy between these values is typical of that between a value obtained from a rotational analysis and from a vibrational analysis. The main reason for the discrepancy is that the rotational and vibrational analyses give zero-point and equilibrium values respectively.

Vibrational analysis of the $\tilde{A}^1B_2$–$\tilde{X}^1A_1$ electronic system of $C_6H_5{\cdot}NH_2$, $C_6H_5{\cdot}ND_2$, $C_6D_5{\cdot}NH_2$, and $C_6D_5{\cdot}ND_2$ showed conclusively [82] that the potential function for the out-of-plane motion of the hydrogen atoms of the

[104] J. C. D. Brand, D. R. Williams, and D. R. Cook, *J. Mol. Spectroscopy*, 1966, **20**, 193.

—NH_2 group, labelled ν_I by Brand *et al.*, has a double minimum in the ground electronic state, with strongly anharmonic vibrational levels. The potential function in the $\tilde{A}^1B_2$ state was shown to be slightly anharmonic and the excited state was called 'quasi-planar'. Figure 11 shows the main vibronic transitions involving ν_I which were observed in undeuteriated aniline. Since no type *C* bands are active in the spectrum, only B_2–A_1 and A_2–B_1 vibronic transitions are observed, *i.e.* transitions which are even–even or odd–odd in quanta of ν_I: therefore, from the electronic spectrum alone, only separations of the even levels or of the odd levels can be obtained. Brand *et al.* assigned a band observed in the i.r. spectrum of undeuteriated aniline [105] at 665 cm^{-1} as the $v_I'' = 0$ to $v_I'' = 3$ transition: this provided the only experimental value for the separation of an even and an odd level. However, Quack and Stockburger have recently shown,[106] from the resonance fluorescence spectrum (see below, p. 111) of aniline, that the i.r. band must represent the $v_I'' = 1$ to $v_I'' = 3$ transition and also that a better value for the separation is 655 cm^{-1}. In addition, Kydd [107] has obtained from the microwave spectrum a value of 41 $\pm$10 cm^{-1} for the $v_I'' = 0$ to $v_I'' = 1$ separation. This value now represents the only experimentally obtained value for an even to odd level separation. It is interesting to note that Brand *et al.*[108] have obtained a value of 41 $\pm$9 cm^{-1} for the $v_I'' = 0$ to $v_I'' = 1$ separation, but in this case the aniline was in the solid phase in an argon matrix at 6—32 K: it appears that the form of the potential function for ν_I is probably very similar in both an argon matrix and the free molecule.

The vibrational energy levels, $G''(v_I)$, for the ground state of undeuteriated aniline can be summarized: $G''(1) = 41$ cm^{-1}, $G''(2) = 423$ cm^{-1}, $G''(3) = 696$ cm^{-1}. The out-of-plane angle of 46° obtained previously [104] is little affected by the new data from the resonance fluoroscence spectra of aniline and deuteriated anilines: it is reduced [106] to 42°, bringing it closer to the microwave value. The height of the barrier to inversion is now estimated [106] to be 450 cm^{-1} compared to the previous value [104] of 565 cm^{-1}.

In the $\tilde{A}^1B_2$ state it was more difficult than in the ground state to obtain reliable vibrational energy levels for ν_I,[82] but the rotational band contour analysis [13] gives a value of 30 $\pm$10° for the out-of-plane angle. Since this is a rotational value it should be compared with the microwave value of 39°21′ for the ground state.[81]

The $\tilde{A}^1B_2$–$\tilde{X}^1A_1$ *System* (279 nm) *of Phenylacetylene and the* $\tilde{A}^1B_2$–$\tilde{X}^1A_1$ *System* (272 nm) *of Phenyl Isocyanide.* Until the 279 nm system of phenylacetylene was investigated by King and So,[31, 65] the strongest band in the system (that at 36 370 cm^{-1}) had been assigned as the O_0^0 band.[109, 110] King

105 J. C. Evans, *Spectrochim. Acta*, 1960, **16**, 428.
106 M. Quack and M. Stockburger, *J. Mol. Spectroscopy*, 1972, in press.
107 R. A. Kydd, personal communication.
108 J. C. D. Brand, V. T. Jones, B. J. Forrest, and R. J. Pirkle, *J. Mol. Spectroscopy*, 1971, **39**, 352.
109 M. R. Padhye and B. S. Rao, *J. Sci. Ind. Res. India, Sect. B*, 1959, **19**, 1.
110 W. W. Robertson, J. F. Music, and F. A. Matsen, *J. Amer. Chem. Soc.*, 1960, **72**, 5260.

and So showed that there are both type A and type B bands in the system. The 36 370 cm^{-1} band is type A and there is a much weaker type B band at 35 879 cm^{-1}, from which there are hot bands separated by known ground-state vibrational intervals. Therefore the 35 879 cm^{-1} band is the O_0^0 band and the 36 370 cm^{-1} band is assigned as 35_0^1, where ν_{35} is the b_2 vibration derived from one component of the doubly degenerate e_{2g} vibration that is the most intensely active in the 260 nm system of benzene.

The gross vibrational structure of the 272 nm system of phenyl isocyanide [66] is very similar to that of the phenylacetylene 279 nm system. Again the O_0^0 band is fairly weak and the b_2 vibration, derived from one component of the benzene e_{2g} vibration, gives rise to the most intense band in the system.

Therefore, of all the substituted benzenes whose electronic spectra have been analysed, phenylacetylene and phenyl isocyanide seem to resemble benzene itself more than any others.

The $\tilde{A}$–$\tilde{X}$ systems of phenylacetylene and phenyl isocyanide join the $\tilde{A}^1B_{2u}$–$\tilde{X}^1A_g$ system of naphthalene [12] as the only known systems which are electronically allowed by electric dipole selection rules but in which the O_0^0 band is weaker than bands involving non-totally symmetric vibrations.

The $\tilde{A}^1A_1$–$\tilde{X}^1A_1$ System (340 nm) *of Phenanthrene.* It has been well known for a long time that symmetry arguments are helpful, but by no means sufficient, in a discussion of Herzberg–Teller intensity stealing [1] in a particular molecule. For example, in benzene there are four e_{2g} vibrations but only one of them is strongly active in the 260 nm system: therefore it is not only the symmetry but the nuclear motions which are important, as well as the proximity to the vibronic state concerned in intensity stealing of an electronic state of the same symmetry as that with which mixing is occurring. These requirements make it more likely that Herzberg–Teller intensity stealing will occur in spectra of molecules with low rather than high symmetry. This is because in molecules of low symmetry there is a greater probability of there being an electronic state of the required symmetry close to the vibronic state with which it is to mix: In addition there is a greater probability of there being a vibration not only of the required symmetry but also of the right form to induce intensity stealing. Therefore in molecules of low symmetry there is a strong possibility also that *totally symmetric vibrations* will be involved in intensity stealing, a possibility which is small in molecules of higher symmetry. This strong possibility is because of the greater likelihood, in molecules of low symmetry, of there being two electronic states close together *and of the same symmetry*, and also because of the larger number of totally symmetric vibrations in the molecule.

These arguments have been put forward by Craig and Gordon [111] in a discussion of the 340 nm system of phenanthrene (C_{2v} point group) in the pure crystal, in the mixed crystal, and in the vapour, although the vapour spectrum is quite diffuse. In this system the totally symmetric vibration with an excited-state wavenumber of 671 cm^{-1} is involved in appreciable intensity

[111] D. P. Craig and R. D. Gordon, *Proc. Roy. Soc.*, 1965, **A288**, 69.

stealing. A characteristic which distinguishes such a vibration, say X, from one which is strongly active in a progression, owing to a change of geometry from the ground to the excited state in the direction of the corresponding normal co-ordinate, is that, although the X_0^1 band is strong, the X_0^2, X_0^3... bands are very weak. Craig and Small [112] have shown that, if a totally symmetric vibration is involved in intensity stealing as well as in a long progression because of an equilibrium geometry change from the ground to the excited state, the Franck–Condon intensity distribution is severely disturbed, and they have also shown that the absorption and fluorescence spectra are affected differently.

As more interest develops in the electronic spectra of less symmetrical molecules it is probable that more examples of intensity stealing by totally symmetric vibrations will be found.

The $\tilde{A}^1A_2$–$\tilde{X}^1A_1$ System (328 nm) *of Cyclopentanone.* Cyclopentanone has a skew C_2 configuration in the ground electronic state and the carbonyl group is coplanar with the two adjacent carbon atoms of the ring. However, the barrier to planarity of the ring is not high and the energy levels can be classified according to the C_{2v} point group. Consequently it might be expected that the 328 nm system would be closely analogous to the $\tilde{A}^1A_2$–$\tilde{X}^1A_1$ system (355 nm) of formaldehyde,[113] in which the molecule is pyramidal in the excited state and the electronically forbidden system is made vibronically allowed principally by the b_1 inversion vibration. Howard-Lock and King [114] have shown that the analogous vibronic mechanism is indeed operative in the 328 nm system of cyclopentanone and that, in the $\tilde{A}^1A_2$ state, the oxygen atom is 35° out-of-plane relative to the three adjacent carbon atoms: this value was obtained by fitting the observed energy levels of the b_1 C=O inversion vibration ν_{25} to a double-minimum potential function which gives a value of 700 cm^{-1} for the height of the barrier to inversion. However, ν_{25} is only quite weakly active in Herzberg–Teller intensity stealing: much more strongly active is the a_2 ring-puckering vibration ν_{18}, of which there is no possible analogue in formaldehyde which has no a_2 vibrations. Howard-Lock and King attribute the fact that the 328 nm system of cyclopentanone is approximately twice as intense as the 355 nm system of formaldehyde to the additional intensity stealing mechanism which occurs in cyclopentanone.

In the cyclopentanone system the hot band structure is relatively simple owing to the fact that the a_2 ring-puckering vibration ν_{18} and the b_1 ring-flapping vibration ν_{26} can be treated, in the ground electronic state, as independent vibrations.[115] However, it is very strange that, although $\nu''_{26} \approx 90$ cm^{-1},[116] this vibration is not active in the 328 nm system, even in sequence bands which are expected to be strong.

[112] D. P. Craig and G. J. Small, *J. Chem. Phys.*, 1969, **50**, 3827.
[113] J. C. D. Brand, *J. Chem. Soc.*, 1956, 858.
[114] H. E. Howard-Lock and G. W. King, *J. Mol. Spectroscopy*, 1970, **36**, 53.
[115] H. Kim and W. D. Gwinn, *J. Chem. Phys.*, 1969, **51**, 1815.
[116] L. A. Carriera and R. C. Lord, *J. Chem. Phys.*, 1969, **51**, 3225.

The cold bands in the system are much more complex and much of the complexity, especially of regions of diffuse absorption which show intensity fluctuations, is attributed [114] to the possible interaction of ν_{18} and ν_{26} in the excited state causing pseudorotation.

The $\tilde{A}^1A_2$–$\tilde{X}^1A_1$ system (330 nm) of cyclobutanone has been investigated by Moule [117] but not yet reported in detail. As in cyclopentanone it was found that, in the excited electronic state, the oxygen atom is not coplanar with the three adjacent carbon atoms.

The 172 nm *System of Cyclopentane and the* 175 nm *System of Cyclohexane.* The spectra of linear saturated hydrocarbons are mostly diffuse, and it is somewhat unexpected that the 172 nm system of cyclopentane and the 175 nm system of cyclohexane both show much sharp vibrational structure.

This has been reported briefly by Bell *et al.*[118] and should be of particular interest in respect of the nature of the ring-puckering vibration in both the ground and excited electronic states of both molecules.

The $^2A_2''$–$\tilde{X}E''$ *System* (338 nm) *of the Cyclopentadienyl Radical.* The cyclopentadienyl radical, C_5H_5, can be produced by flash photolysis of many different molecules including cyclopentadiene.[119] Engelman and Ramsay [120] have observed the 338 nm system at high resolution and shown that the rotational contours strongly resemble those of the 260 nm system of benzene,[8] but a detailed simulation by computed contours has not yet been achieved. A possible complication in the rotational structure is a Jahn–Teller distortion in the electronically degenerate ground state. Although it is expected to be small, the possible effects on the rotational structure are not known.

The vibrational structure is also complex, the dominant feature being the appearance of bands in pairs with a separation of 331 cm^{-1} in C_5H_5, decreasing regularly on successive deuteriation to 233 cm^{-1} in C_5D_5: the interval is not repeated. The authors suggest that the doublets may be due to a Jahn–Teller distortion but point out that the observed intensity pattern would not be consistent with this cause. (It is possible also that they could be due to a symmetrical double-minimum potential in a hydrogenic vibration in either the ground or the excited electronic state, or both.)

The $\tilde{A}^1B_{1g}$–$\tilde{X}^1A_g$ *System* (500 nm) *and the* $\tilde{a}^3A_u$–$\tilde{X}^1A_g$ *System* (536 nm) *of* p-*Benzoquinone.* Following the work of Singh [6] and Anno and Sado,[121] the 477 nm system of *p*-benzoquinone has been assigned [20] as $\tilde{A}^1B_{1g}$–$\tilde{X}^1A_g$. The system is electronically forbidden but made vibronically allowed principally by the activity of the a_u hydrogen out-of-plane vibration ν_1 and also of the a_u ring out-of-plane vibration ν_8 and the b_{3u} vibrations ν_{29} and ν_{30}. The a_u and b_{3u} vibrations give rise to bands with type *A* and type *B* rotational

[117] D. C. Moule, *Canad. J. Phys.*, 1969, **47,** 1235.
[118] S. Bell, R. Davidson, G. D. Gray, and P. A. Warsop, *Chem. Phys. Letters*, 1970, **5,** 214.
[119] B. A. Thrush, *Nature*, 1956, **178,** 155.
[120] R. Engelman, jun. and D. A. Ramsay, *Canad. J. Phys.*, 1970, **48,** 964.
[121] T. Anno and A. Sado, *J. Chem. Phys.*, 1960, **32,** 1602.

contours respectively. The bands were assigned originally [20] using contours computed with the symmetric-rotor approximation, but the assignments were confirmed later [24] with contours computed for an asymmetric rotor.

In the 500 nm system there is an unusual sequence intensity distribution in which the intensity of the parent band is fairly low, then it rises to a maximum at the second or third member, and then falls off. Ross *et al.* have shown [122] that a maximum in the intensity along a sequence may result from three possible circumstances:

(*a*) When there is a change from the ground to the excited electronic state in the origin of the normal co-ordinate concerned with the sequence-forming vibration the $(v \pm 1)$–v bands in this vibration may show an intensity maximum.

(*b*) When the sequence-forming vibration is itself involved in Herzberg–Teller intensity stealing the $(v \pm 1)$–v bands in this vibration may show an intensity maximum.

(*c*) When the sequence comprises $(v+2)$–v transitions these may show an intensity maximum.

In cases (*a*) and (*b*) the intensity $I_v^{v\pm1}$ of $(v \pm 1)$–v bands is given by equation (26), where E_v'' is the vibrational energy in the ground electronic state.

$$I_v^{v\pm1} \propto [\exp(-E_v''/kT)](v+1) \tag{26}$$

In case (*c*) the intensity I_v^{v+2} of $(v+2)$–v bands is given by equation (27).

$$I_v^{v+2} \propto [\exp(-E_v''/kT)](v+1)(v+2) \tag{27}$$

In the $\tilde{A}^1B_{1g}$–$\tilde{X}^1A_g$ system (500 nm) the b_{3u} vibration ν_{30} is involved in Herzberg–Teller intensity stealing and has such a low wavenumber ($\nu_{30}'' = 106$ cm^{-1}) that it forms long sequences and enables the intensity effect to be observed in $(v \pm 1)$–v bands in this vibration. In equation (26) there are two factors determining sequence band intensities, the usual Boltzmann factor and an additional factor of $(v+1)$. Therefore, if the vibration concerned were to have a larger wavenumber, the Boltzmann factor might be dominant even for the lowest value of v'', and the effect might not be observed.

It is only if $\nu'' \neq \nu'$ that $(v+2)$–v transitions have any intensity at all. Since, in the 500 nm system, $\nu_{30}'' = 106$ cm^{-1} and $\nu_{30}' = 140$ cm^{-1}, $(v+2)$–v transitions should be observed weakly, but the rotational contours of bands in this system are such as to cause much overlapping, making it difficult to observe with certainty any weak bands. The $\tilde{a}^3A_u$–$\tilde{X}^1A_g$ system (536 nm) is much less congested since the rotational contours all comprise mainly a single sharp peak. Since, in this system, $\nu_{30}' = 137$ cm^{-1} the possibility of observing $(v+2)$–v bands again arises. They have been observed [20, 122] very clearly and, as equation (27) compared with equation (26) predicts, the intensity maximum is further from the parent band in (*c*) than in (*b*) or (*a*) and therefore the effect is easier to observe.

Another example of case (*b*) has been found in the $\tilde{A}^1A''$–$\tilde{X}^1A'$ system

[122] I. G. Ross, J. M. Hollas, and K. K. Innes, *J. Mol. Spectroscopy*, 1966, **20**, 312.

(372 nm) of benzaldehyde.[123] In this case the a'' vibration involving torsional motion of the CHO group is concerned in Herzberg–Teller intensity stealing and forms long sequences since $\nu'' = 111 \text{ cm}^{-1}$. As $\nu' = 138 \text{ cm}^{-1}$ the situation is even quantitatively similar to that of case (*b*) in the 500 nm system of *p*-benzoquinone, and $(v+1)$–v bands in the torsional vibration in benzaldehyde show an intensity maximum displaced from the parent band.

The $\tilde{A}^1B_1$–$\tilde{X}^1A_1$ *System* (370 nm) *of Quinoxaline.* This system has been known for a long time to be due to a π^*–n transition [124] but, until recently, various attempts to analyse it have led to confusing, and sometimes contradictory, results.

The 370 nm system is rich in vibronic bands, of which there are two types of rotational contour, one sharp and line-like and the other broad (half-width *ca.* 5 cm^{-1}) with a central minimum. Since simple molecular orbital theory predicts two very close-lying π^*n states it was tempting to attribute the two band types to the two π^*–n transitions.[125, 126] However, Glass *et al.*[127] have attributed all the bands to one electronic system $\tilde{A}^1B_1$–$\tilde{X}^1A_1$. The broad bands, including the O_0^0 band, are assigned as type *C* and the sharp bands as type *A* involving b_1 vibrations in single quanta. The identification of the O_0^0 band by Glass *et al.*[127] as being at 27 071 cm^{-1} has been confirmed and put on a much sounder footing by Fischer *et al.*,[128] who have re-examined this system of quinoxaline and have also observed the corresponding system of [$^2H_6$]quinoxaline. Evidence from deuteriation shifts, hot bands, and Franck–Condon calculations have put the assignment beyond doubt. They have also explained a possible third type of band as being due to the fact that, in the excited state, an a_1 and a b_1 vibration have almost exactly the same wavenumber (600 cm^{-1}), giving almost coincident type *C* and type *A* bands. It is surprising, however, that a strong Coriolis interaction does not distort the contours of these two bands.

In connection with the identification of the 370 nm system as involving only one π^*–n transition is the discovery that through-bond interactions between *n*-orbitals can be very important even though the orbitals themselves are quite far apart.[129] As a result the two lowest-energy π^*n states are much more widely split than was previously thought. This kind of through-bond interaction may well explain the observation, in the free molecule, of only one π^*n state where two close-lying ones might previously have been expected, for example in *p*-benzoquinone,† pyrazine, pyrimidine, and pyridazine, and also in *sym*-tetrazine, where four close-lying states might have been expected.

† See note added in proof: p. 112.

[123] J. M. Hollas, E. Gregorek, and L. Goodman, *J. Chem. Phys.*, 1968, **49**, 1745.
[124] R. C. Hirt, F. T. King, and J. C. Cavagnol, *J. Chem. Phys.*, 1956, **25**, 574.
[125] Y. Hasegawa, Y. Amako, H. Azumi, M. Ito, and Y. Kaizu, *Bull. Chem. Soc. Japan*, 1969, **42**, 840.
[126] Y. Kaizu and M. Ito, *J. Mol. Spectroscopy*, 1969, **30**, 149.
[127] R. W. Glass, J. A. Merritt, and L. C. Robertson, *J. Chem. Phys.*, 1970, **53**, 3857.
[128] G. Fischer, A. D. Jordan, and I. G. Ross, *J. Mol. Spectroscopy*, 1971, **40**, 397.
[129] A. D. Jordan, I. G. Ross, R. Hoffmann, J. R. Swenson, and R. Gleiter, *Chem. Phys. Letters*, 1971, **10**, 572.

Resonance Fluorescence Spectra.—Strictly speaking, resonance fluorescence is the process of emission of radiation from molecules at low pressure following monochromatic excitation to a single rotational level in the excited electronic state. The pressure has to be sufficiently low that no rotational relaxation occurs during the lifetime of the excited state. Such fluorescence has been observed frequently in diatomic molecules, using an intense atomic emission line or the light from a laser as the exciting radiation. In large molecules, however, it is not possible to achieve the population of a single rotational level because several rotational transitions will usually lie within the linewidth of the exciting radiation. It is possible to achieve population of a single vibrational level, and the fluorescence process in which large molecules emit from a single vibrational level has also come to be called 'resonance fluorescence'.

The power of resonance fluorescence in trying to analyse the corresponding absorption spectrum has long been realized: indeed, it has been applied in the cases of the 260 nm system of benzene[130, 131] and the 312 nm system of naphthalene.[132] However, the closeness of the O_0^0 bands in these systems to the 254 nm group of lines in the mercury spectrum, in the case of benzene, and to the 313 nm group, in the case of naphthalene, is fortuitous and it would clearly have been desirable to have a tunable source of monochromatic radiation with which to excite the molecules into any excited-state vibrational level at will. The day of the tunable visible and u.v. laser may not be far off but, in the meantime, great advances have been made which are enabling resonance fluorescence spectra of large molecules to be obtained with a tunable source. The experimental technique, which has been developed by Parmenter and co-workers, and also by Stockburger and co-workers, involves a xenon arc source: this is normally a continuous source but the light is passed through a monochromator in which the grating can be rotated to release from the exit slit a narrow band of radiation of any wavelength. The radiation is then passed many times through the vapour to be excited, using a multiple-reflection mirror system. The emitted radiation is collected in a direction perpendicular to that of the exciting radiation, again using a multiple-reflection system. The emitted radiation is passed into a scanning spectrometer with an attached photomultiplier. The best resolution which has been achieved so far is *ca.* 20 cm^{-1}.

Much of the work on resonance fluorescence published so far by Parmenter *et al.* has been concerned with the 260 nm system of benzene.[133–136] Not only has important information on ground-state vibrational levels been obtained

[130] C. K. Ingold and C. L. Wilson, *J. Chem. Soc.*, 1936, 355.
[131] B. R. Cuthbertson and G. B. Kistiakowsky, *J. Chem. Phys.*, 1936, **4**, 9.
[132] J. M. Hollas, *J. Mol. Spectroscopy*, 1962, **9**, 138.
[133] C. S. Parmenter and M. W. Schuyler, *J. Chim. phys.*, 1970, **67**, 92.
[134] C. S. Parmenter and M. W. Schuyler, *J. Chem. Phys.*, 1970, **52**, 5366.
[135] C. S. Parmenter and M. W. Schuyler, *Chem. Phys. Letters*, 1970, **6**, 339.
[136] W. R. Ware, B. K. Selinger, C. S. Parmenter, and M. W. Schuyler, *Chem. Phys. Letters*, 1970, **6**, 342.

but rate constants for radiative and non-radiative decay from various excited vibronic states have been measured.

Blondeau and Stockburger [137] have obtained resonance fluorescence spectra of the 260 nm system of benzene, the corresponding systems of *o*-, *m*-, and *p*-xylene, of aniline, and of toluene, as well as the 312 nm system of naphthalene.[138] Excitation into various excited-state vibronic levels of toluene, aniline, and naphthalene shows that as the energy of the excited vibronic state increases the intensity of a background continuum also increases. These observations show how important it is in such molecules to excite in the O_0^0 band, or close to it, if a background continuum is to be avoided in a fluorescence spectrum. However, in the 260 nm system of benzene the continuum is by no means so pronounced.

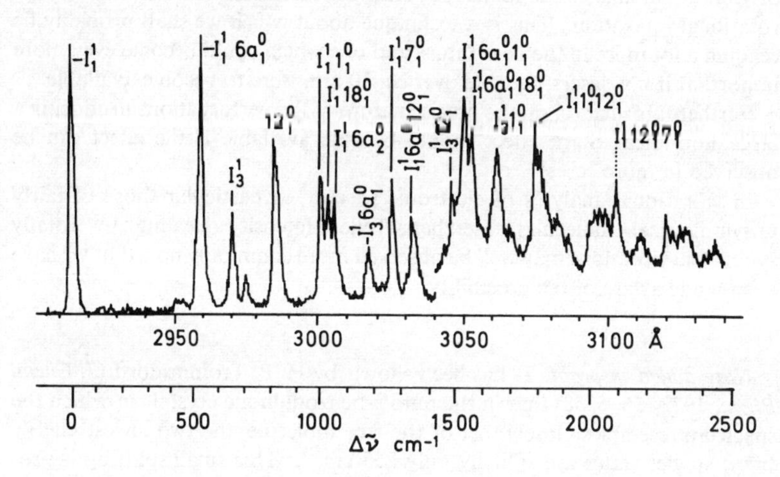

Figure 12 *Resonance fluorescence, showing the* 294 nm *system of aniline excited in the* I_1^1 *band. Emission is plotted upwards on the vertical scale.*
(Reproduced by permission from *J. Mol. Spectroscopy*, 1972, in press)

The resonance fluorescence spectra of aniline (referred to briefly on p. 104) and of many deuteriated anilines have been obtained by Quack and Stockburger,[106] and have led to more information on the vibrational levels in the ground electronic state of the NH_2-inversion vibration ν_I than could be obtained from the u.v. and i.r. absorption spectra.[82] Figure 12 illustrates, for example, how the resonance fluorescence spectrum resulting from excitation in the I_1^1 band shows very clearly the I_3^1 band. The separation of I_3^1 and I_1^1 is 655 cm^{-1}, showing that this is the separation of the I_3 and I_1 levels and not, as was previously thought,[82] that of the I_3 and I_0 levels.

[137] J. M. Blondeau and M. Stockburger, *Ber. Bunsengesellschaft phys. Chem.*, 1971, **75**, 450.
[138] J. M. Blondeau and M. Stockburger, *Chem. Phys. Letters*, 1971, **8**, 436.

5 The Future

It is always dangerous to try to predict what progress might be made even in the next few years but, from what has been achieved in recent years, it is clear that rotational band contour analysis will continue to play an important part in obtaining geometrical and vibrational information about excited electronic states of large molecules. However, the technique is, of course, limited to discrete spectra, a condition which is, unfortunately, not always met in large molecules. The way is open, especially when large computers become available, for progress in the application of the technique to triplet–singlet transitions and to transitions in molecules with internal rotation.

Resonance fluorescence is clearly a very powerful tool for tackling problems of vibrational analysis as well as for studying mechanisms of vibrational and rotational relaxation. This is a technique about which we shall probably be reading a lot more in the near future and one which could become even more important if u.v. lasers, tunable over *ca.* 10 nm, were to become available.

Further information on the precise nature of the perturbations producing a large non-linear Stark effect should become available if the effect can be observed in more cases.

In vibrational analyses of electronic spectra, in particular those of fairly unsymmetrical molecules, Herzberg–Teller intensity stealing by totally symmetric vibrations may well be observed more commonly now that we have been made aware of the possibility.

Note added in proof: It has been shown by H. P. Trommsdorff (*J. Chem. Phys.*, 1972, **56**, 5358) that in the pure *p*-benzoquinone crystal, in which the spectrum resembles closely that of the free molecule, the two lowest energy (π^*n) singlet states are split by only 255 cm^{-1}. This small splitting is probably a result of small through-bond interaction which is limited by symmetry requirements governing mixing between *n*- and σ-orbitals. Such limitations do not apply to *n*-orbitals on nitrogen atoms mixing with σ-orbitals in heterocyclic molecules.

I am very grateful to those authors, particularly Dr. J. R. Lombardi and Dr. M. Stockburger, who have sent copies of their papers to me in advance of publication and also to those who have allowed me to use reproductions of figures from published papers. I am grateful also to Professor I. G. Ross for sending me a preprint of his very illuminating review [40] entitled 'High Resolution Electronic Spectra of Large Polyatomic Molecules'. I have tried to avoid duplicating material in that review and strongly recommend readers of the present Report to read that also.

3

Energy Levels of a Diatomic near Dissociation

BY R. J. Le ROY

1 Introduction

The present chapter is concerned with the influence of the long-range part of the interatomic potential on the distribution of levels near the dissociation limit of a diatomic molecule. This problem received little attention until relatively recently, largely because of the difficulty of making reliable measurements and assignments in the region near dissociation, since the transitions there are usually of low intensity and the levels are very densely packed. However, improvements in experimental resolution and the advent of selective excitation techniques using lasers have combined to make this region much more accessible. These developments added impetus to the increasing theoretical interest in this problem, the first fruits of which are discussed below.

The methods described herein are based on expressions relating level energies to the detailed nature of the long-range potential. Their main application to date has been in the determination of bond dissociation energies and long-range potentials from spectroscopic data, and for the prediction of the number and energies of unobserved levels. Although not yet widely used, they appear to open new routes to these quantities whose accuracy and versatility surpass those of competing methods. In addition, there should be many chemical applications of the understanding of the nature of the distribution of levels obtained in this way. For instance, a knowledge of the density of states is essential for the calculation of unimolecular reaction rates, or for detailed applications of the principle of microscopic reversibility.[1] These examples are suggestive of the rather potent uses of expressions for the level density, and augur well for the methods used herein which describe the region near dissociation where the density of states approaches a maximum. Although such 'second generation' applications of the present results lie in the future, they may eventually overshadow the importance of determining interaction potentials and dissociation energies using the techniques described below.

One of the earliest and most successful applications of modern quantum

[1] See *e.g.* (*a*) D. L. Bunker, 'Theory of Elementary Gas Reactions', Pergamon Press, Oxford, England, 1966; (*b*) R. A. Marcus, *J. Chem. Phys.*, 1970, **53**, 604; (*c*) J. L. Kinsey, *ibid.*, 1971, **54**, 1206.

mechanics has been its explanation of the motion of the nuclei in diatomic molecules. In 1926, immediately after the publication of Schrödinger's [2] differential equation formulation of the quantum theory, Fues [3,4] derived expressions for the energy levels of a diatomic which had the correct dependence on positive powers of $(v+\frac{1}{2})$ and $J(J+1)$, where v and J are, respectively, the (non-negative integer) vibrational and rotational quantum numbers.* The following year Born and Oppenheimer [5] showed how to separate the motions of the electrons and nuclei in Schrödinger's equation, yielding a formal basis for that most ubiquitous concept, the internuclear potential-energy function. From this point on, much of molecular spectroscopy has been concerned with the rationalization of experimentally observed frequencies and intensities as rigorously calculable manifestations of such potential curves. For a long time the intensities resisted analytic treatment and were either discussed qualitatively in terms of the Franck–Condon principle,[6] or examined using exact numerically calculated wavefunctions.[7] However, expressions for matrix elements coupling different vibrational–rotational levels in the lower part of a potential well have recently been obtained as explicit functions of the potential parameters.[8, 9]

The frequencies of transitions involving the lower vibrational–rotational levels have long been understood as quantitative reflections of the nature of the potential near its minimum. From the late 1920's until fairly recently, the determination of an internuclear potential from an energy-level spectrum often involved the use of simple 3- or 5-parameter model potentials for which analytic expressions relate the level energies to the parameters. For a few cases, such as the harmonic oscillator and Morse functions, these expressions give the exact eigenvalues. More often though, they involve approximations which limit their accuracy except near the potential minimum (*e.g.* for the Lennard-Jones, Hulbert–Hirschfelder, and exponential -6 functions). For a wide selection of such forms, comparisons have been made [10] of both

* In the 'old quantum theory' the vibrational and rotational energies were given by $G(v) = \omega_0 v + \omega_0 x v^2 + \ldots$ and $F_v(J) = B_v J^2 + D_v J^4 + \ldots$, the former overlooking the zero-point energy and the latter ignoring terms in odd powers of J.

[2] E. Schrödinger, *Ann. Physik*, 1926, **79**, 361, 489; 1926, **80**, 437; 1926, **81**, 109.
[3] E. Fues, *Ann. Physik*, 1926, **80**, 367; 1926, **81**, 281.
[4] W. Jevons, 'Band Spectra of Diatomic Molecules', The Physical Society, London, 1932.
[5] M. Born and J. R. Oppenheimer, *Ann. Physik*, 1927, **84**, 457.
[6] J. Franck, *Trans. Faraday Soc.*, 1925, **21**, 536; E. U. Condon, *Phys. Rev.*, 1926, **28**, 1182; 1928, **32**, 858; E. U. Condon, *Amer. J. Phys.*, 1947, **15**, 365, gave a general discussion of the 'Principle' and its implications; and R. S. Mulliken, *J. Chem. Phys.*, 1971, **55**, 309, emphasized the oft-neglected role of kinetic energy in the Franck–Condon principle.
[7] R. N. Zare, *J. Chem. Phys.*, 1964, **40**, 1934; D. L. Albritton, A. L. Schmeltekopf, and R. N. Zare, 'Diatomic Intensity Factors', Harper and Row, New York, in preparation.
[8] C. Schlier, *Fortschr. Physik*, 1961, **9**, 455; R. M. Herman and S. Short, *J. Chem. Phys.*, 1968, **48**, 1266, erratum 1969, **50**, 572; R. M. Herman, R. H. Tipping, and S. Short, *J. Chem. Phys.*, 1970, **53**, 595.
[9] W. Benesch, *J. Mol. Spectroscopy*, 1965, **15**, 140; N. Jacobi, *J. Chem. Phys.*, 1970, **52**, 2694; R. H. Tipping, *J. Mol. Spectroscopy*, 1972, **43**, 31.
[10] Y. P. Varshni, *Rev. Mod. Phys.*, 1957, **29**, 664; D. Steele, E. R. Lippincott, and J. T. Vanderslice, *Rev. Mod. Phys.*, 1962, **34**, 239.

their ability to represent sets of experimental level energies and their agreement with the more accurate potentials obtained by RKR techniques (see Section 6). No strong predilection for any particular model was found, other than the expected conclusion that 5-parameter potentials are better than the 3-parameter ones. Furthermore, although the model potentials considered [10] all have the correct overall shape, passing smoothly from a steep short-range repulsion through a minimum and then asymptotically approaching a dissociation limit, they are not sufficiently flexible to represent a real interaction potential with high accuracy except over a fairly narrow region. Thus the present consensus is that this approach is only suitable either when there are too few data to allow a more accurate treatment, or when one is interested only in the rough qualitative properties of the potential.

A partial solution to the limited accuracy of the model-potential approach was given by Dunham.[11] He showed that if the interaction energy is expanded as a power series about its minimum R_e,

$$V(R) = a_0\left(\frac{R-R_e}{R_e}\right)^n\left[1+a_1\left(\frac{R-R_e}{R_e}\right)+a_2\left(\frac{R-R_e}{R_e}\right)^2+\ldots\right] \quad (1)$$

the second-order WKB approximation gives the eigenvalues

$$G(v,J) = \sum_{l,m\geq 0} Y_{lm}(v+\tfrac{1}{2})^l[J(J+1)]^m \quad (2)$$

where the Y_{lm} coefficients are explicitly related to the potential constants (a_i).[12] However, expression (1) diverges at large R values instead of asymptotically approaching a dissociation limit, and hence this approach is not acceptable near the top of the potential well.

Another WKB-based procedure, one which is accurate over the whole attractive potential well, is the RKR method.[7, 13–15] Utilizations of this approach for levels near the dissociation limit will be discussed in Section 6.

Considerable attention has been paid recently to the more limited objective of predicting merely the total *number* of levels bound by a potential, and its dependence on the rotational quantum number J.[16–22] Obvious applications

[11] J. L. Dunham, *Phys. Rev.*, 1932, **41,** 713, 721; J. E. Kilpatrick, *J. Chem. Phys.*, 1959, **30,** 801, later showed that Dunham's eigenvalue expansion equation (2) could be obtained by treating the potential of equation (1) as a perturbed harmonic oscillator.

[12] I. Sandeman, *Proc. Roy. Soc. Edinburgh*, 1940, **60,** 210, later inverted Dunham's expressions to express the turning points at a given energy in terms of the Y_{lm} values, a result which appears to have seen service only under W. R. Jarmain, *Canad. J. Phys.*, 1960, **38,** 217, who used it to prove the formal equivalence of the Dunham [11] and RKR (Section 6) methods.

[13] R. Rydberg, *Z. Physik*, 1931, **73,** 376.

[14] O. Klein, *Z. Physik*, 1932, **76,** 226; R. Rydberg, *ibid.*, 1933, **80,** 514; A. L. G. Rees, *Proc. Phys. Soc.* (*London*), 1947, **59,** 998.

[15] E. A. Mason and L. Monchick, *Adv. Chem. Phys.*, 1967, **12,** 329.

[16] D. E. Stogryn and J. O. Hirschfelder, *J. Chem. Phys.*, 1959, **31,** 1531, 1545, erratum 1960, **33,** 942.

[17] H. Harrison and R. B. Bernstein, *J. Chem. Phys.*, 1963, **38,** 2135, erratum 1967, **47,** 1884; the corrected tables described in the latter were used here.

[18] F. Calogero and G. Cosenza, *Nuovo Cimento*, 1966, **A45,** 867.

of such results are in estimating the equilibrium constant for dimer formation [19] or the bound-state contribution to the second virial coefficient,[16] since at temperatures which are large relative to the molecular binding energy the internal partition function for the dimer is simply the total number of bound levels. Although such results are less general than those sought below, they will yield useful points of comparison with some of the present methods.

In the following, the question of the nature of the distribution of vibrational–rotational levels near dissociation and its dependence on the long-range interaction potential is examined from four points of view. All make use of the asymptotic theoretical inverse-power expansion for interatomic potentials, and hence this material is briefly reviewed in Section 2. Although three of the four methods are based on the semi-classical WKB approximation, each is uniquely characterized by the type and extent of the experimental information required for its application. Hence, their order of presentation is based on the increasing sophistication of the requisite data.

2 A Users' Guide to the Theory of Long-range Intermolecular Forces

General.—Although much of the present theory of long-range intermolecular forces was developed during the early days of quantum mechanics,[23–27] the past few years have seen a resurgence of activity in this field.[28–32] The present section will summarize those parts of the current conventional wisdom which bear on the discussion in the rest of the chapter.

It has long been known that if two atoms are sufficiently far apart that their electron clouds overlap negligibly, their interaction energy may be expanded as (ignoring electronic degeneracy and fine-structure effects):

[19] G. D. Mahan and M. Lapp, *Phys. Rev.*, 1969, **179,** 19.
[20] G. D. Mahan, *J. Chem. Phys.*, 1970, **52,** 258.
[21] A. S. Dickinson and R. B. Bernstein, *Mol. Phys.*, 1970, **18,** 305.
[22] S. Tani and M. Inokuti, *J. Chem. Phys.*, 1971, **54,** 2265.
[23] J. K. Knipp, *Phys. Rev.*, 1938, **53,** 734.
[24] G. W. King and J. H. Van Vleck, *Phys. Rev.*, 1939, **55,** 1165; R. S. Mulliken, *Phys. Rev.*, 1960, **120,** 1674.
[25] H. Margenau, *Rev. Mod. Phys.*, 1939, **11,** 1.
[26] H. B. G. Casimir and D. Polder, *Phys. Rev.*, 1948, **73,** 360.
[27] J. O. Hirschfelder, C. F. Curtiss, and R. B. Bird, 'Molecular Theory of Gases and Liquids', Wiley, New York, 1964, corrected printing.
[28] A. Dalgarno and W. D. Davison, *Adv. Atomic Mol. Phys.*, 1966, **2,** 1; J. O. Hirschfelder, *Adv. Chem. Phys.*, 1967, **12,** vii. H. Margenau and N. R. Kestner, 'Theory of Intermolecular Forces', 2nd. edn., Pergamon Press, London, 1971.
[29] J. O. Hirschfelder and W. J. Meath, *Adv. Chem. Phys.*, 1967, **12,** 3.
[30] E. A. Power, *Adv. Chem. Phys.*, 1967, **12,** 167.
[31] T. Y. Chang, *Rev. Mod. Phys.*, 1967, **39,** 911.
[32] P. R. Certain and L. W. Bruch, in 'MTP International Review of Science, Physical Chemistry Section, Theoretical Chemistry Volume', Medical and Technical Publishing Oxford, England, to be published; also available as University of Wisconsin Theoretical Chemistry Institute Report WIS-TCI-460, 1971.

$$V(R) = D - \sum_{m \geq \tilde{n}} C_m/R^m \tag{3}$$

where D is the dissociation limit, the powers m have positive integer values, and perturbation theory yields formal expressions for the constants C_m.[25,27] The nature of the atoms to which a given molecular state dissociates determines which terms contribute to equation (3), and also sometimes defines their sign. Only the first- and second-order perturbation energies are considered here; the former can yield terms with $m \geqslant 1$, and the latter even-power terms with $m \geqslant 4$. The contributions from third- and higher-order perturbations correspond to $m > \tilde{n}+2$, and hence they would tend to represent a large fraction of $V(R)$ only at small distances where equation (3) is no longer valid (see discussion on p. 119).

If the leading terms in equation (3) all have the same sign, the interaction potential in any subinterval of the long-range region may be accurately approximated by

$$V(R) \simeq D - C_n/R^n \tag{4}$$

where n is some weighted average (in general non-integer) of the powers of the locally important terms. Asymptotically this $n = \tilde{n}$, the (integer) power of the leading term in equation (3); then as R decreases, *i.e.* the magnitude of $V(R)$ increases, this 'local' n will gradually increase owing to the growing relative strength of terms with $m > \tilde{n}$. It will be seen that a theoretical knowledge of the asymptotic power $\tilde{n}$ greatly facilitates utilization of the methods of Sections 3—6. Unless explicitly stated otherwise, the discussion here will always assume that all significant contributions to equation (3) are attractive, and hence that n behaves as described above.

Inverse-power Contributions to the Long-range Potential.—The coulombic $m = 1$ term in equation (3) will be present only if both atoms are charged. For this case the coefficient $C_1 = -Z_A Z_B e^2$, where $Z_A e$ and $Z_B e$ are the charges on atoms A and B, respectively.

The $m = 2$ term would arise classically in the interaction between a dipole moment and an ion. Although no atom has a permanent dipole moment, an electronically excited one-electron atom (*e.g.* excited H, He^+, or Li^{++}) perturbed by the presence of another particle will sometimes behave as if it does. This can occur when eigenstates with different orbital angular momentum quantum numbers but the same principal quantum number are degenerate. The presence of the perturber can then cause a mixing of these states of different symmetry to yield a hybrid atomic orbital which is effectively dipolar.[29] This resulting state will then interact as if it had a permanent dipole, and if its interacting partner is an ion, it will give rise to an $m = 2$ contribution to equation (3).

The $m = 3$ term in equation (3) classically corresponds to the interaction between a pair of dipoles. Thus, as discussed above, it would arise in the interaction between a pair of electronically excited one-electron atoms, each of which is in a hybrid dipolar state. On the other hand, this $m = 3$ term can also arise as a 'resonance' interaction between a pair of atoms of the same species which are in electronic states between which electric dipole transitions are allowed.[24, 29] In practice this means that the total angular momenta of the atomic states must differ by one (S and P, P and D, ... *etc.*). In this case the sign of the C_3 coefficient is determined by the symmetry properties of the particular molecular state; for the interactions between an S- and a P-state atom, results are given in ref. 29.

The $m = 4$ term in equation (3) mainly arises as the second-order charge induced-dipole interaction between an ion and a neutral charge distribution. In this case $C_4 = (Z^2e^2\alpha/2)$ where Ze is the ionic charge and α the polarizability of the neutral atom. Another situation which may also give an $m = 4$ contribution to $V(R)$ is the first-order interaction between an atom with a dipole moment (*e.g.* electronically excited hydrogen, see above) and one with a permanent quadrupole moment (*e.g.* ground-state B, Al, Ga, C, Si, O, S, F, Cl, Br, or I).

Except for particular cases in which the C_5 coefficient is precisely zero for reasons of symmetry, the $m = 5$ term will contribute to the long-range potential whenever *neither* of the atoms is in an S electronic state. It arises from the first-order quadrupole–quadrupole interaction, and accurate general expressions for it have long been known. In 1938 Knipp [23] showed that a C_5 coefficient could be separated into an angular factor which depends only on the symmetry properties of the molecular electronic state formed by atoms A and B, times the product $\langle r_A{}^2\rangle \langle r_B{}^2\rangle$, where $\langle r_X{}^2\rangle$ is the expectation value of the square of the electronic radius of the unfilled valence shell of atom X. Knipp reported values of these angular factors for a number of cases, and Chang [31] extended his tables considerably. Furthermore, relativistic Hartree–Fock–Slater calculations of the expectation values $\langle r_X{}^2\rangle$ have recently been reported for all orbitals of all ground-state atoms with nuclear charge between 2 and 126.[33] Therefore, fairly reliable C_5 constants may be calculated for virtually all molecular states formed from ground-state atoms. On the other hand, the reported [33] values of $\langle r_X{}^2\rangle$ will be lower bounds to those for excited states, so that C_5 values thus calculated for states formed from excited atoms will tend to be lower bounds to the true values.

The $m = 5$ term described above can contribute to all interactions between non S-state atoms. If the atoms are uncharged and have $(np)^\mu$ valence configurations (*i.e.* μ electrons in an unfilled p shell), this is the *only* non-vanishing non-resonant contribution to the first-order interaction. For atoms with $(nd)^\mu$ configurations there will be in addition $m = 7$ and 9 contributions, while $(nf)^\mu$ configurations will cause the appearance of yet higher-power terms in R^{-1}.

[33] C. C. Lu, T. A. Carlson, F. B. Malik, T. C. Tucker, and C. W. Nestor, jun., *Atomic Data*, 1971, **3**, 1; these expectation values appear to be considerably more accurate than Hartree–Fock values reported earlier, the ensuing change in the theoretical C_5 values discussed in Section 3 being as large as 25% (see Table 2).

In addition to the foregoing, a quadrupole resonance contribution to the $m = 5$ term can arise if the interacting atoms are of the same species, but in different atomic states which may be coupled by a quadrupole-allowed transition. This effect may occur on combining like atoms in *S* and *D*, *P* and *F*, ... *etc.* states. In the first of these cases (*S* plus *D*) this resonance energy will be the only non-vanishing $m = 5$ energy, while in the others it will supplement the interaction between the permanent atomic quadrupole moments.

Second-order perturbation theory gives rise to the dispersion terms $m = 6$, 8, and 10 which *always* contribute to the potential. Furthermore, if the interaction is between two uncharged *S*-state atoms, no first-order terms arise, so that they are the leading terms in equation (3) (and hence $\tilde{n} = 6$). The formal expressions for C_6, C_8, and C_{10} yield the general result that these constants are all positive (attractive) for atoms in their ground electronic states.[26, 27] On the other hand, if one or both of the atoms is electronically excited, the coefficients could have either sign. In the past few years considerable progress has been made in developing more accurate ways of calculating C_6 values.[32, 34–38] However, although good results have been obtained for certain types of interactions, a tractable approach of truly general utility is not yet available. For C_8 and C_{10} the situation is somewhat worse, and reliable values have been calculated only for a few species such as H_2 and the diatomic inert gases.[29, 32, 34]

All of the terms discussed above arise from use of a non-relativistic coulombic hamiltonian to describe the system. Additional contributions also arise from magnetic (or relativistic) interactions for which the leading term varies as R^{-3}. However, they are expected to be only a very small fraction of the $m > 3$ coulombic terms in $V(R)$ except at sufficiently large R ($\gtrsim 50$ Å) that their effect on the distribution of spectroscopically observable levels would be negligible.[31]

Validity of the Inverse-power Expansion.—At very long range, where *R* is of the order of wavelengths characteristic of the allowed transitions of the interacting atoms, the form of a number of the terms in equation (3) begins to change owing to 'retardation' effects.[26, 29, 30] For example, in this limit the R^{-6} dispersion terms begin to die off more quickly, asymptotically attaining an R^{-7} dependence. However, such effects are in general negligible for $R \lesssim 100$ Å, and beyond this range the interaction potential is sufficiently weak that it has no significant effect on the level distribution.

At small *R* the increasing overlap of the electron clouds of the two atoms leads both to the addition of 'exchange' terms to the potential, and to the

[34] G. Starkschall and R. G. Gordon, *J. Chem. Phys.*, 1971, **54**, 663.
[35] G. Starkschall and R. G. Gordon, *J. Chem. Phys.*, 1972, **56**, 2102.
[36] G. Starkschall and R. G. Gordon, *J. Chem. Phys.*, 1972, **56**, 2801.
[37] F. E. Cummings, Atlanta University Center Science Research Institute, Research Report AUCSRI-33, 1971.
[38] F. E. Cummings, Ph.D. Thesis, Harvard University, 1972.

increasing inappropriateness of equation (3) for describing the coulombic interaction. The exchange terms increase exponentially with decreasing R, and are attractive or repulsive depending on whether or not the state in question is chemically bound. In addition, the simple inverse-power terms in equation (3) are accurate only at moderate-to-long range, while they overestimate the 'true' coulombic interaction energies by increasingly large amounts at small R. This breakdown has been studied in detail for a number of hydrogen and helium systems.[39—42] One fact clearly demonstrated was the asymptotic nature of expansion (3), which means that although the first terms may yield a fairly good estimate of $V(R)$, the sum to $m = \infty$ will always diverge. The results of refs. 39—42 also suggest that the term of inverse-power m is no longer valid for $R \lesssim (2C_{m+2}/C_m)^{\frac{1}{2}}$.

It seems reasonable to associate the growth of the exchange term and the breakdown of equation (3) with the radii of the valence electron shells on atoms A and B,[33] since both effects are due to electronic overlap. For the cases considered in refs. 39—42, the approximation of representing $V(R)$ by the sum of the first few terms in equation (3) breaks down unless

$$R > 2[\langle r_A{}^2\rangle^{\frac{1}{2}} + \langle r_B{}^2\rangle^{\frac{1}{2}}] \tag{5}$$

This seems a reasonable rule for yielding a lower bound to the region of validity of equation (3). From analogous considerations, Stwalley[43] suggested the less flexible (and often somewhat more severe) rule that equation (3) should not be used for $R < 5$ Å.

Another type of situation in which equation (3) becomes inappropriate is when there are interactions between different molecular states of the same symmetry. This can give rise to avoided curve crossings which manifest themselves as maxima or abrupt changes in the shape of a potential curve. There is no known form for a potential in a region where such interactions arise, and hence methods based on equations (3) and (4) must be used very circumspectly in cases where these effects can occur. Unless stated otherwise, it is assumed that this difficulty does not arise in any of the problems considered below.

3 Distribution of Vibrational Levels near Dissociation

Background: Birge–Sponer Plots.—For almost half a century the Birge–Sponer extrapolation procedure has been essentially the only method used for determining diatomic dissociation limits from experimental vibrational

[39] H. Kreek and W. J. Meath, *J. Chem. Phys.*, 1969, **50**, 2289.
[40] T. R. Singh, H. Kreek, and W. J. Meath, *J. Chem. Phys.*, 1970, **52**, 5565, erratum 1970, **53**, 4121.
[41] H. Kreek, Y. H. Pan, and W. J. Meath, *Mol. Phys.*, 1970, **19**, 513.
[42] Y. H. Pan and W. J. Meath, *Mol. Phys.*, 1971, **20**, 873.
[43] W. C. Stwalley, *Chem. Phys. Letters*, 1970, **7**, 600.

energies.[44–47] It is based on the observation that the spacings between successively higher vibrational levels usually decrease monotonically as the levels approach the dissociation limit. Since level energies $G(v)$ are rarely observed all the way to dissociation, an extrapolation of some sort is required. Birge and Sponer [44] suggested the use of a plot of the level spacing $\Delta G_{v+\frac{1}{2}} = [G(v+1)-G(v)]$ *versus* the vibrational quantum number v, such as that shown in Figure 1. The dissociation energy is then obtained by adding the

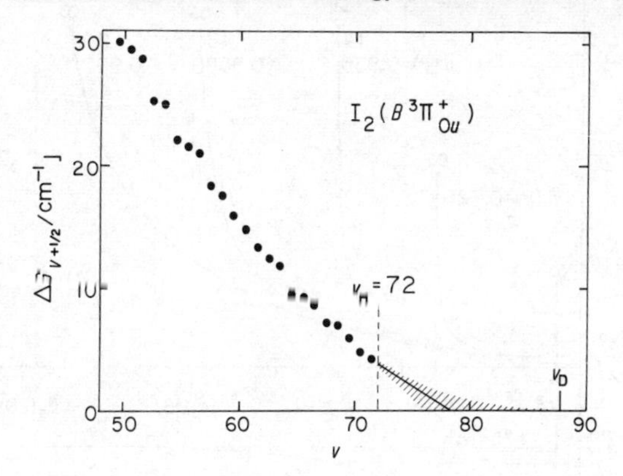

Figure 1 *Birge–Sponer plot of Brown's* [48] *experimental vibrational spacing for* I_2 $(B\,^3\Pi^+_{0u})$

energy of the highest observed level (v_H) to the area under the extrapolated curve between v_H and the intercept. In practice this extrapolation has often been performed linearly, largely because of the lack of a more plausible alternative, but perhaps also because the vibrational spacings of the familiar Morse potential are precisely linear in v. Uncertainty as to the nature of the 'correct' extrapolation is reflected in the area under the extrapolated curve, as is illustrated by the shaded area in Figure 1. The lower limit there corresponds to a straight line through the last few points, the solid line to the intuitive extrapolation by the experimenter who made the measurements,[48] and the upper limit to the 'true' extrapolation (see below).

In spite of its ambiguous nature, the Birge–Sponer procedure has undergone no real change in the past four and a half decades. Alternative extrapolation methods were suggested by Birge [45] and Rydberg,[13] but they were never widely used. Studies have also been made of the ways in which a linear Birge–Sponer approach breaks down for different classes of diatomics,[44–46, 49]

[44] R. T. Birge and H. Sponer, *Phys. Rev.*, 1926, **28**, 259.
[45] R. T. Birge, *Trans. Faraday Soc.*, 1929, **25**, 707.
[46] A. G. Gaydon, 'Dissociation Energies and Spectra of Diatomic Molecules', 3rd. edn., Chapman and Hall, London, 1968.
[47] G. Herzberg, 'Spectra of Diatomic Molecules' Van Nostrand, Toronto, 1950.
[48] W. G. Brown, *Phys. Rev.* 1931, **38**, 709.
[49] C. L. Beckel. M. Shafi, and R. Engelke, *J. Mol. Spectroscopy*, 1971, **40**, 519.

but they were usually impaired by a lack of knowledge of the 'natural' functionality of the distribution of levels near dissociation. This deficiency is removed by the results described below.

Note that throughout this chapter the rotationless vibrational energy $G(v,J = 0)$ is written as $G(v)$, and the turning points (see Figure 2) as $R_1(v) \equiv R_1(v,J = 0)$ and $R_2(v) \equiv R_2(v,J = 0)$.

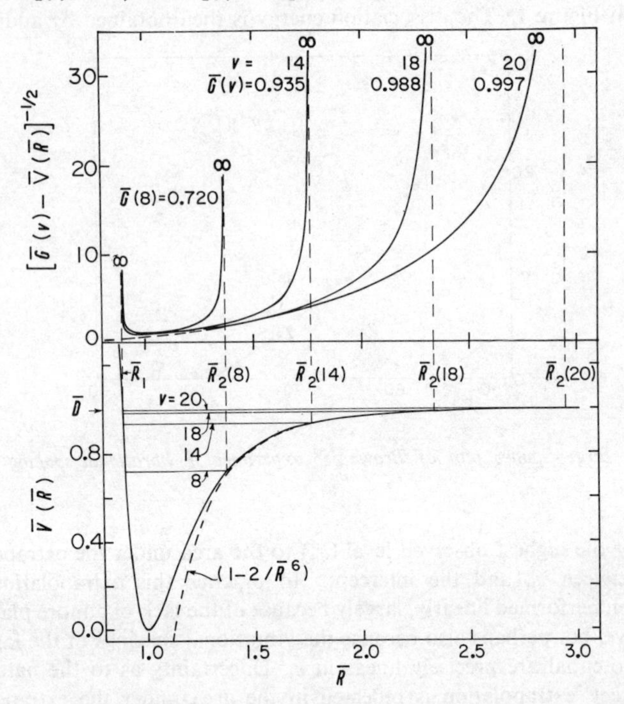

Figure 2 Lower: *eigenvalues and turning points for levels of a model LJ*(12,6) *potential; the dashed curve is the long-range attractive part of the potential. The bars* (e.g. $\bar{R}$) *indicate that energies and lengths are scaled relative to the well depth D and equilibrium distance R_e, respectively*. Upper: *exact integrand of equation* (7) *for the four indicated vibrational eigenvalues; the dashed curve is the approximate $v = 20$ integrand obtained on replacing the exact V(R) by the long-range part* (Reproduced by permission from (*J. Chem. Phys.*, 1970, **52**, 3869)

Derivation and Analysis of the Method.—*For a General R^{-n} Tailed Potential, ignoring Rotation* ($J = 0$). The following presentation is largely based on the work of Le Roy and Bernstein.[50, 51] An alternate derivation of their fundamental result was later reported by Stwalley,[52] but although his approach yields the correct functionality, it is less accurate than that de-

[50] R. J. Le Roy and R. B. Bernstein, *J. Chem. Phys.*, 1970, **52**, 3869.
[51] R. J. Le Roy and R. B. Bernstein, *J. Mol. Spectroscopy*, 1971, **37**, 109.
[52] W. C. Stwalley, *Chem. Phys. Letters*, 1970, **6**, 241.

scribed here. For the particular case of an R^{-4} long-range potential, Tani and Inokuti [22] obtained a more general result which includes the dependence of the level density on the rotational quantum number J; this is described later (p. 128).

Like Dunham's method,[11] the present approach is semi-classical in origin, being based on the first-order WKB quantum condition for the eigenvalues of a potential $V(R)$:

$$v+\tfrac{1}{2} = \frac{(2\mu)^{\frac{1}{2}}}{\pi\hbar}\int_{R_1(v)}^{R_2(v)} [G(v)-V(R)]^{\frac{1}{2}}\,\mathrm{d}R \tag{6}$$

where μ is the nuclear reduced mass, and $R_1(v)$ and $R_2(v)$ are the classical turning points (see Figure 2) at which $G(v) = V(R)$. The allowed eigenvalues are the energies $G(v)$ corresponding to integer values of v. Since we are concerned with the distribution of eigenvalues, it seems natural to differentiate equation (6) with respect to $G(v)$ to obtain the density of levels:

$$\frac{\mathrm{d}v}{\mathrm{d}G(v)} = \frac{(\mu/2)^{\frac{1}{2}}}{\pi\hbar}\int_{R_1(v)}^{R_2(v)} [G(v)-V(R)]^{-\frac{1}{2}}\,\mathrm{d}R \tag{7}$$

Figure 2 shows the integrand of equation (7) for four levels of a model Lennard-Jones (12,6) potential:* $V_{\mathrm{LJ}}(R) = D[(R_e/R)^{12}-2(R_e/R)^6]$. It is immediately clear that for the upper levels the dominant contributions to the integral in equation (7) come from the regions near the turning points where the integrand becomes singular. Furthermore, the anharmonicity of the potential makes the neighbourhood of the outer turning point by far the more important, and this weighting becomes increasingly pronounced as the levels approach dissociation. This suggests the fundamental approximation of replacing the actual potential $V(R)$ with an expression which is accurate at the outer turning points and only approximate elsewhere. Our knowledge of the theoretical inverse-power form of long-range potentials suggests the use of equation (4): $V(R) = D-C_n/R^n$. For the $v = 20$ level of our ($n = 6$) chosen LJ(12,6) potential, the effect of replacing the actual $V_{\mathrm{LJ}}(R)$ by its attractive long-range part $D[1-2(R_e/R)^6]$ is shown by the dashed curve in the lower half of Figure 2. In this case the substitution clearly has a negligible effect on the value of the integral in equation (7). A minor additional approximation is setting $R_1(v) = 0$; its main effect will be to cancel part of the error introduced by ignoring the $R_1(v)$ singularity in the exact integrand of equation (7).

* In the notation of Harrison and Bernstein [17] the parameters of the chosen 24-level model potential correspond to $B_z = 2\mu D R_e^2/\hbar^2 = 10\,000$. The eigenvalues were calculated numerically [50] and are accurate to $10^7/D$.

Substituting equation (4) into equation (7), making a change of variables, and setting $R_1(v) = 0$ yields the desired results:[50]

$$\frac{\mathrm{d}G(v)}{\mathrm{d}v} = K_n[D - G(v)]^{(n+2)/2n} \tag{8}$$

where the factor

$$K_n = \frac{\bar{K}_n}{\mu^{\frac{1}{2}}(C_n)^{1/n}} = \frac{(2\pi)^{\frac{1}{2}}\hbar}{\mu^{\frac{1}{2}}(C_n)^{1/n}} \left[\frac{n\,\Gamma\left(1+\frac{1}{n}\right)}{\Gamma\left(\frac{1}{2}+\frac{1}{n}\right)}\right]$$

and $\Gamma(x)$ is the familiar gamma function.[53] It is often more convenient to use the integrated form of equation (8), which for $n \neq 2$ is

$$G(v) = D - [(v_\mathrm{D} - v)H_n]^{2n/(n-2)} \tag{9}$$

where $H_n = [(n-2)/2n]K_n = \bar{H}_n/[\mu^{\frac{1}{2}}(C_n)^{1/n}]$, and v_D is an integration constant. For various values of n, the numerical constants $\bar{K}_n$ and $\bar{H}_n$ are given in Table 1, together with values of analogous factors which arise in Sections 4 and 5.

Table 1 *Numerical factors from equations* (8), (9), (32), (33), *and* (42) *for various n, assuming units of energy, length, and mass to be* cm^{-1}, Å, *and* amu, *respectively*

n	$\bar{K}_n$	$\bar{H}_n$	$\bar{P}_n$	$\bar{Q}_n$	$\bar{S}_n$
1	16.4234	−8.21171	a	a	a
2	25.7978	0.0	a	a	a
3	34.5429	5.75715	54.8210	60225.244	709.7684
4	43.0631	10.76578	36.8928	4275.947	71.0483
5	51.4763	15.44290	30.6405	1178.333	36.1024
6	59.8301	19.94336	27.4105	546.658	26.6415
8	76.4406	28.66524	24.0754	225.498	20.4229
10	92.9819	37.19274	22.3544	136.330	18.2749

[a] These constants do not exist for $n \leqslant 2$.

For $n > 2$, v_D takes on physical significance as the effective (in general non-integer) vibrational index at the dissociation limit. On the other hand, for $n < 2$, v_D must be smaller than the index v of any of the allowed levels, and hence is negative. For the special case $n = 2$, integration of equation (8) yields

$$D - G(v) = [D - G(v_\mathrm{D})]\exp[-\pi\hbar(v - v_\mathrm{D})(2/\mu C_2)^{\frac{1}{2}}] \tag{10}$$

where v_D is an arbitrarily chosen reference level. Expression (10) is identical

[53] M. Abramowitz and I. A. Stegun, 'Handbook of Mathematical Functions', U.S. Nat. Bur. Stand., Appl. Math. Ser. no. 55, U.S. Dept. of Commerce, Washington, 1964.

to the exact quantal result for a pure R^{-2} potential,[54] except for a small additive correction to the apparent C_2 constant.[50] Results analogous to equations (8)—(10) have also been obtained [50] for exponential-tailed potentials, $V(R) = D - Ae^{-\beta R}$; however, this case is of less physical interest and is not discussed here.

A couple of useful expressions obtained from the derivatives of equation (9) are:

$$\frac{dG(v)}{dv} = \left\{\left(\frac{2n}{n-2}\right)\left(H_n\right)^{2n/(n-2)}\right\}[v_D - v]^{(n+2)/(n-1)} \qquad (11)$$

and*

$$\frac{dG(v)}{dv} \bigg/ \frac{d^2G(v)}{dv^2} = -(v_D - v)\left(\frac{n-2}{n+2}\right), \qquad n \neq 2 \qquad (12)$$

Together with equation (8), equations (11) and (12) suggest that simple plots using derivatives obtained by interpolating over a set of energy levels should yield values of the physically interesting parameters n, v_D, and C_n. Figure 3 shows the plot suggested by equation (12) for levels of the model

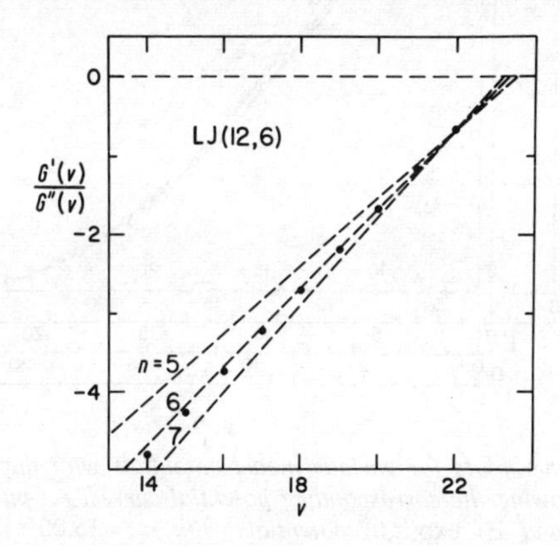

Figure 3 *Plot suggested by equation* (12) *for the highest levels of a model LJ*(12,6) *potential. The broken lines pass through the* $v = 22$ *point and have slopes* $[(n-2)/(n+2)]$ *corresponding to the indicated n*
(Reproduced by permission from *J. Chem. Phys.*, 1970, **52**, 3869)

LJ(12,6) potential discussed above; the highest bound levels clearly exhibit the expected $n = 6$ behaviour.

Equations (8) and (11) imply that Birge–Sponer plots of level spacings for

* For the special case of $n = 2$, the right-hand side of equation (12) becomes $-(\mu C_2/2)^{1/2}/(\pi\hbar)$.

[54] P. M. Morse and H. Feschbach, 'Methods of Theoretical Physics', McGraw–Hill, New York, 1953, vol. 2, p. 1665.

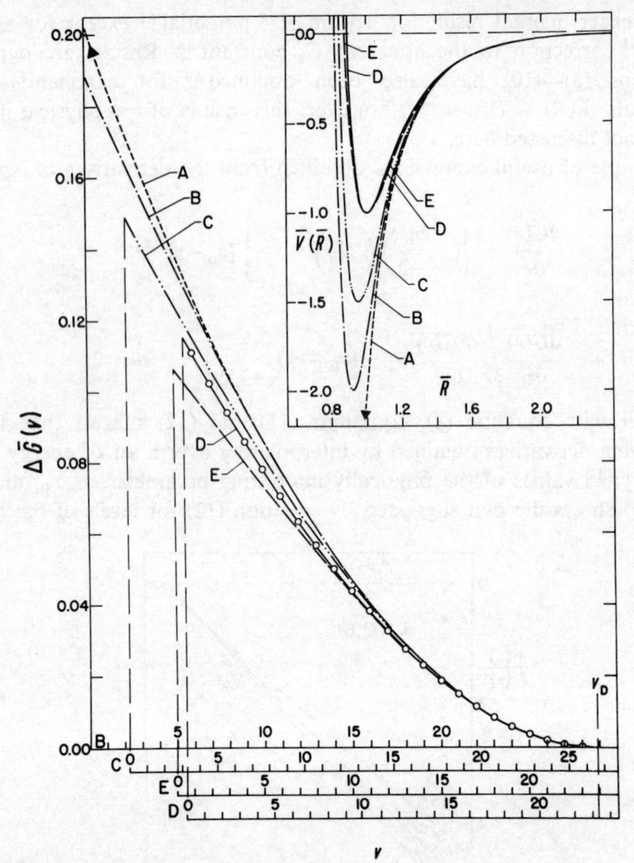

Figure 4 *Birge–Sponer plots for various model potentials with the same long-range tails, the insert showing the corresponding potential curves.* A: *pure* R^{-6} *attractive potential;* B, C, *and* E: exp(α,6) *potentials with* α = 15.00, 13.772, *and* 12.0, *respectively;* D: *LJ*(12,6) *potential. Energies and lengths are scaled relative to the depth and equilibrium distance of this model LJ*(12,6) *curve*
(Reproduced by permission from *J. Chem. Phys.*, 1970, **52**, 3869)

potentials with identical long-range tails, but arbitrarily different short-range behaviour, should be identical near dissociation. This question is examined in Figure 4, which shows plots of $\Delta G(v) = \mathrm{d}G(v)/\mathrm{d}v$ for exp(α,6) potentials

$$V_{\mathrm{exp}}(R) = D\left\{\left(\frac{6}{\alpha-6}\right)\exp\left[-\alpha\left(\frac{R-R_{\mathrm{e}}}{R_{\mathrm{e}}}\right)\right] - \left(\frac{\alpha}{\alpha-6}\right)\left(\frac{R_{\mathrm{e}}}{R}\right)^{6}\right\}$$

with α = 12.0, 13.772, and 15, and for the present model LJ(12,6) curve. The various potentials are shown in the insert on Figure 4; all have the same C_6, and the same μ was assumed in obtaining their vibrational spacings*. The reference curve A was obtained on substituting this C_6 and μ into equation (11), together with $n = 6$; it is the exact semi-classical result for the potential of equation (4) with $n = 6$. The convergence of the other curves to A for levels approaching v_D demonstrates the essential validity of expressions (8)—(10).

A more quantitative test of the present results can be made by fitting exact calculated eigenvalues of a chosen model potential to equation (9), and then comparing the resulting potential parameters with their known values. This has been done for both LJ(12,6) [50] and LJ(12,3) [52] potentials, for which $n = 6$ and 3, respectively. In both cases, least-squares fits of equation (9) to the levels lying closest to D yielded highly accurate results, whereas the addition of more deeply bound levels gradually increased the errors.

An important qualitative implication of equations (8)—(10) is their prediction of positive curvature for Birge–Sponer plots in the region near the dissociation limit where the outer branch of the potential is accurately described by equations (3) or (4). This result should be as universal as the inverse-power form of the attractive long-range potential. Apparent exceptions will occur for molecules with small reduced mass μ whose level density is so low that no eigenvalues lie in the immediate neighbourhood of dissociation. For example, the outer turning points of the two highest vibrational levels of ground-state H_2 are 3.3 and 2.7 Å, distances much too small for the inverse-power expansion (3) to be appropriate. Thus, equations (8) and (9) are not valid for these levels, and it is not surprising that their Birge–Sponer plot shows increasingly negative curvature at D.[55]

Although the region of validity of equations (8)—(12) necessarily reflects itself in positive curvature on a Birge–Sponer plot, observation of such curvature does not necessarily imply that use of these expressions is appropriate. A sufficient condition for the latter is that the outer turning points of the levels considered must lie in the long-range region where the attractive potential is accurately described by equation (3). This criterion appears somewhat nebulous since one is generally dealing only with experimental vibrational energies, and initially has no knowledge of the turning points. However, the parameters obtained on fitting equation (9) to experimental data yield an effective potential of the form of equation (4), and it may be examined retrospectively to determine whether use of this 'near-dissociation' method was justified.

For model potentials consisting of a repulsive part plus a single inverse-power attractive term, errors in equations (8) and (9) arise from approximations to the integrand in equation (7) in the neighbourhood of $R_1(v)$ (see

* More generally, a mass-reduced version of Figure 4 which would be independent of the choice of μ would plot $\mu^{1/2}\Delta G(v)$ *versus* $(v_D - v)/\mu^{1/2}$. The $\Delta G(v)$ results for the exp(a,6) potentials were calculated [50] from the semi-classical results of ref. 17, while the LJ(12,6) $\Delta G(v)$ curve (labelled D) was obtained from exact calculated $\Delta G_{v+1/2}$ values.

[55] G. Herzberg and L. L. Howe, *Canad. J. Phys.*, 1959, **37**, 636.

p. 130). On the other hand, for real molecules this source of error is insignificant compared to the effect on this integral of approximating the real long-range potential of equation (3) by a single inverse-power term. Although the true potential of equation (3) and its derivatives may be accurately represented by equation (4) at any given point, the increasing relative importance of the higher-power terms in equation (3) makes this approximate local potential too weak at smaller distances. Effects of this sort in the region near $R_2(v)$ which dominates the value of the integral in equation (7) are the major sources of error in applications of the present results to real systems (see p. 145).

Rotation-dependent ($J \neq 0$) *Level Density for* R^{-4} *Potentials.* The discussion above considered only the distribution of vibrational levels near the dissociation limit of a non-rotating ($J = 0$) molecule. Tani and Inokuti [22] independently used a similar approach to obtain an analytic expression for the density of levels in an R^{-4} long-range potential as a function of both the binding energy and J. Unfortunately, their derivation appears to be contingent on the fact that the R-dependence of the long-range potential (R^{-4}) is the square of that for the centrifugal potential (R^{-2}), and hence it cannot readily be generalized to any n.

Tani and Inokuti's derivation [22] starts in the same way as that above, except that the total effective potential $V_J(R) = V(R) + (\hbar^2/2\mu)J(J+1)/R^2$ is used in equations (6) and (7) instead of the rotationless potential $V(R)$.* The true $V(R)$ is again replaced by the long-range form, here $(D - C_4/R^4)$, but unlike the Le Roy–Bernstein approach, $R_1(v)$ is held fixed at a constant 'hard-core' value $\underline{R}_1$ instead of being set equal to zero. Changing the variable of integration and defining

$$\Lambda \equiv \left(\frac{\hbar^2}{2\mu}\right) \frac{J(J+1)}{[D - G(,Jv)][R_2(v,J)]^2}$$

and

$$\eta = R_2(v,J)/\underline{R}_1$$

then yields the result

$$\frac{\mathrm{d}G(v,J)}{\mathrm{d}v} = \frac{\pi\hbar[D - G(v,J)]^{\frac{1}{4}}}{(\mu/2)^{\frac{1}{2}} R_2(v,J) T(\eta,\Lambda)} \tag{13}$$

where

$$T(\eta,\Lambda) = (2+\Lambda)^{\frac{1}{2}} E[\varphi|(2+\Lambda)^{-1}] - (1+\Lambda)(2+\Lambda)^{-\frac{1}{2}} F[\varphi|(2+\Lambda)^{-1}]$$

and $F(\varphi|m)$ and $E(\varphi|m)$ are incomplete elliptic integrals of the first and second kind,[53] respectively, with φ being defined by: $\sin\varphi = [1 - \eta^{-2}]^{\frac{1}{2}}$.

* Throughout, the J-dependence of the centrifugal potential, level energies, ... *etc.*, is written in terms of $J(J+1)$ rather than the $(J+\frac{1}{2})^2$ which Langer [56] showed was appropriate for semi-classical analyses using equation (6). For realistic systems, effects associated with the Langer correction are small and, in any case, results obtained herein may be corrected by simply replacing $(J+\frac{1}{2})^2$ by $J(J+1)$.

[56] R. E. Langer, *Phys. Rev.*, 1937, **51**, 669.

Note that the outer turning point $R_2(v,J)$ appearing in equation (13) and in the definition of Λ is given by:

$$R_2(v,J) = \frac{\left[\{4C_4[D-G(v,J)]+[J(J+1)\hbar^2/2\mu]^2\}^{\frac{1}{2}}-J(J+1)\hbar^2/2\mu\right]^{\frac{1}{2}}}{\{2[D-G(v,J)]\}^{\frac{1}{2}}}$$

Equation (13) in general allows the hard core radius $\underline{R}_1$ to be non-zero. However, an increase in η from four to infinity [*i.e.* $\underline{R}_1$ decreases from $R_2(v)/4$ to zero] affects values of $T(\eta,\Lambda)$ by less than 1%.[22] Furthermore, the error introduced by setting $\eta = \infty$ ($\underline{R}_1 = 0$) is in the opposite direction to, and for the R^{-6} model potentials of Figure 4 it is also smaller than, the error due to neglect of the singularity at $R_1(v)$ in the exact integrand of equation (7). Hence it is unlikely that setting $\eta = \infty$ will significantly weaken equation (13). The main effect will be to simplify $T(\eta,\Lambda)$, since $\eta = \infty$ implies $\varphi = \pi/2$, for which case the two incomplete elliptic integrals become the slightly more tractable complete elliptic integrals of the first and second kind.[53]

It is very interesting to note that equation (13) is also valid at energies *above* the dissociation limit, but below the top of the centrifugal potential barrier. The elliptic integrals are continuous over this change from positive to negative Λ, and the $(-1)^{\frac{1}{2}}$ in the factor $[D-G(v,J)]^{\frac{1}{2}}$ is cancelled by one in $(2+\Lambda)^{\frac{1}{2}}$ as long as $\Lambda < -2$. This latter criterion points out the fact that $\Lambda = -2$ corresponds to the maximum of the centrifugal potential barrier for $J \neq 0$.

Equation (13) is clearly a generalization of equation (8) for $n = 4$ to cases where $J \neq 0$, and it is readily shown that the two expressions are identical in the limit $\eta = \infty$ and $J = 0$. Unfortunately, it does not appear to be possible to integrate equation (13) to obtain an expression for $G(v,J)$ as the corresponding analogue of equation (9). This fact, coupled with its relative complexity, effectively precludes the application of equation (13) to the type of spectroscopic analysis for which equations (8) and (9) are used below. However, in chemical applications of a knowledge of the density of levels, its explicit J-dependence makes equations (13) potentially much more useful than equations (8)—(10).

A generalization of equation (13) to the case of any inverse-power potential C_n/R^n is obtained by a rather different route in Section 4. The discussion there shows that the principal source of error in both results arises from the neglect of all but the asymptotic term (here R^{-4}) in the actual long-range potential of equation (3).

Utilization of the Method.—*Distribution of Levels of Simple Model Potentials.* The following discussion is concerned with the distribution of vibrational levels in simple model potentials consisting of a repulsive part plus a single inverse-power attractive term. Those explicitly considered are the generalized LJ(a,n) functions,

$$V_{\text{LJ}}(R) = \varepsilon\left\{\left(\frac{n}{a-n}\right)\left(\frac{R_e}{R}\right)^a - \left(\frac{a}{a-n}\right)\left(\frac{R_e}{R}\right)^n\right\} \qquad (14)$$

the analogous exp(α,n) potentials,

$$V_{\exp}(R) = \varepsilon\left\{\left(\frac{n}{\alpha-n}\right)\exp\left[-\alpha\left(\frac{R-R_e}{R_e}\right)\right]-\left(\frac{\alpha}{\alpha-n}\right)\left(\frac{R_e}{R}\right)^n\right\} \quad (15)$$

and the Sutherland potential which arises from both equations (14) and (15) in the limit $\alpha \to \infty$,

$$V_S(R) = -\varepsilon(R_e/R)^n \quad \text{for } R > R_e \quad (16)$$
$$= \infty \quad \text{for } R \leqslant R_e$$

In each of the above, ε is the well depth, R_e the equilibrium distance, and α and n are constants (in general non-integer) such that $\alpha > n > 0$. For all three types of functions, the long-range potential coefficient is given by the same expression:

$$C_n = \varepsilon(R_e)^n[\alpha/(\alpha-n)] \quad (17)$$

and hence within the context of equation (9) these potentials are precisely equivalent.

Figure 4 shows that equation (9) accurately describes the distribution of levels near the dissociation limits of model LJ(12,6) and exp(α,6) potentials, although it becomes an increasingly poor approximation with increasing binding energy and decreasing α. These observations are readily understood in terms of the two types of error inherent in the derivation of equation (9) for these simple potentials. On the one hand, the approximation of replacing the actual $V(R)$ by equation (4) tends to *underestimate* the integral in equation (7) by ignoring the area under the singularity at $R_1(v)$ (see Figure 2); on the other, replacing $R_1(v)$ by zero tends to *overestimate* this integral by spuriously adding the area under the approximate integrand between $R_1(v)$ and zero.* The fact that the exact model-potential results in Figure 4 all lie below the reference curve A obtained from equation (11) shows that in these examples the first type of error is dominant. However, in the $\alpha \to \infty$ limit where equations (14) and (15) become equation (16), the first kind of error completely disappears and the actual $\Delta G(v)$ curve will tend to lie above that calculated from equation (9). These observations are in agreement with the upward trend with α of the $\Delta G(v)$ curves for exp(α,6) potentials, seen in Figure 4. Furthermore, they imply that for both equations (14) and (15) there will exist some intermediate $\alpha(n)$ values for which the two types of errors roughly cancel. For these cases, equation (9) should describe the overall vibrational spacings with fairly good accuracy.

A more convenient model-potential form of equation (9) is obtained if we introduce Harrison and Bernstein's [17] dimensionless parameter

$$\mathscr{B} = 2\mu\varepsilon(R_e)^2/\hbar^2 \quad (18)$$

* Of course, for real molecules several terms contribute to the attractive part of the potential, and the types of errors discussed here are negligible relative to the effect of the higher-power terms in equation (3) in the region near $R_2(v)$.

to characterize the potential. Substituting it and expression (17) into equation (9) yields:

$$D - G(v) = \varepsilon[(v_D - v)/\mathscr{B}^{\frac{1}{2}} a_0(\alpha,n)]^{2n/(n-2)} \tag{19}$$

where

$$a_0(\alpha,n) = [\alpha/(\alpha-n)]^{1/n}\left[\Gamma\left(\frac{1}{2}+\frac{1}{n}\right)\Big/\Gamma\left(1+\frac{1}{n}\right)\right]/[(n-2)\pi^{\frac{1}{2}}]$$

Since equation (19) is known to be accurate near D (see Figure 4), one can test its utility over the rest of the potential well by seeing whether it correctly predicts the total number of levels supported (*i.e.* the number lying within ε of D).

According to the WKB eigenvalue criterion equation (6), a (hypothetical) vibrational level lying at the potential minimum corresponds to $v = -\frac{1}{2}$, and hence $[D - G(-\frac{1}{2})] = \varepsilon$. Substituting this relation into equation (19) then yields

$$v_D + \tfrac{1}{2} = a_0(\alpha,n)\mathscr{B}^{\frac{1}{2}} \tag{20}$$

as the present method's estimate of the number of levels lying between dissociation and the potential minimum. For LJ(α,n) potentials, an exact semi-classical analysis [21] yields an expression identical to equation (20), but with $a_0(\alpha,n)$ replaced by

$$a_{LJ}(\alpha,n) = n^{\frac{1}{2}}(\alpha/n)^{\lambda}[\Gamma(\lambda-\tfrac{1}{2})/\Gamma(\lambda+1)]/[2\pi^{\frac{1}{2}}(\alpha-n)^{\frac{3}{2}}]$$

where $\lambda \equiv (\alpha-2)/2(\alpha-n)$.* Thus equation (20) will predict the correct number of vibrational levels for those LJ(α,n) potentials for which $a_0(\alpha,n) = a_{LJ}(\alpha,n)$, and equation (19) should yield a good description of the overall level spectra for these cases. For $n = 3$, 4, 5, and 6, respectively, $[a_0(\alpha,n)/a_{LJ}(\alpha,n)]$ grows smoothly from 0.980, 0.961, 0.945, and 0.931 for $\alpha = n+1$, to 1.120, 1.198, 1.253, and 1.294 for $\alpha = \infty$, and equals one at $\alpha = 8.580$, 10.570, 12.563, and 14.557. Thus, equation (19) [or equation (9)] should provide a fairly good description of *all* the vibrational levels of LJ(8.580,3), LJ(10.570,4), LJ(12.563,5), and LJ(14.557,6) potentials. Similarly, Harrison and Bernstein's [17] numerical WKB results for exp(α,6) potentials show that equation (19) will be most accurate for exp(15.6,6) functions. It is also intriguing to note the similarity between the eigenvalue expression obtained for these cases by substituting equation (20) into equation (19):

$$D - G(v) = \varepsilon[(v_D - v)/(v_D + \tfrac{1}{2})]^{2n/(n-2)} \tag{19a}$$

and the exact quantal eigenvalue expression for a Morse potential:[47, 54]

$$D - G(v) = \varepsilon[(v_D - v)/(v_D + \tfrac{1}{2})]^2$$

* Versions of this result for the special cases of LJ(12,6) and LJ(α,6) potentials had been reported earlier.[16,20] For LJ(14,6) potentials Mahan [20] also obtained a semi-classical analogue of equation (20) which is valid for all J. At the appropriate α, $a_{LJ}(\alpha,n)$ is identical to the coefficient of $\mathscr{B}^{1/2}$ in the exact quantum analogues of equation (20) for LJ($2n-2$,n) and Sutherland potentials.[18, 19, 21]

Of course in the latter, $(v_D+\frac{1}{2}) \equiv \mathscr{B}^{\frac{1}{2}}/\beta R_e$, where β is the usual Morse exponential parameter.

Despite the disparaging comments in Section 1, it is often desirable to use simple two-term model potential functions such as equations (14)—(16) in test calculations of physical properties. In this context, equation (19) [or equation (19a)] may prove quite useful as an eigenvalue predictor in the numerical solution of the radial Schrödinger equation to determine exact vibrational eigenvalues and eigenfunctions. This application of the Le Roy–Bernstein expressions may proceed in a number of ways. The simplest approach is to simply substitute equation (20) into equation (19) to obtain

$$D-G(v) = \varepsilon[1-(v+\tfrac{1}{2})/\mathscr{B}^{\frac{1}{2}}a_0(\alpha,n)]^{2n/(n-2)} \tag{21}$$

and use the result equation (21) directly. However, this expression will be adequate for all vibrational levels only if equation (20) is fairly accurate in its own right, *i.e.* for potentials such as the LJ(14.557,6) and exp(15.6,6) functions discussed above. On the other hand, Figure 4 demonstrates that equation (19) will almost always yield a fairly accurate estimate of the vibrational *spacings*. Thus, if one substitutes the exact calculated eigenvalue of any particular level $G(v')$ into equation (19) to first obtain an effective local v_D value, $v_D(v')$, then the use of this constant in equation (19) should yield very accurate predictions of the relative energies of levels near v'. This local $v_D(v')$ may clearly be continuously adjusted as one moves up or down in the well.

For the particular case of an LJ(12,6) potential, Cashion [57] discussed the properties of eigenvalue predictors in some detail. He defined four criteria for an ideal scheme, all of which are reasonably well satisfied by the present approach. Cashion's [57] Dunham expression for LJ(12,6) potentials is certainly much more accurate than the present results, and it has the additional advantage of also being valid for $J \neq 0$. However, equations (19)—(21) may be applied to *any* potential with an inverse-power outer branch, while the restriction to $J = 0$ is not a serious problem since the spacing between levels (v,J) and $(v,J+1)$ is virtually always accurately given by $(1+1/J)[G(v,J)-G(v,J-1)]$.

For Spectroscopic Analyses: Methodology. The present section will describe a few of the ways in which equation (8) and its corollaries may be used to rationalize experimental energy levels and to obtain values of D, n, C_n, and v_D. Equations (8)—(10) describe the distribution of those vibrational levels lying sufficiently near dissociation that the potential at their outer turning points has the long-range inverse-power form of equation (4). In utilizing these expressions for spectroscopic analyses, one must know the energies and relative vibrational numbering of a few such levels. Nothing need be known about the number or distribution of any deeper levels, nor of the rotational properties of the levels under consideration. The latter fact allows the use

[57] J. K. Cashion, *J. Chem. Phys.*, 1968, **48**, 94.

of the unresolved band head measurements which are often the only data available for levels near dissociation, while errors in the absolute vibrational numbering will simply result in an integral shift of the parameter v_D.

There are a number of other, often readily available, facts about the system which can greatly facilitate a spectroscopic analysis based on equations (8)—(10). The simplest of these is the nature of the atoms which the given molecular state yields on dissociation. The discussion of Section 2 shows that this information will yield $\tilde{n}$, the (lower bound) limiting value of the effective power n occurring in equations (4), (8), and (9). It also helps if the given molecular state is properly identified, since this information is usually a pre-requisite to the use of any theoretical knowledge of the $C_{\tilde{n}}$ constant and sometimes also has a bearing on the value of $\tilde{n}$. For example, although states formed from ground-state halogen atoms in general have $\tilde{n} = 5$, the C_5 coefficient for the ground $X\,0_g^+(^1\Sigma_g^+)$ state happens to be precisely zero and hence in this particular case $\tilde{n} = 6$.[23, 31] It will be seen below that a prior knowledge of D or $C_{\tilde{n}}$ can further simplify the application of equations (8) and (9).

The most general way of applying the above results to real systems is simply to fit a set of experimental energies to equation (9) by least squares to obtain values of the four parameters D, n, C_n, and v_D.[50, 51] If the levels considered lie sufficiently near dissociation, this fitted n will be very close to its (usually known) asymptotic value n. In this case, fixing n exactly equal to $\tilde{n}$ and repeating the fit while successively omitting the deepest observed levels from consideration should yield optimal values of D, C_n, and v_D. The constants thus obtained may then be substituted into equation (9) and used to predict the energies of all bound levels lying above the highest one observed.

Figure 5 illustrates the dependence of such least-squares parameter values on the number of data used in the fit.[58] The n values shown there were obtained from fits to equation (9) with all four parameters varying freely. In spite of the indicated statistical uncertainties, the effective local n for the uppermost 5—7 levels had clearly converged to its known asymptotic value of $\tilde{n} = 5$. Therefore the fits were repeated with n fixed at 5 and only three free parameters, yielding the D, C_5, and v_D values shown in the lower parts of Figure 5. Although the standard error of the fit decreases when fewer data are used (*i.e.* as $v_L \rightarrow v_H$), the statistical confidence intervals on the parameters eventually begin to diverge. Thus, although the assumptions on which equations (8) and (9) are based are most accurate if only the four highest levels are considered, the parameter values for this case, corresponding to $v_L = 28$ in Figure 5, are probably less reliable than those corresponding to $v_L = 25$—27.

If the local n for the levels considered is not effectively equal to its asymptotic value, the fitted n and C_n will correspond only to an approximate local potential with the C_n having no theoretical significance. Similarly, the

[58] A. E. Douglas, Chr. Kn. Møller, and B. P. Stoicheff, *Canad. J. Phys.*, 1963, **41**, 1174.

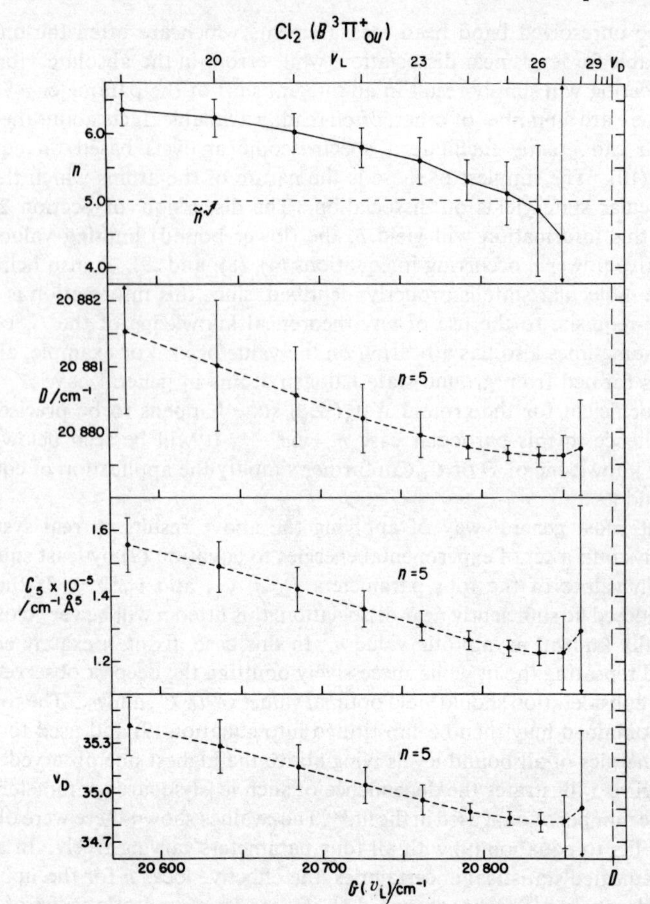

Figure 5 *Parameter values obtained on fitting experimental* Cl_2 $(B\,^3\Pi^+_{0u})$ *energies* [58] *for levels* v_L *to* $v_H = 31$, *to equation* (9)
(Reproduced by permission from *J. Mol. Spectroscopy*, 1971, **37**, 109)

parameter v_D loses its physical significance and becomes merely an integration constant depending on n.* However, the value obtained for D should still be quite accurate. This assertion was tested [50] for the same sample problem considered above [Cl_2 $(B\,^3\Pi^+_{0u})$]. The experimental energies were repeatedly fitted to equation (9) eight adjacent levels at a time, while the highest observed levels were successively omitted from consideration. Figure 6 shows the 'local' D values so obtained, plotted *vs.* the energy of the highest

* The value of v_D obtained in this case would correspond to the vibrational index at D only if the potential varied as R^{-n} all the way to dissociation instead of asymptotically becoming $R^{-\tilde{n}}$ $(n \neq \tilde{n})$.

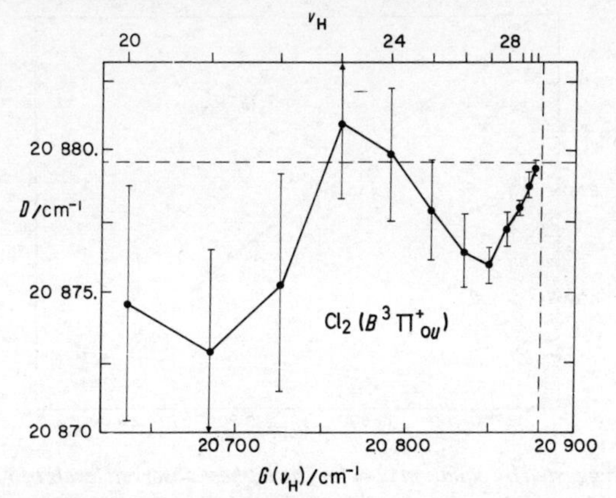

Figure 6 *Estimates of D for Cl_2 ($B\,^3\Pi^+_{0u}$) obtained on fitting equation (9) to the experimental energies* [60] *of levels ($v_H - 7$) to v_H. The broken lines represent the present best estimate of D*
(Reproduced by permission from *J. Chem. Phys.*, 1970, **52**, 3869)

vibrational level included in a given fit, $G(v_H)$. It is striking to note that even if no levels had been observed above $v = 20$, which lies some 244 cm^{-1} below D, the fitted dissociation limit would have been a mere 5.5 cm^{-1} in error! In contrast, a linear Birge–Sponer extrapolation from $v = 20$ yields an estimate of D which is some 70 cm^{-1} too low.

The general least-squares fitting approach discussed above is probably the best way of determining the parameters D, $C_{\tilde{n}}$, and v_D. However, simple graphical methods can often yield virtually indistinguishable results. Such methods presume a knowledge of the asymptotic power $\tilde{n}$ for the state in question, and also that some of the observed levels lie in the asymptotic $R^{-\tilde{n}}$ region. Furthermore, as long as it does not introduce significant error, considerable practical simplification is achieved by approximating derivatives by differences:

$$\mathrm{d}G(v+\tfrac{1}{2})/\mathrm{d}v \approx \Delta G_{v+\frac{1}{2}} = [G(v+1)-G(v)] \qquad (22)$$

$$\mathrm{d}G(v)/\mathrm{d}v \approx \overline{\Delta G}_v \equiv \tfrac{1}{2}[\Delta G_{v+\frac{1}{2}}+\Delta G_{v-\frac{1}{2}}] = \tfrac{1}{2}[G(v+1)-G(v-1)] \qquad (23)$$

Of course, the ensuing expressions become identical to equations (8) and (11) if derivatives obtained by numerical interpolations are used in place of these differences.

Substituting equation (23) into equation (8) and replacing n with $\tilde{n}$ yields

$$(\overline{\Delta G}_v)^{2\tilde{n}/(\tilde{n}+2)} = [D-G(v)](K_{\tilde{n}})^{2\tilde{n}/(\tilde{n}+2)} \qquad (24)$$

An example of the type of plot suggested by equation (24) is shown in Figure 7; it is based on the same experimental data [48] which yielded the Birge–Sponer plot in Figure 1 (for which case [51] $\tilde{n} = 5$). The straightforwardness

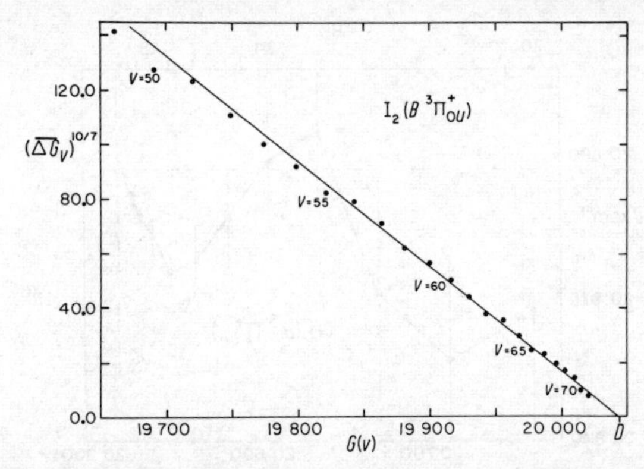

Figure 7 *Plot suggested by equation* (24) *for the highest observed levels of* I_2 $(B\ ^3\Pi^+_{0u})$,[48] *for which* $\tilde{n} = 5$; *energies are in* cm^{-1}
(Reproduced by permission from *J. Mol. Spectroscopy*, 1971, **37**, 109)

of the brief linear extrapolation here is a marked contrast to the uncertainty in Figure 1, the correct result in the latter corresponding to the upper edge of the shaded region. Substituting the D and $C_{\tilde{n}}$ obtained in this way into equation (9) together with experimental $G(v)$ values (and $n = \tilde{n}$) will then yield the remaining parameter, v_D.

The present method's analogue of the Birge–Sponer plot is defined by the expression obtained on substituting equation (22) into equation (11) (with $n = \tilde{n}$):

$$(\Delta G_{v+\frac{1}{2}})^{(\tilde{n}-2)/(\tilde{n}+2)} = (v_D - v - \tfrac{1}{2}) \left\{ \left(\frac{2\tilde{n}}{\tilde{n}-2} \right) (H_{\tilde{n}})^{2\tilde{n}/(\tilde{n}-2)} \right\}^{(\tilde{n}-2)/(\tilde{n}+2)} \qquad (25)$$

For the same problem considered in Figures 1 and 7 [I_2 $(B\ ^3\Pi^+_{0u})$, for which $\tilde{n} = 5$], the plot suggested by equation (25) is shown in Figure 8. As in

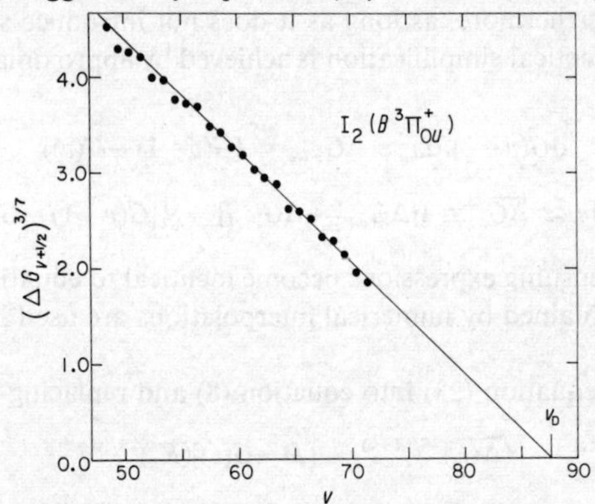

Figure 8 *Plot suggested by equation* (25) *for the highest observed levels of* I_2 $(B^3\ \Pi^+_{0u})$,[48] *for which* $\tilde{n} = 5$; *energies are in* cm^{-1}. *This is the Le Roy–Bernstein method's analogue of the Birge–Sponer plot of Figure* 1

Figure 7, the simple linear extrapolation here involves none of the uncertainty associated with the Birge–Sponer plot in Figure 1.

Once D is accurately known for a single electronic state of a given molecule, its value for all other states of this species may be accurately obtained from the known atomic energy level differences.[59] Thus, a not uncommon situation will be one in which $C_{\tilde{n}}$ and v_D are the only unknown parameters. In this case it is useful to consider a rearranged form of equation (9):

$$[D-G(v)]^{(\tilde{n}-2)/2\tilde{n}} = (v_D - v)H_{\tilde{n}} \tag{26}$$

where, as in equations (24) and (25), the parameters occurring on the left-hand side are known. An example of the type of plot suggested by this expression is shown in Figure 9.

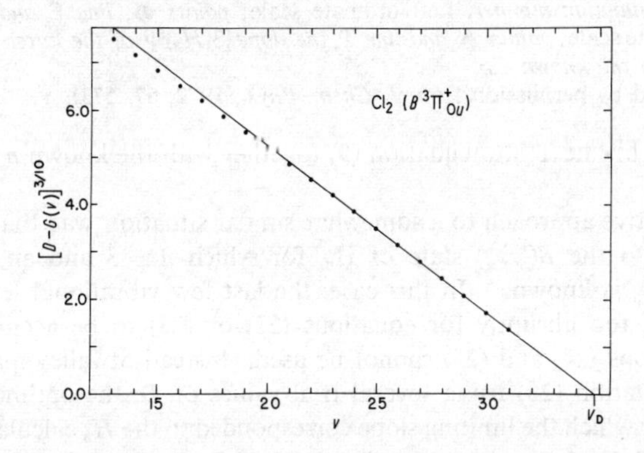

Figure 9 *Plot suggested by equation* (26) *for the highest observed levels of* Cl_2 $(B\,^3\Pi^+_{0u})$,[58] *for which* $n = 5$; *energies are in* cm^{-1}. *The value of D was taken from Table* 2
(Reproduced by permission from *J. Mol. Spectroscopy*, 1971, **37**, 109)

Another type of problem is one in which $\tilde{n}$ and $C_{\tilde{n}}$ (and hence $H_{\tilde{n}}$) are known, but D and v_D are not. This is the case for the ground $(X\,^1\Sigma^+_g)$ state of Ar_2 for which $\tilde{n} = 6$ and an accurate theoretical C_6 has been calculated.[34] Figure 10 shows (triangular points, right ordinate scale) the plot suggested by equation (25) with the true extrapolation (line T) being defined as a straight line through the highest point having a slope $[3(H_6)^3]^{\frac{1}{4}}$ calculated from the known C_6.[60] It differs markedly from the experimenters'[61] linear extrapolation (line E) of a Birge–Sponer plot (round points, left ordinate scale) based on the same data; curve C is the corrected Birge–Sponer extrapolation corresponding to line T. The binding energy of the highest observed level may then be determined by substituting the v_D obtained as

[59] See *e.g.* C. E. Moore, U.S. Nat. Bur. Stand., Circular no. 467, vols. 1–3, 1949, 1952, and 1958.
[60] R. J. Le Roy, *J. Chem. Phys.*, 1972, **57**, 573.
[61] Y. Tanaka and K. Yoshino, *J. Chem. Phys.*, 1970, **53**, 2012.

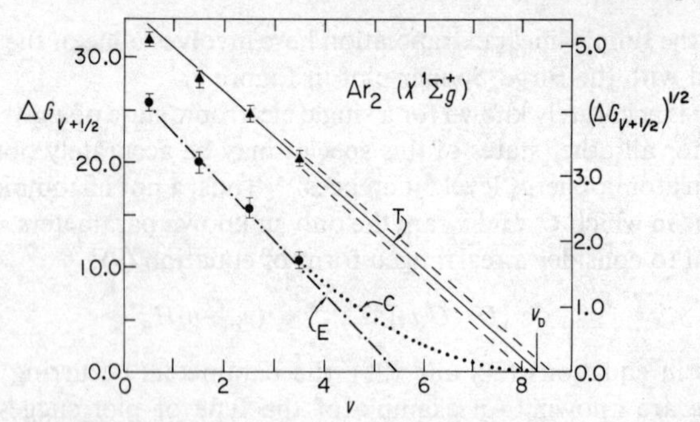

Figure 10 *Observed vibrational spacings*[61] *of* $X(^1\Sigma_g^+)$ Ar_2 (*in* cm^{-1}) *plotted* versus *vibrational quantum number*. Left ordinate scale: *points* ●, *line* E *and curve* C; right ordinate scale: *points* ▲ *and line* T, *the slope* $[3(H_6)^3]^{\frac{1}{6}}$ *of the latter being calculated from the known* C_6
(Reproduced by permission from *J. Chem. Phys.*, 1972, **57**, 573)

the intercept of line T into equation (9) together with the known $n = \tilde{n} = 6$ and C_6.

An alternative approach to a somewhat similar situation was that applied by Stwalley to the $B(^1\Sigma_u^+)$ state of H_2, for which $\tilde{n} = 3$ and an accurate theoretical C_3 is known.[52] In this case, the last few vibrational levels converged much too abruptly for equations (22) or (23) to be accurate, and hence equations (24) and (25) cannot be used. Instead Stwalley made plots based on equation (26) using several trial values of D, the optimum value being that for which the limiting slope corresponded to the H_3 calculated from the known C_3.[52] As is shown in Figure 11, this approach is also capable of yielding very precise values of D and v_D.

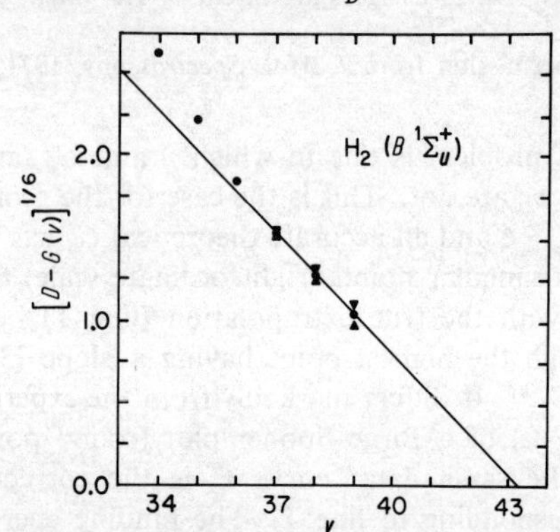

Figure 11 H_2 $(B\,^1\Sigma_u^+)$ *energy levels plotted according to equation* (26), *using three trial D values: points* ● *correspond to* $D = 118337.7$, *points* ▼ *to* 118338.1, *and points* ▲ *to* 118337.3 cm^{-1}. *The slope* H_3 *of the solid line is calculated from the theoretically known* C_3; *energies are in* cm^{-1}
(Reproduced by permission from *Chem. Phys. Letters*, 1970, **6**, 241)

Table 2 *Summary of results obtained using the Le Roy–Bernstein expressions* [equations (8)–(9)]

Species	D/cm^{-1} [a]	v_H [b]	v_D	$\tilde{n}$	$C_{\tilde{n}}$/cm^{-1} Å$^{\tilde{n}}$ Fitted [c]	$C_{\tilde{n}}$/cm^{-1} Å$^{\tilde{n}}$ Theoretical [d]	Ref.
H_2 ($B\,^1\Sigma_u^+$)	118377.7(±0.5)	39	43.26(±0.04)	3	*e*	3.6096×10^4	52
Ar_2 ($X\,^1\Sigma_g^+$)	84.0(±3.5)	4	8.27(±0.21)	6	*e*	$3.26(\pm0.10)\times10^5$	60
$^{35}Cl^{35}Cl$ ($B\,^3\Pi_{0u}^+$)	20879.64(±0.14)	31	34.82(±0.06)	5	$1.22_4(\pm0.04)\times10^5$	1.23×10^5	50,51 [f]
$^{79}Br^{79}Br$ ($B\,^3\Pi_{0u}^+$)	19579.71(±0.27)	52	59.51(±0.3)	5	$1.79(\pm0.2$[illegible]$)\times10^5$	2.00×10^5	51 [f]
$^{81}Br^{81}Br$ ($B\,^3\Pi_{0u}^+$)	19581.77(±0.35)	52	60.28(±0.3)				
$^{127}I^{127}I$ ($B\,^3\Pi_{0u}^+$)	20044.0$_4$(±1.2)	72	87.74$_5$(±0.38)	5	$3.10_6(\pm$[illegible]$16)\times10^5$	3.68×10^5	51

[a] See footnote on p. 140; [b] highest bound level used in the analysis; [c] calculated from the fitted $H_{\tilde{n}}$ using the nuclear reduced mass; [d] with regard to the C_5 values, see footnote on p. 142; [e] in the analysis for this case, $C_{\tilde{n}}$ was held fixed at the known theoretical value; [f] revised here.

Results of Spectroscopic Analyses. At the present time the techniques described above have been applied to only a few systems; the results thus obtained are summarized in Table 2. In all of these cases the highest observed levels lie close enough to dissociation to depend on the theoretically-known asymptotic power $\tilde{n}$, and hence physically significant values of $C_{\tilde{n}}$ and v_D were obtained together with accurate values of D.* This also allowed reliable predictions to be made for the energies of the $(\bar{v}_D - v_H)$ unobserved vibrational levels adjacent to D, where $\bar{v}_D$ is the integer obtained on truncating v_D; the number of such levels ranged from 3, for Ar_2 and Cl_2, to 15, for I_2. Combining the accurate D values in Table 2 with known atomic energy level differences [59] yields the improved ground-state dissociation energies D_0 given in Table 3. Equally good D values are similarly obtainable for all other electronic states of these species.

Table 3 *Diatomic ground* $(X\,^1\Sigma_g^+)$ *state dissociation energies* D_0 (*in* cm^{-1}) *obtained from the results in Table* 2

Species	*Le Roy–Bernstein method*	*Best previous value*
H_2	36118.6($\pm$0.5)	36117.3($\pm$1.0)[a]
Ar_2	84.0($\pm$3.5)	76.9($\pm$1.9)
$^{35}Cl^{35}Cl$	19997.14($\pm$0.14)	19999($\pm$2)
$^{79}Br^{79}Br$	15894.5($\pm$0.4)	15892.2($\pm$0.5)
$^{81}Br^{81}Br$	15896.6($\pm$0.5)	15893.9($\pm$0.5)
$^{127}I^{127}I$	12440.9($\pm$1.2)	12452.5($\pm$1.5)

[a] Herzberg's best estimate was his upper bound, 36118.3 cm^{-1}.[66]

The impact of the present approach on the determination of dissociation limits is perhaps best illustrated by an examination of its effect on the apparent binding energy of the highest observed level, v_H. Table 4 compares the present values of $[D - G(v_H)]$ for the species in Table 2 with the estimates of the experimenters who originally measured the levels. Although neglect of the influence of the long-range potential can evidently lead to large relative errors, the magnitudes of the improvements shown in Table 4 are not very large on a chemical scale. However, Figures 6—9 show that the results obtained in some of these cases would not have been seriously affected if the highest 5—10 reported levels had in fact not been observed. Considerably greater error would almost certainly have arisen in analyses which ignored the characteristic near-dissociation behaviour represented by equations (8) and (9).

The experimental data for H_2 $(B\,^1\Sigma_u^+)$ were analysed using the graphical approach discussed above and illustrated in Figure 11.[52] The improved ground-state dissociation energy thus obtained (see Table 3) is particularly

* Unless specified otherwise, the energy of a dissociation limit D is always expressed relative to the $v = 0, J = 0$ level of the ground electronic state of the particular isotopic species.

Table 4 *Binding energy* $[D - G(v_H)]$ (*in* cm^{-1}) *of the highest observed level* v_H *of each of the species considered in Table* 2

Species	v_H	*Obtained using Le Roy–Bernstein equations* (8) *and* (9)	*Previous estimate*
H_2 ($B\,^1\Sigma_u^+$)	39	1.35($\pm$0.10)	*a*
Ar_2 ($X\,^1\Sigma_g^+$)	4	12.1($\pm$1.7)	5.4
$^{35}Cl^{35}Cl$ ($B\,^3\Pi_{0u}^+$)	31	2.74($\pm$0.14)	3.1($\pm$2)
$^{79}Br^{79}Br$ ($B\,^3\Pi_{0u}^+$)	52	5.24($\pm 0.2_7$)	2.7($\pm$0.5)
$^{81}Br^{81}Br$ ($B\,^3\Pi_{0u}^+$)	52	6.96($\pm 0.3_5$)	4.1($\pm$0.5)
$^{127}I^{127}I$ ($B\,^3\Pi_{0u}^+$)	72	19.6($\pm$1.2)	12.6

[a] Comparison not appropriate; see ref. 52.

interesting in view of the conflict which had existed between experiment [62] and the value of D_0 calculated from the accurate *ab initio* potential of Kolos and Wolniewicz.[63–65] Although Herzberg's new measurements [66] of the photo-dissociation threshold resolved this discrepancy in favour of the theoretical value, Stwalley's analysis [52] yielded a much more precise estimate of D. Unfortunately his improved accuracy was partly compromised by experimental uncertainty in the absolute frequency, as is indicated by the difference between the uncertainty in Table 4 and those in Tables 2 and 3. An additional noteworthy point about the H_2 analysis was the way in which the functionality of equation (9) or equation (26) was exploited in identifying and reassigning lines in the complex experimental spectrum.[52] Similar techniques for making vibrational reassignments were also applied in ref. 51.

Of the species considered in Table 2, the impact of the introduction of the present methods is perhaps most dramatic for ground-state Ar_2 ($X\,^1\Sigma_g^+$). For this case, the molecular dissociation energy was shown [60] to be 9% larger, and the number of bound vibrational levels 50% larger than the experimenters' [61] original estimates. The experimental data for this system were analysed utilizing equation (25) as described in conjunction with the presentation of Figure 10.[60] As with *B*-state H_2, the uncertainty in binding energy of the highest observed Ar_2 level (see Table 4) is considerably smaller than that in the resulting D value (in Tables 2 and 3). For Ar_2 this occurs because the latter also takes account of uncertainties in the experimental $\Delta G_{v+\frac{1}{2}}$ values for $v = 0$—3. It should be noted that the well depth of $D_e = (98.7 \pm 3.5)\ cm^{-1}$ obtained by the present approach is in excellent agreement with both the 98.67 cm^{-1} depth of the current best semi-empirical

[62] G. Herzberg and A. Monfils, *J. Mol. Spectroscopy*, 1960, **5**, 482.
[63] W. Kolos and L. Wolniewicz, *J. Chem. Phys.*, 1964, **41**, 3663.
[64] W. Kolos and L. Wolniewicz, *J. Chem. Phys.*, 1965, **43**, 2429.
[65] W. Kolos and L. Wolniewicz, *J. Chem. Phys.*, 1968, **49**, 404.
[66] G. Herzberg, *J. Mol. Spectroscopy*, 1970, **33**, 147.

potential for this species[67] and the 99.04 cm^{-1} obtained on correcting the vibrational amplitude of an earlier bulk property potential.[68] These values are also in reasonable accord with the 102.2 ± 2.0, 100.4, and 97.81 cm^{-1} obtained from molecular beam scattering measurements.[69, 70]

For the last four cases in Table 2, the desired parameter values were obtained from least-squares fits of experimental energies to equation (9), with n held fixed at the known asymptotic value of $\tilde{n} = 5$.[50, 51] Although C_5 constants for the B-state halogens may be calculated quite readily, such values were not believed to be sufficiently reliable* and hence these coefficients were treated as parameters to be fitted. The resulting experimental C_5 values are seen (in Table 2) to be in reasonable agreement with the corresponding theoretical estimates, but any significant remaining differences are attributed to the latter.

The results presented here for $B(^3\Pi^+_{0u})$-state Cl_2 are slightly different from those given in refs. 50 and 51. The original analysis selected the optimum parameter values as those associated with $v_L = 28$ in Figure 5, since this minimized the influence of deeply bound levels. However, it appears equally desirable to minimize the large statistical uncertainties associated with least-squares fits which have very few degrees of freedom. The results associated with $v_L = 26$ in Figure 5 seem a likely compromise, and they are used in Tables 2—4. The concomitant changes in the predicted binding energies of the three unobserved levels are quite small, the new values being 1.00 ($\pm$0.09), 0.23 ($\pm$0.03), and 0.016 ($\pm$0.004) cm^{-1} for $v = 32$, 33, and 34, respectively.

Within the Born–Oppenheimer approximation,[5] the internuclear potential energy of a given electronic state is precisely the same for all isotopic forms of a particular chemical species. Furthermore, although there are distinct D and v_D values for each isotope, the former merely differ by the differences in the ground-state zero-point energies, whereas the latter are (semi-classically) related through their respective reduced masses:

$$[v_D(i)+\tfrac{1}{2}]/[v_D(j)+\tfrac{1}{2}] = [\mu(i)/\mu(j)]^{\frac{1}{2}} \qquad (27)$$

As a result, data for different isotopes in the same electronic state may be treated concurrently, a rather desirable possibility since it reduces the number of free parameters. This feature was partially exploited in the analysis of experimental data for the $^{79}Br^{79}Br$ and $^{81}Br^{81}Br$ isotopes of $B(^3\Pi^+_{0u})$ Br_2.[51]

* The uncertainty in the theoretical C_5 values is demonstrated by the effect of the recent change in the accepted values of the atomic expectation values $\langle r_X{}^2 \rangle$ required for calculating them (see Section 2).[33] This development is the source of the differences between the theoretical C_5 values for B-state Cl_2, Br_2, and I_2 quoted in Table 2, and the previously reported values[50, 51] of 1.44×10^5, 2.39×10^5, and 4.54×10^5 cm^{-1} Å^5, respectively.

[67] J. A. Barker, R. A. Fisher, and R. O. Watts, *Mol. Phys.*, 1971, **21**, 657.
[68] G. C. Maitland and E. B. Smith, *Mol. Phys.*, 1971, **22**, 861.
[69] M. Cavallini, G. Gallinaro, L. Meneghetti, G. Scoles, and U. Valbusa, *Chem. Phys. Letters*, 1970, **7**, 303.
[70] J. M. Parson, P. E. Siska, and Y. T. Lee, *J. Chem. Phys.*, 1972, **56**, 1511.

As in the preceding case (Cl_2), the levels considered lay close enough to D that the effective n could be held fixed at $\tilde{n} = 5$. Barrow and Coxon [71] recently found that the assumed vibrational numbering of the Br_2 levels used in the analysis of ref. 51 should be reduced by one. Fortunately, the main effect on the published results is an integer decrease in the fitted v_D values (yielding those shown in Table 2), and a corresponding change in the numbering of the unobserved levels whose energies were predicted in Table 2 of ref. 51. Le Roy and Bernstein [51] chose to rely on neither equation (27) nor the known [72, 73] ground-state zero-point energy differences, but fitted the data utilizing five independent parameters (C_5, two D values, and two v_D values) rather than three. Consequently, the statistical uncertainties for their Br_2 results (in Tables 2—4) [51] are considerably larger than the probable errors. This conclusion is attested to both by the very small discrepancy found on substituting the fitted v_D values into equation (27), and by the agreement between the predicted zero-point isotope shift of 2.05 ($\pm$0.12) cm^{-1} and the experimental value of 2.03 ($\pm$0.02) cm^{-1}.[72] It is further confirmed by the excellent agreement between the predicted energies of the $v = 54$—57 levels of $^{79}Br^{79}Br$ of ref. 51 and subsequent measurements of Barrow and Coxon,[71] the $\pm$0.01 cm^{-1} discrepancies being more than an order of magnitude smaller than the statistical uncertainties in the former.

The analysis of B-state I_2 yielded [51, 74] a new value for the ground-state dissociation energy which differed significantly (see Table 3) from that reported by Verma some years earlier.[75] As Figure 7 indicates, a 12 cm^{-1} error in the present value is exceedingly unlikely. In addition, the present method points out an anomaly in the previous analysis which offers a good illustration of the often very compelling qualitative implications of equations (8) and (9). Verma's [75] value of D_0 was based on the well-defined convergence of eighteen vibrational levels lying near the ground-state dissociation limit. Their vibrational spacings $\Delta G_{v+\frac{1}{2}}$ were almost perfectly linear in v, having a maximum possible curvature of $\lesssim 0.001$ cm^{-1}. On the other hand, substituting the known ground-state $\tilde{n} = 6$ and a plausible $C_6 = 3 \times 10^6$ cm^{-1} Å^6 into equation (9) yields a limiting Birge–Sponer curvature of $6(H_6)^3 = 0.054$ cm^{-1} for this case. This conflict, together with the above noted discrepancy in D_0 values, led to a proposed reassignment of these eighteen levels to an excited 0_g^+ state correlating with the same dissociation limit as the ground state, but having a potential maximum at large R.[74] Although this reassignment is still debated,[76] it is evident that the qualitative implications of equation (9) with regard to the shape of Birge–Sponer plots near dissociation also have a useful role to play in spectroscopic analyses.

Results such as those listed in Table 2 were also reported for ground-state

[71] R. F. Barrow, and J. A. Coxon, personal communication.
[72] J. A. Horsley, and R. F. Barrow, *Trans. Faraday Soc.*, 1967, **63**, 32.
[73] J. A. Coxon, *J. Mol. Spectroscopy*, 1971, **37**, 39.
[74] R. J. Le Roy, *J. Chem. Phys.*, 1970, **52**, 2678.
[75] R. D. Verma, *J. Chem. Phys.*, 1960, **32**, 738.
[76] W. C. Stwalley, *J. Chem. Phys.*, 1972, **56**, 680.

$(X\,^1\Sigma_g^+)$ Cl_2, for which $\tilde{n} = 6$;[51] however, they are now believed to be erroneous. In the first place, the fitted C_6 predicts outer turning points of 3.1 and 2.5 Å for the two highest observed levels (bound by 92 and 287 cm^{-1}, respectively). These values are significantly smaller than the minimum distance 4.1 Å corresponding to criterion (5), which implies that the use of equations (8) and (9) was invalid. In addition, this 'experimental' C_6 is an order of magnitude smaller than one which Cummings [37] calculated for this species. These results also cast doubt on the vibrational reassignment for this state proposed in ref. 51, and the reassignment labelled (A,A') now appears more reasonable than the (B,B') assignment adopted there. However, firm conclusions on this point should really await the observation of more levels in the near-dissociation region.

Discussion and Prognosis.—The methods described above provide a hierarchy of different ways of applying equation (8), the basic result of Le Roy and Bernstein,[50] to experimental data. The approach most appropriate in any particular case depends on the nature of the data and our prior knowledge of the system under consideration. The greatest practical distinction is that between the general least-squares fitting techniques and the linear graphical methods. The latter may only be used if the distribution of the levels in question effectively depends on the known limiting asymptotic power $n = \tilde{n}$. Where D is also known (as well as $\tilde{n}$), one may readily determine the remaining parameters $C_{\tilde{n}}$ and v_D using the type of plot suggested by equation (26) (*e.g.* see Figures 9 and 11). If D is not known, but the difference approximations [equations (22) and (23)] for the derivatives $dG(v)/dv$ are known fairly accurately (and again, the effective $n = \tilde{n}$), one may use plots of the type suggested by equations (24) and (25) (*e.g.* see Figures 7, 8, and 10) to obtain $C_{\tilde{n}}$, and D or v_D, respectively. If the approximations of equations (22) and (23) are not accurate, one could still utilize the latter types of plots after first determining derivatives $dG(v)/dv$ by interpolating numerically over the observed energies. However, if numerical methods are to be introduced, one might as well turn to the general least-squares fitting techniques. Of course, if the parameter $C_{\tilde{n}}$ is also known, it defines the slope in all these types of plots and greatly simplifies the determination of the intercept (*e.g.* see the discussion of Figures 10 and 11).

When the effective power n for the levels in question is not known or is not equal to $\tilde{n}$, the data are best treated by fitting them by least squares to equation (9). This approach may also be applied in any of the cases considered above and it will always yield the optimum results, since it does not make use of approximations such as equations (22) or (23).

These methods clearly provide accurate new ways of determining molecular dissociation energies and long-range potentials. In cases to which they have so far been applied, they appear to yield results as good as or better than any obtainable by other means.[50–52, 60] In addition, the use of equation (9) to predict the number and energies of unobserved levels near D obviates the laborious procedure of repeatedly solving the Schrödinger equation to determine their eigenvalues. Another result which may be obtained from equation (9) if $\tilde{n}$ and $C_{\tilde{n}}$ (and hence $H_{\tilde{n}}$) are known is an upper bound $(H_{\tilde{n}})^{2\tilde{n}/(\tilde{n}-2)}$ for the binding energy of the highest bound level. This limit may prove very useful in cases where a lack of data precludes a determination of v_D.

A necessary step in the utilization of the present techniques is an examination of the question of whether or not equation (3) is valid at the outer turning points of the levels under consideration. If it is, one must further determine whether or not the local value of n for these levels is effectively equal to its asymptotic value $\tilde{n}$ for the given state. A test of the latter point for the $B(^3\Pi_{0u}^+)$ halogens [77–79] uncovered a paradox which underlines the importance of an impending sequel to the present results.

The long-range internuclear potential for B-state Cl_2, Br_2, and I_2 has the form

$$V(R) = D - C_5/R^5 - C_6/R^6 - C_8/R^8 - \cdots \quad (28)$$

where the constants C_5, C_6, and C_8 are all attractive (positive).[77] The Table 2 parameter values for these species were obtained from least-squares fits of experimental data to equation (9) with n held fixed at its asymptotic value $\tilde{n} = 5$. For the highest observed levels of Cl_2 and I_2, least-squares fits to equation (9) verified this $n = \tilde{n} = 5$ behaviour, and the extrapolation of this result to Br_2 is very soundly based on the chemical similarity of these molecules and the relative binding energies of the levels concerned.[50, 51] On the other hand, analyses of the RKR potentials for these species (see Section 6) showed that the R^{-6} term in equation (28) also makes a substantial contribution to the potential at the outer turning points of these levels.[77–79] This apparent contradiction raises questions about the significance of the basic results equations (8) and (9). However, the validity of these expressions is attested to by the results of the model-potential analyses (*e.g.* see Figures 3 and 4), by the agreement between the theoretical and experimental C_5 values in Table 2, and by the linearity seen in the plots of Figures 7—9. Further evidence in their favour is the virtually exact agreement between the new data [71] for Br_2 and the energy levels predicted in ref. 51, a result which certainly confirms the accuracy of the present D value for this molecule.

The apparent contradiction is evidently due to a cancellation of errors associated with the influence on equations (8) and (9) of higher-power terms in equation (3) with $m > \tilde{n}$. It has been found that the lowest-order correction to equation (9) arising from inverse-power contributions to equation (3) with $m > n$ is *precisely zero* for $m = n+1$.[80] In this case therefore, equations (8) and (9) with $n = \tilde{n}$ will be 'anomalously accurate' in that they will be valid far into the region where the $R^{-(\tilde{n}+1)}$ term in the potential becomes significant.

[77] R. J. Le Roy, *J. Mol. Spectroscopy*, 1971, **39**, 175.
[78] J. Tellinghuisen, *J. Chem. Phys.*, 1973, **58**, in the press.
[79] R. J. Le Roy and F. E. Cummings, unpublished work.
[80] R. J. Le Roy, to be published.

Thus, a conclusion that n in equations (8) and (9) 'is effectively equal to $\tilde{n}$' does not necessarily carry over to the potential of equation (4) at the outer turning points of the levels considered. At the same time, this result does not affect the accuracy of the parameters D, $C_{\tilde{n}}$, and v_D obtained in such a situation.[80]

Except for the special case of $n = 4$ considered above (p. 128), the results described in this section were restricted to the rotationless $J = 0$ case. As a result, virtually all present applications of equations (8)—(10) have been concerned with spectroscopic analyses and predictions, and the determination of long-range potentials. In contrast, chemical applications of a knowledge of the density of states usually require it to be known for all J. Although equation (13) provides an accurate result for the important case of R^{-4} tailed potentials, its analytic complexity will tend to inhibit its use. A slightly less accurate, but much more useful result, will be obtained below on combining equation (9) with an expression for the v-dependence of the rotational constant B_v near dissociation.

4 Behaviour of the Rotational Constant Bv for Levels near Dissociation

Background.—Understanding the nature of the rotational constant B_v for highly excited vibrational levels was long an unresolved problem in diatomic spectroscopy. For the lower vibrational levels, B_v values may clearly be expressed as

$$B_v = \sum_{l\geq 0} Y_{l1}\,(v+\tfrac{1}{2})^l$$

where the Dunham coefficients Y_{l1} are defined in terms of the properties of the potential at its minimum.[11] Furthermore, for all vibrational levels B_v depends on the expectation value of R^{-2}:[47, 81]

$$B_v = (\hbar/4\pi c\mu)\,\langle v,0|R^{-2}|v,0\rangle \tag{29}$$

However, until recently, no 'natural' behaviour was known for the region near the dissociation limit, other than the generally accepted conclusion that $B_v \rightarrow 0$ as $G(v) \rightarrow D$.

Stwalley considered this question qualitatively in his analysis of the $B \leftarrow X$ spectrum of H_2.[52] He noted that for levels very near dissociation it is reasonable to associate the expectation value in equation (29) with $[R_2(v)]^{-2}$, where $R_2(v)$ is the classical outer turning point of level ($v,J = 0$). Interpreting this conclusion in terms of the H_2 ($B\,^1\Sigma_u^+$) long-range potential ($D-C_3/R^3$) yield the functionality $B_v \propto [D-G(v)]^{2/3}$ for this ($n = 3$) case.[52] Extending Stwalley's approach to the general case of an R^{-n} tailed potential clearly yields:

[81] E. C. Kemble, 'The Fundamental Principles of Quantum Mechanics with Elementary Applications', Dover, New York, 1937, pp. 386–387.

$$B_v = (\hbar/4\pi c\mu)\{[D-G(v)]/C_n\}^{2/n} \tag{30}$$

Although this expression turns out to have the correct dependence on C_n, D, and $G(v)$, it lacks an important numerical factor and hence is considerably in error.

Other precursors of the results described here may be seen in work on the J-dependence of the total number of levels bound by model potential funtions.[16-22] In particular, Dickinson and Bernstein [21] pointed out the dominant effect of the long-range potential on the number of bound and quasi-bound levels at small J. They also evaluated the quantity $\mathrm{d}G(v,J)/\mathrm{d}[J(J+1)]$ at the centrifugal barrier maxima of Sutherland potentials [equation (16)] by utilizing essentially the same techniques used later [82, 83] (see below) in determining an expression for B_v values near D.

Derivation and Analysis of the ***Bv*** **Functionality.**—The following is largely based on the contents of ref. 82; however, the basic equation was also independently obtained by Child.[83] Since the derivation and many qualitative properties of these results have precise analogues in the vibrational problem of Section 3, the present discussion is relatively brief.

Replacing the electronic interaction potential $V(R)$ by the total effective potential

$$V_J(R) = V(R)+(\hbar^2/2\mu)J(J+1)/R^2$$

makes the WKB eigenvalue criterion equation (6) equally valid for any J (see footnote on p. 128). Therefore, one can write

$$B_v \equiv \{\partial G(v,J)/\partial[J(J+1)]\}_{J=0} = -\left\{\frac{\partial(v+\frac{1}{2})}{\partial[J(J+1)]}\bigg/\frac{\partial v}{\partial G(v,J)}\right\}_{J=0}$$

$$= \left(\frac{\hbar}{4\pi\mu c}\right)\int_{R_1(v)}^{R_2(v)} R^{-2}[G(v)-V(R)]^{-\frac{1}{2}}\,\mathrm{d}R\bigg/\int_{R_1(v)}^{R_2(v)}[G(v)-V(R)]^{-\frac{1}{2}}\,\mathrm{d}R \tag{31}$$

where $R_1(v)$ and $R_2(v)$ are the usual classical turning points of level $(v,J=0)$, and $G(v) \equiv G(v,J=0)$. The integral in the denominator of equation (31) is the same one which appeared in the derivation of the vibrational spacing expression (p. 123), while that in the numerator is similar except for a larger weighting of small R values. Figure 12 shows plots of the integrands of the integrals in equation (31) for three levels of the same model LJ(12,6) potential considered in Figure 2 (see also footnote on p. 123); the $k = 2$ curves in the former are the same as the solid curves in the latter. Although the effect is not so pronounced for the $k = 0$ case, Figure 12 demonstrates that both types of integrals are dominated by the influence of the long-range potential and the singularity at the outer turning point $R_2(v)$. Hence, the same approximation used in the vibrational problem may be introduced; the

[82] R. J. Le Roy, *Canad. J. Phys.*, 1972, **50**, 953.
[83] M. S. Child, personal communication, 1971.

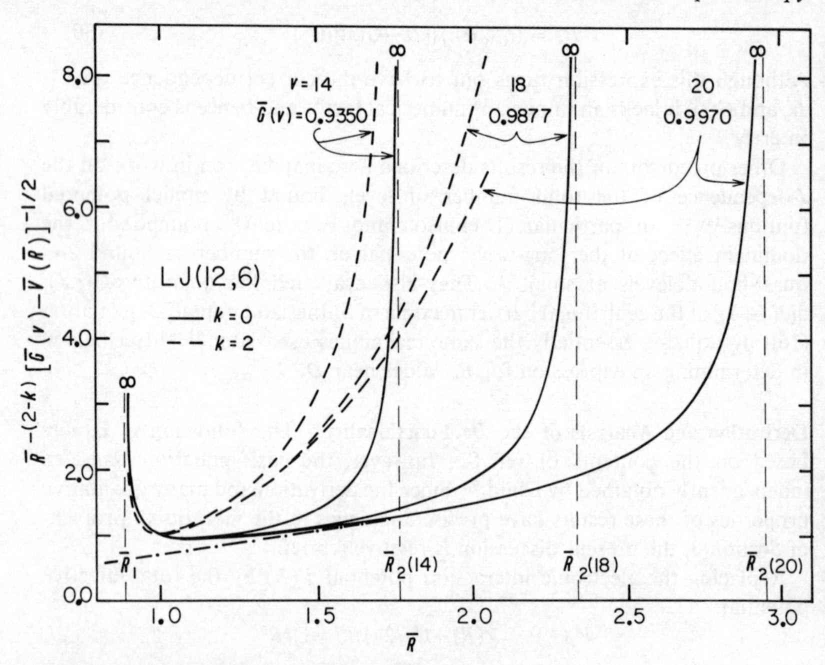

Figure 12 *Integrands of integrals in equation* (31) *for three levels of a model 24-level LJ*(12,6) *potential; $k = 0$ corresponds to the integral in the numerator and $k = 2$ to that in the denominator. The dot-dash curve is the approximate $k = 0$, $v = 20$ integrand obtained on ignoring the repulsive R^{-12} contribution to the potential* (Reproduced by permission from *Canad. J. Phys.*, 1972, **50**, 953)

actual $V(R)$ is replaced by an expression for its long-range part: $D = C_n/R^n$, and $R_1(v)$ is set equal to zero. For $n > 2$ this yields:

$$B_v = P_n[D - G(v)]^{2/n} \tag{32}$$

where the constant

$$P_n = \bar{P}_n/\mu(C_n)^{2/n} = \left(\frac{\hbar}{4\pi c\mu}\right)\left[\frac{\Gamma\left(1+\frac{1}{n}\right)\Gamma\left(\frac{1}{2}-\frac{1}{n}\right)}{\Gamma\left(\frac{1}{2}+\frac{1}{n}\right)\Gamma\left(1-\frac{1}{n}\right)}\right]\Big/(C_n)^{2/n}$$

Combining equation (32) with the vibrational energy expression equation (9) then yields the more useful formula

$$B_v = Q_n(v_D - v)^{4/(n-2)} \tag{33}$$

where $Q_n = \bar{Q}_n/[\mu^n(C_n)^2]^{1/(n-2)} = P_n(H_n)^{4/(n-2)}$, and v_D and H_n are as defined earlier (p. 124). The constants $\bar{P}_n$ and $\bar{Q}_n$ appearing above are collections of numerical factors, and values of them are given in Table 1.

Expressions (32) and (33) are not valid for $n \leqslant 2$ because in this case the integral in the numerator of equation (31) diverges when $R_1(v)$ is replaced by zero. Although the more primitive result equation (30) would appear to be unaffected by this difficulty, it too becomes invalid for $n \leqslant 2$. The physical reason for the breakdown in both cases is that, for $n \leqslant 2$, the expectation value $\langle v,0|R^{-2}|v,0\rangle$ is no longer dominated by contributions from the neighbourhood of the outer turning point. Note that the gamma-function factor in equation (32) which was absent in equation (30) is fairly large, its values ranging from 3.25 to 1.63 as n increases from 3 to 6.

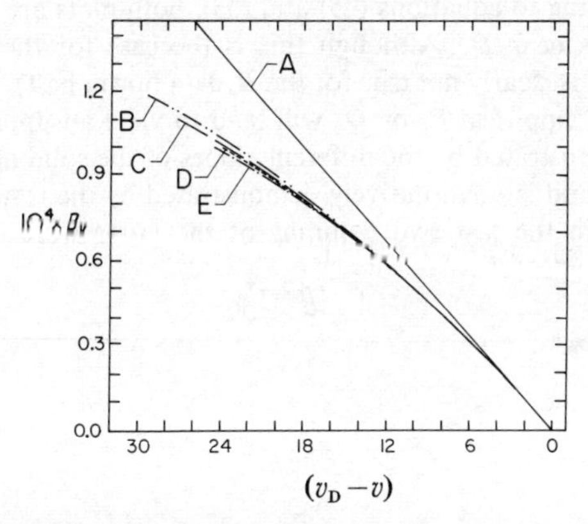

Figure 13 *B_v plots for the diverse model potentials considered in Figure* 4*; the reference curve,* A, *was calculated from equation* (33)

Expressions (32) and (33) imply that for levels lying near dissociation where the approximation of replacing $V(R)$ by equation (4) $(D-C_n/R^n)$ is fairly good, B_v values should depend only on the long-range part of the potential. In Figure 13 this is shown to be the case for LJ(12,6) and exp(α,6) potentials with different well depths and equilibrium distances, but the same long-range tail. These model potentials are the same ones examined in Figure 4. The 'true' B_v values in Figure 13 were obtained by interpolating over Harrison and Bernstein's [17] numerical WKB integrals and substituting the results into equation (31), while the reference curve (A) was obtained from equation (33).

Figure 12 shows that the integrand in the immediate neighbourhood of $R_2(v)$ plays a somewhat less dominant role in the $k = 0$ integral in the numerator of equation (31) than it does in the $k = 2$ integral in the denominator of equation (31) and in equation (7). As a result, errors introduced by the approximations fundamental to the derivation of equations (8), (9), (32), and (33) should be relatively more serious for the latter. A comparison of Figures 4 and 13 demonstrates this fact; the divergence from the theoretical curve A with increasing $(v_D - v)$ is much more rapid in the B_v plot. Of course the source of the error in these examples is quite different from that in real physical problems. As discussed earlier (p. 130), for simple two-term model potentials the error is associated with approximations to the integrands of equation (31) in the neighbourhood of $R_1(v)$. On the other hand, for real systems most of the error is due to the influence of the higher-power terms in the long-range potential (just as in the vibrational problem).

In applications to real molecules the consequence of the relatively lower accuracy of the B_v expressions is that a value of $P_{\bar{n}}$ or $Q_{\bar{n}}$ calculated from a known $C_{\bar{n}}$ will tend to be too large, or in other words, a $C_{\bar{n}}$ obtained by fitting experimental B_v values to equations (32) or (33) will be somewhat high. These effects are illustrated by the data for Cl_2 ($B\,^3\Pi^+_{0u}$) [58] shown in Figure 14. According to equations (25) and (33), both plots are expected to be linear for levels near D. Although this is the case for the vibrational plot (upper half), it is clearly not true for the B_v data (lower half).

The fact that an empirical $P_{\bar{n}}$ or $Q_{\bar{n}}$ will tend to yield an upper bound to the true $C_{\bar{n}}$ is demonstrated by the different slopes of the solid and dot-dash lines in Figure 14 and is quantitatively demonstrated by the results in Table 5. The numbers in the last two columns of the latter were obtained by

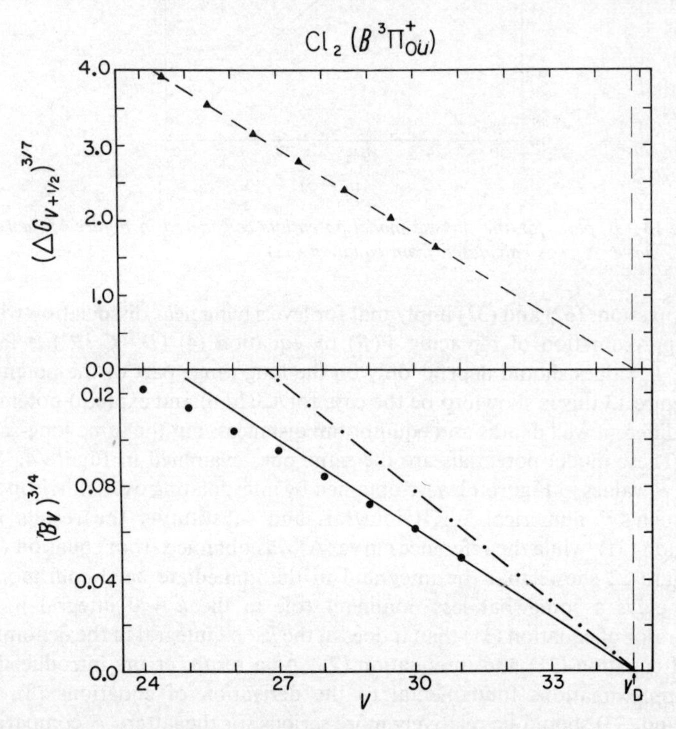

Figure 14 Upper: *experimental vibrational spacings of* Cl_2 ($B\,^3\Pi^+_{0u}$),[58] *for which* $\bar{n} = 5$, *plotted according to equation* (25). Lower: *experimental* B_v *values* [58] *plotted according to equation* (33). *The intercept* v_D *comes from Table* 2, *and the dot-dash line has a slope of* $(Q_5)^{3/4}$ *calculated from the experimental* C_5 *in Table* 2. *Energies are in* cm^{-1}

Table 5 *Test of $C_{\tilde{n}}$ values obtained from the B_v expressions [equations (32) and (33)]*

			$10^{-5} \times C_{\tilde{n}}/\mathrm{cm}^{-1}\,\mathrm{\AA}^{\tilde{n}}$			
Species	$\tilde{n}$	v' [a]	*Theoretical*	*From equations* (8) *and* (9)	*From equation* (33)	*From equation* (32)
$H_2\ (B\,^1\Sigma_u^+)$	3	37	0.36096	[b]	0.431	0.611
$Cl_2\ (B\,^3\Pi_{0u}^+)$	5	31	1.23	$1.22_4(\pm 0.04)$	1.91	2.56
$Br_2\ (B\,^3\Pi_{0u}^+)$	5	52	2.00	$1.79(\pm 0.2)$	3.03	4.31
$I_2\ (B\,^3\Pi_{0u}^+)$	5	62	3.68	$3.10_6(\pm 0.16)$	7.56	13.6

[a] Highest vibrational level for which there exists an experimental B_v; this need not be the same as v_H in Table 2; [b] an independent fitted C_3 was not obtained.

substituting the experimental B_v for level v' into equations (32) or (33), respectively, together with $G(v')$ or v' and the appropriate intercept D or v_D taken from Table 2. The differences among the various estimates of $C_{\tilde{n}}$ in Table 5 may be explained in terms of their dependence on the factor $\bar{P}_{\tilde{n}}$, which effectively contains all the error in the B_v expressions. It is readily seen that $C_{\tilde{n}}$ (32) [*i.e.* the estimate of $C_{\tilde{n}}$ obtained using equation (32)] varies as $(\bar{P}_{\tilde{n}})^{\tilde{n}/2}$, while $C_{\tilde{n}}(33)$ depends on $(\bar{P}_{\tilde{n}})^{(\tilde{n}-2)/2}$ and $C_{\tilde{n}}(8\text{—}9)$ has no $\bar{P}_{\tilde{n}}$ dependence at all. This leads us to expect the observed ordering:

$$C_{\tilde{n}}(32) \geqslant C_{\tilde{n}}(33) \geqslant C_{\tilde{n}}(8\text{—}9) \approx C_{\tilde{n}}(\text{'true'})$$

Utilization of the Bv Expressions.—*Spectroscopic Applications: Prediction of Unobserved B_v Values.* The preceding discussion has shown that parameter values n, C_n, D, and v_D obtained by fitting experimental data to equations (32) and (33) will be less accurate than those obtainable from a vibrational analysis [based on equations (8) and (9)] of the same levels. Fortunately, whenever experimental B_v values are available for levels near D the corresponding vibrational energies $G(v,J = 0)$ will also be known, and hence the desired parameters may be extracted from the latter. As a result, the main spectroscopic role of equations (32) and (33) presently appears to lie in predicting B_v values for levels near D for which they have not been observed.

The expected limiting linearity of plots of $(B_v)^{(\tilde{n}-2)/4}$ *versus* v has not been observed for any of the species considered in Section 3 (see Table 2). However, the foregoing discussion indicates that the achievement of this behaviour is probably an excessively exclusive criterion for the applicability of equation (33). The less restrictive requirement adopted here is that the vibrational energies of the highest observed level(s) must be accurately described by equation (9) with $n = \tilde{n}$. In other words, reliable values of D, $\tilde{n}$, $C_{\tilde{n}}$, and v_D must be known for the state in question; this criterion is satisfied for all the species mentioned in Table 2.

Upper bounds to B_v for levels near D may be obtained from equation (33) using the known $\tilde{n}$ and v_D and a $Q_{\tilde{n}}$ (theory) calculated from the known $C_{\tilde{n}}$. For B-state Cl_2, these upper bounds correspond to points at integer v on the dot-dash line in the lower half of Figure 14. In addition, if an experimental $B_{v'}$ value is known for the highly excited level v', it may be substituted into equation (33) to yield an empirical $Q_v(v')$ which will in turn yield lower bounds to B_v values for levels $v > v'$ and upper bounds to those for $v < v'$. This case corresponds to points at integeral v on the solid line in the lower half of Figure 14. For the species listed in Table 2, these upper- and lower-bound values of $Q_{\bar{n}}$ are given in Table 6. Although the two limits differ by between 25 and 45% of the larger value, the true B_v values will clearly be closer to the lower

Table 6 *Upper and lower bound $Q_{\bar{n}}$ values (in cm^{-1}) for predicting unobserved B_v values*

Species	$\bar{n}$	v_D[a]	v'[b]	$Q_{\bar{n}}(v')$[c]	$Q_{\bar{n}}(theory)$[d]
$H_2\ (B\,^1\Sigma_u^+)$	3	43.26	37	2.54×10^{-4}	3.62×10^{-4}
$Ar_2\ (X\,^1\Sigma_g^+)$	6	8.27	*e*	*e*	1.07×10^{-2}
$^{35}Cl^{35}Cl\ (B\,^3\Pi_{0u}^+)$	5	34.82	31	3.01×10^{-3}	4.06×10^{-3}
$^{79}Br^{79}Br\ (B\,^3\Pi_{0u}^+)$	5	59.51	52	5.71×10^{-4}	8.12×10^{-4}
$^{81}Br^{81}Br\ (B\,^3\Pi_{0u}^+)$	5	60.28	52	5.43×10^{-4}	7.78×10^{-4}
$^{127}I^{127}I\ (B\,^3\Pi_{0u}^+)$	5	87.74_5	62	1.41×10^{-4}	2.55×10^{-4}

[a] From vibrational analyses—see Table 2; [b] highest level for which an experimental B_v value is available; [c] obtained on substituting v_D, v' and the experimental B_v' into equation (33); [d] calculated from the $C_{\bar{n}}$ values given in Table 2; [e] no experimental values of B_v are available for this case.

bound for v near v', but will converge on the upper bound as $v \rightarrow v_D$. Hence a reasonable interpolation between the two limits should yield predicted B_v values with errors considerably smaller than the uncertainties suggested by the difference between the bounds.

Distribution of Vibrational–Rotational Levels near D. One of the more interesting features of the present results [equations (32) and (33)] is that they allow a generalization of the vibrational energy expressions [equations (8) and (9)] to the case of rotating molecules with $J \neq 0$. If J is not too large, the rotational energy of a diatomic molecule in vibrational level v is fairly accurately given by $J(J+1)B_v$ (although this assumption becomes less valid as the levels approach dissociation). In this approximation, equations (9), (32), and (33) allow the total energy of a level near D to be written as:

$$\begin{aligned} G(v,J) &= G(v,0) + J(J+1)P_n[D - G(v,0)]^{2/n} \\ &= D - [(v_D - v)H_n]^{2n/(n-2)} + J(J+1)Q_n(v_D - v)^{4/(n-2)} \end{aligned} \qquad (34)$$

and the inverse of the density of levels at any chosen J is simply:

$$\begin{aligned} \frac{dG(v,J)}{dv} &= \left\{1 - \frac{2J(J+1)\bar{P}_n}{n[(v_D - v)\bar{H}_n]^2}\right\}\left(\frac{2n}{n-2}\right)(H_n)^{2n/(n-2)}(v_D - v)^{(n+2)/(n-2)} \\ &= \frac{dG(v,0)}{dv}\left\{1 - \frac{2J(J+1)\bar{P}_n}{n[(v_D - v)\bar{H}_n]^2}\right\} \end{aligned} \qquad (35)$$

For $n = 4$ and $J = 0$, it is readily shown that both equation (35) and its first derivative with respect to $[J(J+1)]$ are identical to the analogous quantities obtained from the Tani–Inokuti [22] expression [equation (13)] (with $n = \infty$). This equivalence is expected since both were derived from the initial expression (6) using essentially the same approximations. In fact the only real difference between the exact semi-classical result [equation (13)] and the more approximate expression [equation (35)] is that the latter only takes account of a linear term in $J(J+1)$, whereas the former implicitly includes the exact J-dependence. On the other hand, equation (35) has advantages over equation (13) in that it is valid for all n (rather than just $n = 4$), and because its analytic simplicity makes it much easier to use. In addition, there is no integrated form of equation (13) analogous to the simple eigenvalue expression [equation (34)].

For pure R^{-n} potentials, the main deficiency of the present results [equations (34) and (35)] is associated with the fact that they only take account of a linear term in $J(J+1)$ (see also footnote on p. 130). The effect of this approximation will now be examined by comparing expressions for the J-dependence of the total number of bound and quasi-bound levels obtained from equations (34) and (35), with corresponding exact quantal or semi-classical results.[18–21]

Substituting $G(v,J) = D$ into equation (34) and using the definitions of Q_n and H_n one obtains

$$(v_D - v)^2 = J(J+1)\bar{P}_n/(\bar{H}_n)^2$$

where v is the vibrational index (in general non-integral) of the level which lies at D when the molecule has a rotational energy specified by J. The total number of bound levels is therefore

$$N(J) = N(0) - [J(J+1)]^{\frac{1}{2}}(\bar{P}_n)^{\frac{1}{2}}/\bar{H}_n \tag{36}$$

where $N(0) = v_D + \frac{1}{2}$ is the number of bound levels for $J = 0$. Values of $(\bar{P}_n)^{\frac{1}{2}}/\bar{H}_n$ are presented in Table 7, where they are seen to be in reasonable agreement with the analogous coefficients $(n-2)^{-1}$ which occur in both the exact quantal $N(J)$ expressions for LJ$(2n-2,n)$ potentials and in the exact semi-classical $N(J)$ for Sutherland potentials [equation (16)],[21] The discrepancies range from 22% for $n = 3$ down to 3% for $n = 8$. Although additional terms in the Sutherland potential result [21] reflect the present neglect of all but linear $[J(J+1)]$ contributions to equations (34) and (35), it is interesting to note that the linear term $-[J(J+1)]^{\frac{1}{2}}/(n-2)$ represents the *total* J-dependence of the exact quantal LJ$(2n-2,n)$ version of equation (36).[18, 19, 21]

As with the Tani–Inokuti result [equation (13)],[22] the present expressions [equations (34) and (35)] may also be used to describe the quasi-bound levels lying above the dissociation limit. The centrifugal barrier maximum is associated with the zero of $dG(v,J)/dv$. Hence equation (35) suggests that

Table 7 *Coefficients of* $[J(J+1)]^{\frac{1}{2}}$ *in expressions for the J-dependence of the number of levels supported by* R^{-n}*-tailed potentials* [a]

n	Number of bound levels: $[N(J)-N(0)]$		Number of quasi-bound levels: $[N^{\ddagger}(J)-N(J)]$	
	Exact[b,c]	From equation (36)	Exact[c]	From equation (37)
3	1.0000	1.2861	—	0.2360
4	0.5000	0.5642	0.0498	0.1652
5	$0.333\dot{3}$	0.3584	—	0.1317
6	0.2500	0.2625	0.0482	0.1110
8	$0.166\dot{6}$	0.1712	0.044	0.0856

[a] See footnote on p. 128; [b] equal to $(n-2)^{-1}$; [c] see ref. 21.

the total number of levels lying below this maximum, as a function of J, is

$$N^{\ddagger}(J) = N(0) - (2/n)^{\frac{1}{2}}[J(J+1)]^{\frac{1}{2}}(\bar{P}_n)^{\frac{1}{2}}/\bar{H}_n$$

Combined with equation (36), this result yields an expression for the number of quasi-bound levels:

$$N^{\ddagger}(J) - N(J) = [J(J+1)]^{\frac{1}{2}}[1-(2/n)^{\frac{1}{2}}](\bar{P}_n)^{\frac{1}{2}}/\bar{H}_n \qquad (37)$$

Values of the coefficient of $[J(J+1)]^{\frac{1}{2}}$ in equation (37) are given in Table 7, where they are seen to be in relatively poor agreement with the corresponding coefficients from the analogous exact semi-classical expressions for Sutherland potentials.[21] This increased error of equation (37) relative to that of equation (36) is clearly attributable to the fact that the former is concerned with following the vibrational levels to relatively higher J values.

It should be remembered that the exact semi-classical and quantal expressions used to test equations (36) and (37) can only estimate the *total number* of bound or quasi-bound levels. In contrast, the results presented here also describe the distribution of these levels. Furthermore, these tests only examined the effect of cutting-off the $G(v,J)$ expansion [equation (34)] at the linear term in $J(J+1)$, and this is not the dominant source of error in equations (34)—(37) for real molecular systems (see footnote on p. 130).

Discussion and Prognosis.—The main weakness of the results presented in Section 4 is that they are much more sensitive to the effect of additional terms in the long-range potential [equation (3)] than was the case for the vibrational problem discussed in Section 3. For real systems, this sensitivity has much more serious consequences for equations (34)—(37) than does the approximation of including only a linear $[J(J+1)]$ term in equation (34). This weakness must be shared equally by the Tani–Inokuti equation (13) since it rests on precisely the same foundation as the present results. On the other hand, since B_v values are relatively more sensitive to higher-power terms in equation (3) with $m > \tilde{n}$, they may prove to be a useful probe of these interactions.

An additional deficiency of equations (34) and (35) and the results based on them (see Table 7) is their neglect of quadratic and higher-power terms in $J(J+1)$. However, expressions analogous to equations (32) and (33) should be able to be derived for the rotational constant D_v in the near future. In addition to its spectroscopic implications, the inclusion of the ensuing $[J(J+1)]^2$ terms in equations (34) and (35) should lead to significantly improved predictions for $[N(J)-N(0)]$ and $[N^{+}(J)-N(J)]$.

5 Rotational Predissociation and Long-range Potentials

Background: The Limited Curve of Dissociation (LCD).—The total effective internuclear potential of a rotating ($J \neq 0$) diatomic molecule is the sum of the electronic interaction potential plus a repulsive centrifugal term:

$$V_J(R) = V(R) + J(J+1)\hbar^2/2\mu R^2 \tag{38}$$

If the attractive long-range part of $V(R)$ dies off more rapidly than R^{-2}, then $V_J(R)$ will have a potential maximum at large R (for $J>0$). Classically, all levels lying behind the maximum of such a barrier are truly bound, even if they are above the molecular dissociation limit. However, as J increases this centrifugal barrier grows and moves to smaller R; concurrently, the vibrational levels are shifted to higher energies and eventually they move above the potential maximum and become part of the continuum. As this occurs, rotational progressions will suddenly break off at a critical J characteristic of the particular vibrational level, with the last observed line(s) perhaps being measurably broadened. This is the phenomenon known as rotational predissociation.[46, 47]

The more highly excited vibrational levels will clearly be predissociated at relatively smaller J values. Thus, the energy associated with the onset of predissociation will yield an ever decreasing upper bound to D as v increases. Büttenbender and Herzberg [84] noted this trend and proposed the use of plots of the energy of the first predissociating level for each v, $G^{+}(v,J)$ *versus* $J(J+1)$. The $[J(J+1)] = 0$ intercept of a curve of this sort, now known as a 'limiting curve of dissociation' or LCD, is a least upper bound to the dissociation limit D. If $V(R)$ has no rotationless potential maximum this intercept should be D itself; otherwise it is the height of the electronic potential energy barrier.

Schmid and Gerö [85] expanded upon the LCD approach by noting that the tangent to the curve at any point is given by:

$$\mathrm{d}G^{+}(v,J)/\mathrm{d}[J(J+1)] = \hbar^2/2\mu[R^{+}(J)]^2 \tag{39}$$

[84] G. Büttenbender and G. Herzberg, *Ann. Physik*, 1934/35, **21**, 577.
[85] R. Schmid and L. Gerö, *Z. Physik.*, 1937, **104**, 724.

where $R^{\ddagger}(J)$ is the position of the centrifugal barrier maximum. The main application of their result has been as a test of whether the electronic potential $V(R)$ has a potential maximum at some distance R_x. In this case, $R^{\ddagger}(J)$ will be constant and equal to R_x for small J, and the LCD will have a constant finite slope near the $[J(J+1)] = 0$ intercept. On the other hand, if $V(R)$ is monotonically attractive at long-range (with $\tilde{n} > 2$), then $R^{\ddagger}(J)$ smoothly approaches infinity as $J(J+1) \rightarrow 0$ and the LCD will have zero slope at the intercept and positive curvature everywhere.[46, 47] It is important to be able to distinguish between these two cases since the methods described in this section apply only to the latter.

The data requirements for an LCD analysis are somewhat more sophisticated than those for the methods discussed in Sections 3 and 4. First of all, the rotational structure of each vibrational band must be resolved and the correct absolute rotational assignments made. In addition, the observed rotational series for each band must be followed to the critical value of J at which the vibrational level in question predissociates. Unfortunately it is sometimes very difficult to determine whether or not this latter criterion is in fact satisfied, since the sudden breaking-off of a rotational progression can occur for reasons other than rotational predissociation. For instance, the radial wavefunctions of highly excited levels shift to larger R at an accelerating rate as J increases. This could lead to a sharp decrease in Franck–Condon overlap [6] and a concomitant drop in the intensity of the rotational progression so that the latter appeared to break off when the vibrational level was still some distance below the potential maximum. Furthermore, difficulties of this sort are most serious for the highly excited vibrational levels which have the most influence on the determination of D. For example, for the highest bound level ($v = 14$) of ground-state H_2, $\left[\mathrm{d}\langle v,J|R|v,J\rangle/\mathrm{d}J\right]_{J=0}$ is more than 30 times larger than it is for the $v = 0$ level.[86]

In general, errors of the type described above would cause the LCD estimate of D to be too low. An example of a situation in which this appears to have occurred is the $B(^3\Pi^+_{0u})$ state of Br_2. Here the vibrational analysis [51] which superseded the original LCD treatment of this molecule [72] showed that some of the unobserved levels which had been assumed to be rotationally predissociated actually lay *below* the dissociation limit, and hence were not even upper bounds to D.

Doubt about proximity to the barrier maximum is dispelled if a broadening of levels is observed immediately preceding the break-off of a rotational progression. This type of behaviour is due to quantum-mechanical tunneling, and hence can occur only if the level in question lies above the dissociation limit and fairly near the barrier maximum. It is most often observed for species with small reduced mass, since they can tunnel most efficiently.

[86] R. J. Le Roy, *J. Chem. Phys.*, 1971, **54**, 5433; see also University of Wisconsin, Theoretical Chemistry Institute, Report WIS-TCI-387, 1971.

Archetypical examples are found in spectra of AlH and HgH;[87, 88] in both cases rotational progressions are followed from sharp lines, to a reasonably broad line, and on to a very broad (barely discernible) one.

An additional difficulty in any analysis of rotational predissociation data is the problem of determining the precise height of a potential barrier from the discrete, and sometimes widely spaced, levels in a rotational series immediately preceding the onset of predissociation. Fortunately, this problem need not affect the utility of the graphical approach described above, since any broadened or missing levels must lie above the dissociation limit and will necessarily converge to it as the centrifugal barrier shrinks with decreasing J.

Simple graphical LCD plots have come to be widely used for the determination of molecular dissociation limits.[46, 47] However, the quantitative relationship between the LCD and the long-range potential suggested by expression (39) was never really exploited until Bernstein [89] introduced the method described below.

Bernstein's 'Locus of Barrier Maxima' (LBM): Derivation and Analysis.—Bernstein [89] was the first to make use of the dependence of the heights of centrifugal potential barriers on the attractive long-range electronic potential. He noted that at a maximum $R^{\ddagger}(J)$ of the total effective potential expression (38):

$$dV_J[R^{\ddagger}(J)]/dR = 0 = \{dV[R^{\ddagger}(J)]/dR\} - \{J(J+1)\hbar^2/\mu[R^{\ddagger}(J)]^3\} \quad (40)$$

Substituting into equation (40) the long-range potential form [equation (4)], $V(R) = D - C_n/R^n$, then yields an expression for the position of the barrier maximum:

$$R^{\ddagger}(J) = [n\mu C_n/J(J+1)\hbar^2]^{1/(n-2)} \quad (41)$$

Since this maximum corresponds to $V_J[R^{\ddagger}(J)]$, one obtains:[89]

$$E_M(J) = D + S_n[J(J+1)]^{n/(n-2)} \quad (42)$$

for the 'locus of barrier maxima' or LBM, where the constant

$$S_n = \bar{S}_n/[\mu^n(C_n)^2]^{1/(n-2)} = \tfrac{1}{2}(n-2)[\hbar^{2n}/(n\mu)^n(C_n)^2]^{1/(n-2)}$$

Values of the numerical factor $\bar{S}_n$ for various n are listed in Table 1.

There is an important distinction between equation (42) and the expressions (8)—(10) and (32)—(33), which describe the vibrational energies and B_v constants. The latter quantities depend on the potential over the whole of the region between the classical turning points, although for levels near D they are largely dominated by the effect of the attractive long-range part of $V(R)$. In contrast, equation (42) is a purely classical result and it depends only on the nature of $V(R)$ right at the barrier maxima $R^{\ddagger}(J)$. As a result, it will be less seriously affected by errors associated with replacing the actual

[87] L. Farkas and S. Levy, *Z. Physik*, 1933, **84,** 195.
[88] T. L. Porter, *J. Opt. Soc. Amer.*, 1962, **52,** 1201.
[89] R. B. Bernstein, *Phys. Rev. Letters*, 1966, **16,** 385.

potential by the long-range form [equation (4)]. In addition, the fact that the barrier corresponding to the predissociation of a given vibrational level v lies at a distance $R^{\ddagger}(J) = R_2(v,J)$ somewhat larger than the $R_2(v,0)$ turning point makes the use of equation (4) even more appropriate here.

In an analysis based on equation (42) the experimental observables are the energies of the rotational levels preceding the onset of predissociation. Since the levels are sometimes widely spaced, the determination of the energy and J value for which a given vibrational level lies at the centrifugal maximum can be a serious problem. Possible choices are the J and energy of the last sharp observed level, those of any broadened levels which may have been observed, the extrapolated energy and J of the first missing level in the series, or any intermediate values. Thus an experimental LCD may have many definitions and need not coincide with the actual LBM described by equation (42).

This problem has been studied [90] for ground-state H_2, HD, and D_2, for which the exact energies and the widths of *all* quasi-bound vibrational–rotational levels have been calculated [86] from the accurate *ab initio* potential.[63–65] For this case, many of the quasi-bound levels lying behind centrifugal barriers had sufficiently large widths (of from 5 to 80 cm^{-1}) [86, 90] that they would be extremely difficult to observe experimentally. Figure 15 compares the LBM with 'predicted experimental LCDs' defined as the loci of measurably broadened levels with widths

$$0.05 \leqslant \Gamma \leqslant 0.25\ cm^{-1} \tag{43}$$

A mass-reduced abscissa scale is used so that the LBM curves for the three isotopes are identical. Note that the differences ΔE between the LBM and the LCD curves are from 10% to 40% of the LBM energy, with the greatest relative error occurring at small J. This implies that any simple algorithm such as equation (43) will tend to be proportionately most seriously in error at the small J values where Bernstein's expression [equation (42)] should be appropriate. The trend of the results for the different isotopes is in accord with the decreased efficiency of tunneling with increasing reduced mass μ. However, effects of this sort will clearly only become negligible for large $\mu \gg 1$ amu. In conclusion, therefore, the results described above suggest that the best empirical definition of an LCD corresponds to the extrapolated energy and J value of the first unobserved level in the given rotational series. However, for hydrides and deuterides this apparently missing level might still lie well below the centrifugal barrier maximum.

Applications of the LBM Expression: Results and Non-results.—If rotational predissociation is observed for three or more highly excited vibrational levels of a given state, the corresponding energies $E_M(J)$ may in principle be fitted to equation (42) by least squares to yield values of the parameters

[90] R. J. Le Roy and R. B. Bernstein, *J. Chem. Phys.*, 1971, **54**, 5114.

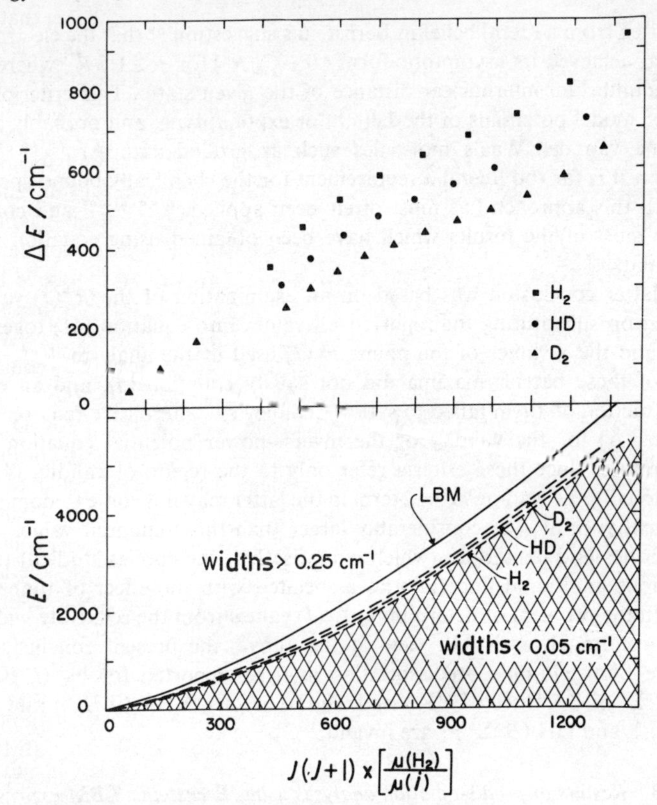

Figure 15 Lower: *comparison of the LBM (solid curve) with 'experimental' LCD values (dashed curves) for ground-state* $(X\,^1\Sigma_g^+)$ H_2, HD, and D_2. Upper: *the error term* $\Delta E \equiv \{E(\text{LBM},J) - E(\text{LCD},J)\}$
(Reproduced by permission from *J. Chem. Phys.*, 1971, **54**, 5114)

D, n, and C_n (from S_n). In practice, however, the small number of data and the uncertainties discussed above seriously hamper any such direct determination of n. These difficulties are compounded by the fact that the power $n/(n-2)$ in equation (42) is only half as sensitive to changes in n as are the powers $2n/(n-2)$ and $4/(n-2)$ appearing in equations (9) and (33)). As a result, it is almost always necessary to assume a value for n, and in practice it is always set equal to its theoretically-known asymptotic value $\tilde{n}$. Plots of $E_M(J)$ *versus* $[J(J+1)]^{\tilde{n}/(\tilde{n}-2)}$ may then be made, and D obtained from the $J = 0$ intercept and $C_{\tilde{n}}$ from the limiting slope. Unfortunately this puts a fairly severe constraint on the present approach, since it may only be used in the asymptotic region where the interaction potential is mostly due to the leading $m = \tilde{n}$ term in equation (3).

A major source of error in past applications of equation (42) appears to have arisen from a literal belief in Bernstein's suggestion [89] that the electronic potential achieved its asymptotic form $(D-C_{\tilde{n}}/R^{\tilde{n}})$ for $R \gtrsim 1.5\ R_e$, where R_e is the equilibrium internuclear distance of the given state. This criterion is valid for model potentials of the LJ(a,n) or exp(a,n) type, and probably also for some van der Waals molecules such as ground-state Ar_2.[60, 61, 67–70] However, it is far too liberal a requirement for the chemically bound species to which this approach has most often been applied,[31, 89, 91–94] and consequently most of the results which have been obtained using equation (42) are invalid.

The latter conclusion was based on an examination of the $R^{\ddagger}(J)$ values obtained on substituting the reported $C_{\tilde{n}}$ values into equation (41) together with $\tilde{n}$ and the J values of the points $E_M(J)$ used in the analyses.[31, 89, 91–94] Many of these barrier maxima did not satisfy equation (5), and an even larger fraction of them failed to satisfy Stwalley's [43] alternative requirement $[R^{\ddagger}(J) > 5\ \text{Å}]$ for the validity of the inverse-power potential equation (3). Furthermore, since these criteria refer only to the region of validity of the expansion (3), the leading $m = \tilde{n}$ term in the latter may not come to dominate the potential until R is considerably larger than this minimum value. For the hydrides and deuterides (which comprise half the species studied) there are serious additional uncertainties associated with the effect of tunneling and with the problem of determining $E_M(J)$ values from the relatively widely-spaced rotational levels. In view of the above, the present conclusion is that the Bernstein-type predissociation analyses reported for N_2 ($C\,^3\Pi$),[89] CO ($B\,^3\Sigma^+$),[31, 89] HgH ($X\,^2\Sigma^+$) and HgD ($X\,^2\Sigma^+$),[89] HF ($X\,^1\Sigma^+$) and DF ($X\,^1\Sigma^+$),[91] and OH ($B\,^3\Sigma^+$)[92] are invalid.

Table 8 *Results of predissociation analyses using Bernstein's LBM expression* [*equation* (42)]

Species	D/cm^{-1}	$\tilde{n}$	$C_{\tilde{n}}$/cm^{-1} Å *Experimental*	$C_{\tilde{n}}$/cm^{-1} Å *Theoretical*	*Ref.*
LiH ($B\,^1\Pi$)	34492.5($\pm$0.5)^a	6	b	0.402×10^6	94
S_2 ($1_u \ldots {}^3P_2 + {}^3P_1$)^c	35999.0($\pm$2.5)^d	5	1.5×10^5	0.803×10^5	93
Se_2 ($1_u \ldots {}^3P_2 + {}^3P_1$)^c	29690 ($\pm$20)^d	5	1.60×10^5	1.171×10^5	91

^{a} See footnote on p. 140; ^{b} C_6 held fixed at the theoretical value; ^{c} These potential parameters characterize the predissociating 1_u state and not the predissociated $B^3\Sigma_u^-$ state; ^{d} expressed relative to the ground-state potential minimum.

The results for the other cases to which equation (42) has been applied are summarized in Table 8. The first, LiH ($B\,^1\Pi$), is an example of Herzberg's [47] case III predissociation involving only a single potential-energy

[91] M. A. Byrne, W. G. Richards, and J. A. Horsley, *Mol. Phys.*, 1967, **12**, 273.
[92] J. A. Horsley and W. G. Richards, *J. Chim. phys.*, 1969, **66**, 41.
[93] J. M. Ricks and R. F. Barrow, *Canad. J. Phys.*, 1969, **47**, 2423.
[94] W. C. Stwalley, and K. R. Way, Paper AA8, Proc. 27th Symposium on Molecular Structure and Spectroscopy, Columbus, 1972.

curve. In this case, only one vibrational level ($v = 2$) predissociates at small enough J that the corresponding centrifugal barrier maximum lies in the long-range region. Therefore Stwalley and Way [94] calculated the asymptotic potential coefficient for this case ($\tilde{n} = 6$) and used it to yield a theoretical value of the constant S_6 in equation (42). Combining the latter with their one datum then yielded the improved dissociation limit of (34492.5 ±0.5) cm^{-1},[94] a value significantly different from the experimenter's [95] estimate of (34495 ±2) cm^{-1}. An additional noteworthy feature of this analysis was the fact that they used trial calculations of quasi-bound level positions and widths for different assumed potentials to estimate the effects of higher-power potential terms (R^{-8}, R^{-10}, ... *etc.*) and of quantum-mechanical tunnelling.[94] The discussion accompanying Figure 15 suggests that for hydrides or deuterides, such model calculations should probably accompany any predissociation analysis.

The other problems to which equation (42) has been applied are examples of Mulliken's [96] case Ib$^+$ predissociation (a subcase of Herzberg's [47] case Ib). Here the observed predissociation arises from the intersection of the attractive outer branches of two potential curves of different but related symmetry which have different dissociation limits. The pure vibrational ($J = 0$) level spectrum of both states will be oblivious to this intersection. However, rotational motion causes a mixing of states such that $J > 0$ levels of the potential with the higher asymptote may be predissociated if they lie above the centrifugal barriers associated with the long-range part of the potential curve with the lower dissociation limit. In this case, the J-dependence of the onset of predissociation for vibrational levels of the first state lying below the asymptote of the second may be used to determine the long-range potential of the latter.

In the absorption and emission spectra of S_2 ($B\,^3\Sigma_u^-$), Ricks and Barrow [93] observed predissociations of this sort, caused by the intersection of the $B\,0_u^+$ curve with a 1_u state which dissociates to yield $^3P_2 + {}^3P_1$ atoms. They analysed the data for the three S_2 isotopes $^{32}S^{32}S$, $^{32}S^{34}S$, and $^{34}S^{34}S$ in terms of equation (42) while assuming $n = \tilde{n} = 5$, and obtained the D and C_5 (experimental) given in Table 8. There are three different 1_u states correlating with these dissociation products, all of which correspond to $\tilde{n} = 5$. However, theory shows [31] that the R^{-5} term is only attractive for the one 1_u state whose calculated C_5 is given in Table 8, while the C_5 values for the other two possible 1_u curves are negative (repulsive).* Note that the theoretical C_5 quoted by Ricks and Barrow [93] erroneously corresponds to one of these latter states.

The disagreement between the theoretical and experimental C_5 for S_2 (see Table 8) is rather large, and the agreement shown in Table 2 implies

* Of course, this does not necessarily mean that the states in question are repulsive, since the dispersion potential terms (R^{-6}, R^{-8}, ... *etc.*) are probably attractive, while the exchange forces could have either sign.

[95] R. Velasco, *Canad. J. Phys.*, 1957, **35**, 1204.
[96] R. S. Mulliken, *J. Chem. Phys.*, 1960, **33**, 247.

that it should not be attributed to error in the former. This appears to leave three possible explanations. If the predissociation is correctly assigned to the 1_u state with an attractive C_5, either some of the data are in error, or the additional attractive R^{-6}, R^{-8}, ... *etc.* terms contributing to the long-range potential are significant over the range of the experimental $R^{\ddagger}(J)$ values. In the latter case, an analysis which assumes *pure* $n = \tilde{n} = 5$ behaviour would be inaccurate. On the other hand, if the predissociating 1_u state is one of those with a repulsive asymptotic R^{-5} tail, then *any* analysis based on equation (42) will be spurious. In any case, there is clearly some uncertainty as to the actual significance of the S_2 results quoted here.

An analysis using equation (42) has also been reported for similar case Ib^+ predissociations of Se_2 ($B\,^3\Sigma^-$).[91] In this case there are two sets of observations corresponding to intersections of the $B\,0_u^+$ curve by 1_u states whose dissociation limits are separated by the atomic $^3P_1 \leftarrow {}^3P_2$ excitation energy.[97] Barrow *et al.*[97] pointed out that there are three possible electronic assignments for this pair of dissociation limits, yielding three different estimates of the ground-state dissociation energy, $D_e(X)$. Using qualitative spectroscopic arguments they chose the middle value of $D_e(X)$, and recent spectroscopic evidence (R. F. Barrow, personal communication, 1972) appears to confirm this assignment. On the other hand, photoionization measurements strongly favour the largest of these three $D_e(X)$ values,[98] which would imply that the observed predissociating 1_u states correlate with $^3P_1 + {}^3P_2$ (upper D value) and $^3P_2 + {}^3P_2$ atoms (lower D value).

Byrne *et al.*[91] analysed the Se_2 predissociation corresponding to the upper dissociation limit ($^3P_2 + {}^3P_1$) using $\tilde{n} = 5$. As in the analogous S_2 situation, there are three different 1_u states correlating with these dissociation products, all corresponding to $\tilde{n} = 5$. If the predissociations are correctly assigned as case Ib^+,[96] they must be due to the one curve with an attractive asymptotic R^{-5} tail. However, the theoretical C_5 for this case (given in Table 8) is significantly smaller than the experimental value of Bryne *et al.*[91] This fact casts further suspicion upon the electron assignment of the predissociation products favoured by the photoionization measurements. On the other hand, Stwalley pointed out that if either of the two lower values of $D_e(X)$ were correct, the 1_u state yielding the higher set of predissociations would correlate with $^3P_1 + {}^3P_1$ or $^3P_1 + {}^3P_0$ atoms, cases for which the $C_5 \equiv 0$.[31] This would have meant that $\tilde{n} = 6$ for the data in question and hence that the reported $\tilde{n} = 5$ analysis was wrong. Note also that Chang's[99] discussion of the theoretical C_5 for this case is erroneous, since he mistakenly assumed that the predissociating potentials had the same 0_u^+ symmetry as the predissociated state.

Discussion and Prognosis.—The dearth of valid applications of equation

[97] R. F. Barrow, G. G. Chandler, and C. B. Meyer. *Phil. Trans.*, 1966, **260**, 395.
[98] J. Berkowitz and W. A. Chupka, *J. Chem. Phys.*, 1969, **50**, 4245.
[99] T. Y. Chang, *Mol. Phys.*, 1967, **13**, 487.

(42) seems surprising in view of the fact that in principle this expression is more accurate than those discussed in Sections 3 and 4. This shortage of results appears to be largely due to the experimental difficulty of observing the predissociation of (highly excited) vibrational levels for which $R^{\ddagger}(J)$ are sufficiently large that equation (42) can be expected to be valid. Compounding these problems are uncertainties as to whether a breaking-off of a rotational series actually corresponds to a barrier maximum, and uncertainties in determining experimental $E_{\mathrm{M}}(J)$ values from the observed energy levels.

Stwalley suggested that problems of the latter sort will diminish if broadened levels are observed and their widths measured, since in this case trial calculations should be able to locate barrier maxima $E_{\mathrm{M}}(J)$ fairly accurately. However, such techniques have as yet only been qualitatively applied,[94] and in any case they would only really be useful for hydrogenic or other light molecules for which tunneling is fairly efficient.

As an alternative to equation (42), Stwalley[100] suggested using Schmid and Gerö's[85] expression [equation (39)] to determine a potential numerically without assuming any analytic form. This would be done by using the derivatives in equation (39) to yield values of $R^{\ddagger}(J)$ which can then be substituted into equation (40) to yield 'experimental' values of $\mathrm{d}V[R^{\ddagger}(J)]/\mathrm{d}R$. Interpolation and integration over the latter would then yield $V(R)$ itself. This type of approach has the advantage of being free from any constraints associated with the validity of the inverse-power form [equation (4)]. However, the determination of accurate derivatives $\mathrm{d}G^{\ddagger}(v,J)/\mathrm{d}[J(J+1)]$ requires both very accurate values for the energies $G^{\ddagger}(v,J) = E_{\mathrm{M}}(J)$ and a fairly high density of such points, both of which are rather difficult criteria to fulfil. These difficulties will be compounded by additional uncertainties arising from the integration of $\mathrm{d}V[R^{\ddagger}(J)]/\mathrm{d}R$. As a result, it seems unlikely that this approach will find very wide usage in the near future.

Very recently Goscinksi and Tapia[100a] derived a generalized version of Bernstein's LBM equation (42) which takes account of higher-power terms ($m > \tilde{n}$) in the long-range potential expansion (3). They represented the first few terms in (3) with the [1,0] Padé approximation:

$$V_{\mathrm{P}}(R) = D - C_{\tilde{n}}/[R^{\tilde{n}}(1 - C_{\tilde{m}}/C_{\tilde{n}}R^{\tilde{m}-\tilde{n}})] \qquad (43a)$$

$$= D - C_{\tilde{n}}/R^{\tilde{n}} - C_{\tilde{m}}/R^{\tilde{m}} - \ldots$$

where $\tilde{m}$ is the power of the first $m > \tilde{n}$ term in (3). Their approach then yielded

$$\tilde{E}_{\mathrm{M}}(J) = D - S_{\tilde{n}}[J(J+1)]^{\tilde{n}/(\tilde{n}-2)} \Big/ \left(1 + \frac{C_{\tilde{m}}(S_{\tilde{n}})^{(\tilde{m}-\tilde{n})/\tilde{n}}\,[J(J+1)]^{(\tilde{m}-\tilde{n})/(\tilde{n}-2)}}{[\frac{1}{2}(\tilde{n}-2)C_{\tilde{n}}]^{\tilde{m}/\tilde{n}}}\right) \qquad (43b)$$

as the generalization of equation (42) for cases in which

$$\alpha(J) = C_{\tilde{m}}/C_{\tilde{n}}[R^{\ddagger}(J)]^{\tilde{m}-\tilde{n}}$$

[100] W. C. Stwalley, Paper AA7, Proc. 27th Symposium on Molecular Structure and Spectroscopy, Columbus, 1972.

[100a] O. Goscinksi and O. Tapia, *Mol. Phys.*, to be published.

is small. The latter requirement does not limit the usefulness of equation (43b), since unless $\alpha(J)$ is significantly less than one, the inverse-power expansion (3) becomes inappropriate for describing the interaction potential (see Section 2: Validity of the Inverse-power Expansion). However, the utility of the improved LBM expression (43b) is still restricted by the difficulty of obtaining experimental data suitable for analysis. In this vein it should be noted that the barrier maxima $R^{\ddagger}(J)$ in the two sample problems considered by Goscinski and Tapia,[100a] $H_2(B'\,{}^1\Sigma_u^+)$ and $HgH(X\,{}^2\Sigma^+)$, do not satisfy criterion (5), and hence their conclusions about these species should be regarded with some suspicion.

In conclusion, therefore, it appears that the methods described in this section are at present of relatively limited applicability. On the other hand, since the difficulties discussed above are partly experimental, the ever-improving techniques of measurement should eventually uncover many more examples in which the LBM equations (42) and (43b) may be profitably applied.

6 Analysis of Long-range RKR Turning Points

Background: Calculating RKR Potentials near Dissociation.—The three methods considered heretofore were all based on expressions relating the long-range interatomic potential directly to the energies or B_v values of levels near dissociation. In contrast, the techniques described below involve the analysis of actual points on the 'true' potential in terms of the theoretically-known long-range form [equation (3)]. Although the latter approach is more straightforward, it is also more restrictive since the calculation of RKR turning points for a given level presumes a knowledge of the energies and B_v values for *all* lower vibrational levels. In practice, however, these extensive data requirements may be partially relaxed for highly excited levels.

The Rydberg–Klein–Rees (RKR) procedure [13–15, 101–109] is based on an exact inversion of the first-order WKB eigenvalue criterion [equation (6)] which involves integrals of the form *

* Kaiser [105] pointed out that the lower limit of integration in equation (44) should actually correspond to the minimum of the potential defined by Dunham's [11] expression (2) as $v \approx -(\frac{1}{2} + Y_{00}/Y_{10})$ rather than simply $v = -\frac{1}{2}$. However, the correction term (Y_{00}/Y_{10}) is usually small and often negligible. This is also true for the 'Langer-correction' $(\hbar^2/8\mu R^2)$ which Howard [106] shows should be subtracted from the calculated potential.

[101] J. T. Vanderslice, E. A. Mason, W. C. Maisch, and E. R. Lippincott, *J. Mol. Spectroscopy*, 1959, **3**, 17; erratum 1960, **5**, 83.
[102] M. L. Ginter and R. Battino, *J. Chem. Phys.*, 1965, **42**, 3222.
[103] W. G. Richards and R. F. Barrow, *Trans. Faraday Soc.*, 1964, **60**, 797.
[104] R. J. Le Roy, *J. Chem. Phys.*, 1970, **52**, 2683.
[105] E. W. Kaiser, *J. Chem. Phys.*, 1970, **53**, 1686.
[106] R. A. Howard, *J. Chem. Phys.*, 1971, **54**, 4252.
[107] J. A. Coxon, *J. Quant. Spectroscopy Radiative Transfer*, 1971, **11**, 443.
[108] W. C. Stwalley, *J. Chem. Phys.*, 1972, **56**, 2485.
[109] A. S. Dickinson, *J. Mol. Spectroscopy*, 1972, **44**, 183; H. E. Fleming and K. N. Rao, *ibid.*, p. 189; J. Tellinghuisen, *ibid.* p. 194.

$$I(k,v) = \int_{-\frac{1}{2}}^{v} (B_{v'})^k [G(v)-G(v')]^{-\frac{1}{2}} \, dv' \qquad (44)$$

where $k = 0$ and 1. The $k = 0$ integral, which utilizes no rotational data, determines the distance between a pair of turning points:

$$R_2(v)-R_1(v) = (\hbar^2/2\mu)^{\frac{1}{2}} I(0,v) \qquad (45)$$

while the B_v-dependent $k = 1$ integral yields

$$1/R_1(v)-1/R_2(v) = (2\mu/\hbar^2)^{\frac{1}{2}} I(1,v) \qquad (46)$$

and effectively determines their means.

The most important contributions to $I(k,v)$ are associated with the integrand at v' near the singularity at v, and this weighting is particularly significant for levels near dissociation for which $G(v')$ approaches $G(v)$ relatively slowly. Thus, it is essential that the interpolation over the discrete experimental $G(v')$ values be highly accurate as v' approaches v. Unfortunately, for the highest observed levels the absence of data for yet higher levels makes accurate interpolations most difficult for precisely those cases which are most sensitive to this type of error. However, such effects may be minimized through use of the natural 'near-dissociation' expressions [9] and [33] (rather than the customary polynomials in v) when interpolating in this region. In contrast to the difficulties described above, the domination of $I(k,v)$ values by the integrand at v' near v also means that errors in the vibrational numbering or in the experimental $G(v')$ or $B_{v'}$ at $v' \ll v$ will have little effect on the turning points of a highly excited level v.

It is often much easier to measure the energies of highly excited vibrational levels than it is to obtain reliable B_v values for them. Fortunately, the distinction between the roles of $G(v)$ and B_v in equations (45) and (46) means that this difficulty need not prevent the determination of reliable $R_2(v)$ values for such levels. The inner turning points for all but the lowest vibrational levels rise so steeply that virtually any reasonable extrapolation beyond the range of the $R_1(v)$ values calculated for the lower levels should yield a fairly accurate repulsive potential wall. The vibrational energies in the region where the B_v values are not known may then be used in equation (45) to yield accurate turning-point differences, which when added to the extrapolated inner branch of the potential should yield reliable outer turning points. This type of RKR-extrapolation, first used by Verma,[75] has proved to be fairly useful[68, 102—104, 107] since the limited availability of rotational data for highly excited levels would otherwise seriously restrict the availability of RKR turning points in the long-range region.

One great strength of the RKR procedure is that it is model-independent, *i.e.* its validity does not depend on the appropriateness of any particular functional form for the potential. However, a concomitant disadvantage is the fact that the method has no built-in way of extrapolating to yield information about the dissociation limit or the long-range potential. To this end, Ginter and Battino[102] suggested that one simply use analytic extrapolations to estimate $G(v)$ and B_v values beyond the range of the experimental data, and then calculate the turning points in this region from these projected data. Although this may yield plausible results over a narrow internal, it merely shifts the uncertainties onto the shoulders of the assumed $G(v)$ and B_v expressions, and hence solves nothing.

Another way in which RKR potentials have been used to estimate dissociation energies involved using the calculated turning points to test the appropriateness of different model potential functions. The analytic potential in closest agreement with the RKR turning points was then assumed to give the best estimate of D.[110—112] This scheme is clearly just a modification of the model-potential approach discussed in Section 1, with the appropriateness of a particular functional form being examined through the intermediary of the calculated turning points rather than by a direct comparison with experimental data. Since no new physical or theoretical information is introduced by such an extrapolation, this approach should not be taken too seriously.

Graphical Methods.—This section describes diverse ways of analysing outer RKR turning points in terms of the long-range potential forms [equations (3) and (4)]. The fundamental assumption of any such treatment is that the use of these expressions is appropriate over the range of the given turning points, and hence a criterion such as equation (5) must always be kept in mind. Of course, turning points satisfying equation (5) should not be expected to depend only on the asymptotic inverse-power potential term $R^{-\tilde{n}}$, since competing higher-power terms in equation (3) with $m > \tilde{n}$ will usually be present also.

Perhaps the simplest way of analysing turning points in the long-range region is by utilizing the expression

$$D - G(v) = C_n/[R_2(v)]^n \qquad (47)$$

obtained on substituting $G(v) = V[R_2(v)]$ into equation (4). If D is accurately known, equation (47) suggests using a log–log plot of binding energy *versus* $R_2(v)$ to determine the power n and coefficient C_n of the effective outer potential over the range of the given turning points. This approach can indicate whether this effective local inverse power approaches its (assumed known) asymptotic value $\tilde{n}$, and when this occurs it will also yield the theoretically interesting coefficient $C_{\tilde{n}}$. Unfortunately, D is usually not known with particularly high accuracy, and relatively small errors in it can significantly affect log–log plots of this sort (*e.g.* see ref. 51). An additional difficulty, shared with the methods of Sections 3—5, is that equation (47) does not offer any natural way of taking account of higher-power contributions to the long-range potential.

Stwalley pointed out that a much more flexible type of RKR analysis

[110] R. B. Singh and D. K. Rai, *Canad. J. Phys.*, 1965, **43,** 1685.
[111] R. B. Singh and D. K. Rai, *J. Quant. Spectroscopy Radiative Transfer*, 1965, **5,** 723.
[112] V. M. Trivedi and V. B. Gohel, *J. Phys.*, (*B*), 1972, **5,** L38.

is possible if one utilizes the first few terms in the long-range potential expansion [equation (3)] for the particular state under consideration.[43] Substituting $G(v) = V[R_2(v)]$ into equation (3) yields:

$$G(v) = D - \sum_{m \geq \tilde{n}} C_m/[R_2(v)]^m \quad (48)$$

where the allowed values of m for the first few terms are assumed to be known (see Section 2). This expression may be used in a number of different ways, depending on whether or not reliable values of D and some C_m are already known. If none are, equation (48) suggests making a plot of $G(v)$ *versus* $[R_2(v)]^{-\tilde{n}}$, which in the limit of large $R_2(v)$ should be linear with intercept D and slope $C_{\tilde{n}}$. However, as with the log–log plots discussed above, this neglect of all $m > \tilde{n}$ terms in equation (48) will tend to make the thus-obtained $C_{\tilde{n}}$ an upper bound to the true value.

If one of the C_m values is already known, say $C_{\tilde{n}}$, then a simple rearrangement of equation (48) suggests the use of a plot of $\{G(v)+C_{\tilde{n}}/[R_2(v)]^{\tilde{n}}\}$ *versus* $[R_2(v)]^{-\tilde{m}}$, where $\tilde{m}$ is the power of the second term in the long-range potential expansion [equation (3)]. In this case, the intercept at $R_2(v) \to \infty$ is again equal to D while the limiting slope should yield the coefficient $C_{\tilde{m}}$. Figure 16 illustrates how Stwalley first used this type of plot in his analysis of the turning points of Mg_2 $(X\,^1\Sigma_g^+)$.[43] In this example, the leading terms in equation (48) have powers $m = 6$, 8, and 10, and a theoretical $C_6 = 3.29$ $(\pm 0.17) \times 10^6$ cm^{-1} Å^6 had been reported.[113]

A somewhat more sophisticated version of such plots was proposed by Cummings[38] for use in cases where reasonably accurate values of D are known. In its simplest form, appropriate when none of the C_m constants

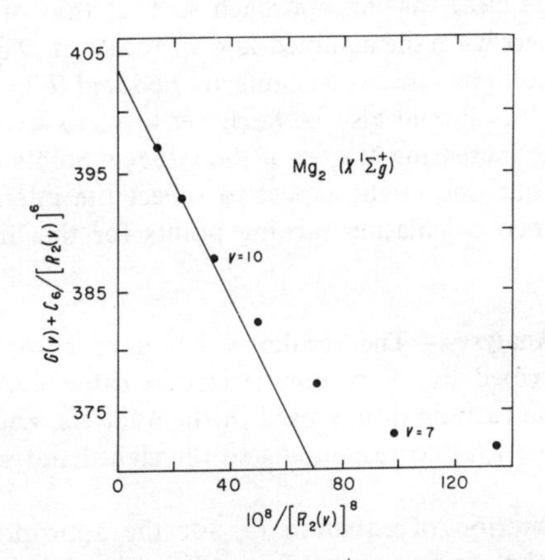

Figure 16 *RKR turning points for* Mg_2 $(X\,^1\Sigma_g^+)$,[114] *for which* $m = 6, 8, 10, \ldots$ etc., *plotted according to equation* (48) *utilizing the theoretically known*[113] *values of* C_6. *Energies are in* cm^{-1} *and lengths in* Å
(Reproduced by permission from *Chem. Phys. Letters*, 1970, **7**, 600)

[113] W. C. Stwalley, *J. Chem. Phys.*, 1971, **54**, 4517.

are known, Cummings' approach suggests plotting $\{[R_2(v)]^{\tilde{n}}[D-G(v)]\}$ *versus* $[R_2(v)]^{\tilde{n}-m}$, where $\tilde{n}$ and $\tilde{m}$ are again the powers of the first and second terms contributing to equation (48). Here, a linear extrapolation as $R_2(v) \rightarrow \infty$ will clearly yield $C_{\tilde{n}}$ as the intercept and C_m as the slope. On the other hand, for the problem to which this approach was first applied, Cl_2 ($B\,^3\Pi^+_{0u}$), for which the leading terms in equation (48) correspond to m = 5, 6, 8, and 10, reliable values of both D and C_5 are known (see Table 2). Thus, Cummings[38] used plots of $[R_2(v)]^6 \times \{D-G(v)-C_5/[R_2(v)]^5\}$ *versus* $[R_2(v)]^{-2}$ such as that shown in Figure 17 to obtain values of C_6 from the intercept and C_8 from the slope.

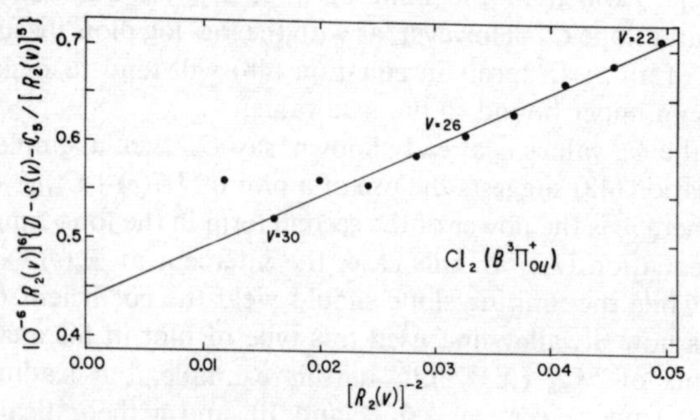

Figure 17 *RKR turning points for* Cl_2 ($B\,^3\Pi^+_{0u}$),[107] *for which* m = 5, 6, 8, 10, . . . etc., *plotted according to equation* (48) *utilizing the known* (*Table* 2) D *and* C_5 *in the manner suggested by Cummings.*[38] *Units as in Figure* 16

It should be quite clear that an approach such as that of Figure 17 will be very sensitive to errors in the assumed-known values of D and C_m. Hence, it should not be used unless these constants are believed to be quite accurate. In addition, such plots should also be fairly sensitive to errors in the RKR turning points. The increasing scatter of the last few points on Figure 17 is the type of behaviour one might expect to reflect the interpolation errors which can arise when calculating turning points for the highest observed levels.

Results of RKR Analyses.—The results which have been obtained using the techniques described above are summarized in Table 9. Also given there are the range of the turning points used in the analysis, and the minimum internuclear distance $R(5)$ corresponding to the right-hand side of equation (5).

Although the criterion of equation (5) for the appropriateness of the long-range potential form equation (48) is satisfied by the turning points for the halogens, this is not so for ground-state Mg_2 ($X\,^1\Sigma^+_g$). The turning-point analysis for this case was performed using a calculated[113] value of C_6 and a plot similar to that shown in Figure 16.[43] It is clear that the D and C_8 values obtained in this way really depend only on the two points corresponding to

Table 9 *Results of graphical analysis of long-range RKR potentials; energies are in* cm^{-1} *and lengths in* Å

Species	$\tilde{n}$	$R(5)$[a]	*Turning Points*[b]	*Assuming*[c]	*Obtain*	*Ref.*
$^{24}Mg^{24}Mg\ (X\,^1\Sigma_g^+)$	6	7.2	$R \leqslant 7.2$	$C_6 = 3.29 \times 10^6$	$D = 403.7(\pm 0.7)$ $C_8 = 48.0(\pm 8) \times 10^6$	43
$^{35}Cl^{35}Cl\ (B\,^3\Pi_{0u}^+)$	5	4.1	$4.5 \leqslant R \leqslant 7.9$	$D = 20879.64$ $C_5 = 1.22_4 \times 10^5$	$C_6 = 0.42(\pm 0.02) \times 10^6$ $C_8 = 5.9(\pm 0.5) \times 10^6$	80
$^{79}Br^{81}Br\ (B\,^3\Pi_{0u}^+)$	5	4.7	$4.7 \leqslant R \leqslant 8.7$	$D = 19580.74$ $C_5 = 1.79 \times 10^5$	$C_6 = 0.84(\pm 0.03) \times 10^6$ $C_8 = 11.8(\pm 1.0) \times 10^6$	80
$^{127}I^{127}I\ (B\,^3\Pi_{0u}^+)$	5	5.5	$5.6 \leqslant R \leqslant 7.9$	$D = 20044.0_4$ $C_5 = 3.10_6 \times 10^5$	$C_6 = 1.63(\pm 0.01) \times 10^6$ $C_8 = 35.1(\pm 0.4) \times 10^6$	117

[a] Internuclear distance corresponding to the criterion of equation (5): $R(5) = 2(\langle r_A^2 \rangle^{1/2} + \langle r_B^2 \rangle^{1/2})$. [b] range of the turning points used in the analysis; [c] for Mg_2, from ref. 113; otherwise, results from Table 2.

$v = 11$ and 12. For these levels, the calculated [114] turning points $R_2(v) = 6.8$ and 7.2 Å come fairly close to satisfying equation (5), and are significantly larger than the minimum distance of 5 Å suggested by Stwalley.[43] Thus, in spite of the violation of the criterion of equation (5), the reported D and C_8 values may be significant. Evidence favouring this possibility is provided by Stwalley's report (W. C. Stwalley, personal communication, 1972) that an additional point corresponding to the recently identified $v = 13$ level lies right on the solid line in Figure 16. In any case, the trend of this plot suggests that the inclusion of additional turning points farther into the long-range region would mainly tend to increase the reported (Table 9) D and C_8 values. Such a change would merely improve the accord between Stwalley's 'experimental' $C_8 = 48\ (\pm 8) \times 10^6\ \text{cm}^{-1}\ \text{Å}^6$ and his approximate theoretical bounds for it:[43]

$$48 \times 10^6 \lesssim C_8 \lesssim 77 \times 10^6\ \text{cm}^{-1}\ \text{Å}^6$$

while slightly increasing the difference between his D value and the experimental estimate [114] of $D_0 = 399\ (\pm 5)\ \text{cm}^{-1}$.

For the $B(^3\Pi^+_{0u})$-state halogens, the first few terms in equations (3) and (48) correspond to $m = 5, 6, 8$, and 10, and the corresponding C_m values are all positive (attractive). Since reliable D and C_5 values are known for B-state Cl_2, Br_2, and I_2 (see Table 2), utilization of their turning points in the manner suggested by Cummings [38] can yield values of C_6 and C_8. The Cl_2 results in Table 9 were thus obtained from a linear least-squares fit to the points for $v = 22-30$ in Figure 17. The point for $v = 31$ was ignored because of its inconsistency with the others, a difficulty probably due to error in the RKR turning point for this highest observed level.

In his related analysis of the outer turning points of Cl_2 ($B\ ^3\Pi^+_{0u}$), Cummings also allowed for variations in the 'assumed-known' D and C_5 values.[38] For a number of trial C_5 values he determined the D for which the $v = 24-30$ points on a plot such as Figure 17 give the best straight line, the line's intercept and slope yielding estimates of C_6 and C_8. The results corresponding to the $C_5 = 1.22 \times 10^5\ \text{cm}^{-1}\ \text{Å}^5$ used here are: $D = 20879.40\ (\pm 0.05)\ \text{cm}^{-1}$, $C_6 = 0.37 \times 10^6\ \text{cm}^{-1}\ \text{Å}^6$, and $C_8 = 6.6 \times 10^6\ \text{cm}^{-1}\ \text{Å}^8$.[38] The differences between Cummings' results and those given in Table 9 are partly due to the different assumed D values, and partly associated with the fact that he used slightly different turning points. Although the theoretical C_6 in Table 10 is in better agreement with the present value (in Table 9) than with Cummings' estimate (above), the difference is not really sufficiently large to determine which of the latter is 'better'. Other Cl_2 analyses by Le Roy [77] and Goscinski [115] are less reliable than that reported here since they ignore or misrepresent the influence of the first (R^{-5}) and third (R^{-8}) terms contributing to equation (48). In addition, in their prior analysis of B-state Cl_2 and Br_2, Byrne *et al.*[91] concluded that the potential tails asymptotically died

[114] W. J. Balfour and A. E. Douglas, *Canad. J. Phys.*, 1970, **48**, 901.
[115] O. Goscinksi, *Mol. Phys.*, to be published.

Table 10 *Test of Cummings'*[38] *theoretical C_6 and C_8 values for the $B(^3\Pi^+_{0u})$-state halogens*

	$10^{-6}\times C_6$/cm^{-1} Å^6		$10^{-6}\times C_8$/cm^{-1} Å^8	
	Experimental[a]	*Theoretical*[b]	*Experimental*[a]	*Theoretical*[b]
Cl_2	0.42(±0.02)	0.45(±0.05)	5.9(±0.5)	2.8(±0.6)
Br_2	0.84(±0.03)	0.83(±0.08)	11.8(±1.0)	6.3(±1.3)
I_2	1.63(±0.01)	1.85(±0.2)	35.1(±0.4)	19(±4)

[a] From Table 9; [b] from ref. 38; the error bounds are only estimates.

off as R^{-6} rather than R^{-5}. However, a re-examination of their data showed that this conclusion was erroneously based on an incorrect assumed value of D, and a neglect of the turning points for the highest observed levels.[51]

Figure 18 shows the Br_2 ($B\,^3\Pi^+_{0u}$) plot analogous to that of Figure 17, the intercept and slope of the solid line yielding the C_6 and C_8 given in Table 9. The RKR turning points used for this case are those reported by Todd, Richards, and Byrne,[116] since the more recent RKR calculation by Coxon [107] was later shown [71] to be based on an incorrect vibrational assignment of levels v = 49—52. The RKR analyses for this state reported by Le Roy [77] and Goscinski [115] should therefore be disregarded, since they were based on Coxon's [107] potential.

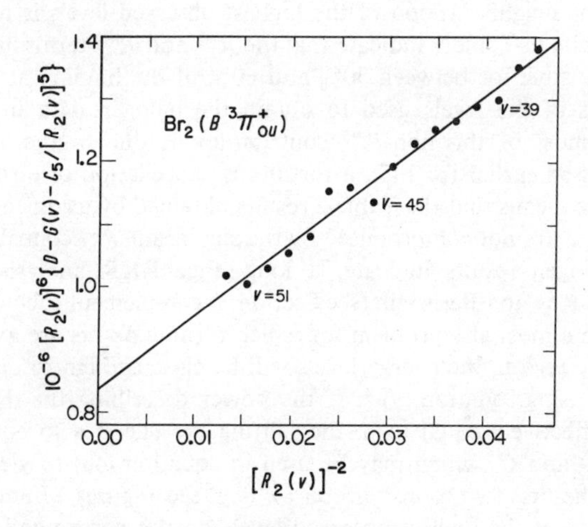

Figure 18 *RKR turning points for* Br_2 ($B\,^3\Pi^+_{0u}$) [116] *plotted as in Figure* 17

A similar analysis of long-range RKR turning points for I_2 ($B\,^3\Pi^+_{0u}$) yielded the C_6 and C_8 for this state shown in Table 9.[117] Prior work by Steinfeld, Campbell, and Weiss [118] had concluded that the v = 45—51 turning points

[116] J. A. C. Todd, W. G. Richards, and M. A. Byrne, *Trans. Faraday Soc.*, 1967, **63**, 2081.
[117] J. D. Brown, G. Burns, and R. J. Le Roy, to be published.
[118] J. I. Steinfeld, J. D. Campbell, and N. A. Weiss, *J. Mol. Spectroscopy*, 1969, **29**, 204.

in the region 4.6—5.0 Å displayed pure R^{-5} behaviour. However, both the RKR potential and the D value used in their analysis have since been shown to be invalid,[51] and in any case the turning points considered lie at too small R for the long-range potential form (3) [or (47)] to be valid. In a more recent study of this state, Tellinghuisen [78] independently noted the lack of pure R^{-5} asymptotic behaviour and re-emphasized the 'analomous accuracy' of the Le Roy–Bernstein results for this case (see discussion on p. 145).

Long-range analyses of outer RKR turning points have been reported for a number of other systems for which the results are believed to be invalid because of violations of the criterion of equation (5). For example, ground-state HF and DF turning points in the region $1.8 < R < 2.2$ Å were reported to display the asymptotic R^{-6} behaviour,[91] a conclusion upon which the $R(5) = 3.1$ Å value for this species casts considerable doubt. In another type of situation, studies of the outer turning points of ground-state I_2 which assumed the asymptotic C_6/R^6 functionality [75, 91, 103] are deemed invalid because the levels whose turning points lie in the long-range region actually belong to a different electronic state.[74]

Discussion and Prognosis.—An important product of the above analyses is the conclusion for the $B(^3\Pi_{0u}^+)$-state halogens that the outer branch of the potential in the neighbourhood of the highest observed levels is not purely R^{-5}. The results in Table 9 indicate that the R^{-6} and R^{-8} terms in equation (48) are responsible for between 30% and 60% of the binding at the outer turning points of the levels used to obtain the halogen data in Table 2. Fortunately, most of this non-R^{-5} contribution is due to the R^{-6} term. As was discussed earlier (p. 145), a fortuitous cancellation of errors in this $\tilde{m} = \tilde{n}+1$ case means that the Table 2 results obtained by using equation (9) with $n = \tilde{n} = 5$ are not compromised by the significant R^{-6} contribution.[80]

As the halogen results indicate, a long-range RKR analysis and the method of Le Roy and Bernstein (see Section 3) complement each other very effectively. In almost any problem for which turning points are available in the long-range region, the energy levels will be close enough to dissociation to be treated using equation (9).* If the power describing the distribution of levels is effectively equal to $\tilde{n}$, then fitting the energies to equation (9) will yield a D and $C_{\tilde{n}}$ which may be used in equation (48) to yield the coefficients of the first two terms with $m > \tilde{n}$ (*e.g.* see Figures 17 and 18). On the other hand, if the highest observed levels do not correspond to $n = \tilde{n}$, a fit to equation (9) will still yield a reliable value of D which when used in equation (48) can yield estimates of $C_{\tilde{n}}$ and C_m.

* An apparent exception is Mg_2 $(X\,^1\Sigma_g^+)$, for which the vibrational spacings' linear dependence [114] on v precludes an analysis using equation (9). However, while the RKR potential was plausibly analysed using equation (48),[43] the fact that the turning points used failed to satisfy the criterion of equation (5) raises questions about the validity of this long-range analysis.

One rather important use of results such as those in Table 9 is as a test of theoretically calculated C_m coefficients. To this end, Table 10 compares the present best C_6 and C_8 values for the $B(^3\Pi^+_{0u})$-state halogens with values calculated by Cummings.[38] The good agreement between the experimental and theoretical C_6 values attests to the usefulness of Cummings [38] methods. The fact that the theoretical C_8 values are roughly twice as large as the experimental values may be due to the approximations involved in the calculation of the former. On the other hand, it might also be a reflection of our neglect of yet higher-power ($m > 8$) contributions to the long-range potential (3).

The discussion in this section has considered only the inverse-power sum representation (3) for the long-range potential. However, for cases in which D is known, it might be preferable to utilize Goscinski's [1,0] Padé approximate representation, equation (43a).[115] In this, setting $G(v) = V[R_2(v)]$ would yield:

$$[R_2(v)]^{-\tilde{n}}[D-G(v)]^{-1} = (C_{\tilde{n}})^{-1}-[C_{\tilde{m}}/(C_{\tilde{n}})^2][R_2(v)]^{-(\tilde{m}-\tilde{n})} \qquad (49)$$

A plot of the left-hand side of equation (49) *vs.* $[R_2(v)]^{-(\tilde{m}-\tilde{n})}$ will clearly yield $C_{\tilde{n}}$ and $C_{\tilde{m}}$ from its intercept and slope. Although the results Goscinski obtained [115] on applying this technique to B-state Cl_2 and Br_2 are less reliable than those presented in Table 9, his approach should prove to be of considerable use in the future.

In summary, therefore, it appears that long-range RKR analyses have a very bright future as a source of information about long-range potential coefficients C_m, particularly when coupled with a prior use of the vibrational eigenvalue expression (9).

7 Concluding Remarks

A feature common to all four of the methods described above is the fact that they readily allow the concurrent treatment of data for different isotopic forms of a molecule in a given electronic state. This is possible because within the Born–Oppenheimer approximation[5] the internuclear potentials for the different isotopes are precisely the same. Furthermore, equations (6) and (38) imply that the vibrational–rotational eigenvalues of the different isotopes may be expressed as a single function of mass-reduced quantum numbers. In view of this, Stwalley [119, 120] has suggested replacing the normal v and J by

$$\eta \equiv (v+\tfrac{1}{2})/\mu^{\frac{1}{2}}$$

and

$$\zeta \equiv J(J+1)/\mu$$

[119] W. C. Stwalley, *Bull. Amer. Phys. Soc.*, 1971, **16**, 1396.
[120] W. C. Stwalley, Paper AA5, Proceedings of the 27th Symposium on Molecular Structure and Spectroscopy, Columbus, 1972.

He demonstrated that all the observed eigenvalues of HgH, HgD, and HgT could be expressed as a single function of these new variables.[120, 121] It is readily seen that the vibrational eigenvalue expressions [equations (8)—(13)] and the LBM equation (42) may be rewritten in terms of this η and ζ. This is also possible for the B_v and RKR-potential expressions, equations (32) and (33) and (44)—(46), respectively, if B_v is replaced by the mass-reduced quantity $\tilde{B}_v \equiv \mu B_v$. Concomitant mass-independent forms of all other expressions described above are also readily obtainable.

The great advantage of the type of 'combined-isotope analysis' which this mass scaling makes possible is that it provides a larger number of independent data in any given region than are available for any single isotopic form. As a result, uncertainties in the results of an analysis of such data would be much smaller than if the data for individual isotopes were analysed separately. This feature is particularly important for a species with a small reduced mass, for which case the density of levels is relatively small. For example, in the case of ground-state HgH the small number of vibrational levels seriously limits the accuracy with which an RKR curve may be obtained. However, using Stwalley's variables η and ζ and treating the three isotopes HgH, HgD, and HgT as a single species allows the calculation of a much more reliable RKR potential.[120, 121] This feature was also exploited in the analysis of the functionality of the energies and B_v values of the highest observed levels of Br_2 ($B\,^3\Pi^+_{0u}$).[51, 82]

Another type of situation in which this 'combined-isotope' approach proves useful is one in which data for different isotopes happen to be available in mutually exclusive regions of the potential well. This occurs for Br_2 ($B\,^3\Pi^+_{0u}$), for which there exist pure isotope ($^{79}Br^{79}Br$ and $^{81}Br^{81}Br$) measurements in the neighbourhoods of both the potential minimum and the dissociation limit,[72, 73] but only mixed-isotope $^{79}Br^{81}Br$ results in the intermediate region.[107] In calculating the RKR potential for this case one should ideally utilize Stwalley's η and ζ to join smoothly the different sets of data.

It is perhaps noteworthy that all of the molecular species studied in this chapter are formed from uncharged atoms. In contrast, for ionic or ionized molecules, for which $\tilde{n} = 1$ and 4, respectively, the asymptotic potential tail will tend to bind a relatively higher density of levels in the immediate neighbourhood of D. Thus, the present techniques, particularly those of Sections 3 and 4, should be even more important there than for the neutral–neutral interactions considered heretofore.

One point which deserves re-emphasis is the importance of verifying the validity of the inverse-power potential expansion [equation (3)] for the data under consideration. The significance of this question is underlined by the discovery in Section 5 that most of the reported applications of Bernstein's[89] LBM expression [equation (42)] were invalid. This problem was first quantitatively considered by Stwalley,[52] who suggested $R > 5$ Å

[121] W. C. Stwalley, Paper T3, Proceedings of the 26th Symposium on Molecular Structure and Spectroscopy, Columbus, 1971.

as a minimum requirement for the validity of equation (3).[43] However, it seems clear that any such criterion should take account of the differing sizes of different atoms, and equation (5) seems to fulfill this requirement admirably. On the other hand, one may obviously formulate any number of analogous criteria for R which utilize other properties of the interacting atoms. In this vein, the RKR analysis of Mg_2 did suggest that the present criterion [equation (5)] might be somewhat pessimistic for atoms of the alkali metals or alkaline earths.

Prior to the development of the methods discussed in this chapter, the best way of studying experimentally long-range interaction potentials was *via* atomic beam scattering.[15, 122–126] It is interesting to note how effectively these two approaches complement one another. In spectroscopic observations, the selection rules for optical transitions readily distinguished between the several molecular states formed by the interaction of a pair of atoms with unfilled valence shells. This type of situation would be quite unsuited to the application of scattering techniques since all the energetically accessible potential curves would concurrently contribute to the observed cross-sections, and it would be virtually impossible to separate them. On the other hand, this latter approach may be very profitably applied to the study of interactions in which at least one of the atoms has a closed-shell electronic configuration, for which case the collision partners effectively interact according to a single potential function. However, this type of species is usually weakly bound and supports few vibrational levels. In addition to being relatively difficult to observe spectroscopically, the small number of levels makes any results that might be obtained quite difficult to analyse using the methods described above.

A particularly interesting connection between beam scattering and spectroscopic measurements is their mutual observation of the broadened rotationally predissociating levels discussed in Section 5.[90] In a scattering experiment these levels, called 'orbiting resonances', manifest themselves as small peaks or dips in the cross-section plotted as a function of collision energy, their position yielding the energy of the particular level above the dissociation limit.[127–130] Since the spectroscopic observation of such a broadened level

[122] H. Pauli and J. P. Toennies, *Adv. Atomic Mol. Phys.*, 1965, **1**, 195.
[123] J. Ross, *Adv. Chem. Phys.*, 1966, **10**, vii.
[124] R. B. Bernstein and J. T. Muckerman, *Adv. Chem. Phys.*, 1967, **12**, 389.
[125] Chr. Schlier, *Ann. Rev. Phys. Chem.*, 1969, **20**, 191.
[126] R. B. Bernstein and R. A. La Budde, University of Wisconsin Theoretical Chemistry Institute Report WIS-TCI-469, 1972; *J. Chem. Phys.*, to be published.
[127] R. A. Buckingham and J. W. Fox, *Proc. Roy. Soc.*, 1962, **A267**, 102; R. A. Buckingham, J. W. Fox, and E. Gal, *ibid.*, 1965, **A284**, 237; J. W. Fox and E. Gal, *Proc. Phys. Soc.*, 1967, **90**, 55.
[128] R. B. Bernstein, C. F. Curtiss, S. Imam-Rahajoe, and H. T. Wood, *J. Chem. Phys.*, 1966, **44**, 4072.
[129] W. C. Stwalley, A. Niehaus, and D. R. Herschbach, Proceedings of the International Conference on the Physics of Electron and Atomic Collisions, Leningrad, 1967, p. 639.
[130] T. G. Waech and R. B. Bernstein, *J. Chem. Phys.*, 1967, **46**, 4905; M. E. Gersh and R. B. Bernstein, *Chem. Phys. Letters*, 1969, **4**, 221.

generally gives its position relative to the potential minimum, combining these two types of results should yield an accurate estimate of the molecular dissociation energy. However, these broad quasi-bound levels are relatively hard to observe by either technique so that this type of result is unlikely to be obtainable for many systems.

In conclusion, we have seen that spectroscopic studies of highly-excited vibrational–rotational levels are capable of yielding much hitherto unobtainable information about long-range intermolecular potentials. The comparisons in Tables 2 and 10 promise that such results will provide a very useful foil in the development of improved theoretical techniques for calculating interatomic forces. Although the utility of the methods described herein clearly depends on the availability of spectroscopic data for levels near dissociation, the growth of selective excitation and fluorescence techniques using lasers has made this region much more open to investigation. At present it appears that the methods of vibrational energy analysis introduced by Le Roy and Bernstein [50, 51] (see Section 3) and the long-range RKR analysis techniques proposed by Stwalley [43] and Cummings [38] (see Section 6) are the most useful. However, future developments [80] may also make B_v values for levels near dissociation (see Section 4) another potent source of information about long-range potential, and improved versions of all the techniques discussed will almost certainly be obtained.

I am very grateful to Professors R. B. Bernstein and W. C. Stwalley for numerous helpful discussions, and to them, Dr. D. M. Gass and Professor W. J. Meath for critical comments on the manuscript. I would also like to thank Drs. F. E. Cummings and O. Goscinksi for sending me results prior to publication.

4

Low-lying Electronic States of Diatomic Halogen Molecules

BY J. A. COXON

1 Introduction

The diatomic halogens constitute an interesting series of molecules about whose electronic states considerable information is available. Their u.v.–visible banded and continuous absorption spectra are particularly well understood, and the role of halogen photodissociation in the development of photochemistry need hardly be mentioned. The many known examples of chemiluminescence from the reverse process, atom combination, now promise to facilitate progress in molecular dynamics.

As in most fields, renewed interest in halogen spectroscopy has recently been encouraged by the introduction of new and penetrating techniques. Thus, for example, they have been found an ideal class of molecule for which the novel properties of lasers operating at visible wavelengths can be utilized. Additionally, the computerization of data reduction has lessened the formidable problems in analyses of the high density of lines in their banded spectra to give comprehensive and accurate molecular constants and energy levels. The use of fast computers in calculating realistic potential energy curves by the Rydberg–Klein–Rees (RKR) method, and in obtaining bound and continuum vibrational wavefunctions by numerical solution of the radial Schrödinger equation, has also been exploited to much advantage for the halogens in describing their detailed molecular properties. A more recent stimulus has been the application of long-range potential theory, which forms the subject of another chapter of this Report. The halogens form a singularly useful series for testing this theory since potential curves are available almost to the dissociation limits of several electronic states.

In preparing this Report, it was evident that a complete review of all the latest developments in halogen spectroscopy would be difficult owing to shortage of space and time. Furthermore, Turner [1] has recently discussed several topics in a review of the physical and spectroscopic properties of the halogens. In the present Report, therefore, an attempt is made to consider in depth some selected problems, where recent work has brought significant progress. Attention is thus confined mainly to the homonuclear halogens

[1] J. J. Turner, MTP International Review of Science, Inorganic Chemistry, Series 1, vol. 3, ed. V. Gutmann, 1972, p. 253.

in their lower electronic states. The spectroscopy of the heteronuclear molecules has advanced relatively little, and since high-lying states are normally of relatively minor chemical importance they will not be discussed here. Tables of spectroscopic constants are not included, since such data are summarized by Turner [1] and in the recent compilation of spectroscopic data.[2] A brief summary of several topics not considered in the main text is given in Section 6.

2 Theoretical Background

Electronic Coupling in Molecular Halogens.—Continuous absorption by halogen molecules in the visible region of the spectrum is well known and is due to electronic transition(s) from low vibrational levels v'' of the ground $X^1\Sigma_g^+$ state to the repulsive limb(s) of the potential curves of one (or more) excited states, which for the moment are labelled $A^3\Pi(1_u)$, $B^3\Pi(0_u^+)$ and $^1\Pi(1_u)$. The A and B states possess bound vibrational levels, whereas $^1\Pi(1_u)$ is entirely repulsive. The g,u symmetry property is relevant only for the homonuclear molecules, F_2, Cl_2, Br_2, and I_2, and is ignored otherwise. Theoretical treatments which explain these transitions, especially their intensities, and which predict the energies of unobserved states have been developed by Mulliken.[3–5] No attempt is made here to describe fully these results; rather, only a summary, relevant to the later sections, is presented.

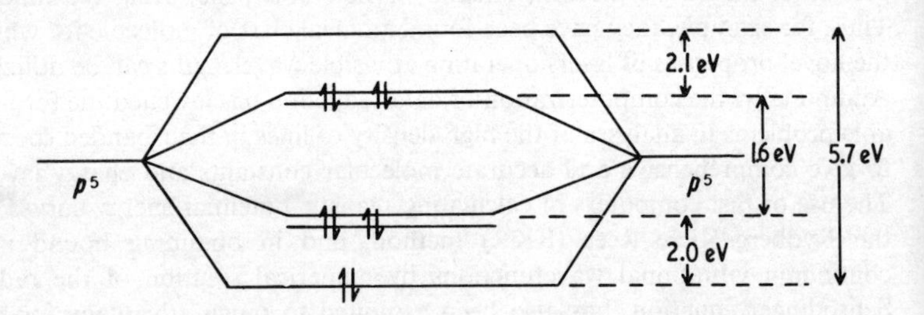

Figure 1 *Molecular orbitals for the valence shell configurations of the homonuclear halogens. The configuration illustrated is the* 2440 $^1\Sigma_g^+$ *ground state, and the energy scale is that estimated by Mulliken* [4] *at* r_e'' *for* I_2. *The photoelectron spectrum* [156] *of* I_2 *gives results in good agreement with Mulliken's estimate*

Consider two ground-state $^2P_{3/2,1/2}$ halogen atoms with np^5 outermost electron configurations; a simple LCAO treatment gives the molecular orbitals shown in Figure 1. The lowest energy arrangement of the ten valence electrons, $(\sigma_g)^2 (\pi_u)^4 (\pi_g)^4 (\sigma_u)^0$, abbreviated to 2440, represents the ground $X^1\Sigma_g^+$ states. Excited valence-shell electronic states are derived by

[2] International Tables of Selected Constants, No. 17, Spectroscopic Data relative to Diatomic Molecules, ed. B. Rosen, Pergamon Press, 1970.
[3] R. S. Mulliken, *Phys. Rev.*, 1930, **36**, 699; 1931, **37**, 1412; 1934, **46**, 549.
[4] R. S. Mulliken, *Phys. Rev.*, 1940, **57**, 500.
[5] R. S. Mulliken, *J. Chem. Phys.*, 1971, **55**, 288.

promotion of one or two electrons from the filled $\sigma_g^2 \pi_u^4 \pi_g^4$ orbitals to the vacant σ_u. The lowest excited electronic state is therefore 2431, $^{1,3}\Pi_u$. Figure 2 shows schematically the three lowest energy configurations obtained in this way with both Hund's case (a) and case (c) notations. In the latter case, only Ω (the electronic angular momentum along the internuclear axis) is defined, and splitting of 0 states into 0^+ and 0^-, and of $^3\Sigma^{+(-)}$ into 1 and $0^{-(+)}$ components is noted.

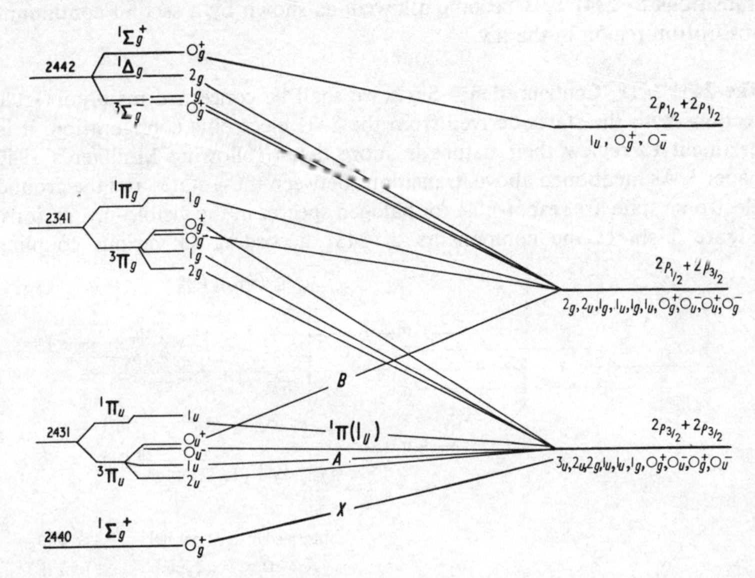

Figure 2 *Schematic correlation diagram for the low-lying halogen valence shell states with their predicted dissociation products. The correlations* [6] *are based on symmetry conservation and the non-crossing rule for states of the same electronic angular momentum*

An important paper by Mulliken [6] enables us to predict which atomic products, X $^2P_{3/2}$+X $^2P_{3/2}$, X $^2P_{3/2}$+X $^2P_{1/2}$, or X $^2P_{1/2}$+X $^2P_{1/2}$ are obtained on dissociation of X_2 for each of the valence shell components. Figure 2 shows that 23 Ω-components are derived from two ^2P atoms, when for large internuclear separations r, J is retained as a good quantum number with component M_J along the internuclear axis. Thus, for example, $^2P_{1/2}+{}^2P_{1/2}$ gives 0_g^+, 0_u^-, and doubly degenerate 1_u components. Using the non-crossing rule that states of the same Ω and same g or u symmetry cannot cross, the correlations shown in Figure 2 are obtained. It is important to note that 0_u^+ from $^2P_{1/2}+{}^2P_{3/2}$ is unique and correlates with 2431 $^3\Pi(0_u^+)$, labelled above as the B state. For heteronuclear halogens, however, with no defined g,u properties, $^3\Pi(0^+)$ could correlate with $^2P_{3/2}+{}^2P_{3/2}$. In addition, it must be

[6] R. S. Mulliken, *Phys. Rev.*, 1930, **36**, 1440.

stressed that Figure 2 is only qualitative. Thus for larger multiplet splittings, the relative positions of 2341 $^3\Pi_g(0_g^+)$ and 2422 $^3\Sigma_g^-(0_g^+)$ could be reversed, which with the non-crossing rule would cause 2341 0_g^+ to correlate with $^2P_{3/2}+{}^2P_{1/2}$ atoms. This situation might arise [5] for I_2 (*cf.* Figure 2 and Figure 4). Finally we note that only 2440 $X^1\Sigma_g^+$ and 2431 $^3\Pi_{0,1,2}$ form bound states. Higher states (see Figure 4) exist, however, but are of ionic $X^+ X^-$ character and correlate with excited atoms. For heteronuclear halogens, transitions to 2341 $^{1,3}\Pi$ become allowed, as shown by a second continuum absorption region in the u.v.

The 2431 $^{1,3}\Pi_u$ Configuration.—Since we shall be concerned mainly in later sections with the states derived from the 2431 electronic configuration, it is pertinent to review their nature in more detail following Mulliken's 1940 paper.[4] As mentioned above, transitions between these states and the ground electronic state are responsible for halogen spectra in the visible–u.v. regions. Figure 3 shows the components of 2431 according to various coupling

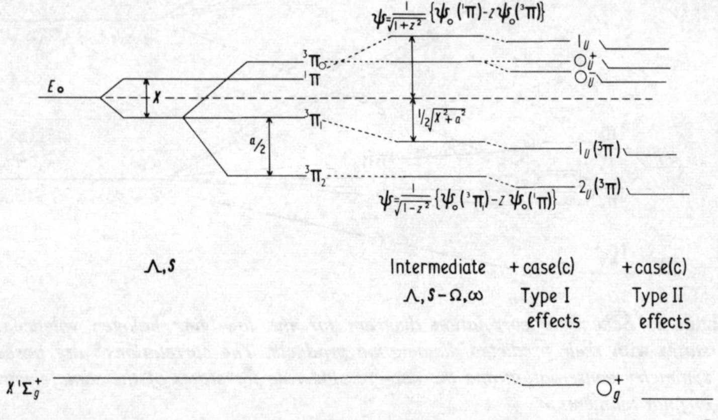

Figure 3 *Components of the* 2431 $^{1,3}\Pi_u$ *configuration for intermediate* $\Lambda,S-\Omega,\omega$ *coupling, with case* (c) *effects.*[4] *Case* (c) *Type I leaves unaffected* $^3\Pi(2_u$ *and* $0_u^+)$, *whereas Type II is expected to depress all components about equally*

schemes. For case (a) coupling, a large splitting of $^1\Pi$ and $^3\Pi$, given by $E = E_0 \pm X/2$ is accompanied by a smaller spin–orbit splitting of the $^3\Pi$ multiplet, for which $E = E_0 - X/2 + a\Sigma/2$, where Σ is the component of $\boldsymbol{S}$ along the internuclear axis. Case (a) coupling is then only a good description if $X \gg a$. In the limit $X \rightarrow 0$, $\boldsymbol{S}$ is not defined, and we have ideal $\boldsymbol{\Omega,\omega}$ coupling, in which for the halogens, the spin of the promoted σ_u electron is weakly coupled to $\boldsymbol{\Omega}$ of the $\pi^3{}_g$, $^2\Pi_{3/2,1/2}$ core. For the heavier halogens, X and a are of similar magnitude, and the coupling is best described as intermediate $\boldsymbol{\Lambda,S}-\boldsymbol{\Omega,\omega}$. In the transition from $\boldsymbol{\Lambda,S}$ to $\boldsymbol{\Omega,\omega}$ coupling, the energies and wavefunctions of the $^3\Pi$(0 and 2) components are unaffected, whereas $^1\Pi_{1u}$ and $^3\Pi_{1u}$ interact, giving the mixed wavefunctions:

$$\psi[^1\Pi(1_u)] = \sqrt{1/(1+z^2)}\,\{\pm z\psi_0(^1\Pi)+\psi_0(^3\Pi_1)\}$$
$$\psi[^3\Pi(1_u)] = \sqrt{1/(1+z^2)}\,\{z\psi_0(^3\Pi_1)\mp\psi_0(^1\Pi_1)\} \quad (1)$$

with eigenvalues, $E = E_0 \pm \frac{1}{2}\sqrt{X^2+a^2}$. ψ_0 are the unperturbed case (a) wavefunctions and z is given by $z = (X/a+\sqrt{X^2/a^2+1})$. This is illustrated in Figure 3. Since $^3\Pi(1_u)$ now takes on some of the properties of $^1\Pi(1_u)$, transitions from the ground state, forbidden in case (a) by the spin selection rule, become partially allowed. Mulliken showed that the ratio of the dipole strengths for the $^1\Pi(1_u) \leftarrow X$ and $^3\Pi(1_u) \leftarrow X$ systems is given by $D(^1\Pi)/D(^3\Pi) = z^2$. The dipole strength D is related to the absorption coefficient k_ν per molecule by

$$g'D = (3hc/8\pi^3e^2)\int k_\nu\, d\nu/\nu$$
$$= (3hc/8\pi^3e^2N_0)\times 2.3\times 1000\times\int \varepsilon_\nu\, d\nu/\nu \quad (2)$$

where g' is the electronic degeneracy of the upper state [2 for $^3\Pi(1_u)$ and $^1\Pi(1_u)$], N_0 is Avogadro's constant and ε_ν is the molar decadic extinction coefficient. For low values of X, tending to pure $\boldsymbol{\Omega},\boldsymbol{\omega}$ coupling, z decreases to unity, and the two transitions have the same dipole strength.

The foregoing discussion has left unaffected $^3\Pi(0_u^+)$ and $^3\Pi(2_u)$, to which transitions from the X state are still forbidden, while experimentally, $^3\Pi(0_u^+) \leftarrow X$ is observed to be considerably more intense than $^3\Pi(1_u) \leftarrow X$. The simple theory, however, neglects the tendencies for case (c) coupling, for which the atomic $\boldsymbol{J}$ coupling partially persists. This has the effect of mixing states of the same symmetry belonging to different molecular electronic configurations, and can be of two types: Type I, interactions between components leading to the same ground state atoms (in our case, the 23 components from $^2P+^2P$ atoms, Figure 2). Thus $A^3\Pi(1_u)$ and $^1\Pi(1_u)$ are lowered in energy by interaction with three higher 1_u components, with similar effects for 0_u^-, 2_u, and the ground 0_g^+ components. $B^3\Pi(0_u^+)$ is exceptional, being still unaffected (Figure 3). Only by interaction with 0_u^+ states of configurations leading to different atomic states can $B^3\Pi(0_u^+)$ be affected [case (c), Type II interactions]. Thus the components of ionic 1342, $^{1,3}\Pi_u$ (see Figure 4 for I_2) can interact with 2431, $^{1,3}\Pi_u$. 1342, $^3\Pi(0_u^+)$ in particular is already mixed by Type I interactions with 2332 $^1\Sigma_u^+$, $^3\Sigma_u^-(0_u^+)$ and 1441 $^1\Sigma_u^+$, and since $^1\Sigma_u^+$ (ionic) $\leftarrow X$ is expected to be exceptionally intense, this indirect route can account, perhaps in part, for the anomalously high intensities of $B^3\Pi(0_u^+) \leftarrow X^1\Sigma_g^+$. The main contribution to the high intensity, however, is probably due[5] to Type I admixture of 2341 $^3\Pi(0_g^+)$ into $X^1\Sigma_g^+$. Unlike Type I, the Type II effects should depress all components of 2431 about equally.

An important application of the preceding considerations lies in interpretation of the visible continuous absorption spectra of the halogens. These arise from overlapping transitions to the repulsive limbs of $A^3\Pi(1_u)$, $B^3\Pi(0_u^+)$ and $^1\Pi(1_u)$ from $X^1\Sigma_g^+$, with relative intensities determined as indicated above. By locating the relative positions of the three states, usually at

$r = r_e''(X)$, and making allowance for case (c) Type I depressions (often difficult exactly), values for X can be found and hence an expectation of the relative intensities.

3 Iodine

Introduction.—Interest in the spectroscopy of the iodine molecule first developed because of its relatively intense absorption of light in the visible and u.v. regions of the spectrum, and the consequent ease with which the molecule can be induced to exhibit fluorescence. Later work revealed many intriguing phenomena, such as the changes from continuous to banded emission spectra on the addition of inert gases, and the quenching of visible fluorescence from the B state by application of a magnetic field. In fact, the results of an impressively large number of spectroscopic investigations over the past 60 years, aided by concurrent development of theoretical models, have been responsible for the unusually large amount of detailed knowledge which is available for the iodine molecule. In recent years, particular attention has been focused on the ground and lowest excited states, and the results of these studies will form the basis of discussion in this section. Higher-lying bound electronic states are known, some of which are of ion-pair type, and are known or predicted to give rise to electronic transitions in the visible and u.v. spectral regions. However, the large mass of iodine and the shallow potential energy curves of most of these excited states also result in large densities of discrete levels. Thus although many of the possible transitions between these excited states, and with the ground state, have now been observed, rotational analyses have so far not been attempted. Even the relatively small errors which are present in band-head measurements have led to several erroneous or ambiguous vibrational analyses, which are responsible for considerable confusion in the literature. The most recent summary of the known spectra with suggested assignments was given by Venkateswarlu,[7] although many of these are now considered unacceptable.[5] Venkateswarlu [7] has also extended the room-temperature absorption spectrum of iodine to 1200 Å in the vacuum-u.v. region, where most transitions are to Rydberg excited states. In view of the great complexity of this spectral region, caution is perhaps also necessary in discussing the Rydberg assignments too finely. As mentioned earlier, Mulliken [5] also has extended recently his useful theoretical procedures with particular regard to I_2, and has discussed in detail many of the existing ambiguities of the absorption and emission spectra in the u.v. region. Clearly though, much more experimental effort is needed in this area before more detailed comparison of the electronic states with Mulliken's estimates can be made.

In recent years, renewed interest in some of the long-standing problems associated with the lower electronic states has been facilitated by the develop-

[7] P. Venkateswarlu, *Canad. J. Phys.*, 1970, **48,** 1055.

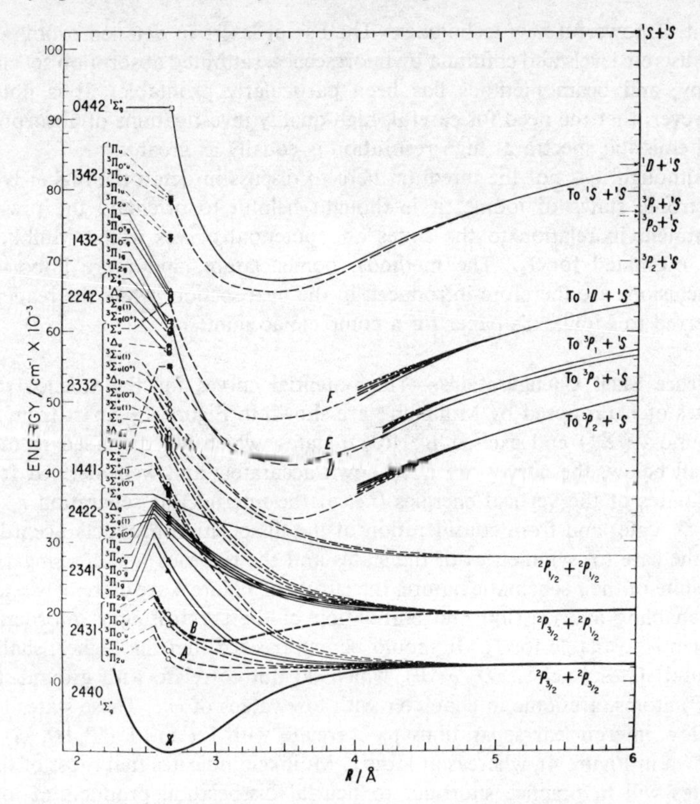

Figure 4 *Estimated potential curves for valence-shell states of iodine. Each curve is drawn to pass through one point at* 2.666 Å (r_e *of state X*) *obtained from an estimated vertical energy at that r value. The curves for states of even* (*g*) *parity are shown by full lines, those for odd* (*u*) *parity by dashed lines. The asymptote of each of the lower set of curves (those which dissociate to two ground-configuration atoms) is reliably known theoretically, except that in few cases, it is not certain which of two curves of the same case-c type and nearly the same estimated vertical energy is really lower. Except for the X and B states (shown by heavy curves), and to some extent the A state, the detailed forms of the curves are not known, but they have been drawn in accordance with qualitative considerations on bonding or antibonding characteristics of the MOs involved. The upper set of states are of ion-pair character, and tend to dissociate to various states of* I^+ *plus* $5p^6$, ^1S *of* I^- *whose energies are indicated at the right of the figure, but actually many of them must shortcut to dissociate to yield one excited-configuration and one ground-configuration atom. For most of these states, only the vertical energy at* r_e *of state X and estimated forms of the curves at large r (based on a* $1/r$ *ion-pair potential modified by some allowance for polarization) are shown. Complete curves are sketched in for a few cases; for the E and F states, the energies of the minima of their potential curves, and some indications as to their* r_e *values, are known. The position for the D curve is uncertain, and its approximate location is indicated*

ment of powerful new techniques. The use of lasers in detailed probing of the discrete levels and continua by fluorescence, saturated absorption spectroscopy, and beam methods has been particularly profitable. It is noted, however, that the need for careful, high-quality investigations of absorption and emission spectra at high resolution is equally as great.

Although it is not the intention here to discuss in detail the higher-lying electronic states of iodine, it is thought helpful to orientate the present treatment in relation to the states and potential curves which Mulliken [5] has estimated for I_2. The methods, nomenclature, and more important conclusions are therefore introduced in the next section, while the reader is referred to Mulliken's paper for a complete account.

Valence Shell Configurations.—The potential curves for the valence shell states of I_2 proposed by Mulliken [5] are shown in Figure 4. Apart from the ground $X(^1\Sigma_g^+)$ and excited $B[^3\Pi(0_u^+)]$ states, which are discussed in more detail below, the curves are not known accurately and were derived from estimates of the vertical energies (*i.e.* at the internuclear separation r_e'' of the X state) and from consideration of the dissociation products according to the case (c) symmetries of the states and the rigorous non-crossing rule. In spite of their schematic nature, the curves of Figure 4 are extremely useful in enabling a discussion and correlation of the spectroscopic information which is available for I_2. It should be emphasized that the higher, shallow bound states (such as D, E, F), which do not correlate with ground-state $I(^2P)$ atoms, are ionic in character with low values of ω_e. These states tend at low internuclear separations to correlate with $I^-(^1S)+I^+(^3P, {}^1D, {}^1S)$ (as shown in Figure 4), whereas at large r, Mulliken indicates that most of these states will in practice short-cut to neutral dissociation products at lower energies than the indicated ionic products.

The vertical energies and approximate forms of the curves are based on a consideration of the MO configurations of the outermost ten electrons, which are distributed over the $\sigma_g\ \pi_u\ \pi_g\ \sigma_u$ set from simple LCAO of I $5p+$ I $5p$. The relative energies of the MOs at $r_e''(X)$ are shown in Figure 4, and were estimated by Mulliken from a combination of the available spectroscopic data on I_2 and I_2^+.

Spin–orbit [or case (c)] splittings are based on experimental values, theory, and analogies with those of O_2. The components shown in Figure 4 are labelled according to both case (a) and case (c) notations. Where states of the same symmetry species are estimated to be at closely similar energies, the true labellings should possibly be reversed (see p. 180).

The $X^1\Sigma_g^+$ Ground State of I_2.—*Rotational and Vibrational Constants; RKR Potential Curve.* Vibrational and especially rotational constants for the ground states of diatomic molecules are usually confined to the few low-lying vibrational levels from which absorption spectra have been observed. For iodine, however, the study of several resonance fluorescence series has resulted in accurate constants for levels almost to the dissociation limit. Rank and Baldwin [8] made interferometric measurements (to 0.001 cm^{-1}) on a single doublet (*PR*) series (Wood's series) excited by the λ 5461 Å Hg green line, and observed levels of the ground state up to $v'' = 22$. Verma [9] later observed several series of u.v. resonance doublets, all excited by the λ 1830.4 Å iodine atomic resonance line by transitions from X, $v'' = 0[P23), R(24), P(47), R(48), R(86)]$ to three vibration-rotation levels ($v' = n, n+1, n+4$; n unknown) of an excited 0_u^+ state. These five series are capable of analysis to give vibrational and rotational constants up to $v'' = 84$. A short, sixth series, of which only the first three high frequency (low v'') members showed resolved doublets, and which occurred at frequencies corresponding to those expected for transitions to levels close to the dissociation limit of $X^1\Sigma_g^+$ to two $^2P_{3/2}$ atoms, was interpreted by Verma as due to transitions to $X^1\Sigma_g^+$; $98 \leqslant v'' \leqslant 115$. Rank and Rao,[10] in the first attempt to combine the data of one of Verma's [9] series ($J_r = 87$; $v'' \leqslant 84$) with the interferometric [8] ($v'' \leqslant 22$) and other [11, 12] ($v'' \leqslant 39$) measurements from the Wood's series, obtained rotational constants for $v'' \leqslant 84$, and vibrational constants which reproduced the fluorescence spectra up to $v'' = 40$. Until recently, these were the accepted constants for $X^1\Sigma_g^+$, even though the available data were only partially utilized. The inadequacy of this treatment of unusually high quality data was noted by Le Roy,[13] who reanalysed them to obtain significantly improved vibrational terms and rotational constants up to $v'' = 82$. For reasons outlined below, Verma's sixth series was not included in Le Roy's analysis, which included the following improvements: (i) data on all blended lines were ignored, (ii) proper account was taken of the rapid increase of D_v with increasing v'' in a reiterative, convergent procedure using trial sets of D_v values, (iii) the stringent condition of fitting the actual frequencies, rather than the doublet separations (which alone are incapable of yielding stretching constants) was adopted, (iv) the whole set of data was exploited, rather than the single u.v. series used by Rank and Rao,[10] (v) allowance was made for the observation that the iodine atomic resonance line and I_2 emission lines were broad, with non-coincident peaks. The final set of constants obtained by Le Roy [13]

$$B_v = 3.7395 \times 10^{-2} - 1.2435 \times 10^{-4}(v+\tfrac{1}{2}) + 4.498 \times 10^{-7}(v+\tfrac{1}{2})^2 - 1.482 \times 10^{-8}(v+\tfrac{1}{2})^3 - 3.64 \times 10^{-11}(v+\tfrac{1}{2})^4$$

$$D_v = 4.54 \times 10^{-9} + 1.7 \times 10^{-11}(v+\tfrac{1}{2}) + 7 \times 10^{-12}(v+\tfrac{1}{2})^2$$

$$G(v) = 214.5481(v+\tfrac{1}{2}) - 0.616\,259(v+\tfrac{1}{2})^2 + 7.507 \times 10^{-5}(v+\tfrac{1}{2})^3 - 1.263\,643 \times 10^{-4}(v+\tfrac{1}{2})^4 + 6.198\,129 \times 10^{-6}(v+\tfrac{1}{2})^5$$

[8] D. H. Rank and W. M. Baldwin, *J. Chem. Phys.*, 1951, **19**, 1210.
[9] R. D. Verma, *J. Chem. Phys.*, 1960, **32**, 738.
[10] D. H. Rank and B. S. Rao, *J. Mol. Spectroscopy*, 1964, **13**, 34.
[11] D. H. Rank, *J. Opt. Soc. Amer.*, 1946, **36**, 239.
[12] R. W. Wood and M. Kimura, *Phil. Mag.*, 1918, **35**, 252.
[13] R. J. Le Roy, *J. Chem. Phys.*, 1970, **52**, 2683.

$$- 2.025\,5975 \times 10^{-7} (v+\tfrac{1}{2})^6 + 3.966\,2824 \times 10^{-9} (v+\tfrac{1}{2})^7$$
$$- 4.634\,6554 \times 10^{-11} (v+\tfrac{1}{2})^8 + 2.933\,0755 \times 10^{-13} (v+\tfrac{1}{2})^9$$
$$- 7.610\,00 \times 10^{-16} (v+\tfrac{1}{2})^{10}$$

revealed the limited accuracy of the vibrational and rotational terms which Verma obtained from the data, *e.g.* differences up to 6.6 cm^{-1} are observed between Le Roy's $G(v)$ and Verma's vibrational terms. Of course, the coefficients of the higher power terms in these expressions, except as an empirical representation of experimental data, have little physical significance. Notwithstanding the sophistication of Le Roy's analysis, however, it appears that small errors must still exist, especially for B_v. The second derivative, d^2U/dr^2, of the repulsive r_{min} limb of Le Roy's RKR potential curve becomes negative above $v'' = 56$. This limb above $v'' = 56$ was therefore estimated by a short extrapolation, which occasioned shifts of the turning points (separations between turning points are known to be relatively more accurate) in the range $56 \leqslant v'' \leqslant 82$ by up to 0.0094 Å at $v'' = 82$. This result, however, seems to be within the limits of error for the derived rotational constants which the total data can afford. It seems unlikely that a curve of greater accuracy (except that [14] derived repeating Le Roy's procedure, but using more recent values for the physical constants,[15] and which also differed in the use of atomic masses rather than nuclear masses) can be obtained until more measurements of greater precision become available or additional resonance series are investigated.

Dissociation Energy of $X^1\Sigma_g^+$ and Le Roy's 0_g^+ State. As mentioned above, the short u.v. resonance doublet series was attributed by Verma [9] to transitions to the uppermost, $98 \leqslant v'' \leqslant 115$, levels of $^1\Sigma_g^+$, and has for several years been the basis for the accepted dissociation energy, $D_0^{0''}$ of $X^1\Sigma_g^+$ (Verma [9] $12\,452.5 \pm 1.5$; Le Roy [16] $12\,452.4 \pm 1.8$ cm^{-1}). This interpretation, however, conflicts with estimates (Brown [17] $12\,439 \pm 8$; Le Roy [16] $12\,440 \pm 1.1$ cm^{-1}) from short extrapolation to the convergence limit of the $B(0_u^+) \leftarrow X^1\Sigma_g^+$ band-head measurements [17] ($48 \leqslant v' \leqslant 72$). In an attempt to reconcile this discrepancy, Le Roy [16] has recently proposed that the short doublet series should be reassigned as transitions to a new shallow-bound state (0_g^+), correlating with two $^2P_{3/2}$ atoms *via* a potential barrier of 13.1 ± 1.4 cm^{-1}. This re-interpretation implies the re-instatement of the dissociation energy from Brown's head measurements. The evidence for the re-assignment [(i) the $\Delta G(v+\frac{1}{2})$ against v plot (Birge–Sponer) does not show the curvature for the highest levels of $X^1\Sigma_g^+$ predicted from long-range attractive C_6/r^6, C_8/r^8... terms, and (ii) the intensities of the members of the resonance series were similar and there would be no reason to expect the Franck–Condon overlap integrals to diminish rapidly below $v'' = 98$] strongly supports LeRoy's assignment to 0_g^+. From a consideration of the selection

[14] J. A. Coxon, *J. Quant. Spectroscopy Radiative Transfer*, 1971, **11**, 443.
[15] B. N. Taylor, W. H. Parker, and D. N. Langenberg, *Rev. Mod. Phys.*, 1969, **41**, 375.
[16] R. J. Le Roy, *J. Chem. Phys.*, 1970, **52**, 2678.
[17] W. G. Brown, *Phys. Rev.*, 1931, **38**, 709.

rules ($g \leftrightarrow u$, $\Delta\Omega = 0$, $+ \nleftrightarrow -$), the new state was assigned as 2341 $^3\Pi(0_g^+)$, which according to Mulliken's predictions[5] (Figure 4) does not correlate with $^2P_{3/2}$ atoms, but with $^2P_{3/2}+^2P_{1/2}$. However, the predicted vertical energy of 2422 $^3\Sigma_g^-(0_g^+)$ lies close to that of 2341 $^3\Pi(0_g^+)$, and it is possible that these two states should be reversed from their positions shown in Figure 4. As Mulliken mentions, however, at the large r values where the new state is observed, the wavefunction would in any case be a case (c) mixture of both states (and with other 0_g^+ states), so that a formal assignment is not really justified. More important than this discussion is the problem which Stwalley[18] has emphasized concerning Le Roy's interpretation, which for the reason outlined above, implies a barrier of around 13 cm^{-1} for dissociation of 0_g^+ to $^2P_{3/2}$ atoms. Stwalley obtains an approximate potential curve for 0_g^+ with minimum near 6 Å, and finds that the potential maximum would occur at $r_{max} \geqslant 10$ Å. The (accurate) calculated C_5 repulsive term for 0_g^+ states correlating with $^2P_{3/2}$ atoms is certainly less, and probably much less, than 2.8 cm^{-1} at $r = 10$ Å. This estimate seems to be based on the value of 2.26×10^{-5} cm^{-1} Å^5 for C_5 given[5] in Mulliken's Table II for 0_g^+ correlating with two $^2P_{3/2}$ atoms. This was obtained using the non-crossing rule and assumed large mixing of all 0_g^+ states at large r. A barrier of 13 cm^{-1} seems untenable on this evidence, and Stwalley concludes that Le Roy's assignment is open to question. Until recently, it has been possible to attempt to reconcile this paradox by postulating systematic error in the $B \leftarrow X$ band-head measurements of Brown[17] at high v'. Such systematic error has been found between band-head and band-origin measurements for the corresponding system of Br_2 near the convergence limit (see below). On this basis, it would be convenient to re-instate the $D_0^{0''}$ for I_2 based on Verma's resonance fluorescence data.[9] However, following a new high-resolution study of the $B \leftarrow X$ band system of I_2 up to $v' = 77$, Yee[19] has established beyond doubt the reliability of Brown's head measurements right up to $v' = 72$. The validity of Le Roy's analysis[16] of the long-range behaviour for $I_2B(0_u^+)$ is thus confirmed, and his value of 12440 ± 1 cm^{-1} for $D_0^{0''}$ ($^{127}I_2$) is likely to be near the true value. A more precise value will of course eventually be obtained from a long-range analysis of Yee's data.[19] The magnitude of possible C_3 spin–spin dipole resonance exchange interaction at very large r, as mentioned by Steinfeld, Campbell and Weiss,[22] will require examination.

In conclusion, therefore, the dissociation energy of iodine is reliably established. Because of the difficulty of accounting for a barrier height of $\sim$13 cm^{-1} for a 0_g^+ state correlating with two I $^2P_{3/2}$ atoms, a re-examination of the resonance fluorescence spectrum of iodine would be highly desirable.

[18] W. C. Stwalley, *J. Chem. Phys.*, 1971, **56**, 680.
[19] K. K. Yee, to be published.
[20] J. Tellinghuisen, *J. Chem. Phys.*, to be published.
[21] R. J. Le Roy and F. E. Cummings, *J. Chem. Phys.*, personal communication.
[22] J. I. Steinfeld, J. D. Campbell, and N. A. Weiss, *J. Mol. Spectroscopy*, 1969, **29**, 204.

Discrete levels of I_2, 2431 $A^3\Pi(1_u)$ and $B^3\Pi(0_u^+)$.—*The $A^3\Pi(1_u)$ State*. From the one study by Brown [23] in 1931 of the $A \leftarrow X$ bands in absorption, the $A^3\Pi(1_u)$ state of I_2 is known to correlate with two $^2P_{3/2}$ atoms in accordance with theoretical predictions (Figure 4). The vibrational numbering and hence the vibrational constants for this state have since remained uncertain, although they are commonly assumed to be correct. Tellinghuisen has recently put forward evidence to suggest that the A state may in fact be much deeper than that obtained by Brown's [23] identification of the lowest observed level with $v' = 0$. New band-head measurements for both $^{127}I_2$ and $^{129}I_2$, together with a rotational analysis of some of the $A \leftarrow X$ bands, would be valuable. An attempt to obtain a vibrational analysis for the banded $A^3\Pi(1_u) \rightarrow X^1\Sigma_g^+$ emission spectrum [24] from $A(1_u)$ populated in $I^2 P_{3/2} + I^2 P_{3/2}$ atom recombination might also serve to provide more information on this state.

The $B^3\Pi(0_u^+)$ State. Although the $B^3\Pi(0_u^+) \leftarrow X^1\Sigma_g^+$ band system of I_2 is one of the most studied molecular spectra, and although a detailed knowledge of the levels of this state are important in many applications, the available data are still very incomplete. They consist of band-head measurements by Mecke [25] and by Brown [17] up to $v' = 72$, rotational analysis [26] of four bands with $v' = 3$ and 4, a high-resolution analysis [27] of the 30–0 band, some unpublished data [28] on $v' = 14$—19, an accurate rotationless term fixed [29] for $v' = 43$ by Ar^+ laser excitation at 5145 Å, and two recent analyses [22, 30] by Steinfeld and co-workers of bands recorded at high resolution with $v' =$ 3—13, 25, 29, and 43. Only in 1965 was the correct vibrational numbering established by measurements [31] on $^{129}I_2$ ($B \leftarrow X$). The previous numbering, in error by 1 unit, led to considerable difficulties in Zare's attempt [32] to compare experimental and RKR-based predicted intensities for several visible fluorescence series. Following Le Roy's [16] report that the band at 19 432.89 cm^{-1}, identified by Steinfeld, Campbell, and Weiss [22] as 49–1, is probably better identified with 57–2 (thus invalidating the latter's rotational constants), Tellinghuisen [20] has recently pointed out some serious inconsistencies between the results reported in the two papers [22, 30] by Steinfeld *et al.* Yee and Miller [33] also mention that their measurements and assignments for the 43–0 band do not agree with those of ref. 22. It appears (with the

[23] W. G. Brown, *Phys. Rev.*, 1931, **38,** 1187.
[24] R. J. Browne and E. A. Ogryzlo, *J. Chem. Phys.*, 1970, **52,** 5774.
[25] R. Mecke, *Ann. Phys.*, 1923, **71,** 104.
[26] F. W. Loomis, *Phys. Rev.*, 1927, **29,** 112.
[27] A. W. Richardson and R. A. Powell, *J. Mol. Spectroscopy*, 1967, **24,** 379.
[28] R. L. Brown and D. Griffiths, unpublished data, 1961.
[29] E. Menke, *Z. Naturforsch.*, 1970, **25a,** 442; Th. Halldorsson and E. Menke, *ibid.*, p. 1356.
[30] J. I. Steinfeld, R. N. Zare, L. Jones, M. Lesk, and W. Klemperer, *J. Chem. Phys.*, 1965, **42,** 25.
[31] R. L. Brown and T. C. James, *J. Chem. Phys.*, 1965, **42,** 33.
[32] R. N. Zare, *J. Chem. Phys.*, 1964, **40,** 1934.
[33] K. K. Yee and G. J. Miller, *J.C.S. Chem. Comm.*, 1972, 1054.

concurrence of Steinfeld) that the assignments and constants of the later (1969) work [22] should be ignored, and that the earlier (1965) data [30] should be reassessed. On this basis, Tellinghuisen [20] has obtained new sets of vibrational ($v' \leqslant 67$) and rotational ($v' \leqslant 60$) constants,

$$\begin{aligned} G(v) = {} & 124.9738\,(v+\tfrac{1}{2}) - 0.618\,385\,(v+\tfrac{1}{2})^2 - 1.624\,880\times10^{-2}\,(v+\tfrac{1}{2})^3 \\ & + 7.027\,188\times10^{-4}\,(v+\tfrac{1}{2})^4 - 2.246\,070\times10^{-5}\,(v+\tfrac{1}{2})^5 \\ & + 3.971\,780\times10^{-7}\,(v+\tfrac{1}{2})^6 - 3.459\,940\times10^{-9}\,(v+\tfrac{1}{2})^7 \\ & + 1.187\,866\times10^{-11}\,(v+\tfrac{1}{2})^8;\ v' < 67 \end{aligned}$$

$$\begin{aligned} B_v = {} & 0.028\,921 - 1.4075\times10^{-4}\,(v+\tfrac{1}{2}) - 2.2212\times10^{-6}\,(v+\tfrac{1}{2})^2 \\ & + 6.5440\times10^{-8}\,(v+\tfrac{1}{2})^3 - 2.4464\times10^{-9}\,(v+\tfrac{1}{2})^4 \\ & + 2.2770\times10^{-11}\,(v+\tfrac{1}{2})^5;\ v' \leqslant 60 \end{aligned}$$

which are the best available at the present time. A new analysis [19] of the absorption bands $B \leftarrow X$ at high resolution with $16 \leqslant v' \leqslant 77$ will eventually be available, however. The above constants reproduce known bands of the $B \leftrightarrow X$ system to 0.07 cm^{-1} for $v' \leqslant 43$, and to 0.5 cm^{-1} for $v' > 43$. Since the RKR curve based on these expressions does not show the expected limiting r^{-5} dependence, Tellinghuisen [20] has raised questions about the validity of the theoretical procedures developed by Le Roy and Bernstein [34] for accurately fixing dissociation limits according to long-range theory. Unfortunately, it appears [21] that when the long-range potential is approximately of the form $C_5/r^5 + C_6/r^6$ (+ negligible C_8, *etc.*), a fortuitous cancellation occurs in the simple theory, causing the vibrational level distribution to appear to depend on C_5 only.

Continuous Absorption, 2431 $\leftarrow X^1\Sigma_g^+$.—Although the visible–i.r. continuum absorption by I_2 has been much investigated,[35–38] many conflicts about its interpretation exist.[20] Of the 23 case (c) components of the 2431 configuration (Figure 2), transitions to only three of them, $A^3\Pi(1_u)$ and $^1\Pi(1_u)$ (correlating with two $^2P_{3/2}$ atoms) and $B^3\Pi(0_u^+)$ (correlating with $^2P_{3/2}+{}^2P_{1/2}$) are allowed. $^1\Pi(1_u)$ is entirely repulsive. It has traditionally been supposed [39–41] that the short wavelength ($\lambda < \sim 5000$ Å) continuum was due entirely to the repulsive limb of the B state. Although studies in the banded region [42–47]

[34] R. J. Le Roy and R. B. Bernstein, *J. Chem. Phys.*, 1970, **52**, 3869; *J. Mol. Spectroscopy*, 1971, **37**, 109; R. J. Le Roy, this Report, Chapter 3.
[35] E. Rabinowitch and W. C. Wood, *Trans. Faraday Soc.*, 1936, **32**, 540.
[36] P. Sulzer and K. Wieland, *Helv. Phys. Acta*, 1952, **25**, 653.
[37] C. A. Goy and H. O. Pritchard, *J. Mol. Spectroscopy*, 1964, **12**, 38.
[38] E. A. Ogryzlo and G. E. Thomas, *J. Mol. Spectroscopy*, 1965, **17**, 198.
[39] J. Franck, *Trans. Faraday Soc.*, 1925, **21**, 506.
[40] L. Mathieson and A. L. G. Rees, *J. Chem. Phys.*, 1956, **25**, 753.
[41] R. S. Mulliken, *J. Chem. Phys.*, 1936, **4**, 620.
[42] E. Wasserman, W. E. Falconer, and W. A. Yager, *J. Chem. Phys.*, 1968, **49**, 1971.
[43] G. E. Busch, R. T. Mahoney, R. I. Morse, and K. R. Wilson, *J. Chem. Phys.*, 1969, **51**, 837.
[44] A. Chutjian and T. C. James, *J. Chem. Phys.*, 1969, **51**, 1242.
[45] A. Chutjian, *J. Chem. Phys.*, 1969, **51**, 5414.

have more recently indicated a larger oscillator strength for ${}^1\Pi(1_u) \leftarrow X^1\Sigma_g^+$ than previously supposed, Kroll's work[48] in 1970, which involved an apparently good matching of the calculated $B \leftarrow X$ continuum spectrum (assuming a constant electronic transition moment) with an experimental curve at 370 K, seemed to confirm the traditional view. The accuracy of the RKR curve used by Kroll, however, is known[20] to be suspect, and the method by which the repulsive r_{min} limb was extrapolated to the continuum region is unknown. The situation was resolved finally only in 1971 by Oldman, Sander, and Wilson,[49] using the technique of photodissociation translational spectroscopy.[50] The experiment consists of crossing a molecular beam with a pulsed monochromatic laser beam. The angular distribution and delay time for appearance of the products (kinetic energy) determined mass spectrometrically serve to identify the final dissociation products and Ω-value of the state(s) responsible for dissociation. Using three laser frequencies well into the region of pure continuum absorption, Oldman *et al.* found for the ratio $P({}^1\Pi/B^3\Pi)$ of absorption probabilities from $X^1\Sigma_g^+$ to ${}^1\Pi$ and $B^3\Pi$, 1.2 ± 0.2 at $\nu = 20\,850$ cm^{-1}, 1.3 ± 0.1 at $\nu = 21\,510$ cm^{-1}, and 2.3 ± 0.5 at $\nu = 22\,230$ cm^{-1}. These results show beyond all doubt that the ${}^1\Pi(1_u)$ state makes in fact a larger contribution to the pure continuum absorption than does $B^3\Pi(0_u^+)$ (see Figure 5), and explain[51] why it was not possible[52] to produce an I_2 photodissociation laser operating on the λ 1.315 μm $I({}^2P_{1/2} \rightarrow {}^2P_{3/2})$ transition. At λ 5310 Å, in the banded $B \leftarrow X$ absorption region, it had not been possible[46c] to resolve the photodissociation spectrum into contributions from ${}^1\Pi(1_u)$ and from predissociation from the $B^3\Pi(0_u^+)$ state. At the longest wavelength laser line used,[46b] however, I atoms were detected, corresponding to much weaker absorption to the expected position of the $A^3\Pi(1_u)$ repulsive limb.

Very recently, Tellinghuisen[20] has made use of the data[49] given above in a new detailed study of absorption by I_2 from λ 4000 to 8000 Å. The experiments were of two types: (i) broad-band ($\sim$26 Å) measurements of extinction coefficients (ε) in the whole spectral region with theory developed to relate (a) apparent ε values for many-line absorption to the electronic transition moment (R_e) and pure Doppler-broadened linewidths ($\Delta\nu_0$), and (b) true ε values for pure continuum absorption to R_e and repulsive potential limbs based on the δ-approximation; (ii) measurements of ε values using the emis-

[46] G. E. Busch, R. T. Mahoney, and K. R. Wilson, *IEEE J. Quantum Electronics*, 1970, QE-6, 171; R. T. Mahoney and K. R. Wilson, *Bull. Amer. Phys. Soc.*, 1969, **14**, 849; K. R. Wilson, in 'Excited State Chemistry', ed. J. N. Pitts, Gordon and Breach, New York, 1970.

[47] C. Jonah, P. Chandra, and R. Bersohn, *J. Chem. Phys.*, 1971, **55**, 1903; J. Solomon, C. Jonah, P. Chandra, and R. Bersohn, *J. Chem. Phys.*, 1971, **55**, 1908.

[48] M. Kroll, *J. Mol. Spectroscopy*, 1970, **36**, 44.

[49] R. J. Oldham, R. K. Sander, and K. R. Wilson, *J. Chem. Phys.*, 1971, **54**, 4127.

[50] G. E. Busch, J. F. Cornelius, R. T. Mahoney, R. I. Morse, D. N. Schlosser, and K. R. Wilson, *Rev. Sci. Instr.*, 1970, **41**, 1066.

[51] G. Hancock and K. R. Wilson, Proceedings of the Esfahan Conference, 1971, Esfahan, Iran; ed. M. Feld, N. Curnit, and A. Javan, Wiley, New York, 1972.

[52] J. V. V. Kasper, J. H. Parker, and G. C. Pimentel, *J. Chem. Phys.*, 1965, **43**, 1827.

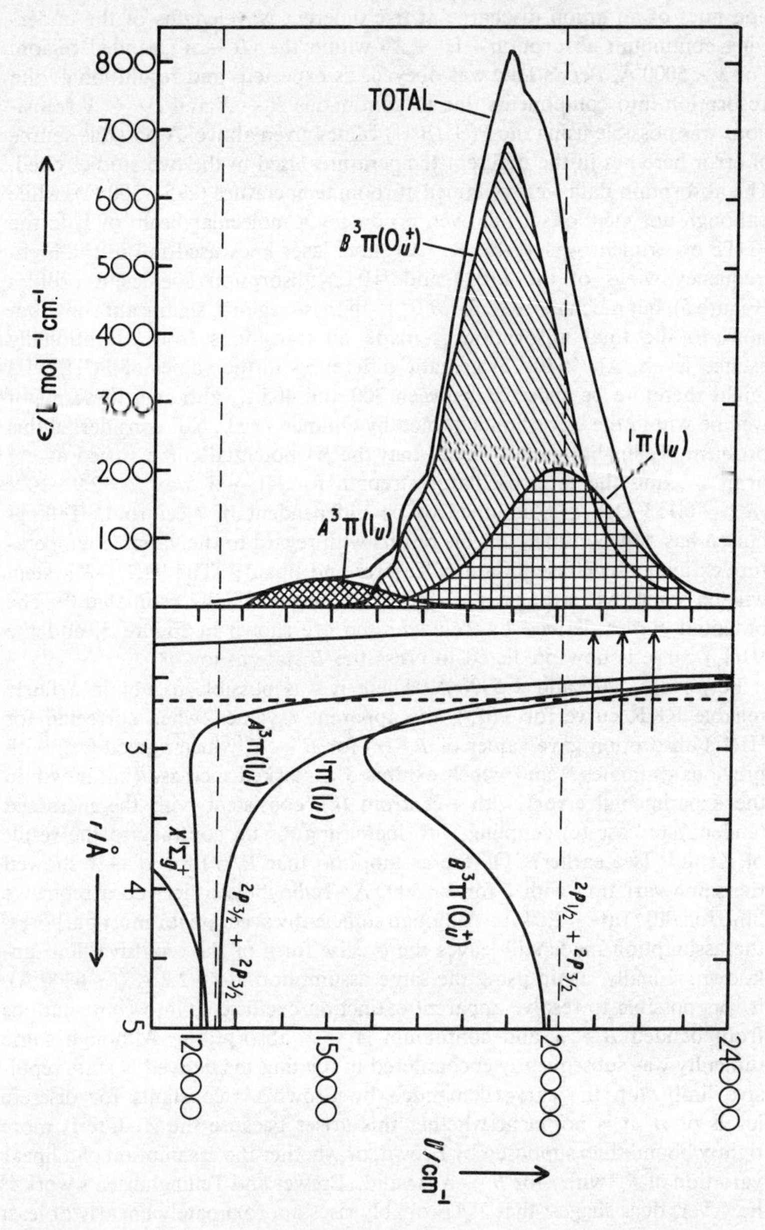

Figure 5 *Potential curves and absorption spectrum of* I_2 *after Tellinghuisen.*[20] *Horizontal arrows are PDTS laser frequencies. See text for full details*

sion lines of an argon discharge at five different wavelengths of the underlying continuum absorption ($^1\Pi \leftarrow X$) within the ($B \leftarrow X$) banded region. For λ < 5000 Å, Beer's Law was obeyed, as expected, and resolution of the absorption into components due to continuous $B \leftarrow X$ and $^1\Pi \leftarrow X$ transitions was possible using the $P(^1\Pi/B^3\Pi)$ values given above. A possible source of error here lies in the different temperatures used in the two studies cited. The absorption data were obtained at room temperature (295—300 K), while although not mentioned, the oven producing a molecular beam of I_2 in the PDTS experiment was at 405 K. All three laser lines used fall in the high-frequency wings of the $B(0_u^+)$ and $^1\Pi(1_u)$ absorption coefficient profiles (Figure 5), but much more so for $B(0_u^+)$. In these regions, significant contributions to the total absorption is made by transitions from vibrationally excited levels, $X(v'' > 0)$. Systematic differences in the values of $P(^1\Pi/B^3\Pi)$ might therefore be expected between 300 and 405 K, although these might well be within the error limits quoted by Oldman *et al.* Not considering this problem, Tellinghuisen concluded that the $^1\Pi$ potential curve varied as r^{-9} near r_e'', and the total oscillator strength for $^1\Pi \leftarrow X$ was $f = 2.9 \times 10^{-3}$ (R_e^2 = 0.153 Debye ^{2}), assumed to be independent of r-centroid. Tellinghuisen has now considered[52a] his results with regard to the variable-temperature extinction-coefficient data of Tamres and Bhat.[52b] The $^1\Pi(1_u)$–X system will be somewhat stronger (< 10%) than that originally estimated.[20] The potential curves derived by Tellinghuisen are shown in Figure 5, and the $^1\Pi(1_u)$ curve is now predicted to cross the B state at low v'.

Between λ 5200 and λ 5900 Å, where it was possible to obtain a fairly reliable RKR curve for $B(0_u^+)$, the apparent ε values, when corrected for $^1\Pi(1_u)$ absorption gave values of R_e^2 ($\bar{r}$) for $B \leftarrow X$, which agreed well with previous estimates,[45] and which exhibited a marked increase (linear within the experimental error) with r-centroid ($\bar{r}$), consistent with the increased tendency to case (c) coupling with increasing $\bar{r}$, and contrary to the result of Kroll[48] (see earlier). On the assumption that $R_e^2(\bar{r})$ for $B \leftarrow X$ showed the same variation with $\bar{r}$ for λ < 5000 Å, Tellinghuisen derived a repulsive limb for $B(0_u^+)$ at r < 2.64 Å. Although sufficiently accurate for most purposes, the assumption made still leaves the precise form of the repulsive limb unknown. Finally, again using the same assumption at $\bar{r}$ > 2.8 Å (λ > 6200 Å), it was possible to resolve apparent extinction coefficients into contributions from banded $B \leftarrow X$ and continuum $A \leftarrow X$ absorption. Although some difficulty was subsequently encountered in relating the derived A state repulsive limb with the curve demanded by Brown's[23] constants for discrete levels of A, it is not clear whether this arises because the A state is more tightly bound than supposed by Brown, or whether the assumption of a linear variation of R_e^2 with r for $B \leftarrow X$ is valid. Brewer and Tellinghuisen's work,[53] however, does suggest that R_e^2 probably rises approximately linearly at least

[52a] J. Tellinghuisen, personal communication.
[52b] M. Tamres and S. N. Bhat, *J. Phys. Chem.*, 1971, **75**, 1057.
[53] L. Brewer and J. Tellinghuisen, *J. Chem. Phys.*, 1972, **56**, 3929.

till ~3 Å, before falling to zero [54] at larger r (corresponding at the limit to the forbidden $5p_z \leftarrow 5p_x$ atomic transition). The oscillator strength for $A \leftarrow X$ is [20] $f = 6.2 \times 10^{-4}$ ($R_e^2 = 0.044$ Debye [2]), about twice that of previous estimates [5, 49] and is likely to be within the quoted 15% error limit.

Radiative Lifetime, Predissociation, and Magnetic Quenching of I_2 $B^3\Pi(0_u^+)$.— *Lifetime and Predissociation.* The lifetime of I_2 $B^3\Pi(0_u^+)$ has been intensively studied during the last few years. From intensities of $B \leftarrow X$ broad-band absorption,[35, 36] as well as from the intensities of a few individual lines,[55, 56] i' is possible to obtain [32, 57] estimates of around $(2 \pm 1) \times 10^{-6}$ s for the purely tadiative lifetime of $B^3\Pi(0_u^+)$ (see Figure 6 of ref. 57). The first direct lifetime measurements [57] by the phase-shift method with population of $B^3\Pi$ by both narrow continuum and sharp-line excitation, showed large variations with vr ot $\tau_{v'}$, the total lifetimes for individual v' levels. Since there existed at that time no firm evidence for a fast predissociation from the B state to a repulsive state correlating with $^2P_{3/2}$ atoms, the interpretation of these results was uncertain. An apparent demonstration [42] of direct predissociation by observation of I^2 $P_{3/2}$ atom formation in the banded region was ambiguous, however, since $^1\Pi(1_u) \leftarrow X$ could also be responsible (and is now known [53] to contribute). Better estimates of the Einstein coefficients ($A_{v'}$) for purely radiative transition to $X^1\Sigma_g^+$ were obtained in 1969 by Chutjian and James,[44] using the method of equivalent widths on rotational lines to the $v' = 12$, 13, 14, 16, 25, and 29 levels. Values of $1/A_{v'}$ were found to vary from 1.8 to 2.2 μs in this range, a slower variation than that more recently determined [53] by Brewer and Tellinghuisen. It was concluded, therefore, that direct predissociation of I_2 $B(0_u^+)$ must occur to explain the much shorter observed $\tau_{v'}$ lifetimes at $v' \sim 5$ and 25. Using these predissociation rates, Chutjian [45] attempted to locate a part of the potential curve of $^1\Pi(1_u)$, which is the only state capable of predissociating $B(0_u^+)$. Since 1969, many more laser-excited fluorescence and phase-shift measurements of $\tau_{v'}$ values have been obtained (see ref. 58 for a summary), all continuing to show large, real variations. The results of Sakurai, Capelle, and Broida,[59] however, cannot be interpreted too finely,[58] on account of their several Å wide dye-laser-excitation and the realization [45, 58] of a pronounced variation of predissociation rate with rotational quantum number J.

Tellinghuisen has recently [58] considered the predicted rate of spontaneous predissociation from $B(0_u^+)$ to $^1\Pi(1_u)$ using a similar approach to that outlined by Chutjian.[45] For a direct, heterogeneous ($|\Delta\Omega| = 1$) predissociation,

[54] S. R. LaPaglia, *J. Chem. Phys.*, 1968, **48**, 537.
[55] R. L. Brown and W. Klemperer, *J. Chem. Phys.*, 1964, **41**, 3072.
[56] F. E. Stafford, University of California, Lawrence Radiation Laboratory, Report UCRL–8854 (1960).
[57] A. Chutjian, J. K. Link, and L. Brewer, *J. Chem. Phys.*, 1967, **46**, 2666.
[58] J. Tellinghuisen, *J. Chem. Phys.*, 1972, **57**, 2397.
[59] K. Sakurai, G. Capelle, and H. P. Broida, *J. Chem. Phys.*, 1971, **54**, 1220.

the first-order rate constant, k_{nr}, for non-radiative transition, $B \rightarrow {}^1\Pi$, is given approximately by

$$k_{nr} = \kappa' q_{cb} J(J+1) \tag{3}$$

with $\kappa' = (h^3/16\pi^2\mu^2)W_e^2(r)$, where $q_{cb} = |<\psi_c|r^{-2}|\psi_b>|^2$ is the square of the matrix element of r^{-2} between the bound ψ_b and continuum ψ_c radial functions of $B(0_u^+)$ and ${}^1\Pi_u$, and $W_e(r)$, the electronic operator mixing the two states, is assumed to be effectively constant in the range of r studied. Thus the rate of predissociation from a rovibrational level $B_{v',J'}$ is predicted to depend not only on the precise form of overlap between the radial functions of the two states [which is negligible for predissociation to $A^3\Pi(1_u)$], but also on $J(J+1)$. The neglect of the J-dependence seems to be responsible for so many apparent discrepancies in $\tau_{v'}$ measurements in the literature. Using equation (3), the total decay rate from $B_{v',J}$ can be written as

$$\tau_{v',J'}^{-1} = A_{v'} + \kappa_{v'} J(J+1) \tag{4}$$

where $\kappa_{v'} = \kappa' q_{cb}$. By numerical integration to give q_{cb} values from the ${}^1\Pi(1_u)$ repulsive curve derived from absorption measurements, Tellinghuisen was able to compare experimental and calculated values of $\kappa_{v'}$. The overall agreement is shown in Figure 6(a), which incorporates all the most recent lifetime data. The small systematic deviation may indicate a finite variation of $W_e(r)$ with r. It should be stressed that although the absolute value of $W_e^2(r)$ is low ($\sim 2 \times 10^{-3}$) compared to its value of unity for a strong predissociation, so that no perturbation of the B-state levels is apparent in absorption, a significant fraction of B-state loss by predissociation still occurs because of the relatively long radiative lifetime of this state. The predissociation rates in Figure 6 are also entirely consistent with estimates [53] of predissociation rates from quantum yields for I formation in photolysis of I_2 in the range λ 5010—6240 Å.

Magnetic Quenching. The $B \rightarrow X$ fluorescence of I_2 is well known to be quenched by a magnetic field (see refs. 60, 61 for a summary of early literature). Following the systematic investigation by Degenkolb *et al.*[61] of the B-state v'-dependence of the quenching, using broad-band ($\sim$15 Å) excitation between λ 5000 and 6000 Å, Chapman and Bunker [62] have recently reported the first study of the phenomenon for 6328 Å He–Ne laser excitation of two single rotational levels ($v' = 6$, $J' = 32$; $v' = 11$, $J' = 128$). The lifetimes (radiative +predissociative) for these two states have already been accurately determined.[63] The observed quenching is due [64] to magnetic mixing of $B(0_u^+)$ with nearby state(s) of 0_u^- character and correlating with ${}^2P_{3/2}$ atoms. In the

[60] L. A. Turner, *Z. Phys.*, 1930, **65**, 464.
[61] E. O. Degenkolb, J. I. Steinfeld, E. Wasserman, and W. Klemperer, *J. Chem. Phys.*, 1969, **51**, 615.
[62] G. D. Chapman and P. R. Bunker, *J. Chem. Phys.*, 1972, **57**, 2951.
[63] K. C. Shotton and G. D. Chapman, *J. Chem. Phys.*, 1972, **56**, 1012.
[64] J. H. Van Vleck, *Phys. Rev.*, 1932, **40**, 544.

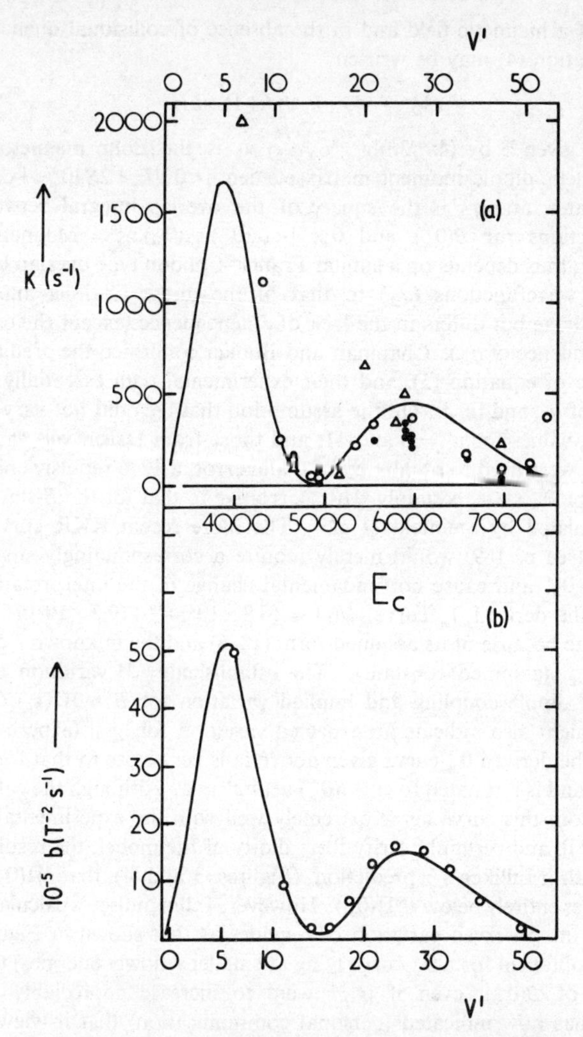

Figure 6 ***Predissociation and magnetic quenching rates for*** I_2 $B^3\Pi(0_u^+)$ ***as a function of B state vibrational level and continuum energy*** (E_c) ***above*** $I\ ^2P_{3/2}+I\ ^2P_{3/2}$. (a) *Figure 3 of Tellinghuisen,*[58] *showing predissociation rate coefficients* $\kappa_{v'}$ (s^{-1}), *equation* (4). *The smooth curve was calculated as described in the text using the potential curves in Figure 5. Experimental points:* ○, *photodissociation yields;* ●, *phase-shift lifetimes,* △, *direct decay lifetimes.* (b) *From data and calculations of Chapman and Bunker,*[62] *showing magnetic quenching coefficients b* ($T^{-2}\,s^{-1}$). *Experimental points for* $v' = 6$ *and* 11 *by Chapman and Bunker;*[62] *remaining points from Degenkolb* et al.[61] *Similar data not mentioned in the text have also been recently obtained by G. A. Capelle and H. P. Broida,* J. Chem. Phys., ***to be*** *published, using broad-band* (5 Å) *laser excitation*

presence of a magnetic field and in the absence of collisional quenching of $B(0_u^+)$, equation (4) may be written

$$\tau_{v',J'}^{-1} = A_{v'} + \kappa_{v'} J(J+1) + bH^2 \qquad (5)$$

where b is given [62] by $(4\pi^2/3h)|g_{eff}|^2\mu_0{}^2q'_{cb}$; μ_0 is the Bohr magneton, $|g_{eff}|$ is the magnetic dipole moment matrix element $|<0_u^-|L_z+2S_z|0_u^+>|$ coupling the two states, and q'_{cb} is the square of the overlap integral between the radial functions for $B(0_u^+)$ and 0_u^-, $|<\psi_c(0_u^-)|\psi_b(0_u^+)>|^2$. Magnetic predissociation thus depends on a similar Franck–Condon type overlap between the radial wavefunctions (q'_{cb}) to that of the direct Coriolis interaction discussed above but differs in the lack of J-dependence (except the expected small dependence of q'_{cb}). Chapman and Bunker confirmed the predicted H^2 dependence of equation (5), and their experimental data essentially reduce to values of q'_{cb} and $|g_{eff}|$. On the assumption that $|g_{eff}|$ did not vary with r, the two q'_{cb} values for $v' = 6$ and 11, and those from Degenkolb *et al.*[61] up to $v' = 48$, were used to obtain, by trial and error, a 12–6 repulsive potential curve for the 0_u^- state, certainly $^3\Pi(0_u^-)$, relative to that for the B-state RKR curve calculated by Steinfeld *et al.*[30] The more recent RKR curves [19, 20] for $B(0_u^+)$ (see p. 189) would merely require a correspondingly small shift in that for 0_u^+, and cause no fundamental change in the interpretation. In any case, the derived 0_u^- curve, $U(r) = (6.8\times10^8)/r^{12} + (9.3\times10^5)/r^6$ is only approximate because of its assumed form (12–6) and the unknown r-dependence of $|g_{eff}|$ (assumed constant). The established [20, 53] variation of $R_e(r)$ for $B \leftarrow X$ dipole coupling and implied variation for $B \leftarrow {}^1\Pi(1_u)$ Coriolis coupling might also indicate an expected variation for $|g_{eff}|$ (experimentally $\sim$0.33). The derived 0_u^- curve given above falls very close to that for $^1\Pi(1_u)$ (Figure 5) and is estimated to cut $B(0_u^+)$ near $v' = 3$. Although the calculated b values from this curve agree extremely well with the experimental points [Figure 6(b)], and certainly verify the validity of the model, the result is out of line with Mulliken's [5] prediction (Figures 3 and 4) that $^3\Pi(0_u^-)$ most probably lies entirely below $B^3\Pi(0_u^+)$. However, Tellinghuisen's calculations [58] show that the observed variation of b values as that shown in Figure 6(b) cannot be obtained for a 0_u^- curve lying at smaller r (lower energies) than the r_{min} limb of $B(0_u^+)$, even if $|g_{eff}|$ were to increase appreciably with r. Mulliken has now indicated (personal communication) that in view of the approximations and assumptions, his conclusion about the relative positions of $^3\Pi(0_u^-)$ and $^3\Pi(0_u^+)$ could be in error. More detailed measurements on more vibrational levels would help to locate the $^3\Pi(0_u^-)$ curve more accurately, especially if the r-dependence of $|g_{eff}|$ could be established.

Laser-excited Fluorescence and Effects of Nuclear Hyperfine Splittings.— *Laser-excited Fluorescence.* The well-known examples of iodine $B \rightarrow X$ fluorescence excited by Cd 5086 Å and Hg 5461 Å atomic line sources have recently been supplemented by several studies using He–Ne, Ar^+, and Kr^+ laser lines at several wavelengths. Yee and Miller,[33] following a re-examina-

Table 1 *Summary*[a] *of resonance fluorescence doublet series of the* $B^3\Pi(0_u^+) \rightarrow X^1\Sigma_g^+$ *system of* $^{127}I_2$

Exciting line	*Molecular assignment*	*Measured frequency*	*Reference*
Ar^+ Laser; 5017 Å, 19 926.04 cm^{-1}	$R(39)$ 64–0	19 925.97	*b, c, d*
	$R(26)$ 62–0	19 925.96	
Cd I; 5086 Å, 19 657.02 cm^{-1}	$R(40)$ 50–0	19 656.90	*e, f*
	$P(9)$ 49–0	19 656.98	
Ar^+ Laser; 5145 Å, 19 429.73 cm^{-1}	$R(15)$ 43–0	19 429.81	*b, c, e, g, h*
	$P(13)$ 43–0	19 429.81	
	(J~100) v'–1	—	*i*
Hg I; 5461 Å, 18 307.48 cm^{-1}	$R(33)$ 25–0	—	*j*
Kr^+ Laser; 5682 Å, 17 594.93 cm^{-1}	$P(96)$ 18–0	17 594.93	*k, l*
	$P(37)$ 17–0	17 594.93	
	(J~125) 26–3 or 21–1	—	
He–Ne Laser; 6382 Å, 15 798.02 cm^{-1}	$R(127)$ 11–5	—	*e, g, i, m, n*
	$P(33)$ 6–3	—	

(*a*) J'' numbering; data mostly taken from K. K. Yee and G. J. Miller, *J.C.S. Chem. Comm.*, 1972, 1054; (*b*) W. Holzer, W. F. Murphy, and H. J. Bernstein, *J. Chem. Phys.*, 1970, **52,** 399, 469; (*c*) Th. Haldorsson and E. Menke, *Z. Naturforsch.*, 1970, **25a**, 1356; (*d*) M. S. Sorem, M. D. Levenson, and A. L. Schawlow, *Phys. Letters* (*A*), 1971, **37**, 33; (*e*) J. I. Steinfeld, J. D. Campbell, and N. A. Weiss, *J. Mol. Spectroscopy*, 1969, **29**, 204; (*f*) J. I. Steinfeld and A. N. Schweid, *J. Chem. Phys.*, 1970, **53**, 3304; (*g*) E. Menke, *Z. Naturforsch.*, 1970, **25a**, 3, 442; (*h*) S. Ezekiel and R. Weiss, *Phys. Rev. Letters*, 1968, **20**, 91; (*i*) D. A. Hatzenbuhler and G. E. Le Roi, 1971, Report TR 11 (36) (AD–711823), Michigan State University, East Lansing, U.S.A.; (*j*) J. I. Steinfeld, R. N. Zare, L. Jones, M. Lesk, and W. Klemperer, *J. Chem. Phys.*, 1965, **42**, 25; (*k*) K. Sakurai and H. P. Broida, *J. Chem. Phys.*, 1969, **50**, 557; (*l*) T. W. Hansch, M. D. Levenson, and A. L. Schawlow, *Phys. Rev. Letters*, 1971, **26**, 946; (*m*) K. Sakurai and H. P. Broida, *J. Chem. Phys.*, 1970, **53**, 1615; (*n*) G. R. Hanes, and C. E. Dahlstrom, *Appl. Phys. Letters*, 1969, **14**, 362.

tion of the $B \leftarrow X$ absorption spectrum at high resolution, have recently been able to clarify the assignments for all excitation sources used so far, and their results with some additions are repeated in Table 1. The fact that for nearly all excitation wavelengths two doublet series are excited, is merely due to fortuitous coincidence of overlap of two absorption lines from low v'' with the laser line. Because of the previous uncertainties for the energy levels of $B(0_u^+)$, some of the earlier assignments are found to be in error. Thus Sakurai and Broida's [65] assignment for Kr^+ 5682 Å excitation is reassigned from (v', J') 16, 38 ± 1; 17, 100 ± 3 to 17, 36; 18, 95. Similarly, for Cd 5086 Å, the earlier assignments 50, 8; 51, 43 [66] and 50, 29; 53, 65 [67] become 50, 41 and 49, 8. Fortunately, the interpretation of energy transfer processes [67] using λ 5086 Å excitation will not require any major revision.

[65] K. Sakurai and H. P. Broida, *J. Chem. Phys.*, 1969, **50**, 557.
[66] P. Pringsheim, 'Fluorescence and Phosphorescence', Interscience, New York, 1948.
[67] J. I. Steinfeld and A. N. Schweid, *J. Chem. Phys.*, 1970, **53**, 3304.

Nuclear Hyperfine Effects. It was recently proposed [68] that λ 6328 Å He–Ne laser absorption by $^{127}I_2$ could be used as a basis for setting a new wavelength standard of length, with a precision of reproducibility more than 2 orders of magnitude better than the present Kr standard. In view of this, the theory and experimental data on I_2 absorption at 6328 Å are considered in some detail.

Because of the non-zero nuclear electric quadrupole moment (eQ) and nuclear spin quantum number ($I = 5/2$) of ^{127}I, each J-rotational level of molecular I_2 is split into several hyperfine components. The two interactions which can cause splittings, (i) interaction of eQ with the internuclear electric field gradient, $q = \partial^2 V/\partial r^2$, and (ii) interaction of the magnetic moment due to molecular rotation with I, have been discussed by several authors.[69–71] Since the latter (spin–rotation) interaction energy, given by $\frac{1}{2}C_I\{F(F+1)-J(J+1)-I(I+1)\}/J(J+1)$ is small and decreases with increasing J, only the nuclear electric quadrupole (NEQ) interaction need be considered in a first approximation for high J levels. The NEQ splittings can also be regarded approximately as due to contributions from the two separate nuclei, when

$$E_{\text{NEQ}} = -\frac{2J+3}{J}\, eqQ\,(\lambda_1+\lambda_2) \tag{6}$$

$$\lambda_i = -\tfrac{1}{2}\left\{\frac{3m_I^2 - I_i(I_i+1)}{4I_i(2I_i-1)}\right\}, \quad m_I = F_i - J$$

and $I_i + J = F_i$. Taking $I = 5/2$ for ^{127}I, equation (6) gives six values of $\lambda_1+\lambda_2$, with different statistical weights. Each J state is thus approximately composed of six degenerate hyperfine components, as shown schematically in Figure 7. Because of the Fermi–Dirac statistics for ^{127}I, only symmetric (anti-symmetric) nuclear spin wavefunctions are possible for odd (even) J rotational states of I_2 $X^1\Sigma_g^+$ $[B(0_u^+)]$. Thus as shown in Figure 7, with the selection rule $\Delta F = \Delta J$, 21 hyperfine components are possible for B (even J) $\leftarrow X$ (odd J), compared with only 15 for B (odd J) $\leftarrow X$ (even J). The six components which are absent in the latter case are shown by broken lines in the diagram. When account is taken of a Doppler broadening at room temperature of $\sim$400 MHz ($\sim$0.012 cm^{-1}) for each component, it is not, of course, usually possible to resolve the hyperfine splittings. Kroll and Innes,[70] however, were able to demonstrate the effect of these splittings by measurements of the composite linewidths for the $B \leftarrow X$ 20–1, 30–0, and 18–0 bands of $^{127}I_2$. The FWHM (full widths at half-maxima) values were found on average to be about 0.005 cm^{-1} more for odd J'' transitions than those for even J'', as well as showing the expected 21:15 intensity ratio. This is readily understood by reference to Figure 7, where four of the extra

[68] G. R. Hanes and C. E. Dahlstrom, *Appl. Phys. Letters*, 1969, **14**, 362.
[69] M. Kroll, *Phys. Rev. Letters*, 1969, **23**, 631; *Chem. Phys. Letters*, 1971, **9**, 115.
[70] M. Kroll and K. K. Innes, *J. Mol. Spectroscopy*, 1970, **36**, 295.
[71] T. W. Hansch, M. D. Levenson, and A. L. Schawlow, *Phys. Rev. Letters*, 1971, **26**, 946.

six hyperfine components for odd J'' lie in the wings of the composite line, and thus broaden the overall Doppler profile.

The most significant investigations of nuclear hyperfine splittings have been obtained using the technique [72] of saturated absorption spectroscopy (inverse Lamb-dip), which enables the hyperfine components of a single $J' \leftarrow J''$ transition to be detected within the overall Doppler absorption profile. A detailed account of an experiment in which a single mode of a He–Ne λ 6328 Å laser was piezo-electrically scanned through its Zeeman-broadened ($\sim$500 G applied field) power-gain curve across two neighbouring lines of $^{127}I_2(B \leftarrow X)$ has been given by Hanes *et al.*[73] With an intracavity absorption cell, absorption of the laser beam travelling in both directions along the laser axis is unique for the class of molecules having zero axial velocity components. Hence absorption at the line centres of the hyperfine components is more easily saturated than for their Doppler profiles. The resulting profile of the emerging laser beam thus shows small dips at frequencies corresponding to the transitions shown in Figure 7. All 21 components of the $P(33)$ 6–3, $B \leftarrow X$ line [73] have been observed, as well as 14 components of $R(127)$ 11–5,[68] which is excited without Zeeman-broadening of the 6328 Å He–Ne laser line. The linewidths (FWHM) were $\sim$9 MHz, and could be located relative to each other with a precision of 0.36 MHz.

Although the simple theoretical model mentioned above is sufficient to describe the nature of the hyperfine spectra, an exact first-order NEQ analysis and allowance for spin–rotation interaction are needed to obtain reasonable agreement between calculated and observed spectra. For the lower J transition, $P(33)$ 6–3, only a second-order treatment of the quadrupole interactions between hyperfine components of the same F but with $\Delta J = \pm 2$ gives a RMS derivation between calculated and observed frequencies similar to the experimental precision.[73] The results of the experiment described above, together with those of other recent studies are summarized in Table 2. Hansch, Levenson, and Schawlow [71] used a tunable Kr^+ 5682 Å laser and observed all 21 hyperfine components of a line assigned as $P(117)$ 21–1. Details of the assignment were not given, and the line is not included in Table 1. Sakurai and Broida's work [65] indicates that the transition is better identified at 26–3, $J' \sim 125$. Sorem, Levenson, and Schawlow [74] used Ar^+ 5017 Å excitation and found the expected 15 components of $R(26)$ 62–0, which is in agreement with the assignment in Table 1. The value $\Delta C_I = 944 \pm 60$ kHz indicates that C_I values increase rapidly for high vibrational levels of $B^3\Pi(0_u^+)$, and it was proposed that this may be due to mixing with other electronic states near the dissociation limit. Knox and Pao [75] used a He–Ne 6328 Å laser with $^{129}I_2$ and observed 38 components: no assignment

[72] R. L. Barger and J. L. Hall, *Phys. Rev. Letters*, 1969, **22,** 4.
[73] G. R. Hanes, J. Lapierre, P. R. Bunker, and K. C. Shotton, *J. Mol. Spectroscopy*, 1971, **39,** 506.
[74] M. S. Sorem, M. D. Levenson, and A. L. Schawlow, *Phys. Letters (A)*, 1971, **37,** 33.
[75] J. D. Knox and Y–H. Pao, *Appl. Phys. Letters*, 1971, **18,** 360.

Table 2 *ΔeqQ (MHz) and ΔC_I (kHz) values for several absorption lines of* $^{127}I_2$ $B^3\Pi(0_u^+) \leftarrow X^1\Sigma_g^+$

Molecular assignment	ΔeqQ	ΔC_I	*Reference*
$P(33)$ 6–3	1938.09 ±0.34	21.69 ±0.96	*a*
	1962 ±2	26.6	*b*
$R(127)$ 11–5	1944.8 ±0.05	28.4 ±0.08	*a, c*
	1948 ±2	29.7	*b*
$J \sim 125$ 21–1 or 26–3	940 ±30	53 ±5	*d*
$R(26)$ 62–0	1878 ±10	944 ±60	*e*

(*a*) G. R. Hanes, J. Lapierre, P. R. Bunker, and K. C. Shotton, *J. Mol. Spectroscopy*, 1971, **39**, 506; (*b*) J. T. La Tourrette and R. S. Eng., VII International Quantum Electronics Conference, Montreal, May, 1972, p. 43; (*c*) revised values, P. R. Bunker, personal communication 1972; (*d*) T. W. Hansch, M. D. Levenson, and A. L. Schawlow, *Phys. Rev. Letters*, 1971, **26**, 946; (*e*) M. S. Sorem, M. D. Levenson, and A. L. Schawlow, *Phys. Letters*, 1971, **37A**, 33.

was possible however, for $^{129}I_2$. Very recently, more precise measurements [76] of some of the relative frequencies of the hyperfine components for $P(33)$, 6–3 and $R(127)$, 6–5 were obtained by measuring the beat frequencies of two lasers simultaneously locked to different hyperfine components. In addition, two extra lines with $\Delta F = 0$, $\Delta J \neq 0$, for $P(33)$ enabled the absolute value, eqQ'' $(v'' = 3) = 2405 \pm 14$ MHz to be found. Although the ΔeqQ and ΔC_I values for $R(127)$ from the two investigations in Table 2 agree reasonably well, those for $P(33)$ do not, the differences being outside the limits of both theory and experiment. However, the difference in ΔeqQ values for the two lines is small in agreement with the expectation that the difference $(q_5'' - q_3'')$ should be similar to $(q_{11}' - q_6')$.

In conclusion, one must mention the anomalously large observed linewidths (~9 MHz) which would correspond to lifetimes of ~0.02 μs, in contrast with directly measured lifetimes [63] for $v' = 6$, $J' = 32$ and $v' = 11$, $J' = 128$ (see p. 196) of 0.31 ±0.01 and 0.41 ±0.015 μs respectively. The possible broadening effects mentioned earlier do not seem capable of such a large effect and the true explanation remains obscure at the present time.

[76] J. L. La Tourrette and R. S. Eng, VII International Quantum Electronics Conference, May 1972, Montreal, p. 43.

Figure 7 *The 21 $J'_{\text{even}} \leftarrow J''_{\text{odd}}$ and 15 $J'_{\text{odd}} \leftarrow J''_{\text{even}}$ hyperfine components of a single $J' \leftarrow J''$ $B^3\Pi(0_u^+) \leftarrow X^1\Sigma_g^+$ transition of $^{127}I_2$. Nuclear spin of ^{127}I is 5/2. Although not to scale, the diagram shows the larger electric quadrupole coupling constant eqQ for the ground state. The frequency scale is relative to the highest frequency component, and is for $P(33)$ of the 6–3 band*

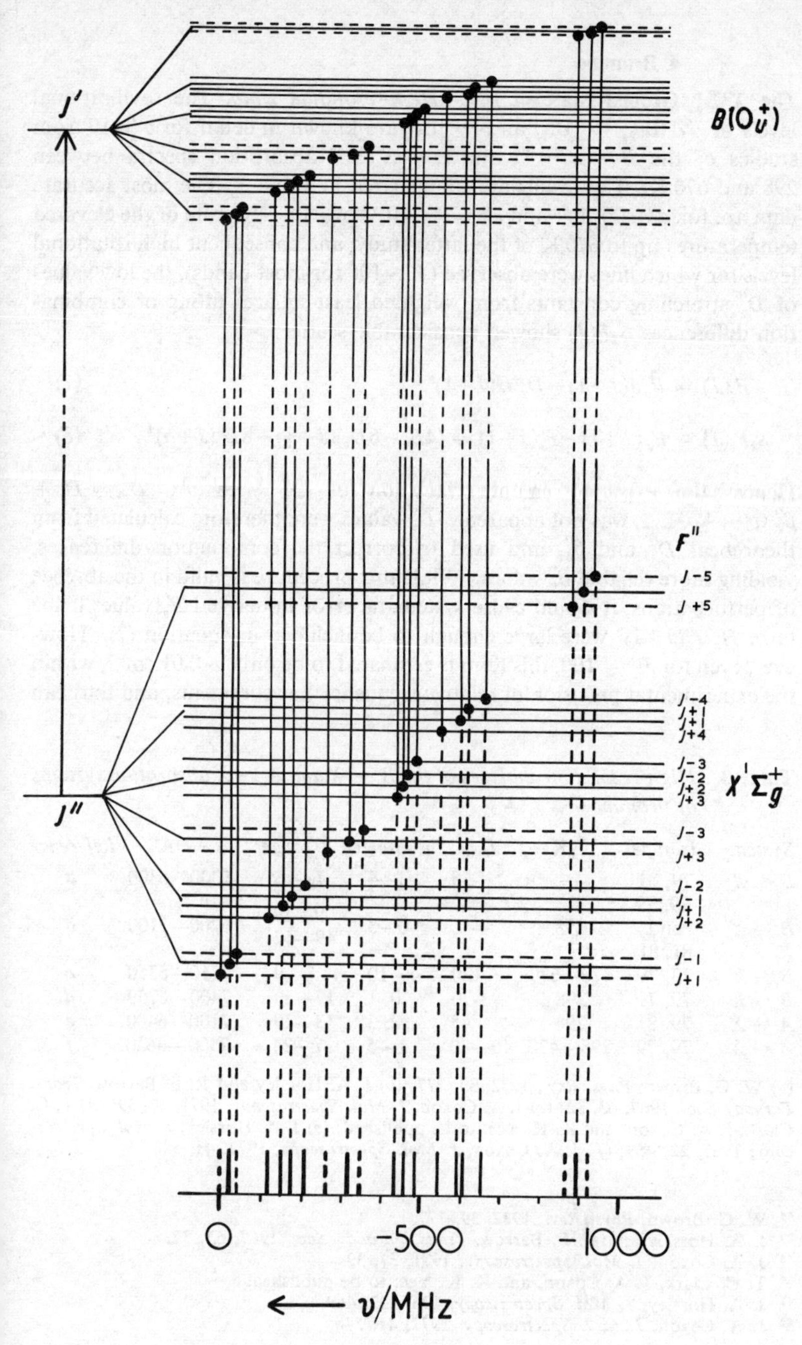
J'
B(0u+)
J''
F''
J-5
J+5
J-4
J+1
J+1
J+4
J-3
J-2
J+2
J+3
X 1Σg+
J-3
J+3
J-2
J-1
J+1
J+2
J-1
J+1
0
-500
-1000
ν/MHz

4 Bromine

The $X^1\Sigma_g^+$ Ground State of Br_2.—*High-resolution Data.* The rovibrational levels of $^{79,79}Br_2$, $^{79,81}Br_2$, and $^{81,81}Br_2$ are known in detail for $v'' \leqslant 10$ from studies of the $B \leftarrow X$ [77–80] and $A \leftarrow X$ [81, 82] absorption spectra between 298 and 670 K. The results are summarized in Table 3. The most accurate data are for $v'' = 0, 1$ [80] and $v'' = 4$—10 [79] of $^{79}Br_2$. In spite of the elevated temperatures up to 670 K of the latter study, and consequent high rotational levels for which lines were observed ($J'' \sim 100$ for most bands), the low values of D_v'' stretching constants from weighted least-squares fitting of combination differences $\Delta_2F(J)$ showed considerable scatter:

$$F_v(J) = B_v J(J+1) - D_v J^2(J+1)^2 \tag{7}$$

$$\Delta_2 F_v(J) = F_v(J+1) - F_v(J-1) = (4B_v - 6D_v)(J+\tfrac{1}{2}) - 8D_v(J+\tfrac{1}{2})^3 \tag{8}$$

Hence, the expected smooth variation of D_v'' with v'', $D_v'' = D_e'' + \beta_e''(v''+\frac{1}{2}) - \ldots$, was not apparent. D_v'' values were therefore calculated from theoretical D_e'' and β_e'', and used to correct the combination differences, yielding more reliable B_v'' values. While this procedure is valid in the absence of perturbations, it would cause systematic error in derived B_v'' values if the term $H_v'' J^3(J+1)^3$ were large enough to be included in equation (7). However, even for $J'' = 100$, this term is estimated to be only ~ 0.01 cm^{-1}, within the experimental precision of relative frequency measurements, and thus can

Table 3 *High-resolution analyses of the $B \leftarrow X$ and $A \leftarrow X$ absorption systems of bromine*

System	*Isotope*	*T/K*	J_{max}	*v″ range*	*v′ range*	*λ/Å*	*Reference*
$B \leftarrow X$	79, 81	—	~55	2—4	7—13	6000—6300	*a*
$B \leftarrow X$	79, 79 and 81, 81	298	—	0—3	9—19, 49—52	6200—5100	*b*
$B \leftarrow X$	79, 79	370—670	~120	4—10	1—9	7710—6350	*c*
$B \leftarrow X$	79, 79	298	~70	0, 1	17—56	7400—6750	*d*
$A \leftarrow X$	79, 81	298	45	2, 3	13—19	7100—6400	*e*
$A \leftarrow X$	79, 79	298—420	40—80	1—5	7—24	7500—6620	*f*

(*a*) W. G. Brown, *Phys. Rev.*, 1932, **39**, 777; (*b*) J. A. Horsley and R. F. Barrow, *Trans. Faraday Soc.*, 1967, **63**, 32; (*c*) J. A. Coxon, *J. Mol. Spectroscopy*, 1971, **37**, 39; (*d*) T. C. Clark, J. A. Coxon, and K. K. Yee, to be published; (*e*) J. A. Horsley, *J. Mol. Spectroscopy*, 1967, **22**, 469; (*f*) J. A. Coxon, *J. Mol. Spectroscopy*, 1972, **41**, 548.

[77] W. G. Brown, *Phys. Rev.*, 1932, **39**, 777.
[78] J. A. Horsley and R. F. Barrow, *Trans. Faraday Soc.*, 1967, **63**, 32.
[79] J. A. Coxon, *J. Mol. Spectroscopy*, 1971, **37**, 39.
[80] T. C. Clark, J. A. Coxon, and K. K. Yee, to be published.
[81] J. A. Horsley, *J. Mol. Spectroscopy*, 1967, **22**, 469.
[82] J. A. Coxon, *J. Mol. Spectroscopy*, 1972, **41**, 548.

be safely neglected. The results of the most recent study [80] are not final at the time of writing. It was undertaken primarily to obtain B'_v and $G(v')$ values for $17 < v' \leqslant 55$ of the $B(0_u^+)$ state, and the data were processed using the term-value method.[83] At the present time, the expressions

$$B''_v = 0.082\,122 - 3.189\times10^{-4}(v''+\tfrac{1}{2}) - 1.14\times10^{-6}(v''+\tfrac{1}{2})^2$$

$$G(v'') = 325.29(v''+\tfrac{1}{2}) - 1.072(v''+\tfrac{1}{2})^2 - 2.6\times10^{-3}(v''+\tfrac{1}{2})^3$$

for $^{79,79}Br_2$ remain the best available for $v'' \leqslant 10$.

Resonance Fluorescence Data. In a study similar to that described above for I_2, Rao and Venkateswarlu [84] have observed a single u.v. resonance fluorescence doublet series from a condensed transformer discharge through flowing bromine vapour. In recent work in this laboratory,[85] the same series has also been observed by excitation with the λ 1575 Å atomic bromine resonance line, which was thus probably the source of excitation in the Rao and Venkateswarlu (RV) experiment also. Since the series provides the only information available for the higher levels ($v'' \leqslant 36$) of $^{79,81}Br_2$ $X^1\Sigma_g^+$, it is unfortunate that the absolute frequencies, and especially the doublet separations, are known only with low accuracy. Thus not only is the uncertainty in the RV measurements at least 1 cm^{-1}, but also the doublet separations appear to decrease with v'' much more rapidly than expected. Furthermore, it is possible

Table 4 *Relative intensities of the fluorescence series of* $^{81}Br_2$ $\{B^3\Pi(0_u^+),\ (v' = 39, J' = 15)\} \rightarrow X^1\Sigma_g^+$, *excited by* 5145 Å *radiation*

v''	$I(39\text{—}v'')^a$	Relative $q_{39,v''}\nu^{4b}$	$\dfrac{I}{q\nu^4}$	$\bar{r}_{39,v''}$/Å^b
1	1.00	1.00	1.00	2.34
2	0.17	0.17	1.0	2.35
3	0.28	0.20	1.4	2.36
4	0.50	0.43	1.2	2.37
5	<0.005	—	—	—
6	0.38	0.29	1.3	2.39
7	0.14	0.13	1.1	2.40
8	0.11	0.07	1.5	2.41
9	0.28	0.23	1.2	2.42
10	<0.002	—	—	—
11	0.26	0.18	1.4	2.44
12	0.07	0.06	1.2	2.45
13	0.12	0.07	1.8	2.46
14	~0.14	0.1	~1.1	2.47
15	<0.02	—	—	—
16	~0.20	0.12	~1.7	2.49

(*a*) W. Holzer, W. F. Murphy, and H. J. Bernstein, *J. Chem. Phys.*, 1970, **52**, 469; (*b*) J. A. Coxon, unpublished data.

[83] N. Åslund, *Arkiv Fysik*, 1965, **30**, 377.
[84] Y. Rao and P. Venkateswarlu, *J. Mol. Spectroscopy*, 1964, **13**, 288.
[85] M. A. A. Clyne and H. W. Cruse, *J. C. S. Faraday II*, 1972, **68**, 1281.

that the measurements could be extended to $v''>36$, although RV gave no information about the intensity dependence on v''. Until the series is re-investigated, its only value lies in providing approximate $G(v'')$ data. As well as this series, however, visible series are also known for Hg 5461 Å [86] and Ar^+ 5145 Å laser excitation.[87] Accurate measurements on these and other series would be valuable.

RV obtained $J_r' = 62$ for the excited state rotational level of $^{79,81}Br_2$ responsible for the u.v. resonance doublet series. A comparison of their doublet separations $[F_v''(63)-F_v''(61)]$ of 19.9, 19.7, 19.9, and 18.7 cm^{-1} for $v'' = 4$, 6, 9, and 10, with values of 19.88, 19.71, 19.48, and 19.41 cm^{-1} calculated from the known [79] spectroscopic constants for $v''\leqslant 10$, seems to be in accord with this assignment. However, with increasing v'', the RV doublet separations decrease too rapidly even if allowance is made for centrifugal stretching. The low B_v'' values derived in this way are clearly demonstrated in the RKR curve derived by Le Roy and Burns [88] (see their Figure 1). These authors therefore calculated B_v'' values from new potential curves using the RKR widths and a reasonable repulsive limb. This enabled refined $G(v'')$ values to be obtained retaining the RV $J_r' = 62$ assignment, and the process was reiterated until convergent. This method unfortunately relies on the low v'' doublet separations. If the RV doublet separations at high v'' were more reliable, this would imply a lower value for J_r', and Le Roy concurs [89] that his $G(v'')$ are in fact quite possibly less reliable than those derived by RV. The latter were used [90] in obtaining a set of Franck–Condon factors and r-centroids for the $B \leftrightarrow X$ system of $^{79,81}Br_2$.

As mentioned earlier, Holzer *et al.*[87] have observed a resonance fluorescence series for Br_2 excited with a Ar^+ 5145 Å laser source. The isotopic species was identified as $^{81}Br_2$ $B^3\Pi(0_u^+)$ with $J' = 15$. However, the vibrational level of the B state was uncertain. The relative intensities of the doublets are given in Table 4, together with values of $q_{39,v''}\nu^4$, where $q_{39,v''}$ are the Franck–Condon factors for the $(39,v'')$ progression of $^{81}Br_2$, $J' = J'' = 0$. The differences between $q_{v',v''}$ values for $J' = J'' = 0$ and $J' = 15$ and $J'' = 14, 16$ transitions are known to be small. The excellent agreement between the relative intensities and $q_{39,v''}\nu^4$ values confirms beyond doubt that the series originates from $v' = 39$, as suggested by Holzer *et al.*[87] The values of $I/q\nu^4$ in Table 4 tend to increase steadily from $v'' = 1$ to $v'' = 16$ ($\bar{r}$ 2.34—2.50 Å), as might be expected for an increase in the electronic transition moment $R_e(r)$ with r, similar to that for I_2 ($B \leftrightarrow X$). However, this important variation can only be established accurately with more experimental data.

Discrete Levels of Br_2, 2431 $A^3\Pi(1_u)$ **and** $B^3\Pi(0_u^+)$.—*The* $A^3\Pi(1_u)$ *state of* Br_2.

[86] H. J. Plumley, *Phys. Rev.*, 1933, **43**, 495; 1934, **45**, 678.
[87] W. Holzer, W. F. Murphy, and H. J. Bernstein, *J. Chem. Phys.*, 1970, **52**, 469.
[88] R. J. Le Roy and G. Burns, *J. Mol. Spectroscopy*, 1968, **25**, 77.
[89] R. J. Le Roy, 1971, personal communication.
[90] J. A. Coxon, *J. Quant. Spectroscopy Radiative Transfer*, 1972, **12**, 639.

In accordance with theory, bound $^3\Pi(1_u)$ states have long been known for I_2 and Br_2. However, accurate data are available only for Br_2. The first high-resolution $A \leftarrow X$ data for nine bands of $^{79,81}Br$ with $13 \leqslant v' \leqslant 19$ were obtained by Horsley,[81] and the vibrational numbering later established by measurements [91] on $^{79}Br_2$. The banded $A \leftarrow X$ spectral region is difficult to examine because of overlapping by the more intense $B \leftarrow X$ band system and the presence of three isotopic species in natural Br_2. Recently, however, the $A \leftarrow X$ system of separated $^{79}Br_2$ was examined [82] in detail between 298 and 420 K, and rotational constants for the 1_u^- component of the Ω-doubled state were found from Q branches for $7 \leqslant v' \leqslant 24$. B_v values for the 1_u^+ component from the weaker P and R branches were found only for the range $11 \leqslant v' \leqslant 18$. The differences, $B_v^+ - B_v^-$, showed a strong dependence on v', similar to that found for the $A^3\Pi(1)$ state of ICl. However, q_v' values were so low that it was assumed [92] that true B_v values would be accurately given as the means of B_v^+ and B_v^-. An extrapolation of B_v from $v' = 7$ to $v' = -\frac{1}{2}$ was made [92] by comparison of experimental α_e values

$$B_v = B_e - \alpha_e(v+\tfrac{1}{2}) + \gamma_e(v+\tfrac{1}{2})^2$$

with those from the Pekeris relationship

$$\alpha_e = 6\{\sqrt{\omega_e x_e B_e^3} - B_e^2\}/\omega_e.$$

For ICl, $A(1)$ and Br_2, $B(0_u^+)$, α_e (expt.) are about 9% lower than α_e (Pekeris). The data available at present for I_2 and Cl_2 $B(0_u^+)$ are not sufficient to make this comparison. Taking α_e for Br_2 $A(1_u)$ to be about 9% lower than α_e (Pekeris), the results $^{79}B_e' = 0.05880\ \mathrm{cm^{-1}}$ and $r_e' = 2.695$ Å were obtained. Vibrational terms for Br_2 $A(1_u)$ were fixed by combining the high-resolution data with those for $v' \leqslant 7$ from the bromine afterglow emission spectrum,[93] and with those for $25 \leqslant v' \leqslant 28$ from Brown's [94] head measurements. It was thus possible to calculate [92] for the first time an RKR curve for the A state of Br_2, together with a set of Franck–Condon factors for the $A \leftrightarrow X$ system. Although these results are subject to the uncertainties described above, they remain the best available until rotational constants for levels near the potential minimum of the A state are determined.

By comparison of the intensities of absorption of two pairs of bands of the $B \leftarrow X$ and $A \leftarrow X$ systems with $\bar{r} \sim 2.45$ Å, it was possible to estimate the relative transition moments of the $B \leftarrow X$ and $A \leftarrow X$ systems. Unfortunately, electronic degeneracy was not taken into account so that the ratio, $[R_e^2(r)]_B : [R_e^2(r)]_A$ should be 28 ± 4, and not 14 ± 2 as previously reported.

The $B^3\Pi(0_u^+)$ *State of* Br_2. Until 1971, the only high-resolution absorption data for the $B(0_u^+)$ state of Br_2 were those of Brown [77] and Horsley and

[91] J. A. Coxon and M. A. A. Clyne, *J. Phys. (B)*, 1970, **3**, 1164.
[92] J. A. Coxon, *J. Mol. Spectroscopy*, 1972, **41**, 566.
[93] M. A. A. Clyne and J. A. Coxon, *J. Mol. Spectroscopy*, 1967, **23**, 258.
[94] W. G. Brown, *Phys. Rev.*, 1931, **38**, 1179.

Barrow [78] (Table 3). Accurate data on the lower levels, $v' \geqslant 1$, were then obtained [79] using Br_2 up to 670 K. From the three studies, which include measurements of four consecutive v',0 bands of $^{79}Br_2$ and $^{81}Br_2$ close to the dissociation limit ($49 \leqslant v' \leqslant 52$), and the band head measurement of Brown,[94] sufficient data were available to obtain an RKR curve for the $B(0_u^+)$ state, and Franck–Condon factors [90] for the $B \leftrightarrow X$ system. It was found,[14, 95] however, that the four bands assigned to $49 \leqslant v' \leqslant 52$ by Horsley and Barrow could only be reconciled with Brown's head measurements if their numbering was in error by unity, and they were accordingly reassigned as $50 \leqslant v' \leqslant 53$. In a new study [80] of the $B \leftarrow X$ absorption bands, $17 \leqslant v' \leqslant 55$, undertaken to provide an improved RKR curve for the analysis of long-range interactions,[96] it has been found that the proposed reassignment was in error, and that Horsley and Barrow's original assignment [78] should be re-instated. Table 5 compares the new band-origin frequencies [80] with Brown's head measurements,[94] together with calculated isotopic displacements. For $v' \leqslant 42$, there is good agreement between the two sets of data (the differences between head and origin frequencies are known to be very low, ~0.1 cm^{-1}). Above $v' = 42$, however, Brown's data systematically diverge from the origin measurements, such that at $v' = 48$ (the highest level observed by Brown), the head measurement coincides closely with the origin for $v' = 47$. It is this anomaly in Brown's head data which was responsible for the proposal that Horsley and Barrow's $49 \leqslant v' \leqslant 52$ bands be reassigned. Diffuse

Table 5 *Comparison of high-resolution origin data* (cm^{-1}) *for* $^{79}Br_2$ *with low-resolution head data for* $^{79,81}Br_2$, $B^3\Pi(0_u^+) \leftarrow X^1\Sigma_g^+$

v'	$^{79,81}Br_2$ ν_{head}^{a}	δG^b	$\nu_{head} + \delta G$	$^{79,79}Br_2$ ν_{origin}^{c}	v'
35	19312.7	7.6	19320.3	19317.21	35
36	344.5	7.1	351.6	350.12	36
37	374.8	6.7	381.5	380.19	37
38	403.4	6.2	409.6	407.51	38
39	429.1	5.7	434.8	432.24	39
40	449.8	5.2	455.0	454.48	40
41	470.3	4.8	475.1	474.33	41
42	486.2	4.4	490.6	491.94	42
43	499.5	3.9	503.4	507.44	43
44	512.5	3.5	516.0	520.97	44
45	524.3	3.0	527.3	532.63	45
46	531.5	2.6	534.1	542.59	46
47	542.2	2.3	544.5	550.98	47
48	551.4	1.9	553.3	557.95	48

(*a*) From $v'' = 0$; W. G. Brown, *Phys. Rev.*, 1931, **38**, 1179; (*b*) Calculated isotopic shift, $\delta G = {}^{79,79}\nu'_{v',0} - {}^{79,81}\nu'_{v',0}$; (*c*) T. C. Clark, J. A. Coxon, and K. K. Yee, unpublished results.

[95] R. J. Le Roy and R. B. Bernstein, *Chem. Phys. Letters*, 1970, **5**, 442.
[96] R. F. Barrow, D. F. Broyd, and L. Pedersen, to be published.

bands at low resolution due to isotopic splitting can be ruled out as the reason for the anomaly since the splittings are lowest where the disagreement is worst.

The main implication of the new data is that the previously reported RKR curve [14] for Br_2 $B^3\Pi(0_u^+)$ is inaccurate for $v' > 42$. A preliminary curve from the new data [80] has been calculated by the Reporter. For the r_{min} limb, the deviations are less than 0.0003 Å, which reflects the accuracy of the variable Morse method used to estimate the r_{min} limb in the absence of a full rotational analysis. Although larger deviations for the r_{max} limb, up to 0.02 Å at $v' = 36$ and 0.05 Å at $v' = 42$ are found, these changes are likely to result in little practical effect on the calculated Franck–Condon factors [90] which, in any case, were reported only to $v' = 44$. A more important consequence is the reconciliation which can be reached between the RKR long-range potential for Br_2 $B(0_u^+)$ with the Le Roy–Bernstein [34, 95] analysis for the dissociation limit for this state. Their method carries the advantage of independence of *absolute* vibrational numbering. Using their values for C_5 and D, Le Roy's [97] analysis of the long-range behaviour of Br_2 $B(0_u^+)$ (to obtain a value for C_6) was one case where unsatisfactory results were obtained. This is now seen to be due certainly to the inaccuracy of the reported [14] RKR turning points near the dissociation limit. From a revised RKR calculation, and using Stwalley's graphical method,[98] the Reporter has obtained a preliminary value of 19 579.63 ± 0.04 cm^{-1} for the convergence limit of $^{79}Br_2$ $B(0_u^+) \leftarrow X^1\Sigma_g^+$, $v'' = 0$. This is in excellent agreement with the result of 19 579.7 ± 0.27 cm^{-1} found by Le Roy and Bernstein.[97] A full analysis will soon be published by Barrow *et al.*[96]

Continuum Absorption, Br_2 $(2431 \leftarrow X^1\Sigma_g^+)$.— The absorption spectrum of Br_2 between λ7000 and 3200 Å has been studied by several investigators.[99–101] In the most detailed study,[101] Passchier, Christian, and Gregory measured extinction coefficients (ε) at seven temperatures in the range 298—713 K. Their data for 298 K, together with segments of the potential energy curves for Br_2 $A^3\Pi(1_u)$, $B^3\Pi(0_u^+)$, and $^1\Pi(1_u)$, are plotted in Figure 8 and are in close agreement with those of Seery and Britton,[100] but are systematically about 10% larger than those of Acton, Aickin, and Bayliss.[99] As shown in Figure 8, the spectrum shows a maximum at ~24 200 cm^{-1}, with a shoulder at lower frequencies, and it has long been recognized that this is due to superposition of at least two systems originating from $X^1\Sigma_g^+$.

Considerable doubt is found in the literature concerning the assignments for the two principal systems. Mulliken [4] suggested two possibilities:

[97] R. J. Le Roy, *J. Mol. Spectroscopy*, 1971, **39**, 175.
[98] W. C. Stwalley, *Chem. Phys. Letters*, 1970, **7**, 600.
[99] A. P. Acton, R. G. Aikin, and N. S. Bayliss, *J. Chem. Phys.*, 1936, **4**, 474.
[100] D. J. Seery and D. Britton, *J. Phys. Chem.*, 1964, **68**, 2263.
[101] A. A. Passchier, J. D. Christian, and N. W. Gregory, *J. Phys. Chem.*, 1967, **71**, 937.

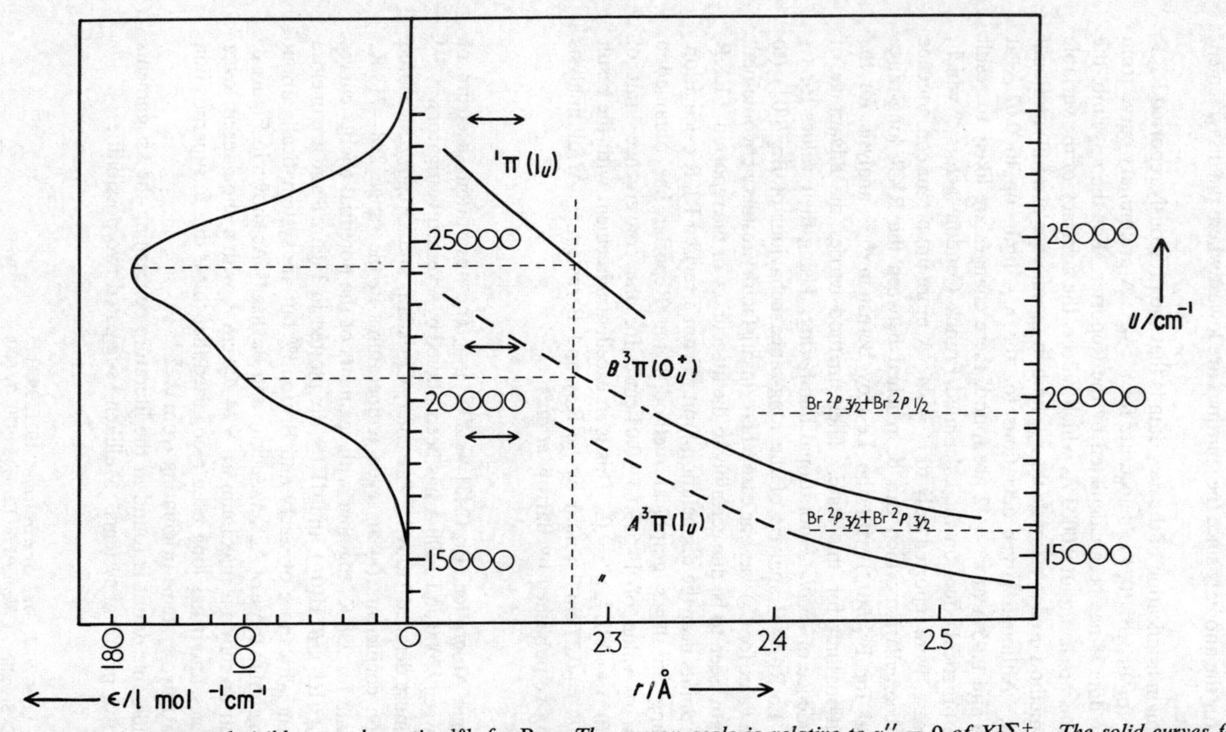

Figure 8 *Potential curves and visible–u.v. absorption*[101] *for* Br_2. *The energy scale is relative to* $v'' = 0$ *of* $X^1\Sigma_g^+$. *The solid curves for* $^3\Pi(1_u)$ *and* $^3\Pi(0_u^+)$ *are RKR potentials (see text). The curve for* $^1\Pi_u$ *and the broken curve for* $B^3\Pi(0_u^+)$ *in the continuum region are those derived by Bayliss*[103] *from extinction coefficient data. Horizontal arrows are PDTS laser frequencies*

	I	II
24 000 cm^{-1} maximum	$^1\Pi(1_u) \leftarrow X$	$^1\Pi(1_u) \leftarrow X$
shoulder	$B^3\Pi(0_u^+) \leftarrow X$	$A^3\Pi(1_u) \leftarrow X$
weak and overlapped	$A^3\Pi(1_u) \leftarrow X$	$B^3\Pi(0_u^+) \leftarrow X$

and Rees [102] resolved the problem in favour of I by showing that the repulsive limb [103] responsible for the shoulder lies close to that predicted by extrapolation of the RKR curve for $B^3\Pi(0_u^+)$. Sulzer and Wieland's 1952 paper,[36] however, caused confusion by the adoption of II. The RKR curve (Figure 8) for $B^3\Pi(0_u^+)$ calculated from the new high-resolution data [80] requires negligible extrapolation and leads almost perfectly into the segment derived by Bayliss [103] in 1937. The discrepancy is only ~0.002 Å. The segment derived by Bayliss for the main absorption peak is due to $^1\Pi(1_u)$, and as Figure 8 shows, absorption to $A^3\Pi(1_u)$ from X, $v'' = 0$ would lie at the lowest frequencies, within the banded $B \leftarrow X$ region. The extrapolation shown for the potential energy curve of $A^3\Pi(1_u)$ was made by retaining the same $^3\Pi(0_u^+) - {}^3\Pi(1_u)$ separation (~1850 cm^{-1}) as that at 2.4 Å. The vertical separation $^1\Pi(1_u) - A^3\Pi(1_u)$ at $r_e'' = 2.281$ Å is then ~5400 cm^{-1}. Using equations (1) and (2) (p. 181) and the spin–orbit splitting constant for Br ^2P of 2457 cm^{-1}, the unperturbed $^1\Pi - {}^3\Pi$ case (a) splitting X is 4809 cm^{-1}, and $z^2 = D(^1\Pi)/D^3\Pi(1_u) = 17$. This result is slightly lower than Mulliken's original estimate [4] of 19, and in reasonable agreement with the experimental result of ~36 (see below), based on $D(^3\Pi_0)/D(^3\Pi_1) = 28$ at $\bar{r}$ ~2.45 Å (see previous section). Some discrepancy between the two values would be expected from error in the $^1\Pi(1_u)$ curve, neglect of case (c) effects, and difference in the variations of D with r for the three systems. A peak extinction coefficient ε~5 l mol^{-1} cm^{-1} is predicted for 298 K continuum absorption $A \leftarrow X$ at 19 000 cm^{-1} within the discrete $B \leftarrow X$ region, and is too low to be manifested in the experimental absorption profile (Figure 8).

As implied above, attempts have been made to resolve the observed spectrum into its two main components, $B \leftarrow X$ and $^1\Pi(1_u) \leftarrow X$. Sulzer and Wieland's [36] approach is empirically valuable but only approximate since the linear dependence of ε on frequency is neglected, leading to the requirement of Gaussian components. Furthermore, in the present example, most of the low-frequency ε measurements refer to $B \leftarrow X$ banded absorption and are thus subject to particular experimental parameters. Thus the frequencies of the Gaussian peaks do not give accurately the vertical energies for the excited-state potential curves. The photodissociation translational spectra of Br_2 have been obtained at three laser frequencies, 28 810, 21 690, and 18 830 cm^{-1}. The results in terms of angular distributions and delay times are entirely in accord with the potential energy curves of Figure 8. When the full details of these studies are known,[104] it may be possible to resolve

[102] A. L. G. Rees, *Proc. Phys. Soc.*, 1947, **59**, 1008.
[103] N. S. Bayliss, *Proc. Roy. Soc.*, 1937, **A158**, 551.
[104] R. J. Oldman, R. K. Sander, and K. R. Wilson, to be published.

accurately the total absorption into the three components, as undertaken for I_2 (see p. 192). However, the reported observation of copious production of Br atoms at 18 830 cm^{-1} may be due in part to predissociation from $B(0_u^+)$ to $^1\Pi(1_u)$.

Lifetime of Br_2 $B^3\Pi(0_u^+)$.—The electronic transition moments (R_e), assumed independent of r, for Br_2 $B \leftrightarrow X$ and $^1\Pi(1_u) \leftrightarrow X$ can be estimated, using equation (2), from the extinction coefficient data described on p. 209. From the early data,[99] Bayliss obtained[103] $(0.107 \times 10^{-8}\ cm)^2e^2$ and $(0.0676)^2e^2$ for $g'R_e^2$ of $^1\Pi(1_u) \leftarrow X$ and $B \leftarrow X$, respectively. Using a correction based on comparison of the Sulzer and Wieland parameters for the early data and those of Passchier *et al.*,[101] we now obtain $R_e^2\ (r_e'') = 0.148$ for $^1\Pi(1_u) \leftarrow X$ and 0.114 Debye [2] for $B \leftarrow X$.

The Einstein coefficient ($A_{v'}$) for spontaneous radiation from a vibrational level v' to a non-degenerate lower state (here $^1\Sigma_g^+$) is given by $A_{v'} = (64\pi^4/3h)\ \overline{\nu_{v'}^3}\ R_e^2$, where $\overline{\nu_{v'}^3}$, is the mean (Franck–Condon weighted) value for transitions to all v'' of the lower state, and R_e is assumed independent of r. Since for transitions from low v' levels most of the contribution to the $\overline{\nu^3}$ term is from a small range of r, this approximation is only subject to small error. $\overline{v_{v'}^3}$ for $B \rightarrow X$ transitions from low v' levels of $B^3\Pi(0_u^+)$ is estimated approximately as $(11\ 000\ cm^{-1})^3$, giving $A \sim 4 \times 10^4\ s^{-1}$ and a radiative lifetime of $\sim 2.5 \times 10^{-5}$ s. If $R_e(r)$ increases with r as found for I_2, the lifetimes would be somewhat lower. The result $\tau \sim 2.5 \times 10^{-5}$ s is considerably less than the previous estimate[105] of 7×10^{-5} s. From the ratio, $R_e^2(B-X)/R_e^2(A-X)$ at $\bar{r} \sim 2.45$ Å, and the lower $\overline{\nu_{v'}^3}$ of $\sim(8000\ cm^{-1})^3$ for $A^3\Pi(1_u) \rightarrow X$, a radiative lifetime $\sim 2 \times 10^{-3}$ s is estimated for bound levels of $A^3\Pi(1_u)$.

The only direct lifetime measurements for Br_2 are those of Capelle, Sakurai, and Broida[106] for $B^3\Pi(0_u^+)$ using broad-band laser excitation. There is some doubt,[107] however, about there liability of these measurements, owing to possible contamination by I_2 impurity. The measured lifetimes appeared to vary randomly with laser frequency between 1.3 and 0.2 μs, much shorter than the radiative lifetime estimated above ($\sim$25 μs). If the shorter lifetimes and their variation with laser frequency (*i.e.* with v' and J'), are confirmed, efficient predissociation of $B^3\Pi(0_u^+)$ to $^1\Pi(1_u)$ would be indicated.

The $E \leftrightarrow 2431\ ^3\Pi$ **system of** Br_2.—Following flash photolysis of bromine, Briggs and Norrish[108] observed a new banded absorption spectrum λ2411—2872 Å, which was assigned as a single progression from Br_2 $B^3\Pi(0_u^+)$, $v = 0$. Wieland, Tellinghuisen, and Nobs[109] have recently found that the band-head

[105] M. A. A. Clyne, J. A. Coxon, and A. R. Woon Fat, *Trans. Faraday Soc.*, 1971, **67,** 3155.
[106] G. Capelle, K. Sakurai, and H. P. Broida, *J. Chem. Phys.*, 1971, **54,** 1728.
[107] H. P. Broida, personal communication, 1971.
[108] A. G. Briggs and R. G. W. Norrish, *Proc. Roy. Soc.*, 1963, **A276,** 51.
[109] K. Wieland, J. B. Tellinghuisen, and A. Nobs, *J. Mol. Spectroscopy*, 1972, **41,** 69.

measurements can be interpreted more satisfactorily as constituting eight progressions from $0 \leqslant v \leqslant 7$, and report the expression

$$\nu = 35\,731.5 + 150.895(v'+\tfrac{1}{2}) - 0.495(v'+\tfrac{1}{2})^2 + 6.5\times10^{-5}(v'+\tfrac{1}{2})^4 - G(v'')$$

where $G(v'')$ are known from the spectroscopic constants [79] for $^{79,81}Br_2$ $B^3\Pi(0_u^+)$. This expression is also found to represent most of the band heads of the strong emission system observed [110] from a discharge through bromine–argon mixtures. Wieland *et al.* therefore suggest that the absorption and emission data may relate to the same system, 1432 $E^3\Pi(0_g^+) \leftrightarrow$ 2431 $B^3\Pi(0_u^+)$. However, it is surprising that the 1432 $^3\Pi(1_g, 2_g) \leftarrow$ 2431 $^3\Pi(1_u, 2_u)$ systems are not also observed in the flash photolysis experiment as 2431 $^3\Pi(1_u, 2_u)$ would be expected to have larger populations than the $B(0_u^+)$ component.

5 Chlorine

The $X^1\Sigma_g^+$ Ground State of Cl_2.—High-resolution $B^3\Pi(0_u^+) \leftarrow X^1\Sigma(0_g^+)$ absorption data [111–113] have defined accurately the rotational constants and vibrational terms for $v'' \leqslant 5$. The whole data were critically assessed in a recent paper,[113] and the expressions

$$B_v'' = 0.243\,99 - 0.001\,49(v''+\tfrac{1}{2}) - 0.000\,007(v+\tfrac{1}{2})^2$$

$$G(v'') = 559.72(v''+\tfrac{1}{2}) - 2.675(v''+\tfrac{1}{2})^2 - 0.0067(v''+\tfrac{1}{2})^3$$

were obtained for $^{35}Cl_2$. These constants were found to give a good computer simulation of the rotational structure of the Raman bands of Cl_2.[114]

Data for higher levels are available from a u.v. resonance fluorescence doublet series,[115] analogous to that described for Br_2. A reconsideration [113] of Rao and Venkateswarlu's data [115] showed that their assignment of J_r', the quantum number of the rotational level of the excited state responsible for the doublet series, was probably in error by unity. The reassignment $J_r' = 21$ gives much closer agreement between the doublet separations and those calculated from the B_v'' expression above. New $G(v'')$ values for $X^1\Sigma_g^+$ have therefore been obtained [14] for $v'' \leqslant 49$, taking account of the revised numbering. Two additional doublets assigned by Rao and Venkateswarlu as $v'' = 52$ and 54 were also observed. Inclusion of these two points in a smooth Birge–Sponer extrapolation of negative curvature yields a dissociation energy for the ground state about 40 cm^{-1} higher than that accurately fixed from the dissociation limit of $B(0_u^+)$ to Cl $^2P_{3/2}$+Cl $^2P_{1/2}$. The suggestion of a potential maximum $\sim$40 cm^{-1} for Cl_2 $X^1\Sigma_g^+$ is untenable from long-range theory,

110 P. Venkateswarlu and R. D. Verma, *Proc. Indian Acad. Sci.* (*A*), 1957, **47**, 251.
111 A. E. Douglas, C. H. Møller, and B. P. Stoicheff, *Canad. J. Phys.*, 1963, **41**, 1174.
112 W. G. Richards and R. F. Barrow, *Proc. Chem. Soc.*, 1962, 297; W. G. Richards, Ph.D. thesis, Oxford, 1962.
113 M. A. A. Clyne and J. A. Coxon, *J. Mol. Spectroscopy*, 1970, **33**, 381.
114 K. Altmann and G. Strey, *Z. Naturforsch.*, 1972, **27a**, 65.
115 Y. V. Rao and P. Venkateswarlu, *J. Mol. Spectroscopy*, 1962, **9**, 173.

which predicts [34] an attractive C_6/r^6 term. The Birge–Sponer plot is thus predicted to show positive curvature for the highest levels, hence implying an even larger discrepancy. Le Roy and Bernstein [116] proposed therefore that the two doublets are better assigned as $v'' = 52$ and 55, although Le Roy now suggests that 53 and 57 are more likely (see Chapter 3 for further details). It is evident that a final interpretation will require more precise experimental work on this and possibly other resonance series. The most recent value [34] of D_0^0 of $^{35}Cl_2$ $X^1\Sigma_g^+$ is 19997.14 ± 0.14 cm^{-1} from the long-range potential of $B^3\Pi(0_u^+)$.

Discrete Levels of Cl_2, **2431** $^3\Pi$.—Rotational analyses [111–113] of the $B^3\Pi(0_u^+) \leftarrow X^1\Sigma_g^+$ system have given vibrational terms and rotational constants for the range $5 \leqslant v' \leqslant 32$ of $B^3\Pi(0_u^+)$. However, the vibrational term for $v' = 32$,[112] which lies only about 1 cm^{-1} from the dissociation limit of $B^3\Pi(0_u^+)$ to $^2P_{3/2} + {}^2P_{1/2}$ atoms is found unreliable [97] from long-range analysis. Vibrational terms for $v' \leqslant 4$ are known less accurately from the chlorine atom recombination emission spectrum.[117] Rotational constants down to $v' = 0$ have been estimated [113] from a quadratic fit in $(v+\frac{1}{2})$ of those for $5 \leqslant v' \leqslant 10$, and were used to derive a new RKR curve [14] for the $^3\Pi(0_u^+)$ state. Although some error undoubtedly exists in the extrapolation, the rotational constants are considered by the Reporter to be more reliable than those used in an earlier RKR calculation.[118] A rotational constant for $v' = 0$ could possibly be obtained from a higher-resolution study of the chlorine atom recombination emission spectrum, and would enable an accurate interpolation between $v' = 0$ and $v' = 5$. The Reporter intends to attempt this experiment. The RKR curves obtained for the B and X states were recently used [119] to obtain a set of Franck–Condon factors for the $B \leftrightarrow X$ system, although extensive intensity data do not yet exist. Holzer *et al.*[87] have observed a few members $(1 \leqslant v'' \leqslant 4)$ of the $B(v' = 22, J' \sim 38) \rightarrow X^1\Sigma_g^+$ resonance fluorescence doublet series of $^{35}Cl_2$ exicted by λ4880 Å radiation. Their intensity data are in good agreement with those expected from the Franck–Condon factors.

In contrast to the spectra of Br_2 and I_2, there is no evidence in the absorption spectrum of Cl_2 for bands which could be assigned to the $A^3\Pi(1_u) \leftarrow X^1\Sigma_g^+$ system. [The $^3\Pi(0_u^+)$ state of Cl_2 is often labelled as the A state in the literature, but to conform with I_2 and Br_2, the B-labelling is adopted here.] The dipole strength of the $A^3\Pi(1_u) \leftrightarrow X^1\Sigma_g^+$ transition is expected (see next section) to be much less than that of $B \leftrightarrow X$. However, some evidence for the $A \rightarrow X$ system of Cl_2 in emission has been found [120] from the chlorine atom recombination spectrum, in which two weak progressions, not part of the $B \rightarrow X$ system, were identified. The same bands also appeared, but

[116] R. J. Le Roy and R. B. Berstein, *J. Mol. Spectroscopy*, 1971, **37**, 109.
[117] M. A. A. Clyne and J. A. Coxon, *Proc. Roy. Soc.*, 1967, **A298**, 424.
[118] J. A. C. Todd, W. G. Richards, and M. A. Byrne, *Trans. Faraday Soc.*, 1967, **63**, 2081.
[119] J. A. Coxon, *J. Quant. Spectroscopy Radiative Transfer*, 1971, **11**, 1355.
[120] J. A. Coxon, Ph.D. thesis, East Anglia, 1967.

with enhanced intensity, in a spectrum of the nitrogen trichloride decomposition flame at pressures near 1 Torr; [121] the heads measured from this spectrum are shown in Table 6, together with the known [14] vibrational intervals for the X state. The two progressions are assigned as $v' = 1$ and 2 for the following reasons: (i) the long radiative lifetime of Cl_2 $A^3\Pi(1_u)$ would ensure extensive vibrational relaxation before radiation; (ii) the single vibrational interval, $\Delta G \sim 245$ cm^{-1}, is similar, as expected, to that of $B^3\Pi(0_u^+)$, $\Delta G(1\frac{1}{2})$ ~ 239 cm^{-1}; (iii) the Franck–Condon factors for $A \rightarrow X$ emission are expected to be closely similar for low v' levels to those of the $B \rightarrow X$ system. Assignment as $v' = 0$ and 1 would be implausible since the Franck–Condon factors for $v' = 0 \rightarrow v'' = 8, 9$ are predicted to be low; (iv) the assignment $v' = 1$ and 2 places the $^3\Pi(1_u)$ state approximately at the energy expected theoretically (see next section). The $A \rightarrow X$ heads were all very weak, and their frequency measurements correspondingly are likely to be subject to considerable error, possibly ~ 5 cm^{-1}. Because of this, and the relationship, $\Delta G(v' + \frac{1}{2}) \sim 0.5\ \Delta G(v'' + \frac{1}{2})$ in this spectral region, various other arrangements of the bands in a Deslandres Table are possible. For example, the 2–12 band of Table 6 could perhaps also be assigned as 0–11. Further measurements at higher dispersion preferably with accurate isotope splittings, would be valuable in obtaining a more certain interpretation.

Continuum Absorption, Cl_2 (2431 $^{1,3}\Pi_u \leftarrow X^1\Sigma_g^+$).—Several detailed studies of the u.v. continuous absorption spectrum of chlorine have been reported. Early data [122, 123] from plate photometry are in good agreement with recent photoelectric determination [100] of the wavelength dependence of the extinction coefficient. From the analogous spectra already described for I_2 and Br_2 for which, in accordance with theory, the coupling tends more to Hund's case (a) with decreasing mass, the spectrum for Cl_2 is expected to be mainly $^1\Pi(1_u)$ $\leftarrow X^1\Sigma_g^+$. The observation of banded $^3\Pi(0_u^+) \leftarrow X^1\Sigma_g^+$ absorption up to the convergence limit would imply, however, that some continuous absorption to the repulsive limb of $B^3\Pi(0_u^+)$ should also be present. Absorption to $^3\Pi(1_u)$ is shown below to be unimportant.

When a continuous absorption spectrum is known to be due to overlapping transitions to more than one upper state, a most useful approach seems to be that adopted by the earlier experimentalists.[122] Briefly, measurements of extinction coefficients as a function of temperature enable the contribution from the $v'' = 0$ level of the ground state to be obtained. Results are then expressed in terms of ε_0, the hypothetical extinction coefficient at 0 K for all molecules in $v'' = 0$. Using this method, Aikin and Bayliss [122] found that the ε_0 *vs.* frequency curve showed a long tail at low frequencies, in accordance with the expectation that a weak $^3\Pi(0_u^+) \leftarrow X$ continuum should be present at lower frequencies than the main $^1\Pi(1_u) \leftarrow X$ system. This

[121] T. C. Clark and M. A. A. Clyne, *Trans. Faraday Soc.*, 1970, **66**, 372.
[122] R. G. Aikin and N. S. Bayliss, *Trans. Faraday Soc.*, 1937, **33**, 1333.
[123] G. E. Gibson and N. S. Bayliss, *Phys. Rev.*, 1933, **44**, 188.

Table 6 *Band heads measured*[a] *from the decomposition flame of nitrogen trichloride,*[b] *and assigned as* Cl_2 $A^3\Pi(1_u) \rightarrow X^1\Sigma_g^+$

v' \ v''	8		9		10		11		12	$\Delta G(v'+\frac{1}{2})$
1	13266	509	12757	502	12255	495	11760			
	238		245		248		248			245
2	13504	502	13002	499	12503	495	12008	492	11516	
$\Delta G(v''+\frac{1}{2})$[c]		510.0		504.2		498.4		492.5		

(*a*) J. A. Coxon, Ph.D. thesis, University of East Anglia, 1967; (*b*) T. C. Clark and M. A. A. Clyne, *Trans. Faraday Soc.*, 1970, **66**, 372; (*c*) J. A. Coxon, *J. Quant. Spectroscopy Radiative Transfer*, 1971, **11**, 443.

interpretation has been adopted generally,[4, 36] although Wilkins [124] assigned the whole continuous absorption to $B(0_u^+) \leftarrow X$.

In recent years, RKR potential energy curves for the $B(0_u^+)$ state of Cl_2 have been calculated,[14, 118] and it is now possible to undertake a more detailed examination of the earlier absorption data. For bromine, Bayliss [103] found a simple empirical approach, in which the ε_0/ν curve is accurately reproduced by reflection of $\psi^2(v'' = 0)$ across the repulsive limb of the upper state(s). Figure 9 shows the low frequency tail of the ε_0 curve for Cl_2, carefully measured from Figure 2 of ref. 122, together with $\psi^2(v'' = 0)$. For $\nu > 25\,000$ cm^{-1}, the rapidly rising ε_0 values (maximum $\varepsilon_0 \sim 70$ at $\sim 30\,300$ cm^{-1}) are due almost entirely to $^1\Pi(1_u) \leftarrow X^1\Sigma_g^+$ absorption. This has recently been clearly demonstrated by photodissociation translational spectroscopy [125, 126] at 28 810 cm^{-1}. In a first approximation, contribution from $B \leftarrow X$ can be neglected, and a rough $^1\Pi(1_u)$ potential curve estimated by Bayliss's method. A short extrapolation of this curve then predicts $^1\Pi(1_u) \leftarrow X$ absorption to be negligible below $\nu \sim 25\,000$ cm^{-1} (at 0 K). The low frequency region can then be assigned with confidence as wholly $B \leftarrow X$, and in fact Bayliss's method qualitatively predicts the observed ε_0 variation in this region. Quantitatively, however, the predicted increase in ε_0 with frequency is a little greater than that observed. The author has, therefore, calculated continuum wavefunctions for Cl_2 $B(0_u^+)$, and the Franck–Condon overlap integrals with $v'' = 0$. The observed ε_0/ν variation between 21 000 and 24 300 cm^{-1} was found to be in excellent agreement with the Franck–Condon factor variation. For this purpose, the RKR curve was extended above $v' = 20$ into the continuum by a Morse curve, $U(r) = D_e(1 - e^{-\alpha(r-r_e)})^2$. A second estimate, shown by a broken line in Figure 9, was obtained using the variable α Morse technique, in which α is allowed to vary with energy in the same way as that for the bound region. This curve differs little at low energies from the simple Morse curve, and predicts closely similar behaviour for ε_0/ν. The integral $\int \varepsilon_0 \, d\nu/\nu$ for $B \leftarrow X$ obtained below is thus subject to little error. Subtracting out the predicted ε_0 values from the total ε_0 above 24 000 cm^{-1} gives ε_0 for $^1\Pi \leftarrow X$ absorption, which itself is then quite satisfactorily in accord with extrapolation of the $^1\Pi(1_u)$ potential curve. The present discussion thus suggests a self-consistent interpretation of the low frequency ε_0 values as contributions from both $^1\Pi \leftarrow X$ and $B \leftarrow X$ absorption. ε_0 for $B \leftarrow X$ is predicted to reach a maximum of ~ 3 l mol^{-1} cm^{-1} at 25 500 cm^{-1}, considerably higher than the values of ~ 1 (23 530 cm^{-1}) [122] and ~ 1.5 (23 600 cm^{-1}) [36] estimated by previous workers.

The electronic transition moment for $B \leftarrow X$ near r_0'' (~ 2.0 Å) was calculated from the ε_0 curve in Figure 9, with the result, $R_e^2 = 0.0059$ Debye ^{2} ($D = 2.56 \times 10^{-4}$ Å^2). This value is about 2.5 times that estimated by

[124] R. L. Wilkins, *J. Quant. Spectroscopy Radiative Transfer*, 1964, **4**, 175.
[125] G. E. Busch, R. T. Mahoney, R. I. Morse, and K. R. Wilson, *J. Chem. Phys.*, 1969, **51**, 449.
[126] R. W. Diesen, J. C. Wahr, and S. E. Adler, *J. Chem. Phys.*, 1971, **55**, 2812.

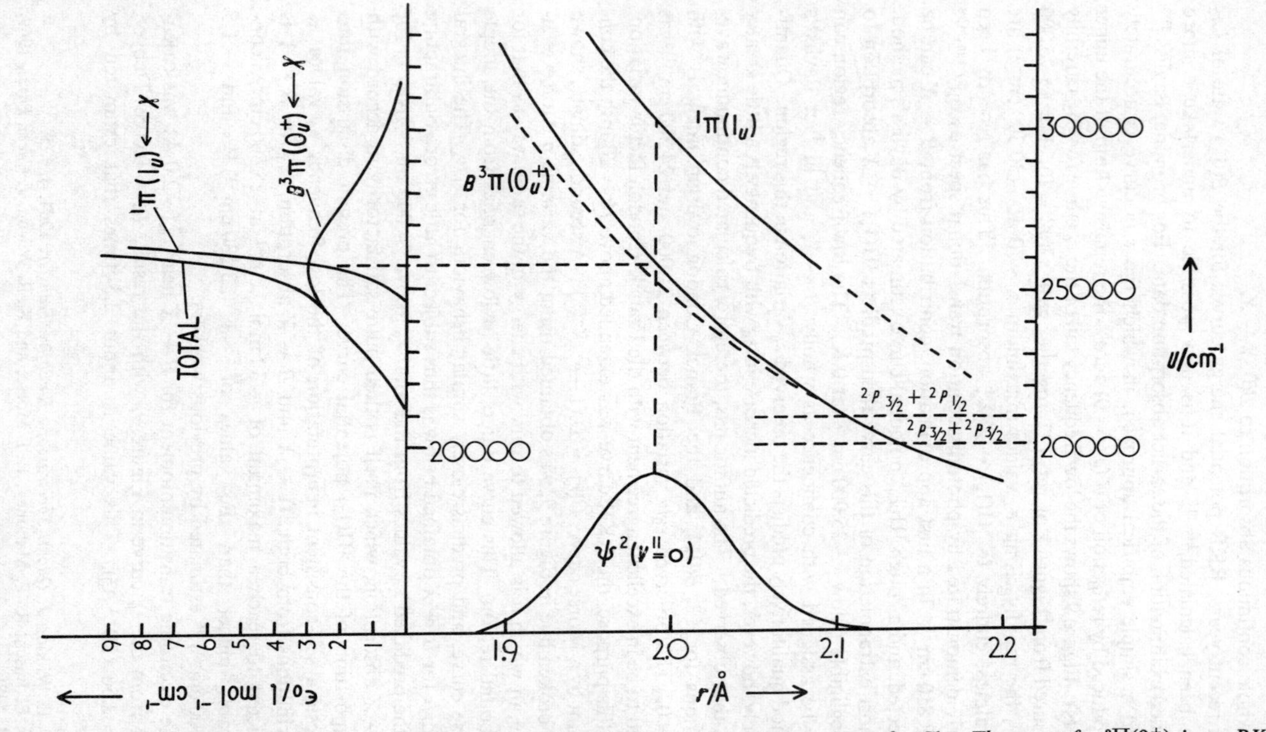

Figure 9 *Potential curves (relative to* $v'' = 0$*) and low-frequency tail of the absorption spectrum for* Cl_2. *The curve for* $^3\Pi(0_u^+)$ *is an RKR potential,*[14] *with two different Morse-based extrapolations above* $v' = 20$. $\psi^2(v'' = 0)$ *was used to calculate the* $^3\Pi(0_u^+) \leftarrow X$ *component of the absorption (see text)*

Mulliken,[4] who used the original estimate of ε_0 (max) ~ 1 l mol^{-1} cm^{-1} of Aikin and Bayliss.[122] The radiative lifetime for bound vibrational levels of Cl_2 $B^3\Pi(0_u^+)$ can now be estimated as for Br_2. $\overline{\nu^3_{0,v''}}$ for transitions from $v' = 0$ to all levels of X was calculated from the Franck–Condon data [119] as (9840 cm^{-1}).[3] This then gives $\tau \sim 0.6$ ms for purely radiative decay of B-state molecules. It has been usual [24, 127] to accept a value of $\tau = 7\times10^{-5}$ s, about an order of magnitude less than the present estimate. This latter value is erroneously attributed to Sulzer and Wieland, but in fact was only estimated (incorrectly) from their data. It should be noted that Carabetta and Palmer's work [128] indicating a lifetime of ~ 2 μs and about equal transition probabilities for $B^3\Pi(0_u^+) \leftarrow X$ and $^1\Pi(1_u) \leftarrow X$ was incorrect [127] since the potential curve was based on the earlier [129] wrong vibrational numbering. In a later assessment,[130] it was not possible to fix an experimental radiative lifetime for Cl_2 $B(0_u^+)$.

Because of the present reassessment of the continuous absorption by Cl_2, Mulliken's estimate [4] for the dipole strength of Cl_2 $(A-X)$ requires revision. The assignment suggested above for the Cl_2 $(A \rightarrow X)$ band heads gives a $B-A$ vertical separation of ~ 350 cm^{-1} and hence a $^1\Pi(1_u) - {}^3\Pi(1_u)$ separation of ~ 4900 cm^{-1}. This gives $X = 4865$ cm^{-1} and $D(^1\Pi)/D[^3\Pi(1_u)] \sim 280$ for the ratio of the dipole strengths of $^1\Pi \leftarrow X$ and $A^3\Pi(1_u) \leftarrow X$. $D[^3\Pi(1_u)]$ is thus estimated as about 10^{-5} Å^2, about twice that given by Mulliken.[4] However, absorption to $^3\Pi(1_u)$ is still expected to be negligible compared with that to the B state, $gD(B)/gD(A) \sim 13$, and the radiative lifetime of bound A state molecules is predicted to be about 15 ms.

The $E \leftrightarrow B^3\Pi(0_u^+)$ System of Cl_2.—In a similar criticism to that given for the absorption bands of flash-photolysed Br_2, Wieland *et al.*[109] have re-interpreted the analogous bands of Cl_2.[108] The previous assignment [108] of a single progression from $B^3\Pi(0_u^+)$, $v = 0$ is now replaced by two progressions from $v' = 0, 1$. With $G(v'')$ known [112] for Cl_2 $B^3\Pi(0_u^+)$, the expression

$$\nu_{v',v''} = 40\,143.9 + \{249.75(v'+\tfrac{1}{2}) - 0.875(v'+\tfrac{1}{2})^2\} - G(v'')$$

accounts quite accurately for all the absorption bands. The same equation is also found [109] to represent most of the emission bands observed by Venkateswarlu and Khanna [131] between λ 2390 and 2600 Å, and which also appear from Cl_2–active-nitrogen systems,[132, 133] and from photodissociation of $COCl_2$.[134]

[127] M. A. A. Clyne and D. H. Stedman, *Trans. Faraday Soc.*, 1968, **64**, 1816.
[128] R. A. Carabetta and H. B. Palmer, *J. Chem. Phys.*, 1967, **46**, 1325.
[129] G. Herzberg, 'Spectra of Diatomic Molecules', Van Nostrand, 1950, 2nd edn.
[130] H. B. Palmer and R. A. Carabetta, *J. Chem. Phys.*, 1968, **49**, 2466.
[131] P. Venkateswarlu and B. N. Khanna, *Proc. Indian Acad. Sci.* (*A*), 1959, **49**, 117.
[132] W. H. B. Cameron and A. Elliott, *Proc. Roy. Soc.*, 1939, **A169**, 463.
[133] G. M. Provencher and D. J. McKenney, *Chem. Phys. Letters*, 1970, **5**, 26.
[134] H. Okabe, A. H. Laufer, and J. J. Ball, *J. Chem. Phys.*, 1971, **55**, 373.

6 Other Topics

Fluorine.—Since neither banded absorption nor emission $^3\Pi_u \leftrightarrow X^1\Sigma_g^+$ spectra, analogous to those described for Cl_2, Br_2 and I_2, are yet known for F_2, a precise, spectroscopically based value for the dissociation energy, D_0^0 (F_2) is not available. However, studies of both photo-ionization [135] and electron impact [136] of F_2, with mass spectrometric detection and kinetic energy analysis of the ionic fragments, now give closely similar values of D_0 (F_2), 36.4 ± 0.7 and 37.2 ± 2.2 kcal mol^{-1}, respectively. The results are also in excellent agreement with the values $D_0^0 = 36.71 \pm 0.13$ and $D_{298}^0 = 37.72 \pm 0.13$ kcal mol^{-1} obtained by Stamper and Barrow [137] from analysis of pressure measurements [138] made on partly dissociated fluorine. The concurrent work of Dibeler *et al.*,[139] interpreted to give a much lower value of 30.9 ± 0.7 kcal mol^{-1}, has been critically assessed by Berkowitz *et al.*,[135] and Dibeler now accepts [140] that his value is in error. Values for the dissociation energy of fluorine have also been based on a short progression of bands observed [141, 142] in absorption near λ870 Å, though some of the assignments have been questioned by Dibeler.[143] Di Lonardo and Douglas [144] have recently observed these bands at high resolution, and conclude that the upper-state levels are violently perturbed. D_0^0 (F_2) values based on a Birge–Sponer extrapolation of these bands are then perhaps only in fortuitous agreement with the mass spectro-

Table 7 *Vibrational intervals and rotational constants* (cm^{-1}) *of* F_2 $X^1\Sigma_g^+$

$\Delta G''(1/2)$	$\Delta G''(3/2)$	B_0''	α	*Reference*
893.95 ± 0.10	870.05 ± 0.10	0.8827 ± 0.001	0.0131 ± 0.0005	*a*
—	—	0.8841 ± 0.0006	—	*b*
891.85 ± 0.4	—	0.8828 ± 0.001	—	*c*

(*a*) G. Di Lonardo and A. E. Douglas, *J. Chem. Phys.*, 1972, **56**, 5185; (*b*) H. H. Claassen, H. Selig, and J. Shamir, *Appl. Spectroscopy*, 1969, **23**, 8; (*c*) D. Andrychuk, *Canad. J. Phys.*, 1951, **29**, 151.

[135] W. A. Chupka and J. Berkowitz, *J. Chem. Phys.*, 1971, **54**, 5126; J. Berkowitz, W. A. Chupka, P. M. Guyon, J. H. Holloway, and R. Spohr, *J. Chem. Phys.*, 1971, **54**, 5165.
[136] J. J. de Corpo, R. P. Steiger, J. L. Franklin, and J. L. Margrave, *J. Chem. Phys.*, 1970, **53**, 936.
[137] J. C. Stamper and R. F. Barrow, *Trans. Faraday Soc.*, 1958, **54**, 1592.
[138] R. N. Doescher, *J. Chem. Phys.*, 1952, **20**, 330; H. Wise, *J. Phys. Chem.*, 1954, **58**, 389.
[139] V. H. Dibeler, J. A. Walker, and K. E. McCulloh, *J. Chem. Phys.*, 1969, **50**, 4592; 1969, **51**, 4230; 1970, **53**, 4414.
[140] V. H. Dibeler, personal communication, 1972.
[141] W. Stricker and L. Krauss, *Z. Naturforsch.*, 1968, **23a**, 486.
[142] R. P. Iczkowski and J. L. Margrave, *J. Chem. Phys.*, 1959, **30**, 403.
[143] V. H. Dibeler, 'Recent Developments in Mass Spectroscopy', ed. K. Ogata and T. Hayakawa, University of Tokyo Press, 1970, p. 781.
[144] G. Di Lonardo and A. E. Douglas, *J. Chem. Phys.*, 1972, **56**, 5185.

metric values given above. Although the work of Di Lonardo and Douglas is not yet fully reported, values [144] for the rotational constant B_0'', and for the two lowest vibrational intervals of $X^1\Sigma_g^+$ have been obtained from 25 absorption bands in the vacuum u.v. Their results are compared in Table 7 with data from the Raman spectrum of F_2.[145, 146] It was also possible to locate in absorption the two vibrational levels of F_2 $C^1\Sigma^+$ responsible for the well-known [141, 147] emission bands, $C^1\Sigma^+ \rightarrow B, B'\ ^1\Pi$.

Some years ago, Rees [148] discussed the intensity variation with wavelength of the λ 4550—2100 Å continuous absorption spectrum of F_2. As is expected from the trend shown by the heavier halogens, this absorption is almost wholly $^1\Pi_u \leftarrow X^1\Sigma_g^+$, the $^1\Pi_u$ state correlating with two F $^2P_{3/2}$ atoms. Since the intensity data [149] are for one temperature only, ε_0 values cannot be obtained other than by the method of Sulzer and Wieland.[36] However, because of the relatively large ground-state vibrational frequency of F_2, only $v'' = 0$ and 1 contribute significantly to the room-temperature absorption. Rees used Bayliss's [103] method, with Morse wavefunctions for $v'' = 0$ and 1, to derive a potential energy curve for the $^1\Pi_u$ state. Convincing evidence was also obtained for a very weak contribution ($\varepsilon_{max} \sim 0.1$ l mol^{-1} cm^{-1} at $\nu \sim 25\,000$ cm^{-1}) in the low-frequency tail of the $^1\Pi_u \leftarrow X$ absorption curve, from which a small segment of the $^3\Pi_{0u}^+$ curve near r_e'' could be determined. Rees's suggestion of the value of a search for the $^3\Pi_{0u}^+ \rightarrow X$ system in emission, which would help to locate this state in more detail, still stands. The dipole strength, $D \sim 5 \times 10^{-6}$ Å^2, implies a radiative lifetime of about 0.3 s for bound levels of F_2 $^3\Pi_{0u}^+$, and it may be possible to observe the system in a F+F fluorine afterglow at $\lambda > 1$ μm from the products of discharge mixtures of CF_4, SF_6, or F_2 with argon.

Heteronuclear Halogens.—*Introduction.* The low-lying states of the heteronuclear halogens are analogous to those of the homonuclear molecules. One important difference, however, arises from the absence of the *g*,*u* symmetry property, which for the homonuclear molecules, allows the 0_g^+ repulsive state from $X\,^2P_{3/2}+X\,^2P_{3/2}$ to cross $^3\Pi(0_u^+)$ from $X\,^2P_{3/2}+X\,^2P_{1/2}$. For the heteronuclear halogens, the interaction between 0^+ and $^3\Pi(0^+)$ appears sufficiently large that the non-crossing rule is a good approximation. In effect, therefore, two new potential curves are formed; the original $^3\Pi(0^+)$ state now correlates with two ground-state atoms *via* a potential maximum, while the new 0^+ state possesses bound vibrational levels. Fuller accounts of the properties of this type of intersection are given by Herzberg,[129] and by Selin [150] in the case of IBr.

[145] D. Andrychuk, *Canad. J. Phys.*, 1951, **29**, 151.
[146] H. H. Classen, H. Selig, and J. Shamir, *Appl. Spectroscopy*, 1969, **23**, 8.
[147] T. L. Porter, *J. Chem. Phys.*, 1968, **48**, 2071.
[148] A. L. G. Rees, *J. Chem. Phys.*, 1957, **26**, 1567.
[149] R. K. Steunenberg and R. C. Vogel, *J. Amer. Chem. Soc.*, 1956, **78**, 901.
[150] L.-E. Selin, *Arkiv Fysik*, 1962, **21**, 529.

ICl. Holleman and Steinfeld [151] have recently reported the first observation of ICl fluorescence excited optically. The fluorescence was examined as functions of pressure and excitation wavelength over the range, λ 5850—6100 Å, where excitation is to bound vibrational levels, $25 \gtrsim v' \gtrsim 16$ of ICl

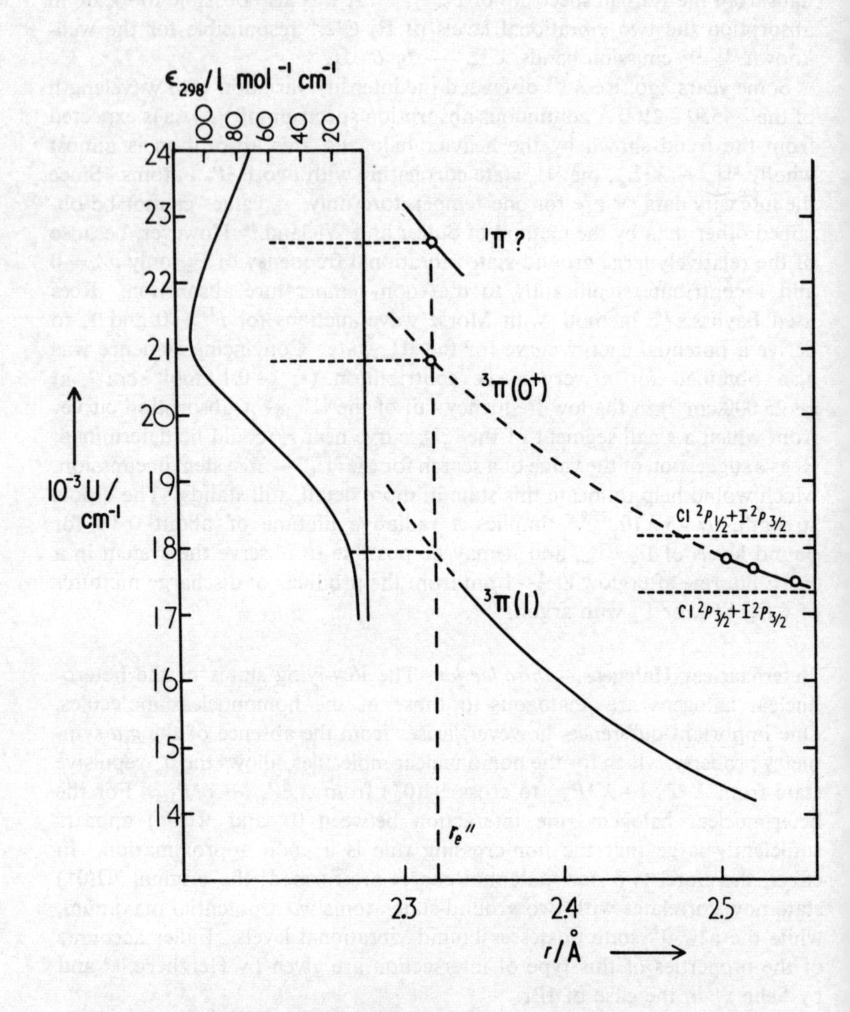

Figure 10 *Potential curves (relative to $v'' = 0$) and extinction coefficients* [100] *at* 298 K *for* ICl. *The solid curve for* $^3\Pi(1)$ *is an RKR potential* (*Table 8*), *as are the three points for* $^3\Pi(0^+)$. *The two points at* r_e'' *are the maxima of the resolved Gaussian components* [100] *and assigned here as* $^3\Pi(0^+)$ *and* $^1\Pi$

[151] G. W. Holleman and J. I. Steinfeld, *Chem. Phys. Letters*, 1971, **12**, 431.

$^3\Pi(1)$. This state is known in detail from analysis of the high-resolution absorption spectrum of ICl by Hulthén *et al.*[152] Figure 10 shows portions of the potential curves for the upper states of ICl. The $^3\Pi(1)$ curve is from RKR turning points (Table 8) calculated by the Reporter from the spectroscopic constants.[152, 153] The two sets of B_v values for the two Ω-

Table 8 *RKR turning points for* $I^{35}Cl\ ^3\Pi(1)$

v	$G(v)$	r_{min}	r_{max}
0	105.42	2.612	2.765
1	312.76	2.565	2.833
2	515.34	2.536	2.884
3	713.03	2.513	2.930
4	905.83	2.494	2.972
5	1093.62	2.478	3.012
6	1276.50	2.464	3.051
7	1453.99	2.452	3.090
8	1626.08	2.441	3.129
9	1792.55	2.430	3.168
10	1953.26	2.421	3.207
11	2107.95	2.413	3.248
12	2256.43	2.405	3.290
13	2398.46	2.398	3.334
14	2533.86	2.392	3.379
15	2662.37	2.386	3.427
16	2783.78	2.380	3.478
17	2897.90	2.375	3.532
18	3004.44	2.371	3.591
19	3103.25	2.367	3.654
20	3194.16	2.363	3.724
21	3277.10	2.359	3.800
22	3351.98	2.356	3.884
23	3418.99	2.354	3.977
24	3478.39	2.351	4.081
25	3530.67	2.349	4.196
26	3576.45	2.347	4.322
27	3616.52	2.345	4.458
28	3651.53	2.344	4.607
29	3681.91	2.343	4.771
30	3708.28	2.342	4.946
31	3731.05	2.341	5.141
32	3750.50	2.340	5.358
33	3766.91	2.339	5.603
34	3780.47	2.339	5.889
35	3791.38	2.338	6.229

[152] E. Hulthén, N. Järlsäter, and L. Koffman, *Arkiv Fysik*, 1960, **18**, 479.
[153] E. Hulthén, N. Johansson, and U. Pilsäter, *Arkiv Fysik*, 1958, **14**, 31.

doubled components were averaged, and a short extrapolation of B_v (mean) from $v' = 3$ to $v' = 0$ was made by the method described in ref. 92. Above $v' = 26$, the r_{min} limb of the RKR curve showed anomalous behaviour, which can be identified with the perturbations known[152] to set in at this energy. The data for $v' \geqslant 27$ in Table 8 assume the calculated RKR widths ($r_{max} - r_{min}$), but r_{min} is from a short extrapolation.

Extinction coefficients at room temperature as a function of wavelength have been obtained by Seery and Britton,[100] and their data are also plotted in Figure 10. Using the Sulzer and Wieland approximation,[36] they resolved the continuous absorption into two Gaussian components, with maximum extinction coefficients (ε_0 at 0 K, l mol^{-1} cm^{-1}) at frequencies (ν_0, cm^{-1}) and with half mean widths ($\Delta\nu_0$, cm^{-1}) respectively of 50.7, 20 818, 1229, and 100.3, 22 598, 2804. Identification of peak frequencies with approximate vertical energies suggests that $\nu_0 = 20\,818$ cm^{-1} can be assigned as $^3\Pi(0^+)$, since a reasonable potential curve for this state is then obtained by interpolation with three points from an RKR calculation. The interpretation of the second, higher frequency peak, $\nu_0 \sim 22\,598$ cm^{-1}, is discussed below with reference to the two continuous absorption peaks also shown by IBr. The much weaker discrete and continuous absorption to $^3\Pi(1)$, evident in the long, low-frequency tail of the absorption curve (Figure 10), was not considered by Seery and Britton, but is likely to cause little change in their parameters given above. Contribution to the low-frequency tail by absorption from excited vibrational levels of $X^1\Sigma^+$ to $^3\Pi(0^+)$ is calculated to be negligible at room temperature. Although data as a function of temperature would be valuable in providing ε_0 values for absorption from $v'' = 0$ alone, one can estimate with fair confidence a value of $\varepsilon_{max} \sim 14$ l mol^{-1} cm^{-1} at $\nu \sim 18\,000$ cm^{-1}.

Holleman and Steinfeld[151] estimate the absorption strength of ICl $^3\Pi(1) \leftarrow X$ to be about one-fiftieth that of I_2 absorption, which is now established to be due mainly to I_2 $B^3\Pi(0_u^+) \leftarrow X^1\Sigma_g^+$. This ratio is in line with the present estimate. Taking account of electronic degeneracy, the experimental lifetime for ICl $^3\Pi(1)$, $v' \sim 21$ of $\sim 10^{-4}$ s is in excellent agreement with the Einstein coefficient from the above absorption data, since $\overline{\nu^3_{v',v''}}$ values for emission to all vibrational levels of the X states are not very different for I_2 $B^3\Pi(0_u^+)$ and ICl $^3\Pi(1)$. Excitation to vibrational levels near the dissociation limit of ICl $^3\Pi(1)$ was found to result in a somewhat shorter lifetime, which Holleman and Steinfeld suggest may be due to a radiationless transition to $X^1\Sigma^+$, high v''.

Precise rotational constants, dipole moments, electric quadrupole coupling constants, and spin–rotation coupling constants for ICl $X^1\Sigma^+$, $v'' = 0$ and 1, have recently been determined[154] from the Stark effect on the $J'' = 2 \leftarrow 1$ microwave spectra of $I^{35}Cl$ and $I^{37}Cl$. The rotational constants, B_0'' and B_1'' are in excellent agreement with those of Hulthén *et al.*[153] (except the value[153]

[154] E. Herbst and W. Steinmetz, *J. Chem. Phys.*, 1972, **56**, 5342.

for B_1'' of $I^{37}Cl$, which seems to be misprinted). The dipole moment of ICl, 1.24 ± 0.02 Debye, is about twice that of previous estimates.

IBr. The low-lying states of IBr are analogous to those of ICl, the discrete levels being well known from the high-resolution absorption data of Selin and co-workers.[150, 155] As for ICl, Seery and Britton[100] have obtained extinction coefficients as a function of wavelength for continuous absorption at $\lambda \geqslant 2200$ Å. For both molecules, as well as BrCl, the higher-wavelength absorption peak was resolved into two Gaussian components; the parameters ε_0^{max}, ν_0, and $\Delta\nu_0$ for IBr are 169.8, 19 715, 1132, and 288.0, 20 951, 2211 respectively. As shown in Figure 10 for ICl, neither of these two components can be assigned as ${}^3\Pi(1) \leftarrow X$. However, by comparison with the electronic states of the homonuclear halogens, and in accordance with Mulliken's predictions,[3] strong absorption to ${}^1\Pi$ repulsive states would also be expected for the heteronuclear molecules. It is tempting therefore, to assign Seery and Britton's two components as ${}^3\Pi(0^+)$ and ${}^1\Pi(1)$. The absorption at higher energies, $\nu_0 = 37\,289$ cm^{-1} for IBr, would correspond to transition to the 2341 ${}^{1,3}\Pi$ configuration (2341 ${}^{1,3}\Pi_g \leftarrow X^1\Sigma_g^+$ is not allowed for the homonuclear molecules). The 0^+ component of ${}^3\Pi$ of this configuration is the state which intersects and mixes with 2431 ${}^3\Pi(0^+)$. This interpretation is supported by data from the photoelectron spectra[156] of ICl and IBr, which give values for the $\Pi - \Pi^*$ separations of 2.68 and 2.14 eV respectively, compared with the values 2.39 and 2.10 eV from Seery and Britton's parameters on the interpretation given above. It is probable, therefore, that the higher energy absorption should also be composed of ${}^1\Pi$ and ${}^3\Pi$ components. It should be mentioned that the 2431, 2341 notations used here refer to $\sigma\pi\pi^*\sigma^*$ orbitals, which for the heteronuclear halogens, are no longer equally associated with the two atoms.

A problem now arises, however, in relating Donovan and Husain's[157] observations on the flash photolysis of IBr, $\lambda \geqslant 3000$ Å, with the assignments suggested above. It was found that only *J*-excited Br ${}^2P_{1/2}$ atoms were produced. Similar observations for ICl were not possible because of overlapping of the vacuum-u.v. Cl atom absorption lines with molecular absorption. For IBr, the major production of Br ${}^2P_{1/2}$ is supported by the work of Giuliano and Hess,[158] who observed Br ${}^2P_{1/2} \rightarrow {}^2P_{3/2}$ 2.7 μm laser action from flash photolysis of IBr. In their discussion, Donovan and Husain imply that the low-frequency absorption of IBr (and ICl) is due wholly to ${}^3\Pi(0^+) \leftarrow X$. Because excitation is predominantly to energies considerably higher than the ${}^3\Pi(0^+) - 0^+$ intersection region, the high velocity of separating I and Br following absorption lowers the degree of mixing, and enables ${}^3\Pi(0^+)$ to

155 L.-E. Selin, *Arkiv Fysik*, 1962, **21**, 479; L.-E. Selin and B. Söderborg, *Arkiv Fysik*, 1962, **21**, 515.
156 A. W. Potts and W. C. Price, *Trans. Faraday Soc.*, 1971, **67**, 1242.
157 R. J. Donovan and D. Husain, *Trans. Faraday Soc.*, 1968, **64**, 2325.
158 C. R. Giuliano and L. D. Hess, *J. Appl. Phys.*, 1969, **40**, 2428.

dissociate to $I(^2P_{3/2})+Br(^2P_{1/2})$ (non-avoided crossing), rather than over the potential maximum to $I(^2P_{3/2})+Br(^2P_{3/2})$. With the interpretation of the absorption data given above, however, absorption to $^1\Pi$ would be stronger than that to $^3\Pi(0^+)$, and production of large amounts of $Br(^2P_{3/2})$ atoms would be expected on the basis that $^1\Pi$ is expected to correlate with ground-state atoms, $I(^2P_{3/2})+Br(^2P_{3/2})$.

The photodissociation translational spectrum [43] of IBr does give large amounts of $Br(^2P_{3/2})$ atoms from a state of 0^+ symmetry $[^3\Pi(0^+)]$; this is entirely reasonable because of the lower excitation energy (18 830 cm^{-1}). The observation of a second peak, also of $\Delta\Omega = 0$ symmetry, and leading to $I(^2P_{3/2})+Br(^2P_{1/2})$, might indicate that the crossing is only partially avoided even at this excitation energy. It seems unlikely that the suggested [43] assignment to 2341, $0^+ \leftarrow X$ could contribute significantly. Since the possible small contribution from a transition of $\Delta\Omega = 1$ symmetry could be due to weak $^3\Pi(1) \leftarrow X$ absorption, here again the apparent absence of $^1\Pi \leftarrow X$ absorption remains paradoxical. One possible explanation is that 2431 $^1\Pi(1)$ correlates like 2431 $^3\Pi(0^+)$ with $I(^2P_{3/2})+Br(^2P_{1/2})$, in which case it would be bound. For reasons given earlier (p. 179), there is no apparent reason why $^3\Pi(0^+)$ itself should not correlate with two ground-state atoms.

BrCl. Until recently, no visible infrared–visible banded spectrum of BrCl was known. A banded emission spectrum was first observed [117] from the decomposition flame of chlorine dioxide in the presence of Br_2 at pressures near 1 Torr, and the excitation was suggested to occur from the reaction $Br+ClOO \rightarrow BrCl(^3\Pi)+O_2$. The excited state of BrCl was assigned as $^3\Pi(0^+)$ since levels were observed above the thermochemically based dissociation limit of BrCl $^1\Sigma^+$ to $Br(^2P_{3/2})+Cl(^2P_{3/2})$. The same system was later observed in absorption,[159] using carefully chosen mixtures of Br_2+Cl_2 between 294 and 690 K. Preliminary results of analyses of a few bands of $^{79}Br^{35}Cl$ and $^{81}Br^{35}Cl$ recorded at high resolution have been reported,[160] and confirm both the assignment as $^3\Pi(0^+) \leftrightarrow X^1\Sigma^+$ and the previously reported vibrational numbering. Plots of $\Delta G(v+\frac{1}{2})$, B_v and D_v against v for the $^3\Pi(0^+)$ state all show large increases in curvature at $v \sim 6$, indicative of a distortion of the electronic potential at this energy. This is consistent with the similar behaviour shown by the $^3\Pi(0^+)$ states of all other interhalogens, which are perturbed as mentioned above by an avoided crossing with the 2341 0^+ repulsive state. A fuller analysis of the absorption bands by the Reporter is currently in progress.

ClF. The most recent analysis of the $B(0^+)$–$X^1\Sigma^+$ system of ClF was published by Schumacher *et al.*[160a] Infrared data for the ground state $v'' = 0$

[159] M. A. A. Clyne and J. A. Coxon, *Nature*, 1968, **217**, 448.
[160] M. A. A. Clyne and J. A. Coxon, *J. Phys.* (*B*), 1970, **3**, L9.
[160a] H. J. Schumacher, H. Schmitz, and P. H. Brodersen, *Anales Asoc. quim. argentina*, 1950, **38**, 98.

and 1 levels are also available.[160b] More recently, the electric quadrupole coupling constants and electric dipole moments for $J = 1$ and 2 of ClF, $X^1\Sigma^+$ have recently been determined [161] by molecular beam electric resonance spectroscopy. The dipole moments are in good agreement with earlier microwave data,[162] and have been found [163] to be in the sense $Cl^- F^+$. However, this finding conflicts with the opposite result, $Cl^+ F^-$, calculated by Green,[163a] a result which is in accord with simple electronegativity considerations.

BrF *and* IF. The $B^3\Pi(0^+)$ emission systems of BrF and IF have recently been observed [164] from the combination reactions of F $^2P_{3/2}$ with Br $^2P_{3/2}$ and I $^2P_{3/2}$ atoms in the presence of metastable electronically excited O_2 ($^1\Delta_g$, $^1\Sigma_g^+$). Since similar chemiluminescence is also known for I_2 [165] and Br_2,[105, 166] it appears that population of halogen $B^3\Pi(0^+)$ states by energy transfer from singlet oxygen is a general and efficient process. In all cases, levels considerably above that of the recombining atoms are populated, and the general mechanism

$$X\,^2P_{3/2} + Y\,^2P_{3/2} + M \rightarrow XY^* + M$$

$$XY^* + O_2(^1\Delta_g, {}^1\Sigma_g^+) \rightarrow XY\,^3\Pi(0^+) + O_2\,X^1\Sigma_g^+$$
$$XY\,^3\Pi(0^+) \rightarrow XY\,X^1\Sigma^+ + h\nu$$

has been suggested.[164] XY* could be one or several of the components of XY $^3\Pi$ correlating with two $^2P_{3/2}$ atoms. For BrF and IF, many new bands to previously unobserved levels of the $X^1\Sigma^+$ states were measured.

It is not yet unambiguously established to which atoms, F $^2P_{3/2}$+I, Br $^2P_{1/2}$ or F $^2P_{1/2}$+I, Br $^2P_{3/2}$, the $^3\Pi(0^+)$ states of IF and BrF correlate. For ICl and IBr, the lighter atom, Cl and Br respectively, is $^2P_{1/2}$. However, the absorption spectrum at low resolution [167] suggests the opposite for BrF, *i.e.* Br $^2P_{1/2}$+F $^2P_{3/2}$. This conclusion is based on the observation of a few bands of an electronic system assigned as $^3\Pi(1) \leftarrow X$. The vibrational levels of the upper state $^3\Pi(1)$ appear to converge at an energy ΔE below that of the 0^+ state, where ΔE corresponds closely with that of the Br $^2P_{1/2} - {}^2P_{3/2}$ separation. The bound 0^+ state is formed above the crossing of $^3\Pi(0^+)$ with repulsive 0^+, just as for ICl and IBr. A re-investigation of the absorption spectrum of BrF would be valuable, since the dissociation energy of BrF $X^1\Sigma^+$ based on the spectroscopic data as interpreted above [167] is at variance with esti-

[160b] A. N. Nielsen and E. A. Jones, *J. Chem. Phys.*, 1951, **19**, 117.
[161] R. Davis and J. Muenter, *J. Chem. Phys.*, 1972, **57**, 2836.
[162] D. A. Gilbert, A. Roberts, and P. A. Griswold, *Phys. Rev.*, 1949, **76**, 1723.
[163] J. J. Ewing, H. L. Tigelaar, and W. H. Flygare, *J. Chem. Phys.*, 1972, **56**, 1957.
[163a] S. Green, personal communication.
[164] M. A. A. Clyne, J. A. Coxon, and L. W. Townsend, *J. C. S. Faraday II*, in press.
[165] R. G. Derwent and B. A. Thrush, *J. C. S. Faraday II*, 1972, **68**, 720.
[166] M. A. A. Clyne, J. A. Coxon, and H. W. Cruse, *Chem. Phys. Letters*, 1970, **6**, 57.
[167] P. H. Broderson and J. E. Sicre, *Z. Phys.*, 1955, **141**, 515.

mates [168] from thermochemical data. A fuller discussion of this problem is given in ref. 164.

For IF, $^3\Pi(1) \leftarrow X$ bands have not yet been observed. In Durie's 1966 paper,[169] the earlier proposal [170] that IF $^3\Pi(0^+)$ correlates with I $^2P_{1/2}$ + F $^2P_{3/2}$, was considered less likely than I $^2P_{3/2}$ + F $^2P_{1/2}$. However, if IF is

Table 9 *Ionization energies and electron affinities of molecular halogens*

	Ionization energy/eV				Electron affinity/eV	
	Adiabatic[a]	*Vertical*	*Technique*	*Ref.*		*Ref.*
F_2	15.7		VUV	*b*	3.08 ± 0.1	*f*
	15.7		PES	*c*		
	15.70 ± 0.01	15.83 ± 0.01	PES	*d*		
	15.69 ± 0.01		PIMS	*e*		
	15.69 ± 0.01		PIMS	*a*		
Cl_2	11.47		VUV	*g*	2.38 ± 0.1	*f*
	11.49		PES	*c*	2.45 ± 0.15	*h*
	11.51 ± 0.01	11.59 ± 0.01	PES	*d*	2.52 ± 0.17	*i*
	11.48 ± 0.01		PIMS	*a*		
Br_2	10.51		PES	*c*	2.51 ± 0.1	*f*
	10.51 ± 0.01	10.56 ± 0.01	PES	*d*	2.55 ± 0.1	*h*
	10.52 ± 0.01		PIMS	*a*	2.87 ± 0.17	*i*
					2.23 ± 0.1	*j*
I_2	9.37		PIMS	*a*	2.58 ± 0.1	*f*
	9.22 ± 0.01	9.35 ± 0.01	PES	*d*	2.55 ± 0.1	*h*
					2.6 ± 0.1	*i*
ClF	12.66 ± 0.01		PES	*k*		
ICl	10.07 ± 0.01		PIMS	*a*		
	10.10 ± 0.02		PES	*d*		
IBr	9.79 ± 0.01		PIMS	*a*	2.7 ± 0.2	*f*
	9.85 ± 0.02		PES	*d*	2.55 ± 0.1	*h*

VUV —Vacuum-u.v. spectroscopy
PES —Photoelectron spectroscopy
PIMS —Photo-ionization mass spectroscopy

(*a*) Based on Table 2 of V. H. Dibeler, J. A. Walker, K. E. McCulloh, and H. M. Rosenstock, *Internat. J. Mass Spectrometry Ion Phys.*, 1971, **7**, 209; (*b*) R. P. Iczkowski and J. L. Margrave, *J. Chem. Phys.*, 1959, **30**, 403; (*c*) A. B. Cornford, D. C. Frost, C. A. McDowell, J. L. Ragle, and J. A. Stenhouse, *J. Chem. Phys.*, 1971, **54**, 2651; (*d*) A. W. Potts and W. C. Price, *Trans. Faraday Soc.*, 1971, **67**, 1242; (*e*) V. H. Dibeler, J A. Walker, and K. E. McCulloh, *J. Chem. Phys.*, 1969, **51**, 4230; (*f*) W. A. Chupka, J. Berkowitz, and D. Gutman, *J. Chem. Phys.*, 1971, **55**, 2724; (*g*) R. P. Iczkowski, J. L. Margrave, and J. W. Green, *J. Chem. Phys.*, 1960, **33**, 1261; (*h*) A. P. M. Baede, *Physica*, 1972, **59**, 541; (*i*) J. J. DeCorpo and J. L. Franklin, *J. Chem. Phys.*, 1971, **54**, 1885; (*j*) R. K. B. Helbing and E. W. Rothe, *J. Chem. Phys.*, 1969, **51**, 1607; (*k*) C. P. Anderson, G. Mamantov, W. E. Bull, F. A. Grimm, J. C. Carver, and T. A. Carlson, *Chem. Phys. Letters*, 1971, **12**, 137.

[168] JANAF Thermochemical Tables, 1st Addendum, ed. D. R. Stull, Dow Chemical Company, Michigan, 1966.
[169] R. A. Durie, *Canad. J. Phys.*, 1966, **44**, 337.
[170] R. A. Durie, *Proc. Roy. Soc.*, 1951, **A207**, 388.

analogous with all other interhalogens, for which intersection of $^3\Pi(0^+)$ with 0^+ affects the levels of $^3\Pi(0^+)$ below the crossing point, Durie's Birge–Sponer extrapolation might merely tend to the crossing-point energy, rather than that of the atomic products at convergence. The failure to observe emission bands from 0^+ levels above the crossing point cannot be regarded as strong evidence that they do not exist, since for BrF, ICl, and IBr, 0^+ levels have to date been observed in absorption only. It seems, therefore, that a firm spectroscopic value for the dissociation energy of IF is not available at the present time.

Ionization Energies and Electron Affinities.—Several investigations of the photoelectron spectra of the halogens have been reported. The more recent high-resolution data can be interpreted to give accurate adiabatic ionization energies, the ordering and energies of the valence molecular orbitals, and properties such as vibrational frequencies and spin–orbit splittings for the ground and excited states of the ionized molecules. Data for F_2 and Cl_2,[171] ClF,[172] and F_2, Cl_2, Br_2, I_2, IBr, and ICl [156] have given ionization potentials accurate to within 0.02 eV. Dibeler *et al.*[173] have recently summarized all available data for halogen ionization energies. The most accurate, adiabatic values, together with the newer results [172, 156] are shown in Table 9. Measurements of photoionization cross-sections for I_2 and Br_2 as functions of wavelength are also available.[174]

Reliable determinations of the electron affinities of molecular halogens have only recently become available. Baede [175] has re-interpreted earlier data from endoergic charge-transfer experiments for Cl_2, Br_2, and I_2. These results, together with values obtained by other workers [176, 177] are also included in Table 9.

The author is greatly indebted to H. P. Broida, P. R. Bunker, G. A. Capelle, G. D. Chapman, V. H. Dibeler, R. J. Donovan, A. E. Douglas, M. Jacon, R. J. Le Roy, J. S. Muenter, R. S. Mulliken, J. Tellinghuisen, J. J. Turner, K. Wieland, and K. K. Yee for helpful correspondence and discussions, and often for providing copies of their work prior to publication.

[171] A. B. Cornford, D. C. Frost, C. A. McDowell, J. L. Ragle, and J. A. Stenhouse, *J. Chem. Phys.*, 1971, **54**, 2651.
[172] C. P. Anderson, G. Mamantov, W. E. Bull, F. A. Grimm, J. C. Carver, and T. A. Carlson, *Chem. Phys. Letters*, 1971, **12**, 137.
[173] V. H. Dibeler, J. A. Walker, K. E. McCulloh, and H. M. Rosenstock, *Internat. J. Mass Spectroscopy Ion Phys.*, 1971, **7**, 209.
[174] J. H. Carver and J. L. Gardner, *J. Quant. Spectroscopy Radiative Transfer*, 1972, **12**, 207; M. Yoshino, A. Ida, K. Wakiya, and H. Suzuki, *J. Phys. Soc. (Japan)*, 1969, **27**, 976; J. A. Myer and J. A. P. Samson, *J. Chem. Phys.*, 1970, **52**, 716.
[175] A. P. M. Baede, *Physica*, 1972, **59**, 541.
[176] W. A. Chupka, J. Berkowitz, and D. Gutman, *J. Chem. Phys.*, 1971, **55**, 2724.
[177] J. J. DeCorpo and J. L. Franklin, *J. Chem. Phys.*, 1971, **54**, 1885.

Apologies are made for any aspects not covered in this Report. Grateful acknowledgement is expressed to M. A. A. Clyne for his encouragement and critical reading of the manuscript. The assistance received from the staff of the Chemistry Department at Queen Mary College is also warmly appreciated.

5
Far-infrared Molecular Spectroscopy

BY M. J. FRENCH

1 Introduction

There is no generally accepted definition of the term 'far-infrared', but this chapter concentrates on absorption processes occurring in the 5—200 cm^{-1} region of the spectrum. Also, the term 'molecular spectroscopy' is not one that permits of a rigid definition, so it is here interpreted to include those areas of spectroscopy which lie predominantly on the chemical physics–physical chemistry interface and to exclude those phenomena, such as magnetic, ferroelectric, super- and semi-conductive, and impurity-induced absorptions, which appear to lie firmly within the province of the solid-state physicist. Research is also excluded which appears to be directed mainly towards the identification and assignment of internal vibration frequencies and subsequent normal co-ordinate analysis.

In a chapter of this length it is not possible to treat in detail every aspect of far-i.r. spectroscopy even in those areas that remain. Therefore, wherever possible, reference is made to reviews and other information sources. In this respect it is fortunate that within the past two years two excellent monographs devoted exclusively to far-i.r. spectroscopy have appeared. The first of these, 'Chemical Applications of Far-Infrared Spectroscopy' by Finch *et al.*,[1] concentrates on what might be called the 'traditional' areas of far-i.r. spectroscopic interest: torsional barriers, weak interactions such as hydrogen-bonding and charge-transfer spectra, low-frequency internal vibrations of inorganic and organic systems, and the spectra of crystal lattices. The second monograph, 'Far-Infrared Spectroscopy' by Möller and Rothschild,[2] extends these areas of basically molecular spectroscopic interest to include collision-induced spectra, the spectra of strained ring systems, and a very comprehensive discussion of the spectra of crystal lattices. Specialist chapters of interest basically to the solid-state physicist on impurity-induced lattice absorptions and ferroelectric, magnetic, semiconductive, and superconductive absorptions are included as appendices to the monograph.

Other general references to far-infrared spectroscopy include the 1969 book

[1] A. Finch, P. N. Gates, K. Radcliffe, F. N. Dickson, and F. F. Bentley, 'Chemical Applications of Far-Infrared Spectroscopy', Academic Press, London, 1970.
[2] K. D. Möller and W. G. Rothschild, 'Far-Infrared Spectroscopy', Wiley-Interscience, New York, 1971.

by Hadni, 'L'Infrarouge Lointain' [3] and the very comprehensive review by Brasch *et al.*[4] covering the literature published up to 1966—67.

In addition, two bibliographies of papers on far-i.r. spectroscopy have been prepared. The Palik bibliography [5, 6] has been extended to late 1969; the complete bibliography forms an appendix to Möller and Rothschild's book,[2] and includes 1512 references in the period 1892—1969. Bloor [7] has prepared a bibliography of 1819 references of which 1752 relate to the period 1960—1969.

These sources provide excellent coverage of the material published up to late 1969 and this review will concentrate, therefore, on research published in the period January 1970 to March 1972.

2 Far-infrared Instrumentation

The components common to all forms of far-i.r. spectrometers, *i.e.* sources, filters, cells, and detectors, will be discussed first, before considering in greater detail the two principal types of far-i.r. spectrometer, the grating instrument and the interferometer.

The problems of far-i.r. instrumentation, and some of the solutions, are discussed in books by Martin,[8] Kimmitt,[9] and Chantry.[10]

Far-infrared Sources.—Traditionally, it has been the energy limitations of the available sources and the consequent detection problems that have led to the isolation of the far-i.r. from other regions of the electromagnetic spectrum. As a result, a great deal of research has been devoted over the years to attempts to increase the available energy of far-i.r. sources. Initially, these efforts concentrated on attempts to extend microwave techniques into the far-i.r.[11] Recently, greater emphasis has been placed on laser sources as potential tunable, narrow-band, far-i.r. sources. Although lasers have, as yet, found limited applications in far-i.r. spectroscopy, it seems appropriate to discuss them in some detail in view of the interesting developments that are taking place in this area.

Comprehensive reviews of far-i.r. sources have been published by Robinson [12] and by Genzel.[13]

Incoherent Sources. Despite many attempts to find more intense, continuous

[3] A. Hadni, 'L'Infrarouge Lointain', Presse Univ. de France, Paris, 1969.
[4] J. W. Brasch, Y. Mikawa, and R. J. Jakobsen, *Appl. Spectroscopy Rev.*, 1968, **1**, 187.
[5] E. D. Palik, *J. Opt. Soc. Amer.*, 1960, **50**, 1329.
[6] E. D. Palik, N.R.L. Bibliography No. 21, U.S. Naval Research Lab., Washington, 1963.
[7] D. Bloor, *Infrared Phys.*, 1970, **10**, 1.
[8] 'Spectroscopic Techniques for Far Infrared, Submillimetre and Millimetre Waves', ed. D. H. Martin, North Holland, Amsterdam, 1967.
[9] M. F. Kimmitt, 'Far-Infrared Techniques', Pion, London, 1970.
[10] G. W. Chantry, 'Submillimetre Spectroscopy', Academic, New York, 1971.
[11] P. N. Robson, ref. 8, Chapter 6.
[12] L. C. Robinson, *Adv. Electronics Electron Phys.*, 1969, **26**, 171.
[13] L. Genzel, in 'Far Infrared Properties of Solids', ed. S. S. Mitra and S. Nudelman, Plenum, New York, 1970, p. 51.

sources of far-i.r. radiation, incoherent sources, such as the medium-pressure mercury arc lamp, are still the most widely used far-i.r. sources. The power output of an incoherent black-body source can be calculated from Planck's equation. The power per unit area, $P(\nu)d\nu$, in a narrow frequency element, $d\nu$, is given by:

$$P(\nu)d\nu = 2\pi hc^2 (\nu^2 d\nu) [\exp(h\nu c/kT) - 1]^{-1}$$

where T is in K, ν in Hz, and P in W.

The effective temperature of the mercury arc has been determined throughout the far-i.r. by a number of workers [9] and there is general agreement that the effective temperature rises from around 1000 K at 250 cm^{-1}, where the emission is mainly thermal emission from the quartz envelope, to around 6000 K at 10 cm^{-1} as plasma radiation assumes an increasing importance. The increasing effective temperature of the Hg arc at low frequencies accounts for it being preferred over other purely thermal sources, such as the Globar, in far-i.r. studies.

Other higher temperature plasmas, such as the argon arc, have been investigated in an attempt to produce more intense far-i.r. sources, though a concomitant increase in the near-i.r. and visible radiation also results. Ukhanov and Filippov [14] showed that the effective temperature of an argon arc (6—14 A discharge current in an argon pressure of 760 Torr) rose from around 2000 K at 200 cm^{-1} to 7000 K at 50 cm^{-1}.

Incoherent thermal sources are frequently used, by virtue of Planck's equation, as primary calibration standards of the energy throughput and frequency response of far-i.r. spectrometers. The use of black bodies as radiation standards has been reviewed by Bedford [15] and by Karoli,[16] and the particular problems encountered in the far-i.r. have been discussed by Lichtenberg and Sesnic.[17] The main problems are to obtain a true black-body source, since the reflectivity of most materials is high in the far-i.r., and to fill completely the entrance slit of the spectrometer. In all practical black-body sources an aperture must be provided for the emission of the radiation, and the implications of this have been considered by Baltes *et al.*[18] who have indicated the corrections that must be applied to Planck's equation when the emission occurs from an open hemisphere or cone rather than a sphere or cube.

Extension of Microwave Techniques to the Far-infrared. Many attempts have been made to extend microwave generators, such as the klystron and magnetron, into the sub-millimetre region of the spectrum.[11] In the far-infrared the most successful of these generators has been the backward-wave oscillator or 'Carcinotron,' [19] which is capable of emitting 1 W at 10 cm^{-1} and 1 mW at

[14] E. V. Ukhanov and O. K. Filippov, *Teplofiz. Vysok. Temp.*, 1970, **8**, 655.
[15] R. E. Bedford, *Adv. Geophys.*, 1970, **14**, 165.
[16] A. R. Karoli, *Adv. Geophys.*, 1970, **14**, 203.
[17] A. J. Lichtenburg and S. Sesnic, *J. Opt. Soc. Amer.*, 1966, **56**, 75.
[18] H. P. Baltes, R. Muri, and F. K. Kenubühl, *Rev. Internat. Hautes Temp. Refract.*, 1970, **7**, 192.
[19] T. Yeou, Proceedings of the 5th International Congress on Microwave Tubes, Academic Press, New York, 1965, p. 151.

30 cm^{-1}. The generation of far-i.r. radiation from microwave sources may also be achieved by conversion of microwave radiation, through a non-linear device, into its higher harmonics.[20] Among the non-linear devices that have been used are crystal diodes and the plasma arc.

This topic has been very adequately reviewed [9, 11, 12, 20] and will not be considered further here.

Generation of Far-infrared Radiation from Laser Sources. The application of laser sources in far-i.r. spectroscopy has, so far, been rather limited. Typical examples include high-resolution studies of gas-phase rotation bands which are fortuitously coincident with a laser emission that is capable of being tuned over a narrow frequency range, studies of the Stark and Zeeman splitting of rotational lines where the absorption is effectively scanned across the fixed-frequency laser emission by application of an external field, and refractive index measurements at discrete wavelengths. Laser sources have, however, considerable potential for far-i.r. spectroscopy, and much research is at present being directed to the generation of far-i.r. radiation from laser sources. In view of this, a somewhat extended treatment of this topic will be given here: it is also discussed in some detail by Chantry.[21]

Fixed-frequency lasers. The first laser to produce appreciable powers in the far-i.r. was the water vapour laser.[22] Over the past eight years a number of other far-i.r. molecular lasers have been developed, including the D_2O,[23] H_2S,[24] HCN,[25] SO_2,[26] and He [27] lasers. Recently considerable work has been directed towards an understanding of the mechanism of the SO_2[28, 29] and H_2S [30] lasers and the assignments of the transitions involved. The properties of far-i.r. molecular lasers have been reviewed by Coleman,[31] and the frequencies of some of the more important far-i.r. laser lines are tabulated by Kimmitt [32] and by Steffen and Kneubuhl.[33] Some of the disadvantages of fixed-frequency lasers as spectroscopic sources could be overcome if a substantial number of far-i.r. lines could be generated from a single lasing medium. A series of lasers have been developed by Chang and his co-workers [34-37] which appear to show

20 J. G. Baker, ref. 8, Chapter 5.
21 G. W. Chantry, ref. 10, Chapter 6.
22 A. Crocker, H. A. Gebbie, M. F. Kimmitt, and L. E. S. Mathias, *Nature*, 1964, **201**, 250.
23 W. W. Müller and G. T. Flesher, *Appl. Phys. Letters*, 1966, **8**, 217.
24 J. C. Hassler and P. D. Coleman, *Appl. Phys. Letters*, 1969, **14**, 135.
25 J. P. Kotthaus, *Appl. Optics*, 1968, **7**, 2422.
26 S. F. Dyubko, V. A. Svich, and R. A. Valitov, *J.E.T.P. Letters*, 1968, **7**, 320.
27 J. S. Levine and A. Javan, *Appl. Phys. Letters*, 1969, **14**, 348.
28 G. Hubner, J. C. Hassler, P. D. Coleman, and G. Steenbeckeliers, *Appl. Phys. Letters*, 1971, **18**, 511.
29 T. M. Hard, *J. Quantum Electronics*, 1970, **6**, 177.
30 J. C. Hassler and P. D. Coleman, *J. Quantum Electronics*, 1970, **6**, 178.
31 P. D. Coleman, *Govt. Report Announcements*, 1971, **71**, 218.
32 M. F. Kimmitt, ref. 9, p. 54.
33 H. Steffen and F. K. Kneubühl, *J. Quantum Electronics*, 1968, **4**, 992.
34 T. Y. Chang and T. J. Bridges, *Optics Comm.*, 1970, **1**, 423.
35 T. Y. Chang, T. J. Bridges, and E. G. Burkhardt, *Appl. Phys. Letters*, 1970, **17**, 249.
36 T. Y. Chang, T. J. Bridges, and E. G. Burkhardt, *Appl. Phys. Letters*, 1970, **17**, 357.
37 T. Y. Chang and J. D. McGee, *Appl. Phys. Letters*, 1971, **19**, 103.

considerable promise in this direction. A molecule with a permanent dipole moment is optically pumped with a near-i.r. laser, such as the CO_2 laser, to produce population inversion in a specific rotational level of the excited vibrational level: far-i.r. emission is then generated by radiative rotational transitions within the vibrationally excited level. Among the molecules in which far-i.r. lasing action has been observed by this technique are MeF (ten far-i.r. lines),[34, 35] $CH_2{=}CHCl$ (three lines),[35] NH_3 (two lines),[36] MeCN (seven lines),[37] and MeC≡CH (eight lines).[37] All these lines lie in the far-i.r. between 150 and 5 cm^{-1}. The 5.51 cm^{-1} line in the MeCN laser is the first molecular laser line to be observed below 10 cm^{-1}. This technique of preferential pumping of a specific rotational level of a molecule should prove a very general method for the production of far-i.r. laser lines.

As an example of the use of fixed-frequency lasers in far-i.r. spectroscopic studies, the magnetic resonance spectrometer of Wagner and Prinz [38] is considered. A pulsed D_2O—H_2O laser is used as a source since this emits over 100 accessible far-i.r. lines in contrast to the CW H_2O laser which emits only 15 lines.[39] The use of a pulsed laser necessitates the use of a fast-response bolometer (Ga doped Ge—see p. 242) as the detector,[40] and a boxcar integrator to average out the pulse-to-pulse amplitude variations in the laser pulse and to increase the signal-to-noise ratio (SNR). Although the laser output itself cannot be tuned, the very large number of fixed-frequency outputs means that frequently one of these is coincident with a cyclotron resonance or other magneto-optic absorption. Effective frequency tuning may then be achieved, not by varying the source frequency, but by variation in the absorption frequency of the sample as a result of changes in the applied magnetic field. Wagner and Prinz investigated the e.p.r. of Dy^{3+} in $DyPO_4$ at 45.5 cm^{-1}. As the magnetic field was varied between 67.2 and 68.2 kG, the central resonance line and the satellite lines resulting from hyperfine interactions were swept across the fixed-frequency laser output. The observed hyperfine splittings corresponded to an effective far-i.r. resolution of 0.01 cm^{-1}. In addition, the intense far-i.r. output of the pulsed laser permitted Wagner and Prinz to observe the cyclotron resonance of almost opaque semiconductors where the optical phonon bands reduced the transmitted intensity by a factor of 1000.

A second example of the use of a fixed-frequency laser in molecular far-i.r. spectroscopy is the observation at high resolution of the Stark spectra of the 6_{24}–6_{15} pure rotational transition in D_2O by Duxbury and co-workers.[41, 42] Effective frequency scanning through the absorption line was achieved by variation of the external electric field applied to the D_2O molecules. Incompletely resolved spectra were also observed for CD_3CN, EtCN, and $CF_2{=}CH_2$.

Other important areas for the use of fixed-frequency far-i.r. lasers are in

[38] R. J. Wagner and G. A. Prinz, *Appl. Optics*, 1971, **10**, 2060.
[39] W. S. Benedict, M. A. Pollack, and W. J. Tomlinson, *J. Quantum Electronics*, 1969, **5**, 108.
[40] R. J. Wagner and G. A. Prinz, *Appl. Phys. Letters*, 1970, **17**, 360.
[41] G. Duxbury and W. J. Burroughs, *J. Phys. (B)*, 1970, **3**, 98.
[42] G. Duxbury and R. G. Jones, *Mol. Phys.*, 1971, **20**, 721.

refractive index studies (see Section 9) and in the establishment of absolute frequency standards. The centre frequency of a rotational absorption in difluoroethylene was measured to an accuracy of 1 part in 10^7 at 29.7 cm^{-1} by Bradley and Knight,[43] using a harmonic mixing technique involving the centre frequency of an HCN laser whose frequency had, itself, been determined to the same order of magnitude by a mixing technique against the twelfth harmonic of a 4 mm klystron.

Optical difference-frequency techniques. Generation of far-i.r. radiation as a difference frequency between two near-i.r. beams was first demonstrated by Zernike and Berman in 1965.[44] If two collinear beams of frequencies ν_1 and ν_2 are mixed in a non-linear crystal, then a difference frequency will be generated at $\nu_d = \nu_1 - \nu_2$. The difference frequency amplitude is greatly enhanced if the two beams are phase-matched within the crystal, and this may be achieved if ν_1 is an ordinary ray (o) and ν_2 is an extraordinary ray (e). There are then two phase-matching conditions:

$$k_1(o) - k_2(e) = k_d(o)$$
$$k_1(o) - k_2(e) = k_d(e)$$

where k represents the wave vector of the radiation. The phase-matching condition, and hence the efficiency of the difference frequency generation, may be controlled by the rotation of the non-linear crystal with respect to the incident radiation.

In the case of two discrete incident frequencies, ν_1 and ν_2, the difference frequency will itself be a single discrete frequency, $\nu_1 - \nu_2$, and the crystal orientation is adjusted so as to meet the phase-matching condition. If, however, one of the incident frequencies is a broad-band source, such as a picosecond mode-locked laser, then the difference frequency is not pre-determined and the output may be frequency-tuned by adjustment of the crystal orientation.

The generation of far-i.r. radiation by the mixing of two discrete frequencies from two ruby laser sources has been described by Faries and co-workers [45, 46] who related their results to the theoretical calculations of Morris and Shen.[47] $LiNbO_3$ and quartz were used as the non-linear crystals and the far-i.r. radiation was detected by a Putley InSb detector (see p. 242). One of the ruby lasers could be operated on the R_1 transition and the other on the R_2 transition. By temperature tuning of the lasers the difference frequency obtained by non-linear mixing of the outputs of the two lasers could be tuned over the range 2—38.8 cm^{-1}. A maximum power of about 1 mW at 8.1 cm^{-1} was achieved.

A more convenient method of tuning and a wider tuning range are achieved by using a broad-band incident light source rather than two discrete frequency sources. A broad-band source may be provided by a picosecond laser pulse.

[43] C. C. Bradley and D. J. E. Knight, *Phys. Letters*, 1970, **32A**, 59.
[44] F. Zernike and P. R. Berman, *Phys. Rev. Letters*, 1965, **15**, 999.
[45] D. W. Faries, K. A. Gehring, P. L. Richards, and Y. R. Shen, *Phys. Rev.*, 1969, **180**, 363.
[46] D. W. Faries, P. L. Richards, Y. R. Shen, and K. H. Yang, *Phys. Rev.*, 1971, **3A**, 2148.
[47] J. R. Morris and Y. R. Shen, *Optics Comm.*, 1971, **3**, 81.

The spectral width, $\Delta\omega$, of a laser pulse, in radian s^{-1}, is related to the pulse duration, τ, in seconds, by the relation

$$\Delta\omega \approx \frac{1}{\tau}$$

Hence a mode-locked laser pulse with a duration of 30 ps would have a spectral width of 100 cm^{-1}. Far-i.r. radiation can be generated by mixing between the various frequency components of the mode-locked pulse in a non-linear crystal, and the radiation generated by the mixing may be continuously tuned over the range 0—100 cm^{-1} merely by adjustment of the orientation of the non-linear crystal to meet the phase-matching condition at different frequencies. Such a scheme has been investigated by Yajima and Takeuchi [48-50] using a Nd-glass laser simultaneously **Q**-switched and mode-locked with a saturable dye. The non-linear crystals used were $LiNbO_3$, $LiIO_3$, ZnTe, ZnSe, or CdS and the far i.r. radiation produced was isolated by means of a mesh filter, crystal quartz, and black polyethylene and detected by a Kinch–Rollin detector (see p. 242). For an input energy at 1.06 μm of 50 mJ, about 1 mW of far-i.r. radiation was observed in the complete bandwidth, 0—150 cm^{-1}, passed by the filter combination. With collinear phase-matching of the crystal the output could be tuned by rotation of the crystal.

A similar scheme using a mode-locked Nd laser and $LiNbO_3$ non-linear crystals has been described by Yang and co-workers.[51] In addition to measuring the power output, they also attempted to measure the bandwidth of the far-i.r. radiation using a Fabry–Perot interferometer. A theoretical bandwidth of about 2 cm^{-1} had been predicted,[47] but this was not verified for the experimental output because of the low resolving power of the interferometer resulting from the high divergence of the far-i.r. beam.

In principle, the tuning range of far-i.r. difference-frequency generation could be extended by the use of other types of laser. For example, Dewey and Hocker [52] have described a near-i.r. source continuously tunable over the range 2500—3500 cm^{-1} by difference-frequency mixing of the radiation from a ruby laser and a dye laser. Other tunable laser sources which might be used in conjunction with difference-frequency-generation techniques to provide tunable far-i.r. sources are the stimulated polariton scattering and the spin-flip lasers described in succeeding paragraphs.

Stimulated polariton scattering lasers. With **Q**-switched laser pulses, stimulated Raman scattering may occur not only from the vibrational modes of the crystal but also from the transverse optical modes or polaritons. Stimulated polariton scattering at near forward scattering has two important properties; not only

[48] T. Yajima and N. Takeuchi, *Jap. J. Appl. Phys.*, 1970, **9**, 1361.
[49] T. Yajima and N. Takeuchi, *Jap. J. Appl. Phys.*, 1971, **10**, 907.
[50] N. Takeuchi, N. Matsumoto, T. Yajoma, and S. Kishida, *Jap. J. Appl. Phys.*, 1972, 11, 268.
[51] K. H. Yang, P. L. Richards, and Y. R. Shen, *Appl. Phys. Letters*, 1971, **19**, 320.
[52] C. F. Dewey and L. O. Hocker, *Appl. Phys. Letters*, 1971, **18**, 58.

is some of the polariton energy carried in the form of an electromagnetic wave so that this energy component can emerge from the crystal as far-i.r. radiation, but also wave-vector conservation requires that

$$k \quad = k_{\text{Stokes}} = k_{\text{polariton}}$$

This implies that the Stokes shift of the stimulated polariton scattering, and consequently the frequency of the far-i.r. emission, is strongly dependent on the scattering angle. In the generation of far-i.r radiation by stimulated polariton scattering, it is arranged that the Stokes radiation propagates orthogonally to the faces of the crystal: the frequency of the far-i.r. emission may then be tuned by varying the angle of incidence of the laser radiation relative to the crystal faces.

The theory of stimulated polariton scattering has been discussed by Barker and Loudon [53] and a tunable far-i.r. source based on this principle constructed by Johnson and co-workers.[54, 55] A **Q**-switched ruby laser was incident on to a $LiNbO_3$ crystal, the opposite faces of which were polished flat and parallel to provide a low-**Q** resonator for the Stokes radiation. By varying the angle of incidence of the incident radiation between 0° and 3° the far-i.r. emission was tuned over the range 50—150 cm^{-1}. For a ruby laser input power of 1 MW, the peak far-i.r. power, measured outside the crystal, was 3 W at 55 cm^{-1} and 0.25 W at 170 cm^{-1}. Conservation of energy for the scattering process requires that the bandwidth of the far-i.r. radiation be equal to, or less than, the bandwidth of the Stokes radiation. The Stokes bandwidth was measured using a Fabry–Perot interferometer and found to be in the range 0.1—0.5 cm^{-1}. It was inferred, therefore, that the far-i.r. radiation had a similar bandwidth, although this was not confirmed experimentally.

Spin-flip lasers. Generation of tunable i.r. radiation by a Raman-scattering process, involving the electron spin-flip of the conduction electrons in InSb, has been reported by Patel [56, 57] and by Smith and co-workers.[58] In a magnetic field the conduction electrons spin about an axis and may be induced by an external magnetic field to flip between higher and lower energy states. The frequency of the scattered radiation, ν_s, is related to that of the incident radiation, ν_i, by

$$\nu_s = \nu_i - g\mu_B B$$

where g is the effective gyromagnetic ratio of the electrons in InSb, μ_B is the Bohr magneton, and B is the external magnetic field intensity.

[53] A. S. Barker and R. Loudon, *Rev. Mod. Phys.*, 1972, **44**, 18.
[54] B. C. Johnson, H. E. Putoff, J. SooHoo, and S. S. Sussman, *Appl. Phys. Letters*, 1971, **18**, 181.
[55] J. M. Yarborough, S. S. Sussman, H. E. Putoff, R. H. Pantell, and B. C. Johnson, *Appl. Phys. Letters*, 1969, **15**, 102.
[56] C. K. N. Patel and E. D. Shaw, *Phys. Rev. Letters*, 1970, **24**, 451.
[57] C. K. N. Patel, *J. Quantum Electronics*, 1971, **7**, 306.
[58] R. L. Allwood, R. B. Dennis, S. D. Smith, B. S. Wherrett, and R. A. Wood, *J. Phys.* (*C*), 1971, **4**, L63.

Tunability of the output frequency is achieved by varying *B*, which determines the spacing between the spin sub-levels. Using a Q-switched CO_2 laser of 2 kW power, Patel and Shaw [56] were able to obtain 1 W of i.r. power that could be tuned over the range 770—850 cm^{-1} by variation of the magnetic field between 48 and 100 kG. Patel [59] has shown that the linewidth of the output of the spin-flip laser is less than 1 kHz, making it by far the narrowest tunable coherent i.r. source. Using a spin-flip laser of this type, Patel, Shaw, and Kerl[60] were able to study the absorption of NH_3 in the 850 cm^{-1} region with a resolution of 0.05 cm^{-1}. The extension of spin-flip laser action to the far-i.r. region of the spectrum could well represent the intense, tunable, coherent source for which far-i.r. spectroscopists have waited so long.

Far-infrared Filters.—The great majority of far-i.r. investigations are carried out using a thermal, incoherent far-i.r. source. This imposes a very great burden on the filtering system if an acceptable spectral purity (say, greater than 99% for practical chemical far-i.r. spectroscopy) is to be obtained. Although this problem is most pronounced in grating spectroscopy, it is normally not entirely absent in interferometry, so that both techniques require the use of low-pass i.r. filters. The standard filtering techniques have been described in detail in the standard far-i.r. books [1–3, 8] as well as in specific descriptions of far-i.r. grating spectrometers.[61–64]

Active interest has continued in the use of metal meshes as far-i.r. filters. Pradhan [65] has developed the theory of reciprocal grids as low-pass transmission filters and has shown [66] that sharper cut-off frequencies may be obtained by the use of a multigrid filter. Demeshina *et al.*[67] have investigated the frequency dependence of both the reflection intensity and the polarization properties of a conventional inductive wire-mesh filter.

The far-i.r. transmission of black polyethylene has been re-studied by Blea *et al.*[68] in order to resolve some of the discrepancies in earlier investigations. The refractive index was measured from the channel spectra (see Section 9) and from the zero-thickness intercept of samples of different thicknesses, and was found to be 1.50 ± 0.02 for all samples in the range 10—170 cm^{-1}. Although the absorption coefficient at a given frequency was found to vary from sample to sample, the frequency dependence of the absorption coefficient was similar for all samples. It was shown that the apparent discrepancies in previous

[59] C. K. N. Patel, *Phys. Rev. Letters*, 1972, **28**, 649.
[60] C. K. N. Patel, E. D. Shaw, and R. J. Kerl, *Phys, Rev. Letters*, 1970, **25**, 8.
[61] T. M. Hard and R. C. Lord, *Appl. Optics*, 1968, **7**, 589.
[62] F. Kneubühl, *Appl. Optics*, 1969, **8**, 505.
[63] I. Iwahashi, K. Matsumoto, S. Matsudaira, S. Minami, and H. Yoshinaga, *Appl. Optics*, 1969, **8**, 583.
[64] M. J. French, D. E. H. Jones, and J. L. Wood, *J. Phys. (E)*, 1969, **2**, 664.
[65] M. M. Pradhan, *Infrared Phys.*, 1970, **10**, 199.
[66] M. M. Pradhan, *Infrared Phys.*, 1971, **11**, 241.
[67] A. I. Demeshina, V. A. Zayats, V. I. Lapshin, and V. N. Murzin, *Zhur. priklad. Spektroskopii*, 1970, **13**, 346.
[68] J. M. Blea, W. F. Parks, P. A. R. Ade, and R. J. Bell, *J. Opt. Soc. Amer.*, 1970, **60**, 603.

measurements were due to differences in carbon particle sizes, carbon form, and carbon concentrations, causing a displacement of frequency dependence of the absorption coefficient curve along the frequency ordinate. Further analyses of the data were consistent with the carbon particles behaving as Rayleigh absorbers.

Wong *et al.*[69] have described the use of disordered solids, such as Ice I and fused quartz, as low-pass far-i.r. transmission filters. These filters are very efficient at low frequencies, they are superior to crossed transmission gratings below 50 cm^{-1} and to inductive mesh reflection filters below 30 cm^{-1}.

Birch and Jones [70] have described a far-i.r. modulator based on the Faraday effect in two polycrystalline ferrites. The modulator will operate at frequencies up to 2 MHz in the 0.1—50 cm^{-1} region of the spectrum. The same authors have described a second modulator based on the control of the free carrier concentration in Ge by the use of an external electric field to control the re-combination effects in the surface layers.[71] This second modulator operates throughout the far- and near-i.r. (up to 1200 cm^{-1}) although the maximum modulation frequency, about 50 kHz, is rather less than for the ferrite modulators. Use of these far-i.r. modulators could result in devices which are common in the microwave region, such as isolators, switches, and rotators, becoming available in the far-i.r.

The far-i.r. transmission and reflection of some common window and filter materials has been redetermined by McCarthy.[72] Tishchenko [73] has described the fabrication of hot-pressed polyethylene echelette transmission grating filters and measured their low-pass characteristics. A band-elimination comb filter, based on the pressure broadening of the HCl rotation lines, has been suggested by Lane *et al.*[74]

Far-infrared Cells.—Scheide and Guilbault [75] have described an interesting cell for the study of gas–solid interfaces in the far-i.r. They used a conventional gas cell with specially machined polythylene windows, and the interacting solid was deposited on the window. Sealing the windows on with paraffin permitted a pressure of 10^{-4} Torr to be achieved in the cell. The reaction of $POCl_3$ vapour with solid $FeCl_3$ was studied, and the P—O deformation band at 191 cm^{-1} and ν_3 of tetrahedral $FeCl_4^-$ at 141 cm^{-1} were observed.

An inexpensive method of making cells of constant path length for liquids has been described by Tsatsas and Risen; [76] polyethylene windows were used. Fleming and Davies [77] have described a cell specifically designed for absorption

[69] P. T. T. Wong, D. D. Klug, and E. Whalley, *J. Opt. Soc. Amer.*, 1972, **62,** 533.
[70] J. R. Birch and R. G. Jones, *Infrared Phys.*, 1970, **10**, 217.
[71] J. R. Birch and R. G. Jones, *J. Phys.* (*D*), 1972, **5**, L18.
[72] D. E. McCarthy, *Appl. Optics*, 1971, **10**, 2539.
[73] E. A. Tishchenko. *Optics and Spectroscopy*, 1971, **30,** 82.
[74] K. P. Lane, C. A. Frenzel and E. B. Bradley, *U.S. Govt. Res. Development Report*, 1970, **70,** 80.
[75] E. P. Scheide and G. C. Guilbault, *J. Phys. Chem.*, 1970, **74,** 3074.
[76] A. T. Tsatsas and W. M. Risen, *Appl. Spectroscopy*, 1971, **24,** 383.
[77] J. W. Fleming and G. J. Davies, *J. Phys.* (*E*), 1971, **4,** 620.

measurements on polar liquids using 3 mm crystal quartz windows and amalgamated lead spacers and sealing gaskets: the precise cell thickness was determined from the channel spectra.

The normal method of obtaining the far-i.r. spectra of powders is by the polyethylene disc technique [78] or, less commonly, as a mull. Peterkin [79] has described an alternative technique in which the powder is cold-pressed into a wax to form a relatively strong and transparent wax disc. Ferraro and Quattrochi [80] have described a diamond anvil cell with platinum or molybdenum steel gaskets for high-pressure investigations on liquids.

A number of studies have been made on light pipes. Pluchino and Möller [81] have described the matching of an $f/4$ Fourier transform spectrometer (FTS) to an $f/8$ multi-pass White gas cell using two conical light pipes, the spectral losses of which were claimed to be frequency-independent in the 40—400 cm^{-1} region. Stroyer-Hansen [82] has described a long-path-length, light-pipe gas cell for the far-i.r. The cell, which is inserted between the interferometer and the detector, consists of a brass tube, of 19 mm internal diameter and 6000 mm length, giving a volume of 1500 ml. This acts not only as the cell but also as a light pipe. Despite its simple construction, the cell has a transmission of 13% at 400 cm^{-1} rising to 30% at 25 cm^{-1}.

With the increasing use of light pipes there has been much interest in the polarization properties of these devices. Frayne [83] has shown that the conventional circular-cross-section light pipe completely depolarizes a linearly polarized incident beam. Poehler and Turner [84] have, however, developed light pipes of square and rectangular cross-section that transmit linearly polarized far-i.r. radiation with low loss and with little deterioration in the degree of polarization.

Far-infrared Detectors.—Far-i.r. detectors may conveniently be divided into two classes: thermal detectors such as the Golay cell and bolometers, and photoconductive detectors such as the InSb detector. Comprehensive reviews of far-i.r. detectors have been given by Putley,[85] by Kimmitt,[9] and by Richards.[86]

In general, thermal detectors have a broad frequency response, their range being limited only by their ability to absorb rather than reflect radiation. Photoconductive detectors tend to have a low-frequency cut-off related to the semiconductor band gap. Thermal detectors tend to be slower (response time 1—10^{-3} s) than photoconductive detectors (response time 10^{-5}—10^{-7} s). The best photoconductive detectors tend to have a greater sensitivity than thermal detectors on account of their reduced spectral range. A disadvantage of photo-

[78] C. Schiele and K. Halfar, *Appl. Spectroscopy*, 1965, **19**, 163.
[79] M. E. Peterkin, *Appl. Spectroscopy*, 1971, **25**, 502.
[80] J. R. Ferraro and A. Quattrochi, *Appl. Spectroscopy*, 1971, **25**, 102.
[81] A. Pluchino and K. D. Möller, *Appl. Optics*, 1971, **10**, 1694.
[82] T. Stroyer-Hansen, *Infrared Phys.*, 1970, **10**, 159.
[83] P. G. Frayne, *J. Phys. (D)*, 1968, **1**, 741.
[84] T. O. Poehler and R. Turner, *Appl. Optics*, 1970, **9**, 971.
[85] E. H. Putley, *Adv. Geophys.*, 1970, **14**, 129.
[86] P. L. Richards, ref. 13, p. 103.

conductive detectors is that they normally require cooling to liquid-helium temperaturcs, whereas thermal detectors can operate satisfactorily in the far-i.r. at room temperature.

Thermal Detectors. It can be shown that for an ideal thermal detector the Noise Equivalent Power (NEP) is given by: [85]

$$\mathrm{NEP} = (16\sigma kT^5 A\Delta f/n)^{\frac{1}{2}}$$

where σ is Stefan's constant, A is the area of the detector, Δf is the bandwidth about the modulation frequency, n is the fractional number of photons absorbed, ($n = 1$ for a black detector) and T is in K. Thus, at 300 K, an ideal thermal detector with an area of 1 cm^2 and an amplifier bandwidth of 1 Hz will have a NEP of 5.5×10^{-11} W Hz$^{-\frac{1}{2}}$. The performance of an actual detector may be compared with that of an ideal detector at the same temperature. Corresponding calculations may be made for the ideal photoconductive detector,[85, 87] showing that the NEP will be greatest at a frequency just in excess of its threshold frequency.

Also frequently encountered in detector terminology is the detectivity, D, which is the reciprocal of the NEP. In those cases where the detectivity is proportional to the square root of the area and amplifier bandwidth, the specific detectivity, D^*, is introduced, defined by

$$D^* = (A\Delta f)^{\frac{1}{2}}/\mathrm{NEP}$$

The Golay cell is too well known to require any further description. A very full account of its properties is given by Hadni.[88] Despite the simplicity of the basic principles underlying its operation, the detectivity of a Golay cell comes within an order of magnitude of that of an ideal room-temperature thermal detector. The Golay cell is, at present, the most widely used room-temperature detector, although it is to be expected that its position will be progressively challenged as the technology of pyroelectric detectors develops.

The NEP of an ideal thermal detector may be very significantly reduced by cooling the detector to liquid-helium temperatures. As this is not feasible with a Golay cell, low-temperature thermal detectors are normally bolometers which provide the highest detectivities of far-i.r. detectors. They are extensively used in preference to photoconductive detectors when speed of response is not of prime importance. A high temperature-coefficient of resistance is required, and among the bolometer elements that have been used are super-conducting Sn,[89] carbon,[90] Ga-doped Ge,[91] Sb-doped Ge,[92] and In-doped Ge.[93] A comparison

[87] S. F. Jacobs and M. Sargent, *Infrared Phys.*, 1970, **10**, 233.
[88] A. Hadni, 'Essentials of Modern Physics Applied to the Study of the Infrared', Pergamon, Oxford, 1967, p. 280.
[89] D. H. Martin and D. Bloor, *Cryogenics*, 1961, **1**, 1.
[90] W. S. Boyle and K. F. Rodgers, *J. Opt. Soc. Amer.*, 1959, **49**, 66.
[91] A. Hadni, P. Strimer, R. Thomas, M. Dugue, J. F. Goullin, M. Gremillet, and M. Moulin, *Phys. Status Solidi*, 1971, **5A**, 707.
[92] S. Zwerdling, R. A. Smith, and J. P. Theriault, *Infrared Phys.*, 1968, **8**, 271.
[93] Y. Nakagawa and H. Yoshinaga, *Jap. J. Appl. Phys.*, 1970, **9**, 125.

of the sensitivities of various bolometers is given by Bertin and Rose.[94,95]

The response time of a bolometer may be controlled to some extent by variation of the thermal conductance and thermal capacity of the detector element and its associated heat sink. The optimization of these parameters is discussed by Deb and Mukerjee.[96]

Impurity doping of Ge bolometers serves to increase the absorption coefficient in the far-i.r. but the transition from 'hopping' to 'metallic' impurity conduction limits the level to which doping may be increased. This transition occurs at a much higher impurity concentration in Si than in Ge, and Kinch[97] has described a semiconducting Sb–B-doped Si bolometer in which a NEP of 2×10^{-14} W $Hz^{-\frac{1}{2}}$ was achieved at a silicon doping level of 2×10^{-18} cm^{-3}.

Thin-film bolometers have much reduced response times. Contreras and Gaddy[98] have described a thin-film bolometer (Bi–Ag on BeO) which they used to monitor CO_2 laser emission: the response time was 15 ns and the NEP 10^{-6} W $Hz^{-\frac{1}{2}}$.

Pyroelectric Detectors. The pyroelectric detector has been developed to meet the need for fast, room-temperature, far-i.r. detectors created by the development of rapid-scan FTS and pulsed far-i.r. lasers. In the ferroelectric effect, the presence of a unique axis along which a permanent dipole exists is revealed by the inability of the stray charges, trapped by the surface normal to the axis, to redistribute themselves on the application of a transient magnetic field. A similar failure of the stray charges to redistribute as a result of a transient thermal change is referred to as the pyroelectric effect. The most commonly used i.r. pyroelectric material is triglycinesulphate (TGS),[99] which has been extensively studied in the far-i.r., although $PbTiO_3$,[100] $PbZrTiO_3$,[101] and polyvinylidene fluoride[102] have also been used. The far-i.r. performance of pyroelectric detectors has been reviewed by Putley,[103] who discusses the sources of noise in pyroelectric detectors, the design of amplifiers for use with these detectors, and the practical construction of TGS detectors. Useful response times down to 1 μs and an NEP of 10^{-7} W $Hz^{-\frac{1}{2}}$ may be achieved.

Blackman and Wright[104] have made a theoretical thermal analysis of a pyroelectric detector linked to an infinite heat sink and subjected to modulated radiation. Putley[105] has described the commercial development of a 128

[94] C. L. Bertin and K. Rose, *J. Appl. Phys.*, 1971, **42**, 163.
[95] C. L. Bertin and K. Rose, 'Treatise on Applied Spectroscopy', ed. B. Newhouse, Academic, New York, 1972.
[96] S. Deb and M. K. Mukherjee, *Infrared Phys.*, 1971, **11**, 195.
[97] M. A. Kinch, *J. Appl. Phys.*, 1971, **42**, 5861.
[98] B. Contreras and O. L. Gaddy, *Appl. Phys. Letters*, 1970, **17**, 450.
[99] J. H. Ludlow, W. H. Mitchell, E. H. Putley, and N. Shaw, *J. Sci. Instr.*, 1967, **44**, 694.
[100] E. Yamaka, T. Hayashi, and M. Matsumoto, *Infrared Phys.*, 1971, **11**, 247.
[101] D. Mahler, R. J. Phelan, and A. R. Cook, *Infrared Phys.*, 1972, **12**, 57.
[102] A. M. Glass, J. H. McFee, and J. G. Bergman, *J. Appl. Phys.*, 1971, **42**, 5219.
[103] E. H. Putley, *Semicond. Semimetals*, 1970, **5**, 259.
[104] H. Blackman and H. C. Wright, *Infrared Phys.*, 1970, **10**, 191.
[105] E. H. Putley, *Opt. Laser Technol.*, 1971, **3**, 150.

element linear-array TGS detector which would appear to have promising applications for multi-channel far-i.r. spectroscopy.

The technology of pyroelectric detectors has progressed to such an extent over the past four years that TGS detectors are now available as standard options on commercial far-i r. Fourier Transform interferometers.

Another fast, room-temperature detector which may soon be applied specifically to the far-i.r. is the Nernst detector. In this the detector element, for example Bi, Bi–Sb,[106] or Cd_3As_2–NiAs,[107] is subjected to an applied magnetic field and the incident radiation induces a measurable voltage gradient, transverse to both the radiation and the magnetic fields. Typically, specific detectivities of 10^8 cm $Hz^{\frac{1}{2}}$ W^{-1} and response times of 10^{-5} s are obtained.

Photoconductive Detectors. Although a continuous range of semiconductors exists with intrinsic energy gaps down to zero, the problems of preparing these materials to the necessary degree of purity have prevented the development of *intrinsic* photoconductors for frequencies less than 1000 cm^{-1}. *Extrinsic* Ge detectors are doped with Ga, B, or Sb to give a peak detectivity at about 100 cm^{-1} and a low-frequency limit, imposed by the ionization energy of the lowest impurity levels, of about 60 cm^{-1}. Enhanced absorption in the tail of the photoconductive curve, permitting the use of Ga–Ge extrinsic photodetectors, at 30 cm^{-1} has been described by Wagner and Prinz.[40] The far-i.r. detectivity at 65 cm^{-1} was increased by a factor of 20 by the interaction with near-i.r. radiation at a frequency around 5800 cm^{-1}.

InSb may function as a far-i.r. detector in two quite distinct modes. In the absence of a magnetic field, or in the presence of a rather small field (less than 5 kG), to increase its resistance the detector functions as a broad-band detector by virtue of free-electron absorptions. It is then referred to as the Kinch–Rollin detector.[108] In the presence of an intense magnetic field, 25—70 kG, the InSb detector becomes a tunable far-i.r. detector dependent on a cyclotron resonance phenomenon of the impurity levels, when it is referred to as a Putley detector;[109] its operation has recently been reviewed by Yoshinaga and Yamamoto.[110] The extension of the Putley-type detector to very low frequencies (less than 50 cm^{-1}) has been considered by Robinson, who proposes a far-i.r. detector analogous to the electron cyclotron maser.[111, 112] Other free-electron-absorption far-i.r. detectors similar to the Kinch–Rollin detector are those based on GaAs.[113]

[106] E. R. Washwell, S. R. Hawkins, and K. F. Cuff, *Appl. Phys. Letters*, 1970, **17**, 164.
[107] H. J. Goldsmid and K. R. Sydney, *J. Phys. (D)*, 1970, **4**, 869.
[108] M. A. Kinch and B. V. Rollin, *Brit. J. Appl. Phys.*, 1963, **14**, 672.
[109] E. H. Putley, *Appl. Optics*, 1965, **4**, 649.
[110] H. Yoshinaga and J. Yamamoto, Optical Instrument Technology Conference Proceedings 1969, Oriel, Newcastle on Tyne, 1970, p. 41.
[111] L. C. Robinson, *Infrared Phys.*, 1970, **10**, 111.
[112] L. C. Robinson and L. D. Whitbourn, *J. Phys. (A)*, 1972, **5**, 263.
[113] G. E. Stillman, C. M. Wolfe, I. Melngailis, C. D. Parker, P. E. Tannenwald, and J. O. Dimmock, *Appl. Phys. Letters*, 1968, **13**, 83.

Point-contact Detectors The use of point-contact and Schottky-barrier microwave detectors in the far-i.r. is limited by their capacitance Becklake *et al.*[114] have reported the first far-i.r. Schottky-barrier GaAs diode, and have compared its performance with those of Ge, Si, and GaAs point-contact diodes. In all cases, the NEP was found to be strongly frequency-dependent, falling from 10^{-9} W at 10 cm^{-1} to 10^{-5} W at 100 cm^{-1}.

Large Area, Two-dimensional Detectors. For military purposes and for the investigation of laser mode structure, there has been considerable interest in processes that convert two-dimensional i.r. images from the i.r. to the visible region of the spectrum (up-conversion) for photoemissive detection. Among the systems that have been studied are parametric up-conversion,[115] Mn-doped Zn_2SiO_4 phosphors,[116] temperature-induced changes in the helical structure of cholesteric liquid crystals,[117] change in reflectivity in the region of total internal reflection,[118] a pyroelectric TV camera,[119] and the differential evaporation of oil films.[120] A number of these have already operated in the far-i.r., and their use in spectroscopy may be foreseen.

Far-infrared Grating Spectroscopy.—Since the introduction of interferometric techniques in the far-i.r., grating spectroscopy has been suffering a progressive decline in popularity, particularly in the general area of molecular spectroscopy, although for the study of intramolecular vibration frequencies, especially of inorganic complexes, it is still preferred by many chemists. The protagonists of each technique have rehearsed their arguments many times [62, 121, 122] and these will not be repeated here. The practice of far-i.r. grating spectroscopy has been reviewed by Kneubühl [62] who provides a very comprehensive bibliography on such topics as filters, gratings, *etc.* Among the single-beam grating spectrometers that have been constructed recently are those of Cole *et al.*[123] and of Silvera and Birnbaum.[124] These are basically similar and are based on conventional, medium-speed (about $f/4$) Czerny–Turner monochromators. A high-pressure Hg arc source, large gratings, and a combination of reflection and transmission filters are used to isolate the far-i.r. radiation. Spectral purities in the region of 98% and resolutions of about 0.2 cm^{-1} are claimed. Double-beam far-i.r. grating spectrometers have been described by Hard and Lord,[61] Iwahashi *et al.*,[63] and by French *et al.*,[64] Loewen [125] has given an interesting historical account of the production of diffraction gratings for spectroscopy,

[114] E. J. Becklake, C. D. Payne, and B. E. Prewer, *J. Phys.* (*D*), 1970, **3**, 473.
[115] R. A. Andrews, *J. Quantum Electronics*, 1970, **6**, 68.
[116] G. A. Condas, *J. Quantum Electronics*, 1971, **7**, 202.
[117] F. Keilman and K. F. Renk, *Appl. Phys. Letters*, 1971, **18**, 452.
[118] W. Ulmer, *Infrared Phys.*, 1971, **11**, 221.
[119] R. M. Logan and R. Watton, *Infrared Phys.*, 1972, **12**, 17.
[120] Y. Takeuchi and S. Kon, *Jap. J. Appl. Phys.*, 1971, **10**, 387.
[121] H. A. Gebbie, *Appl. Optics*, 1969, **8**, 501.
[122] M. F. Kimmitt in ref. 9, p. 121.
[123] A. R. H. Cole, A. A. Green, G. A. Osborne, and G. D. Reece, *Appl. Optics*, 1970, **9**, 23.
[124] I. F. Silvera and G. Birnbaum, *Appl. Optics*, 1970, **9**, 617.
[125] E. G. Loewen, *J. Phys.* (*E*), 1970, **3**, 953.

commencing with the use of a silk handkerchief by Rittenhouse in 1786. Stolen [126] has described the development of laminary as opposed to echelette gratings for use in the far-i.r. In the laminary grating, which has basically a square-wave profile in contrast to the saw-tooth profile of the echelette, the intensity of the even orders is zero, and that for odd orders falls by $1/n^2$. This results in an increased ease of filtering in the far-i.r. which more than compensates for the small reduction $(2/\pi)^2$ in the diffraction efficiency compared with the echelette. Although the advantages of laminary gratings have been appreciated for forty years, the difficulty in their production has limited their use. Stohlen has described a satisfactory photoetching procedure for the production of laminary gratings from printed circuit boards.

Far-infrared Interferometry.—The advent of two-beam Fourier interferometers, particularly the Michelson, and to a lesser extent the lamellar grating, has largely revolutionized the practice of far-i.r. spectroscopy over the past decade. A comprehensive review of the 'state of the art' as it existed in 1970 is given by the Proceedings of the Aspen International Conference on Fourier Spectroscopy held in March 1970.[127]

Space precludes a development of the theory of interferometry but this is, however, very completely covered in a number of papers, outstanding among which are those of Vanasse and Saki,[128] Connes,[129] Richards,[130] Chantry,[131] and Loewenstein.[132] For those unfamiliar with the technique, a number of fairly basic reviews have recently been published, including those by Low,[133, 134] Parrett,[135] Grosse,[136] Hadni,[137] and Peterman.[138]

The interferometer of Sanderson and Scott [139, 140] is an example of an aperiodic Michelson interferometer designed for high-resolution gas-phase studies. To accommodate long cells, 'cat's eye' retroreflectors are used in the long arms of the interferometer. This not only greatly simplifies the optical design but also reduces the effect of pitch and yaw on the motion of the moveable mirror assembly, which is mounted on a 20 cm comparator screw operated by a stepping motor drive. The theoretical resolution is 0.025 cm^{-1} and a value close to this is

[126] R. H. Stolen, *Appl. Optics*, 1970, **9**, 1229.
[127] G. A. Vanasse, A. T. Stair, and D. J. Baker, Editors, Aspen International Conference on Fourier Spectroscopy, A.F.C.R.L. 71–0019 (AD 724 100).
[128] G. A. Vanasse and H. Sakai, *Progr. Optics*, 1967, **6**, 501.
[129] J. Connes, *Rev. Optics*, 1961, **40**, 45, 116, 171, 231.
[130] P. L. Richards in ref. 8, Chapter 2.
[131] G. W. Chantry in ref. 10, Chapters 3 and 4.
[132] E. V. Loewenstein in ref. 127, Lecture 1.
[133] M. J. D. Low, *Naturwiss.*, 1970, **57**, 280; *J. Chem. Educ.*, 1970, **47**, A255, A349, A415.
[134] M. J. D. Low, A. J. Goodsel, and H. Mark, Molecular Spectroscopy 1971, Institute of Petroleum, Applied Science, Essex, 1972, p. 383.
[135] F. W. Parrett, *Lab. Practice*, 1970, **19**, 68, 177.
[136] P. Grosse, Beckman Report, 1970, 3.
[137] A. Hadni, *Opt. Spectra*, 1970, **4**, 57.
[138] B. Peterman, *Obz. Mat. Fiz.*, 1970, **17**, 70.
[139] R. B. Sanderson and H. E. Scott, *Appl. Optics*, 1971, **10**, 1097.
[140] R. B. Sanderson and H. E. Scott, ref. 127, paper 14.

approached. For example, the 105.6 cm^{-1} doublet in the H_2O vapour spectrum, where the separation is 0.052 cm^{-1}, has been resolved.

A far-i.r. reflection Michelson interferometer for measurements of semiconductor reflection spectra at angles close to normal incidence has been described by Birch.[141] A conventional Michelson interferometer was modified by incorporation of a Cassegrain mirror system to reduce the beam diameter to permit the use of small samples. Also, a second beam splitter was used to separate the reflected and incident beams spatially at the sample.

With the increasing availability of small digital computers, the interfacing of FT spectrometers to this type of computer has been receiving increasing attention. Gayles *et al.*[142] give an extensive description of the interfacing of an interferometer to an I.B.M. computer. Curbelo and Foskett [143] and Levy *et al.*[144] describe commercially available interferometer—digital computer packages.

The traditional method of obtaining an acceptable SNR in the interferogram is to advance the moveable mirror in steps and to integrate the signal, for several seconds, at each position. The digital computer and the fast-response TGS detector have made possible the rapid-scan FT spectrometer, in which the interferogram is scanned very rapidly and the data digitized directly into the computer. An acceptable SNR is then achieved by repetitive scanning, each scan lasting for a few seconds. The technique of rapid-scan FTS has been discussed by Loewenstein [145] and by Griffiths.[146]

In certain applications, rapid-scan interferometry can have significant advantages over slow-scan techniques. In the rapid-scan technique it is possible to obtain the spectrum of a source whose intensity varies over a period of a few seconds, but whose spectral content is constant. This makes possible 'hand-held' Fourier spectrometers which can be useful in field applications, such as pollution monitoring. The speed with which a single, fast scan may be accomplished can also be important for the observation of transient phenomena, *e.g.* the monitoring of the effluent of a gas chromatograph.

A number of practical techniques have been described aimed at improving or simplifying FTS. One of the most important of these is the double-beam differencing interferometer, where an optical chopper is used to provide a sample and reference beam which are sampled at identical positions of optical path difference. The sample spectrum may then be computed directly. Interferometers of this type have been discussed by Dowling [147] and by Thorpe *et al.*[148] Mark and Low have described the 'clipping' of the central peak of the inter-

141 J. R. Birch, *Infrared Phys.*, 1972, **12**, 29.
142 J. N. Gayles, W. L. Honzik, and D. O. Wilson, *I.B.M. J. Res. Development*, 1970, **14**, 25.
143 R. Curbelo and C. Foskett, ref. 127, paper 21.
144 F. Levy, R. C. Milward, S. Bras, and R. le Toullee, ref. 127, paper 34.
145 E. V. Loewenstein, ref. 2, Appendix 6.
146 P. R. Griffiths, 'Molecular Spectroscopy, 1971,' Institute of Petroleum, Applied Science, Essex, 1972, p. 371.
147 J. M. Dowling, ref. 127, paper 4.
148 L. W. Thorpe, D. J. Neale, and G. C. Hayward, ref. 127, paper 17.

ferogram [149] and optical cancellation [150] as techniques for obtaining essentially double-beam operation from a single-beam FTS. The same authors [151] have also discussed the use of the path-difference derivative of the interferogram to render weak spectral features more prominent.

Sakai and Murphy [152] have discussed the realization of the multiplex advantage in FTS in which the moving mirror is continuously driven rather than stepped: Bell and Sanderson [153] have discussed the relationship between the spectral error in FTS and the random sampling position errors resulting from imperfections in the mirror drive.

The introduction by the National Physical Laboratory group [154] of phase-modulated rather than amplitude-modulated FT interferometers is likely to represent a significant advance in experimental FTS. In an amplitude-modulated system the use of an optical chopper to modulate the light beam normally prevents 50% of the radiation ever reaching the detector. At best, this represents a $(2)^{\frac{1}{2}}$ decrease in the SNR that may be achieved. In a phase-modulated system the light beam is incident on the detector for the whole experiment so that not only is the SNR maintained at its maximum value but better discrimination against stray radiation is also achieved. The phase modulation is achieved by imparting to one of the mirrors of a Michelson interferometer a sinusoidal movement, whose amplitude is a little less than the mean wavelength of the radiation being detected. In the far-i.r. the requisite amplitude is easily achieved by mounting the interferometer mirror on a commercial loudspeaker coil. The theory has been developed by Chamberlain [155] and its applications to Fourier spectroscopy,[156] distance measurement using sub-millimetre radiation, terametrology,[157] and laser refractometry [158] have been discussed.

There has been considerable interest in the polarization properties of FTS. A polarized FTS using aluminized poly(ethylene terephthalate) as the wire-grid polarizer has been described by Martin and Puplett.[159] Vickers *et al.*[160] have described free-standing wire polarizers for instruments of this type. Fymat has considered in great detail the polarization properties of FTS.[161] He first derived the Jones matrix representation of the beam splitter [162] and the complete Michelson interferometer [163] and used these in conjunction with the formalism of polarization coherency to derive all the polarization characteristics of Fourier

[149] H. Mark and M. J. D. Low, *Appl. Spectroscopy*, 1971, **25,** 605.
[150] M. J. D. Low and H. Mark, *J. Paint Technol.*, 1970, **42,** 265.
[151] M. J. D. Low and H. Mark, *Appl. Spectroscopy*, 1970, **24,** 129.
[152] H. Sakai and R. E. Murphy, *J. Opt. Soc. Amer.*, 1970, **60,** 422.
[153] E. E. Bell and R. B. Sanderson, *Appl. Optics*, 1972, **11,** 688.
[154] J. Chamberlain and H. A. Gebbie, *Appl. Optics*, 1971, **10,** 1184.
[155] J. Chamberlain, *Infrared Phys.*, 1971, **11,** 25.
[156] J. Chamberlain and H. A. Gebbie, *Infrared Phys.*, 1971, **11,** 57.
[157] J. Chamberlain and H. A. Gebbie, *Infrared Phys.*, 1971, **11,** 70.
[158] J. Chamberlain, J. Haigh, and M. J. Hine, *Infrared Phys.*, 1971, **11,** 85.
[159] D. H. Martin and E. Puplett, *Infrared Phys.*, 1970, **10,** 105.
[160] D. G. Vickers, E. I. Robson, and J. E. Beckman *Appl. Optics*, 1971, **10,** 682.
[161] A. L. Fymat and K. D. Abhyankar, ref. 127 paper 39.
[162] A. L. Fymat, *Appl. Optics*, 1971, **10,** 2499.
[163] A. L. Fymat, *Appl. Optics*, 1971, **10,** 2711.

spectrometers.[164] Fymat and Abhyankar [165] have discussed the construction of a far-i.r. interferometric spectropolarimeter based on these considerations.

Burroughs and Harries [166] have given a theoretical analysis of the advantages and disadvantages of apodization in FTS, and it is shown that for gas-phase spectroscopy at relatively high resolutions, better than 0.5 cm^{-1}, there are certain distinct advantages in not apodizing the experimental results. This conclusion is supported by the work of Lightman.[167] Fleming [168] has discussed the convergence corrections that must be applied to the absorption coefficients derived by FTS if the sample is placed in a highly convergent beam.

A second form of two-beam Fourier spectrometer is the lamellar-grating spectrometer developed by Strong and co-workers.[169, 170] In this the two wavefronts are made by wavefront division at the lamellar grating rather than by amplitude division at the beam splitter as in the Michelson interferometer. A lamellar-grating interferometer with a spherical lamellar grating has been described by Hansen and Strong.[171] The spherical grating performs the triple functions of wavefront division, path differencing, and focussing, avoiding the need for collimators or beam splitters. This permits a compact interferometer with a minimum of optical components to be constructed. Baukus and Ballantyne [172] have described the modification of a far-i.r. lamellar-grating interferometer to operate in the 0.27—2.3 cm^{-1} region of the spectrum.

Although a commercial lamellar-grating interferometer is available,[173] the mechanical construction of this type of interferometer is generally considered more complex than for the Michelson, and this has limited the application of the lamellar-grating instrument in the far-i.r. A feature of the lamellar-grating type of interferometer is that its optical layout is similar to that of a Littrow-type grating spectrometer. This has permitted the conversion of Littrow instruments with small gratings and, consequently, low-energy throughputs to the lamellar interferometric mode, to realize not only the 'étendue' but also the 'multiplex' advantage of this form of spectroscopy. The conversion of a Perkin Elmer 301 to the lamellar-grating mode is described by Jones *et al.*[174]

3 Rotational Spectra in the Far-infrared

Pure Rotational Spectra.—The far-i.r. pure rotation spectra of molecules in the gas phase are of importance not only for the determination of the rotational constants of light molecules but also for the calibration of far-i.r. spectrometers.

164 A. L. Fymat, *Appl. Optics*, 1972, **11**, 160.
165 A. L. Fymat and K. D. Abhyankar, *Appl. Optics*, 1970, **9**, 1075.
166 W. J. Burroughs and J. E. Harries, *Infrared Phys.*, 1971, **11**, 99
167 A. Lightman, *Infrared Phys.*, 1971, **11**, 125.
168 J. W. Fleming, *Infrared Phys.*, 1970, **10**, 57.
169 J. Strong and G. A. Vanasse, *J. Opt. Soc. Amer.*, 1960, **50**, 113.
170 J. Strong, *U.S. Govt. Res. Development Report*, 1970, **70**, 196.
171 N. P. Hansen and J. Strong, ref. 127, paper 20.
172 J. Backus and J. Ballantyne, ref. 127, paper 43.
173 R. C. Milward, *Infrared Phys.*, 1969, **9**, 59.
174 B. W. Jones and J. R. Houck, *Appl. Optics*, 1970, **9**, 2582.

The most commonly used calibrant is water vapour, and the far-i.r. spectrum of H_2O is extensively discussed by Möller and Rothschild,[2] who not only illustrate the spectrum in the 380—12 cm^{-1} region but also tabulate the frequencies of the 278 lines in the 13—300 cm^{-1} region as calculated from the rotational constants and centrifugal distortion coefficients.[175]

The pure rotational spectrum of water vapour has been extensively studied by Dowling and co-workers, using a lamellar grating interferometer.[176] The spectrum has been observed with an effective resolution of 0.05 cm^{-1} in the 15—122 cm^{-1} region [177] and the 92—250 cm^{-1} region.[175] The measured frequencies and their assignments are presented, and the r.m.s. deviations from those obtained in other far-i.r. investigations are calculated. In another paper,[178] Dowling extends this comparison to results obtained in the near-i.r., the difference in energy between specific pure rotational energy levels as determined from the pure rotational spectrum in the far-i.r. being compared with those obtained from the vibration–rotation spectra.[179]

At low frequencies, say, less than 50 cm^{-1}, the water vapour spectrum becomes less intense, and fewer isolated well-defined peaks are available for calibration. In this frequency region it becomes advantageous to use small molecules with a significant permanent dipole for the calibration of far-i.r. spectrometers. The pure rotational lines of heteronuclear diatomics and unsymmetric linear triatomics are particularly valuable since, in the absence of hot bands and isotope effects, the pure rotational lines have no overlapping fine structure and the band centre is independent of the resolving power of the spectrometer. Among the molecules commonly employed are HCN, N_2O, and CO. The best available pure rotational frequencies for these molecules have been tabulated by Wilkinson and Martin.[180]

The hydrogen halides would also be more extensively used were it not for the appreciable isotope splittings observed at high resolutions. Even at more moderate resolution, insufficient to resolve the isotope splitting, the frequency of the band centre is dependent on the resolving power of the spectrometer. To permit the use of HCl as a calibrant of medium-resolution spectrometers, Cole and Honey [181] have calculated the contours of the pure rotation spectra of $H^{35}Cl$ and $H^{37}Cl$ from vibration–rotation data and have tabulated the apparent band-centre frequencies of HCl as a function of the resolving power of the spectrometer.

The intensities and linewidths of the pure rotation spectrum of CO have been determined experimentally by Sanderson, Scott, and White,[182] using a far-i.r. Michelson interferometer in the asymmetric mode. An effective re-

[175] R. T. Hall and J. M. Dowling, *J. Chem. Phys.*, 1967, **47**, 2454.
[176] R. T. Hall, D. Vrabec, and J. M. Dowling, *Appl. Optics*, 1966, **5**, 1147.
[177] R. T. Hall and J. M. Dowling, *J. Chem. Phys.*, 1970, **52**, 1161.
[178] J. M. Dowling, *J. Mol. Spectroscopy*, 1971, **37**, 272.
[179] P. E. Fraley and K. N. Rao, *J. Mol. Spectroscopy*, 1969, **29**, 348.
[180] G. R. Wilkinson and D. H. Martin, in ref. 8, p. 104.
[181] A. R. H. Cole and F. R. Honey, *Appl. Optics*, 1971, **10**, 1581.
[182] R. B. Sanderson, H. E. Scott, and J. T. White, *J. Mol. Spectroscopy*, 1971, **38**, 252.

solution of 0.05 cm^{-1} was achieved. The intensity of the $J+1 \leftarrow J$ pure rotational transition may be related to the radial part of the dipole-moment matrix element $\mu(J)$. For a rigid rotor, $\mu(J)$ is just the permanent dipole moment, μ_0. For a real rotor, however, centrifugal stretching gives rise to a J-dependent correction:

$$\mu(J) = \mu_0[1-10^{-4}(J+1)^2]$$

The observed intensities may then be used to calculate the permanent dipole moment and the value obtained for CO, 0.112 Debye, is in good agreement with the value obtained from the Stark splitting of the $J = 0$ line in the microwave spectrum. Dowling [183] had reported that the linewidths of the pure rotational spectrum of CO showed a strong J dependence, which had not been theoretically predicted by Benedict and Herman.[184] The linewidths of Sanderson *et al.*,[182] however, showed little or no J dependence and were in close agreement with the calculated values.

The pure rotation spectrum of the nearly prolate asymmetric top nitrogen dioxide has been studied by Bird *et al.*[185] NO_2, with 17 valence electrons, is the only simple polyatomic free radical stable in the gas phase. The pure rotation spectrum of NO_2 is exceedingly complex on account of the presence of magnetic splitting of the pure rotation lines. This magnetic splitting involves a coupling of the odd electron spin to the magnetic moment generated by the rotation of the molecule. Observation of the magnetic splitting requires a spectrometer of high resolving power, and Bird and co-workers used a large Michelson interferometer with a sample path length of 440 mm, and NO_2 pressure of 226 Torr (3×10^4 N m^{-2}). The pure rotation spectrum was observed in the range 45—104 cm^{-1} with an effective resolution of 0.01 cm^{-1}. The analysis and assignment of the NO_2 pure rotation spectrum is considered in detail in Chapter 6.

Harries *et al.*[186] suggested that several unidentified lines in the emission spectrum of the stratosphere between 18 and 34 cm^{-1} might arise as result of absorption by HNO_3. To test this hypothesis, Harries *et al.*[187] measured the pure rotation spectrum of HNO_3 at 20 Torr using an N.P.L.—Grubb Parsons cube interferometer with an effective resolution of 0.15 cm^{-1}. Over 50 lines in the pure rotation spectrum of HNO_3 were observed between 10 and 30 cm^{-1}. The observed frequencies were in good agreement with those calculated by the rigid-rotor approximation using the microwave rotational constants. There was, however, no correlation between the pure rotation spectrum of HNO_3 and the lines in the stratospheric emission spectrum, so that the origin of the lines in the stratosphere is still uncertain.

Another pure rotational study with a predominantly astrophysical motivation

[183] J. M. Dowling, *J. Quant. Spectroscopy Radiative Transfer*, 1969, **9**, 1613.
[184] W. S. Benedict and R. Herman, *J. Quant. Spectroscopy Radiative Transfer*, 1963, **3**, 265.
[185] G. R. Bird, G. R. Hunt, H. A. Gebbie, and N. W. B. Stone, *J. Mol. Spectroscopy*, 1970, **33**, 244.
[186] J. E. Harries and W. J. Burroughs, *Quart. J. Roy. Met. Soc.*, 1971, **97**, 519.
[187] J. E. Harries, W. J. Burroughs, and G. Duxbury, *Nature Phys. Sci.*, 1971, **232**, 171.

was the study at 248 K of the 23 cm^{-1} absorption in the water vapour spectrum.[188] Atmospheric studies in the far-i.r. are only possible in the gaps between the strong water-vapour lines, and a potentially important gap at 23 cm^{-1} is partially obscured by a weak water-vapour absorption which is more intense in astrophysical studies using the sun as the source than it is in laboratory measurements. Using a 50 m path length of water vapour at 0.4 Torr (50 N m^{-2}) and an argon over-pressure of 2 atm (2×10^5 N m^{-2}), Bohlander and co-workers were able to show that the 23 cm^{-1} feature was significantly more intense at 248 K than at 298 K. This seems to preclude an absorption by monomeric water vapour and it is suggested that the absorption arises from a multimer, possibly the dimer.

The pure rotation spectrum of D_2S has been investigated in the far-i.r. (6.6—33 cm^{-1}) by Osaka and Takahashi.[189] The spectrum was obtained in a 200 mm cell at a pressure of 550 Torr (7×10^4 N m^{-2}) and analysed in terms of the ground-state rotational constants and molecular parameters. The same authors[190] have also measured the far-i.r. rotational spectrum of MeF. Twelve lines, $J = 3$—14, were observed and their frequencies were in good agreement with those calculated from microwave data. Linewidth measurements were also made and although the K dependence of the spectrum was not observed on account of the limited resolving power, the linewidths show a minimum at $J = 10$, in agreement with theory.

The pressure broadening of the pure rotational lines of HCl by high pressures, 100 atm (10^7 N m^{-2}) of argon and helium, has been studied by Van Kreveld and co-workers.[191] The spectra were obtained in the 35—170 cm^{-1} region with a R.I.I.C. FS 720 interferometer. Again, these results are discussed in detail in Chapter 6.

Spectra of Gases Dissolved in Liquids.—Considerable interest has been directed in recent years towards the rotational motion of molecules in liquids and matrices. The theory of pure rotational absorption of polar molecules in non-polar liquids has been considered by Robert and Galatry.[192] They use the rotational autocorrelation function developed by Gordon,[193] while preserving the quantum character of the rotational degrees of freedom of the polar molecule. The observed spectral profile is shown to be determined not only by the position and intensity of the pure rotational absorption of the free molecule but also by the individual widths of these lines, representing the orientational relaxation times.

Experimentally, the far-i.r. spectra of HCl and DCl in non-polar liquids

[188] R. A. Bohlander, H. A. Gebbie, and G. W. F. Pardoe, *Nature*, 1970, **228**, 156.
[189] T. Osaka and S. Takahashi, *Tohoku Daigaku Kagaku Keisoku Kenkyusho Hokoku*, 1970, **18**, 49.
[190] S. Takahashi and T. Osaka, *Tohoku Daigaku Kagaku Keisoku Kenkyusho Hokoku*, 1970, **19**, 43.
[191] M. E. van Kreveld, R. M. van Aalst, and J. van der Elsken, *Chem. Phys. Letters*, 1970, **4**, 580.
[192] D. Robert and L. Galatry, *J. Chem. Phys.*, 1971, **55**, 2347.
[193] R. G. Gordon, *Adv. Magn. Resonance*, 1968, **3**, 1.

were first observed by Datta and Barrow.[194, 195] For HCl and DCl in cyclohexane and carbon tetrachloride they found a broad band envelope, with no resolved rotational fine structure, in the 50—250 cm^{-1} region. The profile of this band was fairly accurately defined by the relative integrated intensities of each of the pure rotational gas-phase transitions. The far-i.r. spectrum of HCl and DCl in liquid SF_6 has been determined by Birnbaum and Ho [196] using a grating spectrometer.[124] The observed spectra were ratioed against the spectrum of an equal path length of SF_6, to eliminate any contribution from the collision-induced spectrum of SF_6 (see Section 4). A total gas pressure of 19 atm (2×10^6 N m^{-2}) was maintained over the liquid at 273 K to ensure that the SF_6 remained in the liquid phase. The DCl spectrum, a broad band envelope with no fine structure, was very similar to that already observed for DCl in C_6H_{12} and CCl_4. In the case of HCl, however, clearly resolved rotational lines were observed on the high-frequency side of the absorption, the frequencies of which were readily identifiable with the pure rotational gas-phase absorption frequencies.

The spectrum of HCl and DCl in SF_6 has been specifically considered by Galatry and Robert [197] as an extension to their general theory of i.r. absorption of polar molecules in liquids.[192] They consider that in the lower quantum states of HCl in SF_6 and for all quantum states of HCl in C_6H_{12} and CCl_4, and of DCl in SF_6, C_6H_{12}, and CCl_4, the orientational action of the solvent is sufficient to ensure that there is a high degree of perturbation of the rotational energy levels and that, consequently, no rotational fine structure can be resolved. In the case of the higher quantum states of HCl, on the other hand, the orientational action of the solvent leaves the energy levels relatively unperturbed and fine structure can be resolved.

These suggestions have received support from the near-i.r. study of the rotation–vibration spectrum of HCl and DCl in SF_6,[198] but contradictory evidence is supplied by the low-frequency Raman spectrum,[199] where the rotational fine structure of DCl in SF_6 was resolved from $J = 5$ to $J = 13$.

The far-i.r. spectrum of HCl in liquid argon and krypton has been measured by van Aalst and van der Elskan.[200] The spectra were recorded using a Beckman-R.I.I.C. FS 720 Fourier Transform spectrometer with HCl in liquid argon at 105 K and 10 atm (10^6 N m^{-2}), and in liquid krypton at 125 K and 6 atm (6×10^5 N m^{-2}). Both spectra showed residual fine structure identifiable with the pure rotational gas-phase transitions, and evaluation of the rotational autocorrelation function showed that, in distinction to the spectra of HCl in gaseous Ar and Kr, the damping of the rotational motion was less in the case of krypton than it was for argon.

[194] P. Datta and G. M. Barrow, *J. Chem. Phys.*, 1965, **43,** 2137.
[195] P. Datta and G. M. Barrow, *J. Chem. Phys.*, 1968, **48,** 4662.
[196] G. Birnbaum and W. Ho, *Chem. Phys. Letters*, 1970, **5**, 334.
[197] L. Galatry and D. Robert, *Chem. Phys. Letters*, 1970, **5**, 120.
[198] P. V. Huong, M. Couzi, and M. Perrot, *Chem. Phys. Letters*, 1970, **7**, 189.
[199] J. P. Perchard, W. F. Murphy, and H. J. Bernstein, *Chem. Phys. Letters*, 1971, **8,** 559; *Mol. Phys.*, 1972, **23,** 499, 519, 535.
[200] R. M. van Aalst and J. van der Elsken, *Chem. Phys. Letters*, 1972, **13,** 631.

Spectra of Gases in Matrices.—The pure rotational spectrum of polar molecules may also be observed when the polar molecule is trapped in an inert-gas matrix. As this normally implies a low-temperature experiment, experimental observation of the rotational transition is not easy as only the lowest rotational quantum levels are significantly populated. For example, in the case of HCl, the far-i.r. absorption occurs in the 10—40 cm^{-1} region of the spectrum. In addition, polar molecules entrapped in rare-gas matrices absorb in two other regions of the far-i.r.: in the 40—100 cm^{-1} region on account of the activation by the impurity of the phonon bands of the matrix itself, and in the 150—200 cm^{-1} region due to polymeric absorptions of the impurity.[201, 202] Von Holle and Robinson [203] have investigated the $J = 1 \leftarrow J = 0$ rotational transition of HCl in argon and krypton matrices, complementing the measurements of Barnes and co-workers.[204] In a krypton matrix, the band was observed as a doublet, with a separation of about 0.8 cm^{-1}, centred at 19 cm^{-1}, whereas in an argon matrix, no splitting was observed and the band was centred at 18 cm^{-1}. Attempts to measure the band in xenon were only partially successful; a wing structure was observed at about 16 cm^{-1} but positive identification was not possible.

In contrast to the situation in argon and krypton matrices, Barnes *et al.*[204] have shown, both by pure rotational and by vibration–rotation studies, that in a nitrogen matrix HCl and DCl are not free to rotate. Katz and Ron [205] were also unable to observe the pure rotational transitions of DCl or HCl in a nitrogen matrix. They did, however, observe two broad and relatively strong absorptions at 154 and 207 cm^{-1} for HCl and at 132 and 162 cm^{-1} for DCl in a nitrogen matrix. Although these absorptions were in the region of the spectrum normally associated with absorptions of dimeric polar molecules,[201] the dependence of the intensity on the concentration and the deposition temperature suggested that in this case the absorption could be ascribed to monomeric polar molecules. This was supported by a study of mixed HCl–DCl systems where no 'mixed' absorptions were observed, but rather a superpositioning of the separate absorptions of HCl and DCl. It is suggested that these absorptions represent either a localized motion of the polar molecule in the matrix, or the motion of the two partners relative to one another in an N_2–HCl complex.

Robinson and co-workers have extended their measurements of the pure rotation spectrum of polar molecules in rare-gas matrices to a study of HF and DF in a series of inert matrices.[206] Using a liquid-cooled Ga–Ge bolometer they were able to improve on their previous data [207] and observe the $J = 1 \leftarrow J = 0$ pure rotational transition of HF at 39.8 cm^{-1} in a neon matrix, rising to 50.5 cm^{-1} in a xenon matrix. Similarly, the DF rotational frequency rose from

[201] B. Katz, A. Ron, and O. Schnepp, *J. Chem. Phys.*, 1967, **46**, 1926.
[202] B. Katz, A. Ron, and O. Schnepp, *J. Chem. Phys.*, 1967, **47**, 5303.
[203] W. G. von Holle and D. W. Robinson, *J. Chem. Phys.*, 1970, **53**, 3768.
[204] A. J. Barnes, J. B. Davies, H. E. Hallam, G. F. Scrimshaw, G. C. Hayward, and R. C. Milward, *Chem. Comm.*, 1969, 1089.
[205] B. Katz and A. Ron, *Chem. Phys. Letters*, 1970, **7**, 357.
[206] M. G. Mason, W. G. von Holle, and D. W. Robinson, *J. Chem. Phys.*, 1971, **54**, 3491.
[207] D. W. Robinson and W. G. von Holle, *J. Chem. Phys.*, 1966, **44**, 410.

19.8 cm^{-1} in an argon matrix to 24.6 cm^{-1} in a krypton matrix. As with HCl in a krypton matrix, the lines were split into a doublet, with a spacing in this instance of about 3 cm^{-1}. It is suggested that this splitting could be accounted for by the polar molecule inducing the rare-gas matrix to crystallize with hexagonal rather than cubic close-packing or that the potential of the polar molecule at the lattice site includes angular-dependent terms. These angular-dependent terms arise from the anisotropy of the polarizability and from the fact that the dispersion forces are not applied to the molecular centre of mass.

4 Collision-induced Spectra in the Far-infrared

Considerable interest has been generated over the past few years in spectroscopic techniques that rely on collision-induced phenomena. These techniques, which include both absorption and light scattering,[208] have proved most powerful in the investigation of intermolecular forces, particularly the quadrupole and higher multipole moments.

A very extensive treatment of collision-induced far-i.r. absorption is given by Möller and Rothschild [2] but some of the salient points will be summarized here.

If the simplest case of a binary collision between two like rare-gas atoms, *e.g.* He–He, is considered, the $D_{\infty h}$ symmetry of the binary 'collision complex' renders it impossible to form an induced dipole moment, and no i.r. absorption is observed. In the case of two unlike atoms, however, the collision leads to the formation of a transient dipole moment which can give rise to i.r. absorption. Two molecular processes could be involved in the absorption, rotation of the collision complex and pure translational motion between two unlike atoms. It may be shown that the translational component is the most important in far-i.r. absorption and, since there are no bound translational states, the absorption is not of a resonant character and its width is determined by the duration of the collisional process. The theory of pure translational absorption was first developed by Poll and van Kranendonk [209] and extended by Sears.[210] The lineshape function and integrated induced absorption coefficients have been derived by Levine and Birnbaum.[211]

In diatomic molecules, however, the collision-induced absorption arises not only from pure translationally induced dipoles but also from rotational motion that is coupled to the translational motion as a result of the dependence of the intermolecular potential on the relative orientation of the collision partners. It is only possible to separate out the rotational and translational motion in the case of very light molecules, *e.g.* hydrogen, where the spacing between the rotational levels is large.

[208] J. P. McTague and G. Birnbaum, *Phys. Rev.*, 1971, **3A**, 1376.
[209] J. D. Poll and J. van Kranendonk, *Canad. J. Phys.*, 1961, **39**, 189.
[210] V. F. Sears, *Canad. J. Phys.*, 1968, **46**, 1163.
[211] H. B. Levine and G. Birnbaum, *Phys. Rev.*, 1967, **154**, 86.

The selection rules for the rotational–translational absorption in the simplest case of the interaction of a homonuclear diatomic molecule with an atom may be deduced by consideration of the transient dipoles induced in the atom by the quadrupole moment of the diatomic molecule. Since in the homonuclear diatom the molecular quadrupole rotates twice as fast as the molecule itself, the collision-induced rotation spectrum is observed at the rotational frequencies of the diatomic molecule with the selection rule $\Delta J = +2$.

Clearly, as the possibility of both colliding molecules possessing quadrupole moments and the anisotropy of the polarizability is taken into account, the theory of collision-induced spectra becomes more complex. For example, in the case of CO_2 the observed collision-induced band is a superposition of contributions from the pure translation band, $\Delta J_1 = \Delta J_2 = 0$, an induced rotational band which is modulated by the translational motion of the molecule, $\Delta J_1 = +2$, $\Delta J_2 = 0$, and a small contribution associated with the anisotropy of the polarizability as a result of double transitions, $\Delta J_1 = \pm 2$, $\Delta J_2 = \pm 2$. The theory of rotational–translational collision-induced spectra has been extensively developed by Colpa and Ketelaar.[212]

Experimentally, the pure translation collision-induced spectra in rare-gas mixtures have been observed by Bosomworth and Gush [213] and more recently by Buontempo and co-workers.[214] The latter authors measured the absorption coefficient of neon in liquid argon using an R.I.I.C. FT spectrometer and correlated the observed lineshapes with the theoretical predictions of Sears [210] to give an intercollisional correlation time of 1.6×10^{-13} s and a pre-exponential induced dipole of 2×10^{-20} e.s.u. At high frequencies, very good agreement was obtained between the experiment and theory, but at lower frequencies, less than 70 cm^{-1}, the agreement was less satisfactory, intercollisional interference effects resulting in a reduction of the experimentally observed absorption. Buontempo *et al.*[215] have also observed the translation-rotation spectra of argon in liquid nitrogen and of nitrogen in liquid argon. In both cases the observed spectra were very similar and in good agreement with the predictions of the Colpa–Ketelaar theory.

Turning to triatomic molecules, there has been considerable interest in the collision-induced spectra of CO_2 at low frequencies, not only because the quadrupole moment of CO_2 is abnormally high, resulting in considerable intensity in the far-i.r. spectrum, but also because the induced absorption of CO_2 has been suggested as a reason for the microwave opacity of Venus. The far-i.r. absorption of compressed CO_2 in the temperature range 200—373 K has been measured by Harries [216] using an N.P.L.–Grubb Parsons cube FTS, a 440 mm path length of CO_2 at 15 atm (1.5×10^6 N m^{-2}), and a Putley or Golay detector. A broad absorption in the 0—120 cm^{-1} region was observed with a maximum

[212] J. P. Colpa and J. A. A. Ketelaar, *Mol. Phys.*, 1958, **1**, 343.
[213] D. R. Bosomworth and H. P. Gush, *Canad. J. Phys.*, 1965, **43**, 751.
[214] U. Buontempo, S. Cunsolo, and G. Jacucci, *Phys. Letters*, 1970, **31A**, 128.
[215] U. Buontempo, S. Cunsolo, and G. Jacucci, *Mol. Phys.*, 1971, **21**, 381.
[216] J. E. Harries, *J. Phys. (B)*, 1970, **3**, 704.

absorption coefficient at about 50 cm^{-1} of 10 neper cm^{-1} atm^{-2}. The quadrupole moment, as calculated from the Colpa–Ketelaar theory, is $7.6 \pm 1.0 \times 10^{-26}$ e.s.u. cm^2. It has, however, been pointed out by Bose and Cole [217] that the Colpa–Ketelaar theory neglects the quadrupole–quadrupole interaction term in the Lennard-Jones potential of the colliding molecules. Because of the very large quadrupole moment of CO_2, neglect of the quadrupole–quadrupole interaction would be expected to lead to a significant error in the calculated value for the quadrupole moment.

Harries [218] has revised his calculation of the quadrupole moment of CO_2 in the light of the comments of Bose and Cole, and as a result the mean corrected value for the quadrupole moment is reduced from 7.6 to $4.5 \pm 0.4 \times 10^{-26}$ e.s.u. cm^2 (4.5×10^{-31} C m^2).

The far-i.r. collision-induced spectrum of CO_2 has also been investigated by Ho, Birnbaum, and Rosenberg [219] in the spectral region 7—250 cm^{-1} using a grating spectrometer [124] The temperature dependence of the spectrum was measured in the range 243—333 K and the CO_2 pressure was 20 amagat.* At all temperatures direct evidence for the separation of the pure translation and the translation–rotation band was obtained. Taking into account the quadrupole–quadrupole interaction terms in the intermolecular potential, Ho *et al.* calculated a quadrupole moment of CO_2 of $4.5 \pm 0.2 \times 10^{-26}$ e.s.u. cm^{-2} (4.5×10^{-31} C m^2).

The intensity of a normal, intramolecular, i.r. absorption varies linearly with pressure, representing the variation of the number density of the absorbing molecules within the sampling volume. One of the most characteristic features of far-i.r. collision-induced spectra is that the intensity varies with the square of the gas pressure, reflecting the collisional nature of the interaction responsible for the absorption. The gas density dependence of the collision-induced far-i.r. absorption in CO_2 has been investigated over the density range 0—85 amagat by Birnbaum, Ho, and Rosenberg.[220] The observed absorption coefficients, $\alpha(\nu,\rho)$, were fitted to an expression of the form:

$$\alpha(\nu,\rho) = \alpha_2(\nu)\rho^2 + \alpha_3(\nu)\rho^3$$

where α_2 represents the induced coefficient due to binary collisions and α_3 represents the induced coefficient due to ternary collisions. The maximum value for α_2 is 7×10^{-5} cm^{-1} $amagat^{-2}$ at 60 cm^{-1}. The bandshape associated with α_3 coincides with the profile of the bar spectrum of the pure rotational transitions and has a maximum value of -2.2×10^{-7} cm^{-1} $amagat^{-3}$ at 40 cm^{-1}.

The far-i.r. absorption of liquid CO_2 was also measured, and although the bandshapes for the gas and liquid were similar, the band centre for the liquid

* Strictly the amagat is a density unit, representing the reciprocal of the molar volume at 273 K and 1 atm. However, for an ideal gas, pressure is linearly related to density.

[217] T. K. Bose and R. H. Cole, *J. Chem. Phys.*, 1970, **52**, 140.
[218] J. E. Harries, *J. Phys. (B)*, 1970, **3**, L150.
[219] W. Ho, G. Birnbaum, and A. Rosenberg, *J. Chem. Phys.*, 1971, **55**, 1028.
[220] G. Birnbaum, W. Ho, and A. Rosenberg, *J. Chem. Phys.*, 1971, **55**, 1039.

was about 25 cm^{-1} higher than in the gas, and the integrated band intensity of the liquid phase was 3.2×10^{-4} cm^{-2} $amagat^{-2}$ compared to a value of 5.11×10^{-3} cm^{-2} $amagat^{-2}$ for the low-density gas. It is not possible, therefore, in considering the far-i.r. collision-induced spectra, to regard the liquid phase of CO_2 merely as a very dense gas. This is in contradistinction to SF_6 (see below) where the liquid- and gas-phase collision-induced spectra are very similar.

In the far-i.r. absorption of dense gases possessing a permanent dipole moment, the collision-induced spectra are superimposed on the allowed pure rotational transitions. Such a situation has been investigated by Baise,[221] who observed the far-i.r. absorption of N_2O in the 8—200 cm^{-1} region using Michelson and lamellar-grating interferometers. It might be expected that the intensity of the allowed rotational absorptions would completely mask the weak, collisionally induced spectrum. The small dipole moment of N_2O (0.17 Debye), however, permits the separation of the pure rotational and collision-induced absorptions and, at high densities, 64 atm (6.4×10^6 N m^{-2}), the intensity of the collision-induced component actually exceeds that of the pure rotational component. Baise measured the far-i.r. spectra at 34.5 and 41.4 atm at 296 K, and at 56.5 atm at 298 K and 64.1 atm at 304 K. The allowed pure rotational component was subtracted out using the integrated intensities derived by Gordon [222] and the residual collision-induced component analysed by the Colpa–Ketelaar theory [212] to give a value of 8.6×10^{-26} e.s.u. cm^2 (8.6×10^{-31} C m^2) for the quadrupole moment of N_2O.

Other triatomic polar molecules studied as compressed gases in the far-i.r. include CS_2 and OCS, which have been investigated by Darmon and co-workers.[223, 224] Using a high-pressure cell in which the liquid and compressed gas were in equilibrium, Darmon *et al.* were able to obtain successive interferograms of each phase. Although a number of bands were observed in the liquid which were assigned to collision-induced transient dipole absorption (see Section 5), no collision-induced bands were observed in the gas phase.

Recently, interest has centred on molecules of high symmetry, such as T_d, which do not possess a quadrupole moment. The far-i.r. spectra may again be approximately decomposed into a low-frequency pure translational band and a high frequency translation–rotation band. These arise predominantly from the octupolar interactions, and Ozier and Fox [225] have extended the formalism of Colpa and Ketelaar [212] to calculate the far-i.r. spectrum of the rotation-translation component of the collision-induced spectrum. The experimental far-i.r. absorption spectra for CH_4,[226] CD_4,[227] and CF_4 [228] were used to calculate the octupole moments of these molecules.

[221] A. I. Baise, *Chem. Phys. Letters*, 1971, **9**, 627.
[222] R. G. Gordon, *J. Chem. Phys.*, 1963, **38**, 1724.
[223] I. Darmon, A. Gerschel, and C. Brot, *Chem. Phys. Letters*, 1970, **7**, 53.
[224] I. Darmon, A. Gerschel, and C. Brot, *Chem. Phys. Letters*, 1971, **8**, 454.
[225] I. Ozier and K. Fox, *J. Chem. Phys.*, 1970, **52**, 1416.
[226] S. Weiss, G. E. Leroi, and R. H. Cole, *J. Chem. Phys.*, 1969, **50**, 2267.
[227] G. Birnbaum and A. Rosenberg, *Phys. Letters*, 1968, **27A**, 272.
[228] A. Rosenberg and G. Birnbaum, *J. Chem. Phys.*, 1968, **48**, 1396.

Ozier and Fox considered only the rotation–translation component of the far-i.r. absorption, neglecting the translational and overlap contributions. The translational contribution to the far-i.r. spectra of tetrahedral molecules has been evaluated by Gray [229] using the formalism of Poll and van Kranendonk.[209] It is shown by Gray that the translational contribution amounts to 8, 12, and 25% of the far-i.r. absorption for CH_4, CD_4, and CF_4, respectively. Although it is surmised that this will lead to errors in the calculated octupole moments, no corrected values are presented.

Of highest symmetry are molecules belonging to the point group O_h, where the lowest non-zero multipole moment is the hexadecapole. The prime example of such a molecule is SF_6 and the far-i.r. absorption of this molecule has been investigated by Rosenberg and Birnbaum.[230] Using a grating spectrometer,[124] the far-i.r. absorption was measured in the 12—250 cm^{-1} region, with a brass light pipe as the cell. The absorption was measured as a function of the temperature (233—273 K) and of pressure (3—19 atm). The gas-phase spectrum exhibited a broad band in the 12—75 cm^{-1} region, the intensity of which was proportional to the square of the gas density: this was assigned to the hexadecapole-induced collision band. In the 70—250 cm^{-1} region two distinct bands were observed, centred at 93 and 173 cm^{-1}, and as their intensities were linearly proportional to the gas pressure they were assigned to intramolecular difference bands. If the whole of the intensity of the low-frequency band is ascribed to dipole induction from the SF_6 hexadecapole (that is, the possibility of 'hard' collisions lowering the O_h symmetry of SF_6 is ignored), a value of between 3.6 and 7.5×10^{-42} e.s.u. cm^4 is obtained for the hexadecapole of SF_6. The range of values arises from the uncertainty in the values to be taken for the parameters of the Lennard-Jones intermolecular potential.

Some multipole moments measured from collision-induced absorption in the far-i.r. are collected in Table 1.

Table 1 *Multipole moments of some molecules as determined by collision-induced absorption in the far-i.r.*

Molecule	*Symmetry*	*Moment*	*Value*	*Ref.*
CO_2	$D_{\infty h}$	Quadrupole	$4.5 \pm 0.4 \times 10^{-26}$ e.s.u. cm^2	218
			$4.5 \pm 0.4 \times 10^{-26}$ e.s.u. cm^2	219
N_2O	$C_{\infty v}$	Quadrupole	8.6×10^{-26} e.s.u. cm^2	221
HCl	$C_{\infty v}$	Quadrupole	5.8×10^{-26} e.s.u. cm^2	*a*
HBr	$C_{\infty v}$	Quadrupole	5.5×10^{-26} e.s.u. cm^2	*a*
CH_4	T_d	Octupole	2.6×10^{-34} e.s.u. cm^3	225
CD_4	T_d	Octupole	2.6×10^{-34} e.s.u. cm^3	225
CF_4	T_d	Octupole	5.6×10^{-34} e.s.u. cm^3	225
SF_6	O_h	Hexadecapole	3.6—7.5×10^{-42} e.s.u. cm^4	230

[a] S. Weiss and R. H. Cole, *J. Chem. Phys.*, 1967, **46**, 644

[229] C. G. Gray, *J. Chem. Phys.*, 1971, **55**, 459.
[230] A. Rosenberg and G. Birnbaum, *J. Chem. Phys.*, 1970, **52**, 683.

The far-i.r. absorption of SF_6 in the liquid phase was also measured by Rosenberg and Birnbaum. The spectra again show a number of intramolecular difference bands and a low-frequency, 50 cm^{-1}, collision-induced band. The general similarity of the results obtained for the liquid- and gas-phase spectra suggests a common mechanism for the collision-induced absorption in both phases.

5 Far-infrared Spectra of Liquids

Far-i.r. absorption has been shown to be a very general property of both polar and non-polar liquids.[231] The absorption is generally very broad, extending over the major part of the far-i.r. spectral region. For non-polar liquids the absorption is rather weak, with a maximum absorption coefficient in the range 1—10 neper cm^{-1}, and for polar liquids it is an order of magnitude greater.[232]

Non-polar Liquids.—It is shown in Section 4 that, in the case of SF_6 and CO_2, the far-i.r. absorption of the liquid closely parallels that of the dense gas, and, as the absorption cannot be considered to arise from allowed translational–rotational motions, a common collision-induced origin is generally accepted for both absorptions.

A distinguishing feature of collision-induced absorption as compared to rotation-allowed absorption is the temperature dependence of the absorption maximum. For collision-induced absorption a decrease in temperature shifts the absorption frequency maximum to lower frequencies, whereas for absorption arising from rotational motion the shift in absorption frequency maximum is to higher frequencies. The temperature dependence of the far-i.r. absorption in non-polar liquids has been investigated by Pardoe [233] for cyclopentane and by Peterman and co-workers [234] for C_6H_6, CCl_4, and CS_2. In all cases the absorption frequency maximum was found to shift to lower frequencies as the temperature was decreased, supporting a collision-induced origin for the bands.

Further supporting evidence is provided by dilution studies in which the far-i.r. absorption is studied as a function of the mole fraction of one non-polar liquid dissolved in a second. If the total absorption were a function of the concentration of each individual component, a linear relationship between the absorption, α, and the mole fraction x would be expected

$$\alpha = x_a\alpha_a + x_b\alpha_b$$

If, however, the absorption arose from a bimolecular collisional process, a non-linear relationship would be expected:

$$\alpha = x_a^2\alpha_a + x_b^2\alpha_b + 2x_a x_b(\alpha_a\alpha_b)^{\frac{1}{2}}$$

[231] G. W. Chantry, H. A. Gebbie, B. Lassier, and G. Wyllie, *Nature*, 1967, **214**, 163.
[232] M. Davies, G. W. F. Pardoe, J. Chamberlain, and H. A. Gebbie, *Trans. Faraday Soc.*, 1970, **66**, 273.
[233] G. W. F. Pardoe, *Trans. Faraday Soc.*, 1970, **66**, 2699.
[234] B. Peterman, B. Borstnik, and A. Azman, *Z. Naturforsch*, 1970, **25a**, 1516.

Such dilution studies have been carried out by Pardoe [233] on cyclohexane in benzene and dioxan, and on cyclopentane in benzene. In all cases a systematic departure from a linear concentration dependence was observed, in good agreement with the proposed bimolecular collision process.

Further support for the collisionally induced origin of the far-i.r. absorption of non-polar liquids comes from the work of Bucaro and Litovitz.[235, 236] They established a collisionally induced mechanism for the Rayleigh scattering of non-polar liquids and showed that the dominant mechanism in the Rayleigh scattering was the change in polarizability anisotropy of the molecular pair. This was related to the repulsive part of the intermolecular potential. Bucaro and Litovitz then compared the Rayleigh scattering profiles of CCl_4 and CS_2 with the far-i.r. absorption of these molecules as determined by Garg *et al.*[237] Very close agreement was observed in the 20—200 cm^{-1} region, suggesting that the induced dipole moments which are responsible for the far-i.r. absorption are also related to the repulsive part of the intermolecular potential. Below 20 cm^{-1} the effect of correlations between successive collisions [238] has the effect of suppressing the low-frequency far-i.r. absorptions while enhancing the low-frequency light scattering.

Savoie and Fournier [239] have investigated the far-i.r. absorption of liquid CH_4, CF_4, and SiF_4 and found that the absorption maximum was approximately dependent on the square root of the molecular moment of inertia. They interpret this as indicating that the motion responsible for the far-i.r. absorption was predominantly rotational. The far-i.r. absorption of liquid O_2, N_2, CH_4, and para and normal H_2 have been investigated by Jones.[240]

The anomalous far-i.r. absorption of 1,4-dioxan has been investigated by Davies *et al.*[232] In contrast to cyclohexane, where the absorption rises to a maximum of 0.5—0.6 neper cm^{-1} at 60—160 cm^{-1}, the far-i.r. absorption of dioxan rises to a maximum of 9 neper cm^{-1} at 75 cm^{-1} before falling to a value of about 5 neper cm^{-1} at 120 cm^{-1}. Davies and co-workers interpret this as reflecting the considerable local polarity of the CH_2—O—CH_2 group (each of which has an oppositely directed electric dipole of 1.8 Debye). Local perturbations of these polar groups are regarded as accounting for the enhanced far-i.r. absorption.

Polar Liquids.—In contrast to non-polar liquids, the far-i.r. absorption of polar liquids has normally been treated in terms of an extension of the microwave, Debye-type, absorption processes. Thus, although collisionally induced absorption processes may still occur, attention has focused on absorption associated with the permanent dipole.

[235] J. A. Bucaro and T. A. Litovitz, *J. Chem. Phys.*, 1971, **55**, 3585.
[236] J. A. Bucaro and T. A. Litovitz, *J. Chem. Phys.*, 1971, **54**, 3846.
[237] S. K. Garg, J. E. Bertie, H. Kilp, and C. P. Smyth, *J. Chem. Phys.*, 1968, **49**, 2551.
[238] J. van Kranendonk, *Canad. J. Phys.*, 1968, **46**, 1173.
[239] R. Savoie and R. P. Fournier, *Chem. Phys. Letters*, 1970, **7**, 1.
[240] M. C. Jones, Nat. Bureau Stand. (U.S.) Tech. Note, 1970, No. 390.

All polar liquids exhibit pronounced dispersions in their permittivities, ε', and an associated absorption or dielectric loss, ε'', in the microwave region of the spectrum, where ε' and ε'' are the real and imaginary parts of the complex dielectric constant. At any frequency, ω, the absorption coefficient, α, and the dielectric loss, ε'' are related:

$$\alpha = \varepsilon''\omega/n$$

where n is the refractive index of the medium.

Debye characterized the relaxation of the polarization contributed by the permanent dipole by a single molecular relaxation time τ. The frequency dependence of the permittivity, the dielectric loss, and the absorption coefficient are then given by [241]

$$\begin{aligned}\varepsilon' &= \varepsilon_\infty + (\varepsilon_0 - \varepsilon_\infty)(1+\omega^2\tau^2)^{-1}\\ \varepsilon'' &= \omega\tau(\varepsilon_0 - \varepsilon_\infty)(1+\omega^2\tau^2)^{-1}\\ \alpha &= \omega^2\tau(\varepsilon_0 - \varepsilon_\infty)(n+\omega^2\tau^2 n)^{-1}\end{aligned}$$

where ε_0 and ε_∞ are, respectively, the limiting low- and high-frequency values between which the dipole dispersion occurs.

Typical values for τ are of the order of 10^{-12} s, so that in the far-i.r. and higher-frequency regions of the spectrum $\omega^2\tau^2$ will greatly exceed unity. The Debye theory predicts, therefore, that in the far-i.r. the absorption coefficient will reach a limiting high-frequency value, α_{FIR}, given by:

$$\alpha_{\mathrm{FIR}} = (\varepsilon_\infty - \varepsilon_0)(\tau n)^{-1}$$

On this basis polar molecules exhibiting dielectric relaxation in the microwave region of the spectrum would be expected to continue to absorb, not only in the far-i.r. but throughout the whole high-frequency region of the spectrum! This physically unrealistic conclusion has been shown to be due to the neglect by the Debye theory of dipole inertial effects,[242] and it is possible to apply corrections to the original Debye equations so that transparency is recovered in the i.r. region of the spectrum.[243] However, the decrease in absorption now predicted in the far-i.r. is, in practice, frequently obscured by the onset of a new absorption mechanism, first considered explicitly by Poley.[244] It was suggested by Hill [245] that the Poley absorption arose as a result of libration of the absorbing molecule contained within a cage of solvent molecules. This concept has been extensively developed by Brot and co-workers,[246] who suggest that the Poley absorption has both a resonant character as the absorbing molecule librates or rotates within a shallow potential well determined by the arrangement of the

[241] N. Hill, W. E. Vaughan, A. H. Price, and M. Davies, 'Dielectric Properties and Molecular Behaviour', Van Nostrand, New York, 1969, Chapter 5.
[242] J. J. O'Dwyer and E. Harting, 'Progress in Dielectrics', 1967, **7**, 1.
[243] G. Birnbaum and E. R. Cohen, *J. Chem. Phys.*, 1970, **53**, 2885.
[244] J. J. Poley, *J. Appl. Sci.*, 1955, **4B**, 337.
[245] N. E. Hill, *Proc. Phys. Soc.*, 1963, **82**, 723.
[246] B. Lassier and C. Brot, *Discuss Faraday Soc.*, 1969, **48**, 39.

nearest neighbour molecules, and a relaxational character as it 'flips' between successive potential wells. The absorption arising from a rotor confined within a potential well with two minima has been considered by Brot and co-workers [246] and by Pourprix and co-workers.[247] The discussion has been extended to potential wells with eight minima by Brot and Darmon.[248]

It was suggested by Kroon and van der Elsken [249] that the rotational contribution to the far-i.r. absorption of polar liquids should be related to the total intensity, A, of the gas-phase rotational spectrum as given by the sum rule of Gordon: [193, 222]

$$A = \frac{\pi}{3\,c^2}\sum \mu_z^2\,(I_x^{-1}+I_y^{-1})$$

where μ_z is the permanent dipole moment along the z axis, I_x and I_y are the principal moments of inertia normal to that axis, and the summation is over the rotational states of the molecule. This equation predicts a total absorption that is proportional to the square of the permanent dipole moment and inversely proportional to the moment of inertia of the molecule.

Attempts have been made to assess the rotational contribution of the far-i.r. absorption of polar liquids by investigating how far they obey the sum rule of Gordon. Jain and Walker [250] investigated the far-i.r. absorption of 27 organic liquids in the 5—70 cm^{-1} region using an R.I.I.C. lamellar-grating interferometer. The liquids studies included mono- and di-substituted benzenes, methyl halides, and long-chain halogenoparaffins. The dipole moments of the liquids varied from 3.97 for benzonitrile and nitrobenzene to zero for the *para*-disubstituted benzenes. The moments of inertia varied between 112 and 1851 $\times 10^{-40}$ g cm^2 and the maximum absorption coefficient between 4 and 100 neper cm^{-1}. Despite these wide variations in molecular properties, Jain and Walker showed that an almost linear relationship existed between the maximum absorption coefficient and μ^2/I, suggesting a very substantial rotational contribution to the far-i.r. absorption.

Davies *et al.*[232] have measured the far-i.r. absorption of simple polar liquids such as methylchloroform, chlorobenzene, benzonitrile, and toluene. It was shown by pressure studies that the far-i.r. absorption was not dependent on a collisional process involving multiple molecules in the liquid state: this supports, indirectly, the assignment of the band to a rotational or librational mode. It was observed that the absorption of pure benzonitrile and of benzonitrile in cyclohexane or benzene solution were essentially the same. This suggests that the local field of the solvent cage is relatively unimportant in determining the far-i.r. absorption, and this would seem to imply that it is the repulsive forces, rather than the attractive ones, which are important in determining the interaction potential.

[247] B. Pourprix, C. Abbar, and D. Decoster, *Compt. rend.*, 1971, **272**, *B*, 1418.
[248] C. Brot and I. Darmon, *Mol. Phys.*, 1971, **21**, 785.
[249] S. G. Kroon and J. van der Elsken, *Chem. Phys. Letters*, 1967, **1**, 285.
[250] S. R. Jain and S. Walker, *J. Phys. Chem.*, 1971, **75**, 2942.

The far-i.r. absorption of *ortho*- and *meta*-dihalogenobenzenes has been reported by Mansingh.[251] It was found that the absorption coefficient for the *ortho*-compound always exceeded that for the *meta*-compound, (*e.g.* 52 neper cm^{-1} for *o*-difluorobenzene and 22 neper cm^{-1} for *m*-difluorobenzene), reflecting the larger permanent dipole of the *ortho*-molecules. Again, a fairly good correlation with the sum rule of Gordon was observed, although the absolute values calculated were too low by about 25%. This suggests that, while the major part of the far-i.r. absorption arises from a rotational mechanism, a significant fraction arises from a relaxational mechanism.

Similar evidence for a predominantly rotational origin for the far-i.r. absorption of polar liquids is provided by Fleming *et al.*[252] who found an approximate correlation between the square of the dipole moment and the integrated intensity for the halogenobenzenes.

Gerschel and co-workers [253] have extended their study of the collision-induced spectra of gases at high pressure [223, 224] to include a study of the dense gas and the liquid along the orthobaric curve for chloroform, chlorobenzene, fluorobenzene, and carbonyl sulphide in the pressure range 0—150 atm ($0—1.5 \times 10^7$ N m^{-2}) and the temperature range 77—659 K. They concluded from an analysis of the far-i.r. absorption in terms of the correlation function for rotational velocity [254] that in the low-density phases, the gas and the liquid to a few tenths of a degree below the critical point, the predominant absorption mechanism was collision-perturbed free rotation. In the liquid phase, on the other hand, it was concluded that the predominant absorption mechanism was strongly perturbed librational motion.

The above considerations apply essentially to rigid polar molecules. In flexible molecules the effective dipole may reorientate without a corresponding movement of the molecule as a whole. Pardoe [255] has considered the far-i.r. absorption of non-rigid molecules such as the alkyl bromides. Using a Grubb Parsons-N.P.L. cube interferometer, the 1-bromo-derivatives of ethane, propane, pentane, and hexane, the 2-bromo-derivatives of butane, pentane, and bromocyclopentane were studied in the far-i.r. The dipole moments of all the liquids were essentially the same, 1.9—2.0 Debye for the 1-bromo-compounds and 2.1 Debye for the 2-bromo-derivatives. It would be expected, therefore, from Gordon's sum rule, that the integrated intensities of the far-i.r. bands would decrease with the increasing moments of inertia of the molecules. Experimentally, it was found that the integrated intensity was essentially independent of the chain length but dependent on the substitution position. It would appear, therefore, that the sum rule of Gordon cannot be applied to the non-rigid alkyl bromides and that the absorption must be understood in terms of the local angular motion of the C—Br dipole. Such limited oscillations are possible even for the longest-chain alkyl bromides, although complete rotational motion seems improbable.

[251] A. Mansingh, *J. Chem. Phys.*, 1970, **52**, 5896.
[252] J. W. Fleming, P. A. Turner, and G. W. Chantry, *Mol. Phys.*, 1970, **19**, 853.
[253] A. Gerschel, I. Darmon, and C. Brot, *Mol. Phys.*, 1972, **23**, 317.
[254] R. G. Gordon, *J. Chem. Phys.*, 1965, **43**, 1307.
[255] G. W. F. Pardoe, *Spectrochim. Acta*, 1971, **27A**, 203.

Warrier and Krimm [256] have placed a different interpretation on the far-i.r. absorption observed in secondary alkyl chlorides. In 2-chloropropane, 2-chlorobutane, 3-chlorohexane, and *meso*-2,4-dichloropentane a relatively sharp band with a half-width of 10 cm^{-1} was observed at 67 cm^{-1}. The band did not shift or sharpen on cooling to 90 K and was absent in the gas-phase spectrum and in solution. It was assigned to a specific intermolecular interaction in the liquid phase, such as a C—Cl ··· H—C hydrogen-bond stretch. A similar specific molecular hydrogen-bond interaction has been proposed by Mierzecki [257] to account for the far-i.r. absorption in chlorobenzene and chlorobenzene–chloroform mixtures.

Clearly, hydrogen-bonding accounts for the very abnormal far-i.r. absorption of liquid water, as measured by Davies *et al.*[232] Experimentally it is observed that the absorption coefficient rises from about 110 neper cm^{-1} at 6 cm^{-1} to 700 neper cm^{-1} at 100 cm^{-1} before reaching a maximum at 193 cm^{-1} as a shoulder on a still stronger band, centred at 685 cm^{-1}. Little progress has been made in analysing this band in terms of the hydrogen-bonding occurring in liquid water and the dipole-allowed rotational transitions. The far-i.r. absorption of liquid water, with especial reference to high temperatures and high pressures, has been reviewed by Andreev and Gal'tsev.[258]

In addition to hydrogen-bonding, torsional modes have also been invoked as a possible mechanism for far-i.r. absorption in pure liquids. To explain the far-i.r. absorption of liquid nitrobenzene, Orville-Thomas *et al.*[259] have suggested that perturbations in the liquid state reduce the C_{2v} point-group symmetry of the nitrobenzene molecule, permitting the $—NO_2$ torsion to become i.r.-active and contribute to the observed absorption. A basically similar mechanism has been proposed by Davies *et al.*[232] to account for the very considerable far-i.r. absorption of aniline. Although the dipole moment of aniline is similar to that of the rigid molecule chlorobenzene, the far-i.r. absorption of aniline rises from 8 neper cm^{-1} at 4 cm^{-1} to reach a limiting value of 80 neper cm^{-1} at 100 cm^{-1}, compared with a maximum value of 16 neper cm^{-1} at 44 cm^{-1} for chlorobenzene. In addition to rotational contributions, it is postulated that the increased absorption in aniline arises (a) from intramolecular $—NH_2$ relaxations in which the $—NH_2$ group rotates or inverts over a wide range of frequencies determined by collisional interactions, and (b) to a lesser extent from N ··· N—H hydrogen-bond deformations.

Electrolyte Solutions.—Another related phenomenon is the far-i.r. absorption exhibited by electrolytes when dissolved in non-aqueous solvents. All salts give rise to a broad absorption of half-width around 100 cm^{-1} whose centre frequency is strongly dependent on the nature of the cation of the salt. In general, Li^+ salts give rise to an absorption around 400 cm^{-1}, NH_4^+ salts at

[256] A. V. R. Warrier and S. Krimm, *J. Chem. Phys.*, 1970, **52,** 4316.
[257] R. Mierzecki, *Acta Phys. Polon.*, 1970, **37A,** 603.
[258] D. V. Andreev and A. P. Gal'tsev, *Probl. Fiz. Atmos.*, 1970, **8,** 59.
[259] W. J. Orville-Thomas, J. A. Ladd, and W. Lüttke, *Chem. Phys. Letters*, 1970, **6,** 629.

200 cm^{-1}, Na^+ salts at 190 cm^{-1}, and K^+ salts at 140 cm^{-1}. It has been established by Edgell *et al.*[260] that, in solvents of low dielectric constant, *e.g.* THF, the absorption frequency is dependent not only on the cation but also on the anion. For example, in THF the $Na^+BPh_4^-$ absorption occurs at 198 cm^{-1}, whereas for Na^+I^- it occurs at 184 cm^{-1}. It is therefore suggested that the absorption arises from the interionic vibration of the cation and anion in a contact ion-pair. In solvents of high dielectric constant, however, such as DMSO, it has been shown that the absorption frequency is now anion-independent but is dependent on the nature of the solvent. Wuepper and Popov[261] have studied $Na^+Al(Bu^t)_4^-$ in DMSO and 1-methyl-2-pyrrolidone, and have shown that the absorption maxima occur at 195 and 206 cm^{-1}, respectively. That the absorption maxima are independent of the nature of the anion was demonstrated by McKinney and Popov[262] for a number of sodium salts in pyridine. The absorption maximum was observed at the same value of 180 cm^{-1}, within the experimental error, for the sodium salts of Cl^-, Br^-, I^-, ClO_4^-, NO_3^-, SCN^- and BF_4^-. It is suggested, therefore, that in solvents of high dielectric constant the absorption arises from the vibration of the cation within a cage of solvent molecules. Wong, McKinney, and Popov[263] have investigated the absorption of Li^+, Na^+, and K^+ salts in acetone, a solvent of intermediate solvating power and dielectric constant. They found that the absorption maximum was anion independent but shifted on deuteriation of the solvent, *e.g.* the Na^+ absorption shifted from 196 to 191 cm^{-1} in deuterio-acetone. It seems, therefore, that the absorption in acetone arises from a cation in a cage of solvent molecules as for solvents of high dielectric constant.

There is, however, some evidence which contradicts the picture presented above. Kalnin'sh *et al.*[264] have investigated the Li^+, Rb^+, Cs^+, and $NH(C_2H_5)_3^+$ salts of heptadecanesulphonic acid in solvents varying from CCl_4 (with a dielectric constant of 2.3) to DMSO (with a dielectric constant of 46). They observe a change of less than 10% in the frequency of the maximum absorption and no apparent correlation between the frequency shift and the dielectric constant of the solvent. They interpret this as indicating that the absorption is arising, in all cases, from contact ion-pairs and that the solvation of the ions and their equilibrium separation is only slightly dependent on the bulk-phase dielectric constant of the solvent.

Edgell and co-workers[260] have measured the pressure dependence of the LiCl vibration in THF over the range 0—20 kbar (2×10^9 N m^{-2}) and found a value for $\delta\nu/\delta P$ of about 1.5 cm^{-1} kbar^{-1}. This value is characteristic of a lattice mode of an ionic solid rather than an internal mode of vibration, where a $\delta\nu/\delta P$ value in the region of 0—0.2 cm^{-1} kbar^{-1} would be expected. The vibra-

[260] W. F. Edgell, J. Lyford, R. Wright, W. Risen, and A. Watts, *J. Amer. Chem. Soc.*, 1970, **92**, 2240.
[261] J. L. Wuepper and A. I. Popov, *J. Amer. Chem. Soc.*, 1970, **92**, 1493.
[262] W. J. McKinney and A. I. Popov, *J. Phys. Chem.*, 1970, **74**, 535.
[263] M. K. Wong, W. J. McKinney, and A. I. Popov, *J. Phys. Chem.*, 1971, **75**, 56.
[264] F. Kalnin'sh, I. A. Brodskii, and B. G. Belen'kii, *Zhur. fiz. Khim.*, 1969, **43**, 2428 (*Russ. J. Phys. Chem.*, 1969, **43**, 1364).

tion of a contact ion-pair would be expected to resemble an internal mode of vibration, so that this result suggests that even in solvents of low dielectric constant the solvent plays an integral part in the absorption mechanism.

Kludt *et al.*[265] have observed the far-i.r. absorption of trialkyl ammonium salts (tri-n-octyl ammonium chloride and triethyl ammonium chloride and bromide) in solvents of very low dielectric constant, CCl_4, $CHCl_3$, and cyclohexane. They observed two absorptions in the far-i.r., a high-frequency band, 133—190 cm^{-1}, which was anion-dependent but occurred at the same frequency in all the solvents, and a low-frequency band, 75—114 cm^{-1}, which was dependent on both the anion and the solvent. The high-frequency band was assigned to the interionic vibration of a contact hydrogen-bonded ion-pair, and the low-frequency absorption was ascribed to the N—H···X bending mode of the ion-pair.

6 Weak Interactions

An area in which far-i.r. spectroscopy has proved to be of exceptional value is in the elucidation of molecular structures involving weak intramolecular interactions such as hydrogen-bonding and charge transference. The application of far-i.r. spectroscopy to these two areas has been considered in detail by Finch *et al.*[1] and by Möller and Rothschild,[2] hence only some of the most recent work will be reported here.

Hydrogen-bonding.—A number of independent studies of the hydrogen-bonding in methanol have been undertaken. Although the O—H···O hydrogen-bond stretching vibration in the higher alcohol homologues has been satisfactorily assigned to absorptions in the 100—200 cm^{-1} region,[266,267] the precise assignment of this mode in methanol remains uncertain. The far-i.r. spectrum of solid methanol has been investigated by Dempster and Zerbi [268] and by Durig and co-workers.[269] Passchier *et al.*[270] have studied the far-i.r. spectra of methanol in both the solid (143—173 K) and the liquid (193—333 K) states in the range 100—1000 cm^{-1}.

Solid methanol exhibits a phase transition from the low-temperature α-form to the β-modification at 159.5 K. The α-form, on which the far-i.r. investigations were performed, has a monoclinic unit cell with two molecules per unit cell and belongs to the space group C_{2h}^2 ($P2_1/m$). Far-i.r. absorption is observed in the 350, 168, 109, and 60 cm^{-1} regions in CH_3OH and in the 325, 164, 100, and 64 cm^{-1} regions in CD_3OD. The low-frequency modes, other than lattice modes, to be assigned are the CH_3 torsion, the hydrogen-bond stretch, and the in- and out-of-plane hydrogen-bond bending vibrations.

[265] J. R. Kludt, G. Y. W. Kwong, and R. L. McDonald, *J. Phys. Chem.*, 1972, **76**, 339.
[266] R. F. Lake and H. W. Thompson, *Proc. Roy. Soc.*, 1966, **A291**, 469.
[267] R. J. Jakobsen, J. W. Brasch, and Y. Mikawa, *Appl. Spectroscopy*, 1968, **22**, 641.
[268] A. B. Dempster and G. Zerbi, *J. Chem. Phys.*, 1971, **54**, 3600.
[269] J. R. Durig, C. B. Pate, Y. S. Li, and D. J. Antion, *J. Chem. Phys.*, 1971, **54**, 4863.
[270] W. F. Passchier, E. R. Klompmaker, and M. Mandel, *Chem. Phys. Letters*, 1970, **4**, 485.

Passchier and co-workers, who did not study the deuteriated analogue, observed in liquid CH_3OH a band at 119 cm^{-1} at 333 K which sharpened and shifted to higher frequencies as the temperature was lowered, reaching 135 cm^{-1} at 193 K. They considered that this band could be correlated with the 165 cm^{-1} band in the solid and assigned it to the O—H···O stretch in agreement with that found in higher alcohols.[266] The band observed at 300 cm^{-1} in the liquid was correlated with the 350 cm^{-1} band in the solid and assigned to the methyl torsion.

Zerbi and Dempster [268] and Durig and co-workers [269] studied the effects of deuteriation on the spectrum of solid methanol and showed that the 350 cm^{-1} band in CH_3OH, assigned by Passchier *et al.* to the methyl torsion, shifted only to 325 cm^{-1} in CD_3OD. This small shift on deuteriation appears to preclude the assignment of this band to the methyl torsion, and both authors conclude that it represents the O—H——O stretching vibration. This is an extremely high value for this vibration compared with that found for other alcohols and may represent a considerable librational contribution to the absorption. The remaining low-frequency bands are assigned to the hydrogen-bond bending vibrations.

Barnes and Hallam [271] have studied the low-frequency absorption of methanol and ethanol in argon matrices and have assigned absorptions at 222 cm^{-1} and 116 cm^{-1} in methanol and 215 cm^{-1} and 125 cm^{-1} in ethanol to the hydrogen-bond stretching and bending vibrations respectively. In methanol, the stretching mode shifts on deuteriation from 222 to 213 cm^{-1}. The small shift and comparable intensity of the two bands assigned to the stretching and bending vibrations suggest that the methanol is present in the matrix in the open-chain dimer configuration (1), where the H-bond stretching motion is translational, rather than the cyclic dimer configuration (2), where the H-bond stretching motion is rotational. A similar open-chain dimer conformation was confirmed in ethanol.

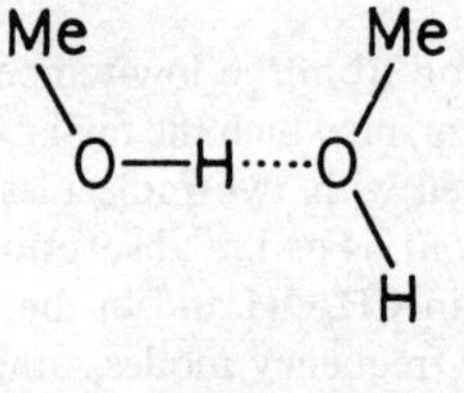

(1) 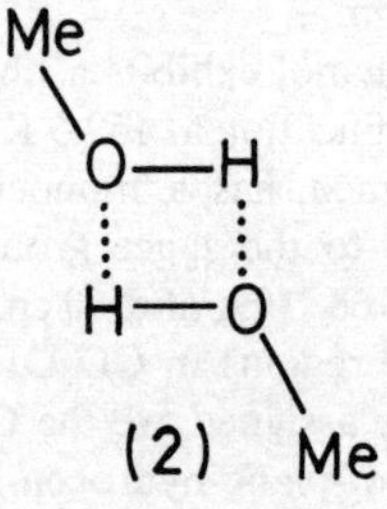 (2)

A similar type of hydrogen-bonded cyclic dimer formation (3) occurs in the carboxylic acids and this has been the subject of extensive far-i.r. investigation.

[271] A. J. Barnes and H. E. Hallam, *Trans. Faraday Soc.*, 1970, **66**, 1920, 1932.

(3)

The hydrogen-bond stretching vibration of these dimers shows a doublet structure in the far-i.r. which has been investigated by Wood and co-workers [272, 273] and by Witkowski and co-workers.[274, 275] It has been shown, both theoretically and by deuteriation studies, that the splitting arises from the non-adiabatic coupling of the hydrogen-bond vibration and the OH stretching vibration, rather than from proton tunnelling. Leviel and Marechal [276] have refined the theoretical calculations to include an anharmonic term in the potential of the hydrogen-bond stretching vibration. This has resulted in very good agreement between the theoretical calculations and the experimentally observed far-i.r. spectrum.

The vapour-phase spectrum of trifluoroacetic acid has been reported by Redington and Lin,[277] who assign a rather weak multiple band at 233—248 cm^{-1} to the H-bond stretching vibration and bands at 146—86 cm^{-1} to the H-bond bending vibrations. All these frequencies are somewhat greater than for the corresponding modes in the acetic acid dimer.

Other hydrogen-bonded systems that have received considerable attention include the hydrogen-bond interactions of amines. Lichtfus and Zeegers-Huyskens [278] studied the far-i.r. spectrum of mixtures of pyridine and aniline with substituted phenols. Foglizzo and Novak [279] studied the hydrogen-bond stretching frequency in pyrimidinium (4) and pyrazinium (5) halides. The hydrogen-bond stretching frequency varied from 107 cm^{-1} in pyrazinium iodide to 198 cm^{-1} in pyrimidinium chloride. The change in the hydrogen-bond

(4) (5)

[272] S. G. W. Ginn and J. L. Wood, *J. Chem. Phys.*, 1967, **46,** 2735.
[273] T. R. Singh and J. L. Wood, *J. Chem. Phys.*, 1968, **48,** 4567.
[274] Y. Marechal and A. Witkowski, *J. Chem. Phys.*, 1968, **48,** 3697.
[275] A. Witkowski, *J. Chem. Phys.*, 1970, **52,** 4403.
[276] J. L. Leviel and Y. Marechal, *J. Chem. Phys.*, 1971, **54,** 1104.
[277] R. L. Redington and K. C. Lin, *J. Chem. Phys.*, 1971, **54,** 4111.
[278] G. Lichtfus and Et. Th. Zeegers-Huyskens, *J. Mol. Structure*, 1971, **9,** 343.
[279] R. Foglizzo and A. Novak, *J. Mol. Structure*, 1971, **7,** 217.

stretching frequency was closely correlated with changes in the NH stretching region of the spectrum.

Charge-transfer Spectra.—The charge-transfer complexes formed between halogens and pyridines have been the subject of extensive far-i.r. interest, which is summarized by Finch *et al.*[1] Yarwood [280] has studied the effect of substituents in the pyridine ring on the frequency and intensity of the far-i.r. donor bands. Pyridine, 1-, 3-, and 4-methylpyridine, and 2,3-, 2,4-, 2,6-, and 3,5-dimethylpyridine were the donors and ICl the acceptor molecule. The complexes were studied in benzene solution and by the polyethylene disc technique. In the solid state, the frequency of the N···I—Cl stretching vibration varied between 137 and 179 cm^{-1} whereas the corresponding frequency range for the bending mode was 75—107 cm^{-1}. It was observed that the frequencies were lower in the presence of α-methyl groups, which also had the effect of reducing the N—I stretching frequency and, presumably, the charge-transfer force constants.

Yamada *et al.*[281] have studied the far-i.r. charge-transfer spectra of I_2, IBr, and ICl with organic sulphides such as chloromethyl sulphide. D'Hondt and Zeegers-Huyskens [282] have studied the far-i.r. charge-transfer spectra of bromine and 16 substituted pyridines, and correlated the observed frequencies with steric effects. Brownson and Yarwood [283] have studied the integrated intensity and frequency of the charge-transfer bands of pyridine with iodine and the interhalogens IBr and ICl.

7 Determination of Torsional Barriers

Direct observation of the far-i.r. torsional frequency provides, along with the microwave splitting method, the most direct technique for obtaining information on the potential barrier to hindered rotation. A very complete discussion of the determination of torsional barriers from far-i.r. data is given by Finch *et al.*[1] and by Möller and Rothschild.[2] A clear introductory account is given by Steele.[284] Lowe [285] provides an introductory account of the determination of torsional barrier heights by a variety of physical techniques.

Three-fold Barriers.—A very large proportion of the current far-i.r. spectroscopic interest in torsional barriers concerns methyl torsions, and the potential barrier for a hindered top possessing a three-fold axis of symmetry, such as the methyl rotor, will be considered first. A two-term torsional potential is assumed:

$$V_\alpha = V_3(1-\cos 3\alpha)+\tfrac{1}{2}V_6(1-\cos 6\alpha)$$

[280] J. Yarwood, *Spectrochim. Acta*, 1970, **26A**, 2099.
[281] M. Yamada, H. Saruyama, and K. Aida, *Spectrochim. Acta*, 1972, **28A**, 439.
[282] J. D'Hondt and Et. Th. Zeegers-Huyskens, *J. Mol. Structure*, 1971, **10**, 135.
[283] G. W. Brownson and J. Yarwood, *J. Mol. Structure*, 1971, **10**, 147.
[284] D. Steele, 'Theory of Vibrational Spectroscopy', Saunders, Philadelphia, 1971, Chapter 9.
[285] J. P. Lowe, in 'Progress in Physical Organic Chemistry', ed. A Streitwieser and R. W. Taft, Interscience, New York, 1968, p. 16.

where α is the angle of deformation from the equilibrium position. Generally V_6 is of the order of 1% or less of V_3, so that no further terms in the Fourier series need be considered. In practice, frequently only one piece of experimental data is available, the far-i.r. frequency of the $1 \leftarrow 0$ torsional vibration, in which case only the V_3 barrier itself can be evaluated. The experimental observation of hot-band vibrations may permit the evaluation of V_6 if it makes a significant contribution to the barrier potential. Substitution of the torsional potential into the vibrational Hamiltonian and solution of the wave equation results in evaluation of the torsional energies, E_v, in terms of the eigenvalues of the Mathieu equation, b_v.

$$E_v = 2.25\ Fb_v$$

where $F = h^2/8\pi^2 I_{\text{red}}$ and I_{red} is the reduced moment of inertia for the torsional oscillation. The experimentally observed far-i.r. frequency, ν_t, is then given by:

$$\nu_t = \Delta E_v/h = 2.25\ F\Delta b_v/h$$

and the torsional barrier, V_3, is given by

$$V_3 = 2.25\ Fs$$

where s is a dimensionless parameter which can be related to Δb_v using tables [286, 287] prepared for the solution of the Mathieu equation.

The observed far-i.r. torsional frequency may then be interpreted in terms of the torsional barrier height, providing that sufficient structural data are available on the molecule to permit the evaluation of the reduced moment of the top.

The main uncertainty in barriers evaluated from far-i.r. torsional data arises from an uncertainty in the magnitude of the reduced moment of inertia of the rotating top. Souter and Wood [288] have shown how the far-i.r. data may be combined with the microwave splitting data to evaluate the torsional barrier without any *a priori* assumptions as to the structural parameters of the molecule. In addition, they have investigated the effect of introducing a non-rigid model which includes first-order coupling of the torsional and intramolecular vibrations into the series CH_3CHO, CD_3CHO, and $CH_3CH{=}CH_2$. Identification of the 'hot-band' transitions and isotopic substitution data allowed the contribution of the higher-order terms in the potential barrier to be evaluated. For CD_3CHO the V_6/V_3 ratio was -0.0195, whereas for propylene a value of 0.007 was obtained.

In highly symmetric molecules, *e.g.* C_{3v}, T_d, the torsional vibration is i.r.-forbidden in the isolated gas-phase molecule, so that the far-i.r. technique for evaluating torsional barriers cannot be used. Frequently, in the solid phase, the effective symmetry of the molecule is reduced, permitting the torsional vibration to be observed, but the very intermolecular forces that 'allow' the

[286] D. R. Herschbach, 'Tables for the Internal Rotation Problem', Department of Chemistry, University of Harvard, 1957.
[287] E. O. Stejskal and H. S. Gutowsky, *J. Chem. Phys.*, 1958, **28**, 388.
[288] C. E. Souter and J. L. Wood, *J. Chem. Phys.*, 1970, **52**, 674.

i.r. absorption perturb the torsional barrier and consequently decrease the applicability of the method. In order to obtain some estimate of the variation of barrier with phase, Durig and co-workers [289] have investigated the effect of crystallization on the far-i.r. torsional absorption, and hence the barrier heights, of some molecules of C_s symmetry where the torsion is i.r.-allowed in both phases (Table 2). The effect of crystallization is to increase the barrier to hindered rotation by about 25%. In order to explain the increased barrier in the solid, several intermolecular potential functions were examined. The most successful was that in which the principal contribution to the torsional barrier in the solid state arose from the repulsion of hydrogen atoms on adjacent molecules.

Table 2 *Comparison of torsional barriers in the solid and gas phases*[289]

Molecule	ν_t *gas* /cm^{-1}	ν_t *liquid* /cm^{-1}	ν_t*solid* /cm^{-1}	ν *gas* / ν *solid*	V_3 *gas* /kJ mol^{-1}	V_3 *solid* /kJ mol^{-1}
CH_3CH_3	289	310	304	1.05	—	—
CH_3CD_3	253	270	267	1.05	—	—
CD_3CD_3	208	216	219	1.05	—	—
CH_3CH_2Cl	251	253	278	1.11	15.4	18.7
CH_3CH_2Br	249	243	276	1.15	15.5	18.9
CD_3CD_2Br	182	—	204	1.12	15.1	19.2
CH_3CH_2I	[230]	—	258	1.12	—	18.5
CD_3CH_2I	—	—	198	—	—	18.8
CH_3COCl	[136]	—	167	1.24	[5.44]	7.55
CD_3COCl	[100]	—	124	1.27	[5.44]	7.72
CH_3COBr	[134]	—	168	1.25	[5.44]	8.22

— : not observed or not calculated
[ ] : calculated from microwave data

The gas-phase torsional vibration of ethyl chloride has also been investigated by Fateley, Kiviat, and Miller.[290] Ethyl chloride and its deuteriated analogues CH_3CD_2Cl, CD_3CH_2Cl, and CD_3CD_2Cl were studied and at least two hot bands were observed for each isotope. This permitted the evaluation of 16 independent values for the three-fold torsional barrier, all of which fell in the range 1270—1298 cm^{-1} (15.1—15.4 kJ mol^{-1}). This is convincing proof that there is no mixing of the torsional mode with other vibration fundamentals, and as the calculated values of V_3 showed no regular trend with increasing quantum number it appears that the contribution of V_6 to the barrier is very small, certainly less than 1% that of V_3.

Molecules, such as $CH_3—O—CH_3$, with two coupled symmetric rotors have

[289] J. R. Durig, C. M. Player, and J. Bragin, *J. Chem. Phys.*, 1971, **54**, 460.
[290] W. G. Fateley, F. E. Kiviat, and F. A. Miller, *Spectrochim. Acta*, 1970, **26A**, 315.

been investigated in the solid state by Durig *et al.*[291] and in the gas phase by Tuazon and Fateley.[292] The coupling of the rotors implies that the torsional potential must be modified by the inclusion of an interaction potential V_{12}, and it is on the magnitude of this interaction potential that interest has centred. The C_{2v} symmetry of the molecule in the gas phase means that in the harmonic-oscillator approximation only the b_1 torsion mode is i.r.-active and evaluation of the interaction constant requires either the frequency of the second a_2 torsion mode from the Raman spectrum or isotopic data on the b_1 torsion mode. Using isotopic data, Tuazon and Fateley evaluated the barrier in both acetone and perdeuterioacetone as 946 ± 15 cm^{-1} (11.3 ± 0.2 kJ mol^{-1}) and the interaction constant as 30 ± 9 cm^{-1} (360 ± 100 J mol^{-1}).

Interestingly, the molecule CH_3OCD_3 belongs to the point group C_s rather than C_{2v}, so that both the torsions are i.r.-active, the CH_3 torsion being observed at 220 cm^{-1} and the CD_3 torsion at 158 cm^{-1}. Evaluation of the torsional barrier leads to a value of 992 cm^{-1} (11.8 kJ mol^{-1}) for the CH_3 torsion and 804 cm^{-1} (9.6 kJ mol^{-1}) for the CD_3 torsion. The large difference in the barriers is due to resonance between the rotors, corresponding to a resonance shift of ± 10.3 cm^{-1}, and a common three-fold barrier of 907 cm^{-1} (10.8 kJ mol^{-1}) for both tops. In the solid phase, Durig *et al.*[292] investigated the series of molecules CH_3OCH_3, CH_3SCH_3, CH_3SeCH_3, and CH_3TeCH_3. Barriers of 14.2—14.6 kJ mol^{-1} were obtained and the interaction constant decreased monotonically from 242 J mol^{-1} for acetone to zero for CH_3TeCH_3.

The torsional barriers in systems with two coupled rotors, such as CH_3CCl_3 and CH_3CF_3, have been determined in the solid state by Durig *et al.*[293] who have also investigated the barriers in three-top methyl rotors, such as $(CH_3)_3CH$ and $(CH_3)_3SiH$ [294] and in four-top methyl rotors, such as $(CH_3)_4C$, $(CH_3)_4Si$, $(CH_3)_4Ge$, and $(CH_3)_4Sn$.[295]

The torsional barriers of other methyl rotors that have been investigated include the molecules *o*-xylene,[296] CH_3GeI_3,[297] $CH_3CH_2GeCl_3$,[298] $CH_2{=}CHGeCl_3$,[299] CH_3PXY_2,[300] *cis-* and *trans*-crotononitrile,[301] 2-cyanopropene,[302] *N*-methylaziridine $CH_3N(CH_2)_2$,[303] and the methyl arsenic torsion in molecules such as CH_3AsCl_2, CH_3AsBr_2, and CH_3AsI_2.[304]

[291] J. R. Durig, C. M. Player, J. Bragin, and Y. S. Li, *J. Chem. Phys.*, 1971, **55**, 2895.
[292] E. C. Tuazon and W. G. Fateley, *J. Chem. Phys.*, 1971, **54**, 4451.
[293] J. R. Durig, S. M. Craven, K. K. Lau, and J. Bragin, *J. Chem. Phys.*, 1971, **54**, 479.
[294] J. R. Durig, S. M. Craven, and J. Bragin, *J. Chem. Phys.*, 1970, **53**, 38.
[295] J. R. Durig, S. M. Craven, and J. Bragin, *J. Chem. Phys.*, 1970, **52**, 2046.
[296] G. W. F. Pardoe, S. S. Larson, H. A. Gebbie, S. J. Strickler, K. C. Ingham, and D. G. Johnson, *J. Chem. Phys.*, 1970, **52**, 6426.
[297] J. R. Durig, C. F. Jumper, and J. N. Willis, *J. Mol. Spectroscopy*, 1971, **37**, 260.
[298] J. R. Durig and C. W. Hawley, *J. Phys. Chem.*, 1971, **75**, 3393.
[299] J. R. Durig and J. B. Turner, *Spectrochim. Acta*, 1971, **27A**, 395.
[300] J. R. Durig and J. M. Casper, *J. Phys. Chem.*, 1971, **75**, 1956.
[301] J. R. Durig, C. K. Tong, C. W. Hawley, and J. Bragin, *J. Phys. Chem.*, 1971, **75**, 44.
[302] J. Bragin, K. L. Kizer, and J. R. Durig, *J. Mol. Spectroscopy*, 1971, **38**, 289.
[303] T. Ikeda, *J. Mol. Spectroscopy*, 1970, **36**, 268.
[304] J. R. Durig, C. F. Jumper, and J. N. Willis, *Appl. Spectroscopy*, 1971, **25**, 218.

Two-fold Barriers.—For torsional barriers possessing two-fold symmetry the first terms of the Fourier expansion of the barrier potential are of the form

$$V_\alpha = \tfrac{1}{2}V_2(1-\cos 2\alpha)+\tfrac{1}{2}V_4(1-\cos 4\alpha)$$

As for the three-fold barrier discussed above, the paucity of experimental data frequently limits the analysis to an evaluation of the V_2 contribution to the barrier. Following the procedure outlined for the three-fold case,

$$V_2 = Fs$$

The torsional vibrations of symmetrically substituted benzaldehydes and phenols possess two-fold symmetry. Campagnaro and Wood [305] have studied the effect of ring substituents on the —CHO torsion in benzaldehyde derivatives. The effect of a substituent *para* to the aldehyde group was shown to be closely correlated with the Hammett σ-values for *ortho*-substituents.[306] They also reported a correlation between the OH torsion frequency in *p*-substituted phenols and the electronic properties of the substituent. The study on *p*-substituted phenols has been extended by Radom *et al.*,[307] who have performed *ab initio* MO calculations of the torsional barriers in these compounds. It was calculated, in agreement with experiment, that *p*-substituents which are π-electron donors lower the torsional barrier and those which are π-electron acceptors raise the barrier height. Carlson *et al.*[308] describe a sample deuteriation procedure to aid in the identification of the —OH and —NH torsional absorptions in far-i.r. spectra.

Torsional Barriers Without Symmetry.—For torsional modes in which the rotor possesses no symmetry the complete Fourier expansion of the torsional potential must be used:

$$V_\alpha = \tfrac{1}{2}V_1(1-\cos \alpha)+\tfrac{1}{2}V_2(1-\cos 2\alpha)+\tfrac{1}{2}V_3(1-\cos 3\alpha)\ldots$$

In general, no single term in the barrier potential is dominant. In addition, the inertial parameters are now functions of the torsional angle, so that evaluation of F requires rather detailed knowledge of the dynamics of motion of the top. A complete evaluation of the significant terms in the torsional potential therefore requires extensive experimental data, including not only 'hot bands' and isotopic substitution data from the i.r. but also microwave and electronic spectral measurements. The problems encountered in the complete analysis of the barrier potential have been discussed by Fateley *et al.*[309]

Ewig and Harris [310] have discussed a co-ordinate transformation of the wave

[305] G. E. Campagnaro and J. L. Wood, *J. Mol. Structure*, 1970, **6**, 117.
[306] R. W. Taft, *J. Phys. Chem.*, 1960, **64**, 1805.
[307] L. Radom, W. J. Hehre, J. A. Pople, G. L. Carlson, and W. G. Fateley, *Chem. Comm.*, 1972, 308.
[308] G. L. Carlson, W. G. Fateley, and F. F. Bentley, *Spectrochim. Acta*, 1972, **28A**, 177.
[309] W. G. Fateley, R. K. Harris, F. A. Miller, and R. E. Witkowski, *Spectrochim. Acta* 1965, **21**, 231.
[310] C. S. Ewig and D. O. Harris, *J. Chem. Phys.*, 1970, **52**, 6268.

equation for the torsional motion of an asymmetric rotor. Providing sufficient experimental data on the 'hot band' frequencies are available, an iterative technique permits the evaluation of the torsional barrier height without any *a priori* knowledge of the moment of inertia of the molecule.

Evaluation of the contribution of V_1, V_2, and V_3 to the barrier potential is also possible in those cases in which two stable conformers exist, normally the *trans* and *gauche* conformers, and spectral data are available on both conformers. In these cases the first three terms in the barrier potential may be evaluated from the experimentally observed far-i.r. torsional frequency of each conformer and the difference in energy between the two conformers, ΔE, as determined by the temperature dependence of the intensity of corresponding vibrational modes. This has been treated theoretically by Cunliffe [311] and its experimental application is illustrated by the determination of the barrier to internal rotation in 1,1,2,2-tetrabromoethane by Carlson *et al.*[312] The spectrum of liquid tetrabromoethane was obtained in the range 33—400 cm^{-1} using a Beckman IR 11 grating spectrometer, and the temperature dependence of the intensity of the Raman bands was measured in the range 273—355 K The far-i.r. torsional mode of the *gauche* conformer was located at 38 cm^{-1}, but the corresponding mode for the *trans*-isomer was not observed. Arguments based on entropy calculations and the product rule were used to justify a frequency of 35 cm^{-1} for the torsional mode of the *trans*-isomer. The Raman measurements were used to show that the *gauche* conformer was 680 ± 150 cal mol^{-1} (2.84 ± 0.63 kJ mol^{-1}) more stable than the *trans*-isomer. The barrier potential was evaluated as

$$V_\alpha = 2.18(1 - \cos \alpha) - 2.81(1 - \cos 2\alpha) + 5.78(1 - \cos 3\alpha) \text{ kcal mol}^{-1}$$

Oskam *et al.*[313] located the *trans* and *gauche* torsional frequencies of N_2F_4 at 131 and 123 cm^{-1} respectively. Although the energy difference between the conformers, 100 cal mol^{-1} (420 J mol^{-1}), was known, the barrier potential was not evaluated.

Frequently, however, the only available experimental datum is the frequency of a single $1 \leftarrow 0$ torsional absorption in the far-i.r. In these cases a number of simplifying assumptions must be made if any numerical value is to be obtained for the barrier height. The cos α terms in the angular dependence of the barrier potential are expanded in a power series of α and terminated after α^2 to give

$$\begin{aligned} V &= \tfrac{1}{4}(V_1\alpha^2 + 4V_2\alpha^2 + 9V_3\alpha^2) \\ &= \tfrac{1}{4}V^*\alpha^2 \end{aligned}$$

where $V^* = V_1 + 4V_2 + 9V_3$.

Using the simple harmonic oscillator approximation, the torsional frequency, ν_t, is related to V^* by

$$V^* = \nu_t^2/F$$

[311] A. V. Cunliffe, *J. Mol. Structure*, 1970, **6**, 9.
[312] G. L. Carlson, W. G. Fateley, and J. Hiraishi, *J. Mol. Structure*, 1970, **6**, 101.
[313] A. Oskam, R. Elst, and J. C. Duinker, *Spectrochim. Acta*, 1970, **26A**, 2021.

where, as before, $F = h^2/8\pi^2 I_{red}$. Normally, the relative magnitudes of the V_1, V_2, and V_3 contributions to V^* are uncertain, so that estimates of V^* are principally of use in inter-comparisons between similar molecules. This is illustrated by the study of glyoxal and its derivatives by Durig and co-workers.[314-318]

(6) (7)

Glyoxal exists as *cis* (6) and *trans* (7) conformers, although the *trans*-conformer is dominant in all phases. Although the torsional motion of the *trans*-conformer occurs in a deep and fairly symmetric potential minimum, the existence of the *cis*-conformer implies significant contributions from V_1, V_2, and V_3 to the barrier potential. In addition to glyoxal itself,[314] the solid-state spectra of oxalyl fluoride,[315] oxalyl chloride,[316] oxalyl bromide,[317] and biacetyl [318] were investigated. The data on the torsional modes of the *trans*-conformer are collected in Table 3.

Table 3 *The torsional vibrations of some glyoxal derivatives*

Molecule	*trans Torsion* (*solid state*) /cm^{-1}	V^* (*solid state*) /kJ mol^{-1}	*trans Torsion* (*gas phase*) /cm^{-1}	*Ref.*
Glyoxal, $(CHO)_2$	192	57.2	128	314, 319, 320
Oxalyl fluoride, $(COF)_2$	94	51.8	54	315
Oxalyl chloride, $(COCl)_2$	55	46.4	–	316
Oxalyl bromide, $(COBr)_2$	40	65.2	–	317
Biacetyl, $(COCH_3)_2$	66	42.2	–	318

The solid-state torsion of glyoxal has also been investigated by Verderame *et al.*[319] Cole and Osborne [320] have observed the gas-phase torsional mode of glyoxal and it deuterio-analogues. Both hot bands and rotational fine structure

[314] J. R. Durig and S. E. Hannum, *J. Cryst. Mol. Structure*, 1971, **1**, 131.
[315] J. R. Durig, S. C. Brown, and S. E. Hannum, *J. Chem. Phys.*, 1971, **54**, 4428.
[316] J. R. Durig and S. E. Hannum, *J. Chem. Phys.*, 1970, **52**, 6089.
[317] J. R. Durig, S. E. Hannum, and F. G. Baglin, *J. Chem. Phys.*, 1971, **54**, 2367.
[318] J. R. Durig, S. E. Hannum, and S. C. Brown, *J. Phys. Chem.*, 1971, **75**, 1946.
[319] F. D. Verderame, E. Castellucci, and S. Califano, *J. Chem. Phys.*, 1970, **52**, 719.
[320] A. R. H. Cole and G. A. Osborne, *Spectrochim. Acta*, 1971, **27A**, 2461.

were observed on the torsional absorption. The band centre of the $1 \leftarrow 0$ torsion mode was located at 126.5 ± 0.8 cm^{-1} and the torsional anharmonicity constant at 0.75 ± 0.05 cm^{-1}. It was concluded that the height of the potential barrier to rotation of the *trans*-conformer was at least 600 cm^{-1} (7.2 kJ mol^{-1}) but analysis of the torsional absorption of the *trans*-conformer did not provide any information on the relative stabilities of the other conformations since the torsional potential did not appear to be significantly perturbed by the energy levels in the adjacent potential minima.

Durig and Harris [321] have observed both the three-fold symmetric CH_3—N torsion and the asymmetric N—N torsion in dimethylhydrazine and its deuterio-derivatives $(CH_3DN)_2$ and $(CD_3DN)_2$. The N—N torsional band was broad and weak in the solid phase, possibly owing to the presence of both *cis*- and *trans*-isomers, and was located at 164 cm^{-1} in $(CH_3NH)_2$ falling to 148 cm^{-1} in the perdeuterio-molecule. The calculated value for V^* was 44.8 kcal mol^{-1} (187 kJ mol^{-1}). Structural considerations suggest that the contributions of V_1 and V_2 to the barrier potential are small, so that the experimental value of V^* leads to a three-fold torsional barrier of about 5 kcal mol^{-1} (21 kJ mol^{-1}). Other similar molecules investigated by Durig and Casper were tetramethyldiarsine,[322] where the As—As torsion was located at 91 cm^{-1}, and dimethylaminodichlorophosphine,[323] where the N—P torsion was observed at 124 cm^{-1}.

8 The Far-infrared Spectra of Strained Ring Systems

The far-i.r. absorption of the following strained ring systems will be considered; (*a*) four-membered ring systems, (*b*) unsaturated five-membered ring systems in which ring puckering occurs, (*c*) the essentially unconstrained five-membered ring systems in which pseudorotation occurs, and (*d*) larger ring systems with ring puckering. The far-i.r. absorption of strained ring systems has been reviewed by Laane.[324]

Ring Puckering in Four-membered Rings.—Cyclobutane is the prototype four-membered ring system. It has a folded ring structure of symmetry D_{2d}, a potential barrier to ring inversion of 518 cm^{-1} (6.18 kJ mol^{-1}), and an equilibrium dihedral angle of 35°. The ring puckering mode is i.r.-inactive and consequently can only be observed in the Raman spectrum [325] or as sum and difference bands in the i.r.[326]

The symmetric double-minimum potential for the ring pucker is of the form:

$$V_x = -6x^2 + x^4$$

321 J. R. Durig and W. C. Harris, *J. Chem. Phys.*, 1971, **55**, 1735.
322 J. R. Durig and J. M. Casper, *J. Chem. Phys.*, 1971, **55**, 198.
323 J. R. Durig and J. M. Casper, *J. Phys. Chem.*, 1971, **75**, 3837.
324 J. Laane, 'Advances in Vibrational Spectroscopy', ed. J. R. Durig, Dekker, New York, 1972.
325 F. A. Miller and R. J. Capwell, *Spectrochim. Acta*, 1971, **27A**, 947.
326 J. M. R. Stone and I. M. Molls, *Mol. Phys.*, 1970, **18**, 631.

where x is the reduced ring-puckering co-ordinate and is a measure of the distance between the ring diagonals. For cyclobutane the exact form of the ring-puckering potential has been determined to be:

$$V_x = 6.932 \times 10^5 z^4 - 3.790 \times 10^4 z^2$$

where z represents the actual distance between the ring diagonals in Å.

Baltagi *et al.*[327] have discussed the choice of internal co-ordinates for the ring puckering mode in four-membered rings. Laane[328] has discussed the solution of the one-dimensional Schrödinger equation for potential functions of the type $V = A(z^4 + Bz^2)$ such as those encountered in ring-puckering studies, and has tabulated the first 17 eigenvalues for 58 separate values of B.

The introduction of a substituent into the cyclobutane ring results in an asymmetric ring-puckering potential in which the double potential minimum has wells of different depths. That is:

$$V_x = -6x^2 + \tfrac{1}{2}x^3 + x^4$$

It is only possible to obtain a complete description of the ring-puckering potential function if a significant number of 'hot bands' are observed. This has been achieved for fluorocyclobutane by Durig *et al.*[329] and for chloro-, bromo- and cyano-cyclobutane by Blackwell *et al.*[330] For the fluoro-derivative, a series of four Q-branches were observed in the far-i.r. at 166.5, 156.0, 145.2, and 130.5 cm^{-1}. These transitions may be considered to arise either from $v = 2$ transitions originating rather deeply in the two alternate wells of a pseudo-symmetric double potential minimum, or from $v = 1$ transitions deep in one well of a very asymmetric potential energy function. For the pseudo-symmetric potential the analysis leads to a ring-puckering barrier of 485 cm^{-1} (5.8 kJ mol^{-1}), a difference in energy between the two conformers of 87 cm^{-1} (1.04 kJ mol^{-1}) and a ring-puckering potential energy function given by:

$$V_x = 2.74 \times 10^6 x^4 - 0.695 \times 10^5 x^2 + 0.307 \times 10^5 x^3$$

For the asymmetric potential function the experimental absorptions lead to a ring-puckering barrier of 803 cm^{-1} (9.9 kJ mol^{-1}), a difference in energy between the two conformers of 722 cm^{-1} (8.6 kJ mol^{-1}), and a ring-puckering potential

$$V_x = 1.72 \times 10^6 x^4 - 0.403 \times 10^5 x^2 + 2.19 \times 10^5 x^3$$

The microwave data[331] on fluorocyclobutane predict a large energy difference between the two conformers, and on this basis the latter potential is believed to be correct.

A similar asymmetric, single-minimum potential function was used to interpret the Q-branch far-i.r. ring-puckering spectrum of the chloro-, bromo-

[327] F. Baltagi, A. Bauder, T. Ueda, and H. H. Günthard, *J. Mol. Spectroscopy*, 1972, **42**, 11
[328] J. Laane, *Appl. Spectroscopy*, 1970, **24**, 73.
[329] J. R. Durig, J. N. Willis, and W. H. Green, *J. Chem. Phys.*, 1971, **54**, 1547.
[330] C. S. Blackwell, L. A. Carreira, J. R. Durig, J. M. Karriker, and R. C. Lord, *J. Chem. Phys.*, 1972, **56**, 1706.
[331] H. Kim and W. D. Gwinn, *J. Chem. Phys.*, 1966, **44**, 865.

and cyano-cyclobutanes.[330] Other monosubstituted cyclobutanes studied include silacyclobutane,[332, 333] cyclobutanol,[334] and cyclobutanone.[335] The available data on the ring-puckering modes are collected in Table 4.

Table 4 *Ring-puckering modes in some cyclobutane derivatives*

Molecule	$1 \leftarrow 0$ *barrier*/cm^{-1}	*Barrier*/cm^{-1}	*Barrier*/kJ mol^{-1}	*Ref.*
Cyclobutane	—	518	6.18	326
Fluorocyclobutane	166.5	803	9.60	329
Chlorocyclobutane	157.7	824	9.80	330
Bromocyclobutane	143.3	—	—	330
Cyanocyclobutane	138.3	—	—	330
Silacyclobutane	159.5 ($2 \leftarrow 1$)	440	5.25	332, 333
Cyclobutanol	170.0	—	—	334
Cyclobutanone	35.3	—	—	335

—, not observable

The other four-membered ring system in which the ring-puckering vibration has been extensively studied in the far-i.r. is trimethylene oxide (8) [336] and its derivatives. Recently, Durig *et al.*[337] have reported a study of the ring puckering vibration of oxetan-3-one (9). A regular progression of *Q*-branches was

(8) (9)

observed, originating at 140 cm^{-1}, which were assigned to the ring-puckering vibration, and the data were fitted to a two-parameter potential of the form:

$$V_z = 13.23\,(z^4 + 27.48\,z^2)$$

Ring Puckering in Unsaturated Five-membered Rings.—In an unsaturated five-membered ring, the ring may assume a planar conformation, if the ring-angle restoring forces are sufficiently strong, or it may assume a non-planar configuration, if the torsional forces about the ring bond are large in relation to

[332] J. Laane, *Spectrochim. Acta*, 1970, **26A**, 517.
[333] J. Laane and R. C. Lord, *J. Chem. Phys.*, 1968, **48**, 1508.
[334] J. R. Durig and W. H. Green, *Spectrochim. Acta*, 1969, **25A**, 849.
[335] J. R. Durig and R. C. Lord, *J. Chem. Phys.*, 1966, **45**, 1961.
[336] S. I. Chan, T. R. Borgers, J. W. Russell, H. L. Strauss, and W. D. Gwinn, *J. Chem. Phys.*, 1966, **44**, 1103.
[337] J. R. Durig, A. C. Morrissey, and W. C. Harris, *J. Mol. Structure*, 1970, **6**, 375.

the ring-angle forces. Ring puckering may be observed in both planar and non-planar rings. In the case of planar rings, however, it will be possible to analyse the observed far-i.r. absorption on the basis of a single minimum in the potential energy function. For non-planar rings, however, the far-i.r. ring-puckering absorptions will be explicable only if a double potential minimum is assumed, with a potential maximum as the ring passes through the planar configuration. The prototype unsaturated five-membered ring is cyclopentene, which has been shown [338] to be non-planar with an inversion barrier of 232 cm^{-1} (2.78 kJ mol^{-1}). On the other hand, 2,5-dihydrofuran (10) [339] and 2,5-dihydrothiophen (11) [340] have been shown to be essentially planar.

(10) (11)

A non-planar ring-puckering mode has been observed in 1-pyrazoline (12) by Durig *et al.*[341] A succession of *Q*-branches were observed below 145 cm^{-1} which were fitted to a symmetric double potential minimum of the form:

$$V_x = 10.16\times10^5 x^4 - 21.46\times10^3 x^2$$

and the barrier to the planar conformation was evaluated as 113 ± 5 cm^{-1} (1.35 ± 0.06 kJ mol^{-1}).

Laane has observed the ring-puckering vibration in the planar molecules silacyclopent-2-ene [342] and silacyclopent-3-ene (13).[343] A series of weak ring-puckering vibrations were observed which could be assigned to the fundamental

(12) (13)

$1 \leftarrow 0$, and hot-band transitions in a single potential minimum. A smooth progression from the $1 \leftarrow 0$ transition at 123.7 cm^{-1} to the $8 \leftarrow 7$ transition at

[338] J. Laane and R. C. Lord, *J. Chem. Phys.*, 1967, **47**, 4941.
[339] T. Ueda and T. Shimanouchi, *J. Chem. Phys.*, 1967, **47**, 4042.
[340] W. H. Green and A. B. Harvey, *J. Chem. Phys.*, 1968, **49**, 177.
[341] J. R. Durig, J. M. Karriker, and W. C. Harris, *J. Chem. Phys.*, 1970, **52**, 6096.
[342] J. Laane, *J. Chem. Phys.*, 1970, **52**, 358.
[343] J. Laane, *J. Chem. Phys.*, 1969, **50**, 776.

199.2 cm^{-1} was observed. The potential function derived for the ring-puckering vibration was of the form:

$$V_x = 18.51 \times 10^5 x^4 + 18.27 \times 10^3 x^2$$

It should be noted that the x^2 coefficient is positive; this is indicative of the ring-puckering vibration of a planar ring system.

In cyclopentanone, the presence of the carbonyl group prevents pseudorotation, and the low-frequency spectrum is best considered in terms of ring puckering. Ikeda and Lord [344] have observed a series of absorptions in the 84—96 cm^{-1} region of the spectrum for cyclopentanone. These were analysed in terms of a two-dimensional potential minimum in which the orthogona x and y co-ordinates represent dimensionless bending and twisting motions respectively. The derived potential was

$$V_{x,y} = 1302x^2 - 14\,531y^2 + 366x^4 + 14\,765x^2y^2 + 70\,352y^4$$

Interconversion between the two twisted conformers occurred through a planar form whose potential energy is about 750 cm^{-1} (8.95 kJ mol^{-1}) greater than the equilibrium forms.

Pseudorotation in Five-membered Rings.—An unconstrained five-membered ring, such as cyclopentane, may assume two extreme conformations: the chair conformation (14) belonging to the C_2 point group and the half-envelope conformation (15) which belongs to the C_s point group. If interconversion between

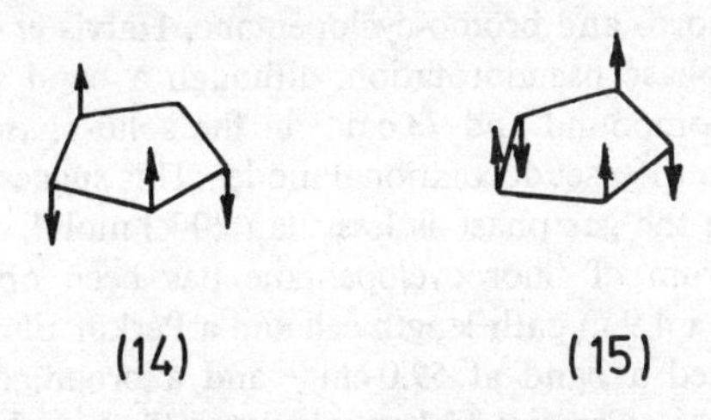

(14) (15)

these two extreme conformations is permitted the maximum out-of-plane amplitude moves around the ring, as a result of which the ring appears to rotate, although in fact the ring atoms have a motion which is perpendicular to the plane of the ring. In addition, the energy levels and the wavefunctions closely resemble those of a plane rotator. A cosine-based potential of the form

$$V_\phi = \tfrac{1}{2}\sum V_n(1 - \cos n\phi)$$

is normally used to describe the pseudorotation, and the numerical solution of the resultant Schrödinger equation has been considered by Ikeda *et al.*[345]

[344] T. Ikeda and R. C. Lord, *J. Chem. Phys.*, 1972, **56**, 4450.
[345] T. Ikeda, R. C. Lord, T. B. Malloy, and T. Ueda, *J. Chem. Phys.*, 1972, **56**, 1434.

The prototype five-membered saturated ring is cyclopentane, but since it does not possess a permanent dipole moment, the pure pseudorotational transition cannot be observed in the far-i.r. Durig and Wertz,[346] however, obtained spectroscopic evidence of pseudorotation in cyclopentane by analysis of the coupling of this mode to the methylene deformation in the mid-i.r. Although pseudorotation has been observed in THF [347, 348] and 1,3-dioxolan,[348] the greatest interest has centred on cyclopentane derivatives. Among the molecules that have been investigated are silacyclopentane,[349, 350] thiacyclopentane,[351] selenacyclopentane,[352] germacyclopentane,[353] chlorocyclopentane,[354] bromocyclopentane,[354] and fluorocyclopentane.[355]

In germacyclopentane, Durig and Willis [353] observed a series of *Q*-branches, commencing at 113.3 cm^{-1} with succeeding transitions at lower frequencies down to 93.7 cm^{-1} with an initial spacing of 2.5 cm^{-1} decreasing to about 1.9 cm^{-1}. The observed pseudorotation bands were quite accurately fitted by a potential of the form:

$$V = -(2043/2)(1-\cos 2\phi)-(21/2)(1-\cos 6\phi)$$

with a barrier to pseudorotation of 23.8 kJ mol^{-1}.

In selenacyclopentane, Green *et al.*[352] again observed a series of converging *Q*-branch spectra in the range 122.2—89.7 cm^{-1}. These were fitted to a pseudorotation potential of the form:

$$V = -(1795.4/2)(1-\cos 2\phi)+(13.45/2)(1-\cos 4\phi)-(86.19/2)(1-\cos 6\phi)$$

with a barrier to pseudorotation of 1881.6 cm^{-1} (22.5 kJ mol^{-1}).

In the cases of chloro- and bromo-cyclopentane, Harris *et al.*[354] were unable to observe the gas-phase pseudorotation, although a band at 96 cm^{-1} in the solid-phase bromo-compound and 73 cm^{-1} in the solid-phase chlorocyclopentane were assigned to the pseudorotational mode. This suggests that the barrier to pseudorotation in the gas phase is less than 20 kJ mol^{-1}.

The far-i.r. spectrum of fluorocyclopentane has been observed by Wertz and Shasky [355] using a 4.9 m path-length cell and a Perkin Elmer FIS-3 spectrometer. They observed a band at 59.0 cm^{-1} and a broad, diffuse absorption rising towards 35 cm^{-1}. Carreira [356] has calculated the pseudorotational potential energy function of fluorocyclopentane by a CNDO/2 quantum-mechanical calculation. He predicted that the potential energy function would have a single

[346] J. R. Durig and D. W. Wertz, *J. Chem. Phys.*, 1968, **49**, 2118.
[347] G. G. Engerholm, A. C. Luntz, W. D. Gwinn, and D. O. Harris, *J. Chem. Phys.*, 1969, **50**, 2446.
[348] J. A. Greenhouse and H. L. Strauss, *J. Chem. Phys.*, 1969, **50**, 124.
[349] J. Laane, *J. Chem. Phys.*, 1969, **50**, 1946.
[350] J. R. Durig and J. N. Willis, *J. Mol. Spectroscopy*, 1969, **32**, 320.
[351] D. W. Wertz, *J. Chem. Phys.*, 1969, **51**, 2133.
[352] W. H. Green, A. B. Harvey, and J. A. Greenhouse, *J. Chem. Phys.*, 1971, **54**, 850.
[353] J. R. Durig and J. N. Willis, *J. Chem. Phys.*, 1970, **52**, 6108.
[354] W. C. Harris, J. M. Karriker, and J. R. Durig, *J. Mol. Structure*, 1971, **9**, 139.
[355] D. W. Wertz and W. E. Shasky, *J. Chem. Phys.*, 1971, **55**, 2422.
[356] L. A. Carreira, *J. Chem. Phys.*, 1971, **55**, 181.

minimum with a barrier to pseudorotation of 4.35 kJ mol^{-1}. The potential energy function calculated was of the form:

$$2V = -340.76(1-\cos\phi)-2.55(1-\cos 2\phi)+15.27(1-\cos 3\phi) +14.28(1-\cos 4\phi)-30.70(1-\cos 5\phi)+22.21(1-\cos 6\phi)$$

Using this potential function, it is possible to evaluate the frequency of the $1 \leftarrow 0$ transition as 63.4 cm^{-1}, in good agreement with the value of 59 cm^{-1} obtained experimentally by Wertz and Shasky.[355]

Durig *et al.*[357] observed the ring-puckering mode at 122 cm^{-1} in solid cyclopentanol, but were unable to identify the pseudorotational mode in the gas phase.

Ring Puckering in Larger Ring Systems.—Dashevsky and Lugovskoy [358] have considered molecular mechanical models for the interconversion of cyclohexane between its boat and half-chair conformers. Laane and Lord [359] have observed the gas-phase ring-puckering of cyclohexa-1,4-diene, and Pickett and Strauss [360] have developed a potential function for ring bending in saturated six-membered rings, such as cyclohexane, dioxan, and trioxan, from the low-frequency spectrum.

9 Intramolecular Absorption, Refractive Index, and Astronomical Studies in the Far-infrared

Intramolecular Absorption.—Space has precluded any discussion of the very large volume of research published on intramolecular absorptions associated with low-frequency vibrational modes. A very high proportion of this far-i.r. work was performed in conjunction with higher-frequency i.r. measurements as a prelude to normal co-ordinate analysis or the calculation of thermodynamic properties. Included in this area is the important topic of the low-frequency vibrations of inorganic complexes. This has been the subject of books by Adams [361] and by Ferraro [362] and of a review by Nakagawa.[363] Maslowsky [364] has reviewed the specific area of the low-frequency vibrations of metal–metal bonds.

Space has also precluded a discussion of either crystal spectra or the optical properties of solids. Crystal spectra are very fully discussed by Möller and Rothschild [2] and by Mitra [365] while for other far-i.r. optical properties of solids,

[357] J. R. Durig, J. M. Karriker, and W. C. Harris, *Spectrochim. Acta*, 1971, **27A**, 1955.
[358] V. G. Dashevsky and A. A. Lugovskoy, *J. Mol. Structure*, 1972, **12**, 39.
[359] J. Laane and R. C. Lord, *J. Mol. Spectroscopy*, 1971, **39**, 340.
[360] H. M. Pickett and H. L. Strauss, *J. Chem. Phys.*, 1970, **53**, 376.
[361] D. M. Adams, 'Metal Ligand and Related Vibrations', Edward Arnold, London, 1967.
[362] J. R. Ferraro, 'Low Frequency Vibrations of Inorganic and Co-Ordination Compounds', Plenum, New York, 1971.
[363] I. Nakagawa, *Coordination Chem. Rev.*, 1969, **4**, 423.
[364] E. Maslowsky, *Chem. Rev.*, 1971, **71**, 507.
[365] S. S. Mitra in ref. 366, p. 333.

reference is made again to Möller and Rothschild [2] and to the books edited by Mitra and Nudelman [366, 367] and by Wright.[368]

Refractive Index.—The optical properties of an isotropic dielectric material at any frequency ν are defined by the real, n', and imaginary, k, parts of the complex refractive index n. The imaginary component is related to the absorption coefficient, α, so that:

$$\begin{aligned} n_\nu &= n'_\nu + \mathrm{i}k_\nu \\ &= n'_\nu + \mathrm{i}\alpha_\nu/4\pi\nu \end{aligned}$$

The absorption coefficient may be determined by the standard techniques of absorption spectrophotometry, but measurement of the real part of the dielectric constant requires the use of specialized experimental techniques.

At a discrete frequency the refractive index may be obtained from the fringe retardation in a two-beam interferometer, such as a Mach–Zender interferometer.[369] Using a discrete-frequency laser as source, the fringe displacement is measured as a function of sample thickness. Alternatively, a single-beam interferometer, such as the Fabry–Perot, may be used and the refractive index related to the mirror displacement required to restore the fringe maximum. The latter technique has been used by Bradley and Gebbie [370] to measure the far-i.r. refractive index of water vapour and nitrogen–water vapour mixtures such as occur in the atmosphere. Measurements were made at the discrete frequncies provided by the HCN and H_2O lasers.

Two techniques are available for measuring the frequency-dependence of the refractive index in the far-i.r., the asymmetric Michelson interferometer and the use of channel spectra. The use of the asymmetric interferometer has been very fully discussed by Chantry [371] and by Chamberlain *et al.*[372] The relative refraction spectra are obtained by taking the full complex Fourier transform of the dispersive interferogram obtained with the specimen in the fixed arm of the interferometer. This is related to the absolute refractive index by comparison measurements at discrete frequencies using laser sources. The application of this technique to tetrahedral molecules such as CCl_4, $GeCl_4$, $GeBr_4$, and $SnBr_4$ has been described by Thomas *et al.*[373]

The alternative method of obtaining the frequency dependence of the refractive index is by the use of channel spectra or Edsel–Butler bands. The channel spectra arise as a result of interference between waves that have made different numbers of passes through the sample because of random internal reflections

[366] S. S. Mitra and S. Nudelman, 'Optical Properties of Solids', Plenum, New York, 1969.
[367] S. S. Mitra and S. Nudelman, 'Far Infrared Properties of Solids', Plenum, New York, 1970.
[368] G. B. Wright, 'Light Scattering Spectra of Solids', Springer-Verlag, Berlin, 1969.
[369] G. W. Chantry, in ref. 10, p. 268.
[370] C. C. Bradley and H. A. Gebbie, *Appl. Optics*, 1971, **10**, 755.
[371] G. W. Chantry, in ref. 10, p. 268.
[372] J. Chamberlain, J. E. Gibbs, and H. A. Gebbie, *Infrared Phys.*, 1969, **9**, 185.
[373] T. E. Thomas, W. J. Orville-Thomas, J. Chamberlain, and H. A. Gebbie, *Trans. Faraday Soc.*, 1970, **66**, 2710.

at the surfaces. The presence of the sample results in the sample interferogram consisting of a series of signatures which replicate the peak at zero path difference. The signatures are inverted and of diminished amplitude relative to the central maximum. The channel spectra are then isolated by subtraction of the cosine FT of the sample from, effectively, that of the reference. The resultant difference spectrum consists of a series of peaks of different amplitudes, corresponding to different orders of the channel spectra, increasing from zero order at zero wavelength. The frequency at which the fringes occur is dependent on the real part of the refractive index, n', and on the sample thickness, h:

$$\nu = m/2n'h \cos \beta$$

where m is the order of the spectrum maximum, ν is the frequency, and β is the angle between the refracted ray and the normal within the sample. The theory has been extensively covered by Randall and Rawcliffe [374] and by Loewenstein and Smith.[375]

The sample thickness must be determined mechanically, *e.g.* by comparison with gauge blocks, but in the far-i.r., where sample thicknesses are of the order of a few hundred microns, this does not introduce any appreciable error. Loewenstein and Smith [375, 376] discuss the measurement of the refractive index of Mylar [poly(ethylene terephthalate)] and Surlyn A (an ionomer resin) in the 50—250 cm^{-1} region. Mylar is a partly crystalline polymer and is optically biaxial. It is only available in rolled sheet form and it was determined that the two directions in the plane of the sheet correspond to the slow and medium directions, the acute bisectrix being normal to the sheet plane.

Randall [377] has measured the refractive index of sodium fluoride from its channel spectra over the range 30—180 cm^{-1} and in the temperature range 10—80 K. The real part of the refractive index rose from 2.2 at 30 cm^{-1} to 2.7 at 180 cm^{-1}. By fitting the measured refractive indices to a Lorentz oscillator model, Randall calculated the low-frequency dielectric constant of sodium fluoride to be 4.70 and the damping parameter to vary between 3.3 cm^{-1} at 80 K and less than 0.025 cm^{-1} at 10 K. The refractive index of gallium phosphide has been measured in the far-i.r. by Parsons and Coleman.[378]

Astrophysical Studies.—A very rapidly developing area of far-i.r. spectroscopy is that associated with astrophysical studies.

Blair *et al.*[379] and Eddy *et al.*[380] have described far-i.r. instrumentation designed for use in airborne and rocket spectroscopy. The far-i.r. absorption

[374] C. M. Randall and R. D. Rawcliffe, *Appl. Optics*, 1967, **6**, 1889.
[375] E. V. Loewenstein and D. R. Smith, *Appl. Optics*, 1971, **10**, 577.
[376] D. R. Smith, in ref. 127, Paper 37.
[377] C. M. Randall, in ref. 127, Paper 38.
[378] D. F. Parsons and P. D. Coleman, *Appl. Optics*, 1971, **10**, 1683.
[379] A. G. Blair, F. Edeskuty, R. D. Hiebert, D. M. Jones, J. P. Shipley and K. D. Williamson, *Appl. Optics*, 1971, **10**, 1043.
[380] J. A. Eddy, R. H. Lee, P. J. Léna, and R. M. MacQueen, *Appl. Optics*, 1970, **9**, 439.

of the atmosphere, using the sun as a source of far-i.r. radiation, has been studied by a number of workers.[381-385] The far-i.r. emission of various parts of the galaxy,[386] of thermal radio sources,[387] and of Jupiter [388] have also been investigated.

The author would like to thank Dr. J. L. Wood for reading a draft of this Report.

381 J. E. Harries and W. J. Burroughs, *Infrared Phys.*, 1970, **10**, 165.
382 I. G. Nolt, T. Z. Martin, C. W. Wood, and W. M. Sinton, *J. Atmos. Sci*, 1971, **28**, 238.
383 W. J. Burroughs and J. Chamberlain, *Infrared Phys.*, 1971, **11**, 1.
384 R. Emery, *Infrared Phys.*, 1972, **12**, 65.
385 A. G. Kislyakov, *Infrared Phys.*, 1972, **12**, 61.
386 D. Muehlner and R. Weiss, *Phys. Rev. Letters*, 1970, **24**, 742.
387 D. A. Harper and F. J. Low, *Astrophys. J.*, 1971, **165**, L9.
388 T. Encrenaz, D. Gautier, L. Vapillon, and J. P. Verdet, *Astronom. Astrophys.*, 1971, **11**, 431.

6

Rotation and Vibration–Rotation Raman and Infrared Spectra of Gases

BY H. G. M. EDWARDS AND D. A. LONG

1 Introduction

This Report aims to give a comprehensive and critical survey of the literature relating to Raman and i.r. studies of the rotation and vibration–rotation spectra of gases for 1970, 1971, and the first six months of 1972. Where necessary, to give proper perspective to the discussion, earlier key papers are also referred to.

2 Experimental Techniques

Raman Spectroscopy.—The development of rotation and vibration–rotation Raman spectroscopy has proceeded in three distinct phases.

First, very soon after the discovery of the Raman effect in 1928, some very elegant studies were made [1] of simple molecules, *e.g.* H_2, O_2, N_2, HCl, NH_3, CO_2, C_2H_2, C_2H_4, and CH_4. After this first phase there was very little work in this field until the early nineteen-fifties. By this time the mercury arc had been brought to a high stage of development as a source for Raman excitation, and in this second phase a number of important studies of rotation and vibration–rotation Raman spectra were undertaken, notably by Welsh and Stoicheff.[1, 2] However, the range of molecules that could be studied was quite restricted, and defined entirely by three major constraints imposed by the mercury arc, which was still the only practicable source for Raman excitation of gases: (i) the natural linewidth of the strongest mercury-arc line at λ 4358 Å is about 0.24 cm^{-1}, so that spectra involving rotational line spacings of less than about 0.24 cm^{-1} cannot be resolved when excited with this source. This severely limited the molecules that could be investigated using mercury-arc excitation. (ii) The maximum power density in the sample that can be achieved with the mercury arc is the relatively low value of about 1W cm^{-2}. It was therefore necessary to observe the Raman scattering from large sample volumes, usually of the

[1] B. P. Stoicheff, in 'Advances in Spectroscopy', ed. H. W. Thompson, Wiley–Interscience, New York, 1959, Vol. I, p. 91.

[2] B. P. Stoicheff, in 'Methods in Experimental Physics', ed. D. Williams, Academic Press, New York, 1962, Vol. III, p. 111.

order of 1—5 litres, and to use mirror-type Raman cells to increase the solid angle over which the scattered radiation is collected. This restricted investigations to commonly available gases of a non-corrosive nature. (iii) The mercury arc provides excitation wavelengths only in the blue and violet regions of the spectrum, at λ 4358 Å and λ 4047 Å, and so the samples to be studied had to be essentially colourless. This second phase came to an end in the early nineteen sixties, when virtually all the molecules accessible to investigation using mercury-arc excitation had been studied.[3, 4]

The development of gas lasers in the nineteen-sixties made available for Raman excitation new sources which, in principle at least, have none of the limitations inherent in the mercury-arc source. With such lasers a good range of excitation wavelengths can be made available, very narrow linewidths can be achieved, and high power densities produced. The gas laser has therefore opened the way to a much wider application of rotation and vibration–rotation Raman spectroscopy to structural problems.[5–8]

Despite the considerable amount of structural information likely to be yielded by such studies, the third phase has been rather slow to develop. All the early work in this phase has been devoted to establishing experimental procedures and to repeating studies made previously with mercury-arc excitation. In 1965, Weber and Porto [9] reported that they had obtained a good photographic record of the pure rotation Raman spectrum of methylacetylene at 0.5 atm with an exposure time of 58 h, using a cell with Brewster-angle windows placed inside the cavity of a helium–neon laser of 20 mW output power. The actual volume of the cell was 340 cm^3 but the volume of gas effective in causing scattering was only 0.59 cm^3, compared with 3000 cm^3 in a conventional mercury-arc set-up. However, with laser excitation the exposure time was about 10 times longer than with mercury-arc excitation. In 1967 Weber, Porto, Cheesman, and Barrett [10] showed that, with a multi-pass Raman cell inside the cavity of the helium–neon laser, the necessary photographic exposure times for rotational spectra were comparable with the classical mercury-arc arrangement. They also made experimental tests of the effectiveness of various sample-illumination arrangements, using a normal cell either inside or outside the cavity, and found that the best arrangement was for the laser beam to be focused into the cell inside the cavity. With this arrangement the illuminated volume effective in scattering is less than 10^{-7} cm^3, and for a

[3] K. S. Rao, B. P. Stoicheff, and R. Turner, *Canad. J. Phys.*, 1960, **38,** 1516.
[4] B. J. Monostori and A. Weber, *J. Mol. Spectroscopy*, 1964, **12,** 129.
[5] H. G. M. Edwards, in 'Essays in Structural Chemistry', ed. D. A. Long, L. A. K. Staveley, and A. J. Downs, Macmillan, London, 1970, Ch. 6, p. 135.
[6] A. Weber, in 'The Raman Effect', ed. A. Anderson, Marcel Dekker, New York, 1971, Vol. II, in the press.
[7] A. Weber, 'Recent Developments in the High-Resolution Raman Spectroscopy of Gases', in 'Developments in Applied Spectroscopy', Vol. 10, in the press.
[8] W. J. Walker and A. Weber, *J. Mol. Spectroscopy*, 1971, **39**, 57.
[9] A. Weber and S. P. S. Porto, *J. Opt. Soc. Amer.*, 1965, **55,** 1033.
[10] A. Weber, S. P. S. Porto, L. E. Cheesman, and J. J. Barrett, *J. Opt. Soc. Amer.*, 1967, **57,** 19.

gas at normal pressure only 10^{11} molecules are involved. This optical arrangement, the allied problems of collection of the scattered radiation, and the superiority of pulse-counting detection systems were fully discussed by Barrett and Adams [11] in 1968. These two papers by Barrett and co-workers present rotation spectra of O_2, N_2, CO_2, and $CH_3C{\equiv}CH$, and vibration–rotation spectra of O_2, N_2, CO_2 which were of excellent quality. They were directly recorded with the sample inside the cavity of an argon-ion laser with an intra-cavity power of 3—5 W at 4880 Å or 5145 Å. In 1970, Barrett and Weber [12] described a cell specially designed for the study of Raman scattering from gases subjected to an electric discharge.

Rich and Welsh [13] have described in some detail the adaptation of a conventional concave mirror-type Raman cell for use with laser excitation. The laser beam is admitted through the slot between the two semicircular front mirrors of the cell, and with careful adjustment continues to traverse the cell for a total of about 50 passes. The front and rear mirrors collect the scattered radiation in the usual way. The authors compare microphotometer traces of photographically recorded spectra of the Q-branch region of the electronic Raman spectrum of the NO molecule (Stokes Raman shift of *ca.* 120 cm^{-1}) obtained with this cell using excitation with (*a*) 4800 Å radiation from an argon-ion laser and (*b*) 4358 Å radiation from mercury arcs. The results show clearly that laser excitation has great advantages over mercury-arc excitation.

In another paper [14] the same authors review some of the problems associated with laser excitation of rotation and vibration–rotation Raman spectra. They discuss *inter alia* the relative merits of recording such spectra photographically and photoelectrically. They conclude that, although direct spectrometric recording, with its intrinsically high sensitivity, is very useful for obtaining preliminary information, frequency measurements of sufficient precision and accuracy can only be made with photographically recorded spectra. These observations are well borne out by the experience of most workers in this field. For example, Barrett and Weber [12] in their study of CO_2 used photographically recorded spectra for frequency measurements, even though they also recorded the spectra directly and could use computer programs to extract weak rotational lines from the background noise. Long and co-workers [15] have investigated in some detail the feasibility of using a Spex 1400 double monochromator for studies of rotation Raman spectra. Although good quality spectra could be obtained, it was not possible to make sufficiently accurate frequency measurements. It was found necessary to use photographic recording for all accurate frequency determinations. However, in a study of the pure rotation Raman spectrum

[11] J. J. Barrett and N. I. Adams, *J. Opt. Soc. Amer.*, 1968, **58**, 311.
[12] J. J. Barrett and A. Weber, *J. Opt. Soc. Amer.*, 1970, **60**, 70.
[13] N. H. Rich and H. L. Welsh, *J. Opt. Soc. Amer.*, 1971, **61**, 977.
[14] N. H. Rich and H. L. Welsh, *Indian J. Pure Appl. Phys.*, 1971, **9**, 944.
[15] D. A. Duddell, D. A. Long, and R. Love, personal communication, 1972.

of fluorine, Claassen *et al.*[16] claimed that a Spex 1400 double monochromator could give frequency measurements of adequate accuracy. They calibrated the instrument using the well-established frequencies of the rotation lines of oxygen, which were scanned alternately with the rotation lines of fluorine. Nevertheless, their experimental results lacked the necessary precision to give a satisfactorily consistent value of the centrifugal stretching constant D_0, values of which differed by more than a factor of two between two experiments. On the other hand, Edwards, Long, and Love,[17] in a more recent study of the rotation spectrum of fluorine using photographic recording, obtained for the first time an acceptable experimental value of D_0 for fluorine.

In the past few years there have appeared the first papers reporting work in which laser excitation of rotation Raman spectra has been used to achieve a degree of resolution and an accuracy of frequency measurement superior to that obtainable with mercury-arc excitation. Several different techniques have been employed to achieve these improvements. Weber and Schlupf[18] have reported the use of a single-moded argon-ion laser in conjunction with a carefully stabilized grating spectrograph to achieve clear resolution of lines with a spacing of 0.07 cm^{-1}. The use of a pressure-scanned Fabry–Pérot interferometer with photon-counting detection for the study of rotation Raman lines has been described by May and co-workers[19, 20] and by Clements and Stoicheff.[21] With this technique the wavenumbers of individual lines could be determined to ± 0.002 cm^{-1}, but only a very limited spectral range, of the order of a few wavenumbers, can be investigated at one time. Butcher, Willetts, and Jones[22] have employed a Fabry–Pérot étalon crossed with a low-dispersion spectrograph to achieve a new level of accuracy in the measurement of rotation lines of oxygen and nitrogen.[23] A typical standard error in a Raman line was 0.003 cm^{-1} and the consequent precision in the rotational constants can be judged from the value for oxygen of $B_0 = 1.437682 \pm 0.000009$ cm^{-1}. These papers represent important advances in technique and will now be considered in more detail.

In Weber and Schlupf's experiments the gas samples were contained in a multiple-pass Raman cell placed within the cavity of the argon-ion laser. A Fabry–Pérot étalon inside the laser cavity reduced the width of the 4880 Å line from 0.15 cm^{-1} (the 'free-running' multi-mode value) to less than 0.001 cm^{-1}. This étalon had to be temperature-stabilized to prevent thermal drifts of the laser frequency owing to changes in room temperature.

[16] H. H. Claassen, H. Selig, and J. Shamir, *Appl. Spectroscopy*, 1969, **23**, 8.
[17] H. G. M. Edwards, D. A. Long, and R. Love, presented at the 3rd International Conference on Raman Spectroscopy, Reims, 1972; to be published in 'Advances in Laser Raman Spectroscopy', ed. J. P. Mathieu, Heyden, London, 1972, Vol. I.
[18] A. Weber and J. Schlupf, *J. Opt. Soc. Amer.*, 1972, **62**, 428.
[19] V. G. Cooper, A. D. May, E. H. Hara, and H. F. P. Knapp, *Canad. J. Phys.*, 1968, **46**, 2019.
[20] V. G. Cooper, A. D. May, and B. K. Gupta, *Canad. J. Phys.*, 1970, **48**, 725.
[21] W. R. L. Clements and B. P. Stoicheff, *J. Mol. Spectroscopy*, 1970, **33**, 183.
[22] R. J. Butcher, D. V. Willetts, and W. J. Jones, *Proc. Roy. Soc.*, 1971, **A324**, 231.
[23] B. P. Stoicheff, *Canad. J. Phys.*, 1954, **32**, 630.

The spectra were recorded photographically, using a grating spectrograph of resolving power *ca.* 700 000 at 4880 Å. Since exposure times ranging from a few hours to several tens of hours were involved, it was necessary to pay particular attention to stabilization of the spectrograph against changes in temperature, atmospheric pressure, and humidity, otherwise shifts in the spectral lines would give rise to asymmetry and consequently to errors in the measured wavenumbers; even total loss of resolution may result. At a temperature of 25 °C, a barometric pressure of 760 Torr, and a relative humidity of 50%, the changes in wavenumber of the 4880 Å argon-laser exciting line owing to changes of temperature, total pressure, and partial water vapour pressure are -0.018 cm^{-1} °C^{-1}, $+0.007$ cm^{-1} Torr^{-1}, and -0.001 cm^{-1} Torr^{-1}, respectively. The temperature and pressure changes decrease slowly across the Stokes Raman region to -0.016 cm^{-1} °C^{-1} and $+0.006$ cm^{-1} Torr^{-1}, respectively, at a displacement of 3000 cm^{-1} from the 4880 Å exciting line, whereas the dependence on humidity is unchanged. The spectrograph was thermostatted to +0.1 °C and the humidity of the laboratory was carefully controlled. Changes in barometric pressure were compensated for in an ingenious way by temperature-tuning of the Fabry–Pérot étalon. A temperature change of 0.05 °C was required to compensate for a change in barometric pressure of 1 Torr. Spectra were photographed using Kodak IIIa-J plates hypersensitized by baking for 20—30 h at 50—60 °C. These plates are suitable for Stokes shifts of up to 1500 cm^{-1} excited by 4880 Å. Special attention had to be paid to the determination of the frequencies of the Raman displacements to achieve the necessary accuracy. Also, the use of the Fabry–Pérot étalon precluded an *a priori* knowledge of the exact value of the exciting frequency. In the study of pure rotation spectra this is of no importance since the uncertainty in the frequency of the exciting line is avoided by averaging the Stokes and anti-Stokes Raman displacements. In the case of vibration–rotation bands this is not possible since, owing to the large dispersion, only a small part of the Stokes Raman spectrum can be photographed at any one time. It was thus necessary to measure independently the frequency of the exciting line at the beginning of each experiment.

Using the apparatus and the techniques described above, Weber and Schlupf have been able to photograph the resolved pure rotation spectra of many heavy polyatomic molecules. Some of these spectra were impossible to obtain previously, either with Hg-arc or He–Ne laser excitation, owing to the intrinsic weakness of the spectra (*e.g.* propane, n-butane, n-pentane). Others, although observed previously, were unresolved (*e.g. trans*-dichloroethylene, carbon suboxide). In addition, the improved resolution has enabled them to observe the resolved rotation Raman spectra of cyanuric fluoride, 1,3,5-trifluorobenzene, hexafluorobenzene, tetramethylallene, and other substances. In the case of hexafluorobenzene a resolution of ~0.07 wavenumbers was achieved. Weber and Schlupf have not yet published any analyses of these spectra but their results are awaited with interest.

Cooper, May, and Gupta [20] and Cooper *et al.* [19] have used an interferometric technique for the measurement of the linewidths and frequencies of the $S_0(0)$ and $S_0(1)$ rotational Raman lines of H_2. Their apparatus consisted of a 150 mW helium–neon laser, a pressure-scanned Fabry–Pérot interferometer, and photon-counting electronics. The lineshapes and the changes in frequency of the rotational lines of H_2 in gas mixtures were measured. The accuracy of the frequency measurements was 0.002 cm^{-1}, an order of magnitude better than previous measurements of this kind which were made without the interferometer. The authors indicate that a further increase in accuracy could be expected if the observations could be made in the forward direction, where the Raman lines would be narrower.

Clements and Stoicheff [21] have also recently used pressure-scanned Fabry–Pérot interferometers with photon-counting detection systems for high-resolution studies of Raman bands. In these investigations they used a specially constructed helium–neon laser which had a total power output of 400 mW in a line of full width at half maximum intensity of 0.025 cm^{-1}. This small linewidth was achieved by designing the laser to operate as a

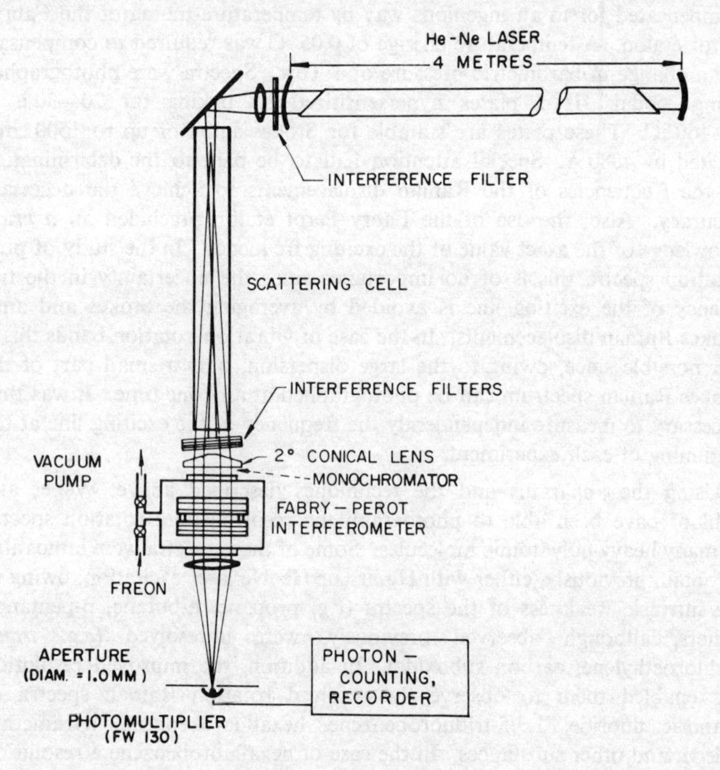

Figure 1 *Diagram of apparatus for high-resolution Raman spectroscopy of gases* (Reproduced by permission from *J. Mol. Spectroscopy*, 1970, **33,** 183)

low-gain oscillator, that is by using a large tube diameter and a 2% transmission mirror at the output end of the cavity. The 6328 Å radiation was isolated with a filter having a band-pass of 10 Å and then focused into the scattering cell by a lens of focal length 1 m (see Figure 1). The Pyrex glass scattering cell was 1 m long and contained the sample gas at pressures up to 2 atm. This was followed by several narrow-band-pass filters which block the laser light and transmit the scattered radiation under study. Radiation scattered at 2° is then collected by a conical lens and analysed with a pressure-scanned Fabry–Pérot interferometer. For investigating rotation Raman lines with small frequency shifts it was found necessary to place a monochromator immediately in front of the Fabry–Pérot interferometer, to be used in conjunction with (or instead of) the narrow-band-pass filters. This not only helped to eliminate the laser light but also to isolate individual rotation lines. In addition, fluorescence originating in the interference filters, lenses, and windows of the scattering cell (with maximum intensity at wavelengths longer than 6800 Å) was suppressed with the use of the monochromator.

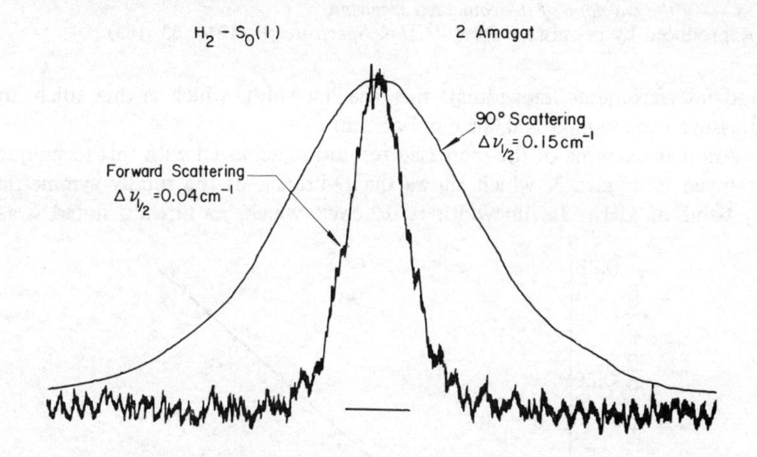

Figure 2 *Profiles of the* $S(1)$ *rotational line of* H_2 *gas at* 2 atm *pressure, observed in forward and* 90° *scattering*
(Reproduced by permission from *J. Mol. Spectroscopy*, 1970, **33,** 183)

Spectra of the $S(1)$ rotational line of H_2, observed with both forward and 90° scattering, are compared in Figure 2. Both spectra were obtained at a gas pressure of 2 atm; the interferometer spectral free range was 0.5 cm^{-1} and the effective resolution 0.02 cm^{-1}. At 90° scattering, the full width at half maximum intensity of the line is 0.15 cm^{-1}, and corresponds to the Doppler breadth at 20 °C modified by motional narrowing. In the forward direction, the observed linewidth is reduced to 0.04 cm^{-1}. However this value represents the total instrumental width (including the laser

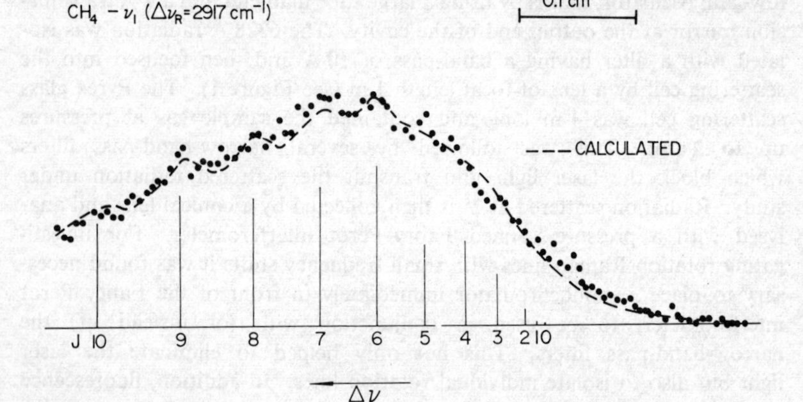

Figure 3 *Q-Branch of the ν_1 band of* CH_4 ($\Delta\nu$ = 2917 cm^{-1}) *at* 2 atm *pressure.* (●), *experimental intensity points obtained with a digital counter using* 500 s *intervals; and* (---), *the envelope of the computed spectrum*
(Reproduced by permission from *J. Mol. Spectroscopy*, 1970, **33**, 183)

and interferometer linewidths); the true linewidth, which is due solely to pressure broadening, is 0.006 cm^{-1} at 2 atm.

Another example of the increased resolution achieved with this technique is given in Figure 3, which shows the *Q*-branch of the totally symmetric ν_1 band of CH_4. Its linewidth is 0.3 cm^{-1} which, as already noted, was

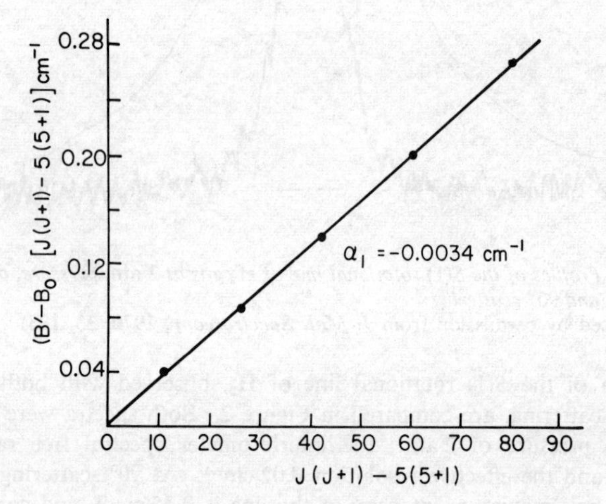

Figure 4 *Frequency separation of observed peaks from the* $J = 5$ *peak plotted against* $J(J+1)$
(Reproduced by permission from *J. Mol. Spectroscopy*, 1970, **33**, 183)

about the limit of resolution in pre-laser work. The spectrum was obtained at 2 atm pressure, with the interferometer spectral free tange being 2.5 cm^{-1} and the total instrumental width 0.07 cm^{-1}. Because of low intensity, a digital counter with a counting interval of 500 s was used, and the observed spectrum is indicated by the solid circles in Figure 3. It is possible to detect some structure within the narrow Q-branch observed with the present techniques. This structure is considered to be due to a small difference in rotational constants of the 0000 and 1000 levels, with the spacing given by the usual expression $(B'-B_0)J(J+1)$. Trial analyses were attempted, with tentative rotational numbering of the observed peaks. The best numbering is that represented by the straight-line graph whose slope gives the value $B'-B_0 = -\alpha_1 = 0.0034$ cm^{-1} (Figure 4). With this value of α_1, the relative positions of the rotational lines within the Q-branch were calculated, and these are indicated in Figure 3. Also, the relative intensities of the rotation lines were computed, and assuming linewidths of 0.07 cm^{-1} a calculated spectrum was obtained. A comparison of this calculated spectrum with the observed spectrum shows satisfactory agreement.

The novel experimental technique used by Jones *et al.*[22] involved crossing a Fabry–Pérot étalon with a low-dispersion spectrograph. The spectrograph served simply to separate the individual rotation lines so that the set of fringes from one line did not overlap and mask the set of fringes from an adjacent line. The general experimental arrangement is shown in Figure 5, and a typical spectrum is shown in Figure 6. The set of fringes observed on each rotation line is a section of a complete set of ring systems emerging from the étalon at angles θ to the normal given by the equation:

$$n\lambda = 2d\cos\theta$$

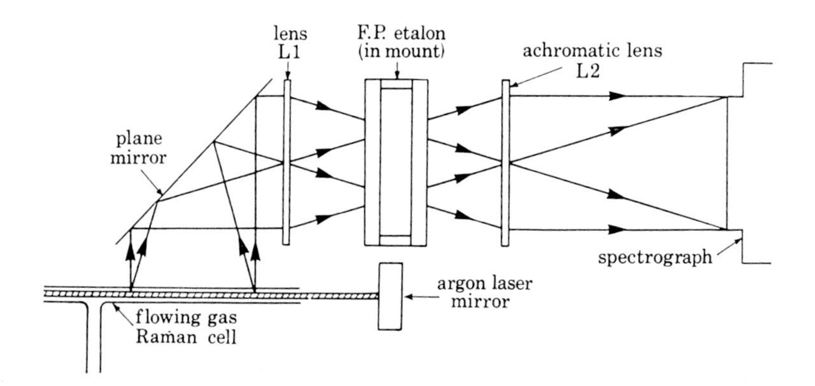

Figure 5 *The optical arrangement used for photographing the rotational Raman spectra of oxygen and nitrogen. Focal lengths of lenses L1 and L2 were about* 250 mm
(Reproduced by permission from *Proc. Roy. Soc.*, 1971, **A324,** 231)

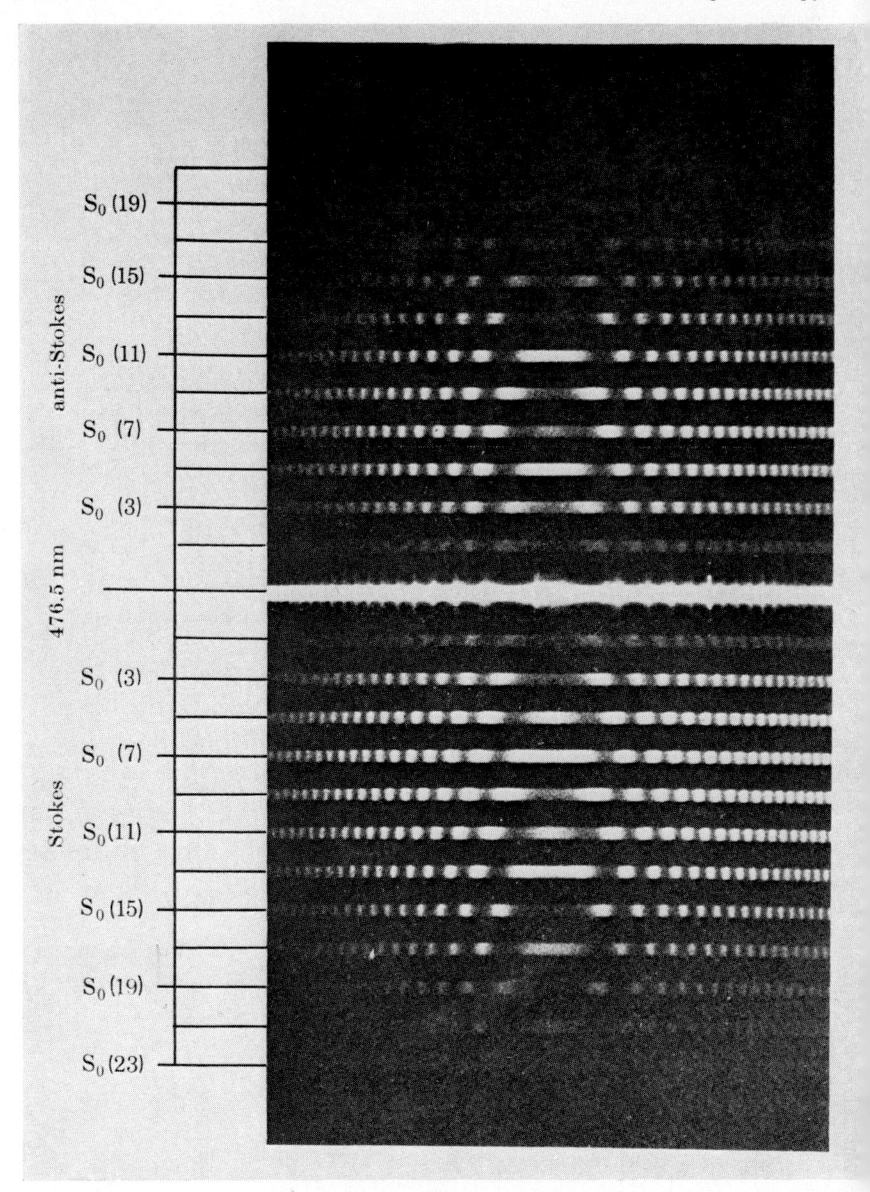

Figure 6 *Pure rotational Raman spectrum of* O_2 *excited by* 476.5 nm *argon laser radiation. Raman radiation dispersed by means of a Fabry–Pérot étalon with plate separation of* 12 mm. *Gas pressure* 1 atm
(Reproduced by permission from *Proc. Roy. Soc.*, 1971, **A324,** 231)

where n is the order of interference, d the étalon spacing, and λ the air wavelength of the line in question. The radiation emerges from the étalon in a series of cones which are imaged on the entrance slit of the spectrograph by lens L2, so giving rise to the characteristic spectral pattern of Figure 6, An analysis based on the twenty or so fringe diameters observed for a given line enabled the wavenumber of a Raman line to be determined to 0.003 cm^{-1}.

It is clear that this interesting technique is capable of providing highly accurate values for the rotational constants of simple molecules. The limitations on the method arise because of the finite spectral widths of the laser lines, typically 0.1 cm^{-1}, as well as the weakness of the Raman scattered lines. For obtaining higher resolution, and consequently greater accuracy, it would be necessary to reduce the width of the laser lines by 'mode selection' and to decrease the pressure of the gas studied to avoid pressure-broadening effects. Both of these processes will undoubtedly decrease the intensity of the scattered light and hence lead to an increase in the exposure time necessary for the observation of the spectra. Such an increase in exposure time would be disadvantageous owing to the difficulties of controlling the temperature and pressure in the vicinity of the étalon. In spite of such difficulties, however, it should prove possible to study rotation Raman spectra at a resolution of *ca.* 0.05 cm^{-1} with simple extensions of the methods reported here.

Infrared Spectroscopy.—Most high-resolution studies in the i.r. involve specially designed grating instruments and interferometers. Recent years have seen continuing improvements in the performance of such instruments. Using interferometric techniques, Connes *et al.*[24a] have achieved a resolution of 0.005 cm^{-1}, and Meyer and co-workers[24b] a resolution of 0.017 cm^{1-} with a Sisam spectrometer. With a grating instrument the best resolution achieved seems to be that reported by Murchison and Overend,[24c] who achieved a resolution of 0.03 cm^{-1}. Tuneable i.r. lasers are now being developed and seem certain to lead to further developments in high-resolution spectroscopy. A recent example of the application of such a laser is the work of Patel and co-workers,[25] who used a CO_2 spin-flip laser tuneable over the 800—900 cm^{-1} region to study the ν_2 band of NH_3. A resolution of 0.05 cm^{-1} was achieved and the estimated laser linewidth was *ca.* 0.03 cm^{-1}.

3 Pure Rotation and Vibration–Rotation Raman Spectra

In this section a review of the pure rotation and vibration–rotation Raman

[24a] P. Connes, J. Pinard, G. Guelachivili, J. P. Maillard, C. Amiot, M. L. Grenier-Besson, C. Camy-Peyret, and J. M. Flaud, *J. Phys.* (*Paris*), 1972, **33**, 77; (*b*) A. Bersellini and C. Meyer, *Compt. rend*, 1970, **270,** *B*, 1672; M. Bellouard and C. Meyer, *ibid.*, p. 1562; (*c*) C. B. Murchison and J. Overend, *Spectrochim. Acta*, 1970, **26A**, 599.

[25] S. K. N. Patel, E. D. Shaw, and R. J. Kerl, *Phys. Rev. Letters*, 1970, **25,** 8.

spectra of simple molecules in the gas phase will be given. The effects of pressure-broadening of the rotation lines in gas mixtures and the investigation of the rotation spectra of gases dissolved in liquids will be reviewed in Sections 4 and 5.

Diatomic Molecules.—*Oxygen.* The most recent study of this molecule is that of Butcher, Willetts, and Jones.[22] Their experimental procedure, involving the use of a Fabry–Pérot étalon, has already been described in detail earlier in this chapter. This procedure leads to more precise values of the rotational constants than have ever been obtained using Raman spectroscopy.

The ground electronic state of the O_2 molecule is $^3\Sigma_g^-$ and, since the oxygen nuclei have zero spin, only the odd rotational levels will be populated in the ground state. An analysis was made of the ten rotational transitions from $N = 3$, $N = 5$ *etc.* to $N = 21$ in the S-branch, where N is the quantum number associated with nuclear rotation (apart from the nuclear spin). Because of the multiplet splitting of each N-level, the rotation lines in the S-branch consist of three overlapping transitions, and it was necessary to apply corrections for this. The final corrected values of the rotational constants were $B_0 = 1.437682 \pm 0.000009$ cm^{-1}, $D_0 = (4.85_2 \pm 0.01_2) \times 10^{-6}$ cm^{-1}.

These constants agree quite closely with those obtained from the recent microwave determination of McNight and Gordy,[26] namely $B_0 = 1.4376807 \pm 0.0000007$ cm^{-1} and $D_0 = (4.910 \pm 0.002) \times 10^{-6}$ cm^{-1}, assuming the velocity of light to be 299792.5 ± 0.1 km s^{-1} for the conversion from MHz into cm^{-1}.

Because of the accuracy of the Raman results, an estimate was made of the H coefficient in the rotational term:

$$F_0(N) = B_0N(N+1) - D_0N^2(N+1)^2 + H_0N^3(N+1)^3.$$

A value for H_0 of $(3 \pm 60) \times 10^{-11}$ cm^{-1} was obtained.

The effect of splitting of the $^3\Sigma_g^-$ ground state by spin–rotation interaction has been known for some time in optical and microwave absorption spectroscopy but it has only recently been observed in the rotation Raman spectrum. The effect of the splitting on the rotation Raman spectrum was first demonstrated by Jammu, St. John, and Welsh,[27] who used Hg 2537 Å radiation for excitation. For the coupling of spin and rotation, the total quantum number J takes the values $(N+1)$, N, and $(N-1)$ for each value of N, the nuclear rotation quantum number. This multiplet splitting of the levels of a particular N value gives rise to transitions in the rotation Raman spectrum that obey the selection rules $\Delta N = 2$, $\Delta J = 0,1,2$. The most intense transitions correspond to $\Delta N = 2$, $\Delta J = 2$ and do not involve any spin reorientation. The transitions governed by $\Delta N = 2$, $\Delta J = 1$ are

[26] J. S. McNight and W. Gordy, *Phys. Rev. Letters*, 1968, **21,** 1787.
[27] K. S. Jammu, G. E. St. John, and H. L. Welsh, *Canad. J. Phys.*, 1966, **44,** 797.

weak, but give rise to observable pairs of satellites on lines in the rotation Raman spectrum. The $\Delta N = 2$, $\Delta J = 0$ transitions are very weak and have not been observed so far.

Renschler, Hunt, McCubbin, and Polo [28] observed the satellites accompanying the S_1 and S_3 Raman lines and the Rayleigh line in the pure rotation spectrum of O_2. They used a Raman cell containing oxygen at three atmospheres pressure, placed in the cavity of a helium–neon laser. The spectra were observed in the thirty-eighth order of an *f*/25 Czerny–Turner double monochromator of focal length 2.5 m, and the detector was an ITT FW130 photomultiplier.

Abe and Shimanouchi [29] have also observed the spin structure in the pure rotation Raman spectrum of O_2 at one atmosphere pressure. Satellites were observed for the rotation lines corresponding to the $N = 1 \rightarrow N = 3$, $N = 3 \rightarrow N = 5$, and the $N = 5 \rightarrow N = 7$ transitions. These workers used an argon ion laser and a Nalumi spectrometer of focal length 2 m and aperture *f*/11.

The most recent work on the spin structure of the O_2 molecule is that of Rich and Lepard,[30] who used the multi-reflection cell described in Section 2 (and ref. 13) and an argon-ion laser with 1 W output at 4880 Å. The rotation spectra were recorded with a Spex double monochromator in the second order of diffraction, using an ITT FW130 photomultiplier and photon counting as the detection system. Accurate relative intensity measurements were made of the satellites and strong central components of the S_3 and S_5 rotation lines. The wavenumber separation of the satellites and the main components is *ca.* 2 cm^{-1}. The intensities determined experimentally by Rich and Lepard, and also by Renschler, Hunt, McCubbin, and Polo, were found to be in good agreement with the theoretical predictions made by Lepard [31] and are shown in Table 1.

Table 1 *Calculated and observed intensity ratios of the satellites and main components of the rotational transitions in the* O_2 *molecule*

N transition	*Rotational line*	*Calc. intensity ratios*[a] (ref. 31)	*Observed intensity ratios*[a] (ref. 28)	(ref. 30)
1→3	S_1	1.3 : 6.0 : 1.5	1.2 : 6.0 : 1.5	not observed
3→5	S_3	1.4 : 33 : 1.0	1 : 30 : 1	1.4 : 35 : 1.0
5→7	S_5	1.2 : 66 : 1.0	not observed	1.5 : 77 : 1.0

[a] The ratios are for satellite:main component:satellite, in that order.

[28] D. L. Renschler, J. L. Hunt, T. K. McCubbin jun., and S. R. Polo, *J. Mol. Spectroscopy*, 1969, **31**, 173.
[29] K. Abe and T. Shimanouchi, *Bull. Chem. Soc. Japan*, 1969, **42**, 3047.
[30] N. H. Rich and D. W. Lepard, *J. Mol. Spectroscopy*, 1971, **38**, 549.
[31] D. W. Lepard, *Canad. J. Phys.*, 1970, **48**, 1664.

Vibration–rotation Raman spectra of the O_2 molecule have been recorded photographically, using Hg 4358 Å excitation, by Weber and McGinnis.[32] From an analysis of ten *O*- and *S*-pairs of rotational lines from $J = 5$ to $J = 21$ they obtained the value of the rotational constant B_1 for the $v = 1$ vibrational state of $B_1 = 1.4220 \text{ cm}^{-1}$, assuming $D_1 = D_0 = 5.6 \times 10^{-6} \text{ cm}^{-1}$. Subsequently, Weber *et al.*[10] obtained a photoelectric recording of the unresolved *Q*-branch of the $v = 1 \leftarrow v = 0$ vibrational transition, but no attendant *O*- and *S*-branch rotational lines were observed. The Raman cell was placed inside the cavity of an argon-ion laser and the 4880 Å line was used for excitation of the spectrum.

Barrett and Adams[11] have also observed the vibration–rotation band of O_2 for the $v = 1 \leftarrow v = 0$ transition. Excitation was with an argon-ion laser at 4880 Å and the sample was placed in the cavity of the laser. The spectrum was recorded directly, using a pulse-counting detection system. Although the Stokes *Q*-branch at 1556 cm^{-1} was unresolved, the rotational lines from S_1 to S_{25} and from O_3 to O_{23} were clearly resolved. However, no measurements of the frequencies of the rotational lines were reported and no rotational constants were given.

Rich and Lepard,[30] in their paper on the spin structure in the rotation Raman spectrum of O_2, also report the observation of multiplet splitting due to spin structure in the fundamental vibration–rotation band. The spectrum was obtained with the same experimental arrangement as was used for the pure rotation spectrum. The *Q*-branch is composed of transitions for which $\Delta N = 0$, $\Delta J = 0, \pm 2$, except for $N = 1$. For $N = 1$, the $\Delta J = \pm 2$ transitions are contained in the satellite *Q*-branch lines for which $\Delta N = 0$, $\Delta J = \pm 1$. These satellites are analogous to those which were observed by Renschler *et al.*[28] accompanying the Rayleigh line in the pure rotation spectrum. Only the high-frequency satellite is observed in the *Q*-branch of the vibration–rotation spectrum since the other is obscured by the main *Q*-branch. The *Q*-branch transitions were resolved from $N = 3$ to $N = 25$, and spin structure similar to that observed in the pure rotation spectrum was also evident on the S_1 and O_3 lines.

Nitrogen. Accurate rotational constants for the N_2 molecule have been known for some time from the Raman-spectroscopic work of Stoicheff,[23] who observed 35 rotational lines in the ground state, using mercury-arc excitation at 4358 Å and photographic recording. The constants obtained were

$$B_0 = 1.9897_3 \pm 0.0003_5 \text{ cm}^{-1} \text{ and}$$
$$D_0 = (6.1 \pm 0.5) \times 10^{-6} \text{ cm}^{-1}.$$

Recently, Butcher, Willetts, and Jones[22] have reinvestigated the pure rotation Raman spectrum of nitrogen, using the experimental procedure involving a Fabry–Pérot étalon and a Czerny–Turner spectrograph, as already described. They analysed 17 Stokes and anti-Stokes pairs of rota-

[32] A. Weber and E. A. McGinnis, *J. Mol. Spectroscopy*, 1960, **4**, 195.

tional lines from $J = 1$ to $J = 18$ (excepting $J = 17$) and obtained the rotational constants $B_0 = 1.98950_6 \pm 0.00002\ \text{cm}^{-1}$ and $D_0 = (5.48 \pm 0.05_5) \times 10^{-6}\ \text{cm}^{-1}$.

These rotational constants agree well with the earlier results of Stoicheff, and the uncertainties in the values have been reduced considerably. However, the value for the constant D_0 is somewhat lower than the value of $5.74 \times 10^{-6}\ \text{cm}^{-1}$ calculated from Dunham's equations.[33] This apparent discrepancy is due to neglect of the term $H_0 J^3 (J+1)^3$ in the rotational term. A revised analysis taking into account this term gave a value of H_0 of $(3.7 \pm 11.0) \times 10^{-10}\ \text{cm}^{-1}$. Introduction of corrections to the observed wavenumber shifts of the rotational lines, based on a value of H_0 of $3.7 \times 10^{-10}\ \text{cm}^{-1}$ followed by a conventional analysis, led to the revised values of $B_0 = 1.98954_8 \pm 0.00002_8\ \text{cm}^{-1}$ and $D_0 = (5.72_3 \pm 0.05_6) \times 10^{-6}\ \text{cm}^{-1}$. The value of D_0 so obtained is almost exactly in agreement with the value calculated from Dunham's expressions.

Stoicheff [23] did not observe rotational structure in the $v = 1 \leftarrow v = 0$ transition of the N_2 molecule. Therefore, to calculate B_e and r_e, he had to use B_0 and r_0 values from his pure rotation spectrum with B_1, r_1 values *etc.* from the electronic spectrum of N_2 in excited vibrational states. Since this earlier work, there have been no measurements of the frequencies of the vibration–rotation transitions in the fundamental band. However, Barrett and Adams[11] have obtained a high-quality, photoelectrically recorded, vibration–rotation spectrum of N_2 for the $v = 1 \leftarrow v = 0$ transition. Although the Q-branch, centred at $2330\ \text{cm}^{-1}$, was unresolved, the rotational lines from S_1 to S_{21} and from O_2 to O_{18} were clearly resolved. Unfortunately, no frequency measurements were reported.

Nitric Oxide. It is well known from i.r. and microwave studies that the NO molecule has a $^2\Pi$ ground electronic state, split by spin–orbit interaction into two substates $^2\Pi_{\frac{1}{2}}$ and $^2\Pi_{\frac{3}{2}}$ which have different components of the total angular momentum J along the internuclear axis. The $^2\Pi_{\frac{3}{2}}$ excited state lies some $120\ \text{cm}^{-1}$ above the $^2\Pi_{\frac{1}{2}}$ state. Since the molecule has an electronic angular momentum, the Raman transitions with $\Delta J = +1$ are allowed, in addition to those with $\Delta J = 0, +2$ in the pure rotation spectrum.

Several papers have recently appeared, reporting the rotation Raman spectrum of the NO molecule. Shotton and Jones [34] have reported the pure rotation spectrum of nitric oxide excited within the cavity of an argon-ion laser (4880 Å line) with an intra-cavity power of 10 W. The S-branch rotational lines from both the $^2\Pi_{\frac{1}{2}}$ and $^2\Pi_{\frac{3}{2}}$ substates were observed, those from the $^2\Pi_{\frac{1}{2}}$ substate being the more intense. The J values are half-integral, with $J = \frac{1}{2}, \frac{3}{2}, \ldots$ for the $^2\Pi_{\frac{1}{2}}$ component and $J = \frac{3}{2}, \frac{5}{2}, \ldots$ for the $^2\Pi_{\frac{3}{2}}$ component. The splitting between corresponding lines of the S-branches increases approximately linearly with J and is a consequence of the slightly different rotational constants in the two substates. The mean of the two

[33] J. L. Dunham, *Phys. Rev.*, 1932, **41**, 721.
[34] K. C. Shotton and W. J. Jones, *Canad. J. Phys.*, 1970, **48**, 632.

rotational levels F_1 (J) and F_2 (J), where the indices 1 and 2 are $^2\Pi_{\frac{1}{2}}$ and $^2\Pi_{\frac{3}{2}}$, respectively, follows rigorously the simple formula:

$$F(J) = B_0J(J+1) - D_0J^2(J+1)^2$$

Shotton and Jones, therefore, determined the rotational constants B_0 and D_0 from the mean of the Raman shifts of the $^2\Pi_{\frac{1}{2}}$ and $^2\Pi_{\frac{3}{2}}$ components by plotting $\Delta\nu_{mean}/(J+\frac{3}{2})$ *versus* $(J+\frac{3}{2})^2$. They based their analysis on the S-branch lines observed for both substates out to $J = \frac{41}{2}$; the values of the rotational constants calculated were $B_0 = 1.6961_4 \pm 0.0001_0$ cm^{-1} and $D_0 = (5.46 \pm 0.20) \times 10^{-6}$ cm^{-1}. These values may be compared with those obtained in the high-resolution i.r. study of Keck and Hause,[35] namely $B_0 = 1.6960_9$ cm^{-1} and $D_0 = 5.4_7 \times 10^{-6}$ cm^{-1}.

An interesting feature of the spectrum obtained by Shotton and Jones was the observation of R-branch rotational lines of the $^2\Pi_{\frac{3}{2}}$ substate for J-values of $\frac{5}{2}$—$\frac{23}{2}$; the intensity of these lines decreased with increasing J value. The presence of weaker R-branch lines arising from $^2\Pi_{\frac{1}{2}}$ was also detected. In addition, at a Stokes shift of *ca.* 120 cm^{-1} a weak unresolved feature was observed, located between the two rotational lines $J = \frac{33}{2}$ and $J = \frac{35}{2}$ of the $^2\Pi_{\frac{1}{2}}$ S-branch. This was attributed to the electronic transition $^2\Pi_{\frac{3}{2}} \leftarrow {}^2\Pi_{\frac{1}{2}}$.

Renschler, Hunt, McCubbin, and Polo[36] have also reported the rotation Raman spectrum of nitric oxide. The spectrum was observed using a He–Ne laser, a Czerny–Turner spectrograph, and a cooled ITT FW130 photomultiplier. The gas pressure was one atmosphere and the cell was placed inside the laser cavity, where the power was 10 W. R-Branch lines were observed for both the $^2\Pi_{\frac{1}{2}}$ and $^2\Pi_{\frac{3}{2}}$ substates from $J = \frac{3}{2}$ to $J = \frac{15}{2}$. S-Branch rotational lines for both substates were observed out to beyond $J = \frac{37}{2}$. These authors analysed their data to give the values for each substate of the ground electronic state of $B_0(^2\Pi_{\frac{3}{2}}) = 1.72$ cm^{-1} and $B_0(^2\Pi_{\frac{1}{2}}) = 1.67$ cm^{-1}. They also discuss fully the intensity distribution in the R- and S-branches and account for the fact that the R-branch lines in the $^2\Pi_{\frac{3}{2}}$ state are 6—9 times more intense than the corresponding lines for the $^2\Pi_{\frac{1}{2}}$ state. This intensity factor is more than enough to overcome the Boltzmann factor and, consequently, in the R-branch the low-displacement line of the pair corresponding to a given J-value is the weaker. They also observed a weak, unresolved feature at ~120 cm^{-1} which they attributed to the $^2\Pi_{\frac{3}{2}} \leftarrow {}^2\Pi_{\frac{1}{2}}$ transition.

Fast, Welsh, and Lepard[37] have recorded photographically the rotation Raman spectrum of the NO molecule with an effective resolution of 0.3 cm^{-1} using Hg 4358 Å radiation. With an exposure time of three hours, the two series of rotational lines in the R-branch were not observed, but the S-

[35] D. B. Keck and C. D. Hause, *J. Mol. Spectroscopy*, 1968, **26**, 163.
[36] D. L. Renschler, J. L. Hunt, T. K. McCubbin jun., and S. R. Polo, *J. Mol. Spectroscopy*, 1969, **32**, 347.
[37] H. Fast, H. L. Welsh, and D. W. Lepard, *Canad. J. Phys.*, 1969, **49**, 2879.

branch lines of the ${}^2\Pi_{\frac{1}{2}}$ and ${}^2\Pi_{\frac{3}{2}}$ substates were clearly resolved out to $J = \frac{41}{2}$. At longer exposures, of up to 12 h, the band attributed to the electronic transition ${}^2\Pi_{\frac{3}{2}} \leftarrow {}^2\Pi_{\frac{1}{2}}$ at *ca.* 120 cm^{-1} was evident. With an exposure time of 120 h, this band was resolved into *O*-, *P*-, *Q*-, *R*-, and *S*-branches. The measured frequencies of the 35 unblended lines in this electronic band were found to be in good agreement with the frequencies calculated from the molecular constants of Keck and Hause.[35] The intensity distribution of the lines in the electronic band was also in excellent agreement with the calculated intensities.

Rich and Welsh [13] have also reported in a short communication some further work which they have carried out on the *Q*-branch of this electronic transition. Spectra were recorded photographically, using both Hg 4358 Å radiation and also the 4880 Å line from an argon-ion laser. The authors comment on the advantage of laser excitation over mercury-arc excitation for the study of rotational Raman scattering. They find that, with the same sample cell and spectrograph, a 2 h exposure using the argon-ion laser produced a more intense rotation spectrum than a 12 h exposure with a battery of Toronto mercury arcs (of total discharge length 7 m). Some assignments of lines in the *Q*- and *R*-branches of the ${}^2\Pi_{\frac{3}{2}} \leftarrow {}^2\Pi_{\frac{1}{2}}$ transition were made.

Fluorine. The first observation of the pure rotation Raman spectrum of fluorine was reported by Andrychuk [38] in 1951, and represented a considerable experimental achievement at that time. Andrychuk used a 1 litre sample of fluorine in a quartz cell illuminated with Toronto mercury arcs. The spectra were recorded photographically with a Zeiss three-prism spectrograph having a reciprocal linear dispersion of 58 cm^{-1} mm^{-1} at 4047 Å. The rotation spectra were excited with Hg 4047 Å radiation and showed the expected intensity alternation consistent with a nuclear spin of $\frac{1}{2}$ for ^{19}F. However, only the strong lines of odd *J* number could be satisfactorily measured, and Andrychuk's analysis of the rotation spectrum was based on measurements of nine Stokes and anti-Stokes pairs (odd *J* values from 7 to 23 inclusive) and four Stokes lines ($J = 5, 25, 27,$ and 29). However, the plot of $\Delta\nu/(J+\frac{3}{2})$ *versus* $(J+\frac{3}{2})^2$ revealed that the accuracy of the measurements was not sufficient to determine the slope (which is equal to $-8D_0$) with any accuracy. Andrychuk therefore calculated a value of $-8D_0 = -0.000027$ cm^{-1} from the *approximate* relationship $D_0 = 4B_0^3/\omega_0^2$, using preliminary values of B_0 and the vibration frequency, ω_0. He then determined the straight line of slope $= -0.000027$ cm^{-1} which best fitted the experimental points. This line had an intercept of 3.5310 whence $B_0 = 0.8828 \pm 0.0010$ cm^{-1} and $r_0 = 1.417_7 \pm 0.001_5$ Å.

In 1969, Claassen, Selig, and Shamir [16] reported that they had recorded directly the pure rotation spectrum of F_2 at 4 atm pressure using a 50 mW helium–neon laser and a Spex 1400 double monochromator. The spectrum they obtained was of good quality, with the weak (even *J*) and

[38] D. Andrychuk, *Canad. J. Phys.*, 1951, **29**, 15.

strong (odd J) rotational lines clearly visible from $J = 1$ to $J = 27$ on both the Stokes and the anti-Stokes sides. They ran the pure rotation Raman spectrum of oxygen immediately after running the fluorine spectrum and used the known frequencies of the pure rotation lines of oxygen, which had been determined by Weber and McGinnis [32] to ± 0.05 cm^{-1}, for calibration. However, plots of $\Delta\nu/(J+\frac{3}{2})$ *versus* $(J+\frac{3}{2})^2$ gave values of D_0 which differed by a factor of more than two for two experiments. Hence Claassen *et al.* assumed a value of $D_0 = 3.47 \times 10^{-6}$ cm^{-1} (or $8D_0 = 2.8 \times 10^{-5}$ cm^{-1}), calculated from the approximate relationship given above. Using this assumed value of D_0, the measured values of the frequency shifts for the seven lines from $J = 19$ to $J = 25$ for both Stokes and anti-Stokes regions were used to calculate independent values of B_0. This procedure gave an average value of $B_0 = 0.8841 \pm 0.0006$ cm^{-1} and a corresponding value of $r_0 = 1.4168 \pm 0.0005$ Å.

Recently, the pure rotation Raman spectrum of F_2 has been reinvestigated by Edwards, Long, and Love.[17] They used an argon-ion laser for excitation, a Czerny–Turner spectrograph of focal length 165 cm and aperture $f/16$, and photographic recording. The gas pressure was one atmosphere and the laser power at the sample was $\sim$3 W at 4880 Å. The pure rotation spectra obtained were of good quality (exposure times 17—36 h) and the weak even-valued J lines and the stronger odd-valued J lines were clearly visible out to $J = 23$ on both the Stokes and the anti-Stokes sides of the exciting line. The analyses were based on measurements of 17 Stokes and anti-Stokes pairs of rotational lines. It was found that the values of D_0 obtained from the analyses did not exhibit the gross variation reported by Claassen *et al.* The best estimates of B_0 and D_0 were $B_0 = 0.8845_0 \pm 0.0001_9$ cm^{-1} and $D_0 = (5.31 \pm 0.32) \times 10^{-6}$ cm^{-1}. This value of B_0 yields a value of $r_0 = 1.4165_7 \pm 0.0001_5$ Å.

Attempts were also made by Edwards, Long, and Love to observe the Stokes vibration–rotation spectrum of F_2, but only the unresolved Q-branch of the $v = 1 \leftarrow v = 0$ transition was observed. There was no trace of features assignable to the O- and S-branches. The frequency of the unresolved Q-branch was determined as $\Delta\nu = 892.02 \pm 0.03$ cm^{-1}, which overlaps with Andrychuk's value [38] of 891.85 ± 0.4 cm^{-1}, but the uncertainty in the measurement has been reduced.

Very recently it has been reported [39] that the vibration–rotation spectrum of F_2 for the $v = 1 \leftarrow v = 0$ transition has been obtained using a scanning spectrometer with digital counting and spectrum accumulation. Rotational features in the O- and S-branches were seen, but the intensity of the strongest O- or S-branch rotational line was seen to be some 80 times weaker than that of the unresolved Q-branch. (It is reported that measurements are in progress).

Tritium. Edwards, Long, and Love [17] have also recently observed the

[39] H. W. Schrötter, personal communication, 1972.

rotation and vibration–rotation Raman spectra of the tritium molecule $^3H_2$. It is believed that this provides the first example of the determination of rotational constants from the laser-excited Raman spectrum of a species which had not been studied previously with mercury-arc excitation.

In the investigation of Edwards *et al.*, the sample volume was 1.5 ml (pressure 600 Torr) and the laser power at the sample was ~3 W at 4880 Å. The spectra were recorded photographically, with exposure times of 30—90 h. Six Stokes–anti-Stokes pairs of rotational lines were measured. As expected for a Fermi gas with a nuclear spin of $\frac{1}{2}$, the odd J-value lines were observed to be three times as intense as the even J-value lines. The values of the rotational constants that were determined were $B_0 = 20.041_9 \pm 0.003$ cm^{-1} and $D_0 = 0.0051_7 \pm 0.00006$ cm^{-1}. From this value of B_0, the bond length of the $^3H_2$ molecule in the ground vibrational state was calculated to be $r_0 = 0.74650 \pm 0.00006$ Å.

The vibration–rotation spectrum of the $^3H_2$ molecule for the $v = 1 \leftarrow v = 0$ transition was also obtained by these authors. Although no traces of O- or S-branch rotational features were discernible, the Stokes Q-branch was resolved and the frequencies of Q_1, Q_2, Q_3, Q_4, and Q_5 were measured; the Q_0 line was not observed on account of its weakness. The Q_1, Q_3, and Q_4 rotational lines were used to derive the rotational constants; Q_2 and Q_5 were too diffuse for good measurement. Using the B_0 and D_0 values obtained from the analysis of the pure rotational spectrum, the rotational constants and molecular parameters for the first vibrational state were calculated to be $B_1 = 19.463_2 \pm 0.002$ cm^{-1}, $D_1 = 0.00500 \pm 0.00006$ cm^{-1}, $\nu_0 = 2464.08_3$ cm^{-1}, and $r_1 = 0.75755 \pm 0.00003$ Å.

The values of the rotational constants and the bond length that were caluculated for the equilibrium state of the $^3H_2$ molecule are $B_e = 20.331_3$ cm^{-1}, $D_e = 0.00526$ cm^{-1}, $r_e = 0.74120$ Å, $\alpha = 0.5787$ cm^{-1}, and $\beta = -0.00017$ cm^{-1}.

Triatomic Molecules.—*Carbon Dioxide.* Barrett and Weber [12] have studied the pure rotation Raman scattering from CO_2 at low pressure (40 Torr), subjected to an electric discharge. Spectra were excited with an argon-ion laser (4880 Å, 1 W power) and were photoelectrically recorded with a Perkin-Elmer E-1 Ebert-mounted spectrometer of focal length 58 cm fitted with a digital grating-drive that was stepped by pulses from an internal clock. The detector was a sensitive photomultiplier and photon-counting system. A computer technique was used for smoothing and re-plotting spectral traces.

With this procedure, high-quality pure rotation Raman spectra of CO_2 were obtained, and rotation bands out to $J = 70$ on both Stokes and anti-Stokes sides of the exciting line could be seen. The Raman spectrum of CO_2 at low pressure, subjected to an electric discharge, contained a series of lines corresponding to the even J-values of the S-branch of the pure rotation spectrum in the $v = 0$ state and, in addition, a new series of lines lying between the *S*-branch lines and having intensity about one-third that of the *S*-branch lines. From measurements of the intensities of the rotational lines, the gas temperature of the discharge was estimated to be approximately 730 K. The authors considered two possible origins of the Raman scattering associated with odd *J*-values. It could arise from the molecule CO_2^+ or from CO_2 in the 010 vibrational state (π_u vibration, $\nu_1 = 667$ cm^{-1}). The concentration of CO_2^+ was considered to be too small to account for the observed Raman intensity. Since the population of the 010 vibrational state is appreciable at 730 K, the authors therefore interpreted their results in terms of the pure rotation transitions of CO_2 in the 010 vibrational state.

The rotational Raman scattering for CO_2 in this π_u state is governed by the selection rule $\Delta J = +1$ (*R*-branch) and $\Delta J = +2$ (*S*-branch). For the 010 (π_u) state the spacing between adjacent rotational lines will be approximately $2B$ for the *R*-branch and approximately $4B$ for the *S*-branch, whereas for the *S*-branch rotational lines in the 000 state ($^1\Sigma_g^+$) of CO_2 the spacing will be approximately $8B$. Thus, the pure rotational lines arising from CO_2 in the 010 state will in general lie between the pure rotational lines arising from CO_2 in the 000 vibrational state.

The photoelectric observations were supplemented by a photographic study of CO_2 at room temperature and at one atmosphere pressure, using a 20 mW helium–neon laser and a multi-pass cell. *S*-Branch rotational lines were measured out to $J = 50$ for the 000 vibrational state and *S*-branch lines from $J = 1$ to $J = 33$ for the 010 vibrational state. Rotational transitions in the *R*-branch were too weak to be measured. The ground- and excited-state rotational constants were obtained by a least-squares fitting of the data to the equation:

$$\Delta\nu/(J+\tfrac{3}{2}) = (4B_v - 6D_J) - 8D_J[(J+\tfrac{3}{2})^2 - l^2]$$

with $l = 0$ for the 000 state and $l = 1$ for the 010 state. The values of the rotational constants which were determined from this analysis are

$$B_{000} = 0.39027 \pm 0.000007\ \text{cm}^{-1},$$
$$D_{000} = (12.9 \pm 0.3) \times 10^{-8}\ \text{cm}^{-1},$$
$$B_{010} = 0.39065 \pm 0.00004\ \text{cm}^{-1},\ \text{and}$$
$$D_{010} = (8.2 \pm 4) \times 10^{-8}\ \text{cm}^{-1}.$$

Carbon Disulphide. Since the ground electronic state of CS_2 is $^1\Sigma_g^+$ and the spin of the ^{32}S nucleus is zero, then, as for CO_2, all rotational levels of odd *J*-value are absent in the ground-state rotational spectrum. The 010 vibrational state is that of the doubly degenerate bending vibration ($\nu_2 = 397$ cm^{-1}), and all *J*-transitions are allowed. Walker and Weber [8] have recently observed the pure rotation Raman spectrum of CS_2 vapour at a pressure of 300 Torr and at room temperature. The spectra were excited by a 20 mW helium–neon laser and recorded photographically using a Czerny-Turner spectrograph in the ninth order of diffraction with a reciprocal linear dispersion of 2.0 cm^{-1} mm^{-1}. The spectrum contained the pure rotational lines from the 000 vibrational state and also pure rotational lines from the 010 vibrational state, which, because of its low vibrational frequency, is appreciably populated at 300 K.

For the 000 vibrational state, 36 pairs of rotational lines (S-branch) were measured ($J = 0$ to $J = 70$); for the 010 vibrational state, 33 pairs of rotational lines in the S-branch were measured ($J = 1$ to $J = 69$). The rotational constants were determined from an analysis of these lines, and are given in Table 2. Table 2 also includes the rotational constants determined from the i.r. measurements of Guenther, Wiggins, and Rank.[40] It is seen that the agreement between these two sets of results is very close.

The R-branch rotational lines of the 010 vibrational state with $J = 3,5$ could be seen on the original spectrograms but were not measurable.

Table 2 *Rotational constants for CS_2 in the* 000 *and* 010 *vibrational states*

Authors	*Method*	*Ref.*	000 *state*		010 *state*	
			B/cm^{-1}	$10^8 D$/cm^{-1}	B/cm^{-1}	$10^8 D$/cm^{-1}
Walker and Weber	Raman	8	0.10912 ±0.000007	0.83 ±0.18	0.010935 ±0.00002	1.5 ±0.6
Guenther, Wiggins, and Rank	i.r.	40	0.109099 ±0.000023	1.05 ±0.36	0.109321	1.1

Tetra-atomic Molecules.—*Acetylene.* The pure rotation Raman spectra and the rotational structure associated with Raman-active vibrations of C_2H_2, C_2HD, and C_2D_2 have been photographed with a spectral resolution of 0.3 cm^{-1} by Fast and Welsh.[41] The Raman spectra were excited using a sample cell that was 1.2 m long, equipped with a four-mirror multiple-reflection system and illuminated by four low-pressure mercury arcs. Both Hg 4047 Å and Hg 4358 Å light were used for excitation of the spectra. The spectra were photographed in the 56th order of diffraction of an f/11 spectrograph with a reciprocal linear dispersion of 3.5 cm^{-1} mm^{-1}.

Extensive rotational structure was recorded for the $\nu_1(\sigma_g^+)$, $\nu_2(\sigma_g^+)$, and $\nu_4(\pi_g)$ bands of C_2H_2 and C_2D_2 and the ν_1 and ν_2 bands of C_2HD. From an analysis of the pure rotation spectrum of each gas, B_0 and D_0 values were obtained. Although these agreed, within the error limits, with the i.r. measurements, they were consistently higher than the i.r. values. This discrepancy probably arises because the pure rotation Raman spectrum is a superposition of the rotation spectra originating in the ground vibrational state and the $v = 1$ states of ν_4 and ν_5. In C_2HD, for example, the relative populations in these three vibrational states are 100 : 16 : 8 respectively.

[40] A. H. Guenther, T. A. Wiggins, and D. H. Rank, *J. Chem. Phys.*, 1958, **28**, 682.
[41] H. Fast and H. L. Welsh, *J. Mol. Spectroscopy*, 1972, **41**, 203.

The B_1 values for the excited vibrational states are all slightly larger than the ground-state values, so that the superposition of the rotation spectra from these states on the ground-state spectrum leads to a distortion of the latter and so to values of the rotational constants which are a little too large. Because of this, these authors considered the i.r. values of B_0 and D_0 to be more accurate than the Raman values. The i.r. values were therefore used in the analyses of the vibration–rotation Raman spectra.

In the vibration–rotation spectra, rotational structure in the following branches was observed and measured:

$$C_2H_2: \quad \nu_1(O, S); \quad \nu_2(O, Q, S); \quad \nu_4{}^1(O, P, R, S)$$
$$C_2D_2: \quad \nu_1(O, S); \quad \nu_2(O, S); \quad \nu_4{}^1(O, P, R, S)$$
$$C_2HD: \quad \nu_1(O, S); \quad \nu_2(O, S).$$

The Q-branch of the ν_3 band of C_2HD was photographed, but attempts to observe the $\nu_4{}^1$ and $\nu_5{}^1$ bands were unsuccessful. The degenerate $\nu_4{}^1$ vibration has P- and R-branches ($\Delta J = \pm 1$) in addition to O- and S-branches, and the resolution was sufficient to show the l-type doubling of the rotational states of the upper level of the $\nu_4{}^1$ band which had hitherto not been observed. The l-type doubling leads to effective B_1 values for the $\nu_4 = 1$ state which are slightly larger for the P- and R-branches than for the Q- and S-branches (Table 3). The spectra of C_2H_2 and C_2D_2 clearly show the expected intensity alternation of 3 : 1 and 1 : 2, respectively, resulting from nuclear-spin statistics. The C_2HD bands show no intensity alternation. The number of unblended lines available for the analysis of a band was 35—45. All five constants B_0, D_0, ν_0, B_1, and D_1 could, theoretically, be obtained from each band analysis; however, as already indicated, the i.r. value [42] of B_0 and D_0 was assumed for each isotopic species (Table 3), and ν_0, B_1, and D_1 were then determined.

The values of α, ν_0, B_1, and D_1 obtained from each vibration–rotation transition are collected together in Table 3. From the Q-branches of hot bands such as $\nu_1+\nu_4{}^1-\nu_4{}^1$ and $\nu_1+\nu_5{}^1-\nu_5{}^1$, the anharmonicity constants were evaluated. The l-type doubling constant, q, was also evaluated for ν_4 of the three species.

The equilibrium value of the rotational constant B_e was obtained from the B_0 and α values for each isotopic species. A combination of i.r. and Raman data was found to give the most consistent set of B_e values, namely $B_e(C_2H_2) = 1.1824_6\ \text{cm}^{-1}$, $B_e(C_2HD) = 0.9956_0\ \text{cm}^{-1}$, and $B_e(C_2D_2) = 0.85069\ \text{cm}^{-1}$.

These rotational constants yield values for the equilibrium bond lengths r_e, of $r_e(\text{C—H}) = r_e(\text{C—D}) = 1.0605 \pm 0.0003$ Å and $r_e(\text{C}\equiv\text{C}) = 1.2033 \pm 0.0002$ Å.

Polyatomic Molecules.—*Methane.* The most recent Raman spectroscopic

[42] W. J. Lafferty and R. J. Thibault, *J. Mol. Spectroscopy*, 1964, **14**, 79; W. J. Lafferty, E. K. Plyler, and E. D. Tidwell, *J. Chem. Phys.*, 1962, **37**, 1981; S. Ghersetti and K. N. Rao, *J. Mol. Spectroscopy*, 1968, **28**, 27.

Table 3 *Molecular constants*[a] *of isotopic species* C_2H_2, C_2HD, *and* C_2D_2

Band	C_2H_2				C_2HD				C_2D_2			
	ν_0	α	B_1	$10^6 D_1$	ν_0	α	B_1	$10^6 D_1$	ν_0	α	B_1	$10^6 D_1$
ν_1	3372.8_2	0.0068_6	1.1697_4	1.5_2	3335.5_1	0.0048_4	0.9867[illegible]	1.3_1	2705.1_2	0.0055_3	0.8423_8	1.3_8
ν_2	1974.3_4	0.0062_1	1.1703_9	1.6_0	1853.8_2	0.0043_2	0.9872[illegible]	1.0_1	1764.7_6	0.0030_3	0.8448_3	0.8_8
$\nu_4^1\ (O, S)$	612.8_2	0.0014_2	1.1751_8	1.5_0	—	—	—	—	511.5_2	-0.0003_0	0.8482_0	0.7_1
$\nu_4^1\ (P, R)$	612.7_9	-0.0039_8	1.1805_8	1.5_9	—	—	—	—	511.5_0	-0.0037_2	0.8516_2	0.9_4
Assumed constants from i.r.[42]												
B_0/cm^{-1}	1.17660 ±0.00003				0.99156 ±0.00004				0.847904 ±0.000009			
$10^6 D_0$/cm^{-1}	1.564 ±0.009				1.17 ±0.07				0.8067 ±0.0038			

[a] Values expressed are ν_0/cm^{-1}, α/cm^{-1}, B_1/cm^{-1}, and $10^6\ D_1$/cm^{-1}.

study of methane has been the investigation of the *Q*-branch of ν_1 at 2917 cm^{-1} by Clements and Stoicheff,[21] using a pressure-scanned Fabry-Pérot interferometer and a helium–neon laser (400 mW) to excite the spectrum. The sample pressure was two atmospheres.

The narrow *Q*-branch, of linewidth ~0.3 cm^{-1}, contained some structure that was due to the small difference in the rotational constants of the 0000 and 1000 levels, and the spacing is given by the usual expression $(B'-B_0)J(J+1)$ if the centrifugal constant D is neglected. The *J*-numbering of the *Q*-branch peaks was subjected to trial analyses, and a value of $\alpha = -(B'-B_0) = 0.0034$ cm^{-1} was determined.

The relative positions and intensities of the *Q*-branch lines were computed, and the calculated band profile was shown to fit the observed *Q*-branch satisfactorily.

Heavy Polyatomic Molecules.—The use of a single-moded argon-ion laser by Weber and Schlupf [18] has already been described. These authors report the successful resolution of the Raman spectra of some heavy polyatomic molecules, but so far they have only published a list of the molecules they have investigated, namely n-butane, n-pentane, *trans*-dichloroethylene, carbon suboxide, cyanuric fluoride, 1,3,5-trifluorobenzene, hexafluorobenzene, and tetramethylallene. In the case of hexafluorobenzene, the odd *J*-valued lines of the *R*-branch were clearly resolved from the adjacent *S*-branch lines: since the *R*-branch rotation-line spacing is 0.07 cm^{-1}, this represents the best resolution yet achieved in Raman spectroscopy when the 90° scattering configuration is employed.

The analyses of the spectra of these molecules have not yet been reported and the results are awaited with interest.

Stimulated Rotational Raman Scattering.—The first observation of stimulated rotational Raman scattering in gases was made by Minck, Hagenlocker, and Rado [43] in 1966. They reported the observation of stimulated rotational Raman scattering from hydrogen and deuterium, using a **Q**-switched ruby laser for excitation. More recently, Mack, Carman, Reintjes, and Bloembergen [44] have obtained stimulated rotational and vibration–rotational Raman scattering from O_2, CO_2, and N_2O. They used a mode-locked ruby laser which produced a train of pulses of 5 ps duration, circularly polarized, with a maximum peak power of 5 GW in a 0.5 cm^2 beam. The pulses were focused into a cell of length 50 cm using a lens of focal length 50 cm. The spectra were recorded photographically, using a Jarrell–Ash 75-152 spectrograph. The authors point out that a particular advantage of producing stimulated Raman scattering by transient excitation with picosecond laser pulses is the discrimination against slower competing

[43] R. W. Minck, E. E. Hagenlocker, and W. G. Rado, *Phys. Rev. Letters*, 1966, **17**, 229.
[44] M. E. Mack, R. L. Carman, J. Reintjes, and N. Bloembergen, *Appl. Phys. Letters*, 1970, **16**, 209.

processes such as stimulated Brillouin scattering. This is especially valuable with gases since, with **Q**-switched pulses, stimulated Brillouin scattering is often the dominant non-linear process.

In N_2O, which has the largest rotational cross-section of the three gases studied, both first- and second-order rotation and vibration–rotation Stokes lines were excited, but accurate identification of the rotational levels was not possible. In carbon dioxide the seven stimulated pure-rotation Raman lines from $J = 12$ to $J = 24$ were observed with a gas pressure of 20 atm in the cell. In oxygen, at a pressure of 60 atm, the $N = 5$ to $N = 13$ rotational lines were stimulated. The wavenumber shifts of the stimulated rotational lines in CO_2 and O_2 were said to agree, within experimental error, with those reported for the spontaneously excited rotation spectra.

Mack *et al.*[44] discuss further the broadening of the rotational lines, which is believed to result from an optical Stark effect.

4 Pure Rotation and Vibration–Rotation Infrared Spectroscopy

The rotation i.r. spectra of gases dissolved in liquids or mixed with other gases will be dealt with in Section 5, under a heading which encompasses molecular interaction and pressure-broadening effects.

Diatomic Molecules.—*Hydrogen.* A relatively new technique for studying the normally inactive i.r. absorption of homonuclear diatomic molecules involves the use of a strong, externally applied electric field. This field distorts the normal symmetry of the molecule, a small dipole moment is induced, and i.r. absorption may occur. The selection rules of Raman spectroscopy apply, *i.e.* $\Delta v = 0, 1, 2, \ldots$; $\Delta J = 0, \pm 2$. The intensity of a transition is proportional to the square of the applied field, so it is important to obtain fields as large as possible. In a recent paper, Boyd, Brannon, and Gailar [45] have reported the construction of a Stark cell, with which they have observed the pure rotation i.r. spectrum of hydrogen by the application of high electric fields to gas samples at ~ 30 atm pressure. The cell utilizes parallel metal plates to apply electric fields of the order of 2×10^5 V cm^{-1} across the sample gas. The pure rotational lines $S_0(1)$, $S_0(2)$, and $S_0(3)$ of H_2 were observed. The measured wavenumbers of these lines and the rotational constants calculated from them are compared in Table 4 with the Raman data obtained by Stoicheff.[46]

Carbon Monoxide. The linewidths and strengths of several lines in the pure rotation spectrum of CO have been determined by Sanderson, Scott, and White,[47] using a Michelson interferometer which had a maximum resolution of 0.05 cm^{-1}. The instrument could be operated in the asymmetric

[45] W. J. Boyd, P. J. Brannon, and N. M. Gailar, *Appl. Phys. Letters*, 1970, **16,** 135.
[46] B. P. Stoicheff, *Canad. J. Phys.*, 1957, **35,** 730.
[47] R. B. Sanderson, H. E. Scott, and J. T. White, *J. Mol. Spectroscopy*, 1971, **38,** 252.

Table 4 *Wavenumbers*/cm^{-1} *of the pure rotation lines of* H_2, *and molecular constants*[a]

	Boyd et al.[45]	*Stoicheff*[46]
$S_0(1)$	587.08	587.055
$S_0(2)$	814.52	814.406
$S_0(3)$	1034.76	1034.651
B_0	59.332	59.339
D_0	0.0450	0.0443
H_0	3.1×10^{-5}	5.2×10^{-5}

[a] All values given are of the constants/cm^{-1}.

or symmetric modes; in the former configuration phase measurements could be made, and in the latter, absorption measurements. From the phase data, line strengths could be calculated.[48] These were then combined with the measurements made in the symmetric mode to yield the linewidths.

The strength of the $(J+1) \leftarrow J$ rotational transition of a diatomic molecule is related to the square of the radial part of the dipole matrix element $\mu(J)$. For a rigid rotor $\mu(J)$ is simply the permanent dipole moment μ_0, but for a real rotor centrifugal stretching gives rise to a J-dependent correction, and the radial dipole-moment matrix element is then, to a second order, given by:

$$\mu(J) = \mu_0[1+\gamma^2(J+1)^2/\theta]$$

where $\gamma = 2\,B_e/\omega_e$ and $\theta = \mu_0/\mu_1 r_e$.

The rotational line strengths determined from the phase data were used to calculate $\mu(J)$. The results confirm that $\mu(J)$ is J-dependent and lead to the expression:

$$\mu(J) = 0.112\,[1-10^{-4}(J+1)^2] \quad \text{Debye.}$$

The value of μ_0, the permanent dipole moment, is 0.112 D, which agrees exactly with the value obtained from the microwave spectrum.

The linewidths determined by Sanderson *et al.* were in good agreement with the theoretical values calculated by Benedict and Herman.[49] The anomalous J-dependence of the rotational linewidths previously reported by Dowling[50] was not confirmed by Sanderson *et al.*

In the vibration–rotation i.r. spectrum of CO, the frequencies, intensities, and linewidths of the vibration–rotation transitions in the $v = 0 \rightarrow v = 3$ transition have been measured by Bouanich, Larvor, and Haeusler.[51] At gas pressures of 1—5 atm there was a shift in band centre with an increase in pressure from 6350.4228 cm^{-1} at 1 atm to 6350.3928 cm^{-1} at 4.75 atm. The half-widths were also calculated, using the Anderson collision

[48] R. B. Sanderson, *Appl. Optics*, 1967, **6**, 1527.
[49] W. S. Benedict and R. Herman, *J. Quant. Spectroscopy Radiative Transfer*, 1963, **3**, 265.
[50] J. M. Dowling, *J. Quant. Spectroscopy Radiative Transfer*, 1969, **9**, 1613.
[51] J. P. Bouanich, M. Larvor, and C. Haeusler, *Compt rend.*, 1969, **269**, *B*, 1238.

theory,[52] which relates the dipole and quadrupole attractive forces and hard-sphere diameters to linewidths in the absorption spectrum. The experimental results of Bouanich *et al.* fitted the theoretical calculations when a quadruple moment of 4.8×10^{-26} e.s.u. was assumed. The intensities of the vibration–rotation lines in the fundamental band of CO have also been measured interferometrically by Hochard-Demolliere.[53]

Hydrogen Fluoride. Meredith [54] has measured the strengths and self-broadened half-widths of eighteen lines in the vibration–rotation spectrum of the first overtone band of HF; the observed linewidths have been compared with those calculated using the Anderson collision theory.

Hydrogen Chloride. Buback and Franck [55] have studied the rotational structure of the fundamental i.r. absorption band of HCl at 2886 cm^{-1} at temperatures between 24 and 400 °C and at pressures between 10 and 1200 atm. A high-pressure multi-reflection cell with a sapphire window was used. Beyond the critical temperature of 51 °C, the gas density could be varied by three orders of magnitude to a maximum of 1.0 g cm^{-3}. With increasing density, the rotational structure was observed to disappear. The frequency of the intense central band decreased with increase in gas density and increased with temperature increase at constant gas density. The behaviour of the bands was interpreted in terms of 'gas-like' and 'liquid-like' states in the dense gas.

Using existing experimental i.r. and microwave results for several isotopically substituted HCl molecules, Bunker [56] has calculated the isotopically independent quantity R_e, which is the equilibrium internuclear distance in the Born–Oppenheimer approximation. Its value was found to be 1.27460 ± 0.00005 Å. The following relationships between the isotopically dependent equilibrium internuclear distances R_e^a in the adiabatic approximation were also obtained for HCl molecules:

$$R_e^a(H^{35}Cl) - R_e^a(D^{35}Cl) = 0.00005 \text{ Å}$$
$$R_e^a(D^{35}Cl) - R_e^a(T^{35}Cl) = 0.000016 \text{ Å}$$

By a similar calculation, Bunker [57] has also determined the isotopically independent R_e value for CO, namely 1.12823 ± 0.00005 Å.

Hydrogen Bromide. Tipping and Herman [58] have derived accurate analytical expressions for the electric-dipole transition moments for the 0—1, 0—2, and 1—2 vibration–rotation bands in an absorption spectrum, using wavefunctions which they had previously reported.[59] The expressions, which include anharmonic effects to the quintic potential constant, are essen-

[52] P. W. Anderson, *Phys. Rev.*, 1949, **76**, 647.
[53] L. Hochard-Demolliere, *Ann. Physique*, 1969, **4**, 89.
[54] R. E. Meredith, *U. S. Clearinghouse Fed. Sci. Tech. Inform.*, 1969, 691900, 34 pp.
[55] M. Buback and E. U. Franck, *Ber. Bunsengesellschaft phys. Chem.*, 1971, **75**, 33.
[56] P. R. Bunker, *J. Mol. Spectroscopy*, 1971, **39**, 90.
[57] P. R. Bunker, *J. Mol. Spectroscopy*, 1971, **37**, 197.
[58] R. H. Tipping and R. M. Herman, *J. Mol. Spectroscopy*, 1970, **36**, 404.
[59] R. M. Herman, R. Tipping, and S. Short, *J. Chem. Phys.*, 1970, **53**, 595.

tially identical with those independently derived by Toth, Hunt, and Plyler.[60] Tipping and Herman report specific calculations for HBr vibration–rotation intensities and compare their results with the observations of Babrov *et al.*[61] and Rao *et al.*[62] They conclude that the data of Babrov *et al.*, which lead to the dipole-moment function:

$$M(x) = (0.824 \pm 0.02) + (0.6438 \pm 0.006)x + (0.167 \pm 0.06)x^2$$

where x is the relative internuclear separation $(R - R_e)/R_e$, are reasonably consistent with the present theory. The more recent higher resolution measurements of Rao *et al.*, which lead to the dipole-moment function:

$$M(x) = 0.826 + 0.482x + 0.037x^2$$

do not agree so well with the present theory. The authors argued that the ultimate resolution of present discrepancies cannot lie in further theoretical refinements.

Hydrogen Iodide. The possibility of utilizing transitions in the hydrogen and deuterium halides for i.r. lasers is currently attracting attention,[63] and a detailed knowledge of the vibration–rotation energy levels of these molecules is thus of particular interest.[64] Hurlock, Alexander, Rao, and Dreska [65] have recently studied the vibration–rotation spectrum of HI and DI, using a high-resolution vacuum spectrograph and a glass absorption cell 1 m long, fitted with sodium chloride or sapphire windows. The 1—0, 2—0, 3—0, and 4—0 bands were observed for HI, and the 1—0 and 2—0 bands for DI. The results are presented in Table 5. Some data on the

Table 5 *Rotational constants*/cm^{-1} *and band origins of* HI *and* DI[65, 66]

	HI	DI
$\nu_0(1—0)$	2229.581 ± 0.006	1599.764 ± 0.004
$\nu_0(2—0)$	4379.225 ± 0.006	3159.354 ± 0.004
$\nu_0(3—0)$	6448.034	4678.529 ± 0.006
$\nu_0(4—0)$	8434.721	—
B_e	6.5111 ± 0.004	3.2839 ± 0.0002
D_e	$(2.07 \pm 0.04) \times 10^{-4}$	$(5.31 \pm 0.08) \times 10^{-5}$
α_e	0.16886 ± 0.00014	0.0612 ± 0.0002
γ_e	$(-9.5 \pm 0.2) \times 10^{-4}$	$(-1.76 \pm 0.14) \times 10^{-4}$

60 R. A. Toth, R. H. Hunt, and E. K. Plyler, *J. Mol. Spectroscopy*, 1969, **32**, 74, 85; 1970, **35**, 110.
61 H. J. Babrov, A. L. Shabott, and B. S. Rao, *J. Chem. Phys.*, 1965, **42**, 4124.
62 B. P. Gustafson and B. S. Rao, *Canad. J. Phys.*, 1970, **48**, 330; B. S. Rao and L. H. Lindquist, *ibid.*, 1970, **46**, 2739.
63 T. A. Cool and R. R. Stephens, *J. Chem. Phys.*, 1970, **52**, 3304.
64 A. W. Mantz, E. R. Nichols, B. D. Alpert, and K. N. Rao, *J. Mol. Spectroscopy*, 1970, **35**, 325.
65 S. C. Hurlock, R. M. Alexander, K. N. Rao, and N. Dreska, *J. Mol. Spectroscopy*, 1971, **37**, 373.
66 L. A. Pugh and K. N. Rao, *J. Mol. Spectroscopy*, 1971, **37**, 375.

3—0 band of DI observed by Pugh and Rao [66] are also included in Table 5. All the results are within the uncertainties of previous measurements.[67]

Triatomic Molecules.—*Water*. The far-i.r. spectrum of the water molecule has been much investigated, largely because of its use to calibrate i.r. spectrometers. Möller and Rothschild [68] have calculated and tabulated the wavenumbers of some 278 rotational lines in the 13—300 cm^{-1} region of the i.r. spectrum, using the rotational constants of Hall and Dowling [69] derived from an analysis of the vibration–rotation spectrum in the near i.r. Using an interferometer with an effective resolution of 0.05 cm^{-1}, Hall and Dowling [70] have also observed the pure rotation spectrum of water vapour in the 92—250 cm^{-1} region with a resolution of 0.14 cm^{-1}. The wavenumbers of the 73 lines in this region assigned to rotational transitions are compared with those calculated from the near-i.r. spectral measurements, and good agreement is observed.

In a later paper, Dowling [71] has made a detailed comparison of the near- and far-i.r. spectral data for the H_2O molecule and has shown that the rotational energy levels obtained from high-resolution near-i.r. data [69, 72] and from high-resolution far-i.r. data [70] are consistent with each other. The internal consistencies of the near-i.r. measurements and of the far-i.r. measurements are themselves very high. The internal consistency of the far-i.r. (pure rotation) analysis was somewhat better than that of the near-i.r. measurements; this was mainly because five microwave transitions were used no fewer than nine times in the calculation of energy-level differences in the far-i.r. An estimate of the consistency between the near- and far-i.r. sets of data was given as $\pm 8 \times 10^{-3}$ cm^{-1}.

Fraley, Rao, and Jones [73] have analysed the rotational structure observed in the ν_1 and ν_3 bands of $H_2{}^{18}O$ occurring in the near-i.r.; their accuracy of wavenumber measurement was estimated at ± 0.005 cm^{-1}. The derived band origins of ν_1 and ν_3 of $H_2{}^{18}O$ are 3649.68_0 cm^{-1} and 3741.58_1 cm^{-1}, respectively. Several vibration–rotation lines were identified as due to $H_2{}^{17}O$ impurity, and the band origins derived for this molecule were $\nu_1 = 3653.15$ cm^{-1} and $\nu_3 = 3748.36$ cm^{-1}.

More recently, Williamson, Rao, and Jones [74] have measured the ν_2 band of $H_2{}^{18}O$, using a Littrow-type vacuum i.r. spectrograph and a germanium–mercury detector operating at liquid-helium temperatures. A spectral resolution of about 0.04 cm^{-1} was achieved. Some 840 vibration–rotation lines occurring between 1330 and 1970 cm^{-1} were measured. In-

[67] C. Haeusler, C. Meyer, and P. Barchewitz, *J. Phys. (Paris)*, 1964, **24**, 289.
[68] K. D. Möller and W. G. Rothschild, 'Far Infrared Spectroscopy', Wiley–Interscience, New York, 1970.
[69] R. T. Hall and J. M. Dowling, *J. Chem. Phys.*, 1967, **47**, 2454.
[70] R. T. Hall and J. M. Dowling, *J. Chem. Phys.*, 1970, **52**, 1161.
[71] J. M. Dowling, *J. Mol. Spectroscopy*, 1971, **37**, 272.
[72] P. E. Fraley and K. N. Rao, *J. Mol. Spectroscopy*, 1969, **29**, 348.
[73] P. E. Fraley, K. N. Rao, and L. H. Jones, *J. Mol. Spectroscopy*, 1969, **29**, 321.
[74] J. G. Williamson, K. N. Rao, and L. H. Jones, *J. Mol. Spectroscopy*, 1971, **40**, 372.

formation from the work of Fraley *et al.*[73] on the ν_1 and ν_3 bands of the same molecule was of help in the assignment of the transitions in the ν_2 band. Values for the energy levels of the first excited state (010) were evaluated.

Varanasi, Tejwani, and Prasad [75] have investigated the linewidths and intensities of 63 vibration–rotation lines in the ν_3 fundamental of H_2O in water–carbon dioxide gas mixtures. A quadrupole moment was computed for CO_2.

Hydrogen Sulphide. The pure rotation i.r. spectrum of D_2S has been reported by Osaka and Takahashi.[76] The spectrum from 6.6 to 33 cm^{-1} was obtained using a cell of length 20 cm and a gas pressure of 550 Torr.

Hydrogen Cyanide. To help in the assignment of laser transitions in HCN, Maki, Olson, and Sams [77] have recently studied the rotational structure associated with a number of vibrational transitions in this molecule. The transitions 11^10—00^00, 04^00—00^00, 12^00—01^10, 05^10—01^10, 13^30—02^20, 13^10—02^00, 12^20—00^00, and 20^00—00^00 were studied for $H^{12}C^{14}N$ and the 11^10—00^00 transitions for both $H^{13}C^{14}N$ and $H^{12}C^{15}N$. Some intensity anomalies were noted. These authors combined these conventional absorption measurements with precise frequency measurements of five HCN laser transitions to characterize the 11^10 and 04^00 levels. Less-precise measurements of four other HCN laser transitions were combined with the absorption measurements to obtain vibration–rotation constants for the 12^00, 12^20, and 05^10 levels.

Infrared absorption due to Δ—Σ transitions (12^00—00^00) was observed for the first time as a series of *P*-branch lines, and *l*-type resonance was invoked to explain their intensities. The energy levels are given by the expression:

$$F_v(J) = \nu_0 + B_vJ(J+1) - D_v[J(J+1) - l^2]^2$$

Inclusion of an H_v term was not necessary since the measurements did not extend to sufficiently high *J* values. The band analyses were carried out by means of a least-squares computer program which fitted the observed transitions to the equations for both the upper and lower energy levels.

The ground-state rotational constants were calculated from a combination of 69 Δ_2F'' values obtained from the vibration–rotation analyses of Maki *et al.*, the i.r. absorption work of Rank, Eastman, Rao, and Wiggins,[78] and from the recent microwave measurements of DeLucia and Gordy [79] of the three *J*-transitions $1 \leftarrow 0$, $2 \leftarrow 1$, and $3 \leftarrow 2$. The values of the

[75] P. Varanasi, G. D. T. Tejwani, and C. R. Prasad, *J. Quant. Spectroscopy Radiative Transfer*, 1971, **11**, 231.
[76] T. Osaka and S. Takahashi, *Tohoku Daigaku Kagaku Keisoku Kenkyusho Hokoku*, 1970, **18**, 49 (*Chem. Abs.*, 1970, **73**, 93 023e).
[77] A. G. Maki, W. B. Olson, and R. L. Sams, *J. Mol. Spectroscopy*, 1970, **36**, 433.
[78] D. H. Rank, D. P. Eastman, B. S. Rao, and T. A. Wiggins, *J. Opt. Soc. Amer.*, 1961, **51**, 929.
[79] F. DeLucia and W. Gordy, *Phys. Rev.*, 1969, **187**, 58.

Table 6 *Lower-state rotational constants used in the analysis of the vibration–rotation spectra of hydrogen cyanide* [77]

Species	*Vibrational level*	B_v/cm^{-1}	$10^6 D_v$/cm^{-1}
$H^{12}C^{14}N$	00^00	1.47822162 ±0.00000001	2.9093 ±0.0012
	01^10	1.481768	2.981
	02^00	1.485826	3.04
	02^20	1.485014	3.04
$H^{13}C^{14}N$	00^00	1.439995	2.786 ±0.033
$H^{12}C^{15}N$	00^00	1.435248	2.73 ±0.11

ground-state rotational constants for the three isotopic species, and, in addition, the values of the constants calculated in a similar manner for the 01^10, 02^00, and 02^20 levels of $H^{12}C^{14}N$ are shown in Table 6.

In Table 7 are given the values of the band constants ν_0, ΔB, ΔD, and q_v (the *l*-type doubling constants) for the vibration–rotation transitions, analysed using the constants of Table 6.

Hougen, Bunker, and Johns [80] have derived an expression for the vibration–rotation Hamiltonian of a triatomic molecule (linear or bent) which allows for the large amplitude of the bending vibration. They have calculated the energy levels associated with the bending vibrations of HCN and DCN in the first overtone ($2\nu_2$) and for the third overtone ($4\nu_2$) of HCN only. Their results agree quite closely with the experimentally observed wavenumbers:

HCN: $2\nu_2$ (observed) (ref. 77) 1412.0 cm^{-1}; (calculated) 1412.0 cm^{-1}
$4\nu_2$ (observed) (ref. 77) 2803.0 cm^{-1}; (calculated) 2816.3 cm^{-1}
DCN: $2\nu_2$ (observed) (ref. 81) 1130.0 cm^{-1}; (calculated) 1123.7 cm^{-1}

Cyanogen Chloride. Murchison and Overend [82] have reported the vibration–rotation spectrum of the Fermi doublet, ν_1 and $2\nu_2$, of ClCN with a resolution of better than 0.03 cm^{-1}. The spectra were recorded using a 2.5 m vacuum grating spectrograph and a germanium detector cooled to liquid-helium temperatures. The sample was prepared from natural chlorine and contained $^{35}Cl^{12}C^{14}N$ (75%) and $^{37}Cl^{12}C^{14}N$ (25%). The ν_1 band was recorded using a sample pressure of 6 Torr and a sample cell of effective path-length 6 m; for the $2\nu_2$ band a path-length of 3 m and a pressure of 3 Torr was used. The Fermi doublet was found at *ca.* 783 cm^{-1} and 714 cm^{-1} and was accompanied by hot bands from the ν_1 and $2\nu_2$ transitions.

The rotational constants were obtained from an analysis of the spectra, with the lower-state rotational constants constrained at the microwave

[80] J. T. Hougen, P. R. Bunker, and J. W. C. Johns, *J. Mol. Spectroscopy*, 1970, **34**, 136.
[81] A. G. Maki, E. K. Plyler, and R. Thibault, *J. Opt. Soc. Amer.*, 1964, **54**, 869.
[82] C. B. Murchison and J. Overend, *Spectrochim. Acta*, 1970, **26A**, 599.

Table 7 *Band constants for* HCN [77]

Species	*Transition*	ν_0/cm^{-1}	$\Delta B/cm^{-1}$	$\Delta D/cm^{-1}$	q_v/cm^{-1}
$H^{12}C^{14}N$	20^00—00^00	4173.10(07)	−0.02019(16)	0.0	—
	11^10—00^00	2805.5843(10)	−0.006650(6)	0.070(45)	q_{110} = 0.007483(6)
	12^20—01^10	2804.894(2)	−0.007100(9)	0.065(9)	—
	13^30—02^20	2804.308(7)	−0.00757(10)	0.05(30)	—
	04^00—00^00	2802.9616(12)	0.15636(1)	0.296(24)	—
	12^00—01^10	2790.143(15)	−0.006292(7)	0.045(8)	—
	13^10—02^00	2789.797(3)	−0.006712(36)	0.00(7)	—
	05^10—01^10	2783.21(10)	−0.01590(15)	0.30	q_{050} = 0.007921(12)
$H^{12}C^{15}N$	11^10—00^00	2772.223(3)	−0.00638(3)	0.06(6)	q_{110} = 0.007072(30)
$H^{13}C^{14}N$	11^10—00^00	2765.306(15)	−0.006696(15)	0.062(30)	q_{110} = 0.007196(30)

Table 8 *Band origins and rotational constants for* ^{35}ClCN *and* ^{37}ClCN

Species	*Transition*[a]	ν_0/cm^{-1}	B'/cm^{-1}	$10^8 D'$/cm^{-1}
^{35}Cl^{12}C^{14}N	10^00—00^00	714.018 ±0.004	0.199169(5)	6.99 ±0.14
	02^00—00^00	782.825 ±0.004	0.199440(6)	4.43 ±0.19
	11^10—01^10	700.694 ±0.009	0.19976(1)	—
	03^10c—01^10c	795.911 ±0.005	0.200076(10)	4.63 ±0.6
	03^10d—01^10d	795.919 ±0.005	0.199669(10)	4.41 ±0.5
^{37}Cl^{12}C^{14}N	10^00—00^00	708.322 ±0.005	0.194960(8)	6.68 ±0.2
	02^00—00^00	779.241 ±0.007	0.19541(1)	4.72 ±0.3
	03^10c—01^10c	791.656 ±0.005	0.19604(1)	4.33 ±0.2
	03^10d—01^10d	791.652 ±0.005	0.19578(2)	4.95 ±0.8

[a] c and d refer to the levels of the 01^10 state, which is affected by *l*-type doubling.

values;[83] namely, ^{35}ClCN, $B_{000} = 0.199165$ cm^{-1} and ^{37}ClCN, $B_{000} = 0.195043$ cm^{-1}. The rotational constants calculated for the upper vibrational states of the molecules ^{35}ClCN and ^{37}ClCN are given in Table 8.

The *l*-type doubling constants q' and q'' were also determined from splittings in the *P*- and *R*-branches of the $03^10 \leftarrow 01^10$ transition in ^{35}ClCN; q_{010} was found to be $(2.51 \pm 0.1) \times 10^{-4}$ cm^{-1}, in excellent agreement with the microwave value [83] of 2.4899×10^{-4} cm^{-1}, and q_{03^10} was found to be $(3.979 \pm 0.005) \times 10^{-4}$ cm^{-1}. There were insufficient data for the calculation of the analagous *l*-type doubling constants for the ^{37}ClCN isotopic species. In addition, the Fermi-resonance operators were estimated for the 10^00, 02^00 and 11^10,03^10 diads of ^{35}ClCN and the 10^00,02^00 diad of ^{37}ClCN.

Hypochlorous Acid. Ashby [84] has re-examined the i.r. spectrum of HOCl in the gas phase. The spectra were recorded with a Jarrell–Ash Czerny–Turner spectrometer of focal length 2 m, using glass cells fitted with sodium chloride windows. Spectra were recorded at a spectral slit width of 0.24 cm^{-1}. The ν_2 band, with a wavenumber of *ca.* 1240 cm^{-1}, was parallel in character, and lines up to $J \approx 40$ in both the *P*- and *R*-branches were measured. Because of the existence of two isotopic species in the sample, *viz.* HO^{35}Cl and HO^{37}Cl, and because of considerable overlapping of sub-bands, rotational constants could not be evaluated with any high degree of accuracy. Analysis of the *Q*-branch lines Q_1, Q_2, and Q_3 of HO^{35}Cl gave ν_2(HO^{35}Cl) $= 1239.9$ cm^{-1} and $\Delta A = 1.56$ cm^{-1}. Q_1 of HO^{37}Cl was found at 1239.2 cm^{-1}, and assuming ΔA(HO^{37}Cl) $= \Delta A$(HO^{35}Cl), this gives a value for ν_2(HO^{37}Cl) $= 1237.7$ cm^{-1}.

The parallel ν_3 band of HOCl was obscured by absorption due to Cl_2O; this factor, together with the presence of two isotopic molecules and overlapping of sub-bands, prevented acceptable rotational data being obtained. The band centre was observed, however, at 725 cm^{-1}, and not at 739 cm^{-1} as previously reported.[85]

Ozone. Because of the conflicting values of the rotational constants of

[83] W. J. Lafferty, D. R. Lide, and R. A. Toth, *J. Chem. Phys.*, 1965, **43**, 2063.
[84] R. A. Ashby, *J. Mol. Spectroscopy*, 1971, **40**, 639.
[85] K. M. Hedberg and R. M. Badger, *J. Chem. Phys.*, 1951, **19**, 508.

previous i.r. work and those obtained from a recent microwave investigation,[86] Tanaka and Morino [87] have re-examined the rotational structure in the ν_2 fundamental of the O_3 molecule. The band envelope, *i.e.* the profile of each individual vibration–rotation line, was computed using the rotational constants determined in the microwave investigation for the ground and excited states. Contributions from all the transitions with $J \leqslant 40$ were taken into account and the computed spectrum was compared with the i.r. spectrum. Excellent agreement between the calculated and observed spectra is observed. The band origin was re-assigned to 700.93 cm^{-1}, which is ~0.5 cm^{-1} lower than the previous value of 701.42 cm^{-1}.

Sulphur Dioxide. Barbe and Jouve [88] have obtained for the first time the vibration–rotation i.r. spectrum of $^{32}S^{18}O_2$. Nine fundamental, harmonic, and combination bands were recorded, using a Beckman IR12 double-beam spectrometer and a multi-pass absorption cell of path length 4 m.

Four perpendicular bands, ν_1, ν_2, $2\nu_1$, and $2\nu_3$, were sufficiently well resolved to permit the assignment of Q-branch lines. Since the two principal moments of inertia I_B and I_C are nearly equal, the molecule can be considered as a symmetric top; the Q-branches were analysed by the method of combination differences to yield the values of the rotational constants shown in Table 9.

Table 9 *Band constants for* $S^{18}O_2$

Band	*Band centre*/cm^{-1}	$[A''-(B''+C'')/2]$	$[A'-(B'+C')/2]$	$10^5 D_K''$/cm^{-1}	$10^5 D'_K$/cm^{-1}
ν_2*	496.7	1.632 ± 0.004	1.665 ± 0.004	7.0 ± 0.5	7.3 ± 0.5
ν_1	1100.65	1.636 ± 0.006	1.639 ± 0.006	7.8 ± 0.7	7.8 ± 0.7
ν_3	1317.9	—	—	—	—
$\nu_2+\nu_3$	1811.5	—	—	—	—
$2\nu_1$	2195.0	1.630 ± 0.010	1.585 ± 0.010	—	—
$\nu_1+\nu_3$	2406.0	—	—	—	—
$2\nu_3$	2426.5	1.637 ± 0.010	1.639 ± 0.010	—	12.0 ± 1.2
* *Cf.* microwave[89]		1.629	1.665	7.6	8.3

The band centres of ν_3 and certain combination bands were reported (Table 9), and the zeroth-order band wavenumbers were calculated, namely $\omega_1 = 1113.98$ cm^{-1}, $\omega_2 = 505.30$ cm^{-1}, and $\omega_3 = 1335.18$ cm^{-1}. The following Coriolis coupling constants were obtained, $\zeta_{13} = 0.30_1$ and $\zeta_{23} = 0.95_3$. Inertia defects, Δ, were calculated using the Coriolis coupling constants and the zero-order frequencies. A comparison of the Δ values obtained from the i.r. and microwave[89] spectra is given for $S^{18}O_2$:

[86] T. Tanaka and Y. Morino, *J. Mol. Spectroscopy*, 1970, **33**, 538.
[87] T. Tanaka and Y. Morino, *J. Mol. Spectroscopy*, 1970, **33**, 552.
[88] A. Barbe and P. Jouve, *J. Mol. Spectroscopy*, 1971, **38**, 273.
[89] R. van Riet, *Ann. Soc. sci. Bruxelles*, 1963, **77**, 164; 1964, **78**, 237; 1968, **82**, 405.

State	*Microwave* Δ (ref. 87)	*I.r.* Δ (ref. 86)
000	0.231	0.229
100	—	0.289
010	0.6909	0.699
001	—	0.169

Carbon Dioxide. The i.r. chemiluminescence spectrum of CO_2 excited by activated N_2 molecules has enabled transitions involving the higher vibrational levels of the molecule to be observed, and so has resulted in a further refinement of the molecular constants. Val[90] has reportd the i.r. chemiluminescence spectrum of CO_2, identifying thirteen vibrational transitions and deriving values for the rotational constants B and D for each of the vibrational levels involved. The molecular constants and the populations of the vibrational levels were calculated and used to build up 'computed spectra' which were compared with the experimentally observed spectra. There is quite good agreement between the experimental and the computed spectra; the rotational constants obtained from the analysis of the spectra are shown in Table 10.

Table 10 *Band constants*[90] *of* CO_2

Vibrational level	*Vibrational energy*/cm^{-1}	B (*Literature value*)/cm^{-1}	B (*from chemiluminescence*)/cm^{-1}	$10^8 D$/cm^{-1}
00^00	0	0.390210	—	13.40
00^01	2349.142	0.38714044	—	13.25
00^02	4673.311	0.3840352	0.3840750(5)	13.3
00^03	6972.555	0.380985	0.3810184(13)	13.3
00^04	9246.920	0.377860	0.3779326(24)	13.3
01^12c	5315.700	0.384546	0.3845561(14)	13.5
01^12d	—	—	0.3851232(13)	13.5
01^13c	7602.460	0.381480	0.3815622(36)	13.05
01^13d	—	0.382040	0.3820828(34)	13.55
$(10^01, 02^01)_{II}$	3612.844	0.387490	0.3874976(30)	15.84
$(10^01, 02^01)_{I}$	3714.782	0.387060	0.3870409(38)	11.4
02^20c	1335.129	0.391657	0.3916343(94)	13.56
02^20d	—	0.391657	0.3916405(98)	13.56
02^21c	3659.271	0.388634	0.3886101(70)	14.2
02^21d	—	0.388634	0.3886207(80)	14.2

Toth, Hunt, and Plyler[91] have measured the line intensities of the five Σ—Σ bands of CO_2 in the 1.43—1.65 μm region at low sample pressures and at resolutions in the range 0.35—0.06 cm^{-1}. All the transitions studied were relatively weak; four of the bands studied belonged to the Fermi tetrad $(3\nu_1+\nu_3)$, $(2\nu_1+2\nu_2+\nu_3)$, $(\nu_1+4\nu_2+\nu_3)$, and $(6\nu_2+\nu_3)$, and the fifth band was the $3\nu_3$ overtone of the antisymmetric C—O stretching fundamental. Overlapping rotational lines and lines blended with the hot bands of CO_2

[90] J. L. Val, *J. Mol. Spectroscopy*, 1971, **40**, 367.
[91] R. A. Toth, R. H. Hunt, and E. K. Plyler, *J. Mol. Spectroscopy*, 1971, **38**, 107.

which occurred in this region of the spectrum were excluded from the analysis. Line intensities were measured with a planimeter and were corrected for wing and base losses.

The intensity of the $3\nu_3$ band at 296 K is found to be 3.62×10^{-2} cm^{-2} atm^{-1}. The intensities of the four bands in the Fermi tetrad in order of increasing frequency are 1.15×10^{-3}, 1.14×10^{-2}, 1.12×10^{-2}, and 1.27×10^{-3} cm^{-2} atm^{-1} at 296 K. The analogous total band intensity values for a temperature of 273 K were calculated by changing the total number of molecules per unit volume in accord with the Ideal Gas law and by using the vibrational partition function at 273 K to find the number of molecules in the ground vibrational state. Boese *et al.*,[92] and Burch *et al.*[93] in similar investigations in 1968, converted their band intensity results at room temperature into ones for 273 K by correcting for only the number of molecules per unit volume. The results of Toth *et al.* agree very closely with those of Birch *et al.* and to within 2—15% of those of Boese *et al.*, when the results of the latter two groups of workers are adjusted to include the effect of reducing the vibrational partition function to the appropriate one at 273 K.

Carbonyl Sulphide. Eleven i.r. absorption bands of $O^{13}CS$ have been measured, seven for the first time, by Fayt and Vandenhaute.[94] The spectra were obtained using a 50% enriched ^{13}C sample, and the pressures were from 5 to 50 Torr in a cell of path-length 11—50 m. The transitions observed were from the 00^00 state to 40^00, 24^00, 20^01, 12^01, 04^01, 10^02, 14^01, and 02^02. Transitions from the 01^10 state to 21^11, 13^11, and 11^12 were also observed.

Because the number of measurements on $O^{13}CS$ was insufficient for a full independent vibration–rotation analysis to be made, the band constants of least significance were constrained to their values in OCS, which had previously been analysed by Fayt.[95] The B_{000} value of 0.2022024 cm^{-1} was taken from the microwave measurements of Strandberg *et al.*;[96] the D_0 value of 4.3×10^{-8} cm^{-1} was taken from the i.r. work of Triaille,[97] and appropriate values for B and D for the $00^01 \leftarrow 00^00$, $10^01 \leftarrow 00^00$, and $02^01 \leftarrow 00^00$ transitions were taken from the work of Maki *et al.*[98]

Fayt and Vandenhaute determined the ΔB values for the transitions under investigation and, from the values of $10^5\alpha_1 = 65.865$ cm^{-1}, $10^5\alpha_2 = -34.495$ cm^{-1}, and $10^5\alpha_3 = 117.967$ cm^{-1}, they calculated a B_e value of 0.20272 ± 0.00003 cm^{-1}. The zeroth-order wavenumbers are $\omega_1 = 870.805$ cm^{-1}, $\omega_2 = 509.134$ cm^{-1}, and $\omega_3 = 2038.797$ cm^{-1}.

[92] R. W. Boese, J. H. Miller, E. C. Y. Inn, and L. P. Giver, *J. Quant. Spectroscopy Radiative Transfer*, 1968, **8**, 1001.
[93] D. E. Burch, D. A. Gryvnak, R. R. Patty, and C. E. Bartky, *J. Opt. Soc. Amer.*, 1968, **59**, 267.
[94] A. Fayt and R. Vandenhaute, *Ann. Soc. sci. Bruxelles*, 1971, **85**, 105.
[95] A. Fayt, *Ann. Soc. sci. Bruxelles*, 1970, **84**, 69.
[96] M. W. P. Strandberg, T. Wentink, and R. L. Kyhl, *Phys. Rev.*, 1949, **75**, 270.
[97] E. A. Triaille, *Ann. Soc. sci. Bruxelles*, 1965, **79**, 193.
[98] A. G. Maki, E. K. Plyler, and E. D. Tidwell, *J. Res. Nat. Bur. Stand., Sect. A*, 1962, **66**, 163.

An accidental Coriolis resonance was found between the 12^01 and 41^10 levels, the effective coupling term being of the order of 0.1 cm^{-1}.

Carbon Disulphide. Smith and Overend [99] have reported the i.r. vibration–rotation analysis at a resolution of 0.03 cm^{-1} of the ν_3 fundamental of the CS_2 molecule at 1535 cm^{-1}. The principal transition in the ν_3 band is $00^01 \leftarrow 00^00$ of $^{12}C^{32}S_2$, but a number of weaker transitions due to hot bands and isotopic molecules were also found; the $00^01 \leftarrow 00^00$ and $01^11 \leftarrow 01^10$ transitions of both $^{12}C^{32}S_2$ and $^{13}C^{32}S_2$ were assigned, and the rotational constants in each vibrational state were determined. The ν_3 fundamental would be obscured completely by the 6 μm absorption band of water if the spectrometer were not evacuated. The absorption lines of atmospheric water vapour were eliminated from the spectrum by enclosing the short path between the exit window and the detector within a polythene bag, inside which was maintained a dry nitrogen atmosphere. A liquid-helium-cooled copper-doped germanium detector was used with a cooled dielectric filter of band-pass 4—7.4 μm. Sample pressures in a cell of length 13 cm were 0.2 Torr for ν_3 of $^{12}C^{32}S_2$ and 20 Torr for ν_3 of $^{13}C^{32}S_2$.

The ν_3 band is dominated by the intense vibration–rotation spectrum of the $00^01 \leftarrow 00^00$ transition of $^{12}C^{32}S_2$, but near the band centre the spectrum is more complex because of the appearance of rotational features belonging to the hot band, ($01^11 \leftarrow 01^10$) transition. At 1485 cm^{-1} another band system, similar in appearance to, but much weaker than, the 1535 cm^{-1} system, appears that is due to $^{13}CS_2$ (ν_3 fundamental). The relative intensities are 100 : 1, in agreement with the natural abundance of carbon isotopes. The 1535 cm^{-1} band contains as components the ν_3 bands of $^{12}C^{32}S_2$, $^{12}C^{32}S^{34}S$, and $^{12}C^{34}S_2$ in the relative intensities 450 : 40 : 1, this being the ratio of the natural abundances of the sulphur isotopes.

Since the ^{32}S nucleus has a nuclear spin of zero, the antisymmetric vibration–rotation levels of the symmetric molecules are unpopulated, *i.e.* the levels of odd J in the ground state. In the degenerate 01^10 states, which are the lower states of the hot bands, all J levels are populated. Band origins were located as gaps of $6B$ in the rotational lines. About 40 values of J were observed in both the P- and the R-branches of the ν_3 fundamental of $^{12}C^{32}S_2$ and $^{13}C^{32}S_2$ for the $00^01 \leftarrow 00^00$ and $01^11 \leftarrow 01^10$ transitions. The rotational constants obtained are included in Table 11. The table has been restricted to include the rotational constants of the fundamental vibrational levels and the 01^11 level only. There is good agreement between the results of Smith and Overend and the results of some previous work by Agar *et al.*[100]

Finally, Smith and Overend used the following expressions to derive values for q_v, the l-type doubling constants:

[99] D. F. Smith and J. Overend, *Spectrochim. Acta*, 1970, **26A,** 2269.
[100] D. Agar, E. K. Plyler, and E. D. Tidwell, *J. Res. Nat. Bur. Stand., Sect. A*, 1962, **66,** 259.

Table 11 *Rotational constants of* CS_2

Isotopic species	*Ref.*	B_{00^00}/cm^{-1}	B_{00^01}/cm^{-1}	B_{01^10}/cm^{-1}	B_{10^00}/cm^{-1}	B_{01^11}/cm^{-1}	$10^8 D_{00^00}$/cm^{-1}	$10^8 D_{01^10}$/cm^{-1}
$^{12}C^{32}S_2$	99	0.109110 ±0.00002	0.108399 ±0.00002	0.10931 ±0.00003	—	0.10860 ±0.00003	1.35 ±0.4[a]	0.7 ±0.5[b]
$^{12}C^{32}S_2$	100	0.10912	—	—	0.10940	—	1.38	2
$^{13}C^{32}S_2$	99	0.109118 ±0.00002	0.108432 ±0.00002	0.10935 ±0.00005	—	0.10867 ±0.00005	1.49 ±0.2[a]	1.7 ±2.0[b]
$^{12}C^{32}S_2$	40	0.109099 ±0.000023	—	0.109321	—	—	1.05 ±0.36	1.1
$^{12}C^{32}S_2$	101	0.109123 ±0.000010	—	—	—	—	1.23 ±0.12	—
$^{12}C^{32}S_2$ (Raman)	103	0.10910 ±0.00005	—	—	0.108945	—	1.0	—
$^{12}C^{32}S_2$ (Raman)	8	0.10912 ±0.000007	—	0.10935 ±0.00002	—	—	0.83 ±0.18	1.5 ±0.6
$^{12}C^{32}S_2$	102	0.109114 ±0.00005	0.108399	—	0.108958 ±0.0000009	—	1.4	—
$^{13}C^{32}S_2$	102	—	0.108432	—	0.108972 ±0.0000014	—	1.15[a]	—

[a] D_{00^01} value; [b] D_{01^11} value.

$$\Delta E/hc = q_v J(J+1)$$
$$q_v = 2\,B_e^{\,2}/\omega_2 + 8\,B_e^{\,2}\omega_2/(\omega_3^{\,2} - \omega_2^{\,2})$$

The values of the *l*-type doubling constants were calculated to be:

$$q_{010} = 7.5 \times 10^{-5}\ \text{cm}^{-1}\ \text{for}\ ^{12}C^{32}S_2$$
$$(= 5.6 \times 10^{-5}\ \text{cm}^{-1}\ \text{for}\ ^{13}C^{32}S_2)$$

and

$$q_{011} = 7.3 \times 10^{-5}\ \text{cm}^{-1}\ \text{for}\ ^{12}C^{32}S_2$$
$$(= 6.0 \times 10^{-5}\ \text{cm}^{-1}\ \text{for}\ ^{13}C^{32}S_2)$$

cf.

$$q_{010} = 6.7 \times 10^{-5}\ \text{cm}^{-1}\ \text{(Guenther et al.}^{40}\text{)}.$$

Blanquet and Courtoy[101] have also published the results of their work on the vibration–rotation spectrum of $^{12}C^{32}S_2$ in the 3000—6000 cm^{-1} region of the i.r. Four new combination bands have been studied in the gas phase, namely $(4\nu_2+\nu_3)$ [ν_0 = 3129.983 cm^{-1}], $(2\nu_1+2\nu_2+\nu_3)$ [ν_0 = 3597.047 cm^{-1}], $(3\nu_1+\nu_3)$ [ν_0 = 3478.386 cm^{-1}], and $(\nu_1+3\nu_3)$ [ν_0 = 5201.151 cm^{-1}]. The resolution achieved was 0.04—0.07 cm^{-1} and the gas pressure was 2.5—21 cmHg, with a path length of 22—49 m. The rotational constants B and D were determined for each of the vibration–rotation transitions; the B_{00^00}, D_{00^00} values derived from the combination differences are given in Table 11.

Smith, Chao, Lin, and Overend[102] have studied the fundamental Fermi diad 10^{0}0,02^{0}0 of CS_2 through the i.r.-active transitions between these states and the 00^{0}1 state. The $(\nu_3-\nu_1)$, $(\nu_3-2\nu_2)$, and $(\nu_3+4\nu_2)$ bands have been analysed to yield values of B and D in these vibrational states. Although the 10^{0}0—00^{0}0 transition had previously been observed in the Raman spectrum by Stoicheff,[103] and the rotational constants had been evaluated for the excited state, the other component of the Fermi diad was not resolved. Smith *et al.* have now measured, under conditions of high resolution, the vibration–rotation spectra of the difference bands 00^{0}1—10^{0}0 and 00^{0}1—02^{0}0; in addition, the weak transition in the Fermi triad, 04^{0}1—00^{0}0, was also analysed. The 00^{0}1—10^{0}0 transition occurred at 877 cm^{-1} for $^{12}C^{32}S_2$ and 828 cm^{-1} for $^{13}C^{32}S_2$. The 00^{0}1—02^{0}0 transition in $^{12}C^{32}S_2$ was found at 733 cm^{-1} but was too weak to be observed for $^{13}C^{32}S_2$. The rotational constants obtained in this work were compared with those obtained by Stoicheff[103] for the 00^{0}0 and 10^{0}0 vibrational states and by Walker and Weber[8] for the 00^{0}0 and 01^{1}0 states. Both sets of data are included in Table 11.

Nitrogen Dioxide. Blank, Olman, and Hause[104] have analysed the vibration–rotation i.r. spectra of the 003 and 103 vibrational states of $^{14}N^{16}O_2$ and $^{15}N^{16}O_2$. Previously, the 101 and 201 vibrational states were analysed in the near-i.r.[105]. NO_2 is nearly an accidentally symmetric

[101] G. Blanquet and C. P. Courtoy, *Ann. Soc. sci. Bruxelles*, 1970, **84**, 293.
[102] D. F. Smith, T. Chao, J. Lin, and J. Overend, *Spectrochim. Acta*, 1971, **27A**, 1979.
[103] B. P. Stoicheff, *Canad. J. Phys.*, 1958, **36**, 218.
[104] R. E. Blank, M. D. Olman, and C. D. Hause, *J. Mol. Spectroscopy*, 1970, **33**, 109.
[105] M. D. Olman and C. D. Hause, *J. Mol. Spectroscopy*, 1968, **26**, 241.

prolate top; the electronic angular momentum is quenched so that the major effect of the unpaired electron in the absence of external fields is an N–S spin–rotation interaction, which leads to doubling of the low-N lines. The 003 band was analysed with a sample pressure of 15 Torr and an absorbing path-length of 6.3 m; the weaker 103 band was observed with a pressure of 50 Torr and a 16 m path-length in the cell. For both bands the spectra were obtained with an effective resolution of 0.03 cm^{-1}. A detailed account of the theoretical expressions for the slightly-asymmetric-top molecule has been given by Olman and Hause.[105] The rotational constants for the ground state were fixed at the values obtained from the earlier work, and the upper-state constants were calculated using these ground-state constants. The 103 band is located in the 5970 cm^{-1} region for $^{14}NO_2$ and in the 5860 cm^{-1} region for $^{15}NO_2$. Sub-bands for the $K_{-1} = 0$ and $K_{-1} = 1$ transitions were observed, but the Q-branch lines were too weak to be measured. Transitions were identified with $K_{-1} = 5$ and $N = 46$. Spin splitting was not observed in the low-N lines because of lack of intensity and the presence of overlapping bands in this region. The 003 band is a strong band in the 4730 cm^{-1} region for $^{14}NO_2$ and the 4650 cm^{-1} region for $^{15}NO_2$. Analysis of the bands was successfully accomplished out to $K_{-1} = 7$ and $N = 48$. The Q-branch lines were well resolved, and spin splitting was observed and analysed; it was appreciable in the Q-branches, large in the R-branches, and small in the P-branches of the sub-bands. Spin–rotation coupling constants were obtained for the excited vibrational states.

Blank and Hause[106] have reported a re-measurement of the 301 vibrational band of NO_2 under high resolution (0.05 cm^{-1}) in the near-i.r., using similar conditions of pressure and cell path-length to those described above. The 301 band is at 5430 cm^{-1} for $^{14}NO_2$ and 5350 cm^{-1} for $^{15}NO_2$. P-, Q-, and R-branches were observed in both bands; the P- and R-lines were analysed and the molecular constants derived therefrom were used to predict positions for the Q-sub-branches. Owing to the weak absorption and the overlapping of band lines, spin splitting was not observed. Improved values of the rotational and vibrational constants were obtained.

Bird *et al.*[107] have studied the pure rotation spectrum of NO_2. The spectrum was observed and measured interferometrically (0.10 cm^{-1} resolution) over the wavenumber range 45—104 cm^{-1}, with a gas pressure of 226 Torr and a cell path-length of 44 cm. It should be pointed out that, under the conditions of the experiment, NO_2 is actually an equilibrium mixture, $2NO_2 \rightleftharpoons N_2O_4$, but, since N_2O_4 lacks a permanent dipole, it does not interfere with the far-i.r. spectrum of NO_2. Previous analysis of the far-i.r. pure rotation spectrum located the Q-branches, but the lack of instrumental resolution limited the analysis for rotational constants. In the work of Bird *et al.*, a resolution sufficient to allow the observation of the magnetic splittings has

106 R. E. Blank and C. D. Hause, *J. Mol. Spectroscopy*, 1970, **34**, 478.
107 G. R. Bird, G. R. Hunt, H. A. Gebbie, and N. W. B. Stone, *J. Mol. Spectroscopy*, 1970, **33**, 244.

been achieved. NO_2 is the first simple polyatomic molecule for which magnetic splitting has been observed in the far-i.r. *P*-, *Q*-, and *R*-branch lines were observed for values of *N* and *K* up to 40, and a detailed theoretical structure of the $K = 4 \rightarrow 5$ *Q*-branch (*N* values $5 \rightarrow 35$), showing the magnetically split sub-branches, is given. The *Q*-branches themselves were not resolved under these experimental conditions but the shapes of the band envelopes are consistent with the theoretical prediction.

The rotational constants of NO_2 that were obtained from the work described above are collected together in Table 12.

Table 12 *Rotational constants*[a] *of* NO_2

Ref.	*Vibrational level*		*Isotopic species* $^{14}NO_2$	$^{15}NO_2$
107	000	$(B+C)/2$	0.422018 ±0.000020	—
		$A-(B+C)/2$	7.579139 ±0.00039	—
		D_N	$(2.812 \pm 0.016) \times 10^{-7}$	—
		D_{NK}	$(1.904 \pm 0.042) \times 10^{-5}$	—
		D_K	$(2.516 \pm 0.034) \times 10^{-3}$	—
104	003	ν_0	4754.209	4655.228
		A	7.3427 ±0.0012	7.0217 ±0.0019
		B	0.425457 ±0.000032	0.425781 ±0.000046
		C	0.402916 ±0.000024	0.402214 ±0.000036
	103	ν_0	5984.705	5874.951
		A	7.4140 ±0.0087	7.0791 ±0.0057
		B	0.423211 ±0.000039	0.42355 ±0.00016
		C	0.399358 ±0.000017	0.398728 ±0.000065
106	301	ν_0	5437.540	5367.316
		A	8.0140 ±0.0056	7.5328 ±0.0090
		B	0.42384 ±0.00011	0.42452 ±0.00027
		C	0.399519 ±0.000079	0.39911 ±0.00019

[a] All values quoted are of the constants/cm^{-1}.

Nitrous Oxide. In 1968, Pliva [108] reported a thorough vibrational and rotational analysis of $^{14}N_2{}^{16}O$ and was able to derive a set of rotational constants capable of reproducing the experimental data to $\pm 0.006\ cm^{-1}$. Some uncertainty in some interaction constants was due to the fact that the values of these constants depended largely on measurements in the photographic i.r. region, the absolute accuracy of which was only $\pm 0.08\ cm^{-1}$. Pliva [109] has now re-measured, with considerably increased resolution and accuracy, the overtone ($5\nu_3$) and combination ($4\nu_2{}^0 + 4\nu_3$) bands of $^{14}N_2{}^{16}O$ which are most influenced by these interactions. He used a 7.3 m vacuum Ebert spectrograph and a multi-reflection cell of length 33.3 m, which was used to attain path-lengths of up to 6 km at pressures of N_2O as low as 40 Torr. The previously derived vibrational and rotational constants were adjusted to fit the new, accurate data. The ground-state constants

[108] J. Pliva, *J. Mol. Spectroscopy*, 1968, **27**, 461.
[109] J. Pliva, *J. Mol. Spectroscopy*, 1970, **33**, 500.

Table 13 *Rotational constants of* $^{14}N_2{}^{16}O$

Vibrational state	ν_0/cm^{-1}	$10^3(B'-B_0)$/cm^{-1}	$10^6(D'-D_0)$/cm^{-1}	$10^9(H'-H_0)$/cm^{-1}
00^05	10815.249	−14.651	1.643	0.405
	±0.001	±0.007	±0.012	±0.006
04^04	10820.130	−14.335	−1.629	−0.538
	±0.001	±0.012	±0.033	±0.025

$B_0 = 0.4190113$ cm^{-1}, $D_0 = 1.795\times10^{-7}$ cm^{-1}, and $H_0 = 1\times10^{-12}$ cm^{-1} were accurately known from previous work [110] and were kept fixed in the analysis. The $(B'-B_0)$, $(D'-D_0)$, and $(H'-H_0)$ rotational constants obtained for $^{14}N_2{}^{16}O$ are shown in Table 13. The large values of the distortion constants $(D'-D_0)$ and $(H'-H_0)$ are ascribed to the strong interaction between the 00^05 and 04^04 states. This is also supported by the observation that these constants are of approximately equal magnitude but of opposing sign for the interacting states.

Toth [111] has measured the line-strengths of N_2O in the 2.9 μm region (3400 cm^{-1}) of the i.r., using low sample pressures so as to minimize pressure-broadening effects, and a resolution of 0.03 cm^{-1}. The two Σ—Σ bands (10^01—00^00 and 02^01—00^00) and the Π—Π band (11^11—01^10) were studied with a sample-cell path-length of 16 m and pressures of 0.3—9 Torr. The line strengths were determined by the method of equivalent widths.[111] The data were analysed to determine the dipole matrix elements. Five lines of the 06^00—00^00 band and two lines of the 06^20—00^00 band were also observed in the spectrum of the 10^01—00^00 band. The results were used to determine the Fermi interaction term between the 06^00 and 10^01 levels.

Toth [112] goes on to report the self- and N_2-broadened linewidths of N_2O, which were obtained from direct measurements of the 10^01—00^00 and 02^01—00^00 bands at 297 K. The experimental results were compared with the linewidths calculated from Anderson's impact theory, as amplified by Tsao and Curnutte, in order to determine the hard-sphere diameters and quadrupole moments of N_2O and N_2. These best-fit values were re-inserted into the theory to calculate the linewidths of N_2O and N_2 as a function of temperature for the temperatures 297 K, 260 K, 220 K, and 180 K. The quadrupole moments determined from nuclear-spin-relaxation data are in excellent agreement with the values obtained by Toth.

Jansson, Hunt, and Plyler [113] have described a method for correction of high-resolution i.r. spectra for the distortion introduced by the spectrometer. The method was applied to a portion of the Q-branches of N_2O near 3.3 μm at a sample gas pressure of <1.0 Torr. The spectra, after deconvolution, show an improvement of resolution which approaches the limit set by the

[110] J. Pliva, *J. Mol. Spectroscopy*, 1968, **25**, 62.
[111] R. A. Toth, *J. Mol. Spectroscopy*, 1971, **40**, 588.
[112] R. A. Toth, *J. Mol. Spectroscopy*, 1971, **40**, 605.
[113] P. Jansson, R. H. Hunt, and E. K. Plyler, *J. Opt. Soc. Amer.*, 1970, **60**, 596.

Doppler widths of the lines. Line separations measured from the deconvoluted spectra are within 0.001 cm^{-1} of the calculated values.

Tetra-atomic Molecules.—*Hydrogen Fluoride Dimer*. The i.r. spectrum of the polymers of HF has been investigated in the past decade by a number of observers; Kuipers [114] and Smith [115] have shown that the i.r. absorption around 3965 cm^{-1} previously attributed to the monomer in the region from the rotation lines R_1 to P_3 in the 1—0 vibrational band of the monomer was due to the dimer H_2F_2. The absorption spectrum of the dimer in the null-gap region of the 1—0 vibration–rotation band of HF has recently been reinvestigated by Himes and Wiggins.[116] These authors used a specially constructed sample-cell of monel metal, 1.6 m long, with sapphire windows. The cell was maintained at 0 °C. The optimum pressure for observing the H_2F_2 spectrum was found to be 80 Torr.

The spectrum was observed under high resolution, and problems with atmospheric water vapour absorption were reported. The observed spectrum of H_2F_2 was similar to that reported by Herget *et al.*;[117] an intensity alternation was found and also an intense absorption region which was assignable to a Q-branch. Since the band origin could not be located, the observed spectrum could not be analysed unambiguously. The intensity alternation was discussed, and the obvious explanation, that the molecule possessed a centre of symmetry, was discounted for the following reasons. The presence of a centre of symmetry in H_2F_2 would mean that the molecule would be planar and have zero dipole moment; but a large dipole moment is in fact observed for this molecule. Also, the symmetric model would not be amenable to hydrogen-bonding, nor to an analysis of the spectrum in terms of a single branch. The spectrum has been explained by Himes and Wiggins as the overlapping of two bands and analysed on this basis. They have predicted a wavenumber of 3966 cm^{-1} for the band origin and have suggested a $(B'+B'')$ value of 0.449 cm^{-1}. Other possible explanations of the intensity alternation are also suggested, including a double set of lines being produced by rotational inversion. Additional work is needed before the spectrum can be analysed quantitatively.

Disulphane. The HSSH molecule has been the subject of several structural studies, including a recent microwave investigation.[118] Winnewisser [119] has made a high-resolution i.r. study of the ν_1 and ν_5 vibrational bands to determine molecular parameters of excited vibrational states which were not accessible to the microwave measurements. The i.r. spectra were recorded between 2490 and 2650 cm^{-1} at a resolution of 0.045 cm^{-1}. The

114 G. A. Kuipers, *J. Mol. Spectroscopy*, 1958, **2**, 75.
115 D. C. Smith, *J. Mol. Spectroscopy*, 1959, **3**, 473.
116 J. L. Himes and T. A. Wiggins, *J. Mol. Spectroscopy*, 1971, **40**, 418.
117 W. F. Herget, N. M. Gailar, R. J. Lovell, and A. H. Nielsen, *J. Opt. Soc. Amer.*, 1960, **50**, 1264.
118 G. Winnewisser, M. Winnewisser, and W. Gordy, *J. Chem. Phys.*, 1968, **49**, 3465.
119 B. P. Winnewisser, *J. Mol. Spectroscopy*, 1970, **36**, 414.

sample cell was a 1 m Pyrex tube with calcium fluoride windows, and the gas pressures used were 5—50 Torr. H_2S is a product of decomposition of H_2S_2 and its spectrum in the same frequency range was recorded under similar conditions, so that lines due to H_2S could be identified. The spectrum of H_2S_2 consisted of a superposition of the ν_1 band (symmetric S—H stretch) and the ν_5 band (antisymmetric S—H stretch). The spectrum was very dense and the resolution of single lines proved to be almost impossible. The molecule is nearly an accidentally symmetric top, and the largest calculated difference from symmetric-top behaviour was a splitting of $<0.005\ cm^{-1}$ in the P- and R-branches of the $K = 1$ sub-band. Since this deviation is much less than the linewidth observable, the spectrum could be analysed without considering asymmetry. The analysis was carried out, using the expression:

$$\nu = \nu_0 + (\bar{B}' + \bar{B}'')m + (\bar{B}' - \bar{B}'')m^2 - (4D_J)m^3$$

for the $K = 0$ sub-band in the ν_5 band, where $m = J+1$ for the R-branch and $m = -J$ for the P-branch and where $\bar{B}$ is $(B+C)/2$. The molecular constants obtained from the Q-branches and $K = 0$ sub-band were used to generate a simulated spectrum with a computer. Improved constants were obtained by utilizing the $K = 1$ sub-band measurements. The extent of splitting in the excited states due to torsional doubling is also discussed.

Boron Trifluoride. Ginn, Johansen, and Overend [120] have measured the perpendicular ν_4 bands at $480\ cm^{-1}$ in the i.r. spectra of isotopically enriched $^{10}BF_3$ and $^{11}BF_3$ with a resolution of $0.03\ cm^{-1}$. The ν_4 system had previously only been measured under low resolution; the Q-branches of the individual K sub-bands are almost coincident, and they appeared as a single feature under the low-resolution conditions. The ν_2 and ν_3 fundamentals had earlier been analysed by Ginn *et al.*[121] The ν_4 bands of each species were similar in appearance. The Q-branch was broad and featureless, but on the low-wavelength side there was a series of lines spaced by about $0.07\ cm^{-1}$. The series was repeated at intervals of $0.7\ cm^{-1}$ ($\sim 2B$), and it appears as though each series represents the K components of a single J transition in the R-branch; the $J = 6$ group is discussed in detail. At higher values of J, the K-multiplet structure of adjacent J-lines begins to overlap and complicate the assignment. The P-branch lines could be assigned on a similar basis. l-Type resonance was considered and judged to be of little importance in this analysis. The rotational constants were determined and are given in Table 14.

Ginn *et al.*[122] have published a paper on the vibrational anharmonicity in BF_3, using the results of the above, and other, vibration–rotation analyses.

[120] S. G. W. Ginn, D. Johansen, and J. Overend, *J. Mol. Spectroscopy*, 1970, **36**, 448.
[121] S. G. W. Ginn, C. W. Brown, J. K. Kenney, and J. Overend, *J. Mol. Spectroscopy*, 1968, **28**, 509; S. G. W. Ginn, J. K. Kenney, and J. Overend, *J. Chem. Phys.*, 1968, **48**, 1571.
[122] S. G. W. Ginn, S. Reichman, and J. Overend, *Spectrochim. Acta*, 1970, **26A**, 291.

Table 14 *Rotational constants of* $^{10}BF_3$ *and* $^{11}BF_3$ *(from analysis of* ν_4*)* [120]

$^{10}BF_3$	ν_0	$= 481.13\ cm^{-1}$
	B'	$= 0.34520 \pm 0.0002\ cm^{-1}$
	B''	$= 0.34471 \pm 0.0002\ cm^{-1}$
	D_J'	$= (4.5 \pm 0.3) \times 10^{-7}\ cm^{-1}$
	D_J''	$= 4.2 \times 10^{-7}\ cm^{-1}$
	D_K'	$= (3.3 \pm 0.6) \times 10^{-7}\ cm^{-1}$
	D_K''	$= 3.4 \times 10^{-7}\ cm^{-1}$
$^{11}BF_3$	ν_0	$= 479.36\ cm^{-1}$
	B'	$= 0.34524 \pm 0.0002\ cm^{-1}$
	B''	$= 0.34455 \pm 0.0002\ cm^{-1}$
	D_J'	$= (4.1 \pm 0.6) \times 10^{-7}\ cm^{-1}$
	D_J''	$= 4.2 \times 10^{-7}\ cm^{-1}$
	D_K'	$= (2.7 \pm 0.8) \times 10^{-7}\ cm^{-1}$
	D_K''	$= 3.4 \times 10^{-7}\ cm^{-1}$

Nitrogen Trifluoride. The NF_3 molecule has previously been studied by Popplewell, Masri, and Thompson [123] in a medium-resolution (0.2 cm^{-1}) i.r. investigation, and by Otake, Hirota, and Morino [124] in the microwave. The rotational constants obtained from the i.r. and microwave analyses are in good agreement, apart from data concerning the rotational levels in the ν_4 state. It has been shown that the rotational levels in this state are likely to be severely perturbed by *l*-type resonance. Reichman and Ginn [125] have recently reported the results of a reinvestigation of the vibration–rotation i.r. spectrum of the ν_4 band of NF_3 at 500 cm^{-1} under high-resolution conditions. Spectral resolutionw as 0.025—0.05 cm^{-1}; the sample was contained in a multi-pass cell of path-length 3 m at 10 Torr pressure.

The *P*-, *Q*-, and *R*-branches were obtained, but it was found that a computed spectrum could not be fitted to the observed results unless *l*-type resonance effects were considered. Cartwright and Mills [126] have discussed theoretically the effect of the *l*-type perturbation on individual lines within a perpendicular band, and Stone and Mills [127] have recently computed the ν_4 band contour in NF_3 using the constants obtained by Otake *et al.*[124] and allowing for the effect of *l*-type resonance. The effect of *l*-type resonance in ν_4 is most clearly seen in the recent work of Reichmann and Ginn,[125] where the two *Q*-branches are separated; the ^RQ-branch has been weakened in intensity and the ^PQ-branch has been strengthened by *l*-type resonance. This intensity pattern confirms conclusively that the *l*-type doubling constant q_4, which had been left undetermined in the microwave investigation, must be positive. The band constants obtained in this work are $\nu_0 = 493.43 \pm 0.02\ cm^{-1}$ and $q_4 = 1.71 \times 10^{-3}\ cm^{-1}$.

[123] R. J. L. Popplewell, F. N. Masri, and H. W. Thompson, *Spectrochim. Acta*, 1967, **23A**, 2797.
[124] M. Otake, E. Hirota, and Y. Morino, *J. Mol. Spectroscopy*, 1968, **28**, 316.
[125] S. Reichman and S. G. W. Ginn, *J. Mol. Spectroscopy*, 1971, **40**, 27.
[126] G. J. Cartwright and I. M. Mills, *J. Mol. Spectroscopy*, 1970, **34**, 415.
[127] J. M. R. Stone and I. M. Mills, *J. Mol. Spectroscopy*, 1970, **35**, 354.

Arsenic Trifluoride and Phosphorus Trifluoride. Reichman and Overend [128] have reported the vibration–rotation analyses of high-resolution i.r. spectra of AsF_3 and PF_3. Spectra of ν_1 of AsF_3 and ν_1 and ν_2 of PF_3 have been measured with a resolution of 0.03—0.07 cm^{-1}.

In the case of the very reactive arsenic trifluoride, silver chloride windows had to be used in the sample cell. The loss of energy was significant and the spectral resolution was only 0.07 cm^{-1}. The *P*- and *R*-branch lines do not show evidence of *K*-splitting. In addition to the main band, there is another series of lines from the 'hot' transiton $(\nu_1+\nu_4) \leftarrow \nu_4$. From the relative intensities of the *P*- and *R*-branches, it is concluded that there is a significant Coriolis interaction between $\nu_1(A_1)$ and $\nu_3(E)$, although there is no indication of any frequency perturbation. The band constants of AsF_3 (ν_1) are given in Table 15. It is interesting that $\alpha_1{}^B$ for AsF_3 is positive because of a particularly large Coriolis resonance term that couples ν_1 and ν_3, whereas $\alpha_1{}^B$ for PF_3 (see below) is negative.

For phosphorus trifluoride, the ν_1 band at 890 cm^{-1} and the ν_2 band at 488 cm^{-1} were measured. In the ν_1 band, as in the case of AsF_3, the *P*- and *R*-lines are sharp and not significantly degraded. A smearing out of the *Q*-branch structure makes the *J* assignment somewhat ambiguous; and B_0 has therefore been constrained towards the results of previous microwave investigations. The ν_2 band of PF_3 was observed at a resolution of 0.03 cm^{-1}, and signs of degradation were observed in the *P*- and *R*-branch lines. In contrast to the ν_1 analysis, the *J* assignment of ν_2 is straightforward. Values of the band constants are given in Table 15.

Table 15 *Band constants for* AsF_3 *and* PF_3

Compound	AsF_3	PF_3	
Band	ν_1	ν_1	ν_2
ν_0/cm^{-1}	740.546 ± 0.002	891.919 ± 0.002	487.718 ± 0.001
B_0/cm^{-1}	0.19645 ± 0.00007	0.26100 ± 0.00009	0.26091 ± 0.00002
D_J/cm^{-1}	$(2.1 \pm 0.3) \times 10^{-7}$	—	—
D_J'/cm^{-1}	—	$(4.2 \pm 0.7) \times 10^{-8}$	$(2.3 \pm 0.2) \times 10^{-7}$
D_J''/cm^{-1}	—	$(4.7 \pm 0.7) \times 10^{-8}$	$(2.2 \pm 0.2) \times 10^{-7}$
α_1^B/cm^{-1}	$(1.57 \pm 0.02) \times 10^{-4}$	$-(1.072 \pm 0.006) \times 10^{-3}$	—
α_2^B/cm^{-1}	—	—	$(2.61 \pm 0.03) \times 10^{-6}$

Sulphur Trioxide. The rotational fine structure of several parallel and perpendicular bands of SO_3 has been partially resolved and analysed by Thomas and Thompson.[129] The spectral resolution was 0.1—0.3 cm^{-1} and the absorption cells were of length 10 cm with windows of Teflon, arsenic sulphide, or silver chloride for different spectral ranges. The ν_3 band (S—O

[128] S. Reichman and J. Overend, *Spectrochim. Acta*, 1970, **26A**, 379.
[129] R. K. Thomas and H. W. Thompson, *Proc. Roy. Soc.*, 1970, **A314**, 329.

asymmetric stretch) revealed a series of Q-branches superposed upon unresolved P- and R-branch structure. Nine ^{R}Q-branches and nine ^{P}Q-branches were measured and yielded a value for ζ_3, the Coriolis coupling coefficient, of +0.47. α^A and α^B values were derived from the expression:

$$^{R}Q_K + {}^{P}Q_K = 2\nu' + 2K^2[(A'-A'')-(B'-B'')]$$

where $$(A'-A'') = \alpha^A,\ (B'-B'') = \alpha^B,$$

and in which $$\nu' = \nu_0 + [A'(1-\zeta)^2 - B']$$

A method of correcting for the displacement of the Q-branch unresolved contours from the true band origins is described and is used to yield α values more accurate than the above method could provide.

Table 16 *Band constants for* SO_3

	Band			
	ν_3	$2\nu_3$	ν_2	ν_4
B''/cm^{-1}	0.353	0.353	0.353	0.353
α^A/cm^{-1}	0.0005	0.001	0.001	0.0015
α^B/cm^{-1}	0.001	0.0024	0.002	0.003
ν_0/cm^{-1}	1389.86	2777.72	529.65[a]	498.5[a]
ζ	+0.46	−0.92	+0.52	−0.46
A″/cm^{-1}	—	—	—	0.176

[a] These values of ν_2 and ν_4 should be compared with the work of Stopperka (ref. 130); ν_2 = 484 cm^{-1}, ν_4 = 530 cm^{-1}.

The $2\nu_3$, ν_2, and ν_4 bands of SO_3 were also partially resolved. ν_2 and ν_4 were re-assigned (see Table 16). In a detailed analysis of ν_2 and ν_4, Thomas and Thompson measured thirty-eight lines in the R-branch of ν_2 ($J = 4$ to $J = 41$) and obtained B'' from the relationship:

$$R(J+1) - R(J) = 2(B'-B'')(J+1) + 2B'$$

neglecting centrifugal distortion terms (which could well be significant at higher J values). Using the data for $J = 4$ to $J = 30$, B'' was found to be 0.38 cm^{-1}. The difference in this B'' and that calculated from an electron-diffraction investigation arises from Coriolis interaction between ν_2 and ν_4. The value of ζ_{24} was calculated to be 0.5 and B'' was found to be 0.353 ±0.005 cm^{-1}, giving an S—O bond length of 1.41 ±0.01 Å, which is significantly smaller than the electron-diffraction value of 1.43 Å.

A summary of the band constants is given in Table 16.

Ammonia. Shimizu and Shimizu [131] have analysed the ν_2 band of isotopically substituted ammonia, $^{15}NH_3$, at 10 μm. Many of the absorption lines

[130] K. Stopperka, *Z. anorg. Chem.*, 1966, **345,** 277.
[131] F. O. Shimizu and T. Shimizu, *J. Mol. Spectroscopy*, 1970, **36,** 94.

coincide in frequency with CO_2 and N_2O laser lines, and the analysis of the high-resolution i.r. spectrum has therefore been a useful adjunct to laser spectroscopy. 170 Absorption lines in the wavenumber range 842—1154 cm^{-1} were observed using a 2 m grating vacuum spectrometer. The absorption cell was 10 cm long and the gas pressure was 11 Torr. The spectral resolution attained was 0.15 cm^{-1}. All observed spectral lines could be divided into two groups, '*Group s*' consisting of the transitions from the symmetric levels of the inversion doublets in the ground state to the asymmetric levels of the doublets in the excited state, and '*Group a*' consisting of all transitions which start from the asymmetric levels in the ground state. The inversion frequencies were eliminated from the transition frequencies by averaging the transitions $\nu_s\ (J, K)$ and $\nu_a\ (J, K)$. From these frequency values, the band origin and the rotational constants could be determined using the expression for a symmetric-top molecule without inversion doubling:

$$E = BJ(J+1)+(C-B)K^2-D_J J^2(J+1)^2-D_{JK}J(J+1)K^2-D_K K^4$$

Some of the rotational constants obtained from the analysis are $\nu_0 = 945.321\ cm^{-1}$; $B'' = 9.918825\ cm^{-1}$ from microwave;[132] $D''_J = 7.95 \times 10^{-4}\ cm^{-1}$; $D''_{JK} = -15.55 \times 10^{-4}\ cm^{-1}$; $B' = 9.9582\ cm^{-1}$; $D'_J = 8.84 \times 10^{-4}\ cm^{-1}$; and $D'_{JK} = -18.65 \times 10^{-4}\ cm^{-1}$.

Shimizu and Shimizu also derived the inversion frequencies in the ground state for both *s* and *a* transitions using a power-series expansion in $J(J+1)$ and K^2, namely:

$$\nu''_{\text{inv}} = \nu''_i + b''J(J+1)+(c''-b'')K^2-d''_J J^2(J+1)^2-d''_{JK}J(J+1)K^2-d''_K K^4$$

and similarly for the excited state. Values of ν'_i and ν''_i, *b*, *c*, and *d* are given. It was found that the discrepancies between the observed and calculated inversion frequencies were large for weak transitions and for the transitions with high *J* and *K* values. By the adoption of another method of calculation, the frequencies of the i.r. transitions could be calculated to within the estimated experimental error as follows. The energies of the initial levels of the *s* transitions and the *a* transitions were calculated, and the transition frequencies and constants in the excited state which best fitted the observed frequencies by a least-squares method were derived. The results, two sets of 'modified' rotational constants and band origins, are given in Shimizu and Shimizu's paper.[131]

Kelley, Francke, and Feld [133] have reported that the 949 cm^{-1} vibration–rotation transition of $^{14}NH_2D$ can be Stark-tuned into resonance with the P_{14} line of CO_2. An estimate of the inversion splitting in NH_2D is given as 18.30 cm^{-1} and the band origins $(0_a$—$1_s)$ and $(0_s$—$1_a)$ are at 875.9 and 894.6 cm^{-1}, respectively.

[132] P. Helminger and W. Gordy, *Phys. Rev.*, 1969, **188,** 100.
[133] M. J. Kelley, R. E. Francke, and M. S. Feld, *J. Chem. Phys.*, 1970, **53,** 2979.

Abouaf-Marguin, Dubost, and Legay [134] have obtained the i.r. spectrum of NH_3, trapped in an argon matrix at liquid-helium temperatures and irradiated with a CO_2 laser at 975 cm^{-1}. The peak intensity of the 1638 cm^{-1} absorption band was decreased because of the depopulation of the fundamental level. The appearance of *P*- and *R*-branch rotational lines was noted and was explained as follows: as the vibrationally excited NH_3 molecules return to the ground vibrational level, the vibrational energy is transferred as heat to the surrounding matrix, so enabling the molecules to rotate. These authors also noted that the inversion splitting, 23 cm^{-1}, was not greatly decreased relative to the gas-phase value of 36.3 cm^{-1}.

The inversion–rotation emission spectrum of thermally excited NH_3 in the 60—200 cm^{-1} region has been measured, using interference spectroscopy, by Walker and Hochheimer.[135] Previously, Lowenstein [136] had observed the far-i.r. (10—70 cm^{-1}) inversion and inversion–rotation spectra of the ν_2 vibrational level. Walker and Hochheimer have extended the investigation of inversion rotation spectra to other vibrationally excited levels of NH_3, namely the $2\nu_2$ and ν_4 levels. These two levels are nearly in resonance and the Coriolis interactions in the doubly degenerate ν_4 vibration are hence accentuated. The ground-state inversion–rotation spectrum was strongly absorbed by cold gas near the window of the sample cell which was heated to 230 °C. The ν_2 level is the lowest-lying vibrational level and is therefore the most populated. The inversion levels are separated by 35.6 cm^{-1}, so the inversion-doubled rotation spectra are well separated. The ν_2 inversion–rotation spectrum will have frequencies which are given by the expression:

$$\nu = 2\,BJ - 4\,D_J J^3 - 2\,D_{JK} JK^2 \pm \nu_0$$

where ν_0 is the inversion frequency. The ν_4 inversion–rotation spectrum was found to be too limited in extent and resolution to yield any refined molecular constants. The *K*-splitting was largely unresolved except for those lines which were strongly perturbed by Coriolis interactions. In the $2\nu_2$ vibrational level, the inversion levels are separated by 284.7 cm^{-1}. The incomplete inversion–rotation spectra from the $2\nu_2$ and ν_4 vibrational levels do, however, agree with those predictions based on earlier results.

Acetylene. Baldacci, Ghersetti, and Rao [137] have re-measured the vibration–rotation bands of $^{12}C_2HD$ in the 6—10 μm region of the i.r. The $2\nu_4$, $(\nu_4+\nu_5)$, and $2\nu_5$ bands were analysed, but difficulties in interpretation were caused by band overlap from C_2H_2 and C_2D_2 bands. The spectral resolution achieved was 0.05 cm^{-1}. Errors in the *J*-assignment of previous work were noted, and the new ground-state combination differences $R(J-1)-P(J+1)$ were shown to be in agreement with similar data for other bands in $^{12}C_2HD$. The effect of perturbation on the upper vibrational levels of this molecule

[134] L. Abouaf-Marguin, H. Dubost, and F. Legay, *Chem. Phys. Letters*, 1970, **7**, 61.
[135] R. E. Walker and B. F. Hochheimer, *J. Mol. Spectroscopy*, 1970, **34**, 500.
[136] E. V. Lowenstein, *J. Opt. Soc. Amer.*, 1960, **50**, 1163.
[137] A. Baldacci, S. Ghersetti, and K. N. Rao, *J. Mol. Spectroscopy*, 1970, **36**, 358.

was discussed briefly and a comparison made with C_2H_2 and C_2D_2. The molecular constants obtained arc:

	ν_0/cm^{-1}	$10^3 (B'-B'')/\mathrm{cm}^{-1}$
$2\nu_4$	1033.934 ± 0.002	5.48 ± 0.02
$(\nu_4+\nu_5)$	1200.493 ± 0.002	3.99 ± 0.02
$2\nu_5$	1342.231 ± 0.002	3.01 ± 0.01

Ghersetti, Pliva, and Rao [138] have reported the results of improved measurements of the $^{12}C_2D_2$ bands which occur in the 2—2.5 μm and 5—10 μm regions of the i.r. Ten combination bands were observed and measured and also one overtone band, namely $3\nu_5$. The effective spectral resolution achieved was 0.03 cm^{-1} at low wavelengths and 0.04—0.05 cm^{-1} at the higher wavelengths. The data from this extensive investigation were combined with the results of earlier work [139, 140] to obtain an optimum set of molecular constants for all the observed vibration–rotation states of $^{12}C_2D_2$.

Formaldehyde. Yamada *et al.*[141] have recently measured the ν_1 and ν_5 bands of H_2CO in the C—H stretching-frequency region in the i.r., with a resolution of 0.3 cm^{-1}. Rotational structures of these bands were analysed as asymmetric-top bands by means of band-contour calculations and least-squares fits. The band origins and rotational constants of the upper states were determined. The band origin of ν_5, *viz.* 2843.24 $\pm$0.03 cm^{-1}, is in good agreement with the result of an earlier determination which used a symmetric-top approximation. The ν_1 band-origin, 2782.40 $\pm$0.07 cm^{-1}, is significantly higher than the value of 2766.4 cm^{-1} previously assigned to it. Coriolis interaction of ν_1 with other vibrational states was discussed. Equilibrium-value rotational constants were calculated to be $A_e = 9.5795 \pm 0.0231$ cm^{-1}, $B_e = 1.3033 \pm 0.0016$ cm^{-1}, and $C_e = 1.1462 \pm 0.0016$ cm^{-1}.

Assuming r_e (C—O) = 1.203 $\pm$0.003 Å, the equilibrium values of bond length and bond angle derived from the above rotational constants were r_e (C—H) = 1.099 $\pm$0.009 Å and CH_2 bond angle = 116.5 $\pm$1.2°.

Nakagawa and Morino [142] have also investigated the Coriolis interactions in the ν_4 and ν_6 bands of H_2CO. I.r. spectra were measured in the 800—1440 cm^{-1} region and the rotational assignments were made on the asymmetric-top assumption. Strong Coriolis interaction was observed between ν_4 and ν_6 and taken into account in a simultaneous least-squares analysis of 460 i.r. and microwave transitions. Two sets of calculations were carried out; in set (*a*), the band origins were constrained to ν_6 = 1249.14 cm^{-1} and ν_4 = 1167.26 cm^{-1}, values which had been obtained in a preliminary analysis. The four rotational parameters $A_6-\frac{1}{2}(B_6+C_6)$, (B_6-C_6), $A_4-\frac{1}{2}(B_4+C_4)$, and (B_4-C_4) were regarded as independent variables,

[138] S. Ghersetti, J. Pliva, and K. N. Rao, *J. Mol. Spectroscopy*, 1971, **38**, 53.
[139] S. Ghersetti and K. N. Rao, *J. Mol. Spectroscopy*, 1968, **28**, 27, 353.
[140] K. F. Palmer, S. Ghersetti, and K. N. Rao, *J. Mol. Spectroscopy*, 1969, **30**, 146.
[141] K. Yamada, T. Nakagawa, K. Kuchitsu, and Y. Morino, *J. Mol. Spectroscopy*, 1971, **38**, 70.
[142] T. Nakagawa and Y. Morino, *J. Mol. Spectroscopy*, 1971, **38**, 84.

Table 17 *Band constants* [a] *of* CH_2O [142]

	Set (a)	*Set (b)*
ν_6	1249.14 fixed	1249.08 fixed
$A_6-\bar{B}_6{}^b$	8.1773 ±0.021	8.2001 ±0.011
$\bar{B}_6{}^b$	1.212965 fixed	1.214085 fixed
B_6-C_6	0.170646 ±0.000082	0.167978 ±0.000078
ν_4	1167.26 fixed	1167.21 fixed
$A_4-\bar{B}_4{}^b$	8.1706 ±0.021	8.1516 ±0.010
$\bar{B}_4{}^b$	1.209050 fixed	1.211466 fixed
B_4-C_4	0.146206 ±0.000082	0.150972 ±0.000082
$2A\zeta_{64}\approx\xi_{64}$	10.3401 ±0.084	10.3039 ±0.043

[a] All values given are of constants/cm^{-1}, except ζ; [b] $\bar{B}_v = \frac{1}{2}(B_v+C_v)$.

whereas $\frac{1}{2}(B_6+C_6)$ and $\frac{1}{2}(B_4+C_4)$ were taken from microwave measurements. In set (*b*), the energy shifts due to the Coriolis interactions between ν_3 and ν_6 and between ν_3 and ν_4 were estimated. The observed frequencies of the rotational transitions corrected for these shifts were used in the least-squares analysis. In both sets of calculations, the Coriolis interaction between ν_6 and ν_4, ξ_{64}, was considered as a variable parameter. The results of the sets of calculations for cases (*a*) and (*b*) are given in Table 17.

The systematic deviations in the case of set (*b*) are significantly smaller than those in set (*a*) because the perturbations of the ν_6 level by ν_3 have been corrected in the former calculations. The Coriolis interaction term $\xi_{64} = 10.30 \pm 0.04$ cm^{-1}. Theoretical band-contours of ν_4 and ν_6 were computed, taking into consideration the intensity perturbation due to the Coriolis interaction, and good agreement was obtained with the observed spectra.

Fulminic Acid. The microwave spectrum of the HCNO molecule has recently been investigated.[143] The first high-resolution i.r. measurements were reported by Winnewisser and Winnewisser,[144] who made a rotational analysis of the C—H stretching vibration, ν_1. Winnewisser [145] has now reported the further analysis of five hot bands in the i.r. spectrum in the region 3200—3400 cm^{-1}. The *l*-type doubling constants *q* for the $(\nu_1+\nu_5-\nu_5)$ and $(\nu_1+\nu_4-\nu_4)$ transitions were evaluated and compared with the microwave values. The rotational constants obtained from the analysis were also compared with the more accurate microwave determinations; the rotational constants obtained by Winnewisser are given in Table 18.

Sheasley *et al.*[146] have carried out a rotational analysis of the ν_1 vibration–rotation band in DCNO occurring at 2620 cm^{-1} in the i.r. region. 39 Lines

[143] M. Winnewisser and K. K. Bodenseh, *Z. Naturforsch.*, 1967, **22a**, 1724; H. K. Bodenseh and M. Winnewisser, *ibid.*, 1969, **24a**, 1966; M. Winnewisser and B. P. Winnewisser, *ibid.*, 1971, **26a**, 128.
[144] B. P. Winnewisser and M. Winnewisser, *J. Mol. Spectroscopy*, 1969, **29**, 505.
[145] B. P. Winnewisser, *J. Mol. Spectroscopy*, 1971, **40**, 164.
[146] W. D. Sheasley, C. W. Mathews, E. L. Ferretti, and K. N. Rao, *J. Mol. Spectroscopy* 1971, **37**, 377.

Table 18 *Rotational Constants of* HCNO *and* DCNO

Ref.	*Molecule*	*Band*	*Band origin*/ cm^{-1}	B''/cm^{-1}	$10^3(B'-B'')$/ cm^{-1}	$10^7(D'')/cm^{-1}$	$10^8(D'-D'')$/ cm^{-1}	*Microwave value*, B''/cm^{-1} (*ref.* 143)
145	HCNO	ν_1	3336.1165 ±0.0008	0.38251 ±0.00002	−1.033 ±0.002	1.05 ±0.07	0.67 ±0.11	0.38256632(20)
		$\nu_1+\nu_5{}^1-\nu_5{}^1$	3308.0046 ±0.0009	0.38357 ±0.00003	−0.972 ±0.001	1.67 ±0.11	—	0.38357100(20)
		$\nu_1+\nu_4{}^1-\nu_4{}^1$	3332.9354 ±0.0019	0.38267 ±0.00006	−1.009 ±0.003	2.03 ±0.29	—	0.383045518(22)
		$\nu_1+2\nu_5{}^0-2\nu_5{}^0$	3294.3700 ±0.0028	0.38382 ±0.00004	−1.123 ±0.005	0.95 ±0.75	—	0.384122728(50)
		$\nu_1+2\nu_5{}^2-2\nu_5{}^2$	3284.4554 ±0.0013	0.38407 ±0.00003	−0.918 ±0.002	0.72 ±0.13	—	0.384226445(54)
146	DCNO	ν_1	2620.727 ±0.002	0.34334 ±0.00006	−1.048 ±0.006	1.4 ±0.2	1.3 ±0.4	0.3433211

in the R-branch and 41 lines in the P-branch were measured with a spectral resolution of 0.05 cm^{-1}. As in the investigations of Winnewisser,[145] the hot band $(\nu_1+\nu_5-\nu_5)$, with its characteristic l-type doubling, was also observed for the DCNO molecule. A value for B_0 was determined and it is in good agreement with that from the microwave determination. The band constants of DCNO are given in Table 18.

Polyatomic Molecules.—*Methane.* The ν_3 i.r. absorption band of the CH_4 molecule has been re-measured with a spectral resolution of 0.02 cm^{-1} by Henry *et al.*[147a] The absorption cell was of length 7 m and the methane pressure was 0.5 Torr. A feature of the experimental arrangement was the laser-illuminated goniometric system which was used for precise wave-number interpolation between calibration lines. The spectrum recorded covered the range 2884—3141 cm^{-1}, corresponding to the vibration–rotation lines from P_{12} to R_{12}. 323 Vibration–rotation lines were measured; the assignments were assisted by information available on the relative intensities of the lines. Other recent investigations of CH_4 are of ν_3, $2\nu_3$, ν_4, and $(\nu_2+\nu_3)$.[147b]

The $3\nu_3$ band of CH_4 at 9050 cm^{-1} has been investigated by Margolis.[148a] An unexpected J-dependence of the half-width of the rotational lines is discussed.

Ozier, Ho, and Birnbaum[148b] have observed the pure rotation spectrum of CH_3D in the ground vibrational and electronic states, using a far-i.r. grating spectrometer. The spectrum is of particular interest since the electric dipole moment arises entirely from an isotopic substitution. The spectrum was observed using an $f/3.8$ spectrometer operating at a spectral slit width of 0.67—2.31 cm^{-1}. Ten lines were observed in the wavenumber range 40—120 cm^{-1}; these were identified with the $J+1 \leftarrow J$ transitions for the rotational levels $J = 5$—14. For the $J = 6 \rightarrow 7$ transition, the absolute intensity and linewidth were determined by a curve-of-growth method. It was found that, for this line, the dipole moment μ has the value $(5.68 \pm 0.30) \times 10^{-3}$ D; a broadening parameter was also evaluated. This measurement constitutes the first correct experimental determination of the dipole moment of CH_3D; an earlier determination of μ by a molecular-beam method had been based on an incorrect assignment of the lines in the Stark spectrum. Further intensity measurements on other rotational lines showed that, to within 6%, μ is independent of J in the range $5 \leqslant J \leqslant 12$. A detailed account of the experimental intensity measurements is given. The rotational constants that were evaluated were $B_0 = 3.882 \pm 0.002$ cm^{-1} and $D_0 = (7.7 \pm 0.3) \times 10^{-5}$ cm^{-1}.

[147] (*a*) L. Henry, N. Husson, R. Andia, and A. Valentin, *J. Mol. Spectroscopy*, 1970, **36**, 511; (*b*) N. Husson and M. Dang Nhu, *J. Phys.* (*Paris*), 1971, **32**, 627; N. Husson and G. Poussigne, *ibid.*, p. 859; J. C. Hilico, *ibid.*, 1970, **31**, 289; R. Brégier, *ibid.*, p. 301; B. Bobin, *ibid.*, 1972, **33**, 345.

[148] (*a*) J. S. Margolis, *J. Quant. Spectroscopy Radiative Transfer*, 1970, **10**, 165; (*b*) I. Ozier, W. Ho, and G. Birnbaum, *J. Chem. Phys.*, 1969, **51**, 4873.

Methyl Fluoride. Osaka and Takahashi [149] have measured the far-i.r. rotation spectrum of methyl fluoride. Twelve rotational transitions for $J = 3$—14 were observed; the frequencies were in good agreement with those from the microwave determinations. The K-splitting was not observed because of limited resolution, but linewidth measurement indicated a maximum splitting at $J = 10$, in agreement with the theoretical prediction.

Methyl Chloride. Peterson and Edwards [150a] report a high-resolution i.r. investigation of CD_3Cl; the ν_4 (2300 cm^{-1}) and $2\nu_4$ (4600 cm^{-1}) vibrational bands were analysed at a resolution of 0.04—0.05 cm^{-1}. The rotational isotope effects due to the ^{35}Cl and ^{37}Cl isotopes were not resolved, and average ground-state rotational constants from a microwave determination [150b] were used in the analysis. The molecular constants obtained from an analysis of ν_4 and $2\nu_4$ are given in Table 19 . Morillon-Chapey *et al.*[151a] have examined ν_3, $2\nu_5$, and $(\nu_2+\nu_5)$ of CH_3Cl, and ν_1 of CD_3Cl has been recorded [151b] at 0.035 cm^{-1} resolution.

Table 19 *Molecular constants*[a] *of* CD_3Cl *and* CD_3I

	CD_3Cl	CD_3I	
	Ref. 150	*Ref.* 154	*Ref.* 155
$\nu_0(\nu_4)$	2283.445 ± 0.008	2298.526 ± 0.003	—
$\nu_0(2\nu_4)(\perp)$	4551.273 ± 0.006	4580.494 ± 0.005	—
$A_0 - \frac{1}{3}\eta_{44}$	2.5930 ± 0.0006	2.5788 ± 0.0004	2.5792 ± 0.0003
$D_0{}^K$	$(23 \pm 2) \times 10^{-6}$	$(35 \pm 2) \times 10^{-6}$	$(36 \pm 1) \times 10^{-6}$
$\alpha_4{}^A$	$(12.53 \pm 0.06) \times 10^{-3}$	$(12.9 \pm 0.04) \times 10^{-3}$	$(12.87 \pm 0.03) \times 10^{-3}$
$\alpha_4{}^B$	$(114 \pm 5) \times 10^{-6}$	$(83 \pm 2) \times 10^{-6}$	$(86 \pm 4) \times 10^{-6}$
B_0	0.360127^b	0.2014822^c	0.2014822^d
$D_0{}^{JK}$	$3.39 \times 10^{-6\ b}$	$1.612 \times 10^{-6\ c}$	$1.612 \times 10^{-6\ d}$
$D_0{}^J$	$3.59 \times 10^{-7\ b}$	$1.197 \times 10^{-7\ c}$	$1.19 \times 10^{-7\ d}$

[a] All values/cm^{-1}; [b] microwave, ref. 151; [c] microwave; [d] microwave.

Methyl Bromide. The ν_3 parallel band of CH_3Br in the i.r., at 610 cm^{-1}, has been measured at 0.03—0.06 cm^{-1} resolution by Anderson and Overend.[152] The isotopic splitting of the molecules $^{12}CH_3{}^{79}Br$ and $^{12}CH_3{}^{81}Br$ was resolved for the lines R_{12} to R_{39} in the R-branch. There was no evidence of K-structure in the P- and R-branch lines; the lines P_3—P_{52} and R_1—R_{48} of $^{12}CH_3{}^{79}Br$ and P_2—P_{51} and R_1—R_{50} of $^{12}CH_3{}^{81}Br$ were assigned. The band constants agreed well with recent microwave [153] measure-

[149] S. Takahashi and T. Osaka, *Tohoku Daigaku Kagaku Keisoku Kenkyusho Hokoku*, 1970, **19**, 43 (*Chem. Abs.*, 1971, **74**, 132 630t).

[150] (*a*) R. W. Peterson and T. H. Edwards, *J. Mol. Spectroscopy*, 1971, **38**, 524; (*b*) A. K. Garrison, J. W. Simmons, and C. Alexander, *J. Chem. Phys.*, 1966, **45**, 413.

[151] (*a*) M. Morillon-Chapey and G. Graner, *J. Mol. Spectroscopy*, 1969, **31**, 155; M. Morillon-Chapey, G. Graner, C. Alamichel, C. Betrencourt-Stirnemann, M. Betrencourt, and J. Pinard, *J. Phys.* (*Paris*), 1970, **31**, 519; (*b*) C. Betrencourt-Stirnemann and C. Alamichel, *ibid.*, p. 285.

[152] D. R. Anderson and J. Overend, *Spectrochim. Acta*, 1971, **27A**, 2013.

[153] Y. Morino and C. Hirose, *J. Mol. Spectroscopy*, 1967, **24**, 204.

ments. The slight difference is accounted for by errors in the i.r. analysis that are due to the inability to resolve the K fine-structure of the J lines. For $^{12}CH_3{}^{79}Br$, $\nu_0 = 611.088 \pm 0.002$ cm^{-1}, $B_0 = 0.31947 \pm 0.00006$ cm^{-1}, and $D_0 = (4.2 \pm 0.6) \times 10^{-7}$ cm^{-1}. For the $^{12}CH_3{}^{81}Br$ species, $\nu_0 = 609.903 \pm 0.003$ cm^{-1}, $B_0 = 0.31817 \pm 0.00009$ cm^{-1}, and $D_0 = (3.7 \pm 0.6) \times 10^{-7}$ cm^{-1}. The microwave values are $B_0 = 0.319161(1)$ cm^{-1} and $D_0 = 3.3 \times 10^{-7}$ cm^{-1} for $CH_3{}^{79}Br$, and $B_0 = 0.317948(1)$ cm^{-1} and $D_0 = 3.2 \times 10^{-7}$ cm^{-1} for $CH_3{}^{81}Br$.

Methyl Iodide. Peterson and Edwards [154] have analysed the i.r. spectra of the ν_4 and $2\nu_4$ bands of CD_3I at a resolution of 0.04—0.05 cm^{-1}. As in the case of CD_3Cl described above, over 200 rotational lines in ν_4 and $2\nu_4$ were simultaneously assigned and analysed using the microwave values of ground-state constants. They obtained a value for $[A_0 - \frac{1}{3}\eta_{44}]$ of 2.5788 ± 0.0004 cm^{-1}. The results of the simultaneous analysis of ν_4 and $2\nu_4$ are shown in Table 19. The $(\nu_1 + \nu_5)$, $(\nu_2 + \nu_4)$, and $(\nu_4 + \nu_5)$ bands of CD_3I have been observed and measured by Kurlat *et al.*[155] in the region 3100—3400 cm^{-1} with a resolution of 0.06 cm^{-1}. The fine-structure data from the $(\nu_4 + \nu_5)$ and $(\nu_2 + \nu_4)$ bands have been combined with data from $2\nu_4$ and analysed. The $(\nu_1 + \nu_5)$ band is thought to be subject to strong l-type resonance effects and was, thus, not included in the analysis. The rotational constants (given in Table 19) obtained from the simultaneous analysis agree closely with those of Peterson and Edwards.[154] Kurlat *et al.*[155] found that $[A_0 - \frac{1}{3}\eta_{44} + \frac{1}{3}\eta_{42}] = 2.5792 \pm 0.0003$ cm^{-1}.

Matsuura and Overend [156] have measured the ν_5 and $(\nu_3 + \nu_6)$ bands of CH_3I with a resolution of 0.03 cm^{-1}. The excited states of these bands form a Fermi diad and, from an analysis of 600 rotational lines in the spectrum, the matrix element of the Fermi-resonance operator has been determined. In addition to the Fermi resonance, a further perturbation due to Coriolis resonance between the $v = 5$ and $v = 3$, $v = 6$ states was discovered.

Connes *et al.*[24a] have recently reported a high-resolution study (0.005 cm^{-1}) of ν_4 of CH_3I using a Fourier Transform interferometer.

Trifluoromethane. Hoskins [157] has examined the vibration–rotation spectra of the perpendicular bands of CHF_3 and CDF_3. The ν_4 band of CHF_3 and the ν_5 band of CDF_3 were resolved and Coriolis coupling constants were determined. The measured band profiles of the ν_6 bands also enabled Coriolis constants to be calculated and used to determine the E-species force-constants of the harmonic potential function.

Bürger and Ruoff [158] and Ruoff *et al.*[159] have investigated the vibration spectra of symmetric-top molecules, and report the i.r. spectra of CHF_3

[154] R. W. Peterson and T. H. Edwards, *J. Mol. Spectroscopy*, 1971, **38**, 1.
[155] H. Kurlat, M. Kurlat, and W. E. Blass, *J. Mol. Spectroscopy*, 1971, **38**, 197.
[156] H. Matsuura and J. Overend, *J. Chem. Phys.*, 1971, **55**, 1787.
[157] L. C. Hoskins, *J. Chem. Phys.*, 1970, **53**, 4216.
[158] H. Bürger and A. Ruoff, *Spectrochim. Acta*, 1970, **26A**, 1449.
[159] A. Ruoff, H. Bürger, and S. Biedermann, *Spectrochim. Acta*, 1971, **27A**, 1359.

and CDF_3 measured with a resolution of 0.2—1.0 cm^{-1}. The rotational structure was observed in all fundamental bands, $\nu_1 \rightarrow \nu_6$. The bands were analysed and compared with computer-simulated profiles. Coriolis coupling coefficients ζ_4, ζ_5, and ζ_6, and band origins, were calculated. In Table 20 are shown the ν_0 values, ζ values, and some α constants for CHF_3 and CDF_3.

Table 20 *Rotational constants, band origins, and* ζ *constants for* HCF_3 *and* DCF_3[158]

	HCF_3			DCF_3		
	ν_0/cm^{-1}	ζ	α/cm^{-1}	ν_0/cm^{-1}	ζ	α/cm^{-1}
ν_1	3034.2	—	—	2261.11	—	—
ν_2	1140.7	—	—	1111.2	—	—
ν_3	700.0	—	—	694.61	—	—
ν_4	1377.27	—	—	1212.6	—	—
ν_5	1152	—		$\left\{\begin{matrix}975.05\\975.15\end{matrix}\right.$	—	—
ν_6	508.1	—	—	502.4	—	—
ζ_4	—	0.993 $\pm$0.005	—	—	—	—
ζ_5	—	—	—	—	0.73 $\pm$0.02	—
ζ_6	—	−0.79 $\pm$0.01	—	—	−0.73 $\pm$0.02	—
α_1^B	—	—	—	—	—	11.0×10^{-4}
α_3	—	—	$(6.5\pm1.0)\times10^{-4}$	—	—	7.2×10^{-4}
α_4^A	—	—	$(3.8\pm0.9)\times10^{-4}$	—	—	—
α_6^A	—	—	$(2.8\pm0.9)\times10^{-4}$	—	—	$(1.9\pm1.0)\times10^{-4}$

Methylacetylene. Duncan, Wright, and Ellis[160] have reinvestigated the Fermi- and Coriolis-interacting band system ν_4, ν_7, $(\nu_8+\nu_{10})$ of methylacetylene with a resolution $\leqslant$0.3 cm^{-1}. The previous analysis of Thomas and Thompson[161] was shown to be in error. The band system is complicated by a localized perturbation with $(\nu_5+\nu_9)$ at 1540 cm^{-1} and by a series of hot transitions. Identification and analysis of the *Q*-branches arising from ν_7 and $(\nu_8+\nu_{10})$ leads to the molecular parameters $\nu_7 = 1450.88$ cm^{-1}, $\zeta_7 = -0.311$, $(\nu_8+\nu_{10}) = 1371.08$ cm^{-1}, $\zeta = -1.296$. By taking the Coriolis interaction between ν_4 and ν_7 into account, the complicated structure in the region of 1390 cm^{-1} was assigned, and enabled the following molecular parameters to be determined: $\nu_4 = 1390.55 \pm 0.02$ cm^{-1}, $\zeta_{47} = 0.55 \pm 0.05$. During the course of the analysis, it was shown that the *K*-structure of the very weak fundamental ν_8 had been misassigned by three units in *K*.

[160] J. L. Duncan, I. J. Wright, and D. Ellis, *J. Mol. Spectroscopy*, 1971, **37**, 394.
[161] R. K. Thomas and H. W. Thompson, *Spectrochim. Acta*, 1968, **24A**, 1337.

Hence, the band centre of ν_8 was revised from the original value of 1053.2 cm^{-1} to the new value of 1036.03 cm^{-1}.

Dimethylacetylene. Olson and Papousek [162] have measured the i.r. spectra at high resolution (0.03 cm^{-1}) of the perpendicular ν_9, ν_{13} C—H stretching band of MeCCMe. The *Q*-branches of the band are grouped into bunches. The interval between the bunches is 9.38 cm^{-1} and between neighbouring *Q*-branches is 0.22 cm^{-1} ($2B$ cm^{-1}). The overall appearance of the band confirms the intensity predictions in the bunches of *Q*-branches. The *Q*-structure in the bunches of *Q*-branches could only be partially resolved, which indicated a very small value for $(B'-B'')$, perhaps as low as 10^{-5} cm^{-1}. A value was obtained of $B_0 = 0.11156 \pm 0.00006$ cm^{-1}, which is in excellent agreement with the B_0 value of 0.1116 ± 0.0002 cm^{-1} obtained from an earlier electron-diffraction investigation.

Allene. Stone [163] has presented some new experimental and theoretical results for the ν_{10} fundamental of allene. Previous work had shown the presence of a strong Coriolis interaction between the $v = 9$ and $v = 10$ vibrational levels. In the recently observed spectrum of the ν_{10} band the $^rQ_2$ branch is broader than the *Q*-branches in the rest of the band, and an explanation of this has been suggested in terms of an *l*-type doubling of the $K = 1$ levels in the $\nu_{10} = 1$ state. A reassignment of the branches associated with the $(\nu_{10}+\nu_{11}-\nu_{11})$ hot band is also given.

Duncan *et al.*[164] have obtained spectra of the ν_9 and ν_{10} fundamentals of allene with a resolution of 0.8 cm^{-1} in the 700—1100 cm^{-1} region of the i.r. Comparison between the observed and the calculated band contours was made to determine the vibrational transition-moment ratio for the Coriolis-interacting ν_9 and ν_{10} fundamentals. The transition-moment ratio $|M_{10}/M_9| = 3.8 \pm 0.8$ differs from all previous estimates, and the discrepancies are discussed.

Vinyl Halides. A re-analysis of the vibration–rotation spectra of ν_7, ν_{10}, ν_{11}, ν_8, and ν_{12} of vinyl chloride, vinyl bromide, and vinyl iodide has been carried out by Elst and Oskam,[165] and by the same authors for vinyl fluoride.[166] All these fundamentals involve motion of the C and H atoms with the exception of ν_8, which is the carbon–halogen bond stretching frequency. The literature assignment of the CH twist, ν_{12}, in vinyl iodide was shown to be incorrect, and a reassignment was suggested on the basis of the vibration–rotation analysis. Vibration–rotation analyses of the fundamentals were carried out and Coriolis interactions are discussed. The ν_8 fundamental of vinyl iodide was found at 545.3 cm^{-1} and ν_{12} at 539.4 cm^{-1}.

[162] W. B. Olson and D. Papousek, *J. Mol. Spectroscopy*, 1971, **37**, 527.
[163] J. M. R. Stone, *J. Mol. Spectroscopy*, 1971, **38**, 503.
[164] J. L. Duncan, D. Ellis, I. J. Wright, J. M. R. Stone, and I. M. Mills, *J. Mol. Spectroscopy*, 1971, **38**, 508.
[165] R. Elst and A. Oskam, *J. Mol. Spectroscopy*, 1971, **40**, 84.
[166] R. Elst and A. Oskam, *J. Mol. Spectroscopy*, 1971, **39**, 357.

Glyoxal. The rotational structure of the fundamental ν_9 of glyoxal has been recorded under low resolution;[167] analysis proved difficult on account of the asymmetry of the molecule and overlap of the hot bands due to molecules in excited states of the torsion (ν_7 = 127 cm^{-1}). Cole and Osborne [168] have recently studied the ν_9 band under high resolution, and 636 vibration–rotation lines have been assigned to the *K* sub-bands. Many lines have been assigned in the central region of the band, where the sub-bands are split by asymmetry. The rotational constants were determined and the ground-state constants are in good agreement with those obtained from the visible electronic spectrum.[169] The constants obtained from the i.r. vibration–rotation analysis are given in Table 21.

Table 21 *Rotational constants of glyoxal* ($C_2H_2O_2$) *from an analysis of* ν_9 (*ref.* 168) *and electronic spectra* (*ref.* 169)

Ref.	*Vibrational state*	ν_0/cm^{-1}	*A*/cm^{-1}	*B*/cm^{-1}	*C*/cm^{-1}	$10^4 D_K$/cm^{-1}
168	Ground state	—	1.8445 ±0.0002	0.15993 ±0.00007	0.14723 ±0.00007	0.199 ±0.014
168	$\nu_9 = 1$	2835.07_4 ±0.002	1.8379 ±0.0002	0.15982 ±0.00007	0.14707 ±0.00007	0.209 ±0.012
169	Ground state	—	1.8441 ±0.0002	0.16008 ±0.00004	0.14738 ±0.00004	—

N-*Methylaziridine.* Ikeda [170] has reported for the first time the far-i.r. spectrum of *N*-methylaziridine, $MeN(CH_2)_2$, in the range 60—250 cm^{-1}, with a resolution of 0.2—1.6 cm^{-1}. Transitions involving torsion–vibration levels were found and the spectrum was explained as a composite of several type *B* bands. The band envelopes were not resolved. The potential function for hindered internal rotation of the methyl group was described. The barrier height is 1252 cm^{-1}. The effect of uncertainty in the moment of inertia of the methyl group on determining the barrier height is discussed.

Diborane. Lafferty *et al.*[171] have studied the two i.r.-active terminal B—H stretching bands ν_8 and ν_{16} of $^{10}B_2H_6$ and $^{11}B_2H_6$ with a resolution of 0.04—0.05 cm^{-1}. The following ground-state rotational constants were determined:

$$
\begin{aligned}
{}^{10}B_2H_6\colon A_0 &= 2.65550 \pm 0.00056 \text{ cm}^{-1}\\
B_0 &= 0.642190 \pm 0.000080 \text{ cm}^{-1}\\
C_0 &= 0.587372 \pm 0.000066 \text{ cm}^{-1}\\
{}^{11}B_2H_6\colon A_0 &= 2.65569 \pm 0.00026 \text{ cm}^{-1}\\
B_0 &= 0.606463 \pm 0.000068 \text{ cm}^{-1}\\
C_0 &= 0.557279 \pm 0.000056 \text{ cm}^{-1}
\end{aligned}
$$

[167] R. K. Harris, *Spectrochim. Acta*, 1964, **20**, 1129.
[168] A. R. H. Cole and G. A. Osborne, *J. Mol. Spectroscopy*, 1970, **36**, 376.
[169] F. W. Birss, J. M. Brown, A. R. H. Cole, A. Lofthus, S. L. N. G. Krishnamachari, G. A. Osborne, J. Paldus, D. A. Ramsay, and L. Watmann, *Canad. J. Phys.*, 1970, **48**, 1230.
[170] T. Ikeda, *J. Mol. Spectroscopy*, 1970, **36**, 268.
[171] W. J. Lafferty, A. G. Maki, and T. D. Coyle, *J. Mol. Spectroscopy*, 1970, **33**, 345.

These constants were used to derive the structural parameters B···B = $1.762_8 \pm 0.0026$ Å, B—H$_t$ = $1.200_5 \pm 0.0036$ Å, B—H$_b$ = $1.320_4 \pm 0.0010$ Å, $\angle H_tBH_t = 121.0 \pm 0.6°$, and $\angle H_bBH_b = 96.2 \pm 0.2°$. Upper-state band constants were also reported.

Nitric Acid. The pure rotation i.r. spectrum of HNO_3 has been observed by Harries, Burroughs, and Duxbury.[172] The absorption spectrum of HNO_3 at 20 Torr was observed with a spectral resolution of 0.15 cm^{-1}, and the observed spectrum agreed with that calculated from the known rotational constants.

Tetramethyltin. As part of a review of the assignments of the C—H stretching modes of all the Group IV tetramethyls, Graham [173] has made some vibrational reassignments on the basis of high-resolution i.r. studies. The broad band at 3047 cm^{1-} in the i.r. spectrum of $SnMe_4$ possesses fine structure in the form of poorly resolved maxima with a spacing of 10 cm^{-1}. The maxima are interpreted as *Q*-branch bunches which are affected by Coriolis coupling between rotation and the C—H vibrations.

5 Rotation Spectra of Gases in Solution

Galatry and Robert [174] have considered two factors which determine the general appearance of the far-i.r. spectrum of a diatomic polar molecule dissolved in a non-polar solvent: the average value of the interaction energy between the dissolved molecule and its neighbourhood, and the mean square of the fluctuations of that interaction. A quantitative analysis was made for HCl dissolved in CCl_4 and in SF_6. The same authors [175] have extended the treatment, using the rotational autocorrelation function, and they have shown that the observed spectrum was determined not only by the position and intensities of the pure rotation lines of the free molecule, but also by the individual linewidth. Robert and Galatry found that the molecular interaction was so low that the highest rotational quantum states of HCl in SF_6 were relatively unperturbed, *i.e.* the HCl molecules are freely rotating. Conversely, the low *J* states of HCl in SF_6 were highly perturbed, and no rotational structure could be resolved.

The above work has been confirmed by the investigations of Huong *et al.*,[176] who observed the vibration–rotation spectra of HF (3700—4250 cm^{-1}), DF (2700—3100 cm^{-1}), HCl (2550—3150 cm^{-1}), and DCl (1900—2250 cm^{-1}), each dissolved in SF_6.

Perchard, Murphy, and Bernstein [177] have studied the pure rotation Raman

172 J. E. Harries, W. J. Burroughs, and G. Duxbury, *Nature Phys. Sci.*, 1971, **232,** 171.
173 S. C. Graham, *Spectrochim. Acta*, 1970, **26A,** 345.
174 L. Galatry and D. Robert, *Chem. Phys. Letters*, 1970, **5,** 120.
175 D. Robert and L. Galatry, *J. Chem. Phys.*, 1971, **55,** 2347.
176 P. V. Huong, M. Couzi, and M. Perrot, *Chem. Phys. Letters*, 1970, **7,** 189.
177 J. P. Perchard, W. F. Murphy, and H. J. Bernstein, *Chem. Phys. Letters*, 1971, **8,** 559.

spectra of HCl, DCl, and HBr dissolved in SF_6 and C_2F_6 at 296 K and excited with the 4880 Å line of an argon-ion laser. Rotational lines of the low J-value transitions were clearly resolved from $J = 2$ to 9 for HCl in SF_6, $J = 1$ to 7 for HBr in C_2F_6, and from $J = 5$ to 13 for DCl in SF_6. A solvent-dependent broadening of the Q-branches of the fundamentals was also observed. Observation of the $J = 2 \rightarrow 4$ transition for HCl and $J = 1 \rightarrow 3$ transition for HBr close to their respective gas-phase frequencies indicates that the low rotational energy levels for these molecules are not perturbed as much as predicted by Galatry and Robert. No J-dependence of the rotational lines was found by Perchard *et al.*, in contrast to the observations of Huong *et al.* on the vibration–rotation transitions of HF in SF_6.

Birnbaum and Ho [178] have observed the pure rotation i.r. spectra of HCl and DCl in SF_6. Only a broad band envelope was obtained for DCl, but in the case of HCl, clearly-resolved rotational fine structure was seen at higher values of J. The band contour in both cases was similar to that which was obtained in the gas phase.

Pure rotational transitions have also been observed in the i.r. spectrum of HCl in liquid argon and krypton [179] at 105 K and 125 K, respectively.

Bratoz *et al.*[180] have proposed a theory to explain the vibrational i.r. spectra of inert solutions of diatomic molecules. The vapour–solution-phase band-shifts were shown to depend on the difference between the solvent–solute interaction energies of the two vibrational states involved in the transition. Band profiles with P- and R-type rotational wings and their relationships with the dominating relaxation process are discussed.

Afanas'eva *et al.*[181] have reported the results of temperature and solvent variation on the 2—0 and 3—0 i.r. vibrational band profiles of CO and HCl dissolved in C_5F_{12}, CF_2Cl_2, C_2Cl_4, and CS_2. Unlike the spectrum of the fundamental band in the gas phase, the general appearance of the solution-phase spectra is that of a low-intensity Q-branch and an increase in intensity of the R-branch relative to the P-branch. The temperature range used in these studies was from −150 to +96 °C. Increase in temperature was observed to decrease the Q-branch intensity, leaving the P- and R-branch intensities unaffected. The overall increase in intensity of the R-branch relative to the P-branch was attributed to vibration–rotation interaction.

The band moments of CO, $CHCl_3$, $CHBr_3$, and CHI_3 molecules dissolved in CCl_4 and CS_2 have been evaluated experimentally by Rossi-Sonnichsen *et al.*[182] The band moments differed markedly from the calculated values owing to asymmetry of the bands. The rotation of the molecules in the liquids was discussed in terms of the Gordon theory.

[178] G. Birnbaum and W. Ho, *Chem. Phys. Letters*, 1970, **5**, 334.
[179] R. M. van Aalst and J. van der Elskan, *Chem. Phys. Letters*, 1971, **9**, 631.
[180] S. Bratoz, J. Rios, and Y. Guissani, *J. Chem. Phys.*, 1970, **52**, 439.
[181] N. I. Afanas'eva, M. O. Bulanin, and N. D. Orlova, *Optika i Spektroskopiya*, 1971, **30**, 669.
[182] I. Rossi-Sonnichsen, J. P. Bouanich, and N. V. Thanh, *Compt. rend.*, 1971, **273**, *C*, 19.

6 Pressure-induced Rotation Spectra and the Rotation Spectra of Gas Mixtures

Collison-induced i.r. spectra in the far-i.r. are treated fully in Chapter 5. In this section, therefore, only the Raman-spectroscopic work and the i.r. investigations which have direct relevance to rotational spectroscopy will be considered.

Most effort has been concentrated on the study of the rotation spectra of simple diatomic molecules under pressure or in admixture with other polar or non-polar gases. Diatomic molecules will first be considered; then the recent work carried out with polyatomics will be reviewed.

Diatomic Molecules.—*Hydrogen and Deuterium.* Cooper, May and Gupta [19,20] have measured interferometrically (resolution 0.03 cm^{-1}) the linewidths and frequencies of the S_0 and S_1 rotational lines in the pure rotation Raman spectrum of H_2, excited by a 150 mW helium–neon laser. The S_0 and S_1 lines were measured in pure H_2, and the S_1 lines in 1 : 10 mixtures with helium and argon as a function of the total density. Gas densities were from 2 to 38 Amagat in the case of pure hydrogen and from 7 to 85 Amagat for the hydrogen mixture with argon or helium. It was observed that, at lower densities, the effect of collisions was to suppress the linewidth; with further increase in gas density, the effect of the collisions was increasingly to broaden the line by perturbation of the molecules (collision-broadening contribution to the linewidth). Line-broadening coefficients B were calculated, and the collisional broadening characterized by the experimentally derived parameter B was compared with the theoretical predictions of Van Kranendonk *et al.* Values of the Raman frequencies of the S_0 and S_1 rotational lines extrapolated to zero density are calculated to be $S_0 = 354.390$ cm^{-1} and $S_1 = 587.060$ cm^{-1}, with an accuracy of ± 0.002 cm^{-1}.

The contour of the depolarized Rayleigh line has been measured for hydrogen (and N_2, CO_2) at various pressures by Hess.[183] The linewidth was determined from two additive contributions, collisional broadening and diffusional broadening. The latter contribution depended on the scattering angle between incident and detected light. It was found that, for scattering of less than 90°, collisional broadening dominated if the pressure was sufficiently high that the molecular mean free path was very short compared with the wavelength of the incident light.

Varghese and Reddy [184] have studied the collision-induced i.r. absorption of the H_2 fundamental in hydrogen–oxygen and hydrogen–xenon mixtures at pressures up to 250 atm. Although the band profiles in the oxygen mixtures showed the usual features of collision-induced absorption, the xenon mixtures exhibited some interesting features; the peak separation between the two components of the Q-branch remained almost constant with increas-

[183] S. Hess, *Z. Naturforsch.*, 1969, **24a**, 1852.
[184] G. Varghese and S. P. Reddy, *Canad. J. Phys.*, 1969, **47**, 2745.

ing density of the mixture, and the lines of the quadrupolar branches Q and S were more pronounced than those in any other binary mixture of H_2 previously studied.

The pressure-induced i.r. spectrum of H_2 in the first overtone region (1.1—1.25 μm) has been recorded at various temperatures and pressures by Watanabe, Hunt, and Welsh.[185] A spectrum observed at 24 K with a gas density of 31 Amagat is analysed on the basis of the Van Kranendonk theory of quadrupole-induced absorption. The relative intensities of the 15 single and double transitions which compose the band and the integrated intensity of the band are in satisfactory agreement with the theory. James [186] has applied the pressure-narrowing theory of Galatry to the calculations of equivalent widths of lines in the quadrupole vibration–rotation spectrum of H_2. A value for the pressure-broadening coefficient of 0.0015 cm^{-1} atm^{-1} at 300 K is obtained for the S_1 line of the 0—1 vibrational transition.

The linewidths of $Q_0 \rightarrow Q_4$ in the electric-field-induced i.r. fundamental band of H_2 have been measured by Hunt *et al.*,[187] with 0.027 cm^{-1} resolution for pressures of 4.4—35 atm. The widths vary linearly with density, with negligible contributions from the Doppler effect because of collisional narrowing. At 300 K, the values obtained for the broadening coefficients are 2.78, 1.83, 3.05, 4.60, and 3.26×10^{-3} cm^{-1} $Amagat^{-1}$ in order of increasing J value ($0 \rightarrow 4$). These results are some 20—30% larger than those of previous Raman data.

Reddy and Kuo [188] have studied the collision-induced first overtone i.r. absorption band of D_2 in the pure gas and in binary mixtures of argon and nitrogen at pressures $\leqslant$800 atm. The band profiles in the nitrogen mixtures are complicated by $Q+Q$ and $Q+S$ double transitions from colliding pairs of D_2 molecules and also of N_2 molecules. Analysis of the single transitions of the D_2–Ar band confirmed that the contribution of overlap forces to the band intensity is negligible.

Oxygen. An interpretation of the visible and near-i.r. absorption spectra of compressed O_2 as collision-induced electronic transitions has been carried out by Tabisz, Allin, and Welsh [189] over a range of pressures and temperatures. Assumption of an appropriate rotational structure and broadening of each rotational transition reproduced the observed band profiles with only minor discrepancies. No specific effects of $(O_2)_2$ species were identifiable in the spectra.

Hydrogen Fluoride. Shaw and Lovell [190] have measured the shifts caused by the presence of CO_2 in the P_4—R_6 lines of the fundamental vibration–rotation band of HF. The pattern of the shifts in the R-branch

[185] A. Watanabe, J. L. Hunt, and H. L. Welsh, *Canad. J. Phys.*, 1971, **49,** 860.
[186] T. C. James, *J. Opt. Soc. Amer.*, 1969, **59,** 1602.
[187] R. H. Hunt, W. L. Barnes, and P. J. Brannon, *Phys. Rev. (A)*, 1970, **1,** 1570.
[188] S. P. Reddy and C. Z. Kuo, *J. Mol. Spectroscopy*, 1971, **37,** 327.
[189] G. C. Tabisz, E. J. Allin, and H. L. Welsh, *Canad. J. Phys.*, 1969, **47,** 2859.
[190] B. M. Shaw and R. J. Lovell, *J. Opt. Soc. Amer.*, 1969, **59,** 1598.

was the same as that which was found by Jaffe *et al.*[191] for the 1—0 band of HF broadened by Kr and other noble gases. The variation in line shift closely followed that which was previously reported for the 1—0 band of HCl broadened by CO_2. Wiggins *et al.*[192a] have measured the rotational linewidths in the 1—0 and 2—0 bands of HF in the presence of argon, krypton, and xenon at one atmosphere pressure. Results for the 1—0 band are similar to those seen for HCl. In the 2—0 band, the widths of the rotational lines increase at high J values for the heavier perturbers and show a significant decrease with increase in temperature. Atwood *et al.*[192b] have examined the integrated intensity of the vibration–rotation bands in HF, HCl, and HBr as a function of foreign gas broadening.

Hydrogen Chloride. The pressure-broadening of the $J = 1$, $J = 2$, and $J = 3$ transitions of the pure rotation i.r. spectrum of HCl have been studied by Lane *et al.*[193] The observed half-widths are all smaller than those predicted on the basis of Lindholm's pressure-broadening theory. The pressure effect of noble gases on the pure rotation spectrum has been discussed by Van Kreveld *et al.*[194] and calculated by Cattini.[195] In the former work, Van Kreveld *et al.* have compared the experimentally observed pure rotation spectra under 10—100 atm argon and helium with the predictions of a simple perturbation theory.

Linewidths of the 0—2 band vibration–rotation lines of gaseous $H^{35}Cl$ and $H^{37}Cl$ alone, and perturbed by argon, have been measured by Levy *et al.*[196] The half-widths of the P_1, P_3, and R_5 lines were observed to increase linearly with argon pressure up to 5—6 atm and then to level off; all half-widths were greater in the presence of argon gas than in its absence. Toth, Hunt, and Plyler[197] have compared their measurements of the linewidths in the 1—0 band of HCl broadened by H_2, D_2, and HCN with the calculated linewidths from Anderson's impact theory. Values for the quadrupole moments and hard-sphere diameters of HCl, H_2, D_2, and HCN are listed.

A theory for the rare-gas-induced Q-branch in the HCl fundamental vibration–rotation band at high pressures has been developed by Herman.[198] The absorption intensity distributions of HCl–rare gas mixtures were computed and show substantial agreement with experiment, especially in the overall magnitude of the Q-branch intensity and its dependence on the rare-

[191] J. H. Jaffe, A. Rosenberg, M. A. Hirschfeld, and N. M. Gailar, *J. Chem. Phys.*, 1965, **43**, 1525.

[192] (*a*) T. A. Wiggins, N. C. Griffen, E. M. Arlin, and D. L. Kerstetter, *J. Mol. Spectroscopy*, 1970, **36**, 77; (*b*) M. R. Atwood, H. Vu, and B. Vodar, *J. Phys.* (*Paris*), 1972, **33**, 495.

[193] K. P. Lane, C. A. Frenzel, and E. B. Bradley, *Ky. Univ. Off. Res. Eng. Serv. Bull.*, 1970, **93**, 26 pp. (*Chem. Abs.*, 1970, **73**, 30 249y).

[194] M. E. Van Kreveld, R. M. Van Aalst, and J. Van der Elsken, *Chem. Phys. Letters*, 1970, **4**, 580.

[195] M. Cattini, *Phys. Letters* (*A*), 1970, **31**, 106.

[196] A. Levy, E. Piolet-Mariel, and C. Haeusler, *Compt. rend.*, 1969, **269**, *B*, 427.

[197] R. A. Toth, R. H. Hunt, and E. K. Plyler, *J. Chem. Phys.*, 1970, **53**, 4303.

[198] R. M. Herman, *J. Chem. Phys.*, 1970, **52**, 2040.

gas species and temperature. The principal discrepancy lies in the underestimation of the component linewidths. Williams *et al.*[199] have proposed a simple fitting procedure for the observed linewidths in the 0—1 and 0—2 vibration–rotation bands of HCl (and CO). The collision process between hard spheres and the transitions between rotational levels are considered in detail and collision parameters are described. The procedure was successfully applied to the foreign-gas broadening of HCl (and CO).

Triatomic Molecules.—*Water*. Varanasi and Prasad[200] have reported the measurement of rotational half-widths and line intensities of the ν_2 fundamental of H_2O in H_2O–CO_2 mixtures. The experiments were performed with small amounts of water vapour (2—10 Torr) and 2—5 atmospheres of CO_2 in a cell of adjustable path-length. The CO_2-broadened half-widths are approximately 0.25 cm^{-1} atm^{-1}, which are about half as large as the self-broadened half-widths of 0.5 cm^{-1} atm^{-1}. In a later high-resolution study of the same fundamental, Varanasi[201] has measured 12 rotational lines. The half-widths of the experimentally observed CO_2-broadened lines are compared with the widths computed using the Anderson–Tsao–Curnutte theory; agreement between theory and experiment is excellent.

Nitrous Oxide. Chalaye *et al.*[202] have determined the i.r. rotational correlation functions, second band moments, and intermolecular mean-squared torques from the observed ν_3 vibration–rotation band profiles of C_2H_2 and N_2O in admixture with helium, argon, hydrogen, oxygen, nitrogen, and methane. Toth[112] has measured the self-broadened and N_2-broadened linewidths of N_2O for the 101 and 021 bands with a resolution of 0.03 cm^{-1}. Quadrupole moments and hard-sphere diameters of N_2O and N_2 were calculated. Experimentally observed linewidths at various temperatures agree well with the predictions of the Anderson impact theory, as modified by Tsao and Curnutte.

Polyatomic Molecules.—Ozier and Fox[203] have extended the binary collision theory of far-i.r. pressure-induced spectra and applied it to tetrahedral molecules such as CH_4, CD_4, and CF_4. Octupole moments were deduced from the experimentally observed translation–rotation spectrum.

A well-resolved vibration–rotation transition in the ν_3 i.r. fundamental of SF_6 has been studied by Djeu and Wolga[204] at low pressures, and its optical saturation has been studied. Further details of the far-i.r. collision-induced spectra of polyatomic molecules, including SF_6, are given in the appropriate section of Chapter 5.

[199] D. Williams, D. C. Wenstrand, R. J. Brockman, and B. Curnutte, *Mol. Phys.*, 1971, **20**, 769.
[200] P. Varanasi and C. R. Prasad, *J. Quant. Spectroscopy Radiative Transfer*, 1970, **10**, 65.
[201] P. Varanasi, *J. Quant. Spectroscopy Radiative Transfer*, 1971, **11**, 223.
[202] M. Chalaye, E. Dayan, and G. Levi, *Chem. Phys. Letters*, 1971, **8**, 337.
[203] I. Ozier and K. Fox, *J. Chem. Phys.*, 1970, **52**, 1416.
[204] N. Djeu and G. J. Wolga, *J. Chem. Phys.*, 1971, **54**, 774.

7 Contours of Vibration–Rotation Bands

Currently, much use is made of band-contour methods for analysing the i.r. bands of large symmetric- and asymmetric-top molecules in which the rotational structure is unresolved or only partly resolved. In these methods, computer programs are used to synthesize band profiles by calculating the frequencies and intensities of rotational lines from the rotational constants of the combining states; the latter are varied until the calculated and experimental profiles match each other. Kidd and King [205] have recently critically surveyed the types of band-contour analysis which have been used by various authors over the past six or seven years to obtain rotational constants. Particular regard has been paid to the matching of intensities of the contour. A brief summary of the available programs for the theoretical calculation of contours is given.

Jalsovszky and Nemes [206] have described a computer program for simulating the rotational structure of the perpendicular vibrational transitions of slightly asymmetric prolate-top molecules. Accounting for the Coriolis interactions, the symmetric-top transitions are exactly matched to $\geqslant K_{-1} = 4$. Calculated and experimental vibration–rotation spectra of the ethylene ν_7 band (880—1000 cm^{-1}) agree well. Paal and Varsanyi [207] have commented on the accuracy of calculation of i.r. band contours of the planar asymmetric tops C_6H_5Cl and C_5H_5S.

Cartwright and Mills [208] have reviewed the effects of *l*-type resonance on vibration–rotation bands in i.r. spectra. Observed spectra were compared with the computer-simulated spectra obtained by solving the Hamiltonian matrix and calculating the true (perturbed) wavenumber and intensity of each line in the band. The most obvious effects in the spectra were shown to result from intensity perturbations rather than line-shifts. For example, in oblate symmetric tops the *Q*-branch structure near the band centre may show anomalies due to *l*-type resonance, even at low resolution. Numerical values of the *l*-doubling constants were obtained for several cyclopropane bands by comparing computed contours with those observed at about 0.2 cm^{-1} resolution. Further details of the analysis of the perpendicular fundamentals ν_{11}, ν_{10}, ν_9, and ν_8 of C_3H_6 are given by Mills.[209]

Seth-Paul [210] has reviewed some classical and modern procedures for calculating the *PR* and *QQ* separations of band envelopes of symmetric- and asymmetric-top molecules. Fletcher [211] has investigated the general theory of contours of unresolved Raman bands for spherical tops and symmetric tops and has written high-speed programs for plotting Raman bands at any

[205] K. G. Kidd and G. W. King, *J. Mol. Spectroscopy*, 1971, **40**, 461.
[206] G. Jalsovszky and L. Nemes, *Acta Chim. Acad. Sci. Hung.*, 1971, **68**, 65.
[207] E. Paal and G. Varsanyi, *Acta Chim. Acad. Sci. Hung.*, 1969, **62**, 51.
[208] G. Cartwright and I. M. Mills, *J. Mol. Spectroscopy*, 1970, **34**, 415.
[209] I. M. Mills, *Pure Appl. Chem.*, 1969, **18**, 285.
[210] W. A. Seth-Paul, *Mededel. vlaam. chem. Ver.*, 1969, **31**, 117.
[211] W. H. Fletcher, *U.S. Govt. Res. Develop. Reports*, 1971, **71**, 62.

chosen resolution. Masri and Williams [212] have described a program in Fortran IV language for the calculation of degenerate vibration–rotation bands of symmetric tops in the i.r. or Raman spectrum. Where the rotational structure is partially resolved, the program can be used to assist in the assignment of band features. Masri and Fletcher[213] have performed numerical calculations of the *PR*-branch separation of the three-fold degenerate vibrations of spherical-top molecules, taking into consideration the Coriolis interaction. Aliev [214] has extended the band analysis of Masri and Fletcher and has derived expressions for the relative intensities and frequency separations of the branches. Aliev applied the expressions to the analysis of the Raman spectra of CF_4, SF_6, WF_6, OsF_6, IrF_6, and ReF_6.

The contour of the ν_4 band of CF_4 ($\nu = 631$ cm^{-1}) consists of three maxima,[215] the interval between the peaks of the side maxima (Δ) being 14 cm^{-1}. Assuming a B value of 0.19 cm^{-1} and $\zeta_4 = -0.35$, Aliev calculated $\Delta(R^+, P^-)$ to be 16 cm^{-1} and $\Delta(R^-, P^+)$ to be 17 cm^{-1}. The analysis of the ν_3 band was more complicated as this band has a single central branch and six side maxima. Despite this complication, Aliev was able to assign the side maxima to the constituent *P*-, *Q*-, *R*-, and *S*-branches. In the case of the hexafluorides (SF_6, WF_6) only the triply degenerate ν_5 was considered. Because of the small Δ values and a sufficiently uniform intensity distribution between the seven branches, individual branches may be unresolved. This has been confirmed experimentally; in octahedral molecules a broad ν_5 band is observed with unresolved fine structure. The widths of the bands (δ) calculated by Aliev for SF_6, WF_6, OsF_6, IrF_6, and ReF_6 according to the relationship $\delta \approx 8(BkT)^{1/2}$ agree very closely with the experimentally observed bandwidths.[216]

Gaufres and Sportouch [217] have obtained band constants from an analysis of the contour of the intense *Q*-branch of the Σ^+—Σ^+ Raman band of natural Cl_2 ($\Delta v = 1$). The *Q*-branches for the $^{35}Cl_2$, $^{37}Cl^{35}Cl$, and $^{37}Cl_2$ molecules were identified. The α_e value was determined to be 0.0015 cm^{-1}.

Vibration–Rotation Coupling Constants.—It is well known that the Coriolis coupling constants of symmetric-top molecules influence the shapes of the i.r. band contours for degenerate vibrations. Müller *et al.*[218] have derived expressions for the Coriolis coupling constants ζ_1, ζ_2, and ζ_3 of the *E*-type vibrations of XY_3Z molecules and have explained the signs of the ζ-constants by means of the *L*-matrix. Reasonably good agreement between theory and experiment was given for a large number of molecules, including CCl_3H and $SiCl_3H$.[158]

212 F. N. Masri and I. R. Williams, *Computer Phys. Comm.*, 1970, **1**, 349.
213 F. N. Masri and W. H. Fletcher, *J. Chem. Phys.*, 1970, **52**, 5759.
214 M. R. Aliev, *Optika i Spektroskopiya*, 1971, **31**, 376.
215 B. Monostori and A. Weber, *J. Chem. Phys.*, 1960, **33**, 1867.
216 H. H. Claassen and H. Selig, *Israel J. Chem.*, 1969, **7**, 499.
217 R. Gaufres and S. Sportouch, *Compt. rend.*, 1971, **272**, *B*, 995.
218 A. Müller, R. Kebabcioglu, S. J. Cyvin, and H.-J. Schumacher, *J. Mol. Spectroscopy*, 1970, **36**, 551.

Jacobi [219] has made a classical treatment of the vibration–rotation interaction on the line intensities of diatomic molecules for the fundamental and first and second overtone bands (*e.g.* CO). The use of vibration–rotation coupling constants for calculating the force fields in simple molecules is illustrated by the recent work of Kredentser and Sverdlov [220] on BHF_2. Coriolis coupling constants have also been evaluated for NO_2F and NO_2Cl (ref. 221), NF_3 (ref. 222), and COF_2 (ref. 223).

A general theoretical treatment of the intensity perturbations due to centrifugal distortion for the vibration–rotation lines of asymmetric rotors has been given by Ben-Aryeh.[224] An example was provided by the analysis of the ν_2 band of H_2O. Tarrago [225] has calculated the rotational contributions to the vibration–rotation energies of molecules with C_{3v} symmetry; frequency expressions as functions of J and K for pure rotational transitions in the ground state and in excited vibrational states were obtained.

Watson [226] and Mills *et al.*[227] have pointed out that some of the pure rotational transitions of polyatomic molecules that are forbidden according to the rigid-rotor selection rules may acquire intensity by a vibration rotation interaction mechanism involving centrifugal distortion; the intensity is dependent on the dipole derivatives and on the normal-co-ordinate displacements produced by the centrifugal distortion. The structures of the predicted spectra are discussed for non-polar molecules which belong to the point groups D_{3h}, D_{2d}, and T_d. Some of the calculated R-branch lines of CH_4 are predicted to be stronger than some of the rotational lines already observed for HD in the same frequency range. There is, thus, a possibility that the far-i.r. forbidden rotational transitions of molecules such as CH_4, CH_2CCH_2, BF_3, NH_3, and the methyl halides may be observed in the near future.

[219] N. Jacobi, *J. Chem. Phys.*, 1970, **52**, 2694.
[220] E. I. Kredentser and L. M. Sverdlov, *Izvest. V.U.Z. Fiz.*, 1971, **14**, 26.
[221] E. V. Rao, *Indian J. Pure Appl. Phys.*, 1969, **7**, 762.
[222] I. W. Levin, *J. Chem. Phys.*, 1970, **52**, 2783.
[223] E. I. Kredentser and L. M. Sverdlov, *Optika i Spektroskopiya*, 1971, **30**, 990.
[224] Y. Ben-Aryeh, *J. Opt. Soc. Amer.*, 1970, **60**, 1469.
[225] G. Tarrago, *J. Mol. Spectroscopy*, 1970, **34**, 23.
[226] J. K. G. Watson, *J. Mol. Spectroscopy*, 1971, **40**, 536.
[227] I. M. Mills, J. K. G. Watson, and W. L. Smith, *Mol. Phys.*, 1969, **16**, 329.

7
Vibrational Spectroscopy of Macromolecules

BY V. FAWCETT AND D. A. LONG

1 Introduction

This chapter reviews recent work on the i.r. and Raman vibrational spectroscopy of polymers, including biopolymers. Spectroscopic studies of a number of individual polymers are described and the section on biopolymers includes accounts of work on polypeptides, proteins, nucleic acids, and polynucleotides. The literature for the period 1970—mid 1972 is covered in detail and, where appropriate, some earlier papers are dealt with. Reference is also made to other structural techniques where necessary.

Polymers have been the subject of spectroscopic study since the early days of i.r. and Raman spectroscopy, although progress had tended to be rather slow until recently, because spectra were hard to obtain and difficult to interpret. However, as this Report will show, in recent years there have been very considerable advances in the vibrational spectroscopy of polymers, both from the experimental and theoretical standpoints. Vibrational spectroscopy is now able to make substantial contributions to our knowledge of the structure of polymers, often complementing information obtained from other structural investigations.

The improvements in experimental techniques have been most substantial in Raman spectroscopy, and stem particularly from the special properties of gas lasers, now almost invariably used for Raman excitation. Thus the high power densities available make it possible to obtain spectra from very small samples, and the polarization properties and low divergence of the laser beam enable the scattering from oriented crystalline polymers and oriented fibres to be studied as a function of illumination direction and scattering angle. Double and triple monochromators and special post-sample filters discriminate very effectively against stray scattering, and enable low-frequency Raman shifts to be observed as readily as those of higher frequency. Improvements in detection systems and the use of systems which average the signal over a series of spectral scans enable very weak Raman scattering to be recorded. In i.r. spectroscopy developments in instrumentation for the far-i.r. region have made it possible to obtain low-frequency i.r. spectra quite readily. The spectra shown in Figures 1—5 give an indication of what can now be achieved.

On the theoretical side there has been a considerable increase in our understanding of the theory of vibrations in large molecules. With the

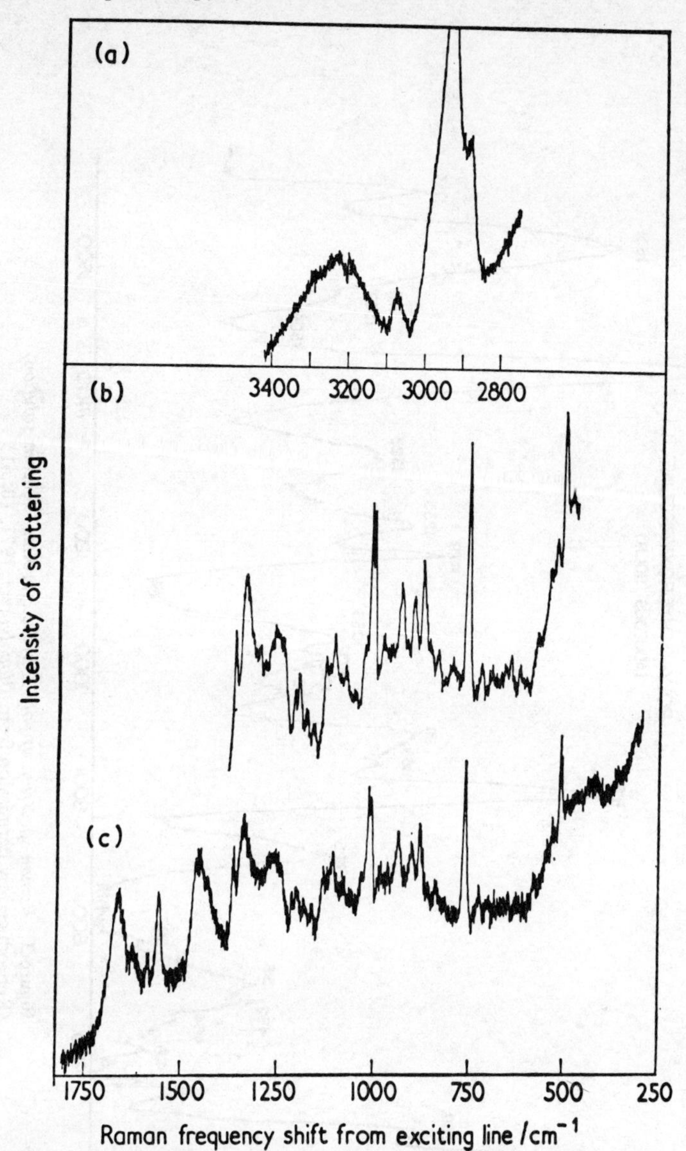

Figure 1 *Raman spectrum of lysozyme in water. The concentration was* 12 mg *in* 40μl *of solution, pH* 5.2, 28 °C (a) *Spectral slit width,* $\Delta\sigma$, 15 cm^{-1}; *sensitivity, s,* 1.6×1000; *rate of scan, r,* 0.10 cm^{-1} s^{-1}; *period,* τ, 21 s. *Because of drop-off in photomultiplier sensitivity at high frequency shifts, the apparent intensity of the spectrum above* 2800 cm^{-1} *is greatly reduced compared to the region below* 1750 cm^{-1}. (b) $\Delta\sigma$, 5 cm^{-1}; s, 2.2×1000; *r,* 0.025 cm^{-1} s^{-1}; τ, 35 s. (c) $\Delta\sigma$, 5 cm^{-1}; *s,* 1.2×1000; *r,* 0.10 cm^{-1} s^{-1}; τ, 13 s
(Reproduced by permission from *J. Mol. Biol.*, 1970, **50**, 509)

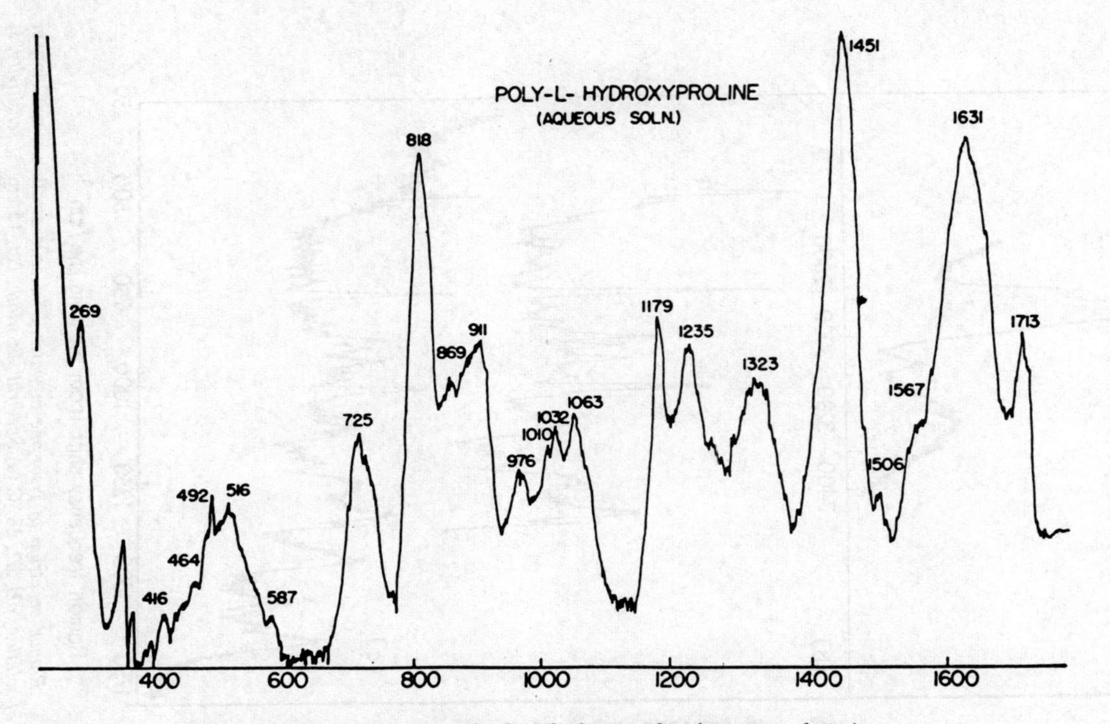

Figure 2 *Raman spectrum of poly-L-hydroxyproline (aqueous solution)*
(Reproduced by permission from *Biopolymers*, 1971, **10**, 615)

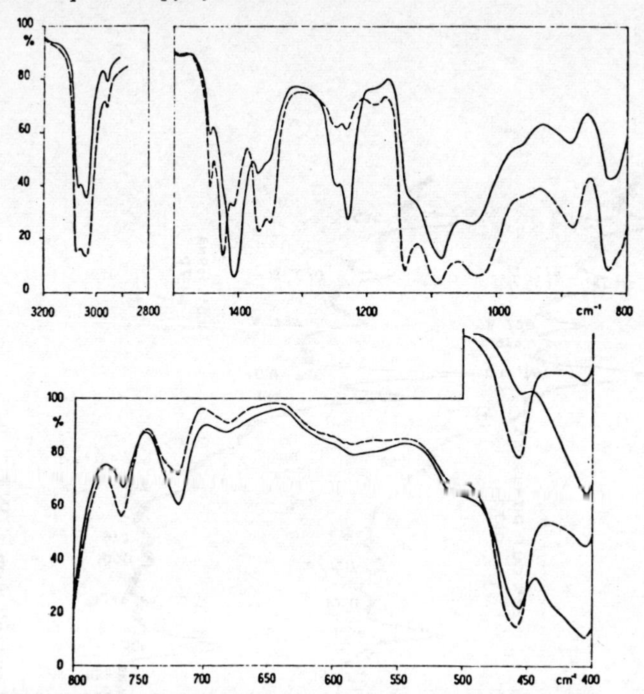

Figure 3 *Infrared spectrum of an oriented sample of poly(vinyl fluoride) in polarized light;* ——— *electric vector parallel to the drawing direction;* – – – – *electric vector perpendicular to drawing direction*
(Reproduced by permission from *Spectrochim. Acta*, 1970, **26A**, 733)

availability of large computers, it has been possible to carry out detailed calculations on very large systems and so identify frequencies or frequency regions associated with particular types of motion. Such calculations have been extended to include the effect on the vibrational spectra of the introduction of various kinds of structural defects in a polymer.

The structural information derived from vibrational spectroscopy frequently cannot be obtained using other techniques. Vibrational spectroscopy, particularly Raman spectroscopy, is especially valuable for the study of biopolymers in aqueous solutions; the effect of temperature, pH, and ionic strength may then be investigated. It should be noted too that vibrational spectra tend to reflect the detailed fine structure of large molecules, whereas many other optical techniques give information only on overall gross structure. A particularly subtle example is provided by the studies of Raman intensities of certain bands of the nucleic acids. The intensities of these bands are controlled, through the pre-resonance Raman effect, by the u.v. absorption characteristics of these molecules. Since the u.v. absorption is sensitive to the stacking of the purine and pyrimidine bases, the Raman intensities form a sensitive probe of the base-stacking arrangements.

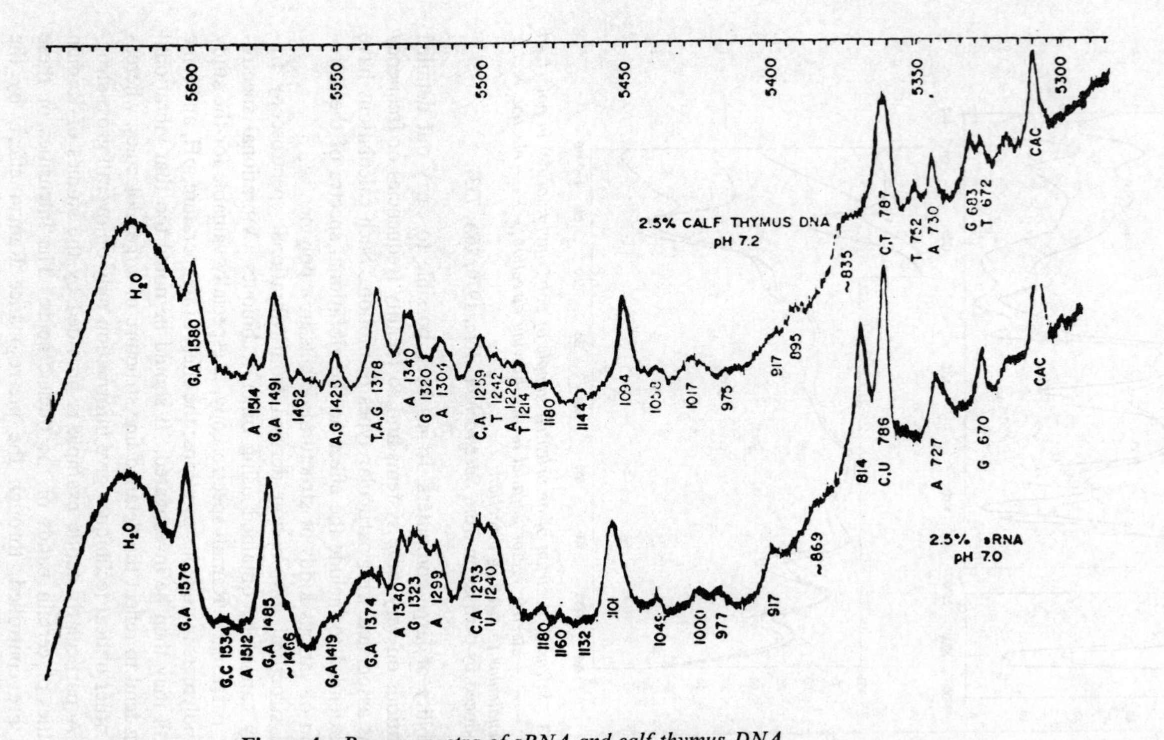

Figure 4 *Raman spectra of sRNA and calf thymus DNA.*
(Reproduced by permission from *Biopolymers*, 1971, **10**, 1377)

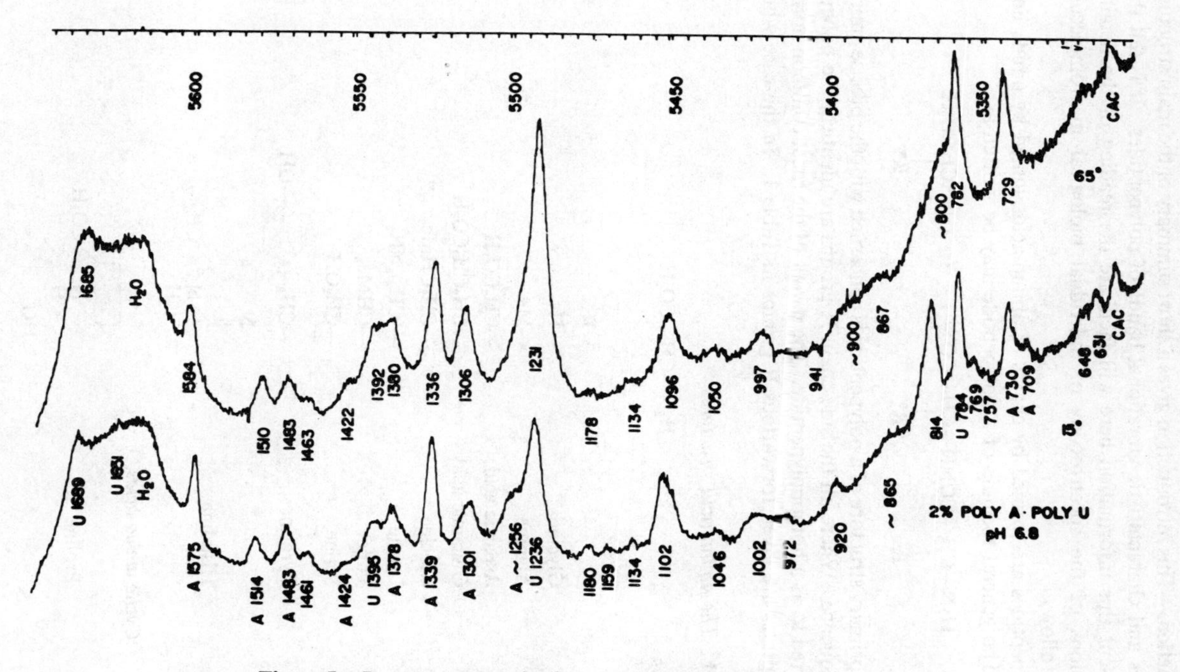

Figure 5 *Raman spectra of poly A.U at 15 and 65 °C*
(Reproduced by permission from *Biopolymers*, 1971, **10**, 1377)

2 Biological Macromolecules

Introduction.—This introduction gives a short summary of the main structural features and characteristic vibrational bands of polypeptides. It is felt that to present this information here will facilitate the reading of the detailed discussions of the spectroscopy of individual biological macromolecules which follow.

Polypeptides are formed by chains of amino-acids joined by amide linkages. The general formula of a polypeptide may be written as:

$$H_2N\text{—}\underset{\displaystyle R^1}{\underset{|}{CH}}\text{—}CONH\text{—}\underset{\displaystyle R^2}{\underset{|}{CH}}\text{—}CO\text{—}\cdots\cdots\cdots\cdots NH\underset{\displaystyle R^n}{\underset{|}{CH}}CO_2H$$

The primary structure of a polypeptide is concerned with the precise amino-acid sequence. Where all the side-chain R groups are identical the polymer is referred to as a homopolypeptide. The names of the most common homopolypeptides are given for various R groups in Table 1. In the case where

Table 1 *The amino-acid structures*

$$H_2N\text{—}\underset{\displaystyle R}{\underset{|}{CH}}\text{—}CO_2H$$

	R
Glycine	—H
Alanine	—Me
Aspartic acid	—CH_2CO_2H
Glutamic acid	—$CH_2CH_2CO_2H$
Leucine	—CH_2CHMe_2
Lysine	—$(CH_2)_4NH_2$
Valine	—$CHMe_2$
Serine	—CH_2OH
Tyrosine	—CH_2—C_6H_4—OH
Histidine	(imidazolyl: N, NH ring)—CH_2—
Cyclic amino-acids	
Proline	(pyrrolidine ring, NH) CO_2H
Hydroxyproline	HO (pyrrolidine ring, NH) CO_2H

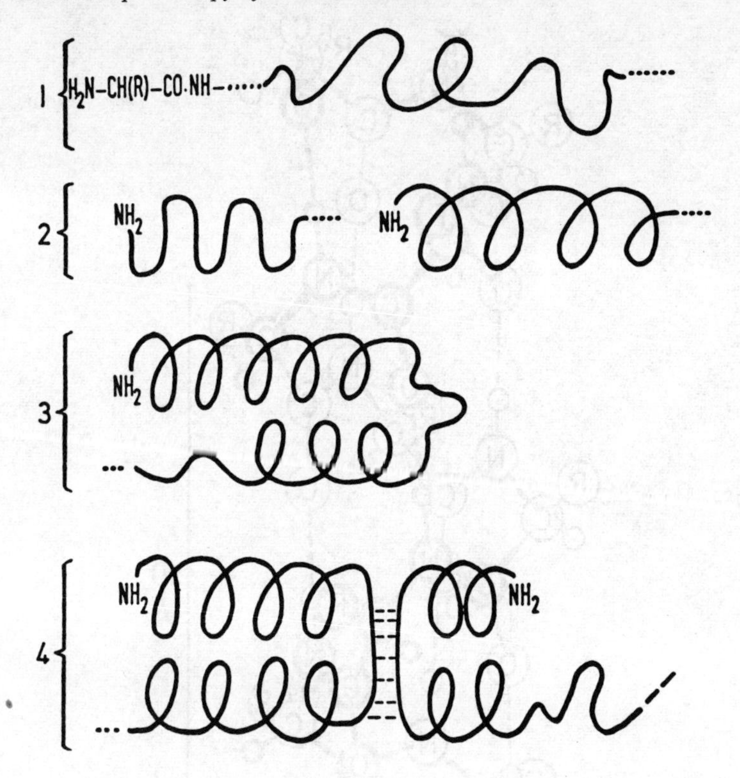

Figure 6 *Schematic representation of primary* (1), *secondary* (2), *tertiary* (3), *and quaternary* (4) *structures in proteins*
(Reproduced by permission from 'Optical Rotatory Dispersion of Proteins and Other Macromolecules', Springer–Verlag, Berlin 1969, p. 2)

a number of different R groups occur in a sequential manner the polymer is referred to as a regular copolymer. If there is no definite sequence then the polymer is called a random copolymer. A block copolymer occurs where two or more homopolymers are linearly linked together to form a single chain. Each homopolymeric sequence is referred to as a block. The secondary structure refers to the detailed arrangement of the polypeptide in space (see Figure 6). The conformation adopted by a particular polypeptide is often dependent upon its physical state. Two main types of secondary interactions occur which influence the conformation adopted by a polypeptide: hydrogen bonding between the oxygen of the carbonyl groups and the hydrogen of the amide groups, and non-polar hydrophobic bonding between hydrocarbons in the polypeptide side-chains. There are three main secondary structures: the α-helical conformation, the β-planar sheet or extended conformation, and the random-coil conformation.

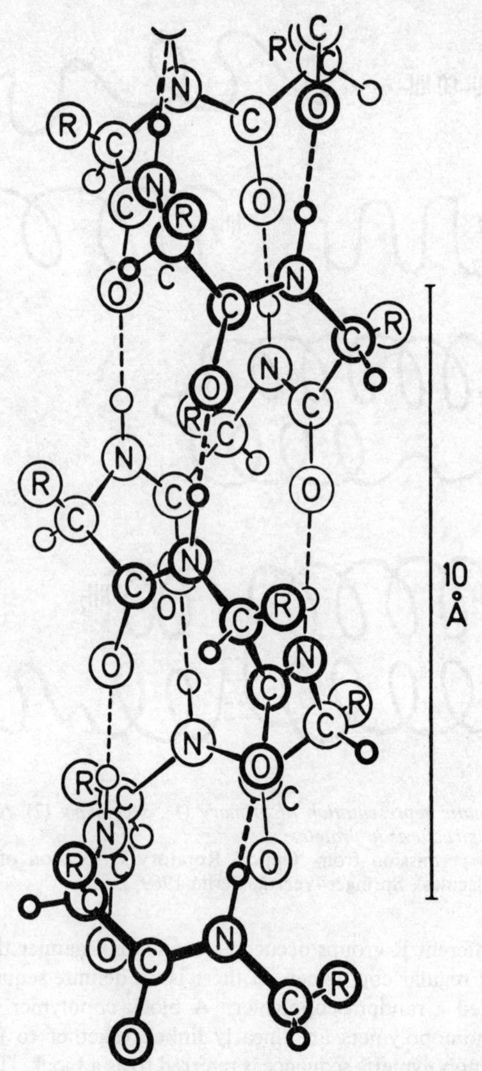

Figure 7 *The Pauling–Corey α-helix*
(Reproduced by permission from *Proc. Roy. Soc.*, 1953, **B141,** 21)

Figure 7 shows a typical α-helix structure. It has been established that the L-enantiomorph of polyalanine is a right-handed α-helix. In general, the screw-senses of other polypeptides have been linked with poly-L-alanine by optical rotation measurements.

The β-planar sheet or extended conformation was first proposed for the structure of silk and steam-stretched hair keratin. In this form, hydrogen

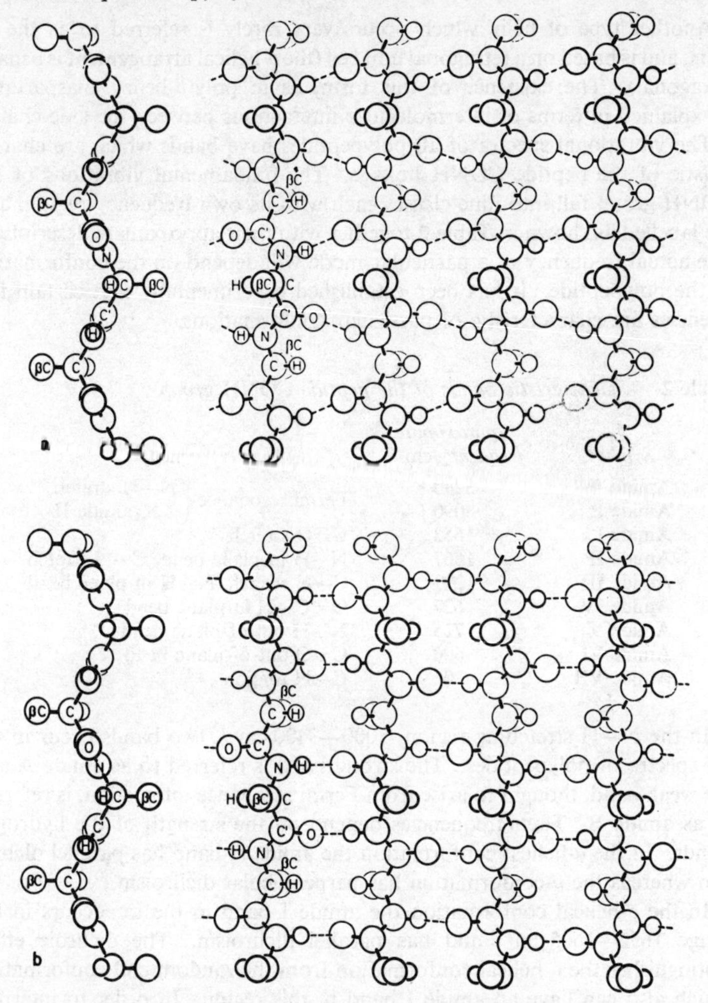

Figure 8 *Parallel* (a) *and antiparallel* (b) *pleated sheet β-structures of extended polypeptide chains*
(Reproduced by permission from *Proc. Roy. Soc.*, 1953, **B141**, 21)

bonding occurs between the NH and CO groups of one chain and the NH and CO groups of adjacent chains on both sides, thus forming a sheet-like structure as shown in Figure 8. Two types of β-conformation are often distinguished and are referred to as the parallel β-structure or antiparallel β-structure depending on the arrangement of the chains in the sheet structure.

The random-coil conformation describes a form of the polypeptide in which the degree of ordering is very small.

Another type of helix which occurs very rarely is referred to as the ω-helix, and is based on a tetragonal unit cell (the α-helical arrangement is usually hexagonal). The existence of this form, as in poly(β-benzyl-L-aspartate), is explained in terms of intermolecular interactions between the side-chains.

The vibrational spectra of all polypeptides have bands which are characteristic of the peptide CONH linkage. The fundamental vibrations of the CONH group fall into nine classes each with its own frequency region, and are labelled as shown in Table 2 together with their approximate description. The actual frequency of a particular mode will depend on the conformation of the polypeptide. It has been established experimentally that certain frequencies are characteristic of particular conformations.

Table 2 *Characteristic bands of the peptide* CONH *group*

Symbol	*Approximate frequency*/cm^{-1}	*Assignment*
Amide A	3280	Fermi resonance {N—H stretch
Amide B	3090	Fermi resonance {2 × amide II
Amide I	1653	C=O stretch
Amide II	1567	N—H in-plane bend, C—N stretch
Amide III	1299	C—N stretch, N—H in-plane bend
Amide IV	627	O=C—N in-plane bend
Amide V	725	N—H out-of-plane bend
Amide VI	600	C=O out-of-plane bend
Amide VII	206	C—N torsion

In the N—H stretching region, 3000—3400 cm^{-1}, two bands occur in the i.r. spectra of polypeptides. The strong band is referred to as amide A and the weak band, thought to arise from Fermi resonance interaction, is referred to as amide B. Their frequencies depend on the strength of the hydrogen bonds. In the α-helical conformation the amide A band has parallel dichroism whereas the β-conformation has perpendicular dichroism.

In the α-helical conformation the amide I band in the i.r. occurs in the range 1652—1665 cm^{-1} and has parallel dichroism. The dichroic effect distinguishes the α-helical conformation from the random-coil conformation which also can have an amide I band in this region. In order to interpret correctly the amide I region for the antiparallel β-conformation, the interactions of four residues (a pair of residues from each of two adjacent chains) must be considered. Taking into account both intrachain and interchain interactions, four spectroscopically active bands arise characterized by their phase relationships, and are described as $\nu(0,0)$, $\nu(0,\pi)$, $\nu(\pi,0)$, and $\nu(\pi,\pi)$ where the first phase relationship is intrachain and the second is interchain. Each of these modes has, in general, a different frequency. The $\nu(0,0)$ mode is Raman-active but i.r.-inactive, the $\nu(0,\pi)$ mode is Raman- and i.r.-active and has parallel dichroism, and $\nu(\pi,0)$ and $\nu(\pi,\pi)$ are Raman- and i.r.-active and have perpendicular dichroism.

In the typical antiparallel β-structure a strong band in the i.r. spectrum occurs in the region of 1632 cm^{-1}, assigned to the $\nu(\pi,0)$ mode, together with a weak band in the region of 1685 cm^{-1}, assigned to the $\nu(0,\pi)$ mode. In general the $\nu(\pi,\pi)$ mode, although i.r.-active, is too weak to be observed but has been calculated to lie at about 1668 cm^{-1}. In the Raman spectrum a strong band occurs in the region of 1674 cm^{-1} and is assigned to the $\nu(0,0)$ mode.

In the typical parallel β-structure a strong band occurs in the i.r. spectrum in the region of 1630 cm^{-1} together with a weak band at 1645 cm^{-1}. The random-coil structure appears to exhibit no distinct features in either the Raman or i.r. spectrum but rather a broad band centred around 1665 cm^{-1}.

In the amide II region of the i.r. spectrum, a typical α-helical structure has a strong band at about 1546 cm^{-1} and a weak band at 1516 cm^{-1}. The β-structures both have strong bands at 1530 cm^{-1} and a weak band at 1550 cm^{-1}. The random-coil structure has a band at about 1535 cm^{-1}. The amide II region in the Raman spectrum tends to be diffuse.

In the amide III region of the i.r. the α-helical conformation has a strong band at 1262 cm^{-1} whereas the β conformation has weaker bands at 1240 cm^{-1} and 1220 cm^{-1}. In the Raman spectrum the α-helical conformation has a strong band in the region of 1265 cm^{-1} whereas the β-conformation has a strong band at 1234 cm^{-1}. The random-coil form has a broad band centred in the vicinity of 1250 cm^{-1}. All the preceding points are summarized in Table 3.

In some cases other amide bands have been used as an indication of the conformation of a polypeptide as will emerge in the following sections.

Table 3 *Typical Raman and i.r. band frequencies* (cm^{-1}) *characteristic of polypeptides in the various conformations.*

Conformation	*Amide I*		*Amide II*	*Amide III*	
	Raman	*I.R.*	*I.R.*	*Raman*	*I.R.*
α-helical	1660 (s)	1657 (s)	1546 (s)	1265 (s)	1262 (s)
			1516 (w)		
β-antiparallel		1685 (w) $\nu_{\parallel}(0,\pi)$	1530 (s)	1234 (s)	1240 (m)
	1674 (s) $\nu(0,0)$			1550 (w)	1220 (w)
		1632 (s) $\nu_{\perp}(\pi,0)$			
β-parallel		1645 (w)	1530 (s)		
			1550 (w)		
		1630 (s)			
random-coil	1665 (br)	1665 (br)	1535 (w)		1250 (br)

Individual Polypeptides.—*Polyglycine.* The simplest of all polypeptides is polyglycine. It is known to exist in two crystalline modifications which are designated I and II. Polyglycine I has the extended planar β-conformation whereas polyglycine II has a 3-fold helical conformation. In both forms of the polypeptide, interchain hydrogen bonding of the type N—H$\cdots$O occurs.

In the case of polyglycine II hydrogen bonding of the type C—H···O also occurs similar to that found in collagen structures.

Normal mode analyses of the homopolypeptides are essential for the understanding of their vibrational spectra, neutron inelastic scattering data, and thermodynamic properties. There have been several attempts to calculate the phonon dispersion curves for both forms of polyglycine. The various models that have been used yield basically the same results for the high-frequency modes. It is in the low-frequency region that the greatest differences occur. Most of these normal-co-ordinate analyses are based on isolated chain models.[1-3] This approach, in the main, ignores the part played by intermolecular interactions. Some caution should therefore be exercised when using the results of such calculations to interpret spectra obtained from the solid-state homopolypeptides. The calculations for the case of helical homopolypeptides must also include terms allowing for intramolecular hydrogen bonding. Peticolas,[4] for instance, in his calculations on α-helical poly-L-alanine, showed that the intramolecular hydrogen-bonding terms had a profound effect on the low-frequency modes.

Polyglycine I. The antiparallel β-sheet structure of polyglycine I in the crystalline state contains four chemical residues per unit cell. The space-group operations are: $\bar{C}_2(z)$ a two-fold screw axis along the chain axis; $\bar{C}_2(y)$ a two-fold screw axis parallel to the interchain hydrogen bond, and $C_2(x)$ a two-fold rotation. Together with the identity operation these symmetry operations form a group isomorphous with point group D_2. The line-group symmetry of the isolated chain is C_{2v} and the site symmetry is C_2. The correlation table for this is shown in Table 4.

Table 4 *Correlation table between line group C_{2v}, site group C_2 and point group D_2 for polyglycine I*

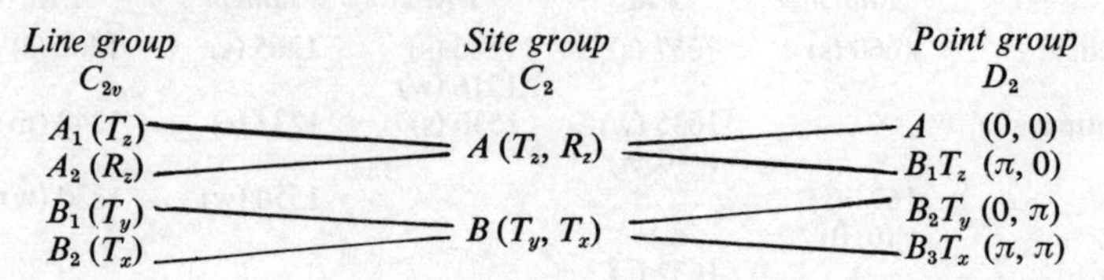

The spectroscopically active modes are those corresponding to all unit cells moving in phase. In this instance it corresponds to the conditions of $\phi = 0,\pi$ and $\theta = 0,\pi$. The spectroscopically active modes are then labelled as (ϕ,θ), that is (0,0), (0,π), (π,0) or (π,π). The symmetry of each of these species is shown in Table 4. In this instance, the modes of type (0,0) are Raman active only while the other modes are both i.r. and Raman active.

[1] V. D. Gupta, A. K. Gupta, and M. V. Krishnan, *Chem. Phys. Letters*, 1970, **6**, 317.
[2] R. D. Singh and V. D. Gupta, *Chem. Phys. Letters*, 1971, **8**, 294.
[3] E. W. Small, B. Fanconi, and W. L. Peticolas, *J. Chem. Phys.*, 1970, **52**, 4369.
[4] B. Fanconi and W. L. Peticolas, *Biopolymers*, 1971, **10**, 2223.

Fanconi[5] has made sophisticated calculations of the normal modes of polyglycine I which take into account intermolecular interactions. The dynamical matrices for the two-dimensional lattice were expressed in terms of an interchain phase angle ϕ and an intrachain phase angle θ. The calculations for this fully extended form of polyglycine were based on two different models: the first was a five-mass model of the repeat unit and included interchain forces directed perpendicular to the chain axis; the second was based on a single-mass model for the repeat unit and was used to determine the effects of torsion about the hydrogen bond and hydrogen-bond bending on the low-frequency vibrations.

The initial set of Urey–Bradley force-field constants used for the five-mass model were taken from a calculation by Fukushima *et al.*[6] A slight

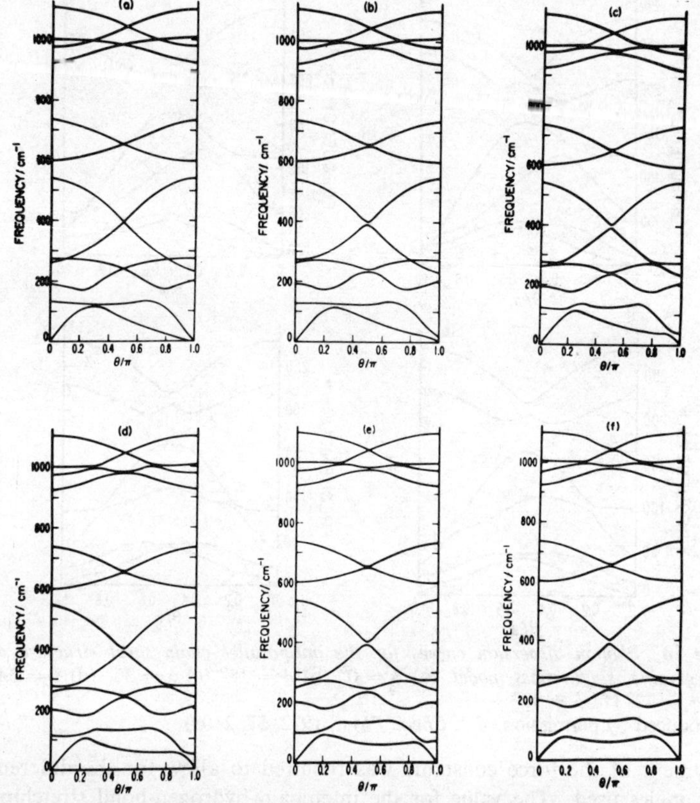

Figure 9 *In-plane phonon dispersion curves for the antiparallel chain sheet structure of polyglycine, five-mass model.* (a) $\phi = 0°$ (b) $\phi = 18°$ (c) $\phi = 36°$ (d) $\phi = 54°$ (e) $\phi = 72°$ (f) $\phi = 90°$

(Reproduced by permission from *J. Chem. Phys.*, 1972, **57**, 2109)

[5] B. Fanconi, *J. Chem. Phys.*, 1972, **57**, 2109.

[6] K. Fukushima, Y. Ideguchi, and T. Miyazawa, *Bull. Chem. Soc. Japan*, 1963, **36**, 1301.

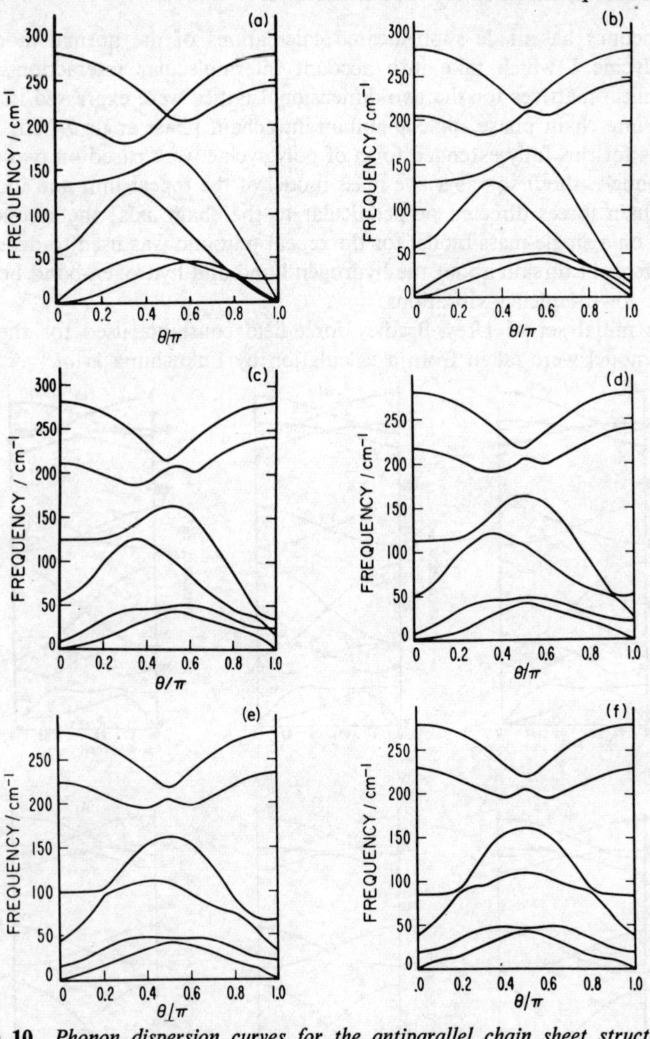

Figure 10 *Phonon dispersion curves for the antiparallel chain sheet structure of polyglycine, single-mass model.* (a) $\phi = 0°$ (b) $\phi = 18°$ (c) $\phi = 36°$ (d) $\phi = 54°$ (e) $\phi = 72°$ (f) $\phi = 90°$
(Reproduced by permission of *J. Chem. Phys.*, 1972, **57**, 2109)

adjustment of the force constants was required to allow for the different bond angles used. The value for the interchain hydrogen-bond stretching force constant was taken from a calculation on poly-L-alanine.[7] The valence force-field constants for the single-mass model were found by fitting the frequencies of the spectroscopically active modes calculated from the five-mass model.

[7] B. Fanconi, E. W. Small, and W. L. Peticolas, *Biopolymers*, 1971, **10**, 1277.

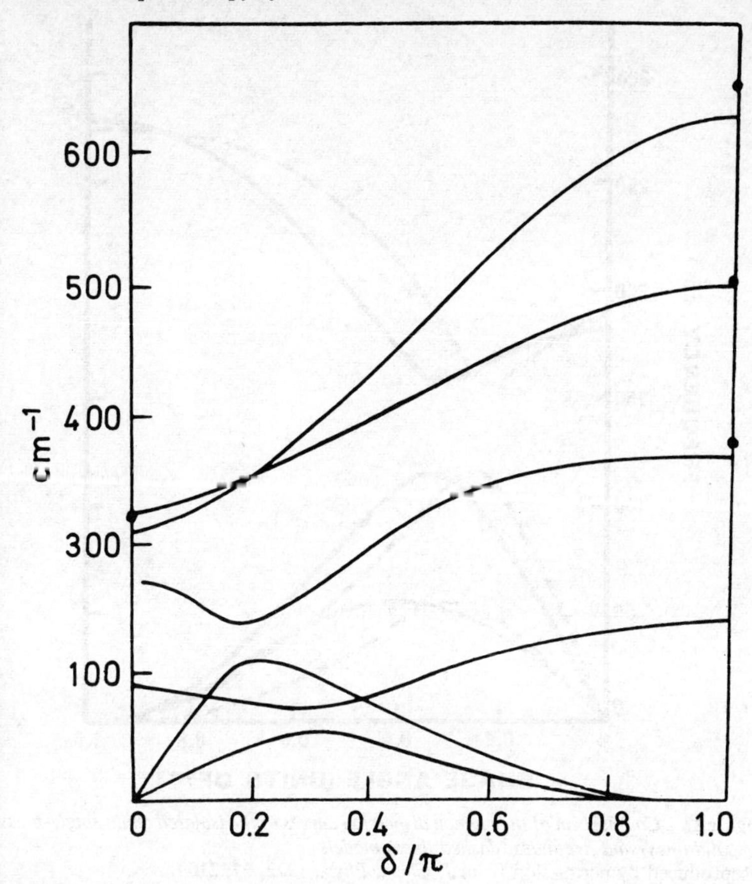

Figure 11 *Dispersion curves for acoustical and low-frequency optical modes for polyglycine I*
(Reproduced by permission from *Chem. Phys. Letters*, 1970, **6**, 317)

The results of Fanconi's calculations for the five-mass model are shown in Figure 9 and for the single-mass model in Figure 10. In each case the frequencies are plotted as a function of θ/π for values of ϕ at intervals of 18°. The higher frequency modes are not included in the Figures since they showed very little dispersion.

By comparison with the phonon dispersion curve calculated for a single-chain model by Gupta *et al.*[1] shown in Figure 11, it can be seen that the inclusion of an intermolecular hydrogen-bonding term has a profound effect on the low-frequency modes.

For the single-mass model the masses were placed so that the interchain bonds were collinear with the hydrogen bonds. The calculated values for the in-plane phonon dispersion curves for an isolated chain are compared

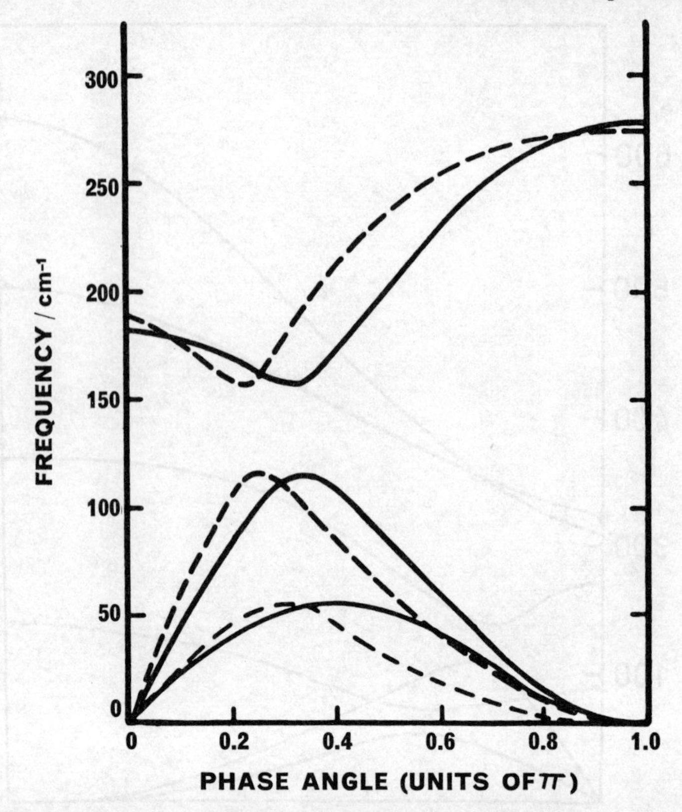

Figure 12 *Comparison of the phonon dispersion curves of the isolated chain single-mass (solid lines) and five-mass (dashed lines) models*
(Reproduced by permission from *J. Chem. Phys.*, 1972, **57**, 2109)

with those obtained from the five-mass model in Figure 12. The main differences occur at the zone boundary. The phonon dispersion curves for a two-dimensional lattice, determined from the single-mass model, were also plotted as a function of intrachain phase angle and are shown in Figure 13. The lowest two branches result from out-of-plane motions and the other four arise from in-plane motions. A histogram was plotted for the frequency distribution as determined from the single-mass model.

The neutron inelastic scattering spectrum of polyglycine I has been determined.[8] Gupta *et al.*[1] have compared this spectrum with their calculations for a polyglycine I isolated chain and conclude that the 90—160 cm^{-1} region contains contributions from hydrogen-bond bending and stretching. The results of Fanconi,[5] however, show that although the hydrogen-bond stretch does contribute to the density of states in the 90—160

[8] Molecular Dynamics and Structure of Solids, N.B.S. Publications No. 301, 1969, p. 559.

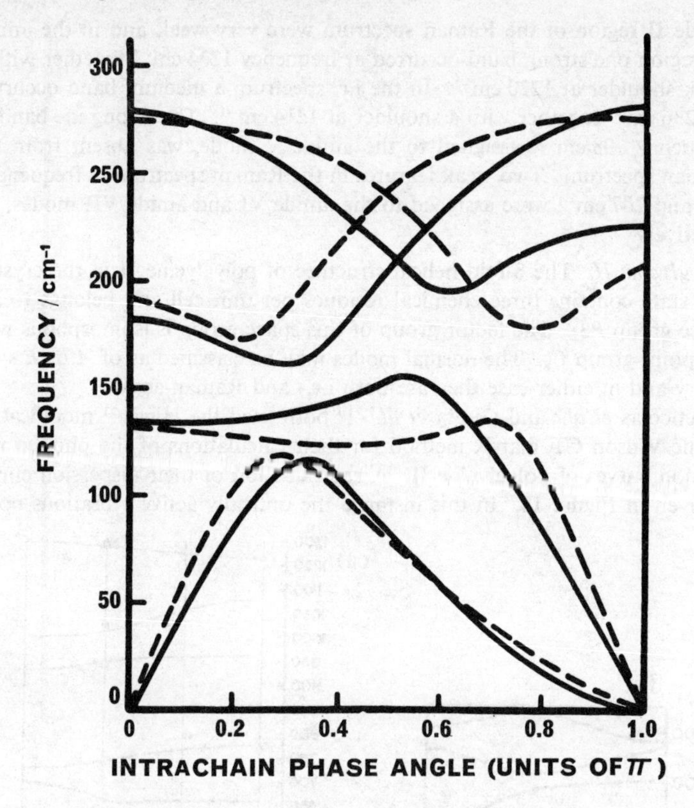

Figure 13 *Comparison of the in-plane phonon dispersion curves of the two-dimensional lattice from the single-mass model (solid line) and five-mass model (dashed line)* (Reproduced by permission from *J. Chem. Phys.*, 1972, **57**, 2109)

cm^{-1} region, the contribution of the hydrogen-bond bending modes to the density of states occurs at considerably lower frequencies. Fanconi suggested that a weak feature in the neutron inelastic scattering spectrum at 50 cm^{-1} may correspond to this contribution. The higher frequency neutron inelastic scattering data is in good agreement with the calculated values of all the various models.

Peticolas *et al.*[3] observed the Raman spectrum of polyglycine I in the solid state and compared their results with the i.r. data obtained by Shimanouchi *et al.*[9] In the amide I region, a single strong band was observed in the Raman spectrum with frequency 1674 cm^{-1} compared with two bands in the i.r. spectrum with frequencies 1685 and 1636 cm^{-1}. These amide I frequencies are typical for a polypeptide in the β-conformation. The bands in the

[9] S. Suzuki, Y. Iwashita, and T. Shimanouchi, *Biopolymers*, 1966, **4**, 337.

amide II region of the Raman spectrum were very weak and in the amide III region one strong band occurred at frequency 1234 cm^{-1} together with a weak shoulder at 1220 cm^{-1}. In the i.r. spectrum a medium band occurred at 1236 cm^{-1} together with a shoulder at 1214 cm^{-1}. The strong i.r. band at frequency 708 cm^{-1}, assigned to the amide V mode, was absent from the Raman spectrum. Two weak features in the Raman spectrum at frequencies 601 and 207 cm^{-1} were assigned to the amide VI and amide VII modes, respectively.

Polyglycine II. The 3-fold helical structure of polyglycine II in the crystalline state contains three chemical residues per unit cell and belongs to the space group $P3_1$. The factor group of this space group is isomorphous with the point group C_3. The normal modes may be classified as of *A* or *E* symmetry and in either case they are both i.r.- and Raman-active.

Peticolas *et al.*[3] and Gupta *et al.*[2, 10] both used the Higgs [11] modification of the Wilson GF matrix method for their calculations of the phonon dispersion curves of polyglycine II. A reproduction of their dispersion curves is given in Figure 14. In this instance the optically active vibrations occur

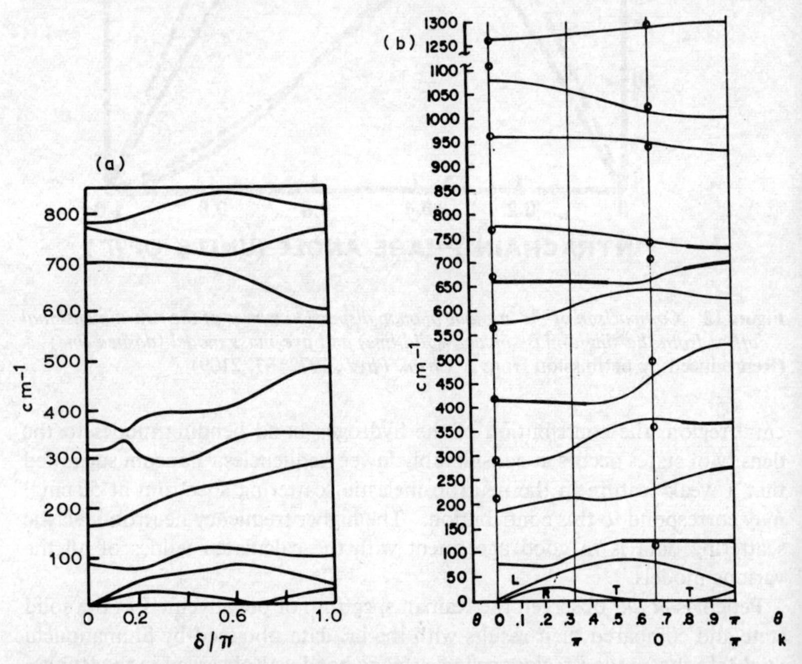

Figure 14 *Comparison of the phonon dispersion curves for polyglycine II as calculated by* (a) *Singh and Gupta* (b) *Small, Fanconi, and Peticolas*
(Reproduced by permission from *Chem. Phys. Letters*, 1971, **8**, 294, and *J. Chem. Phys.*, 1970, **52**, 4369)

[10] R. D. Singh and V. D. Gupta, *Spectrochim. Acta*, 1971, **27A**, 385.
[11] R. Higgs, *Proc. Roy. Soc.*, 1953, **A220**, 472.

at phase differences of 0° and 120° on the phonon dispersion curve. It can be seen that the results of both workers are substantially the same.

The neutron inelastic scattering spectrum of polyglycine II has been determined.[12] There was reasonably good agreement between this spectrum and the frequency distribution histogram for polyglycine II calculated by Gupta *et al.*[2, 10] They considered that a more extensive calculation on a three-dimensional structure which included interchain interactions would have given better agreement.

Peticolas *et al.*[3] have observed the Raman spectrum of polyglycine II in the solid state. These results were compared with the i.r. spectrum obtained by Shimanouchi *et al.*[9] In the amide I region a single strong band was observed in the Raman spectrum with frequency 1654 cm^{-1} compared with a single band in the i.r. at 1644 cm^{-1}. These amide I frequencies are thought to be typical of a polypeptide in the 3-fold helical conformation. A band at 1283 cm^{-1} in the amide II region of the Raman spectrum was coincident in frequency with a band in the i.r. spectrum. Two strong bands in the i.r. spectrum at frequencies 698 and 363 cm^{-1} had no strong counterpart in the Raman spectrum and were both assigned to amide IV bands.

Peticolas *et al.*[3] have also obtained the Raman spectrum of *N*-deuteriated polyglycine II in the solid state. The i.r. spectrum of this compound was obtained by Shimanouchi *et al.*[9] As would be expected there are differences in the amide frequencies of this compound compared with the non-deuteriated species. The amide I′ band occurs with frequency 1640 cm^{-1} in the Raman spectrum and at 1639 cm^{-1} in the i.r. spectrum. The amide III′ band occurs with frequency 995 cm^{-1} in the Raman spectrum and 987 cm^{-1} in the i.r. spectrum, where amide I′ and III′ refer to the frequencies of the COND group.

Glycine oligomers. Dwivedi and Gupta[13] compared the i.r. spectra of several glycine oligomers in the solid state with the i.r. spectra of polyglycine I and polyglycine II. A summary of their results is shown in Table 5. It can be seen that the i.r. spectrum of dodecaglycine is closely similar to that of polyglycine II and suggests that this compound also exists in a 3-fold helical conformation. The i.r. spectra of the oligomers up to and including hexaglycine strongly resemble that of polyglycine I. It would appear then that the lower oligomers tend to favour the extended β-conformation.

The Raman and i.r. spectra of the glycine oligomers up to the pentamer have been studied by Koenig *et al.*[14] The results appeared to agree closely with those expected for these oligomers in the antiparallel extended chain conformation.

Poly-L-alanine. It is well known that poly-L-alanine exists in two distinct crystalline forms with differing chain conformation. In the α-form of poly-L-alanine the chains form right-handed α-helices which are arranged in an

[12] V. D. Gupta, *Proc. Nat. Acad. Sci. India*, 1969, **35A**, 864.
[13] A. M. Dwivedi and V. D. Gupta, *Chem. Phys. Letters*, 1971, **8**, 220.
[14] M. Smith, A. G. Walton, and J. L. Koenig, *Biopolymers*, 1969, **8**, 29.

Table 5 *Amide bands and methylene group frequencies* (cm^{-1}) *in the oligomers of glycine and in the polyglycines*

Bands	*Di*	*Tri*	*Tetra*	*Penta*	*Hexa*	*Dodeca*	*poly I*	*poly II*
Amide I	1645	1675	1685	1683	1682	1650	1685	1644
		1635	1635	1630	1632		1636	
Amide II	1548	1550	1522	1520	1525	1555	1517	1554
		1515			1545			
Amide III	1243	1282	1240	1243	1290	1280	1236	1283
	1233	1245	1220	1230	1235	1250		1244
			1215	1214				
Amide IV	615	645	640	633	635	700	628	698
Amide V	708	718	700	712	705	745	708	740
				695				
Amide VI	588	585	575	585	582	568	589	573
Amide VII						365	217	365
CH_2 bend	1445	1429	1435	1432	1432	1419	1432	1420
CH_2 wag	1408	1395	1402	1403	1400	1375	1408	1384
Skeletal stretch	1007	1026	1006	1011	1012	1029	1016	1028

hexagonally close-packed lattice. The β-form contains nearly fully extended chains arranged in an antiparallel pleated-sheet structure.

α-Helical poly-L-alanine. Several normal-co-ordinate calculations of α-helical poly-L-alanine have been made. The earlier calculations did not determine the full phonon dispersion curve. The treatment by Miyazawa *et al.*[15] for instance gave only the i.r.-active modes below 800 cm^{-1} and the model used neglected intrachain hydrogen bonding. The more recent work of Peticolas *et al.*[7] and Gupta *et al.*[16] calculated the full phonon dispersion curves of poly-L-alanine in the α-helical conformation, using the Higgs[11] modification of the Wilson GF matrix method. Both sets of workers used a Urey–Bradley force-field for their calculations. Peticolas *et al.* in the main used force constants transferred from normal-co-ordinate analyses of polyglycine with slight modifications to the torsional values. The hydrogen-bond force constant was a modified value of that used for the formic acid dimer. Gupta *et al.* used principally the force constants transferred from their calculation on the β-form of poly-L-alanine with considerable modifications to the values for the torsional, wagging, and other skeletal modes. The results of both workers are substantially the same and those of Peticolas *et al.*[7] are shown in Figure 15. The i.r.-active bands are found at phase differences of 0° and 100° corresponding to the A and E_1 modes respectively. The Raman-active bands are found at phase differences of 0°, 100°, and 200° corresponding to the A, E_1, and E_2 modes respectively. The frequencies at phase difference 200° are identical with those found at 160° because of the periodicity of the Brillouin zone.

There is particular interest in the low-frequency vibrations of α-helical

[15] T. Miyazawa, K. Fukushima, and S. Sugano, 'Conformation of Biopolymers', ed. G. N. Ramachandran, Academic Press, New York, 1967, Vol. 2, p. 557.

[16] M. V. Krishnan and V. D. Gupta, *Chem. Phys. Letters*, 1970, **6**, 231.

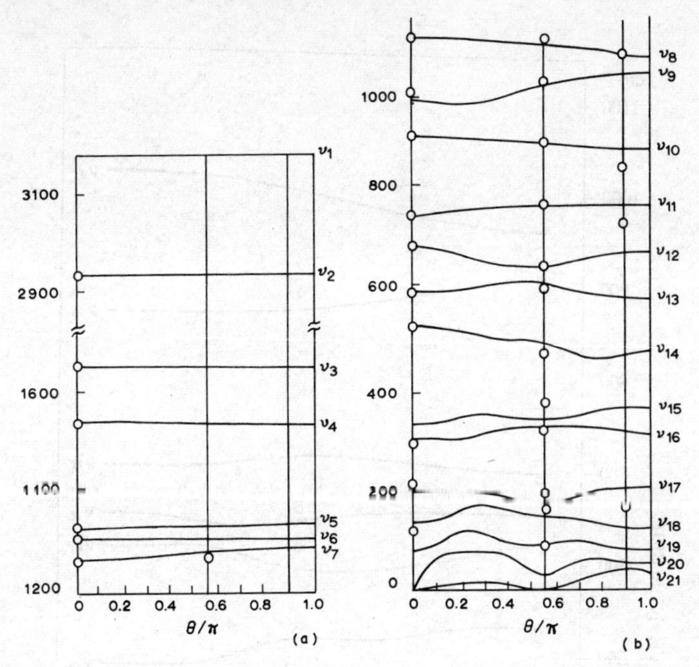

Figure 15 *Phonon dispersion curves of* α*-helical poly*-L*-alanine*: (a) 1200—3200 cm^{-1} *and* (b) 0—1200 cm^{-1}. (*Vertical lines represent phase differences of* 100° *and* 160°. *Open circles are spectral observations*)
(Reproduced by permission from *Biopolymers*, 1971, **10**, 1277)

polymers and several workers have made calculations of these modes for α-helical poly-L-alanine. Shimanouchi and Itoh [17] calculated the low-frequency phonon dispersion curves for α-helical poly-L-alanine using a model in which the CH_3 group was treated as a single dynamic entity. The force constants were transferred from *N*-methylacetamide and appropriate hydrocarbons. The results in Figure 16 show that the modes below 300 cm^{-1} have frequencies largely dependent on the phase difference.

The α-helix of poly-L-alanine having a finite number of residues is expected to have band progressions indicative of the length of the helix. The frequency of this longitudinal accordion-like vibration should be inversely proportional to the chain length as with n-paraffin, its extended planar zig-zag chain analogue.[18] Shimanouchi and Itoh[17] calculate these accordion-like frequencies for different values of *N*, the number of amino-acid residues in the chain. These results are shown in Table 6.

Peticolas *et al.*[7] have also made some preliminary calculations of the acoustical phonon dispersion curves. They found that the value of the hydrogen-

[17] K. Itoh and T. Shimanouchi, *Biopolymers*, 1970, **9**, 383.
[18] R. F. Schaufele and T. Shimanouchi, *J. Chem. Phys.*, 1967, **47**, 3605.

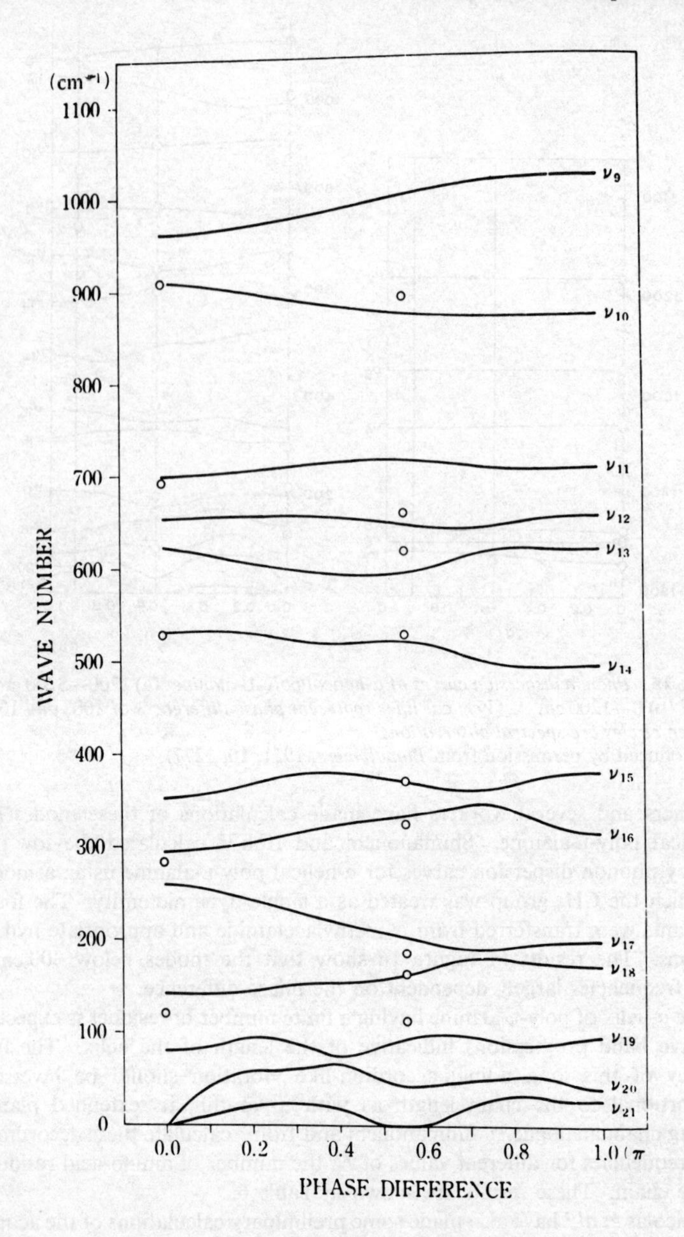

Figure 16 *Dispersion curves of the vibrations for the poly-L-alanine α-helix.* (*The circles at* 0.0 *and* 0.57 *give the observed frequencies for the A and E species vibrations, respectively*)
(Reproduced by permission from *Biopolymers*, 1970, **9,** 383)

Table 6 *Frequencies of accordion-like vibrations of the poly-L-alanine α-helix with N amino-acid residues.*

N	ν/cm^{-1} (*ref.* 7)	ν/cm^{-1} (*ref.* 17)
5	—	56
10	67	40
15	50	29
20	40	23
25	32	—
30	28	15
40	24	12
50	20	10
60	—	8
70	—	7
75	15	—
80	—	6
90	—	5
100	11	—

bond force constant had a significant influence on the low-frequency dispersion curves. For the calculations a model was chosen where the entire chemical repeat unit was treated as a single dynamical mass with nearest-neighbour and third-nearest-neighbour hydrogen-bonded interactions. The results in Figure 17 indicate that there are considerable differences between the calculated dispersion curves of Peticolas *et al.*[7] and those of Shimanouchi and Itoh,[17] especially for the longitudinal accordion-like mode. Peticolas *et al.* pointed out that the differences could probably be attributed to the following factors: (a) the force constant for the hydrogen bond used in their calculation was twice as large as that used by Shimanouchi and Itoh, (b) a formula for a polymer with free ends was used in their calculation whereas Shimanouchi and Itoh used a formula for a polymer with fixed ends. It is to be noted that the values for the chain-length-dependent accordion-like vibrations are also different (see Table 6).

In the light of these results, Peticolas and Fanconi[4] have made further simplified force-field calculations of the low-frequency motions of the α-helix. Three models of the chemical repeat unit were studied and compared. The first model treated each atom in the chemical repeat unit as a single dynamical entity except for the α-carbon methyl group which was counted as a single mass. This was called the seven-mass model. The second model treated the whole chemical repeat unit as a single dynamical unit. The third model treated the whole chemical repeat unit, minus the mass of the amide hydrogen, as a single point mass located at the oxygen position in the α-helix. The hydrogen atom was taken as a point mass located at the position of the amide hydrogen and was considered to be hydrogen-bonded to the mass at the third-nearest-neighbour oxygen position. The low-frequency motions of the α-helix were known to be very sensitive to the hydrogen-bond force constant and the phonon dispersion curves were calculated for several values of this constant. The curves for the seven-mass model are given in

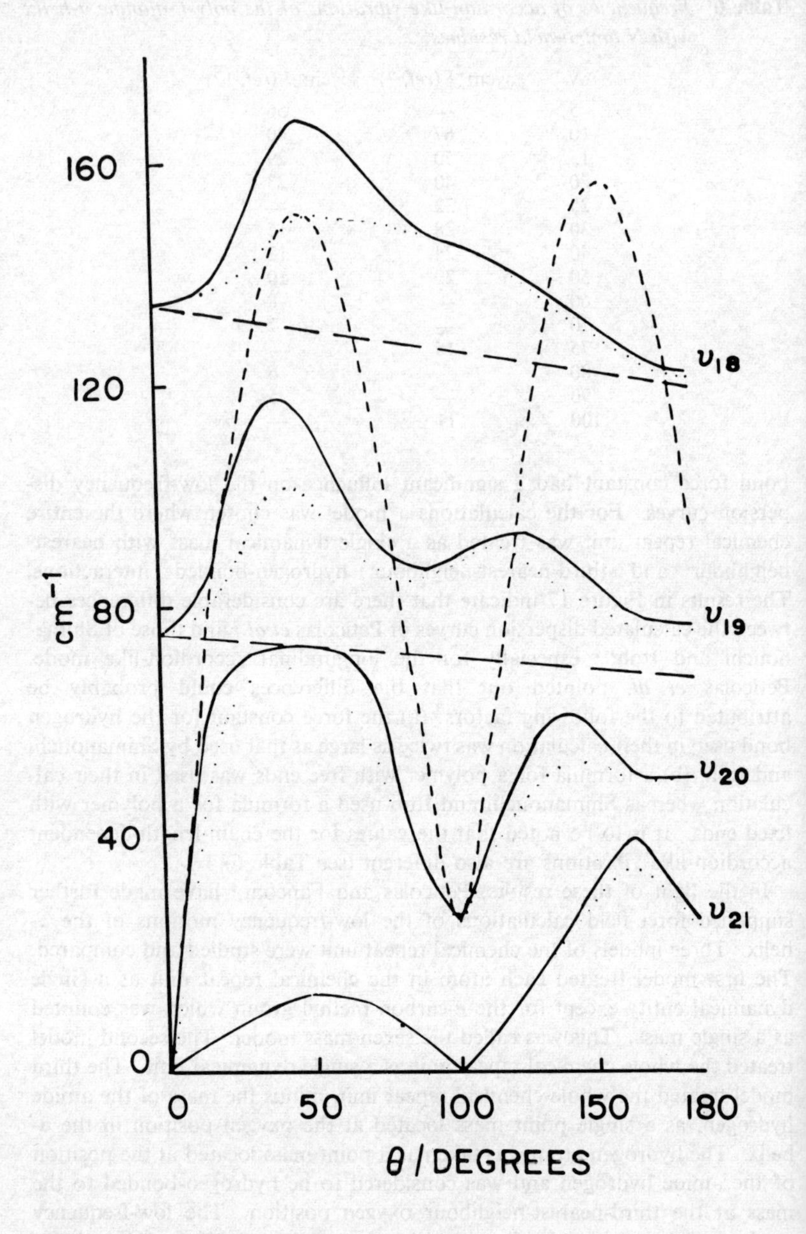

Figure 17 *Low-frequency phonon dispersion curves for α-helical poly-L-alanine (solid lines)*
(Reproduced by permission of *Biopolymers*, 1971. **10**, 1277)

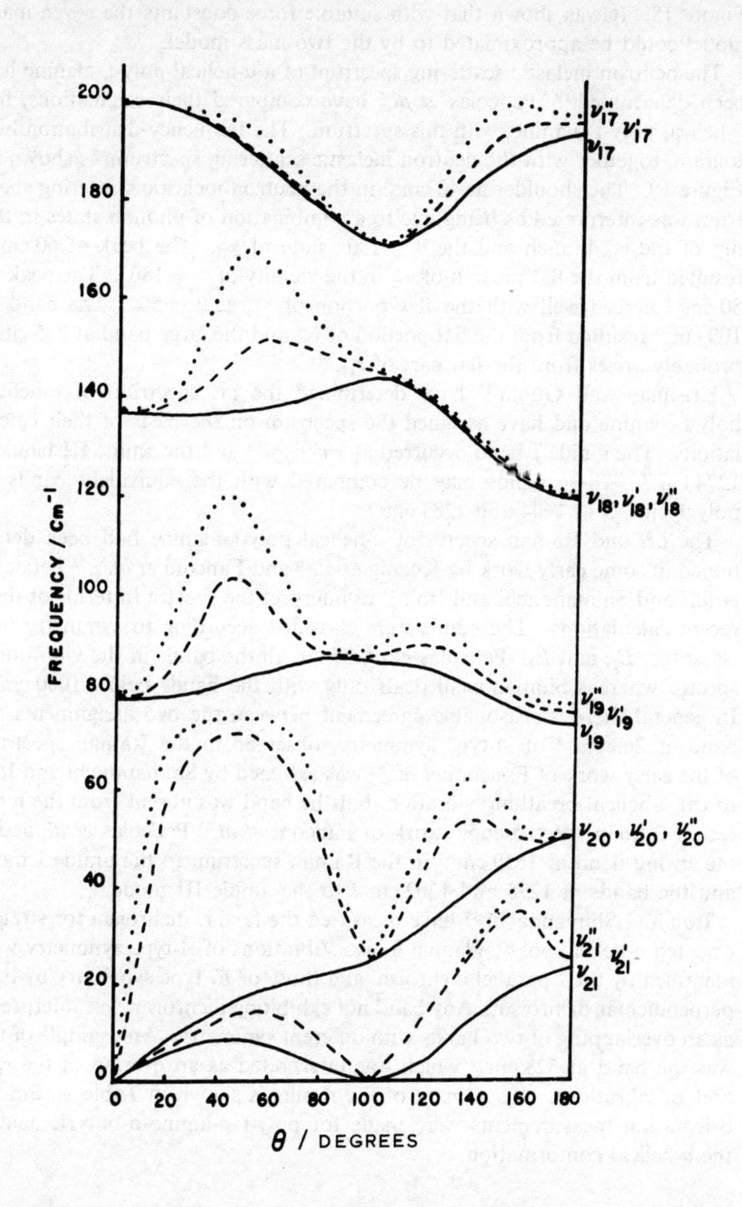

Figure 18 *Low-frequency phonon dispersion curves of the seven-mass model of α-helical poly-L-alanine for three values of the hydrogen-bond force constant:* $F_{N-H\cdots O} = 0.0$ mdyn Å^{-1} (*solid line*); $F_{N-H\cdots O} = 0.15$ mdyn Å^{-1} (*dashed line*); $F_{N-H\cdots O} = 0.30$ mdyn Å^{-1} (*dotted line*)
(Reproduced by permission from *Biopolymers*, 1971, **10**, 2223)

Figure 18. It was shown that with suitable force constants the seven-mass model could be approximated to by the two-mass model.

The neutron inelastic scattering spectrum of a α-helical poly-L-alanine has been determined.[19] Peticolas *et al.*[7] have compared their calculations, for α-helical poly-L-alanine, with this spectrum. The frequency-distribution histogram, together with the neutron inelastic scattering spectrum, is shown in Figure 19. The shoulder at 40 cm^{-1} in the neutron inelastic scattering spectrum was interpreted as being due to a combination of phonon states in the dip of the ν_{20} branch and the $\theta = 180°$ state of ν_{21}. The peak at 60 cm^{-1} resulted from the flat portion of ν_{20} in the vicinity of $\theta = 180°$. The peak at 80 cm^{-1} agreed well with the flat portion of ν_{20} at $\theta = 50°$. The band at 100 cm^{-1} resulted from the flat portion of ν_{19} and the large band at 235 cm^{-1} probably arises from the flat part of ν_{17}.

Krishnan and Gupta [16] have determined the i.r. spectrum of α-helical poly-L-alanine and have assigned the spectrum on the basis of their calculations. The amide I band occurred at 1659 cm^{-1} and the amide III band at 1274 cm^{-1}. These bands may be compared with the equivalent bands in polyglycine II at 1644 and 1283 cm^{-1}.

The i.r. and Raman spectra of α-helical poly-L-alanine had been determined in some early work by Koenig *et al.*[20] and Fanconi *et al.*[3, 21] Peticolas *et al.*[7] and Shimanouchi and Itoh [22] re-interpret the spectra in terms of their recent calculations. The bands were classified according to symmetry into classes A, E_1, and E_2. Peticolas *et al.* assign all the bands in the vibrational spectra whereas Shimanouchi deals only with the bands below 1000 cm^{-1}. In general there is reasonable agreement between the two assignments. A band at 268 cm^{-1} of A-type symmetry, observed in the Raman spectrum of the early work of Fanconi *et al.*,[20] was assigned by Shimanouchi and Itoh to the α-helical breathing vibration, but the band was absent from the more recent Raman spectroscopic work of Fanconi *et al.*[3] Peticolas *et al.* assign the strong band at 1659 cm^{-1} in the Raman spectrum to the amide I mode and the bands at 1275 and 1262 cm^{-1} to the amide III mode.

Itoh and Shimanouchi [17] have measured the far-i.r. dichroism for stretch-oriented α-helical poly-L-alanine films. Vibrations of A-type symmetry were identified by their parallel dichroism and those of E_1-type symmetry by their perpendicular dichroism. Any band not exhibiting dichroism was interpreted as an overlapping of two bands with different symmetry. An example of this was the band at 528 cm^{-1} which was interpreted as an overlap of the $\nu_{14}A$ and E_1 vibrations. A summary of the results is shown in Table 7. Similar orientation measurements were made for poly-L-α-amino-n-butyric acid in the α-helical conformation.

[19] V. D. Gupta, H. Boutin, and S. Trevino, *Nature*, 1967, **214,** 1325.
[20] J. L. Koenig and P. L. Sutton, *Biopolymers*, 1969, **8,** 167.
[21] B. Fanconi, B. Tomlinson, L. A. Nafie, E. W. Small, and W. L. Peticolas, *J. Chem. Phys.*, 1969, **51,** 3993.
[22] K. Itoh and T. Shimanouchi, *Biopolymers*, 1971, **10,** 1419.

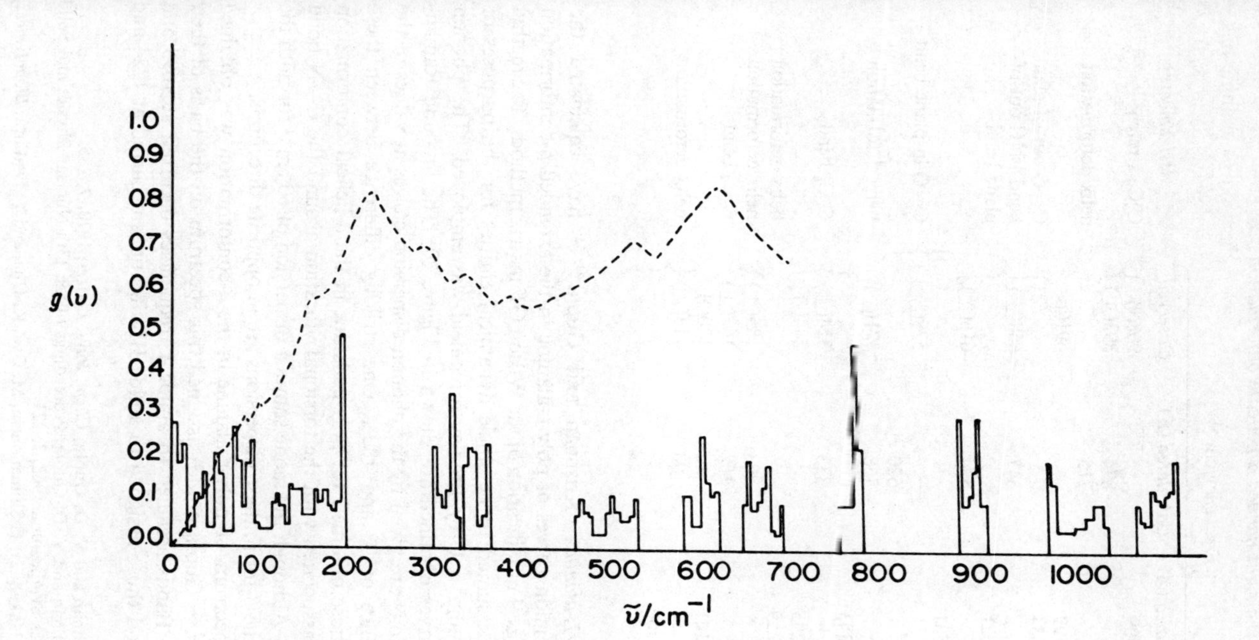

Figure 19 *Comparison of the frequency distribution with the neutron inelastic scattering spectrum (dashed line) of α-poly-L-alanine*
(Reproduced by permission from *Biopolymers*, 1971, **10**, 1277)

Table 7 *Observed and calculated frequencies, and assignments of poly-L-alanine with the right-handed α-helical conformation*

	Frequency/cm^{-1}			
	Calculated (A)	*Calculated (E)*	*Observed*	*Assignments*
ν_{10}	909		908(∥)	$C_\alpha C_\beta$ stretch
		874	893(⊥)	
ν_{11}		715		helix deformation
	699		691(∥)	
ν_{12}	654			C=O out-of-plane bend+NH out-of-plane bend
		647	658(⊥)	
ν_{13}	621			
		594	616(⊥)	
ν_{14}	531			
			528(∥, ⊥)	C=O in-plane bend
		520		
ν_{15}		368	371(⊥)	helix deformation
	353			
ν_{16}		333	324(⊥)	$C_\alpha C_\beta$ bend
	295		284(∥)	
ν_{17}	259			helix deformation
ν_{18}		203	185(⊥)	helix deformation+ $C_\alpha C_\beta$ bend
		162	163(⊥)	
	141		120(∥)	
ν_{19}		93	113(⊥)	helix deformation
	80			
ν_{20}		44		

β-Sheet poly-L-alanine. Krishnan and Gupta [23, 24] have calculated the phonon dispersion curves of poly-L-alanine in the extended β-conformation using the Higgs [11] modification of the Wilson GF matrix method. The methyl groups were assumed to be single dynamical entities for the purposes of computation and a Urey–Bradley force-field was employed. The phonon dispersion curves obtained are shown in Figure 20. The i.r.- and Raman-active vibrations are given by the phonon dispersion curves at values of the phase difference of 0° and 180°. One striking difference between these curves and those obtained for poly-L-alanine in the α-helical conformation is the curve associated with the torsional vibration around the C—N bond (amide VII). Although the mode shows a dispersion of 60 cm^{-1} for both the α-form and the β-form, these dispersions are in opposite directions.

The i.r. spectrum of poly-L-alanine in the β-conformation was obtained by Itoh *et al.*[25] Gupta *et al.*[23] assigned this spectrum on the basis of their calculations. Bands assigned to amide I frequencies occurred at 1695 and 1634 cm^{-1} and those assigned to amide III vibrations occurred at 1224 and

[23] M. V. Krishnan and V. D. Gupta, *Chem. Phys. Letters*, 1970, **7**, 285.
[24] M. V. Krishnan and V. D. Gupta, Proceedings of the 13th Nuclear Physics and Solid State Physics Symposium, 1968, **3**, 221.
[25] K. Itoh, T. Nakahara, T. Shimanouchi, M. Oya, K. Una, and Y. Iwakura, *Biopolymers*, 1968, **6**, 1759.

1241 cm⁻¹. These frequencies are clearly quite different from those of the α-form but are closely similar to those values for the β-conformation of polyglycine.

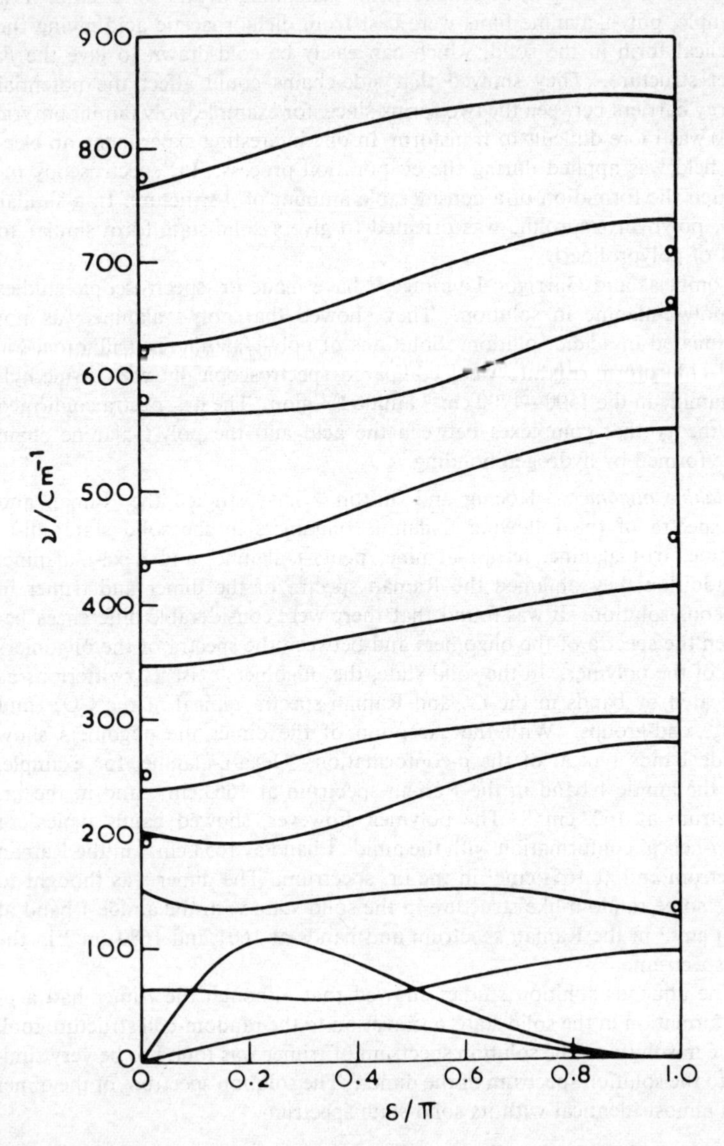

Figure 20 *Dispersion curves for poly-L-alanine in the β-form* (*circles represent observed frequences*)
(Reproduced by permission from *Chem. Phys. Letters*, 1970, 7, 285)

Bensing and Pysh [26] indicated the need for further understanding of the potential energy barriers separating various conformational forms of polypeptides. They prepared several polypeptides whose chain conformations were known and could be altered while remaining in the solid state. For example, poly-L-alanine films were cast from dichloroacetic acid giving the α-helical form in the solid, which can easily be cold-drawn to give the β-sheet structure. They showed that side-chains could affect the potential energy barriers between the two forms since, for example, poly (aminobutyric acid) was more difficult to transform. In one interesting experiment, an electric field was applied during the evaporation process. I.r. spectroscopy indicated the formation of a considerable amount of β-structure. In a similar way, polyhydroxyproline was oriented to give a solid-state form similar to that of polyproline I.

Combelas and Garrigou-Lagrange [27] have made i.r. spectroscopic studies of poly-L-alanine in solution. They showed that poly-L-alanine was not protonated in acidic solution. Solutions of poly-L-alanine in trifluoroacetic acid–chloroform mixture were compared spectroscopically with *N*-methylacetamide in the 1500—1700 cm^{-1} amide I region. The i.r. spectra confirmed the theory that complexes between the acid and the poly-L-alanine chain were formed by hydrogen bonding.

L-Alanine oligomers. Koenig and Sutton [28] have studied the Raman and i.r. spectra of the following L-alanine oligomers in the solid state: di-L-alanine, tri-L-alanine, tetra-L-alanine, penta-L-alanine, and hexa-L-alanine. In addition they obtained the Raman spectra of the dimer and trimer in aqueous solution. It was found that there were considerable differences between the spectra of the oligomers and between the spectra of the oligomers and of the polymer. In the solid state, the oligomers exist as zwitterions as indicated by bands in the i.r. and Raman spectra typical of the CO_2^- and NH_3^+ end-groups. With the exception of the dimer the oligomers show amide bands typical of the β-conformation. Hexa-L-alanine, for example, has the amide I band in the Raman spectrum at 1663 cm^{-1} and in the i.r. spectrum at 1623 cm^{-1}. The polymer, however, showed bands typical of the α-helical conformation with the amide I band at 1654 cm^{-1} in the Raman spectrum and at 1651 cm^{-1} in the i.r. spectrum. The dimer was thought to have some random-like structure in the solid state with the amide I band at 1681 cm^{-1} in the Raman spectrum and bands at 1601 and 1680 cm^{-1} in the i.r. spectrum.

The aqueous solution studies showed that although the trimer had a β-conformation in the solid state, a transition to the random-coil structure took place in solution. The solution spectrum of trimer was found to be very similar to the solution spectrum of the dimer. The solution spectrum of the dimer was almost identical with its solid-state spectrum.

[26] J. L. Bensing and E. S. Pysh, *Biopolymers*, 1971, **10**, 2645.
[27] P. Combelas and C. Garrigou-Lagrange, *Compt. rend.*, 1971, **272**, *C*, 153.
[28] P. Sutton and J. L. Koenig, *Biopolymers*, 1970, **9**, 615.

Poly-L-valine. Poly-L-valine in the solid state occurs in the antiparallel β-sheet conformation and presents an ideal case for the study of this structure.[29] The line group of this polymer is isomorphous with the point group D_2. In this group the A modes are Raman-active only, and the B_1, B_2, and B_3 modes are both i.r.- and Raman-active.

Koenig and Sutton[30] have studied both the i.r. and Raman spectra of poly-L-valine in the solid state. In the i.r. spectrum the amide I bands show splitting characteristic of the β-conformation with frequencies 1687 and 1625 cm^{-1}. Similar splitting occurs in the amide II region with bands at frequencies 1538 and 1555 cm^{-1} typical of the β-conformation. The amide III mode was found to be split into four components as discussed earlier. Of the three i.r.-active components, two were observed with frequencies 1287 and 1224 cm^{-1} and were assigned to the $\nu\|$ $(0,\pi)$ and $\nu\perp$ $(\pi,0)$ modes, respectively. The third i.r.-active component of $\nu\perp$ (π,π), normally a weak feature, was not observed.

In the Raman spectrum a strong band at 1666 cm^{-1} was observed in the amide I region. This band was considered to be comparable with bands of similar frequency in the Raman spectrum of the β-oligomers of glycine and the β-oligomers of alanine. Only a diffuse Raman scattering was observed in the amide II region. Three bands were observed in the amide III region with frequencies 1291, 1276 and 1231 cm^{-1}. The bands at 1291 and 1231 cm^{-1} were assigned to the $\nu\|$ $(0,\pi)$ and $\nu\perp$ (π,π) amide III modes, respectively. The 1276 cm^{-1} band, inactive in the i.r., was assigned to the ν $(0,0)$ amide III mode. The $\nu\perp$ $(\pi,0)$ mode, which should also be Raman active, was not observed. The two strong Raman lines at 1462 and 1455 cm^{-1} were assigned to deformations of the *gem*-dimethyl group.

Poly-L-leucine. Poly-L-leucine exists in the α-helical conformation in the solid state.[31] Koenig and Sutton[30] have determined the i.r. and Raman spectra of this compound in the solid state. In the i.r. spectrum the amide I band occurs at 1650 cm^{-1} and the amide II band at 1540 cm^{-1}. These bands are comparable with the equivalent bands in the α-helical form of poly-L-alanine which occur at 1652 and 1546 cm^{-1}. Three i.r. bands occur in the 600—700 cm^{-1} region at frequencies 694, 657, and 614 cm^{-1}. These bands were thought to indicate the presence of the extended β-form, the random-coil form, and the α-helical form, respectively; the 614 cm^{-1} band was the strongest.

In the Raman spectrum the amide I band occurs at 1652 cm^{-1} and the amide III band at 1320 cm^{-1}, typical of the α-helical conformation. A weak band was observed at 1547 cm^{-1} similar to that found in α-helical poly-L-alanine at 1549 cm^{-1}. The bands in the region of 1450 cm^{-1} were assigned to

[29] D. B. Fraser, B. S. Harrap, T. P. MacRae, F. H. C. Stewart, and E. Suzuki, *J. Mol. Biol.*, 1965, **12**, 482.
[30] J. L. Koenig and P. L. Sutton, *Biopolymers*, 1971, **10**, 89.
[31] E. R. Blout, C. Loze, S. M. Bloom, and G. D. Fasman, *J. Amer. Chem. Soc.*, 1960, **82**, 3787.

methylene vibrations and those in the 930—970 cm^{-1} region were assigned to symmetric methyl group vibrations.

Poly-L-serine. Poly-L-serine is a high molecular weight polypeptide possessing a β-conformation in the solid state, as determined by *X*-ray diffraction.[32] Koenig and Sutton [30] have determined the i.r. and Raman spectra of this compound in the solid state. In the i.r. spectrum the amide I region shows splitting characteristic of the β-conformation with bands at 1695 and 1621 cm^{-1}. A weak i.r. band at 1322 cm^{-1} was assigned to a methylene wagging vibration.

In the Raman spectrum a band at 1668 cm^{-1} was assigned to the i.r.-inactive ν (0,0) mode. The frequency of this band is typical of the β-conformation. A strong Raman line at 1466 cm^{-1} was assigned to a deformation mode of the methylene group.

Poly-L-lysine. The i.r. and Raman spectra of poly-L-lysine has been studied by Wallach and Graham.[33] For the Raman studies they used a one watt argon laser in conjunction with a Spex double monochromator and a photon-counting detection system linked to a computer-of-average-transients. The Raman spectra of poly-L-lysine in the antiparallel β-conformation, α-helical conformation, and random-coil form were studied. The spectra were signal-averaged over four runs and normalized with respect to the 1638 cm^{-1} line. The concentration of poly-L-lysine used was about 1 mg ml^{-1}. Prominent bands in the amide I region were observed at 1631 and 1672 cm^{-1}. These were thought to be typical of the antiparallel β-sheet structure. The equivalent bands in the i.r. spectrum of the antiparallel β-structure were observed at 1635 and 1687 cm^{-1}. The random-coil form of poly-L-lysine had Raman bands at 1653, 1665, and 1683 cm^{-1}; the α-helical form had one intense band at 1647 cm^{-1}. For comparison purposes the Raman spectra of poly-L-glutamate in its α-helical and random-coiled forms were also measured. Again, the α-helical form exhibited a strong band at 1647 cm^{-1} considered to be typical of the α-helical form of polypeptide chains.

Koenig and Sutton [34] determined the Raman spectrum of solid-state poly-L-lysine hydrochloride and its precursor poly-(ε-benzoxycarbonyl-L-lysine). The aqueous solution properties of the hydrochloride had already been studied extensively by *X*-ray, i.r., and u.v. spectroscopy and by circular dichroism measurements. All of these techniques indicated that poly-L-lysine hydrochloride formed the α-helical conformation in solution.

The chemical structure of the polymers is shown in Figure 21. In the solid state they were both found to have a strong band in the region of 1650 cm^{-1} considered typical of α-helical polymers. This band was observed in the Raman spectrum of the α-helical forms of poly-L-alanine at 1654 cm^{-1}, in poly-L-leucine at 1652 cm^{-1}, and in poly-(γ-benzyl-L-glutamate) at 1652 cm^{-1}. The amide I region thus suggests predominance of the α-helical form

[32] Z. Bohak and E. Katachalski, *Biochemistry*, 1963, **2**, 228.
[33] D. F. H. Wallach and J. M. Graham, *F.E.B.S. Letters*, 1970, **7**, 330.
[34] J. L. Koenig and P. L. Sutton, *Biopolymers*, 1970, **9**, 1229.

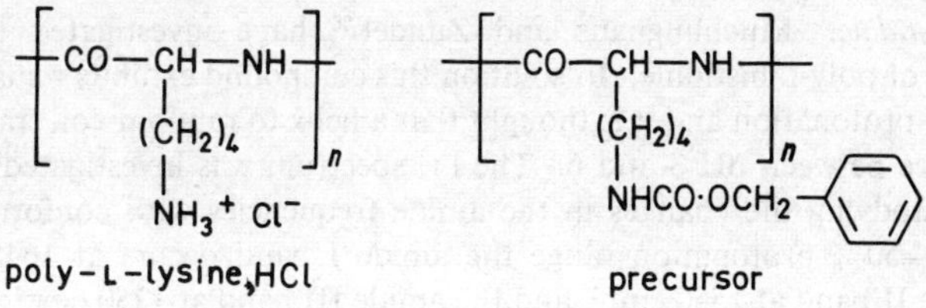

$-[CO-CH(-(CH_2)_4-NH_3^+\,Cl^-)-NH]_n-$ poly-L-lysine,HCl

$-[CO-CH(-(CH_2)_4-NHCO{\cdot}OCH_2-C_6H_5)-NH]_n-$ precursor

Figure 21 *Diagram of the structure of poly-L-lysine hydrochloride and its precursor poly-(ε-benzoxycarbonyl-L-lysine)*

although, since the i.r. spectrum in the amide V region has bands at 610, 650, and 690 cm⁻¹, there may be some random coil and β-structure present. In aqueous solution the weak Raman line of water at 1640 cm^{-1} obscures the amide I region but, using D_2O as the solvent, a band was observed at 1659 cm^{-1} characteristic of a random-coil structure for poly-L-lysine hydrochloride.

Poly-L-tyrosine. Conio *et al.*[35] have studied the conformational transitions of poly-L-tyrosine in mixed water–ethanol solvents, as a function of solvent composition, using i.r. spectroscopy, optical rotatory dispersion, and potentiometric titration. Bands dependent on conformation were observed in the 1600—1700 cm^{-1} region of the i.r. spectrum.

Poly-L-tyrosine in 20% deuterioethanol at pD 13.0 gives a broad band in the i.r. centred at 1648 cm^{-1}, corresponding to the ionized random-coil configuration. Slow titration down to pD 11.6 yielded two new intense bands at 1683 and 1623 cm^{-1}, which were thought to correspond to the anti-parallel β-sheet conformation. Fast titration down to pD 11.6 produced very little change in the i.r. spectrum. This supported the view that under these conditions poly-L-tyrosine cannot assume the β-conformation. In 40% deuterioethanol, after slow titration down to pD 11.4, it was found that only a little β-conformation was present. In 60% deuterioethanol after slow titration down to pD 11.4 it was found that no β-conformation was present. It can thus be seen that increasing ethanol concentration decreases the stability of the β-conformation.

The 1400—1500 cm^{-1} region of the i.r. spectrum was also investigated. The results strongly indicated the presence of an α-helical form in 20% ethanol solution in the fast titration experiments and also in the alcohol-rich mixtures after fast or slow titration. For instance, the i.r. spectrum of

[35] G. Conio, E. Patrone, and F. Salons, *Macromolecules*, 1971, **4**, 283.

poly-L-tyrosine in 60% ethanol titrated down to pH 10.96 was typical of an α-conformation.

Poly-L-histidine. Muehlinghaus and Zundel [36] have investigated the i.r. spectrum of poly-L-histidine. In solution this compound exhibits remarkable effects on protonation and it is thought that a helix to random-coil transition takes place between pH 3 and 6. The i.r. spectrum was investigated with a view to studying the changes in the amide frequencies with conformation. In the 0—50% protonation range the amide I band occurs at 1645 cm^{-1}, the amide II band at 1550 cm^{-1}, and the amide III band at 1250 cm^{-1}. These frequencies agree well with those expected for a polypeptide in the α-helical configuration. At 100% protonation the amide I band occurs at 1659 cm^{-1} and the amide II band at 1528 cm^{-1}. These results are characteristic of the random-coil structure. There are some indications that there is also present a β-form characterized in the spectra by a shoulder at 1689 cm^{-1}, a weak shoulder at 1550 cm^{-1}, a more intense band at 1630 cm^{-1}, and a weaker feature at 1680 cm^{-1}.

Poly-L-proline. This polypeptide occurs in two distinct ordered helical configurations: a right-handed helix with ten chemical residues in three turns of the helix [37] and a left-handed helix with three residues per turn. The first form, poly-L-proline I, is only slightly soluble in salt-free water and the second form, poly-L-proline II, is soluble at low temperatures but precipitates at higher temperatures.

Johnston and Krimm [38] have studied the i.r. spectrum of both forms of poly-L-proline. In the solid state, for both forms, the amide I band occurs at 1643 cm^{-1} and the methylene bending vibrations occur at 1447 and 1421 cm^{-1}. In aqueous solution both forms gave a strong band at 1619 cm^{-1} and a less intense band at 1458 cm^{-1} in the i.r. spectrum. In 6M-$CaCl_2$ solution a new band appears in the amide I region at 1641 cm^{-1}. This band increased in intensity with increasing calcium chloride concentration. A broadening of the features in the methylene bending region also occurred. These results suggested that the spectral changes occurring in the solution were primarily due to alteration in the hydrogen bonding between the carbonyl groups of the polypeptide and the solvent. The increasing disorder of the chain with increasing salt concentration is indicated both by i.r. spectroscopy and c.d. measurements. The results were interpreted in terms of an increasing range of accessible C^{α}—C' rotation angles with increasing salt concentration rather than the random introduction of *cis*-imide bonds in the chain.

Swenson [39] has also studied the i.r. absorption spectrum of poly-L-proline in concentrated aqueous salt solutions. In this work several salts other than calcium chloride were used, like lithium bromide, potassium bromide, sodium

[36] J. Muehlinghaus and G. Zundel, *Biopolymers*, 1971, **10**, 711.
[37] P. M. Cowan and S. McGavin, *Nature*, 1965, **176**, 1062.
[38] N. Johnston and S. Krimm, *Biopolymers*, 1971, **10**, 2597.
[39] C. A. Swenson, *Biopolymers*, 1971, **10**, 2591.

bromide, sodium iodide, and sodium thiocyanate. Further, the spectra were measured at three different temperatures, 15, 30, and 50 °C. The results were interpreted in terms of a *cis–trans* isomerization about the peptide bond as indicated by the frequency of the amide I absorptions. The aqueous salt-free solution spectrum showed only single bands in the amide I and methylene bending regions. In contrast, the salt solutions showed two bands in each region exactly as observed by Krimm and Johnston.[38] It was concluded that in salt solutions there were present both *cis* and *trans* peptide linkages. The temperature studies showed that as the temperature was increased the proportion of *cis*-form present also increased.

Koenig *et al.* [40, 41] have studied both the i.r. and Raman spectra of poly-L-proline I, the right-handed 10_3 helical form, and poly-L-proline II, the left-handed 3_1 helical form. The amide I band of poly-L-proline I occurs at 1647 cm^{-1} in the Raman spectrum and that of poly-L-proline II occurs at 1650 cm^{-1}. The 1647 cm^{-1} band in poly-L proline I has almost the same intensity as the 1446 cm^{-1} methylene bending mode, whereas in poly-L-proline II the 1650 cm^{-1} band is only about half the intensity of the 1446 cm^{-1} band. Table 8 summarizes the Raman and i.r. bands which may be used to distinguish between the two forms.

Table 8 *Summary of the Raman and i.r. bands* (cm^{-1}) *which may be used to distinguish the two forms of poly-L-proline*

Raman		*Infrared*	
Poly-L-proline I	Poly-L-proline II	Poly-L-proline I	Poly-L-proline II
		1355	
	1198		
1187	1187		
	1176		
	1000		
957		960	
781			
	722		
662			670
540			
	530		
	400		400
363			

The Raman spectra of a series of proline oligomers have also been obtained.[40] It was concluded that in aqueous solution and lyophilized solids, the oligomers from the trimer upward were capable of existing in a helical form similar to that of poly-L-proline II. The spectra of dried solids indicated that the tetramer was the lowest oligomer to have a stable helical conformation.

[40] W. B. Rippon, J. L. Koenig, and A. Ghalt, *J. Amer. Chem. Soc.*, 1970, **92,** 7455.
[41] M. Smith, A. G. Walton, and J. L. Koenig, *Biopolymers*, 1969, **8,** 173.

Poly-L-hydroxyproline. The conformations of biopolymers containing the pyrrolidine ring are important in biochemistry. The collagen structure, for instance, is thought to be partly due to the high pyrrolidine ring content. Recent theoretical calculations by Walton and Hopfinger [42] suggested that the only stable structure of poly-L-hydroxyproline in the solid state was a left-handed *trans* helix similar to that of poly-L-proline II.

Koenig *et al.* [43] have investigated the Raman spectra of a number of compounds containing the pyrrolidine ring. They have studied the Raman spectrum of poly-L-hydroxyproline in the solid state and in aqueous solution. They also report the Raman spectra of the imino-acids, proline and hydroxyproline, as dipolar ions, hydrochlorides, and as sodium salts. The imino-acids exist as zwitterions in the solid state, and bands characteristic of the ionized carboxy-group are observed in the Raman spectrum, together with bands associated with the NH_2^+ moiety. In the case of the hydrochlorides, a band was observed in the region 1700—1750 cm^{-1}, characteristic of a non-ionized carboxy-group. In solution the Raman spectrum of hydroxyproline was found to be very dependent on pH as would be expected.

The main differences in the Raman spectrum of a solid hydroxyproline polymer, as compared with that of a solid hydroxyproline monomer, are the absence of those bands associated with the NH_2^+ and CO_2^- moieties and the appearance of strong bands at 556 and 520 cm^{-1} associated with the polymer skeleton. The most intense band in the spectrum of the polymer was at 1628 cm^{-1} and this was assigned to the carbonyl stretching frequency. Its low frequency was thought to arise from hydrogen bonding with the hydroxy-group of the pyrrolidine ring.

The Raman spectrum of poly-L-hydroxyproline in aqueous solution was found to be very insensitive to change in pH. A variation from pH 1 to 13 was found to affect only one band in the spectrum. This band, assigned to a methylene deformation mode, shifted only 4 cm^{-1} over this wide pH range. In general, the aqueous solution spectrum was very little different from the solid-state spectrum, indicating that the conformation in both phases was essentially the same. Minor differences could usually be attributed to environmental influences on the pyrrolidine group.

Poly-(γ-benzyl-L-glutamate) and Related Compounds. In general, the stable conformation of poly-(γ-benzyl-L-glutamate) depends on such parameters as molecular weight, physical state, and temperature.

High molecular weight poly-(γ-benzyl-L-glutamate) has been shown by *X*-ray diffraction to assume the α-helical conformation in the solid state.[44] This conformation persists as the stable form on dissolving the polymer in weakly interacting solvents such as chloroform. In strongly interacting sol-

[42] A. J. Hopfinger and A. G. Walton, *J. Macromol. Sci.*, 1969, **B3**, 195.
[43] J. L. Koenig, M. J. Deveney, and A. G. Walton, *Biopolymers*, 1971, **10**, 615.
[44] M. F. Perutz, *Nature*, 1951, **167**, 1053.

vents such as dichloroacetic acid or trifluoroacetic acid, the polymer changes its conformation to the random-coil form.[45]

Combelas and Garrigou-Lagrange [46] have studied the i.r. absorption spectrum of poly-(γ-benzyl-L-glutamate) in solution in a mixture of trifluoroacetic acid and chloroform. In pure chloroform the amide I band was at 1651 cm^{-1}, a typical value for the α-helical form, the amide II band was at 1550 cm^{-1}, and the ester carbonyl frequency was at 1733 cm^{-1}. The ester carbonyl frequency disappeared almost immediately on addition of acid to the chloroform solution, the amide I band shifted to 1641 cm^{-1}, and the amide II band also shifted to 1542 cm^{-1}. There were also marked intensity changes in the 3200—3400 cm^{-1} region. These spectroscopic results were interpreted in terms of a helix to random-coil transition of the polymer at a mole fraction of trifluoroacetic acid in chloroform between 0.08 and 0.10.

Maschka *et al.*[47] determined a relation between the electric-field strength and the degree of orientation for molecules possessing a permanent dipole moment. They measured the degree of orientation in samples of poly-(γ-benzyl-L-glutamate) subjected to an electric field. The directions of the transition moments of the amide A and amide I bands as found by i.r. dichroism studies were in good agreement with the theory.

Koenig and Sutton [30] have obtained the Raman spectra of α-helical poly-(γ-benzyl-L-glutamate) in the solid state. To assist in the interpretation of the spectrum they used the i.r. data of Masuda and Miyazawa.[48] The Raman spectrum, as is to be expected, was found to be dominated by the bands due to the benzene ring vibrations. Nevertheless, these intense features did not prevent observation of the weaker amide frequencies. In the Raman spectrum the amide I band occurs at 1652 cm^{-1}, typical for an α-helical conformation, and may be compared with the band observed at 1653 cm^{-1} in the i.r. spectrum. The amide II region is fairly diffuse in the Raman spectrum and a single band at 1336 cm^{-1} occurs in the amide III region. There is a certain amount of ambiguity regarding the assignment of the 1336 cm^{-1} band, because C—H deformations are also expected to occur in this region. The amide V band observed at 614 cm^{-1} in the i.r. spectrum was absent from the Raman spectrum, as is characteristic of the α-helical conformation.

Goodman *et al.*[49] have made some interesting studies of the i.r. spectra of γ-ethyl-L-glutamate oligopeptides in the solid state and in solution. The amide I and amide II bands were carefully investigated and conformational assignments were based on their frequencies. In the solid state the amide V band proved invaluable as a diagnostic frequency for the β-structure. In trimethyl phosphate solutions the oligomers up to the pentamer were shown to have random-coil configurations. Above the heptamer it was suggested

[45] J. R. Parrish and E. R. Blout, *Biopolymers*, 1971, **10**, 1491.
[46] P. Combelas and C. Garrigou-Lagrange, *Compt. rend.*, 1971, **272**, *C*, 1537.
[47] A. Maschka, G. Bauer, and Z. Dora, *Monatsh.*, 1971, **102**, 1516.
[48] Y. Masuda and T. Miyazawa, *Makromol. Chem.*, 1967, **103**, 261.
[49] M. Goodman, Y. Masuda, and A. S. Verdini, *Biopolymers*, 1971, **10**, 1031.

that the molecules could be present in a folded structure, but no spectral indications of the β-conformation were observed. In chloroform solution for oligomers higher than the pentamer the amide I band occurred at 1625 cm^{-1}, indicative of the extended β-structure. Poly(γ-methyl-L-glutamate) in chloroform solution has a band at 1650 cm^{-1} in the i.r. spectrum, indicative of the α-helical conformation.

Watanabe *et al.*[50] studied the conformation of γ-polyglutamic acid by i.r. spectroscopy and optical rotatory dispersion. The acid form of the γ-polyglutamic acid gave amide I bands at 1630 (strong) and 1650 cm^{-1} (weak) and amide II bands at 1520 (strong) and 1550 cm^{-1} (weak). This indicated that the polymer was in the β-form. The sodium salt of the acid appeared to be random-coil with diffuse bands at 1655 and 1535 cm^{-1} in the i.r. spectrum.

Poly-(β-benzyl-L-aspartate) and Related Compounds. Poly-(β-benzyl-L-aspartate) is well known to form a left-handed α-helix which is converted to an ω-helix on heating.[51] This ω-helix is characterized by the tetragonal stacking of the benzene rings around the central helix giving a more ordered side-chain conformation than in the case of the α-helical form. Clearly this stacking variability of the side-chain is of great interest in relation to polypeptide molecules. Obata and Kanetsuna [52] used i.r. spectroscopy to monitor this phenomenon. The amide I and amide II bands of the α-form were at frequencies 1667 and 1557 cm^{-1}, respectively. For the ω-form the equivalent bands were at 1677 and 1537 cm^{-1}. Another interesting feature was the variation of the band at 756 cm^{-1} associated with the benzene rings. The band at 740 cm^{-1} in the i.r. spectrum of the α-form was replaced almost completely by the 756 cm^{-1} band of the ω-form. This shift to higher frequency was thought to be due to the increased interaction of the benzene rings in the ω-form.

Bradbury *et al.*[53] used polarization measurements in i.r. spectroscopy to study the side-chain orientation in a number of poly-L-aspartate esters. They showed that poly-(β-benzyl-L-aspartate) has different side-chain conformations depending on whether it is in the right-handed or left-handed α-helical configuration. It was shown that all right-handed poly-L-aspartate esters have the same side-chain conformation as poly-(β-benzyl-L-aspartate). Bradbury *et al.* also studied the effect of the length of the side-chain on the type of α-helix formed. It appeared that *para*-substitution of such groups as NO_2, Cl, Me, or CN resulted in the formation of a right-handed helix.[54] Poly-(β-*p*-chlorobenzyl-L-aspartate) was found, for instance, to be in the right-

[50] T. Watanabe, T. Ina, K. Ogawa, T. Matsumo, S. Sawa, and S. Ono, *Bull. Chem. Soc., Japan*, 1970, **43**, 3939.
[51] E. M. Bradbury, A. R. Downie, A. Elliott, and W. E. Hanby, *Proc. Roy. Soc.*, 1960, **A259**, 110.
[52] H. Obata and H. Kanetsuna, *J. Polymer Sci., Part A-2, Polymer Phys.*, 1971, **9**, 1977.
[53] E. M. Bradbury, B. G. Carpenter, and R. M. Stephens, *Macromolecules*, 1972, **5**, 8.
[54] E. H. Erenrich, R. H. Andreatta, and H. A. Scheraga, *J. Amer. Chem. Soc.*, 1970, **92**, 1116.

handed helical conformation whereas the *ortho*- and *meta*-substituted analogues were in the left-handed configuration.

Aragao and Loucheux [55] have made similar studies of the i.r. spectrum of *ortho*-, *meta*-, and *para*-nitrobenzyl-L-aspartate esters. Homopolymers and corresponding copolymers of the nitro-derivatives with benzyl-L-aspartate were studied with a view to determining the influence of the nitro-group substituent on the stability and conformation of the polymer. The stability of the α-helix was found to vary with the solvent and the position of the nitro-group. I.r. bands whose frequencies depended on the handedness of the helix were observed. In the case of the copolymers plots of the frequencies of amide A, amide I, and amide II bands were made as a function of NO_2 concentration.

Other physical factors can influence the handedness of the copolymers. For instance, random copolymers of β-ethyl-L-aspartate and β-benzyl-L-aspartate can undergo a temperature transition from right-handed to left-handed helical forms. Malcolm [56] has studied the polarized i.r. spectra of orientated specimens of high molecular weight polymer cast in the form of a monolayer from chloroform–dichloroacetic acid (90:10). A band was observed in the amide I region at frequency 1658 cm^{-1} with parallel dichroism, together with a band at 1552 cm^{-1} with perpendicular dichroism. These figures were in good agreement with those expected for a right-handed α-helical conformation. The values expected for a left-handed α-helical conformation were given in an earlier paper [57] as 1668 and 1557 cm^{-1}. Malcolm showed that the right-handed α-helical form could be converted to the left-handed α-helical form by exposure of the film to the vapour of chloroform–dichloroacetic acid (90:10). Heat treatment of the specimen was thought to produce the ω-conformation characterized by bands at frequencies 1675 and 1536 cm^{-1}. Malcolm also showed that the spectra of films cast from low molecular weight material varied with time. The intensity of the low-frequency component of the amide I region at 1635 cm^{-1} increased with time and was thought to indicate that a β-conformation was developing.

Polypeptide Copolymers. The precise order of the amino-acid sequence in a protein has an important influence on the conformation adopted by various parts of the molecule. Since polypeptide copolymers contain specific amino-acid sequences they serve as useful model compounds for the study of the influence of sequence on conformation.

Collagen contains a glycyl residue at every third position along its chain and thus the polypeptide copolymer poly-(L-alanyl-L-alanylglycine) is an appropriate model. Doyle *et al.*[58] investigated the i.r. spectra of this compound in the solid state and in solution. After exposure to water or formic

[55] J. B. Aragao and M. H. Loucheux, *J. Chim. phys.*, 1971, **68**, 1578.
[56] B. R. Malcolm, *Biopolymers*, 1970, **9**, 911.
[57] E. M. Bradbury, B. G. Carpenter, and R. M. Stephens, *Biopolymers*, 1968, **6**, 905.
[58] B. B. Doyle, W. Traub, G. P. Lorenzi, F. R. Brown, and E. R. Blout, *J. Mol. Biol.*, 1970, **51**, 47.

acid, this compound gave bands characteristic of the β-conformation; a small shoulder at 1700 cm^{-1} suggested that the hydrogen-bonded chains were in the antiparallel conformation. Dichroism studies on oriented films indicated that the amide A and amide I bands had parallel dichroism and the amide II band had perpendicular dichroism. The film orientation was obtained by stroking and the chains became aligned perpendicular to the stroke direction. Films and powders obtained from hexafluoropropan-2-ol solution gave a spectrum different from that associated with the β-structure. The amide I band at 1650 cm^{-1} could not be assigned to a definite structure and the spectrum lacked the high amide A frequency thought to be typical of the collagen structure. It was thought that the structure was random-coil together with some modified form of the α-helical conformation.

Iio and Takahashi [59] have made similar measurements on the conformation adopted in solution by polypeptides containing ordered sequences of L-alanyl and glycyl residues. They studied the i.r. spectra of solutions of poly-tri-L-alanylglycine $(A_3G)_n$, poly-di-L-alanylglycine $(A_2G)_n$, and poly-L-alanyldiglycine $(AG_2)_n$. Each sample gave two bands in the amide I region at 1630 and 1655 cm^{-1}. The intensity of the two bands varied with the sample. The intensity of the band at 1630 cm^{-1} was taken as an indication of the amount of β-structure, whereas the intensity of the band at 1655 cm^{-1} was taken as an indication of the amount of α-structure. The band at 1655 cm^{-1} was strong in $(A_2G)_n$ but weak in the other two compounds. The overall indication was that the stability of the α-helix decreased in the order $(A_2G)_n$, $(A_3G)_n$, $(AG_2)_n$. This was not quite in the order of glycine content as had been expected. Consequently, the α-helical content could not be explained merely in terms of chain flexibility at the glycyl residues, and consideration has to be given to sequence and side-chain interactions.

Poly(glycylalanylglutamate-OEt) has also been studied in some detail. *X*-Ray diffraction gave reflections typical of a β-sheet structure [60] and this, together with results from electron microscopy, led to an antiparallel chain structure being postulated. The i.r. spectral information from oriented films was thought [61] to be consistent with a doubly oriented, cross β-conformation, superfolded structure with its molecular axis perpendicular to both the fibre axis and the film plane. A weak band at 1700 cm^{-1} was thought to be typical of this structure. Helical forms normally give parallel dichroism for the amide I band and perpendicular dichroism for the amide II band; β-structures give lower frequencies and reverse dichroism. This structure gave frequencies typical of the β-conformation but with α-helical type dichroism. Orientation measurements were made in three directions by careful manipulation of the sample. The results are shown in Table 9.

Poly(glycylalanylglutamate-OEt) might be expected to form an α-helix since both the alanyl and the glutamyl ester residues are predominantly

[59] T. Iio and S. Takahashi, *Bull. Chem. Soc. Japan*, 1970, **43**, 515.
[60] J. C. Andries and A. G. Walton, *J. Mol. Biol.*, 1971, **56**, 515.
[61] W. B. Rippon, J. M. Anderson, and A. G. Walton, *J. Mol. Biol.*, 1971, **56**, 507.

Table 9 *Infrared orientation data for poly(glycylalanylglutamate-*OEt)

Frequency/cm^{-1}	*Dichroism*	*Assignment*
3290 (s)	‖	Amide A
1730 (s)	‖	CO stretch of γ-ester
1695 (w)	⊥	Amide I, $\nu(0, \pi)$
1628 (s)	‖	Amide I, $\nu(\pi, 0)$
1545 (sh)	‖	Amide II, $\nu(\pi, \pi)$
1518 (s)	⊥	Amide II, $\nu(0, \pi)$
1440 (m)	‖	CH_2 bend
1370 (m)	‖	CH_2 wag
1330 (w)	⊥	Amide III
1250 (m)	⊥	CH_2 twist
1170 (m)	‖	CO stretch of γ-ester
697 (m)	⊥	Amide V

α-formers. However, polyglycine, polyalanine, and poly-(γ-benzyl-L-glutamate) can all form β-structures. This polytripeptide does to some extent reflect the structure of collagen although there are charged species in collagen which induce further modifications to the conformation.

Itoh, Oya and Shimanouchi [62] studied the copolymers of L-alanine with glycine, L-alanine with L-valine, L-alanine with L-leucine, and L-alanine with L-phenylalanine. I.r. spectroscopy of the copolymers in the solid state was used to determine their conformation. Attempts were made to correlate the conformational structure of the copolymer with its constituent amino-acids.

In the L-alanine–L-valine copolymer a band at 1630 cm^{-1} was found to increase in intensity relative to the 1655 cm^{-1} band as the L-valine content was increased. Since the band at 1633 cm^{-1} in poly-L-valine is characteristic of the β-conformation, and the band at 1656 cm^{-1} in poly-L-alanine is typical of the α-helical structure, these results imply that as the amount of L-valine in the copolymer was increased the proportion of β-conformation also increased. Other distinct differences between the α- and β-forms were found in the 200—700 cm^{-1} region.

Spectroscopic studies also showed that the L-alanine–L-leucine copolymer and the L-alanine–L-phenylalanine copolymer always take the α-conformation. This suggested that copolymers made up of amino-acid residues whose homopolymers adopt the α-conformation also form the α-helical conformation.

Lewis and Scheraga [63] applied the results of previous spectroscopic conformational studies of polypeptides to the interpretation of the Raman spectra of various water-soluble block copolymers in the solid state and in solution. For example, they studied the solid-state Raman spectrum of the block copolymer poly-L-alanine–poly-DL-lysine and found a band at 238 cm^{-1} which was not observed in the Raman spectrum of poly-L-alanine. This band was assigned to a conformationally dependent mode of the DL-lysine

[62] K. Itoh, M. Oya, and T. Shimanouchi, *Biopolymers*, 1972, **11**, 1137.
[63] A. Lewis and H. A. Scheraga, *Macromolecules*, 1971, **4**, 539.

blocks. A band at 390 cm^{-1} in the block copolymer was assigned to a helix deformation mode of the poly-L-alanine block. This occurs at 375 cm^{-1} in the spectrum of poly-L-alanine homopolymer. A band at 579 cm^{-1} in the block copolymer was assigned to an ordered backbone structure of the poly-L-lysine component. Indeed, most of the bands in the spectrum could be assigned to one or other of the constituent homopolymers. The low-frequency spectrum, together with bands in the 900—1300 cm^{-1} region, suggested that the poly-L-alanine block of the copolymer was predominantly in the α-helical conformation. Other features indicated that the poly-DL-lysine block may be in the α-helical conformation in the solid state, suggesting that D and L residues may not be randomly distributed.

The band at 238 cm^{-1} disappeared when the block copolymer was dissolved in salt-free water and two new bands appeared with frequencies 250 and 345 cm^{-1}. In solution the helix deformation mode shifted from 390 to 376 cm^{-1}, the values observed in previous work on solid-state poly-L-alanine. It was thought that poly-DL-lysine was exclusively in the random-coil conformation, whereas poly-L-alanine was a mixture of random-coil and α-helical conformations. Addition of salt to the copolymer in solution tended to broaden the features and was thought to indicate the presence of alanine–alanine hydrophobic bonds. This shielding of the charges of the poly-DL-lysine end blocks permitted the central poly-L-alanine block to adopt a hairpin-like conformation stabilized by side-chain to side-chain hydrophobic bonding.

Lewis and Scheraga [64] have also studied the laser-excited Raman spectra of random copolymers of hydroxybutylglutamine (HBG) and glycine in the solid state. The spectra were considered in terms of the effect of the glycine content on the helical content and also in terms of the behaviour of an isolated glycyl residue among the hydroxybutylglutamine residues in helical and coil sequences. It is known that as the glycine content increases so the structure transforms from a predominantly α-helical conformation to a random-coil form, and it was found that the Raman bands at 205, 260, 313, and 382 cm^{-1} assigned to helix deformation modes disappeared as the glycine content was increased. In the random copolymer of high α-helical content two bands at 1005 and 1040 cm^{-1} were observed, the most intense band being that at 1040 cm^{-1}. These were assigned to skeletal stretching modes of glycine. As the α-helical content decreased the band at lower frequency increased in intensity with concomitant decrease in intensity of the high frequency band. In the totally random-coil conformation the 1040 cm^{-1} band disappeared.

Perhaps the most interesting and novel feature of the work was the band found at 61 cm^{-1}. This band shown conclusively to be a Raman line was extremely sharp and was thought to arise either from a specific interaction of the glycyl residue with an HBG residue in an α-helix, or from a localized mode similar to those observed in doped solids. If this latter explanation is

[64] A. Lewis and H. A. Scheraga, *Macromolecules*, 1972, **5**, 450.

correct then it would be the first example of an observed localized mode in a helical biopolymer. If such localized modes proved to be a general feature then much information could be obtained from them on the coupling and energetics of biopolymers.

Proteins.—*Introduction.* Proteins are classified broadly into two groups: simple proteins, which yield only amino-acids on hydrolysis, and conjugated proteins, which contain non-amino-acid constituents. The conjugated proteins themselves do not appear to have been the subject of spectroscopic investigation.

The simple proteins include the albumins, the globulins, the glutelins, the prolamines, the protamines, the histones, and the scleroproteins like collagen, keratin, and elastin. The conjugated proteins include the nucleoproteins which contain nucleic acid, the glycoproteins which contain carbohydrate, the lipoproteins which contain lipids, the mucoproteins which contain hexosamines, the chromoproteins which contain chromophores, the metalloproteins which contain metals, and the phosphoproteins which contain phosphorus. In many cases a conjugated protein may be classified under more than one heading.

The simple proteins are generally classified into two groups according to their macromolecular shape: the globular proteins and the fibrous (or linear) proteins. There are some proteins whose structure can be considered as intermediate between these two cases. Within each group the proteins can be further classified according to their conformation (see Figure 22).

Fibrous Proteins. Fibrous proteins can be roughly divided into four classes (a) keratins (α-helical proteins), (b) silk (β-structure), (c) collagens, (d) elastin, resilin, and the amorphous proteins.

Recent spectroscopic studies have been concerned only with the fibrous

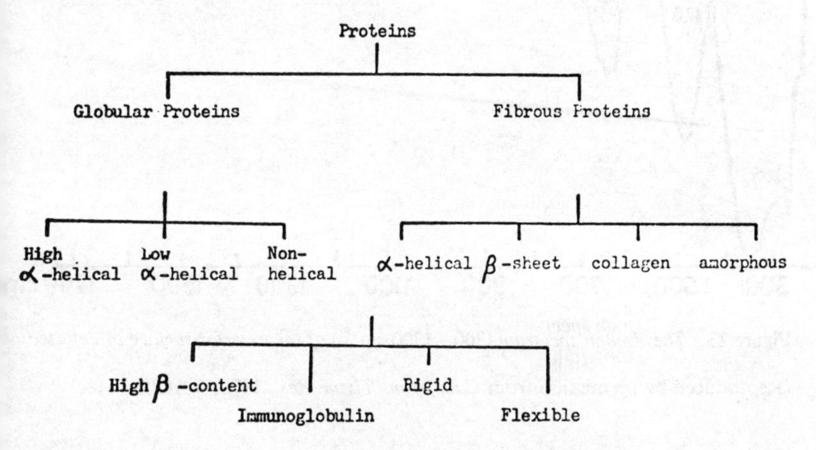

Figure 22 *Diagram of the approximate protein classification scheme*

proteins in the collagen class. Collagen is the major protein constituent of the tissues of both vertebrates and invertebrates. Its role is primarily structural, achieving its variety by alteration in its amino-acid sequence. It exists within the body as a constituent of skin, cartilage, myofibril, bone, tendon, and cornea. Tissue collagen is almost insoluble although some soluble collagens do exist containing small amounts of carbohydrate and nucleic acid.

X-Ray and physicochemical studies have provided information on the structure of collagen and it is known to be a triple-stranded polypeptide chain twisted into a triple helix stabilized by hydrogen bonding. The collagen chains have a glycyl residue at every third amino-acid position and the structure is not unlike that of polyglycine II.

The i.r. spectra of collagen obtained from rat, pike, and cod are very similar and appear to indicate the presence of one hydrogen bond per tripeptide.[65] The Raman spectra of ox tibia and rat tail tendon have been obtained by

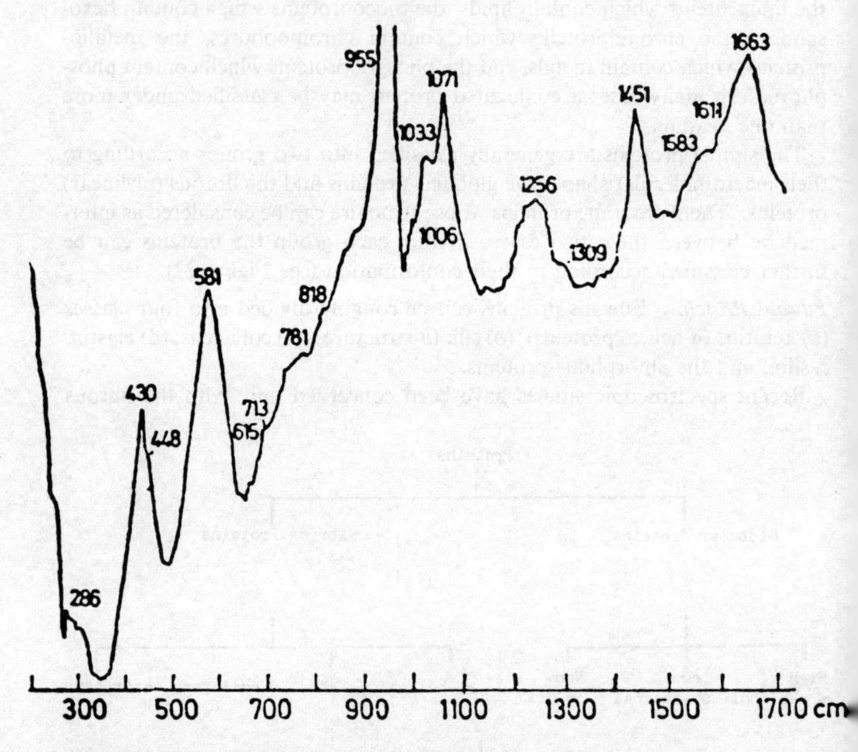

Figure 23 *The Raman spectrum* (200—1700 cm^{-1}) *of the exterior surface of defatted ox tibia*
(Reproduced by permission from *Calciferous Tissue Res.*, 1970, **6**, 162)

[65] N. G. Esipova and Y. A. Lazarev, Proceedings of the 1st European Biophysics Congress, 1971, **1**, 155.

Koenig *et al.*[66] and compared with the i.r.-spectra of calcified tissue and collagen (see Figure 23). The not unusual problem of high background emission was encountered and was responsible for the poor quality spectra. However, after exposure to the laser radiation for one hour the emission background had decayed sufficiently for a Raman spectrum to be obtained. This spectrum was of better quality than that previously reported for the transmission and total reflection i.r. spectrum of human bone. The bands of the Raman spectrum were assigned to calcium phosphate and collagen. In view of the splitting of the 1633 cm^{-1} band the carbonyl group was thought to exist in a variety of environmental conditions. No orientation studies like the previous i.r. orientation studies [67] on the C—O and N—H bonds were made. The published Raman spectra did not extend to the region of the amide A band (N—H stretch). Previous i.r. studies had shown that an amide A band at 3330 cm^{-1} seemed to be characteristic of the collagen structure, but Koenig *et al.* expressed the view that this band by itself should not be regarded as unequivocal evidence for the triple-helix conformation.

Globular Proteins. Globular proteins fall into three classes: those with a high content of the α-helical form like serum albumin and myoglobin; those with only a low α-helical content like lysozyme, and those with a non-helical conformation. There are, of course, borderline cases within this classification and often instances arise where the classification of a particular protein will depend upon its physical state. No globular protein has an α-helical content greater than 90% and the majority are found to have an α-helical content in the 15—75% range.

An example of a globular protein with a high α-helical content is serum albumin. There have been several studies of this compound both by i.r. [68–70] and Raman spectroscopy.[71–74] Far-i.r. spectra of the solid down to 50 cm^{-1} have been obtained using a Michelson interferometer.[69, 70] The bovine serum albumin sample was examined as a solid film, or cast on crystalline quartz, or in a solid polyethylene pellet. The far-i.r. spectrum of such an α-helical polypeptide would be expected to show narrow bands in the 50—400 cm^{-1} region,[75] characteristic of the peptide linkage in the α-helical configuration. The observed spectrum of bovine serum albumin contained only a single broad intense band centred around 150 cm^{-1} with no fine structure. The far-i.r. spectra of some other globular proteins of low α-helical content, in-

[66] A. G. Walton, A. G. Deveney, and J. L. Koenig, *Calciferous Tissue Res.*, 1970, **6**, 162.
[67] J. D. Termine and A. S. Posner, *Nature*, 1966, **211**, 268.
[68] M. A. Semenov, G. N. Kostoglodova, and V. Y. Maleev, *Dopovidi Akad. Nauk Ukrain R.S.R., Ser. B*, 1970, **32**, 1022.
[69] U. Buontempo, G. Careri, and P. Fasella, *Phys. Letters (A)*, 1970, **31**, 543.
[70] U. Buontempo, G. Careri, P. Fasella, and A. Ferraro, *Biopolymers*, 1971, **10**, 2377.
[71] G. Careri, V. Mazzacurati, M. Sampoli, G. Signorelli, and P. Fasella, *Phys. Letters (A)*, 1970, **32**, 495.
[72] G. Careri, V. Mazzacurati, and G. Signorelli, *Phys. Letters (A)*, 1970, **31**, 435.
[73] A. M. Belocq, R. C. Lord, and R. Mendelsohn, *Biochim. Biophys. Acta*, 1972, **257**, 280.
[74] J. P. Biscar, P. K. Dhall, and J. L. Pennison, *Phys. Letters (A)*, 1972, **39**, 111.
[75] G. H. Vineyard, *J. Phys. and Chem. Solids*, 1957, **3**, 121.

cluding ribonuclease and lysozyme, also showed this single broad band. No fine structure in this feature was revealed even at a resolution of 2.5 cm^{-1} and liquid nitrogen temperatures. It was suggested that the fine structure, which is usually associated with well-ordered homopolymers, was lost because the peptide chains of a globular protein assume many different conformations and contain many different amino-acids. Further broadening would result because of the non-periodical intramolecular interactions. It was also suggested that the broad band arose from anharmonic coupling among the soft modes, chiefly through hydrogen-bonded groups.[76] Presumably this broad intense band obscures the low-frequency peptide bands. It should be noted, however, that the neutron inelastic scattering spectrum does show a number of relatively sharp features in this region, as expected.[77]

The near-i.r. spectrum of bovine serum albumin has been studied briefly.[70] Intensity changes with temperature in the 1600—1750 cm^{-1} region of the i.r. spectrum of human serum albumin in D_2O have been investigated.[68] The far-i.r. spectrum of myoglobin, another protein of high α-helical content, was also observed by Buontempo *et al*;[70] the results were similar to those already discussed for bovine serum albumin. Chirgadze and Ovesepyan [78] have also studied the far-i.r. spectrum of sperm whale myoglobin and have found conformationally dependent bands in the 300—500 cm^{-1} region.

For bovine serum albumin in both the solid state [71] and in aqueous solution, Careri *et al.*[71, 72] claim to have observed a Raman band over 3000 cm^{-1} wide with a depolarization ratio of $\frac{3}{4}$. They suggested this was the Raman spectrum of the substance arising from the juxtaposition of many narrow Raman lines. Some very recent work [74] claims to support this interpretation. However, Lord *et al.*[73] have determined a much more credible Raman spectrum of aqueous bovine serum albumin. They observed some thirty discrete Raman bands in the 500—1600 cm^{-1} region, many of which they assign to the functional groups of the constituent amino-acids. Some other bands were attributed to the disulphide bridges, of which there are sixteen in bovine serum albumin. However, the intense broad water peak at 468 cm^{-1} merges with the bands of the S—S stretching region, making interpretation difficult in this region, but there does appear to be a low-intensity band at 650 cm^{-1} attributable to a C—S vibration. The amide III region gave some indication of the presence of an ordered structure. It would seem that the so-called Raman spectra observed by earlier workers was in fact simply high background emission typical of so many polymeric samples.

Bernstein *et al.*[79] have made some interesting resonance Raman studies of the interaction between a 3% aqueous bovine serum albumin solution and methyl orange (10^{-5} mol l^{-1}). The spectra were obtained using up to 800 mW of radiation at 4880 Å and a rotating sample cell. All the bands observed

[76] H. Frohlich, *Phys. Letters* (*A*), 1968, **26**, 402.
[77] M. Hirata, M. Hirata, and H. Saito, *J. Phys. Soc. Japan*, 1969, **27**, 405.
[78] Y. N. Chirgadze and A. M. Ovesepyan, *Doklady Akad. Nauk S.S.S.R.*, 1971, **201**, 744.
[79] P. R. Carey, H. Schneider, and H. J. Bernstein, *Biochem. Biophys. Res. Comm.*, 1972, **47**, 588.

under these conditions were attributable to the dye (ligand). In the presence of bovine serum albumin there was a general shift of frequencies of the methyl orange to lower values compared with those of free methyl orange. These shifts were explained in terms of methyl orange being essentially removed from its aqueous environment and buried in the protein structure. Although no further conclusions were drawn, the authors thought that intensity changes in the spectrum on complex formation might provide information relating to ligand protein interaction and protein conformation.

The detailed molecular structure of lysozyme, a globular protein with a low α-helical content, has been determined by *X*-ray diffraction.[80] The amino-acid sequence is known but depends on the source of the lysozyme. As already stated, the i.r. spectrum of lysozyme in the solid state has been studied by Buontempo *et al*,[69, 70] who related changes in the amount of β-component in chicken egg-white lysozyme to intensity changes in the 1630—1660 cm^{-1} region. The i.r. spectrum of lysozyme treated with n-propanol indicated that the protein had undergone a conformational change resulting in an increase of the β-structure. The band at 1630 cm^{-1} showed an increase in intensity whereas the band in the 1650—1660 cm^{-1} region decreased in intensity.

The Raman spectrum of chicken egg-white lysozyme in aqueous solution has also been investigated by Lord and Yu.[81] The spectra were obtained using a helium–neon laser with a power as low as 30 mW but the signal-to-noise ratio was not very good. The main features of the spectra were interpreted in terms of the constituent amino-acids. To assist in the interpretation of the Raman spectrum of lysozyme, Lord and Yu carried out a systematic study of the Raman spectra of the constituent amino-acids and oligopeptides in aqueous solution. They also determined the Raman spectrum of an aqueous solution containing the constituent amino-acids of lysozyme in approximately the same proportions as in the protein. The spectrum of this amino-acid mixture looks like the Raman spectrum of lysozyme (see Figure 24). Lord and Yu discussed features in the Raman spectrum which could be related to conformation, and suggested that the frequencies and intensities of the Raman bands assigned to the disulphide crosslinks could be related to the environment of the bridging group. They also suggested that bands in the amide I region and amide III region, together with bands in the 800—1150 cm^{-1} range, could provide information on the conformation of the molecule.

Mendelsohn and Lord [82, 83] investigated the Raman spectrum of aqueous lysozyme denatured by lithium bromide. Although the disulphide band with frequency 509 cm^{-1} was broadened and became less intense, the approximate constancy of area under the band indicated that the number of disul-

[80] C. C. F. Blake, D. F. Koenig, G. A. Mair, A. C. T. North, D. C. Phillips, and V. R. Sarma, *Nature*, 1965, **206**, 757.
[81] R. C. Lord and N. T. Yu, *J. Mol. Biol.*, 1970, **50**, 509.
[82] R. C. Lord and R. Mendelsohn, *J. Amer. Chem. Soc.*, 1972, **94**, 2133.
[83] R. C. Lord, *Pure Appl. Chem. Suppl.*, 1971, **7**, 179.

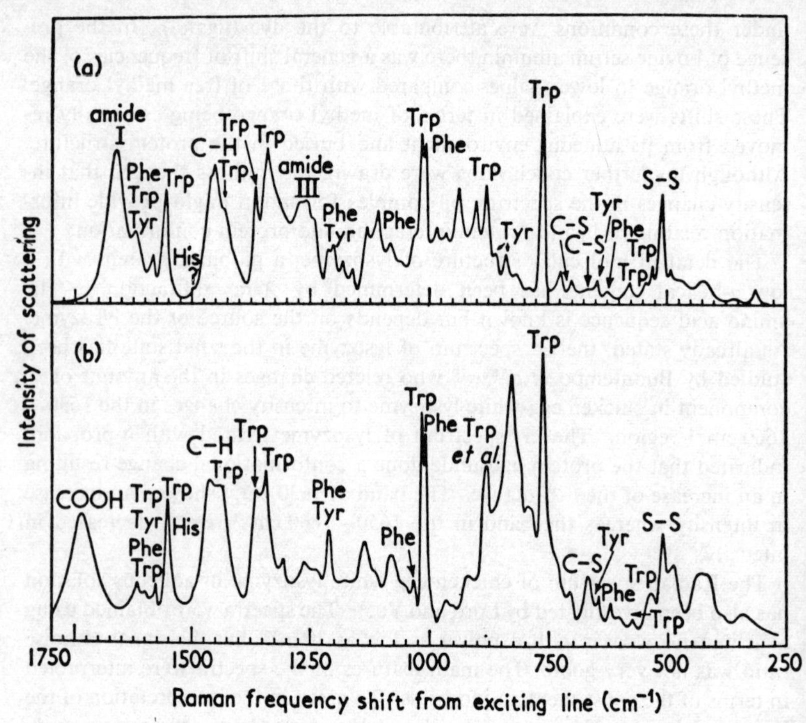

Figure 24 *The Raman spectrum of* (a) *lysozyme in water* pH 5.2 (b) *superposition of the Raman spectra of constituent amino acids at* pH 1.0
(Reproduced by permission from *J. Mol. Biol.*, 1970, **50**, 509)

phide bonds had not changed. The amide III region, however, resembled that of a denatured enzyme in which the disulphide linkages were broken. The region also indicated that there was little alteration in the skeletal structure on thermal denaturation of lysozyme.

The third class of globular proteins, those without helical form, fall roughly into four groups: those with a high content of the antiparallel β-structure such as β-lactoglobulin and α-chymotrypsin, the immunoglobulins, those with a rigid structure such as ribonuclease, and those with flexible chains.

Lord *et al.*[73] have investigated the Raman spectrum of the globular protein β-lactoglobulin in aqueous solution. This protein is known to contain two disulphide linkages and to have substantial pleated-sheet structure. The S—S stretching vibrations appeared as a broad band centred at 507 cm^{-1} (see Figure 25). In the amide III region a band was found at 1242 cm^{-1} with a shoulder at 1262 cm^{-1}. The latter peak was assigned to an amide III band of the β-structure and the main band was assigned to a random-coil structure. From the relative intensities it appeared that β-lactoglobulin had

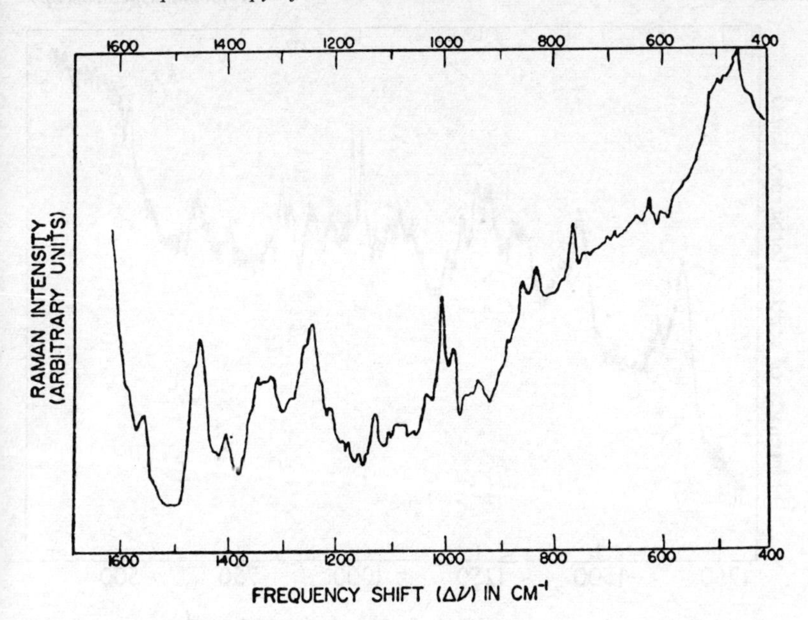

Figure 25 *Raman spectrum of β-lactoglobulin*; 48 mg ml^{-1}, pH 5.8, 0.1M-NaCl. *Spectral slit width* 7 cm^{-1}, 4880 Å Ar^+ *excitation*, 200 mW
(Reproduced by permission from *Biochim. Biophys. Acta*, 1972, **257,** 280)

predominantly a random-coil structure. A recent i.r. study of α-casein in methanol [84] has shown that a band at 1625 cm^{-1} thought to be characteristic of the β-conformation was observed only at high concentration.

The Raman spectrum of α-chymotrypsin in aqueous solution has been determined by Lord and Yu [85] (see Figure 26). *X*-Ray diffraction has shown that α-chymotrypsin contains a high amount of the extended β-structure.[86] In the region 500—750 cm^{-1} the Raman lines are very weak, although the line at 511 cm^{-1} was clearly the S—S stretching frequency. The intensity of the 511 cm^{-1} line was difficult to measure because of the steep slope of the background arising from the solvent, but it was neither as strong, nor as sharp as one would expect on the basis of the observed S—S line intensity in the Raman spectrum of lysozyme. These spectral features suggested that in contrast to lysozyme all the C—S—S—C groups in α-chymotrypsin do not have the same conformation. The C—S stretching frequencies due to the two methionine groups were difficult to locate because of the poor spectral quality. The amide I band was observed at 1669 cm^{-1} and was much sharper than the band observed in this region for random-coil poly-L-glutamic acid. In the amide III region two bands were observed with frequencies

[84] K. Shimazaki, S. Sugai, R. Niki, and S. Arima, *Agric. Biol. Chem.*, 1971, **35,** 1995.
[85] R. C. Lord and N. T. Yu, *J. Mol. Biol.*, 1970, **51,** 203.
[86] B. W. Matthews, P. B. Sigler, R. Henderson, and D. M. Blow, *Nature*, 1967, **214,** 652.

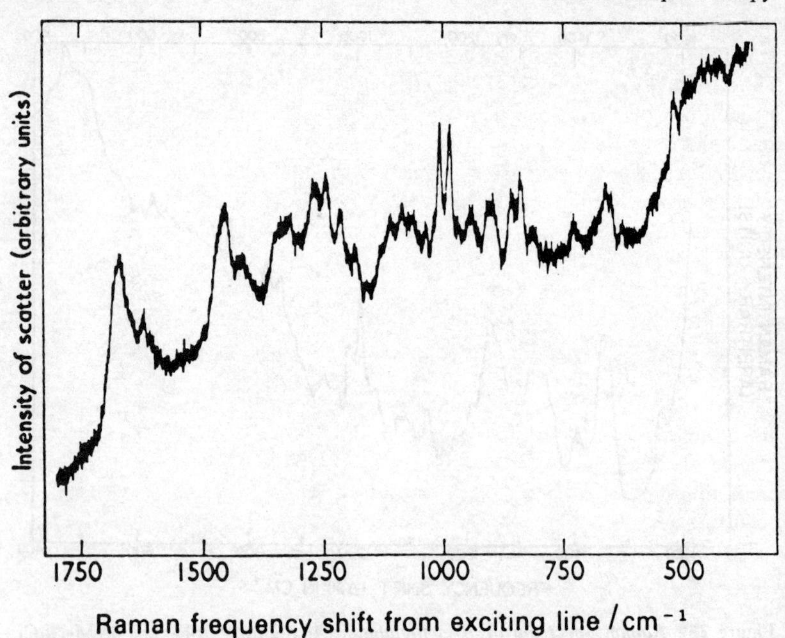

Figure 26 *Raman spectrum of α-chymotrypsin at* pD 5.0
(Reproduced by permission from *J. Mol. Biol.*, 1970, **51**, 203)

of 1245 and 1260 cm^{-1}. This separation of 15 cm^{-1} was smaller than that found in ribonuclease. The lines characteristic of the aromatic rings were readily located by comparison of the Raman spectrum of α-chymotrypsin with a mixture of the constituent amino-acids under the same conditions.

Optical rotatory dispersion and circular dichroism studies have shown that immunoglobulin M does not have an α-helical structure. Miller [87] has shown that the principle features of the i.r. spectrum of this compound in D_2O solution were the 1644 cm^{-1} (amide I) band and a second less intense shoulder at 1635 cm^{-1}. A number of weaker bands were observed including a shoulder at 1685 cm^{-1}. The band at 1644 cm^{-1} was considered to be typical of a random-coil protein in aqueous solution. The band at 1635 cm^{-1} was thought to be typical of compounds containing the β-conformation. The very weak shoulder at 1685 cm^{-1} was tentatively assigned to the antiparallel β-structure. However, the authors did not consider the spectroscopic evidence for the existence of β-structures to be conclusive. For instance, soya bean trypsin inhibitor which is known to have a random-coil structure has a band at 1635 cm^{-1}.[88] This serves as an illustration of the limitations of this type of approach to spectral assignments; often only tentative conclusions can be arrived at.

[87] J. N. Miller, *Biochim. Biophys. Acta*, 1971, **236**, 655.
[88] B. Jirgensons, M. Kawabata, and S. Capetillo, *Makromol. Chem.*, 1969, **125**, 126.

Termine *et al.*[89] have also investigated the i.r. spectra of a number of immunoglobulin samples and immunoglobulin derivatives in the solid state. They found that, although these compounds did not normally exhibit β-structure, most of them formed this structure when cast from 50% formic acid solution. One compound particularly well studied was human amyloid fibril. It was recognized that amyloid fibrils might also be derived from proteins other than immunoglobulins in some circumstances. I.r. spectroscopy was used to confirm the presence of antiparallel β-pleated sheet conformation in isolated human amyloid fibrils. A strong absorption band in the 1630 cm^{-1} region together with a weak but discernible shoulder at 1695 cm^{-1} was taken as diagnostic for the presence of antiparallel β-sheet conformation. It was found that the amount of β-structure increased with decreasing pH.

Lord and Yu [85] have studied the Raman spectrum of the rigid-chain globular protein, ribonuclease, in aqueous solution. Since the amino-acid sequence of ribonuclease has been fully established, mixtures of the constituent amino-acids in the correct proportions were used to simulate the ribonuclease spectrum. This enabled tentative assignments of the ribonuclease spectrum to be made. The bands at 1240 and 1262 cm^{-1} were thought to be indicative of two distinct structural components of the ribonuclease backbone. The band at 1262 cm^{-1} was associated with an α-helical structure containing a strongly hydrated peptide group and the band at 1240 cm^{-1} was associated with a random-coil structure. Attempts to obtain spectra of denatured samples were unsuccessful.

Intermediate Proteins. The borderline between fibrous and globular proteins is rather indefinite. Fibrinogen, which is converted into fibrin during the blood-clotting process, can be considered as either a globular or a fibrous protein. I.r. spectroscopy has been used to study the dichroism of bovine fibrinogen and fibrin.[90] It supports the view that no major conformational modification takes place in fibrinogen during the clotting process. The identification of the structure was made in this instance from the amide frequencies in the i.r. spectrum. In the case of fibrinogen the bands were weak and broad with the amide A band at 3306 cm^{-1}, the amide B band at 3070 cm^{-1}, the amide I band at 1654 cm^{-1}, and the amide II band at 1539 cm^{-1}. In the case of soluble fibrin the equivalent bands were at 3306, 3070, 1657, and 1518 cm^{-1} in the aqueous solution i.r. spectrum. Addition of calcium ion to the solution affected the hydrogen bonding of the protein, as exemplified by a shift in frequency of the amide A band from 3307 to 3299 cm^{-1}; this was thought to indicate either a strengthening of the hydrogen bonds and a concomitant weakening of some of the N—H bonds, or a competition between hydrogen-bond acceptors and the calcium ions. The amide II band also appeared to shift to lower frequencies on addition of calcium ion, but the diffuse nature of this spectral region made interpretation difficult.

[89] J. D. Termine, E. D. Eanes, D. Ein, and G. G. Glenner, *Biopolymers*, 1972, **11**, 1103.
[90] C. R. Kahn, R. M. Huseby, and M. Murray, *Life Sci.*, 1970, **9**, 1125.

Infrared dichroism measurements were also made on oriented fibrin films.[90] The amide A and amide I bands both showed parallel dichroism whereas amide II showed perpendicular dichroism. These results, together with the band positions, strongly indicate an α-helical conformation for the fibrin film.

Another example of an intermediate protein is insulin. This protein can be induced to make reversible globular to fibrous structure transformations. Yu *et al.*[91, 91a] assigned the amide I bands at 1662 and 1685 cm^{-1}, in the Raman spectrum of native insulin, to the α-helical and random-coil conformations, respectively. In a more recent paper, Yu *et al.*[92] studied the Raman spectrum of the crystalline hormone glucagon and found a strong amide I band at 1660 cm^{-1}. Since o.r.d. studies [93] of this compound had shown that it contained a 75% α-helical structure, these observations supported the

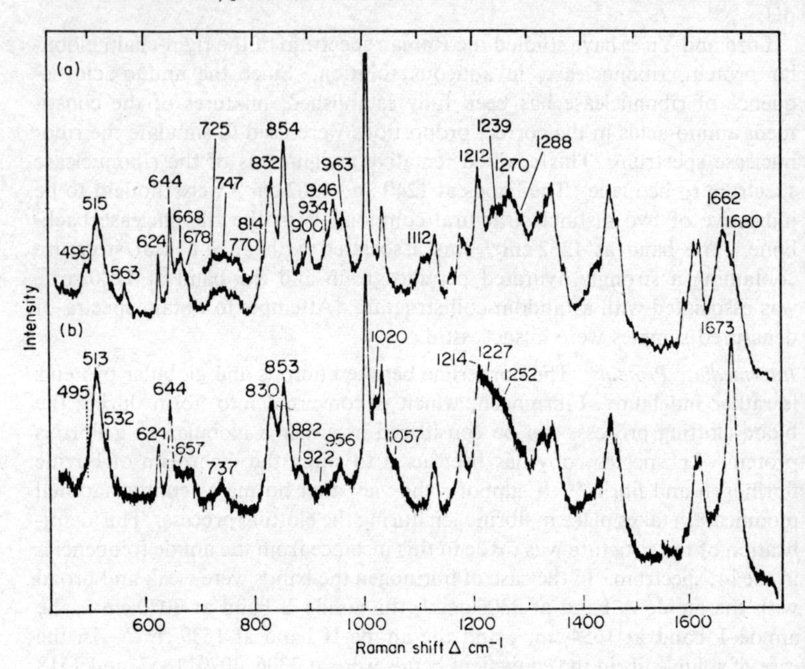

Figure 27 *Raman spectra of native and denatured insulin (bovine) in the solid state.* (a) *Native zinc–insulin crystalline powder. Spectral slit width,* $\Delta\sigma$, 4 cm^{-1}; *sensitivity, s,* 1000 cts s^{-1} *full scale; rate of scan, r,* 10 cm^{-1} min^{-1}; *standard deviation, d,* 1%. (b) *Denatured* insulin (*heat precipitated*). *The solution from which the sample was prepared had a concentration of* 10 mg insulin ml^{-1} *at* pH 2.42. $\Delta\sigma$, 5 cm^{-1}; *s,* 5000 cts s^{-1}; *r,* 10 cm^{-1} min^{-1}; *d,* 1%. *The laser power, p, at both samples was* 200 mW *at* 514.4 nm

(Reproduced by permission from *J. Mol. Biol.*, 1972, **70,** 117)

[91] J. Culver, N. T. Yu, C. S. Liu, and D. C. O'Shea, *Biochim. Biophys. Acta*, 1972, **263,** 1.
[91a] N. T. Yu and C. S. Liu, *J. Amer. Chem. Soc.*, 1972, **94,** 3250.
[92] N. T. Yu, C. S. Liu, and D. C. O'Shea, *J. Mol. Biol.*, 1972, **70,** 117.
[93] M. Schiffer and A. B. Edmundson, *Biophys. J.*, 1970, **10,** 293.

assignment in insulin of the band at 1662 cm^{-1} to an α-helical structure. The same authors [92] also observed the Raman spectrum of native and denatured insulin in the solid state (see Figure 27). It can be seen that interesting spectral changes occurred upon denaturation. The amide I band shifted from 1662 to 1673 cm^{-1} and became considerably sharper. This type of shift had also been observed in lysozyme upon chemical denaturation. The sharpening of the amide I band in denatured insulin was interpreted as being due to the presence of a greater uniformity arising from the now weaker hydrogen bonds. A comparison of the amide III region of denatured insulin with polyglycine I leads to the conclusion that denatured insulin, like polyglycine I, is in the β-conformation.

Denaturation also lead to significant changes in the S—S and C—S stretching regions of the disulphide bridges. The S—S stretching region about 515 cm^{-1} increased in intensity whereas the band at 668 cm^{-1}, in the C—S stretching region, decreased in intensity and shifted to 657 cm^{-1}. The C—S stretching frequency at 678 cm^{-1} remained unaffected. This was thought to arise because the geometry of the two interchain disulphide links in the denatured form was different from that in the native form but the intrachain disulphide link remained in the same conformation.

Other bands in the region below 800 cm^{-1} assigned to skeletal bending frequencies changed upon denaturation as expected. The changes that occurred in the 800—1200 cm^{-1} region were interpreted in terms of the unfolding of the insulin backbone.

Yu *et al.*[92] also studied the Raman spectrum of insulin in aqueous solution. They showed, by observations on the amide I and amide III frequencies, that the β-structure of the fibrillar form of insulin remained in solution. They also determined the effect of pH change on native insulin in solution. The Raman spectroscopic results indicated that no conformational change took place in insulin over the pH range 2.40—8.30.

Yu *et al.*[92] also compared the solid-state Raman spectra of insulin and proinsulin. The comparison provided good evidence that the insulin moiety in both compounds existed in the same conformation.

Nucleic Acids and Polynucleotides.—Nucleic acids are important constituents of all living cells; they are polynucleotides built up from mononucleotides through phosphodiester linkages. Nucleotides consist of purine or pyrimidine bases joined by glycosidic linkages to ribose or 2-deoxyribose, which is esterified with phosphoric acid. Nucleic acids containing ribose are called ribonucleic acids (RNA) and those containing deoxyribose are called deoxyribonucleic acids (DNA). RNA and DNA contain cytosine, thymine, adenine, guanine, and small amounts of other bases. The Watson and Crick model for DNA consists of a double-stranded helix held together by hydrogen bonds between bases in different chains. Three main types of RNA occur: ribosomal RNA, which is a high molecular weight polynucleotide; transfer RNA which has, in comparison to ribosomal RNA, a relatively low molecular weight; and messenger RNA which has a composition which is believed to reflect the composition of a particular section of DNA. Important aspects of the structure of both RNA and DNA are the nucleotide sequence along the polynucleotide chain and the stacking interactions of the constituent bases.

The i.r. and Raman spectra of these compounds have been studied with some thoroughness. Several recent reviews [94–97] have been written on the vibrational spectroscopy of these compounds. Tsuboi [98] discusses the i.r. work on the nucleic acids published before 1969, and Susi [99] covers in detail the i.r. spectra published before 1969 of proteins, polypeptides, nucleosides, nucleotides, and nucleic acids. Koenig [94, 95] has reviewed the techniques and applications of Raman spectroscopy using laser excitation in the study of biological macromolecules and in particular those with a helical conformation.

The main spectroscopic interest in the nucleic acids lies in their conformational behaviour in solution. Almost any change in the physical environment of DNA produces a change in its secondary structure. C.d. measurements, the main optical method used to study these conformational changes, give information on the bulk structural changes only. Vibrational spectroscopy, and especially Raman spectroscopy, provides a method for studying conformational changes taking place within specific parts of the molecule.

Infrared spectroscopy has been used to study the base composition of healthy and diseased human epidermal RNA.[100] Parker and Khare [101] have investigated hydrogen bonding in DNA and related compounds using i.r. spectroscopy. They studied the structure of bands in the N—H stretching region. These bands are broad but the authors claim that the 'structure' of these bands is not consistent with existing models of hydrogen bonding in DNA. They appear to have neglected possible contributions from Fermi resonance. The contribution of Fermi resonance in the N—H stretching region of the i.r. spectrum has been stressed by Zundel *et al.*[102]

Shie *et al.*[103] claim that the i.r. spectra of DNA and RNA in aqueous solution in the 950—1350 cm^{-1} region are characteristic of the sugar–phosphate skeleton. They studied the temperature dependence of the i.r. spectrum of calf thymus DNA, yeast RNA, and tobacco mosaic virus. In calf thymus DNA, in aqueous solution at 28 °C, a band at 1228 cm^{-1}, assigned to a phosphate vibration, shifted to 1231 cm^{-1} at 90 °C. This was interpreted

[94] J. L. Koenig, *Appl. Spectroscopy Rev.*, 1971, **4**, 233.
[95] W. B. Rippon, J. L. Koenig, and A. G. Walton, *J. Agric. Food Chem.*, 1971, **19**, 692.
[96] G. P. Zhizhina and E. F. Oleinik, *Uspekhi Khim.*, 1972, **41**, 474.
[97] H. Fritzsche, *Z. Chem.*, 1972, **12**, 1.
[98] M. Tsuboi, *Appl. Spectroscopy Rev.*, 1969, **3**, 45.
[99] H. Susi, *Struct. Stabil. Biol. Macromol.*, 1969, **2**, 575.
[100] N. M. Rumen and W. Pieretti, Proceedings of the 1st European Biophysics Congress, 1971, **1**, 275.
[101] B. R. Parker and G. P. Khare, *J. Mol. Spectroscopy*, 1972, **41**, 195.
[102] G. Zundel, W. D. Lubos, and K. Kölkenbeck, *Canad. J. Chem.*, 1971, **49**, 3795.
[103] M. Shie, I. G. Karitonenkov, T. I. Tikhonenko, and Y. Chirgadze, *Nature*, 1972, **235**, 386.

as a weakening of the hydrogen bonding between the phosphate group and water molecules. The heating of yeast RNA caused all the bands in the spectrum to broaden and decrease in intensity. This was interpreted as a denaturation of the RNA structure. On cooling the solution the original spectrum was restored and was interpreted as a renaturation of the RNA. Heating of the tobacco mosaic virus above 60 °C was thought to rupture the virus protein coat and release RNA into the solution. Changes occurring in the amide III region supported this theory. Pitha [104] has made similar measurements on these compounds and has discussed the results in terms of changes in the degree of base pairing. Thomas and Hartman [105] have used bands in the 900—1450 cm^{-1} region as a quantitative measure of the degree of base pairing.

Fritzsche [106] has discussed the NH out-of-phase bending vibrations and NH_2 wagging vibrations of base-paired molecules of DNA and related substances. Assignments were made on the basis of frequency shifts caused by deuteriation, change of temperature, and changes in complementary base pairing. Maleev *et al.*[107, 108] have discussed the effect of temperature on the i.r. spectra of D_2O solutions of DNA, RNA, and synthetic polynucleotides in the 1550—1750 cm^{-1} region. Bands in this region associated with the ring vibrations of the bases were shown to increase in intensity with temperature increase. These bands have also been shown to be sensitive to pH changes.[109] These changes were interpreted in terms of the unstacking of bases in various parts of the molecules.

Raman spectroscopy has been quite extensively used for the study of conformational changes of nucleic acids and related compounds. Tsuboi *et al.*[110] have studied the Raman spectrum in aqueous solution of transfer RNA obtained from *Escherichia coli* and the polynucleotides; polyriboguanylic acid, polyribocytidylic acid, polyriboadenylic acid, and polyribouridylic acid. The data obtained from the homopolymeric polynucleotides were used to synthesize a Raman spectrum of the transfer RNA. This synthesized Raman spectrum was found to be in quite good agreement with the observed spectrum of RNA. Tsuboi *et al.* interpreted the spectral differences in terms of the environmental differences of the base residues. Frequency shifts were thought to be due to changes in the hydrogen bonding, and intensity changes were attributed to differences in the stacking of the bases. Work of this kind up to 1969 has been reviewed by Peticolas *et al.*,[111] and since then a

[104] J. Pitha, *Biochim. Biophys. Acta*, 1971, **232**, 607.
[105] K. A. Hartman and G. J. Thomas, *Science*, 1970, **170**, 740.
[106] H. Fritzsche, *Experentia*, 1972, **28**, 391.
[107] V. Y. Maleev and M. A. Semenov, *Biofizika*, 1971, **16**, 389.
[108] V. Y. Maleev, M. A. Semenov and L. N. Blok, *Dopovidi Akad. Nauk Ukrain. R.S.R.*, *Ser. B*, 1970, **32**, 448.
[109] B. I. Sukhorukov, L. A. Kozlova, and M. A. Mazo, *Zhur. fiz. Khim.*, 1972, **46**, 548.
[110] M. Tsuboi, S. Takahashi, S. Muraishi, T. Kajiura, and S. Fishimura, *Science*, 1971, **174**, 1142.
[111] W. L. Peticolas, B. Fanconi, B. Tomlinson, A. L. Nafie, and E. W. Small, *Ann. New York Acad. Sci.*, 1970, **168**, 564.

number of other papers on the subject have appeared. Peticolas and co-workers have published several important papers reporting studies of the effect of change of temperature on the Raman spectra of nucleic acids. They have interpreted their results in terms of conformational alterations.

The first of these studies by Peticolas and Tomlinson [112] concerned the temperature dependence of the band at 720 cm^{-1} in the Raman spectrum of polyriboadenylic acid. This band was shown to decrease in intensity when the temperature was lowered from 26 to 3 °C. The effect was interpreted as being due to changes in base stacking with temperature leading to intensity changes among the various electronic levels and hence changes in the polarizability which controls the Raman intensity.

The second of these studies by Peticolas and Small [113] concerned the temperature dependence of the Raman scattering from adenosine 5′-monophosphate and a variety of polynucleotides in aqueous solution: polyriboadenylic acid (poly A), polyribocytidylic acid (poly C), polyribouridylic acid (poly U), and polyriboinosinic acid (poly I). Each sample was studied in a 12 μl cell specially designed for accurate temperature control. Care was taken to ensure that the pH of the solutions did not change appreciably over the 10—80 °C temperature range by using a 0.01 mol l^{-1} cacodylate buffer. Intensity measurements of bands due to the polynucleotides were made relative to a strong band of the cacodylate buffer at 610 cm^{-1} which was assumed to remain invariant in intensity with temperature. Most of the Raman bands observed for polynucleotides are related to the constituent bases attached to a sugar–phosphate polymer backbone chain. Significant intensity increases with temperature increase were found for 725, 1252, 1303, 1377, 1424, and 1508 cm^{-1} bands of polyriboadenylic acid. The largest changes occurred in the 725, 1303, and 1508 cm^{-1} bands which were assigned to ring vibrations of the bases. Poly C and poly I behaved in a manner similar to poly A, but poly U showed no intensity changes at all in the temperature range 30—80 °C.

The marked decrease in intensity of the 725, 1303, and 1508 cm^{-1} bands of poly A with decreases in temperature was interpreted as due to a stacking of the bases and an increase in the ordering of the poly A structure. At the higher temperatures the spectrum of poly A was very similar to the room temperature spectrum of its monomer adenosine 5′-monophosphate and it was concluded that the amount of base stacking must be very similar in the two substances under these conditions. The absence of intensity changes in poly U lead to the conclusion that there was little change in conformation with temperature in agreement with other studies.[114]

In this paper a theory of these intensity changes was developed in terms of the preresonance Raman effect. The normal resonance Raman effect occurs when the frequency of the exciting radiation approaches closely a region of

[112] B. L. Tomlinson and W. L. Peticolas, *J. Chem. Phys.*, 1970, **52**, 2154.
[113] W. L. Peticolas and E. W. Small, *Biopolymers*, 1971, **10**, 69.
[114] E. G. Edwards, C. P. Flessel, and J. R. Fresco, *Biopolymers*, 1963, **1**, 431.

electronic absorption of the molecule; Raman intensities are then greatly enhanced. In the preresonance Raman effect the exciting radiation is relatively far removed in frequency from a region of electronic absorption of a molecule but is still near enough to produce some intensity enhancement. The colourless nucleic acids have electronic absorptions which occur in the u.v. region. These u.v. absorptions show marked changes on alteration of the polynucleotide conformation which is thought to be due to the variation in the stacking of the purine and pyrimidine bases. The broad features of the u.v. spectrum make a detailed analysis impossible, but the much sharper features of the Raman spectrum yield information on the environmental conditions of individual parts of the molecule. The relationship between the changes in the u.v. absorption spectrum and the intensity changes in the Raman spectrum is very complex. If certain empirical assumptions are made a rough working hypothesis emerges which qualitatively explains the results from a wide variety of compounds. The most important generalization is that some vibrations appear to be more strongly coupled to the first u.v. absorption band than others. It would appear that given a better theoretical understanding, more definite assertions could be made on conformational changes taking place within biopolymers.

Peticolas and Small [115] have also studied the intensity dependence of Raman scattering from polynucleotides during order–disorder phase changes of the helical systems. They investigated the temperature dependence of the Raman intensity of RNA, DNA, poly A.U complex, poly(dAT) (alternating copolymer of deoxyadenosine and deoxythymidine), guanosine 5′-monophosphate, and guanosine 3′-phosphate. The complex poly A.U was studied because at certain ionic strengths it gives a double-stranded helix which had been shown to give rise to a co-operative melting transition.[116] The Raman spectrum of this complex at pH 6.8 between 15 and 65 °C showed appreciable intensity variations. These spectral changes were divided into three categories: (a) bands arising from ring vibrations of the nucleotide bases which increased in intensity with decreased base stacking, (b) a band at 814 cm^{-1} associated only with the α-helical conformation which decreased in intensity with decreased base stacking, and (c) bands in the carbonyl stretching region, 1600—1700 cm^{-1}, which change markedly with temperature variation. Examples from each of these categories will now be considered.

An example of the effect of temperature on a ring vibration is given in Figure 28 which shows a plot of the intensity of the 1236 cm^{-1} band, due to a uracil ring vibration, as a function of solution temperature. The helix to random-coil transition temperature of this compound under these conditions was at 59 °C. The sharp increase in intensity of the 1236 cm^{-1} band at this temperature was thought to be typical for a co-operative transition from the helix to random-coil conformation. The gradual increase in intensity of the 1236 cm^{-1} band from 20 to 58 °C was thought to be an indication

[115] E. W. Small and W. L. Peticolas, *Biopolymers*, 1971, **10**, 1377.
[116] H. T. Miles and J. Frazier, *Biochem. Biophys. Res. Comm.*, 1964, **14**, 129.

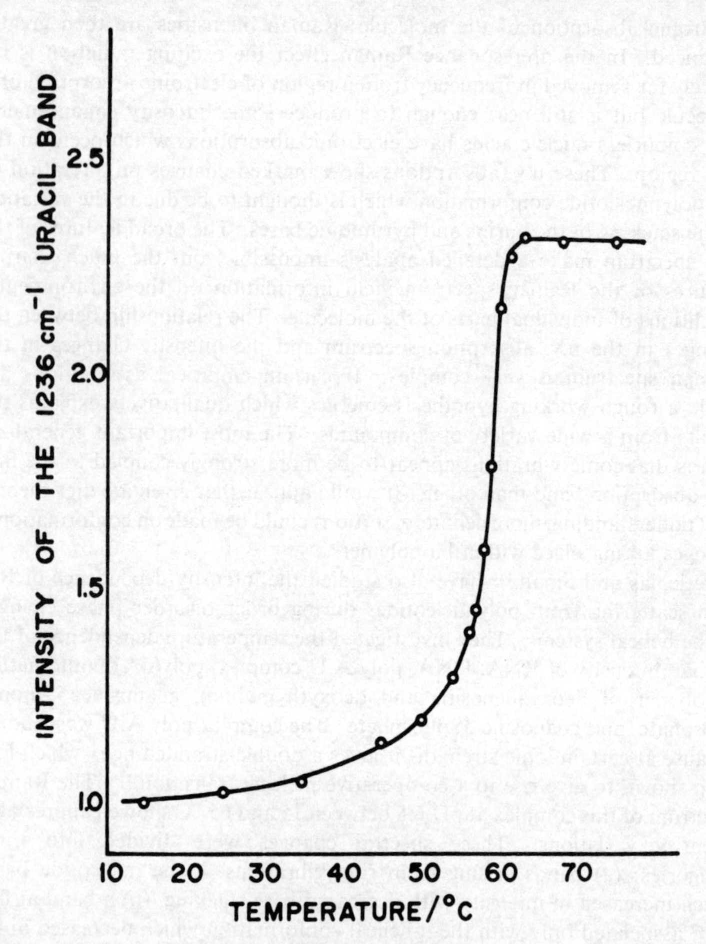

Figure 28 *Plot of the intensity of the uracil band at* 1236 cm^{-1} *in poly A.U vs. temperature* (*ratio to intensity at* 15 °C)
(Reproduced by permission from *Biopolymers*, 1971, **10**, 1377)

of a weakening of the poly U structure on approaching the transition temperature. The flatness of the curve above the transition temperature indicated that no further disordering of the poly U chain took place. A further example of the effect of temperature on a ring vibration is given in Figure 29 which shows a plot of the intensity of an adenine ring vibration at 730 cm^{-1} as a function of temperature. The intensity is approximately constant until the transition temperature at 59 °C is reached. Again at the transition temperature there is a sudden increase in the intensity of the 730 cm^{-1} band indicative of the co-operative phenomenon which leads to a sudden uncoiling

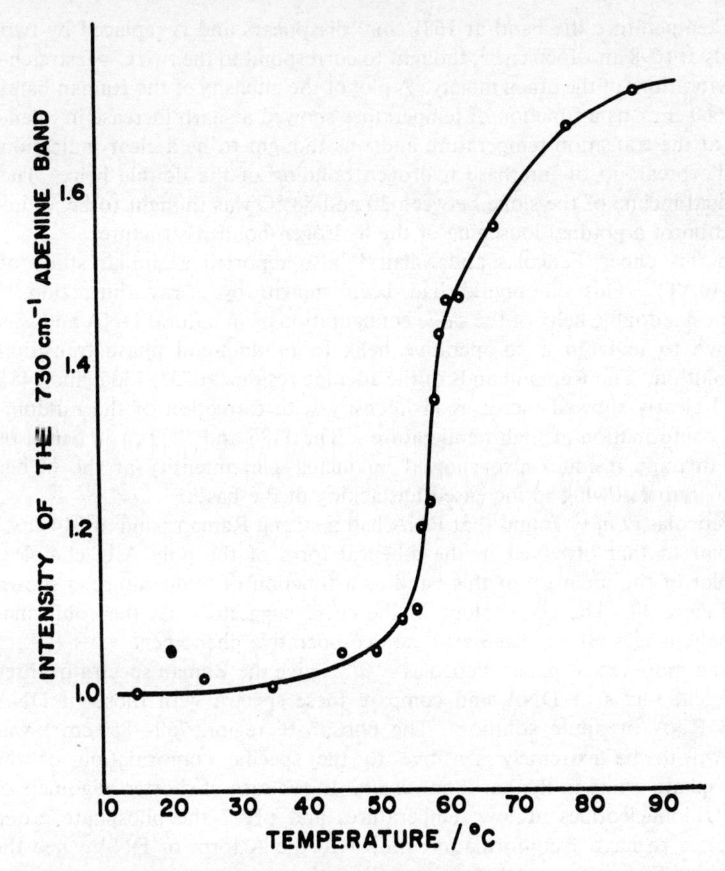

Figure 29 *Plot of the intensity of the adenine band at* 730 cm^{-1} *in poly A.U vs. temperature* (*ratio to intensity at* 15 °C)
(Reproduced by permission from *Biopolymers*, 1971, **10**, 1377)

of the helix at this temperature. In this instance, there is a further gradual increase in the intensity of the band above the transition temperature. This was thought to be an indication of further unstacking of the poly A chain above the transition temperature.

The band at 814 cm^{-1} which is characteristic of the α-helical conformation was tentatively assigned to a symmetric stretch of the phosphate diester because of its polarisation characteristics. At the transition temperature the intensity of the band drops sharply to almost zero indicating its association only with the α-helical conformation.

In the carbonyl stretching region, 1600—1700 cm^{-1}, one strong band at 1681 cm^{-1} was observed at low temperatures, thought to be due to one of the C=O stretching vibrations of the hydrogen-bonded uracil. Above the transition temperature the band at 1681 cm^{-1} disappears and is replaced by two bands at 1698 and 1660 cm^{-1}, thought to correspond to the two C=O stretching vibrations of the uracil moiety. A plot of the intensity of the Raman band at 1660 cm^{-1} as a function of temperature showed a sharp increase in intensity at the transition temperature and was thought to be a clear indication of the break-up of interbase hydrogen bonding of the double helix. The gradual incline of the slope between 20 and 58 °C was thought to be an indication of a gradual loosening of the hydrogen-bonded structure.

In this paper, Peticolas and Small [115] also reported a similar study of poly(dAT). This compound had been shown, by *X*-ray diffraction,[117] to form a double helix of the same configuration as in natural DNA and was known to undergo a co-operative helix to random-coil phase transition in solution. The Raman bands of the adenine residue at 733, 1306, and 1485 cm^{-1} clearly showed increases in intensity with formation of the random-coil conformation at high temperatures. The 1186 and 1242 cm^{-1} bands of the thymine residues also showed an increase in intensity at the higher temperatures owing to increased unstacking of the bases.

Peticolas *et al.*[115] found that RNA had a strong Raman band at 814 cm^{-1} similar to that observed in the α-helical form of the poly A.U complex. A plot of the intensity of this band as a function of temperature is shown in Figure 30. The gentle slope of the curve suggested that the conformational changes taking place were non-co-operative phenomena.

In a more recent paper, Peticolas *et al.*[118] give the Raman spectra of three different forms of DNA and compare these spectra with those of DNA and RNA in dilute solution. The phosphate region 750—850 cm^{-1} was shown to be extremely sensitive to the specific conformation of the phosphate group in the backbone chain. In the case of shorter oligomers of the ribonucleotides at low temperatures and pH 7 the phosphate group appears to have a conformation similar to the A form of DNA when the stacking forces between bases are sufficiently strong.

Peticolas *et al.*[119] have also studied the temperature dependence of the Raman scattering from transfer RNA. The spectra were obtained from an aqueous solution at pH 7.2 in the temperature range 10—70 °C. A band at 814 cm^{-1} was observed to decrease in intensity as the temperature was raised and was interpreted as a disordering of the sugar–phosphate backbone. A band at 725 cm^{-1} assigned to an adenine in-plane ring vibration was found to increase in intensity with increase in temperature and this observation was interpreted as an unstacking of the adenine bases. A band at 1230 cm^{-1} assigned to a uracil ring vibration was found to increase in intensity with increase in temperature indicating an unstacking of the uracil bases.

[117] D. R. Davies and R. L. Baldwin, *J. Mol. Biol.*, 1963, **6**, 251.
[118] S. C. Erfurth, E. J. Kiser and W. L. Peticolas, *Proc. Nat. Acad. Sci. U.S.A.*, 1972, **69**, 938.
[119] E. W. Small, K. G. Brown, and W. L. Peticolas, *Biopolymers*, 1972, **11**, 1209.

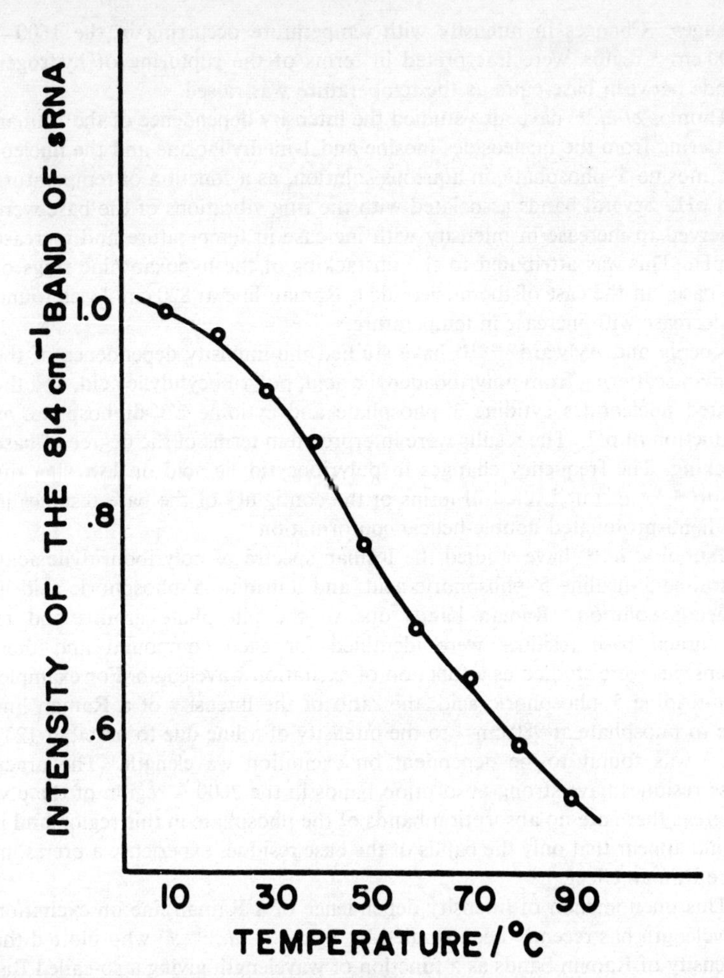

Figure 30 *Plot of the intensity of the sRNA band at* 814 cm^{-1} *as a function of temperature* (*ratio to intensity at* 10 °C)
(Reproduced by permission from *Biopolymers*, 1971, **10**, 1377)

Thomas *et al.*[120, 121] have measured Raman intensity changes with temperature and ionic strength for nucleic acids, in particular ribosomal RNA obtained from *Escherichia coli*. A band at 814 cm^{-1} was found to decrease in intensity with either increase in temperature or decrease in ionic strength and was thought to indicate a change in conformation of ribose–phosphate

[120] G. J. Thomas, *Biochim. Biophys. Acta*, 1970, **213**, 417.
[121] G. J. Thomas, G. C. Medeiros, and K. A. Hartman, *Biochem. Biophys. Res. Comm.*, 1971, **44**, 587.

linkages. Changes in intensity with temperature occurring in the 1600—1700 cm^{-1} region were interpreted in terms of the rupturing of hydrogen bonds between base-pairs as the temperature was raised.

Thomas *et al.*[122] have also studied the intensity dependence of the Raman scattering from the nucleosides inosine and 1-methylinosine and the nucleotide inosine 5′-phosphate, in aqueous solution, as a function of temperature and pH. Several bands associated with the ring vibrations of the base were observed to increase in intensity with increase in temperature and increase in pH. This was attributed to the unstacking of the hypoxanthine rings of the base. In the case of the nucleotide a Raman line at 820 cm^{-1} was found to decrease with increase in temperature.

Koenig and Aylward [123, 124] have studied the intensity dependence of the Raman scattering from polyriboadenylic acid, polyribocytidylic acid, and the related nucleotides cytidine 3′-phosphate and cytidine 2′3′-diphosphate as a function of pH. The results were interpreted in terms of the degree of base stacking. The frequency changes in polyribocytidylic acid on lowering the pH to 5.5 was interpreted in terms of the contiguity of the base residues in the hemi-protonated double-helical conformation.

Tsuboi *et al.*[125] have studied the Raman spectra of polyribouridylic acid, β-uridine-3′-uridine 5′-phosphoric acid, and β-uridine 5′-phosphoric acid in aqueous solution. Raman bands due to the phosphate groups and to the uracil base residues were identified for each compound and their intensities were studied as a function of excitation wavelength. For example, in β-uridine 5′-phosphoric acid, the ratio of the intensity of a Raman line due to phosphate at 980 cm^{-1} to the intensity of a line due to uracil at 1233 cm^{-1} was found to be dependent on excitation wavelength. The uracil base residues have strong absorption bands in the 2600 Å region of the u.v. whereas there are no absorption bands of the phosphate in this region and it would appear that only the bands of the base residues experience a preresonance Raman effect.

This phenomenon of intensity dependence of a Raman line on excitation wavelength has recently been studied by Rimai *et al.*,[126, 127] who plotted the intensity of Raman bands as a function of wavelength giving a so-called Raman profile. The results showed that, for some molecules of biological importance such as retinal, retinol, and β-carotene, the resonance Raman effect can arise from spin-forbidden transitions which are not observable in the absorption spectrum in the visible region 4000—7000 Å.

Other Molecules.—The metalloproteins rubredoxin,[128] haemoglobin, and their

[122] G. C. Medeiros and G. J. Thomas, *Biochim. Biophys. Acta*, 1971, **247,** 449.
[123] J. L. Koenig and N. N. Aylward, *Macromolecules*, 1970, **3,** 583.
[124] J. L. Koenig and N. N. Aylward, *Macromolecules*, 1970, **3,** 590.
[125] M. Tsuboi, S. Takahashi, S. Muraishi, and T. Kajiura, *Bull. Chem. Soc. Japan*, 1971, **44,** 2921.
[126] L. Rimai, *Chem. Phys. Letters*, 1971, **10,** 207.
[127] L. Rimai, R. G. Kilponen, and D. Gill, *J. Amer. Chem. Soc.*, 1970, **92,** 3824.
[128] T. V. Long and T. M. Loehr, *J. Amer. Chem. Soc.*, 1970, **92,** 6384.

related compounds [129, 130] have been studied by Raman spectroscopy. In the case of rubredoxin only two bands were observed with frequencies 365 and 311 cm^{-1}, but in the case of haemoglobin a resonance Raman effect occurred and many bands were observed. The solutions of haemoglobin and its related compounds were very dilute (10^{-4} mol l^{-1}), and the exciting radiation used was at 5682 Å, which falls in the middle of an absorption envelope of the compounds. Spectra of excellent quality were observed for oxyhaemoglobin, deoxyhaemoglobin, carboxyhaemoglobin, and cytochrome *c*. All the bands observed were assigned to fundamental vibrations of the haeme groups. In the case of carboxyhaemoglobin in the i.r., the C=O stretching frequency had been assigned to a band in the 1950—2000 cm^{-1} region but no band was observed in this region in the resonance Raman spectrum.

Infrared spectra [131] of whole cells have been determined in an attempt to characterize their genera. It has been shown [132] that the i.r. spectra of cell membranes originate mainly from the amide I and amide II bands of proteins and other bands due to such constituents as lipids. The bands in the methylene stretching region were thought to be characteristic of protein–lipid interactions. This type of interaction has been the subject of study by Peticolas *et al.*[133] using Raman spectroscopy.

There has been only a modest amount of interest in the vibrational spectroscopy of biological macromolecules other than proteins and polypeptides. There has been a comparison between the laser-excited Raman spectrum of native cellulose and the cell walls of an alga *V. ventricosa*.[134] These data were used together with the earlier i.r. data to assign the cellulose vibrational spectrum. It was shown that the CH_2OH side-chains existed in only one conformation. Other spectroscopic studies of cellulose have stemmed from its importance in the textile industry.[135]

There has been some interest in the chain-folding properties of sugars, and Koenig and Vasko [136] have made i.r. studies on chain-folding in amylose. A band at 1295 cm^{-1} was assigned to a regular tight-loop fold. This study confirmed that amylose had a fold-chain structure in the crystalline state with helix axes oriented normal to the lamellar faces. A similar study was made of amylopectin.[137] The spectroscopic results were thought to suggest that 'swelling agents' attacked films of amylopectin at the folded portions of the chain.

[129] T. C. Strekas and T. G. Spiro, *Biochim. Biophys. Acta*, 1972, **263**, 830.
[130] S. McCoy and W. S. Caughley, *Biochemistry*, 1970, **9**, 2387.
[131] N. A. Krasil'mikov, G. I. El-Registan, V. B. Il'yasova, and N. S. Agre, *Microbiology (U.S.S.R.)*, 1971, **40**, 58.
[132] D. H. Green and M. R. J. Salton, *Biochim. Biophys. Acta*, 1970, **211**, 139.
[133] J. L. Lippert and W. L. Peticolas, *Proc. Nat. Acad. Sci. U.S.A.*, 1971, **68**, 1572.
[134] J. Blackwell, P. D. Vasko, and J. L. Koenig, *J. Appl. Phys.*, 1970, **41**, 4375.
[135] F. Grass, H. Siesler, and H. Kraessig, *Melliand Textilber. Internat.*, 1971, **52**, 1001.
[136] J. L. Koenig and P. D. Vasko, *J. Macromol. Sci.*, 1970, **B4**, 347.
[137] P. D. Vasko and J. L. Koenig, *J. Macromol. Sci.*, 1972, **B6**, 117.

3 Polymers

Introduction.—Several reviews [138-144] and three books [145-147] on the spectro-spectroscopy of polymers have appeared recently. Hendra *et al.*[138] review the current techniques of Raman spectroscopy as applied to polymers and discuss the potential of this form of spectroscopy. Willis *et al.*[139] review possible industrial applications of Raman spectroscopy to the study of polymers. The review of Schaufele [142] covers the techniques for measuring the Raman spectra of polymers using laser excitation. This review includes typical Raman spectra for $CH_3[CH_2CHCl]_nCH_3$ and $CH_3[CH_2C(Me)_2]_n$-CH_3. Schaufele has also reviewed experimental and theoretical work on the so-called symmetric longitudinal accordion-like mode characteristic of extended planar zig-zag carbon chains. The accordion mode was first recognized by Mizushima and Shimanouchi in 1949.

General Considerations.—There have been several important developments in the theory of the vibrational spectroscopy of polymers. Linear polymers are highly anisotropic and their spectra are complicated and difficult to simulate, as pointed out in a recent review.[143] Broadhurst and Mopsik [148] have given a method for calculating the vibrational frequencies of model linear polymers. They used for their calculations a system of N harmonically coupled oscillators of mass M, with bending and stretching force constants. Each mass was also considered to be quasiharmonically coupled to a Debye lattice through a lattice force constant k_L corresponding to a cut-off frequency $\omega_L = (2k_L/M)^{\frac{1}{2}}$. The $3N$ free-chain eigenfrequencies ω_j were found by diagonalizing the appropriate force constant matrices. Since the lattice is randomly fluctuating, the effect is to smear out the frequencies ω_j. Each of these frequencies forms a band with a low frequency cut-off at ${\omega_j^2}_{min} = \omega_j^2$ and a high frequency cut-off at ${\omega_j^2}_{max} = \omega_j^2 + \omega_L^2$. Broadhurst and Mopsik used these results to calculate the frequency spectra of the n-alkanes.[149] Their results were found to be in good agreement with the results of other workers [18] and in fair agreement with the Wilson GF matrix calculations for polyethylene.[150]

[138] M. J. Gall, P. J. Hendra, C. J. Peacock, and D. S. Watson, *Appl. Spectroscopy*, 1971, **25**, 423.
[139] M. E. A. Cudby, H. A. Willis, P. J. Hendra, and C. J. Peacock, *Chem. and Ind.*, 1971, 531.
[140] H. Tadokoro and M. Yokoyama, *Sen-i To Kogyo*, 1970, **3**, 603.
[141] Y. Tanaka, *Nippon Gomu Kyokaishi*, 1970, **43**, 966.
[142] R. F. Schaufele, *Macromol. Rev.*, 1970, **4**, 67.
[143] B. Wunderlich and H. Bauer, *Adv. Polymer Sci.*, 1970, **7**, 151.
[144] R. Jankow and J. N. Willis, *J. Mol. Spectroscopy*, 1972, **41**, 412.
[145] W. L. Peticolas, 'Polymer Characterisation, Interdisciplinary Approaches', Plenum, New York, 1971.
[146] P. J. Hendra, 'Polymer Spectroscopy', Verlag Chemie, Weinheim, 1971.
[147] T. R. Gilson and P. J. Hendra, 'Laser Raman Spectroscopy', Wiley, London, 1970.
[148] M. G. Broadhurst and F. I. Mopsik, *J. Chem. Phys.*, 1971, **55**, 3708.
[149] M. G. Broadhurst and F. I. Mopsik, *J. Chem. Phys.*, 1970, **54**, 4239.
[150] T. Kitagawa and T. Miyazawa, *Reports Progr. Polymer Phys., Japan*, 1968, **11**, 219.

Gribov [151] has also given a method for calculating the vibrational spectrum of two-dimensional and three-dimensional structures of interacting polymer chains. The calculations were based on a system of parallel periodic chains located equidistant from each other and coupled together in a simple manner. Identical interactions of adjacent chains along their length were assumed. It was found that most of the results of the calculations based on these more complicated two- and three-dimensional structures were similar to those obtained from the simple one-dimensional model. Gribov concluded that, on the basis of his theory, qualitative conclusions could be drawn about the behaviour of amorphous polymers but quantitative calculations which included polymer defects were not possible.

Most of the earlier calculations of polymer chain vibrational frequencies have been based on models of ideal, isolated, infinitely long, and perfectly regular chains. Modified Wilson GF matrix methods were used to calculate the $k = 0$ modes. It was generally recognized that many features in the i.r., Raman, and neutron inelastic scattering of polymeric materials could not be accounted for by the $k = 0$ modes or by multiphonon processes. However, polymers are not ideal systems and exhibit various defects. They are not infinite chains and they bend, fold, and twist out of the stereoregular conformation. Polymer defects can be classified into four types: (a) chemical defects, (b) conformational defects, (c) steric defects, and (d) mass defects.

Zerbi [152, 153] has considered in some detail the vibrational spectrum of chain molecules with conformational disorder. He gives a method for the exact calculation of the dynamical matrix of a polymer chain which contains a random distribution of conformational defects. The main difficulty lies in the solution of the eigenvalue equation of the very large dynamical matrices that result. The negative eigenvalue theorem was used to calculate the density of vibrational eigenstates $g(\nu)$; $g(\nu)$ is a function defined such that $g(\nu)\,d\nu$ gives the number of frequencies in the interval $(\nu, \nu + d\nu)$ and is plotted as a histogram. A generalization of the results may be given as follows: (a) the translational symmetry is removed and the shape of the phonon waves is slightly perturbed resulting in the activation of the whole of the density of states of the perfect lattice; (b) if the concentration of defects is small the localized defect modes will also give their contribution to the observed spectra; (c) if the concentration of defects is high then the modes of the defects may interact and the phonons of the host lattice will be strongly perturbed. The whole density of states will then be perturbed and because of the lack of symmetry all the modes will be optically active.

Zerbi applied this theoretical treatment to polyethylene and also re-investigated its vibrational spectrum. Many, but not all, of the features of the observed i.r. and Raman spectra could be explained in terms of a $k = 0$ one

[151] L. A. Gribov, *Optika i Spektroskopiya*, 1970, **29**, 876.
[152] G. Zerbi, L. Piseri, and F. Cabassi, *Mol. Phys.*, 1971, **22**, 241.
[153] G. Zerbi, *Pure Appl. Chem.*, 1971, **26**, 499.

phonon spectrum, which has been characterized with a fair degree of certainty as a result of previous extensive experimental work and theoretical calculations on single isolated chains and three-dimensional crystals. However, bands which could not be explained in terms of $k = 0$ modes occur in the solid-state i.r. spectrum at 198, 252, 386, 538, 1075, 1128, 1300, 1350, 1367, and 1440 cm^{-1} and in the Raman spectrum of crystalline polyethylene at 1465 and 720 cm^{-1}.

Zerbi adopted a model of polyethylene consisting of 200 methylene units joined in an all-*trans* conformation. Kinks and fold defects were introduced into the chain in both a regular and a random way. The values of density of states were calculated and compared with the experimental anomalies listed above. The whole of the density of states of the perfect crystal become spectroscopically active. The i.r. bands at 1440, 1300, 1128, and 1075 cm^{-1} coincided with singularities in the density of states. The broad features at 538 and 252 cm^{-1} in the i.r. were accounted for as activation of the ν_5 and ν_9 acoustical branches. The Raman bands at 1465 and 720 cm^{-1} were accounted for in terms of singularities which were normally only i.r. active for the ideal chain but became Raman active for the imperfect chain. The bands at 1350 and 1367 cm^{-1} were thought to be localized defect modes. However, not all the observed bands could be accounted for in this way.

Luongo [154] has discussed the diverse uses of i.r. spectroscopy in the study of polymers. He showed that, although most of the observed features in the i.r. spectrum are attributable to functional groups, there are also a num-

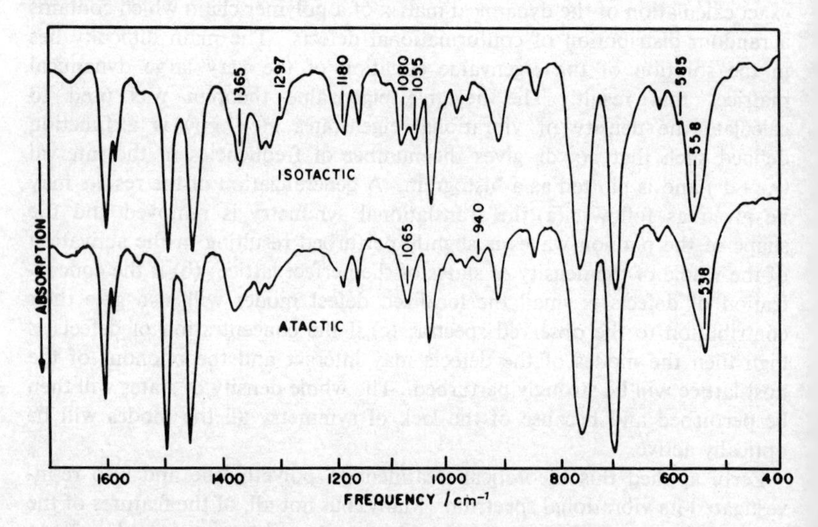

Figure 31 *Infrared spectra of isotactic and atactic polystyrene*
(Reproduced by permission from *Appl. Spectroscopy*, 1971, **25**, 76)

[154] J. P. Luongo, *Appl. Spectroscopy*, 1971, **25**, 76.

ber of weaker bands which can be related to conformation. Thus, for example, when the solid-state i.r. spectra of isotactic and atactic polystyrene are compared (see Figure 31), although the general features of the spectra are the same, the isotactic form has a doublet with frequencies 1080 and 1055 cm^{-1} whereas the atactic form has only a single band at 1065 cm^{-1} in this region. A further example shows how polymorphism can be followed by i.r. spectroscopy. Isotactic polybut-1-ene can exist in three different crystalline forms. Two of these modifications are produced by hot-moulding a film of the material and allowing it to age at room temperature for several weeks. During this time a crystalline phase transition occurs involving the expansion of a four-fold helix modification to a three-fold helix modification. This can be monitored by i.r. spectroscopy. Typical spectra obtained by Luongo are shown in Figure 32. A third example shows how i.r. spectroscopy can be used to study the surface region morphology of polymer films. The degree of crystallinity in the surface region of polyethylene films is found to be dependent on the nature of the nucleating substrate. A fourth example shows how structural rearrangements in for example polyethylene, induced by electron irradiation, can be monitored by i.r. spectroscopy.

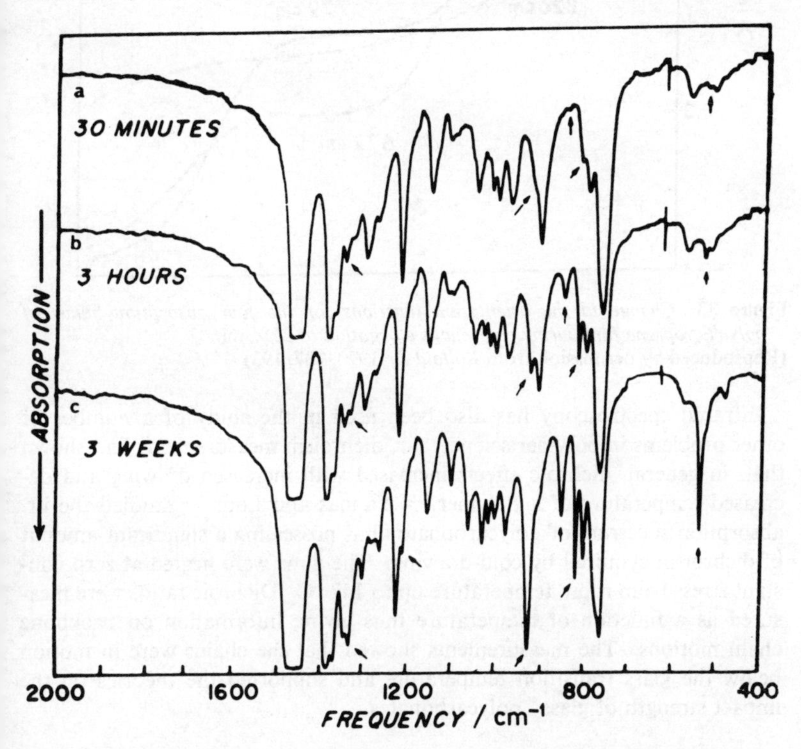

Figure 32 *Infrared spectra of polybut-1-ene during the II → I transformation* (Reproduced by permission from *Appl. Spectroscopy*, 1971, **25**, 76)

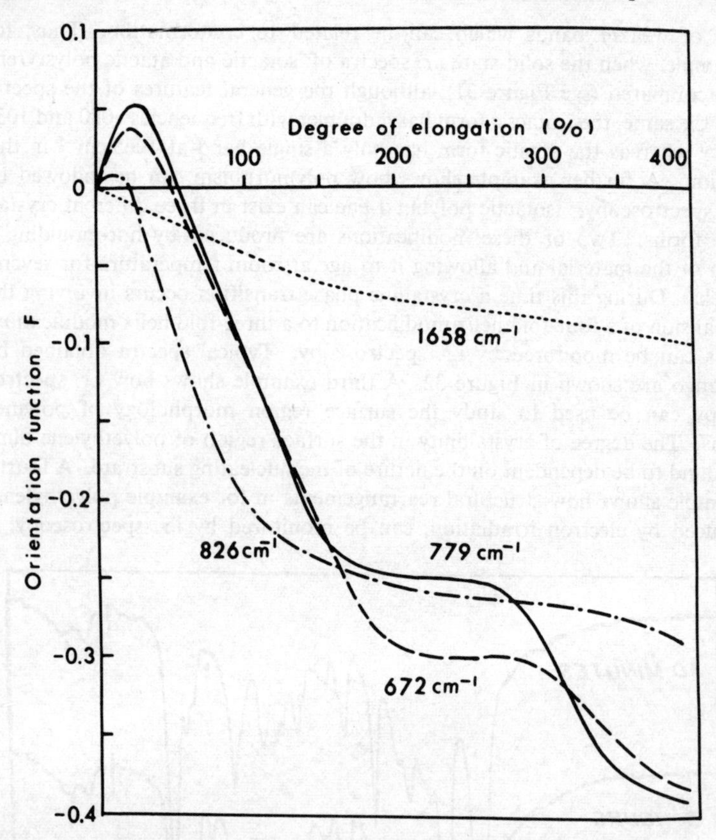

Figure 33 *Change in the orientation functions for the four absorption bands of polychloroprene film during continuous elongation at* 25% min^{-1}
(Reproduced by permission from *Kolloid Z.*, 1970, **237**, 193)

Infrared spectroscopy has also been used in the study of a number of other problems in polymer science. I.r. dichroism measurements had shown that, in general, dichroic effects increased with increased drawing and decreased temperature of a polymer.[155] Yannas and Lunn [156] studied the i.r. absorption spectrum of polycarbonate films possessing a significant amount of dichroism acquired by cold-drawing. The films were heated at zero constant stress from room temperature up to 147 °C. Dichroic ratios were measured as a function of temperature thus giving information on backbone chain motions. The measurements showed that the chains were in motion below the glass transition temperature and supported the theories on the impact strength of glassy polycarbonates.

[155] G. M. Bartenev and A. A. Valishin, *Mekh. Polimerov*, 1970, **6**, 979.
[156] V. I. Yannas and A. C. Lunn, *J. Polymer Sci., Part B, Polymer Letters*, 1971, **9**, 611.

Tanenaka *et al.*[157] studied the relation between stress and i.r. dichroism. In this instance they studied continuous elongation and stress relaxation in polychloroprene films at room temperature. They assumed uniaxial orientation and defined a function F such that

$$F = \frac{A_{\parallel} - A_{\perp}}{A_{\parallel} + 2A_{\perp}}$$

where $A_{\parallel}$ and $A_{\perp}$ are absorbances of the sample for radiation polarized parallel and perpendicular to the stretch direction, respectively. By studying bands sensitive to crystallinity, the authors claimed that it was possible to follow changes in crystallinity during mechanical treatment. Three bands with frequencies 672, 779, and 826 cm^{-1} were used to determine the crystallinity. Polychloroprene is orthorhombic with space group D_2^4 and four molecules per unit cell. The bands at 672 and 779 cm^{-1} are the CH_2 rocking mode of B_2 or B_3 symmetry and the C—Cl stretching mode of B_2 or B_3 symmetry, respectively. The transition moments of these bands can be thought of as parallel to the crystal *ab* plane. Figure 33 shows the change in orientation function F as a function of percentage elongation at a continuous elongation rate of 25% min^{-1} for the three bands. Included in this plot is the function F for the band at 1658 cm^{-1} which was considered to be insensitive to changes in crystallinity. The difference in behaviour of the 826 cm^{-1} line compared with the 779 cm^{-1} and 672 cm^{-1} lines was interpreted in terms of an overlap of two bands at 826 cm^{-1} due to out-of-plane =C—H deformations, one characteristic of the amorphous state and the other sensitive to crystallinity.

Figure 34 shows the change in value of F/F_0 as a function of time where F_0 is the instantaneous orientation function. The polychloroprene film was stretched to four times its original length and then held at this constant elongation while relaxation of the polymer was allowed to take place. The value of F/F_0 for each of the bands, including the band at 1658 cm^{-1}, can be seen to increase rapidly at first and then to settle to a constant value. Although the F/F_0 value for the crystalline-insensitive band at 1658 cm^{-1} was slightly larger than for the crystalline sensitive bands, the difference between these values was as great as in the case of vulcanized rubber.[158] This result suggested that in contrast to the results for vulcanized natural rubber the sample orientation in polychloroprene in the crystalline phase was almost the same as that in the amorphous phase. In the case of rubber,[158] orientation of the crystalline phase was completed almost immediately upon elongation, whereas in the amorphous phase the molecular chains oriented gradually during the course of stress relaxation.

Zhurkov [159] has used i.r. spectroscopy to study the formation of sub-

[157] T. Takenaka, Y. Shimura, and R. Gotoh, *Kolloid Z.*, 1970, **237**, 193.
[158] R. Gotoh, T. Takenaka, and N. Hayama, *Kolloid Z.*, 1965, **205**, 18.
[159] S. N. Zhurkov, V. A. Zahrevskii, V. E. Korsukov, and V. S. Kuksenko, *Fiz. Tverd. Tela*, 1971, **13**, 2004.

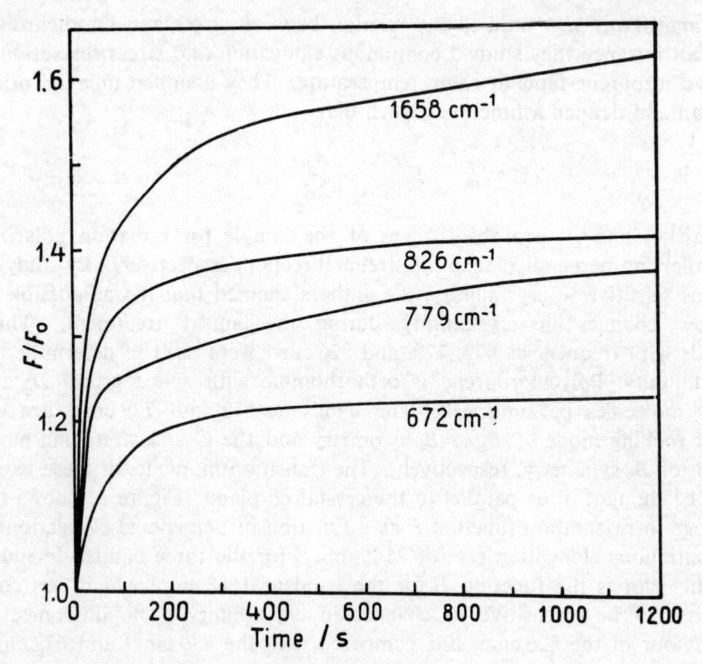

Figure 34 *Change in F/F_0 value for the four absorption bands of polychloroprene film during stress relaxation at 400% elongation*
(Reproduced by permission from *Kolloid Z.*, 1970, **237**, 193)

microcracks in polymers under stress and Tanabe [160] has studied the changes in the i.r. spectrum of a number of polymers at pressures up to 5000 atm. For example, he obtained the i.r. spectra of the three modifications of poly-(vinylidene fluoride). The three modifications labelled I, II, and III were obtained as follows: modification II by crystallization of the melt at atmospheric pressure; modification I by drawing modification II at 50 °C; modification III either by crystallization of the melt at high pressure or by heat treatment of modification II under high pressure. In general, high pressures were found to significantly increase the crystallinity of the samples. It was found that the degree of molecular orientation of polymer samples also had an effect on the polymorphism. For instance, the crystalline modification of poly(vinylidene fluoride) produced by treatment under a given set of conditions depended upon the degree of orientation of the sample.

Savitskaya *et al.*[161] investigated the relation of tensile strength of a polymer to molecular orientation of chains within the amorphous part of oriented fibres. They studied poly(vinyl alcohol) and its copolymer with

[160] R. Hasegawa, Y. Tanabe, M. Kobayashi, H. Tadokoro, A. Sawaoka, and N. Kawai, *J. Polymer Sci., Part A-2, Polymer Phys.*, 1970, **8**, 1073.

[161] A. N. Savitskaya, I. B. Klimenko, L. A. Vol'f, and V. F. Androsov, *Polymer Sci.* (*U.S.S.R.*), 1970, **12**, 894.

N-vinylpyrrolidine. The dichroic ratio of a band occurring at 916 cm^{-1} in both polymers, thought to arise from the amorphous regions, was found to be a linear function of the degree of polymer elongation at several different temperatures. In contrast, the band at 1141 cm^{-1}, sensitive to crystallinity, gave an infinitely large dichroic ratio at three-fold elongation. The breaking strengths of the polymers were also a linear function of elongation. It follows that dichroic ratio is also a linear function of breaking strength and could therefore be used to estimate the fracturing strength of the polymers.

Koenig *et al.*[162, 163] have described an improved method for measuring the dichroic ratio in the i.r. of polymers. Normally, samples have to be tilted in a special way with respect to the radiation and the spectrometer entrance slit, and as a result corrections for refractive index and tilt angles have to be applied. Such corrections are often uncertain. Koenig *et al.* developed a method which used a constant refractive index correction and the precision of measurement of i.r. dichroism was thus improved by at least 50%. They illustrated their technique with polyacrylonitrile and showed that for their sample it was only partially uniaxially orientated.

Snyder[164] has given a theoretical treatment of the Raman activity of vibrations of partially oriented molecules which is directly applicable to oriented polymer fibres. The Raman activities were calculated for two cases of restricted molecular orientation: (a) one axis of the molecule is maintained in a fixed orientation while the other two axes are free to rotate, (b) one axis of the molecule is confined to a particular plane while the other two axes are free to rotate. Examples of case (a) are a drawn fibre, or an n-paraffin in a urea clathrate. An example of case (b) is a matte of oriented fibres. Snyder derived expressions for the scattering activities in terms of the elements of the derived polarizability tensors for each possible molecular point group. The results are tabulated for three orthogonal orientations of the sample and for 90° and 180° observation. Examination of these tables reveals how carefully axes must be chosen in experiments intended to lead to unambiguous distinction of symmetry species.

There have been very few papers devoted to the investigation of half-widths of polymers, in contrast to the extensive study of the half-widths of bands in liquids and crystals. Although there had been no satisfactory theory to explain half-widths, it had been found that they could be related in an empirical way to polymer interactions.[165] For instance, half-widths of i.r. bands in polypropylene, poly(ethylene terephthalate), and polyoxymethylene films were found to increase with increasing temperature and decreasing crystallinity.[166]

[162] J. L. Koenig and M. Itoga, *Appl. Spectroscopy*, 1971, **25**, 355.
[163] L. E. Wolfram, J. G. Grasselli, and J. L. Koenig, *Appl. Spectroscopy*, 1970, **24**, 263.
[164] R. G. Snyder, *J. Mol. Spectroscopy*, 1971, **37**, 353.
[165] J. M. Konarski, *Chem. Phys. Letters*, 1971, **9**, 54.
[166] V. I. Vettegren, I. V. Dreval, V. E. Korsukov, and I. I. Novak, *Vysokomol. Soedineniya, Ser. B*, 1970, **12**, 680.

Vettegren and Kosobukin [167, 168] explain the half-widths of crystalline polymers in terms of anharmonic interactions. The anharmonicities were derived for one-dimensional crystal models of both linear and zig-zag chains. These calculations predicted a particular relationship between the half-width of a band and the temperature. The experimental values for the half-widths of the 975 cm^{-1} poly(ethylene terephthalate) band, the 809, 899, and 975 cm^{-1} polypropylene bands and the 1146 cm^{-1} band of poly-(vinyl alcohol) determined by Vettegren *et al.* as a function of temperature were used to test the predicted temperature dependence. A plot of log $[\Gamma - \Gamma_0]/\Gamma$ against $1/T$, where Γ is the half-width of the band at T K and Γ_0 is the extrapolated value for the half-width at 0 K, was almost linear. This linear dependence indicated that the calculations were valid and that the half-widths were related to third-order anharmonicity as predicted. Deviation from linearity at higher temperatures indicated the participation of even higher order anharmonicities.

Melveger [169] has compared the effectiveness of intensities and half-widths of Raman lines as measures of crystallinity in polymers. A plot against sample density of the intensity ratio of the band at 1096 cm^{-1} to the band at 632 cm^{-1} in poly(ethylene terephthalate) showed a large scatter of points although a rough downward trend was indicated as the density decreased. It also seemed that heat-crystallized samples having no orientation did not fall on the same line as that for drawn fibre samples. Melveger also measured the bandwidth at half peak height for the 1730 cm^{-1} band in the Raman spectra of a variety of samples of poly(ethylene terephthalate) including powders, unoriented heat-crystallized filaments, drawn yarns, and high-pressure crystallized material. The bands were measured at a constant resolution of about 2 cm^{-1}. The 1730 cm^{-1} band is attributed to the C—O stretching vibration. A plot of the band half-width against density gave a remarkably good straight line considering the diversity of the samples investigated. However, as the author rightly points out, the assumption of constant amorphous density in oriented samples may not be valid and therefore caution should be exercised in inferring a relation between Raman bandwidths and sample crystallinity. The variation in band half-width is explained in terms of a rotation of the carbonyl groups out of the plane of the benzene ring in the amorphous phase. This increase in the number of environmental conditions of the carbonyl group would tend to increase the bandwidth.

Individual Polymers.—*Polyethylene.* Several reviews have been written on the vibrational spectroscopy of polyethylene.[94, 170–172] The review by Koe-

[167] V. A. Kosobukin, *Mekh. Polimerov*, 1971, **7**, 579.
[168] V. I. Vettegren and V. A. Kosobukin, *Optics and Spectroscopy*, 1971, **31**, 311.
[169] A. J. Melveger, *J. Polymer Sci., Part A-2, Polymer Phys.*, 1972, **10**, 317.
[170] T. Shimanouchi, *Kagaku No Ryoiki*, 1971, **25**, 377.
[171] T. Shimanouchi, Colloquium on N.M.R., Aachen, 1971, Springer-Verlag, Berlin, vol. 4, p. 287.
[172] P. J. Hendra and M. A. Cudby, *Polymer*, 1972, **13**, 104.

nig [94] covers various aspects of Raman spectroscopy as applied to polymers. Included in this review are references to the Raman spectroscopy of polyethylene in the molten state and in the crystalline state. The review by Shimanouchi [171] gives a summary of the work on polyethylene and the n-paraffins. He discusses their spectra in terms of the phonon dispersion curves and conformation. Hendra *et al.*[139, 172] comment on the structural and analytical aspects of the Raman spectrum of polyethylene.

In spite of the considerable amount of experimental and theoretical work on this polymer, its vibrational spectrum is still not completely understood. It is true that most of the major features in the spectrum have been unambiguously assigned but the weaker features of the spectrum still present problems. The general approach to the interpretation of the more dominant features of the spectrum has involved detailed calculations on both polyethylene itself and the n-paraffin oligomers.[173, 174]

As discussed earlier, Zerbi [152, 153, 175] has made considerable advances in the theory and interpretation of the weaker features. In a short note, Hölzl and Schmid [176] make observations on the defect-induced changes in the density of phonon states for highly crystalline polyethylene. Using a single chain approximation, an integration method was employed to obtain smooth frequency distribution curves rather than histograms. It was shown that conformational defects could induce localized modes and resonance modes.

The conformational defect produced in polyethylene by chain-folding has been the subject of many i.r. and Raman spectroscopic investigations. It is thought that the fold in polyethylene may have a *ggtgg* (*gauche-gauche-trans-gauche-gauche*) tight-fold conformation. Since the crystal structure [177] of the cyclic paraffin $(CH_2)_{34}$ indicates that it also has a tight-fold conformation, this molecule has been used as a model compound for the study of chain-folding and longitudinal acoustic modes of extended planar zig-zag segments in polyethylene crystals. This cyclic paraffin has been the subject of a number of spectroscopic studies,[142, 178–180] and recently Krimm and Jakes [181] re-investigated this molecule. They carried out a normal-co-ordinate analysis using a force field which had been derived for the normal paraffins and also applied to some polyamides.[182] They determined the i.r. spectrum of $(CH_2)_{34}$ in the solid state but used the Raman frequencies determined by Shimanouchi *et al.*[179] On melting, the i.r. band at 1342 cm^{-1} disappeared and was replaced by a strong band at 1350 cm^{-1};

[173] J. H. Schachtschneider and R. G. Snyder, *Spectrochim. Acta*, 1963, **19**, 117.
[174] R. G. Snyder and J. H. Schachtschneider, *Spectrochim. Acta*, 1963, **19**, 85.
[175] G. Zerbi, *Pure Appl. Chem.*, 1971, **26**, 499.
[176] K. Hölzl and C. Schmid, *J. Phys.* (*C*), 1972, **5**, L185.
[177] H. F. Kay and B. A. Newman, *Acta Cryst.* 1968, **24B**, 615.
[178] R. F. Schaufele and M. Tasumi, *Polymer J.*, 1971, **2**, 815.
[179] M. Tasumi, T. Shimanouchi, and R. F. Schaufele, *Polymer J.*, 1971, **2**, 740.
[180] T. Shimanouchi, *Discuss. Faraday Soc.*, 1970, no. 49, p. 60.
[181] S. Krimm and J. Jakes, *Macromolecules*, 1971, **4**, 605.
[182] J. Jakes and S. Krimm, *Spectrochim. Acta*, 1971, **27A**, 19.

there was also an increase in intensity of the 1365 cm^{-1} band. These changes were thought to be due to the release of strain in the *tggt* regions of the molecule. Assignments were not given for bands below 600 cm^{-1} because of strong mixing between modes of the planar chain region and those of the fold region. Their analysis supported the association of a band at 1342 cm^{-1} with the fold region. However, Krimm and Jakes emphasized that when using these results for $(CH_2)_{34}$ in the interpretation of chain-folding in polyethylene, the following points should be taken into consideration. The presence of a band at 1342 cm^{-1} does not necessarily indicate a specific structure but rather arises from a distorted *gg* conformation; the absence of a band at 1342 cm^{-1} in the spectrum of crystalline polyethylene may be taken as an indication of the probable absence of the *ggtgg* fold; however, it is not certain that the tight-fold vibration occurs at this frequency in polyethylene. Krimm and Jakes concluded that an analysis of the vibrational spectrum of $(CH_2)_{34}$ does not conclusively show whether folding in polyethylene is tight or not. Further they showed that the calculated values for the longitudinal acoustical modes, based on all-*trans* paraffin chains, did not agree with the spectral observations for the cyclic paraffin. They concluded that caution must be exercised when applying the Schaufele–Shimanouchi* relationship to folded paraffin chains, since coupling cannot be neglected.

Peticolas *et al.*[183] have studied the Raman-active low-frequency longitudinal acoustical vibrations of a variety of crystalline polyethylene samples. These modes, which have been likened to the motion of an accordion, have A_g symmetry in the case of polyethylene, whose single crystals have space group D_{2h}^{16}. The Raman spectrum at these low frequencies was observed using a single-moded argon laser operating at 5145 Å and an iodine filter.

Polyethylene forms single crystals in the form of platelets about 100 Å thick which double in thickness on annealing. At the pre-annealing stage the samples have a broad band centred at 35 cm^{-1} which is attributed to the longitudinal acoustical mode. This band shifts to 12 cm^{-1} on annealing indicating an increase in length of the all-*trans* chain in the crystal. If the Schaufele–Shimanouchi curve is extrapolated to low frequencies, 35 cm^{-1} corresponds to an all-*trans* carbon backbone of about 70 carbon atoms

* On the basis of a least-squares treatment for the observed frequencies of the longitudinal acoustic modes of normal paraffin chains in the planar zig-zag configuration Schaufele and Shimanouchi show that the frequencies depend on the order of the mode and the number of carbon atoms in the chain and closely fit the expression:

$$\Delta\nu/\text{cm}^{-1} = A\left(\frac{m}{n}\right) + B\left(\frac{m}{n}\right)^2 + \ldots + F\left(\frac{m}{n}\right)^6$$

where $A = 2495\ \text{cm}^{-1}$, $B = -5.867\times10^3\ \text{cm}^{-1}$,
$C = 6.253\times10^4\ \text{cm}^{-1}$, $D = -3.485\times10^5\ \text{cm}^{-1}$,
$E = 7.329\times10^5\ \text{cm}^{-1}$, $F = -4.724\times10^5\ \text{cm}^{-1}$.

[183] W. L. Peticolas, G. W. Hibler, J. L. Lippert, A. Peterlin, and H. Olf, *Appl. Phys. Letters*, 1971, **18**, 87.

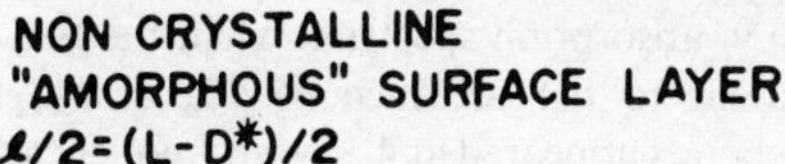

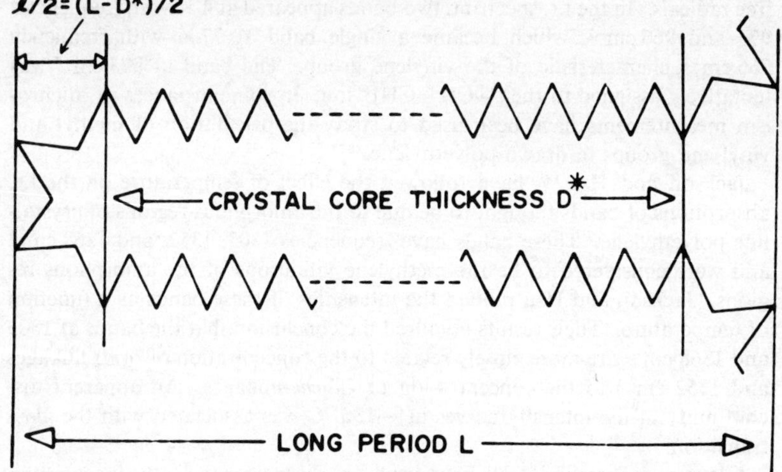

Figure 35 *Model of chain conformation in a polyethylene single crystal of thickness* L *and all-trans conformation length* D*. *The 'amorphous' surface layer contains the chain loops*

(Reproduced by permission from *J. Polymer Sci., Part B, Polymer Letters*, 1971, **9**, 583)

(87 Å), whereas *X*-ray analysis indicates 95 Å. This discrepancy can be explained in terms of end-effects. The nature of the groups at the ends of the all-*trans* segments determines whether the segment ought to be considered as a chain with free ends or fixed ends. This in turn can influence the frequency of the longitudinal acoustic mode. Peticolas [184] proposed a sandwich type of structure for the polymer in which a crystalline core is contained inside two amorphous surface layers (see Figure 35). The low-frequency Raman spectrum did not support the notion that in oriented single crystals of polyethylene there is no coherent crystal lattice extending from one fold-containing surface to the next. The data suggest a less dense 'amorphous' surface layer in the basal planes of the crystal.

Various types of chemical defect in polymers have also been the subject of spectroscopic study. Thus i.r. spectroscopy has been used to determine the degree of unsaturation and chain-branching in polyethylene [185] and to estimate the extent of chemical defect caused by γ-irradiation.[186] Waterman and Dole [187] measured the u.v. and i.r. absorption spectrum at 4 K of high-density polyethylene after irradiation with 1 MeV electrons. A

[184] A. Peterlin, H. G. Olf, W. L. Peticolas, G. W. Hibler, and J. L. Lippert, *J. Polymer Sci., Part B, Polymer Letters*, 1971, **9**, 583.
[185] S. Badilescu, M. Toader, H. Opea, and I. I. Badilescu, *Mater. Plast.*, 1970, **7**, 468.
[186] J. Y. Kim and S. S. Lee, *New Phys., S. Korea*, 1970, **10**, 107.
[187] D. C. Waterman and M. Dole, *J. Phys. Chem.*, 1971, **75**, 3988.

shoulder appeared at 258 nm in the u.v. absorption spectrum when the temperature was raised to 77 K, equivalent to the presence of 4% of alkyl free radicals. In the i.r. spectrum two bands appeared at 4 K with frequencies 973 and 966 cm^{-1}, which became a single band at 77 K with frequency 966 cm^{-1}, characteristic of the vinylene group. The band at 973 cm^{-1} was tentatively assigned to the $[\text{—CH=CH}]^+$ ion. In some instances i.r. dichroism measurements have been used to study the orientation of methyl and vinyl end-groups in drawn polyethylene.[187a]

Jackson and Hsu [188] have followed the effect of temperature on the i.r. absorptions of bands thought to be due to the amorphous regions of crystalline polyethylene. These bands have frequencies 1303, 1352, and 1368 cm^{-1} and were considered to be the methylene vibrations of the amorphous regions. Jackson and Hsu plotted the intensities of these bands as a function of temperature. Their results favoured the conclusion that the bands at 1303 and 1368 cm^{-1} are more closely related to the concentration of *trans*-linkages and 1352 cm^{-1} to the concentration of *gauche*-linkages. An apparent discontinuity in the intensity curves at −130 °C was associated with the glass transition.

Glenz and Peterlin [189–191] have used i.r. dichroism to study the orientation of chain molecules in linear and ethyl-branched polyethylene for both amorphous and crystalline regions of the polymer during the drawing and annealing process. The bands chosen as typical of the crystalline and amorphous regions of the spectrum had freqencies of 1894 and 1368 cm^{-1}, respectively. The degree of orientation as determined by measurement of i.r. dichroism of these bands was plotted as a function of draw ratio and annealing temperature. The results favoured a two-phase model for partially orientated crystalline polyethylene. Glenz and Peterlin further showed that estimates could be made of the contribution to the overall orientation of the molecules and chain ends and the number of methylene groups involved in chain-folding. Several interesting conclusions were drawn which agreed with observations from quite independent sources. For instance, it was shown that chain molecules in the crystalline phase become oriented earlier than those in the amorphous phase, in agreement with models of plastic deformation in polymers. It was also concluded that the vinyl and methyl end-groups have become incorporated in the non-crystalline phase, in agreement with earlier theoretical predictions.[192] From the mean orientation of the chain molecules in the non-crystalline regions it was estimated, for one sample of polyethylene, that about 25% of the chains form tie molecules, 52% form folds, and 23% contain end-groups. It was also estimated that one chain-fold contained twenty-four methylene groups.

[187a] W. Glenz and A. Peterlin, *Makromol. Chem.*, 1971, **150**, 163.
[188] J. F. Jackson and T. S. Hsu, *Polymer Preprints*, 1971, **12**, 726.
[189] W. Glenz, A. Peterlin, and W. Wilke, *J. Polymer Sci., Part A-2, Polymer Phys.*, 1971, **9**, 1191.
[190] W. Glenz and A. Peterlin, *Kolloid Z.*, 1971, **247**, 786.
[191] W. Glenz and A. Peterlin, *J. Macromol. Sci.*, 1970, **4B**, 473.
[192] H. G. Kilian, *Kolloid Z.*, 1969, **231**, 534.

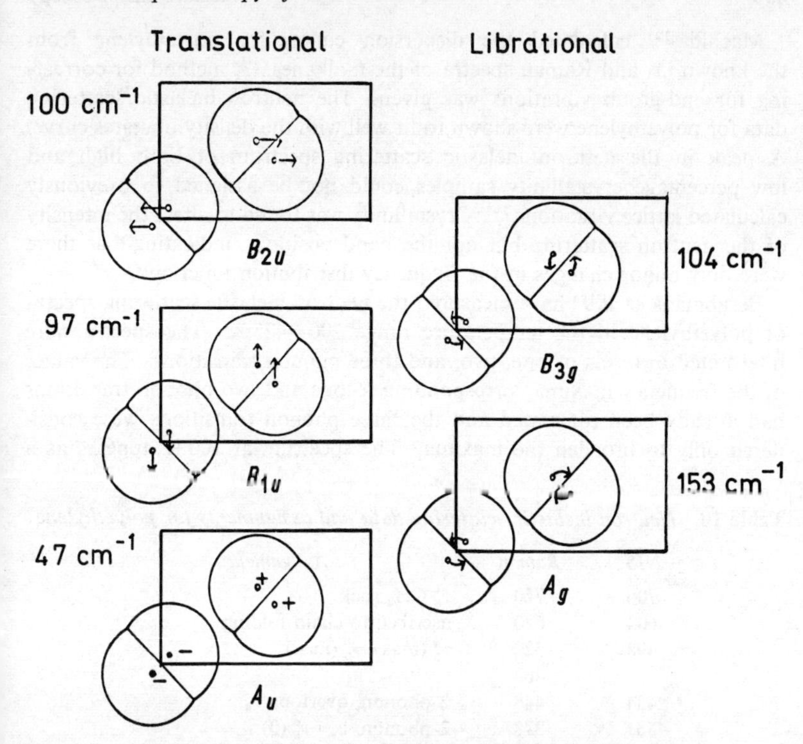

Figure 36 *Lattice vibrations* ($\delta = 0$) *of orthorhombic polyethylene*
(Reproduced by permission from *Bull. Chem. Soc. Japan*, 1970, **43,** 372)

There have been several theoretical calculations of the phonon dispersion curves of polyethylene in the low-frequency region because of its relevance to specific heats and neutron inelastic scattering. Kitagawa and Miyazawa [193] have calculated the frequency distribution, specific heat, and Young's modulus for orthorhombic polyethylene. Since specific heats are dependent on the frequency distribution in crystals, the calculated values are dependent on the force field selected. Kitagawa and Miyazawa present a method of calculating the frequency distribution of chain molecules in polymer crystals. Methylene groups were approximated to point masses since their internal frequencies all lie above 700 cm^{-1}. The force field was divided into intrachain and interchain interactions. The intrachain force field was of the Urey–Bradley type and the interchain force field was expressed in terms of the three types of nearest-neighbour interactions. In the orthorhombic case, in the limit, there are three acoustic vibrations equivalent to translations along the *a*, *b*, and *c* axes and five $k = 0$ vibrations of the lattice altogether (see Figure 36). Dispersion curves were given for the five lattice modes for phase differences $\delta = (0,0,0) \rightarrow (0,0,\pi)$.

[193] T. Kitagawa and T. Miyazawa, *Bull. Chem. Soc. Japan*, 1970, **43,** 372.

Machida[194] calculated the dispersion curves for polyethylene from the known i.r. and Raman spectra of the n-alkanes. A method for correcting for end-group vibrations was given. The neutron inelastic scattering data for polyethylene were shown to fit well with the density of states curve. A peak in the neutron inelastic scattering spectrum of both high and low percentage crystallinity samples could not be assigned to previously calculated lattice vibrations.[195] Crystallinity was found to affect the intensity of the neutron scattering but not the band positions, indicating that there were only minor changes in the frequency distribution function.

Berghmans *et al.*[196] have measured the neutron inelastic scattering spectra of polyethylene in the temperature range 200—425 K. The spectra were interpreted in terms of one, two, and three phonon transitions.. The values of the frequency maxima corresponding to one and two phonon transitions had already been calculated and the three phonon transitions were considered only to broaden the maxima. The spectrum at 200 K appears as a

Table 10 *Neutron inelastic scattering data and assignments for polyethylene*

NIS	*Raman*	*Assignment*
706	720	ν_8^b CH_2 rock
604	620	ascribed to chain-folding
498	520	ν_5^a (max), ν_5^b (max)
	468	
434	448	2-phonon, overtone ν_9
351	378	2-phonon, $\nu_9+\nu_5^a$ (0)
	325	
277	284	2-phonon, ν_5^a (0)$+\nu_5^b$ (π)
247	254	
214	228	co-operative chain torsion ν_9^a (max)
193	180	co-operative chain torsion ν_9^b (max)
163		antiparallel rotary lattice, ν_5^a (0)
130	143	parallel rotary lattice ν_5^b (0)
118	125	antiparallel translation ∥ *b* axis, ν_5^b (π)
104	104	
92		
83		
71	74	antiparallel translation ∥ *a* axis ν_5^a (π)
63	52	antiparallel translation ∥ *c* axis ν_9^b (0)
45		
38	36	
33		
24		

[194] K. Machida, Nippon Genshiryoku Kenkyusho Kenkyu Hokoku, 1970, JAERI-1197, 101.
[195] K. Doi, M. Sakamoto, M. Izumi, N. Masaki, and H. Motohashi, Nippon Genshiryoku Kenkyusho Kenkyu Hokoku, 1970, JAERI 1197, 109.
[196] H. Berghmans, G. J. Safford, and P. S. Leung, *J. Polymer Sci., Part A-2, Polymer Phys.*, 1971, **9**, 1219.

broad band centred at 163 cm^{-1} with a number of minor features on either side of this central maximum. In all, sixteen features were thought to be significant and many of these were in good agreement with the calculated values. The assignments are shown in Table 10 together with the Raman results of Frenzel *et al.*[197] for comparison. The experimental values for the neutron inelastic scattering data tended to vary with the morphology of the sample and this was used to explain the variations with temperature. The general broadening of the spectral maxima and apparent loss of resolution with increased in temperature were interpreted in terms of segmental rotation which occurred first in the amorphous regions of the polymer and then spread to the ordered regions.

Koenig and Boerio [198] have studied crystal-field splitting effects in the Raman spectra of polyethylene and polyperdeuterioethylene at low temperatures. Crystal-field effects have been observed in the i.r. spectra of these polymers at room temperatures. The evidence for crystal-field splitting in the Raman spectra at room temperature is much less certain. The pair of bands at 1418 and 1441 cm^{-1} in the Raman spectrum of polyethylene and the pair of bands at 977 and 992 cm^{-1} in polyperdeuterioethylene have been considered as arising from crystal-field effects but the splitting is very large. At -160 °C, Koenig and Boerio observed splitting of the bands at 1296 cm^{-1} (1295 and 1297 cm^{-1}) and 1066 cm^{-1} (1065 and 1068 cm^{-1}) in polyethylene and at 828 cm^{-1} (827 and 830 cm^{-1}) in polyperdeuterioethylene and considered that these smaller splittings definitely arose from the crystal field arising from repulsive forces between hydrogen atoms on neighbouring chains. A similar explanation was used by Koenig and Boerio [199] to explain band splitting in polytetrafluoroethylene.

Mixed crystals of polyethylene and polyperdeuterioethylene were the subject of study by Bank and Krimm.[200] They showed that these crystals provided evidence of chain segregation associated with crystallization rates and molecular weight. Evidence was obtained from band splitting in the i.r. in the methylene rock region which suggested that similar chains associate with each other rather than mixing with chains of the opposite species.

Snyder [201] has determined the Raman spectrum of polyethylene and polyperdeuterioethylene in an attempt to clarify the rather confused earlier interpretation of orientated polyethylene fibres. The experimental difficulties in obtaining meaningful orientation data were carefully enumerated. Among the more difficult problems were the retention of polarization of the incident beam as it entered the polymer fibre, since a bunch of narrow fibres tends to randomize the polarization of the incident radiation, and retention of the polarization of the scattered radiation. The work of Carter [202] indi-

[197] C. A. Frenzel, E. B. Bradley, and M. S. Mathur, *J. Chem. Phys.*, 1968, **49**, 3789.
[198] F. J. Boerio and J. L. Koenig, *J. Chem. Phys.*, 1970, **52**, 3425.
[199] F. J. Boerio, *Diss. Abs. (B)*, 1971, **32**, 1766.
[200] M. I. Bank and S. Krimm, *J. Polymer Sci., Part B, Polymer Letters*, 1970, **8**, 143.
[201] R. G. Snyder, *J. Mol. Spectroscopy*, 1970, **36**, 222.
[202] V. B. Carter, *J. Mol. Spectroscopy*, 1970, **34**, 356.

cated that, by using single fibres, this problem of scrambling of polarization could be overcome. As a result of careful orientation measurements on oriented single polyethylene fibres she was able to show that the band at 1133 cm^{-1} had A_g symmetry and not B_{2g} symmetry as previously thought. She suggested that a weak band at 1380 cm^{-1} should be assigned to the B_{2g} wagging mode. The calculated intensity of this band is in fact very low. The theory of polarization data derived from oriented polymer fibres has been given by Snyder [164] and has been discussed earlier.

Polypropylene. Zerbi [203] has calculated the normal vibrations of polypropylene based on a single-chain model, but the results were found to be in disagreement with the earlier neutron up-scattering data. Yasukawa *et al.*[204, 205] have studied the down-scattering of pulsed neutrons and have plotted the time-of-flight distribution of neutrons scattered at right angles from isotactic polypropylene. The results agreed with those of Safford;[206] the improved resolution showed clearly two bands centred at 200 and 170 cm^{-1}, assigned to ν_{24} and ν_{25} respectively.

Roylance and De Vries [207] have used i.r. spectroscopy to study stress in the backbone of an orientated polymer. Highly orientated polypropylene was subjected to stress while its i.r. spectrum was measured. The results were obtained as rapidly as possible after the application of the stress. A band at frequency 975 cm^{-1} was observed to shift to lower frequency accompanied by an appearance of a shoulder at 905 cm^{-1}. The authors obtained a linear plot of frequency against stress but the ten points covered only a 2 cm^{-1} frequency range. No indication was given as to what spectrometer was used or how it was calibrated so accurately. These changes in the C—C stretching region were considered to be due to an increase in length of the C—C bond and a concomitant change in force constant.

The Raman spectra of monoclinic, smectic, and molten isotactic polypropylene have been observed by Koenig and Vasko.[208] Polarization data were obtained from transparent films of the smectic phase and from the molten phase. The bands were assigned in the light of the polarization data and the helical conformation was discussed in relation to the melt structure. In the low-temperature spectrum at −160 °C, a weak band observed at 417 cm^{-1} was thought to be a characteristic of the smectic isotactic polypropylene structure.

Poly(ethylene terephthalate). Danz and co-workers [209] have made calculations

[203] G. Zerbi, *J. Chem. Phys.*, 1968, **49**, 3840.
[204] M. Kimura, N. Watanabe, M. Oyanada, T. Yasukawa, and T. Suzuki, Research Report Lab. Nuclear Sci., Tokyo University, 1969, **2**, 89.
[205] T. Yasukawa, N. Kimura, N. Watanabe, and Y. Yamada, *J. Chem. Phys.*, 1971, **55**, 983.
[206] G. J. Safford, H. R. Danner, H. Boutin, and M. Berger, *J. Chem. Phys.*, 1964, **40**, 1426.
[207] D. K. Roylance and K. L. DeVries, *J. Polymer Sci., Part B, Polymer Letters*, 1971, **9**, 443.
[208] J. L. Koenig and P. D. Vasko, *Macromolecules*, 1970, **3**, 597.
[209] R. Danz, J. Dechant, and C. Ruscher, *Faserforsch. Textiltech.*, 1970, **21**, 503.

of the normal modes of a planar model of poly(ethylene terephthalate). The bands observed in the i.r. with frequencies 973, 437, 387, and 138 cm^{-1} were assigned to vibrations of the $—O—CH_2CH_2—O—$ part of the molecule. The frequencies of these bands were considered to be sensitive to conformational changes within this group. Changes in the frequencies of these bands are observed during crystallization of poly(ethylene terephthalate).[210] Manley and Williams [211] have studied the i.r. spectra of a variety of compounds related to poly(ethylene terephthalate). They showed that the rigid crystalline model for poly(ethylene terephthalate) was not wholly satisfactory for explaining the i.r spectrum; the model does not allow for the extent of rotation about the glycol fragment. They therefore based their assignments on the existence of *trans* and *gauche* isomers (symmetries V_h and C_{2v} respectively) produced as a result of rotation about the glycol fragment. The spectral changes observed on drawing poly(ethylene terephthalate) were attributed to *trans–gauche* isomerization.

Hoffman and co workers [212] have calculated phonon dispersion curves for poly(ethylene terephthalate), discussing these curves in terms of the i.r. spectra of this polymer. The far-i.r. absorption spectra are given for both amorphous and crystalline samples. Huang and Koenig [213] have followed the low-temperature phase transitions in poly(ethylene terephthalate) and polystyrene by i.r. spectroscopy. It had already been shown that the location of the transitions was dependent on mechanical treatment and orientation. Theoretical considerations by Ovander have shown that in a condensed phase the integrated band intensity should have the following linear relationship: $I = I_0 + aT$, where a the temperature coefficient is generally negative. Since thermoexpansion measurements show abrupt changes in the region of transition temperatures one would expect the intermolecular force field to show similar abrupt changes. The coupled oscillator theory of Hester shows that band intensities in the i.r. spectrum will be sensitive to changes in the intermolecular force field. Thus if abrupt changes occur in the force field then kinks will appear in the plots of integrated band intensity as a function of temperature. For the sake of clarity it was the derivatives of the integrated band intensities that were plotted as a function of temperature. Changes in the slope of such plots were investigated down to 13 K and transitions were found at 50, 120, 180 and 210 K, approximately, depending on sample condition. Two transitions were found in polystyrene at 50 and 230 K.

McGraw [214] investigated the polyester structure by Raman spectroscopy, studying the intensities of several bands as a function of temperature. The preliminary results appeared to correlate the Raman intensity of certain bands with sample density as determined on a lithium bromide density gra-

[210] O. F. Schonherr, *Faserforsch. Textiltech.*, 1970, **21**, 246.
[211] T. R. Manley and D. A. Williams, *Polymer*, 1969, **10**, 339.
[212] V. Hoffman, B. Heise, and W. Frank, *Kolloid-Z.*, 1971, **247**, 795.
[213] Y. S. Huang and J. L. Koenig, *J. Appl. Polymer Sci.*, 1971, **15**, 1237.
[214] G. E. McGraw, *Polymer Preprints*, 1970, **11**, 1122.

dient column, but the work of Melveger,[169] discussed earlier, appears to have superseded this.

Polytetrafluoroethylene (*PTFE*). Twisleton and White [215] have used neutron inelastic scattering to determine the phonon dispersion curve for the longitudinal lattice vibrations perpendicular to the chain axes in hexagonal PTFE at 25 °C.

Boerio and Koenig [216] determined a nineteen-parameter force field for PTFE using the method of damped least-squares. The model for the polymer used was the conformation stable at 25 °C, involving a helix with fifteen CF_2 groups in seven turns of the helix. The factor group of this line group is isomorphous with the point group $D_{\frac{14}{15}\pi}$ and the vibrations are classified as follows: $4A_1$ (R, p), $3A_2$ (i.r., ‖), $8E_1$ (R, dp; i.r., ⊥) and $9E_2$ (R, dp). Boerio and Koenig calculated the phonon dispersion curves and frequency distribution histogram. Several bands in the vibrational spectrum were reassigned in the light of these calculations. A band at 1450 cm^{-1} previously assigned [217] to an antisymmetric CF_2 mode of A_2 symmetry was shown not to be a fundamental. The antisymmetric CF_2 mode was reassigned to the intense i.r.-active band at 1210 cm^{-1}. The Raman-active band at 1215 cm^{-1} was assigned to the ν_{asym} (CF_2) mode of E_2 symmetry because of its probably proximity to the A_2 symmetry mode. The ν_{asym} (CF_2) mode of E_1 symmetry was assigned to the Raman active band at 1298 cm^{-1}, although it was not observed in the i.r.

Three papers on the Raman spectrum of PTFE have appeared recently. The first, by Koenig and Boerio,[218] concentrates on the phase transition in PTFE at 19 °C. This first-order phase transition involves a change in form of the helix. The temperature dependence of two bands with frequencies 575 and 595 cm^{-1} were particularly carefully studied. The assignment of the bands as fundamentals of each type of helix was rejected on the grounds of the persistence of the 575 cm^{-1} band well above 19 °C. Studies [217] on the fluorocarbon oligomers C_5F_{12}, C_6F_{14}, C_7F_{16}, C_8F_{18}, $C_{14}F_{30}$, and $C_{18}F_{38}$ suggest that the 575 cm^{-1} band is probably characteristic of a crystalline fluorocarbon below or just above its transition temperature, while the 595 cm^{-1} band is characteristic of an amorphous or crystalline fluorocarbon above its transition temperature. The 595 cm^{-1} band was assigned to a CF_2 deformation mode adjoining a *gauche* linkage in the backbone, whereas the 575 cm^{-1} band was assigned to a CF_2 deformation mode of a group adjoining a *trans*-linkage in the chain.

This view has now been revised by Koenig and Boerio [219] and the effect is thought to be merely due to crystal-field splitting. The authors discuss the splitting of the 575—595 cm^{-1} doublet and other features in the spectrum

[215] J. F. Twisleton and J. W. White, *Polymer*, 1972, **13**, 40.
[216] J. L. Koenig and F. J. Boerio, *J. Chem. Phys.*, 1970, **52**, 4826.
[217] M. J. Hannon, F. J. Boerio, and J. L. Koenig, *J. Chem. Phys.*, 1968, **50**, 2829.
[218] J. L. Koenig and F. J. Boerio, *J. Chem. Phys.*, 1970, **52**, 4170.
[219] F. J. Boerio and J. L. Koenig, *J. Chem. Phys.*, 1971, **54**, 3667.

at low temperatures. They found that the bands in the vicinity of 1215 and 389 cm^{-1} also have a marked temperature dependence. At 100 °C the band at 389 cm^{-1} has only a weak shoulder on the low-frequency side. As the temperature is decreased this shoulder increases its intensity relative to the 389 cm^{-1} band and shifts to lower frequency, reaching 383 cm^{-1} at −180 °C. The previous assignment [217] of the band and the shoulder to CF_2 deformation fundamentals of A_1 and E_2 symmetry, respectively, does not account for these intensity observations. Koenig and Boerio now propose that this splitting of 6 cm^{-1}, although quite large, is due to crystal-field splitting. A similar problem arises with the doublet in the region of 1215 cm^{-1}, which shows a splitting of 4.4 cm^{-1} at −180 °C. It was thought probable that the low-temperature phase of PTFE was monoclinic and that the observations on the lattice modes supported this theory.

Hendra *et al.*[220] have studied the Raman spectrum of PTFE under a variety of conditions of sample crystallinity and temperature. To assist in the assignment of the PTFE Raman spectrum they determined depolarization ratios for bands in the Raman spectrum of perfluoropropylene–perfluoroethylene copolymer. On the basis of somewhat doubtful depolarization ratios taken from the solid copolymer, three A_1 modes of PTFE were assigned to bands with frequencies 732, 335, and 290 cm^{-1}. A band at frequency 1381 cm^{-1} was also assigned to an A_1 mode, even though it had a depolarization ratio well within the range expected for a depolarized band. The other bands in the spectrum were assigned to modes of A_2, E_1, and E_2 symmetry according to their activity in the i.r. and Raman spectra and their dichroism in the i.r. The two bands mentioned earlier at frequencies 576 and 597 cm^{-1} were discussed by Hendra *et al.* The band at 576 cm^{-1} was assigned to a CF_2 rocking mode. The intensity of the band at 597 cm^{-1} was shown to be dependent on crystallinity and was assigned to a CF_2 rocking mode at a reversal point, in the screw sense, of the helical chains in the crystalline phase.

Polydifluorosilylene, $(SiF_2)_n$, the silicon analogue of PTFE, has been studied by Raman spectroscopy by Margrave and Wilson.[221] They give the frequencies together with the assignments.

Poly(vinyl chloride) (*PVC*). The intermolecular interactions in crystalline PVC have been studied by i.r. spectroscopy.[222] A band with frequency 67 cm^{-1}, thought to be due to CH · · · · Cl interaction in crystalline syndiotactic PVC, was observed at room temperature and at liquid nitrogen temperature. The band is not observed in commercial samples of PVC, indicating that it is almost certainly due to vibrational modes of the crystalline phase. On the grounds of symmetry the band could not be assigned to a translational mode. It was suggested that this band was due to a localized mode.

[220] C. J. Peacock, P. J. Hendra, H. A. Willis, and M. E. A. Cudby, *J. Chem. Soc.* (*A*), 1970, 2943.
[221] J. L. Margrave and P. W. Wilson, *European Polymer J.*, 1971, **7**, 989.
[222] S. Krimm and A. V. R. Warner, *Macromolecules*, 1970, **3**, 709.

Bands similar to this are observed in other liquid secondary chlorides and are thought to arise from intermolecular interaction.

Other papers on the i.r. spectroscopy of PVC deal with such properties as orientation in PVC plastics [223] and the study of ordering in the amorphous regions of PVC.[224] Solution properties of PVC have been studied, with the conclusion that PVC solubility has nothing to do with the rupturing of hydrogen bonds.[225]

The Raman spectrum of PVC has been studied by Koenig and Druesedow [226] and later by Liebman *et al.*[227] The latter pointed out the spectral differences obtained with varying sample treatment, and attempted to determine whether these differences were due to a decomposition process or to more subtle conformational changes within the polymer. The samples were heated and cooled in a variety of ways with careful temperature control. Heating at a rate of 5 °C per minute to 180 °C, for example, gave rise to two new bands with frequencies 1124 and 1511 cm^{-1} which were thought to be due to a folded syndiotactic structure. The work of Rimai [127] suggested that these bands arise from decompsition products.

Shimanouchi [171] summarizes the spectroscopic results for highly syndiotactic PVC. He gives the i.r. dichroism data together with results from a normal-co-ordinate calculation. The i.r. bands are classified into three groups according to their dichroic properties: those that show only perpendicular dichroism of A_1 symmetry; those that show parallel dichroism for the fully stretched film of B_1 symmmetry; those that show a dichroism dependent on the degree of elongation of the film mainly of B_2 symmetry. Bands at 1426, 1333, 1102, and 638 cm^{-1} belong to the first class, bands at 1379, 1229 and 833 cm^{-1} belong to the second class, and bands at 1354, 1254, 957, and 603 cm^{-1} belong to the third class. The Raman results of Koenig and Druesedow [206] appeared to confirm these assignments.

Poly(vinylidene difluoride) (PVF_2). This compound exists in two crystalline states, one form containing planar PVF_2 and the other a non-planar form with a *tgtg* conformation.[228] The Raman spectrum of the non-planar form was studied by Koenig and Boerio [229] in 1969. In a later paper the same authors considered the planar form.[230] The factor group of the line group of planar PVF_2 is isomorphous with the point group C_{2v} and the normal modes are distributed as $5A_1+3B_1+2A_2+4B_2$. The non-planar

[223] Y. Kazama and O. Yamamoto, *Zairyo*, 1971, **20**, 665.
[224] Y. V. Glazkovskii, A. N. Zav'yalov, N. M. Bakardzhiev, and I. I. Novak, *Vysokomol. Soedineniya, Ser. A*, 1970, **12**, 2697.
[225] W. Lesch and L. Ulbrich, *Makromol. Chem.*, 1970, **140**, 229.
[226] J. L. Koenig and D. Druesedow, *J. Polymer Sci., Part A-2, Polymer Phys.*, 1969, **7**, 1075.
[227] S. A. Liebman, C. R. Flotz, J. F. Reuwer, and R. J. Obremski, *Macromolecules*, 1971, **4**, 134.
[228] W. W. Doll and J. B. Lando, *J. Macromol. Sci.*, 1970, **4B**, 309.
[229] F. J. Boerio and J. L. Koenig, *J. Polymer Sci., Part A-2, Polymer Phys.*, 1969, **7**, 1489.
[230] F. J. Boerio and J. L. Koenig, *J. Polymer Sci., Part A-2, Polymer Phys.*, 1971, **9**, 1517.

form of lower symmetry has a factor group isomorphous with point group C_s. Assignments of the planar form were fairly straightforward and assignment of the non-planar form was made with the aid of i.r. dichroism measurements and by analogy with the planar form. Force constant calculations were made for both forms of PVF_2. The initial force constants were transferred from polyethylene and PTFE, and subsequently refined until a set of force constants were obtained which gave good results for both forms of PVF_2.

In solution the configuration of PVF_2 depends on the solvent.[231] In acetone, the configuration can be planar or non-planar depending on the starting configuration; in dimethylformamide, the configuration changes until all the molecules are in the non-planar form; in dioxan, tetrahydrofuran, amyl acetate, or cyclohexane the final form is planar. These variable observations explain, as suggested by Zerbi,[232] why the n.m.r. results appear to be in conflict with *X*-ray results. In the analysis of the planar syndiotactic configuration by Zerbi, two monomer units were taken as the repeat unit. The line group was isomorphous with the point group C_{2v} and the modes were distributed as $9A_1+7A_2+7B_1+9B_2$.

Polyoxymethylene (POM) and Poly(propylene oxide) (PPO). Polyoxymethylene exists in two distinct crystalline forms. The most common form is that of a helix with nine residues in five turns of the helix. I.r. studies have shown that the configuration is dependent on the conditions of preparation.[233] The unit cell of one form is hexagonal, containing one molecule per unit cell space group $P3_1$ or $P3_2$. The other form is orthorhombic $P2_12_12_1$ with two molecules per unit cell.

Tanaka *et al.*[234] have studied the i.r. spectrum of POM, and Kitagawa [235] has examined the i.r. spectroscopic evidence for orientation of end-groups in POM. Dichroism measurements suggest that polyamide additives are aligned in uniaxially oriented POM specimens.

Boerio and Cornell [236] have studied the Raman-active lattice vibrations of orthorhombic POM to gain information about the intermolecular forces in the crystal. Boerio and Cornell also calculated the phonon dispersion curve and a frequency-distribution histogram for POM. The vibrational spectrum was assigned with the aid of the i.r. results obtained by Zerbi.[237] The band observed in the Raman spectrum at 37 cm^{-1}, calculated to occur at 37 cm^{-1}, was assigned to the out-of-phase translational lattice mode

[231] L. I. Taruntina, *Vysokomol. Soedineniya, Ser. B*, 1930, **12**, 780.
[232] G. Zerbi and G. Contili, *Spectrochim. Acta*, 1970, **26A**, 733.
[233] M. Mikhailov and L. Tarlemezyan, *Otd. Khim. Nauk. Bulg. Akad. Nauk.*, 1970, **3**, 267.
[234] A. Tanaka, S. Uemura, and Y. Ishida, *J. Polymer Sci., Part A-2, Polymer Phys.*, 1970, 1585.
[235] T. Kitagawa, A. Tanaka, and M. Nishii, *J. Polymer Sci., Part B, Polymer Letters*, 1971, **9**, 579.
[236] F. J. Boerio and D. D. Cornell, *J. Chem. Phys.*, 1972, **56**, 1516.
[237] G. Zerbi and G. Masetti, *J. Mol. Spectroscopy*, 1967, **22**, 284.

along the crystal *c* axis. The band at 130 cm^{-1} in the i.r. spectrum, calculated to occur 123 cm^{-1}, was assigned to a rotational lattice mode about the crystal *c* axis. The band at 83 cm^{-1} in the i.r., judged to be coincident with a band at 81 cm^{-1} in the Raman spectrum, was assigned to an out-of-phase translational lattice mode along the *b* axis, calculated to occur at 85 cm^{-1}. The band at 89 cm^{-1}, calculated to occur at this frequency, was assigned to an out-of-phase translational lattice mode along the crystal *a* axis. These assignments differ from those given by Zerbi.[237] The good agreement between the calculations of Boerio and Cornell and the observed vibrational spectra were thought to indicate that the intermolecular potential consists mainly of repulsion forces between non-bonded hydrogen atoms and other non-bonded atoms because of the nature of the force field used.

Hendra *et al.*[238] have studied the vibrational spectra of polythiomethylene and polyselenomethylene. The former exists in a hexagonal form and the latter in both hexagonal and orthorhombic forms. Tentative assignments were made for these polymers and compared with assignments for POM.

The neutron inelastic scattering spectra of oriented and non-oriented samples of poly(ethylene glycol) and poly(ethylene glycol) in solution have been measured by Assarsson *et al.*[239] The results were shown to be in good agreement with the normal-co-ordinate calculations and i.r. and Raman spectra of Tadokoro *et al.*[240] and Miyazawa *et al.*[241]

Kumpanenko *et al.*[242] have used the low-temperature i.r. spectrum of PPO to determine the tacticity of the polymer. It was shown theoretically and confirmed experimentally that the number of *trans* PPO units was equal to the number of *gauche* PPO units in dilute solutions at low temperatures. This was due to the untwining of PPO chains. The i.r. band intensities of PPO crystals were the same as those of a 2% solution in carbon disulphide at −96 °C.

Poly(vinyl alcohol), Poly(vinyl formate), and Poly(vinylsiloxane). An excellent review of the i.r. and n.m.r. spectroscopy of poly(vinyl alcohol) and the related compounds poly(vinyl acetate) and poly(vinyl formate) has been written by Fujii.[243] Optical methods for studying orientation within poly(vinyl alcohol) have been reviewed and compared.[244]

The i.r. band at 1141 cm^{-1}, which is assigned to a C—O stretch, has been the subject of extensive study. Recently [245] the transition moment angle of this band has been determined for a uniaxially oriented stretched poly-

[238] P. J. Hendra and D. S. Watson, *Spectrochim. Acta*, 1972, **28A,** 351.
[239] P. G. Assarsson, P. J. Leung, and G. J. Safford, *Polymer Preprints*, 1969, **10,** 1241.
[240] T. Yoshihara, H. Tadokoro, and S. Murahaski, *J. Chem. Phys.*, 1964, **41,** 2902.
[241] H. Matsuura and T. Miyazawa, *Bull. Chem. Soc. Japan*, 1968, **41,** 1798.
[242] I. V. Kumpanenko and K. S. Kazanskii, *Vysokomol. Soedineniya, Ser. A*, 1971, **13,** 719.
[243] K. Fujii, *J. Polymer Sci., Part D, Macromol. Rev.*, 1971, **5,** 431.
[244] M. L. Gulrajani and M. R. Padhye, *Indian J. Technol.*, 1971, **9,** 211.
[245] S. Hibi, M. Maeda, S. Makino, S. Nomura, and H. Kawai, *Sen-i Gakkaishi*, 1971, **27,** 246.

(vinyl alcohol) film by both i.r spectroscopy and by *X*-ray diffraction. The angles found were 81°15′ and 83°25′ for samples annealed at 160 and 90 °C, respectively.

Poly(vinyl formate) [246] and poly(vinylsiloxane) [247] have been studied recently by i.r. spectroscopy. The latter polymer has been shown to form stable chemical bonds with glass fibres through Si—O—Si linkages.

Poly(methyl methacrylate) and Related Compounds. No *X*-ray data are available for poly(methacrylic acid), but some structural information has been obtained by vibrational spectroscopy. I.r. dichroism studies are useless as it is not possible to obtain oriented fibres, but depolarization ratios can be obtained from Raman measurements on solutions.

The Raman spectrum of syndiotactic poly(methacrylic acid) was investigated in the solid state and in aqueous solution,[248] but the similarity of the solid-state and solution spectra suggested that only minor conformational changes take place. Polarization data and coincidences with i.r. bands suggested that the polymer structure corresponded to D_n symmetry. This requires that the poly(methacrylic acid) helix has a pitch greater than 3_1. The Raman spectrum of the sodium salt of the acid was also obtained. The band at 772 cm^{-1}, strongly polarized in the Raman spectrum of the acid, shifts to 832 cm^{-1} for the salt in solution. It was suggested that this large shift was due to an uncoiling of the helix in the solution of the salt. Changes in the spectra on neutralization suggest quite complex processes taking place in solution. Small amounts of base do not alter the helical conformation. Further addition of base broadens the spectral features, suggesting an increase in the number of conformational isomers present. Further neutralization once more simplifies the spectrum to that characteristic of the salt in solution. The results suggested that the helix to random-coil transformation was not co-operative as has been suggested for the denaturation of globular proteins, but rather a progressive structural randomisation.

Manley and Martin [249] have studied the low-frequency vibrations of poly(methacrylic acid) using far-i.r. and Raman spectroscopy. A model was chosen for syndiotactic poly(methacrylic acid) which in the planar zig-zag form gives a line group having a factor group isomorphous with the point group C_{2v}. The frequencies of the skeletal vibrations were calculated using Zbinden's method [250] for a planar zig-zag chain. The calculated values were in extraordinary good agreement considering the approximations involved in this approach, and were used as a guide to the assignment of the modes.

[246] J. K. Haken and R. L. Werner, *Spectrochim. Acta*, 1971, **27A**, 343.
[247] J. L. Koenig and T. K. P. Shih, *J. Colloid Interface Sci.*, 1971, **36**, 247.
[248] J. L. Koenig, A. C. Angood, J. Semen, and J. B. Lando, *J. Amer. Chem. Soc.*, 1969, **91**, 7250.
[249] T. R. Manley and C. G. Martin, *Polymer*, 1971, **12**, 524.
[250] R. Zbinden, 'Infrared Spectroscopy of High Polymers', Academic Press, London, 1964.

Haken and Werner[251] have given the i.r. spectrum of poly(methyl methacrylate) together with assignments for the range 150—4000 cm^{-1}. Trapeznikova and Belopol'skaya[252] studied the i.r. spectrum of poly(methyl methacrylate) with special regard to its temperature characteristics and solution properties. The disappearance of a band at 1172 cm^{-1} as the temperature is decreased was interpreted as an equilibrium shift between two rotational isomers of the ester group. Further solution studies of polyacrylates in trifluoroethanol were made by Lucas and co-workers.[253] Schneider[254] has investigated the structure of poly(methyl methacrylate) by i.r. and Raman spectroscopy. The dimethyl ester of 2,2,4,4-tetramethylglutaric acid can be regarded as a very simple model of this compound and the interpretation of the spectrum of the polymer was based on comparison with this model compound. If only staggered conformations for the model compound are considered and rotation of the ester group is neglected, then only four different isomeric forms can occur. I.r. dichroism data on oriented crystals of the model compound indicated that only one conformation was present in this state. The structure of this conformation was ascertained by comparison with the i.r. spectrum of oriented crystals of 2,4-dichloro-2,4-dimethylpentane.[254a] The i.r. spectrum of the liquid state indicated the presence of other conformations in this phase. If staggered conformations only of the polymer are considered and rotation of the ester groups is neglected, then six different conformations can occur in both isotactic and syndiotactic poly(methyl methacrylate) dyads $—CR^1R^2—CH_2—CR^1R^2—$. The i.r. spectrum indicated that the dyad conformation of the polymer corresponded closely with the crystalline conformation of the model.

Infrared spectroscopy has been used to study the conformation of the α-chloroacrylate polymers.[255] The earlier results of Furukawa on spectral dependence on tacticity were supported. Major differences occur in the carbonyl stretching region for the isotactic and syndiotactic forms. The far-i.r. spectra of cross-linked triethyleneglycol dimethacrylate polymers have also been interpreted in terms of degree of crystallization, cross-linking, and other forms of order.

Polyacrylonitrile (*PAN*). Thermal transitions in polyacrylonitrile have been studied by i.r. spectroscopy.[256] Many transitions of this type occur in PAN in the temperature range 80—145 °C. The bands in the C—N stretching region were found to be the most sensitive to temperature changes.

[251] J. K. Haken and R. L. Werner, *British Polymer J.*, 1971, **3**, 263.
[252] O. N. Trapeznikova and T. V. Belopol'skaya, Tel. Dvizhenie Mol. Mezhmol. Vzaimodeistvie Zhidk Rastrorakh, 1969, 355.
[253] T. Lucas, B. Sebille, and C. Quivoron, *J. Chim. phys.*, 1971, **68**, 1278.
[254] B. Schneider, J. Stokr, S. Dirilikov, and M. Mikailov, *Macromolecules*, 1971, **4**, 715.
[254a] J. Stokr, S. Dirlikov, and B. Schneider, *Coll. Czech. Chem. Comm.*, 1971, **36**, 1923.
[255] B. Wesslen and R. W. Lenz, *Macromolecules*, 1971, **4**, 21.
[256] K. Ogura, S. Kawamura, and H. Sobue, *Macromolecules*, 1971, **4**, 79.

Huang and Koenig [257] have obtained the Raman spectrum of polyacrylonitrile and have proposed a structural model for the compound. They suggest a generalized division of the i.r. and Raman lines which is valid for any polymer. Each line can be classified into one of the following eight categories: [p,0], [d,0], [0,π], [0,σ], [p,π], [p,σ], [d,π], and [d,σ] where 0 means inactive, p means polarized, d means depolarized, π means parallel, and σ means perpendicular. The precise distribution of bands within this scheme will naturally depend on the symmetry of the polymer. The absence of a band in a spectrum is a negative result which may only be due to its inherent weakness. Thus the classification of any band to one of the first four classes will always be at best tenuous. In the case of PAN, neglecting the negative results, the modes can all be placed in the categories [p,σ], [d,σ], and [d,π]; no modes belong to the category [p,π]. This classification rules out the following structures: the atactic model since [p,π] would then be present; the two-fold syndiotactic helix since [p,σ] is present; the three-fold isotactic helix since [p,σ] and [d,π] are present and [p,π] is absent. The planar syndiotactic model is consistent with the above results, although some bands in the category [d,0] would be expected but were not found. Thus the results suggest a predominantly planar structure.

Nylons. Krimm and Jakes [182] have given a method by which the normal vibrations of polyamides may be calculated. Many examples are given which include a variety of nylons.

Normally, i.r. dichroism measurements give only two-dimensional information, but recently techniques have been developed allowing three-dimensional information to be obtained. Kablan *et al.*[258] have used such a technique for studying cross-linking in epoxy resins. Sibilia [259] obtained three-dimensional i.r. dichroism information for a nylon 6 film. Nylon 6 can exist in several different conformations: an extended chain form, a twisted or pleated-chain form, an amorphous form, and a smectic hexagonal form. Dichroic measurements of the 835 and 935 cm^{-1} bands in the i.r. were used to study the orientation of nylon 6 in its extended form. The dichroism results proved that the hydrogen-bonded sheets, made up of extended chains, were parallel to the film surface in a biaxially drawn film. In the case of uniaxial drawing, at 100 and 150 °C, a high degree of chain alignment was found in the draw direction for the extended form, especially for draw ratios greater than 2.50. A small degree of planar orientation of both forms was observed for high draw ratios.

Frayer *et al.*[260] have shown that the i.r. bands in α-nylon 6 with frequencies 1210, 1233, and 1288 cm^{-1} are caused by a unique conformation in

[257] Y. S. Huang and J. L. Koenig, *Appl. Spectroscopy*, 1971, **25**, 620.
[258] J. Kablan, S. Lunak, and J. Tamchyna, *Analyt. Fys. Metody Vyzk. Plastu Pryskyric*, (Proc. Conf.), 1971, **1**, 203.
[259] J. P. Sibilia, *J. Polymer Sci., Part A-2, Polymer Phys.*, 1971, **9**, 27.
[260] P. D. Frayer, J. L. Koenig, and B. Lando, *J. Macromol. Sci.*, 1972, **B6**, 129.

the fold of a tightly folded chain with adjacent re-entry. The bands at frequencies 1210 and 1288 cm^{-1} were thought to arise from a *gauche* N—CH_2 group in the fold and the 1233 cm^{-1} band from a *gauche* CO—CH_2 group in the fold. The band at 1198 cm^{-1} arose from the *trans*-conformation of the amide group in the normal planar zig-zag configuration.

The Raman spectra of nylon 6, nylon 610, and nylon 11 were determined by Hendra *et al.*[261] Wunderlich and Bauer [262] also obtained the spectrum of nylon 66 and discussed its interpretation. Most of the bands in the vibrational spectrum were interpreted as arising from the amide groups and the zig-zag methylene backbone. The spectra vary according to the actual methylene sequence length and appear to increase in complexity for longer sequences. Several general points emerge, however. For example, the NH stretching amide A modes are comparatively weak in the Raman spectrum as are also the amide II bands and sometimes the amide I bands. The methylene groups give rise to bands which dominate the Raman spectrum. The strong band at 1440 cm^{-1} was assigned to a methylene symmetric deformation, the strong band at 1295 cm^{-1} to a methylene twist, and two strong bands near 1110 cm^{-1} to skeletal stretching vibrations These assignments for nylon were compared to those for polyethylene.

Poly(*ethylene glycol*).* Koenig and Angood [263] have studied the Raman spectrum of poly(ethylene glycol) in the solid, molten, and solution phases using an argon-ion laser. The spectral data were compared with those obtained from Raman spectra using helium–neon laser [264] and mercury arc sources.[265] Several new features were observed in the spectrum of the solid-state polymer using the argon laser. For instance, the bands with frequencies 1484, 1237, and 1064 cm^{-1} were found to contain more than one component. New weak features were also observed at 1286, 811, and 231 cm^{-1}. Depolarization ratios were obtained from the molten polymer; however, the poly(ethylene glycol) Raman spectrum changes dramatically on change of phase. The band splitting disappears and frequency shifts occur. The solid-state structure consists of a helix with seven residues in two turns of the helix, and four chains per unit cell. The spectral changes were considered to be due to loss of order in the molten phase with the appearance of new rotational isomers arising from rotation around the C—O bonds.

The spectrum of poly(ethylene glycol) in chloroform solution is, as might be expected, similar to the melt but quite different from the aqueous solution spectrum. The latter more closely resembles the solid-state spectrum although the features are not as sharp, and the splittings are no longer ob-

* See also polyoxymethylene and poly(propylene oxide) on p. 437.

[261] P. J. Hendra, D. S. Watson, M. E. A. Cudby, H. A. Willis, and P. Holliday, *Chem. Comm.*, 1970, 1048.

[262] B. Wunderlich and H. Bauer, *Adv. Polymer Sci.*, 1970, **7**, 151.

[263] J. L. Koenig and A. C. Angood, *J. Polymer Sci., Part A-2, Polymer Phys.*, 1970, **8**, 1787.

[264] R. F. Schaufele, *Trans. New York Acad. Sci.*, 1967, **30**, 69.

[265] Y. Matsui, T. Kubota, H. Tadokoro, and T. Yoshihara, *J. Polymer Sci., Part A*, 1965, **3**, 2275.

served. It was thought that the destruction of the helical backbone was not complete in the aqueous solution. This was in agreement with previous n.m.r. observations.[266]

Polybutadiene and Polyisoprenes. I.r. and Raman spectra have proved of value to the rubber industry, for instance in the monitoring of structural changes during the curing process and for the study of vulcanized networks in *cis*-1,4-polybutadiene.[267] Observations on *cis*-1,4-polybutadiene showed how the intensity of Raman lines, due to dialkenyl and cyclic sulphides together with conjugated triene structures, displayed asymptotic trends when plotted as a function of time.[268] The ratio of i.r. absorption bands at 1347 and 1337 cm^{-1} has been used [269] to determine the relative amounts of *trans* and *gauche* structure in 1,4-polybutadiene in solution. As the temperature was increased there appeared to be a decrease in the number of *trans* units present.

Woodward and co-workers [270] have obtained well-resolved i.r. spectra of *trans*-1,4-polybutadiene single crystal mats in the 1000—1400 cm^{-1} region. The mats were grown from six different solvents or solvent mixtures and were annealed near or above the crystal transition temperature. Other samples were melt-cooled or cast from solution. A band at 1350 cm^{-1}, thought to be associated with the amorphous phase, was compared with a band at 1335 cm^{-1} thought to be associated with the crystalline region. The intensity ratio of these bands varied with previous history of the sample. Information was obtained on the amount of amorphous material present in various preparations. The results supported the view that for certain methods of preparation of the mats the amorphous regions can be predominantly on the sample surface. There also seemed to be a correlation between the extent of internal amorphous contribution and the temperature of crystal growth.

Cornell and King [271] studied the i.r. and Raman spectra of *cis*-1,4-polybutadiene. At room temperature this compound is a rubber which, because of its stereoregularity, can be made to crystallize by stretching at room temperature or by cooling below the glass transition temperature as for *cis*-1,4-polyisoprene. The spectra at room temperature and at −77 °C, which is below the glass transition temperature, were compared. There are several new bands in Raman spectrum of the crystalline phase which have no counterpart in the i.r. spectrum. An assignment of these frequencies was attempted and it was suggested that a rule of mutual exclusion was operating in the crystalline phase. As a consequence it was suggested that the

[266] K. Liu and J. L. Parsons, *Macromolecules*, 1969, **2**, 529.
[267] P. H. Starmer, J. L. Koenig, M. Coleman, and J. R. Shelton, *Rubber World*, 1971, **164**, 47.
[268] J. R. Shelton and J. L. Koenig, *Rubber Chem. Technol.*, 1971, **44**, 904.
[269] P. V. Mikhailova and V. N. Nikitin, Sin. Strukl. Svoistva Polim Tr. Nauch. Konf. 15th, 1970, 197.
[270] C. Hendrix, D. A. Whiting, and A. E. Woodward, *Macromolecules*, 1971, **4**, 571.
[271] S. W. Cornell and J. L. King, *J. Polymer Sci., Part B, Polymer Letters*, 1971, **8**, 137.

crystalline form of *cis*-1,4-polybutadiene involved two residues connected by a centre of symmetry.

Infrared spectroscopy has been used to characterize various polyisoprenes at room temperature. *cis*-1,4-Polyisoprene and *trans*-1,4-polyisoprene,[272] and also polycycloisoprene and 1,4-polyisoprene,[273] can be readily distinguished. The i.r. spectra of a number of partially deuteriated polyisoprenes have also been obtained.[274] These spectra lead to major reassignments of the vibration frequencies of the polyisoprene unit.

Copolymers. Gardner *et al.*[275] have investigated the i.r. spectra of the ethylene–propylene copolymers. Differing sequence distributions were studied using the absorptions at 1155, 1378, and 720 cm^{-1}. Somewhat similar studies have been made of styrene–propylene copolymers [276] and of flame-resistant copolymers.[277] A study of the variation in hydrogen bonding between the NH and CO groups was made on *N*-methylmethacrylamide–styrene copolymer.[278] Folkes *et al.*[279] have investigated the i.r. spectra of a three-block copolymer, polystyrene–polybutadiene–polystyrene. They found no evidence of dichroism in any of the samples. This suggests that the structure is isotropic. I.r. dichroism measurements have also been used to study the relaxation of ethylene–methacrylic acid copolymers [280] and their salts and the changes in morphology under stress of copolymers like the polyurethane elastomers.[281, 282] These measurements provided information on molecular orientation within various domains. This in turn gave information on the role of the domain structure in determining the elastic properties of these segmented elastomers. A model consistent with the i.r. dichroism data was proposed in which the domains consisted of 'hard' ordered bundles of urethane linkages imbedded in a 'soft' polyester or polyether matrix. I.r. spectra [283] of crystalline urethane copolymers revealed polymorphic phase transitions probably due to conformational changes in the various polymer segments.[284] The degree of hydrogen-bonding has also been studied by i.r. spectroscopy in substituted urethanes.[285]

[272] Y. Tanaka, Y. Takouchi, M. Kobayashi, and H. Tadokoro, *J. Polymer Sci., Part A-2, Polymer Phys.*, 1971, **9**, 43.
[273] K. J. Clark, J. Lal, and J. N. Henderson, *J. Polymer Sci., Part B, Polymer Letters*, 1971, **9**, 49.
[274] M. A. Golub, *Spectrochim. Acta*, 1970, **26A**, 1883.
[275] I. S. Gardner, C. Cozewith, and G. Verstrate, *Rubber Chem. Technol.*, 1971, **44**, 1015.
[276] A. Dankovich and Y. V. Kissin, *Polymer Sci.* (*U.S.S.R.*), 1970, **12**, 908.
[277] K. A. Makarov, S. P. Shenkov, L. N. Mashlyakovskii, and N. A. Lipatnikov, *Vysokomol. Soedineniya, Ser. B*, 1971, **13**, 675.
[278] N. A. Kuznetsov and A. L. Smolyanskii, *Zhur. priklad. Spektroskopiya*, 1971, **15**, 299.
[279] M. J. Folkes, A. Keller, and F. P. Scalisi, *Polymer*, 1971, **12**, 793.
[280] Y. Uemura, R. S. Stein, and W. J. MacKnight, *Macromolecules*, 1971, **4**, 490.
[281] G. M. Ester, R. W. Seymour, and S. L. Cooper, *Polymer Preprints*, 1970, **11**, 516.
[282] G. M. Ester, R. W. Seymour, and S. L. Cooper, *Macromolecules*, 1971, **4**, 452.
[283] S. V. Laptii, V. N. Vatulev, Y. Y. Kercha, and N. A. Lipatnikov, *Zhur. priklad. Spektroskopiya*, 1971, **15**, 498.
[284] V. N. Vatulev and S. V. Luptii, *Vysokomol. Soedineniya, Ser. B*, 1971, **13**, 475.
[285] T. Tanaka, T. Yokoyama, Y. Yamaguchi, S. Naganuma, and M. Furukawa, *Kogyo Kagaku Zasshi*, 1971, **74**, 171.

Koenig and Meeks[286] have studied the Raman spectrum of the vinyl chloride–vinylidene chloride copolymers. The Raman spectrum of the constituent homopolymers had already been carefully studied.[287, 288] The Raman band at 2926 cm^{-1}, assigned to a methylene asymmetric stretching frequency and occurring at this frequency in the spectra of both homopolymers and in the spectra of all copolymers, was used as an internal standard for comparing Raman band intensities. The intensity of a band at 2906 cm^{-1}, known to be associated only with vinyl chloride, gave a linear plot as a function of vinyl chloride concentration of the copolymer. The intensity of a band at 1320 cm^{-1}, known to be associated with a vinyl chloride dyad (two adjacent vinyl chloride units), gave a linear plot as a function of dyad concentration. The dyad concentration was deduced from kinetic measurements. In a similar way the band at 1167 cm^{-1}, associated with vinyl chloride triads, also gave a linear plot. Hence it would appear that a quantitative relationship exists between the Raman scattering intensity of a unit and its concentration in the copolymer microstructure.

[286] M. Meeks and J. L. Koenig, *J. Polymer Sci., Part A-2, Polymer Phys.*, 1971, **9**, 717.
[287] J. L. Koenig and D. Druesedow, *J. Polymer Sci., Part A-2, Polymer Phys.*, 1969, **7**, 1075.
[288] P. J. Hendra, J. R. Mackenzie, and P. Holliday, *Spectrochim. Acta*, 1969, **25A**, 1349.

8

Vibrational and Vibrational–Rotational Spectroscopy of the Cyanide Ion, the Cyano-radical, the Cyanogen Molecule, and the Triatomic Cyanides XCN (X = H, F, Cl, Br, and I)

BY B. M. CHADWICK AND H. G. M. EDWARDS

1 Introduction

Our objective in this chapter is to review critically the recent work and developments in the application of vibrational and vibrational–rotational spectroscopy to some of the simpler cyano-species, *viz.* the parent cyano-species, the diatomic ion CN^-, and the radical CN; the tetra-atomic molecule, C_2N_2; and the molecular triatomic cyanides XCN, where X = H, F, Cl, Br, and I. These species have only been briefly mentioned in recent books on inorganic vibrational spectroscopy.[1] The detailed review presented here is not available elsewhere as far as we are aware.

Cyano-compounds in general were the subject of some of the earliest infrared[2, 3] and Raman spectroscopic studies.[4, 5] The cyano-species considered in this chapter contain only a few atoms. Most of the constituent atoms have relatively low atomic masses and have more than one stable isotopic form. As a consequence a great amount of detailed spectroscopic information can be obtained for these species. It is not surprising therefore that these species have continued to be much investigated spectroscopically. Indeed it seems almost that each refinement in technique has been followed by further spectroscopic studies of some of these species and the consequent elucidation of further spectroscopic details. Thus we feel this chapter will prove of interest not only as a detailed account of the spectroscopic properties of these simpler cyano-species but also as an illustration of the high state of development of the art and science of spectroscopy.

Our primary aim has been to discuss all the relevant work published on

1 (*a*) S. D. Ross, 'Inorganic Infrared and Raman Spectra,' McGraw-Hill, London, 1972; (*b*) L. H. Jones, 'Inorganic Vibrational Spectroscopy,' M. Dekker, New York, 1971, Vol. I.
2 W. Burmeister, *Verhandl. deutsch. phys. Gesellschaft*, 1913, **15**, 589.
3 H. Rubens and H. von Wartenberg, *Verhandl. deutsch. phys. Gesellschaft*, 1911, **13**, 796.
4 A. Petrikaln and J. Hochberg, *J. Phys. Chem.*, 1929, **3**, 217; 1930, **3**, 217, 440.
5 A. Dadieu and K. W. F. Kohlrausch, *Monatsh.*, 1930 **55**, 58; A. Dadieu, *ibid.*, 1931, **57**, 437.

the species listed above in the recent period January 1969—July 1972, as defined by *Chemical Titles*, 1972. Where relevant, we have also made reference to work published before this period. We have preferred to leave i.r. absorptions and Raman shifts in units of cm^{-1} rather than transfer them to the S.I. unit of wavenumber, m^{-1}. Likewise, internuclear separations are reported in Å rather than nm. On the other hand, force constants, where appropriate, have been given in N m^{-1}. When rotational constants measured in frequency units have been converted into units of wavenumber, the I.U.P.A.C. recommended value of the velocity of light, $c = 2.997925 \times 10^{10}$ cm s^{-1}, has been used. We use ν to represent an observed vibration, and ω to signify an observed vibration corrected for anharmonicity; ν_0 and ω_0 are the corresponding vibrations corrected for rotation. In the case of triatomics, XCN, we have adopted the convention: $\nu(CN) \equiv \nu_1$, $\nu(XC) \equiv \nu_2$, and $\delta(XCN) \equiv \nu_3$.

2 The Cyanide Ion

The vibrational spectrum of the cyanide ion has been studied in aqueous solution, as a substitutional impurity in various alkali-metal halide single crystals, and in the polycrystalline and single-crystal bulk material (alkali-metal salts).

In aqueous solution the stretching fundamental, $\nu(CN)$,[6–8] is observed in the i.r. at 2080 ± 2 cm^{-1} ($\varepsilon_M = 29 \pm 1$) for KCN and in the Raman at 2079 cm^{-1} for NaCN. The corresponding $\nu(CN)$ values for the solid state were reported [9–11] many years ago as 2080 cm^{-1} (i.r.) and 2085 cm^{-1} (Raman) for NaCN and 2081 cm^{-1} (Raman) for KCN. A somewhat more recent Raman measurement [12] of $\nu(CN)$ in a KCN crystal gave 2076 cm^{-1}. [Although, as will be seen later, the first overtone region of $\nu(CN)$ has been much investigated in studies of doped crystals, this overtone region does not appear to have been studied at all in the bulk solid or in aqueous solution.] These values, especially those in aqueous solution, provide the basis for the almost invariable assumption that $\nu(CN)$, the stretching vibration of the free, non-complexed cyanide ion, is about 2080 cm^{-1}.

Solid potassium cyanide is, in fact, one of a broad group of polymorphic compounds $M^+(XY)^-$ whose crystalline phase immediately below the melting point exhibits a cubic (usually rock salt) structure. Such cubic structures must necessarily be disordered either with individual linear XY^- ions attaining effective cubic symmetry by rotation or with a random distribution of ion orientations among equivalent directions in the unit cell, pro-

[6] L. H. Jones and R. A. Penneman, *J. Chem. Phys.*, 1954, **22**, 965.
[7] R. A. Penneman and L. H. Jones, *J. Chem. Phys.*, 1956, **24**, 293.
[8] G. W. Chantry and R. A. Plane, *J. Chem. Phys.*, 1960, **33**, 736.
[9] W. Gordy and D. Williams, *J. Chem. Phys.*, 1935, **3**, 664; 1936, **4**, 85.
[10] P. Krishnamurti, *Indian J. Phys.*, 1930, **5**, 633.
[11] N. N. Pal and P. N. Sen Gupta, *Indian J. Phys.*, 1930, **5**, 13.
[12] J. P. Mathieu, *Compt. rend.*, 1954, **238**, 74.

viding long-range cubic symmetry. Despite extensive investigation, however, the orientation and dynamical behaviour of the cyanide ions in the cubic phase of KCN are still in doubt [as is the case with the linear ions in virtually all similar $M^+(XY)^-$ compounds]. In the latest study [13] neutron-diffraction measurements were made on single crystals of KCN. The results rule out free rotation of the CN^- ions, but do not suggest a preferred direction of orientation. The analysis indicated that the K^+ and CN^- ions in KCN crystals suffer large and temperature-independent (180—295 K) displacements from their equilibrium lattice sites. A tentative explanation of this phenomenon is that the disordering of the CN^- ions in the crystal involves a low activation energy (*ca.* 110 cm^{-1}) for rotation of CN^- ions between neighbouring equivalent positions. It is envisaged that these rapid reorientations of the CN^- ions would produce local strains in the crystal to which the K^+ ions could respond. Some circumstantial evidence as to the plausibility of this mechanism is provided by the report [14] that preliminary Raman measurements on KCN single crystals at room temperature using laser excitation suggest a broad rotational band centred at about 60 cm^{-1} shift from the exciting line. This appears to be the only recent vibrational spectroscopic study of an alkali-metal cyanide.

More interest has centred on the i.r.[15] and Raman [16] spectra of the cyanide ion as a substitutional impurity in various alkali-metal halide single-crystal matrices. These investigations were aimed at establishing the rotational motion of the cyanide ion in these crystals. Quite apart from the intrinsic value of these studies, some of the findings may have a bearing on the situation in KCN. The i.r. studies involved the fundamental and first-overtone regions; the maximum doping was 0.5 mol % and measurements were made over the temperature range 1.36—295 K. The Raman investigations were concerned with two frequency regions, 0—400 cm^{-1} and frequencies in the neighbourhood of ν(CN). The doping was in the range 0.3—0.72 mol % and the temperature range 4.2—300 K. The view has been expressed that, in general, the observation of the Raman spectra of isolated species in doped alkali-halide matrices does not require cryogenic conditions.[17] However, these studies of the cyanide ion illustrate the information that can be obtained when investigations are conducted over a range of temperatures down to liquid-helium temperatures.

The i.r. observations have been discussed in terms of a model in which the rotational motion of the cyanide is determined by a Devonshire-type three-dimensional potential function with six potential minima along the $\langle 100 \rangle$ directions. (Uniaxial stress measurements indicate that these are the equilibrium directions and that reorientation can take place down to 1.36 K.)

[13] D. L. Price, J. M. Rowe, J. J. Rush, E. Prince, D. G. Hinks, and S. Susman, *J. Chem. Phys.*, 1971, **56**, 3697.
[14] J. J. Rush and R. Khanna, personal commuication in ref. 9.
[15] W. D. Seward and V. Narayanamurti, *Phys. Rev.*, 1966, **148**, 463.
[16] R. Callender and P. S. Pershan, *Phys. Rev.* (*A*), 1970, **2**, 672.
[17] G. A. Ozin, *The Spex Speaker*, 1971, **XVI**, No. 4.

In this model it is assumed that the centre-of-mass system of the cyanide ion is tied to a lattice point (vacancy and entropy studies indicate that CN^- generally substitutes for the halide) and that the matrix ions are at rest so that the effect of the matrix is to produce a static potential, totally symmetric under O_h symmetry, of the form $V(\theta,\varphi) = -\frac{1}{8}\bar{K}(3-30\cos^2\theta+35\cos^4\theta+5\sin^4\theta\cos^4\varphi)$, where $\bar{K}$ is a constant. At low temperatures, if kT is very much less than the barrier height, V_B, the i.r. spectrum in the vicinity of ν(CN) should consist of a strong Q branch [ν(CN)] with weak satellites separated from the fundamental by multiples of the librational vibration ν_0^{lib} (similarly for the first overtone). In this situation reorientation can be visualized as taking place *via* a classical thermally activated jump process or *via* quantum-mechanical tunnelling. At high temperatures ($kT \gg V_B$) the spectrum should approximate to that of 'free rotation' and should consist of P and R branches with a missing central Q branch. For intermediate temperatures the spectrum should consist of P, Q, and R branches of comparable intensity, corresponding to a situation of hindered rotation. The KCl and KBr data indicate that below 20 K the motion of the cyanide ion is best described as librational, with ν_0^{lib} of 11—12 cm^{-1}. At 2 K this frequency is clearly observed as a sum satellite [$f_{1u}(J = 1) \rightarrow f_{2g}(J = 2)$] of the fundamental ν(CN), at 2089 cm^{-1} in KCl and 2078 cm^{-1} in KBr. It is also observed as a sum satellite of the first overtone of ν(CN) at 8 K. Between 20 and 50 K the cyanide ions perform hindered rotational motions. Above 60 K the Q branch disappears, indicating that the ions occupy energy states lying above the barrier. Moreover, as expected, the separations of the P and R maxima increase with temperature according to a $T^{\frac{1}{2}}$ law, reaching 40 and 35 cm^{-1} for KCl and KBr respectively at 295 K. This P–R separation for KCl leads to a B value of 1.25 cm^{-1} (recalculated [16] to be 1.0 cm^{-1}) for the KCl matrix, where the P and R branches are more clearly resolved, and a V_B value of 25 cm^{-1} in the $\langle 110 \rangle$ direction (based on a $\bar{K}$ of 20 cm^{-1}). However, tunnelling transitions at low temperature and the rotational fine structure of the P and R branches at high temperature were not resolved in either KCl or KBr. In NaCl and NaBr the i.r. spectrum in the fundamental region consists of only one narrow band (*ca.* 2106 and 2088 cm^{-1} at 300 K) whose half-width (*ca.* 20 cm^{-1} at 300 K) is proportional to the square of the temperature, with the result that the instrumental resolution (0.6 cm^{-1} in this region) did not allow the determination of the lineshape below 80 K. It was concluded from the absence of P and R branches, the extreme narrowness of the vibrational transition, and the fact that the half-width does not follow a $T^{\frac{1}{2}}$ law, that the barrier to rotation in NaCl and NaBr is extremely high and almost certainly greater than 100 cm^{-1} and that no free rotation is observed even at 300 K.

As the more recent Raman analysis [16] represents a significant extension of that provided by the i.r. data and as it seems clear that the Raman effect will be equally useful in studying systems of an analogous nature, it seems worthwhile to give a brief and didactic summary of the theory used in analysing the scattering of alkali-halide matrices containing molecular ion impurities. Because the energy of the stretching vibration of the cyanide ion is large compared with the energy of the rotational states of the cyanide ion or the vibrational states of the impure matrix, two fundamentally different and clearly definable regions can be anticipated in the Raman spectrum of cyanide-doped alkali-metal halide crystals: a low-frequency-shift region, covering approximately 0—400 cm^{-1}, and a high-frequency-shift region in the vicinity of ν(CN). Now, the cyanide ion is likely to occupy a site of O_h symmetry in the alkali-metal halide. According to semi-classical theory there are two contributions to the scattering in the vicinity of ν(CN) from such a 'point'. The first, variously called the trace scattering, the δ function, and the '$Q(0)$' branch, is totally polarized, is parallel to the electric field of the incident light, is independent of crystal orientation, and corresponds to an a_{1g} component. As it is centred at ν(CN) and should not be broadened by rotational motion it can be used to locate ν(CN) precisely and to give a measure of the extent to which the theory holds. The second contribution, which has components parallel and perpendicular to the incident electric field, arises from the traceless components of the polarizability tensor and is modulated by the rotational motion of CN^-; it can be expressed as an on-diagonal scattering function $\langle\beta^{\nu}_{xx}(t)\beta^{\nu}_{xx}(0)\rangle$ (corresponding to an e_g component) and an off-diagonal one $\langle\beta^{\nu}_{xy}(t)\beta^{\nu}_{xy}(0)\rangle$ (corresponding to an f_{2g} component), where $\langle\beta^{\nu}(t)\rangle$ is a second-rank tensor invariant under all operations of O_h symmetry. These two scattering functions* provide a measure of the rotational motion of the cyanide ion, the most important experimental manifestation of which is the side-band structure of ν(CN). No matter what form the interaction between matrix and impurity takes and how complex the resultant rotational motion of the cyanide ion is, the Raman cross-section in the vicinity of ν(CN) is a projection of the rotational motion of the cyanide ions. The trace scattering and the on- and off-diagonal functions can be isolated by choosing appropriate scattering geometries. Though the trace scattering can never be completely isolated, in practice it is quite narrow (half-band width of *ca.* 1 cm^{-1} at 300 K) and can therefore be easily identified.

In contrast, the scattering in the low-frequency-shift region is much harder to interpret. First, the cyanide ion will cause induced first-order scattering (due to a relaxation of the wave-vector selection rule in the presence of the impurity) from phonon-like modes which do not include the rotational motion of the cyanide ion. Secondly, there will be matrix-rotational scattering which involves rotational motion of the cyanide ion and a simultaneous motion of alkali-metal and halide ions. This is in contrast to the side-band structure of ν(CN), which only involves the rotational motion of the cyanide ion. These two scatterings have to be separated from

* There is some traceless $Q(0)$ scattering both parallel and perpendicular to the laser electric field; in the case of the free rotor these have been calculated to be at least an order of magnitude less intense than the trace scattering.

one another (assuming that there are separate rotational and modified matrix phonon states), and also from the rather large normal second-order Raman scattering of the pure alkali halide. Certain simplifications can be made, however. It should be possible to freeze out a large amount of the second-order scattering at low temperature, and the induced first-order scattering can be calculated, at least if it is assumed that the cyanide ion is a spherical ion defect. Moreover, because the change induced in the polarizability tensor by ν(CN) has exactly the same symmetry as the static polarizability tensor of the rigid ion, the shape and form of the side-band spectra and the low-frequency Raman spectra due to rotational-like motion of the rigid ion will be identical.

The scattering behaviour observed in the vicinity of ν(CN) was as expected. The values of ν(CN), as established by the narrow trace scattering, are listed in Table 1. The sharpness of the band at room temperature compared very favourably with the broad band envelopes in the i.r.

Table 1 *Raman trace scattering in various* CN^-*-doped alkali-metal halides*

Matrix	ν(CN)/cm^{-1}	$\Delta\nu_{\frac{1}{2}}$/cm^{-1}
KCl (300 K)	2084.4	0.9
KCl (8.5 K)	2087.1	0.4
KBr (300 K)	2075.4	$\sim$1.0
KBr (9 K)	2077.5	$<$0.6
NaCl (300 K)	2102.4	0.85
NaCl (10 K)	2106.2	0.4
RbCl (300 K)	2077.1	$<$1.0
RbBr (300 K)	2068.5	$<$1.5

It can be seen that ν(CN) varies over a range of about 35 cm^{-1} in these alkali-metal halides. At room temperature the depolarized scattering of symmetry $\langle\beta^{\nu}_{xx}(t)\beta^{\nu}_{xx}(0)\rangle$ and of symmetry $\langle\beta^{\nu}_{xy}(t)\beta^{\nu}_{xy}(0)\rangle$ both look very much alike for the KCl and KBr matrices. For each symmetry it consists of a reasonably well-defined peak on each side (*S* and *O* branches) of the trace position (at which there is a small amount of *Q* scattering, some of which at least may arise from spurious trace scattering). This is consistent with the cyanide ion being a free rotor. However, the peak separations for the two scattering symmetries are not equal, indicating that even at room temperature the crystalline environment of the cyanide ion significantly modifies the free rotational states. The peak separations in the scattering of symmetry $\langle\beta^{\nu}_{xx}(t)\beta^{\nu}_{xx}(0)\rangle$ (62 and 60 cm^{-1} for KCl and KBr respectively) at 300 K are consistent with the room-temperature i.r. *P–R* separations. The *S–O* separations also follow a $T^{\frac{1}{2}}$ relationship (down to $\sim$100 K for KCl and $\sim$60 K for KBr), yielding mean *B* values of 0.75 $\pm$0.15 cm^{-1} (KCl) and 0.55 $\pm$0.15 cm^{-1} (KBr) which agree reasonably well with the i.r. *B* values (1.0 and 0.8 cm^{-1} respectively). However, just as in the i.r., no rotational fine structure was observed. On cooling, reason-

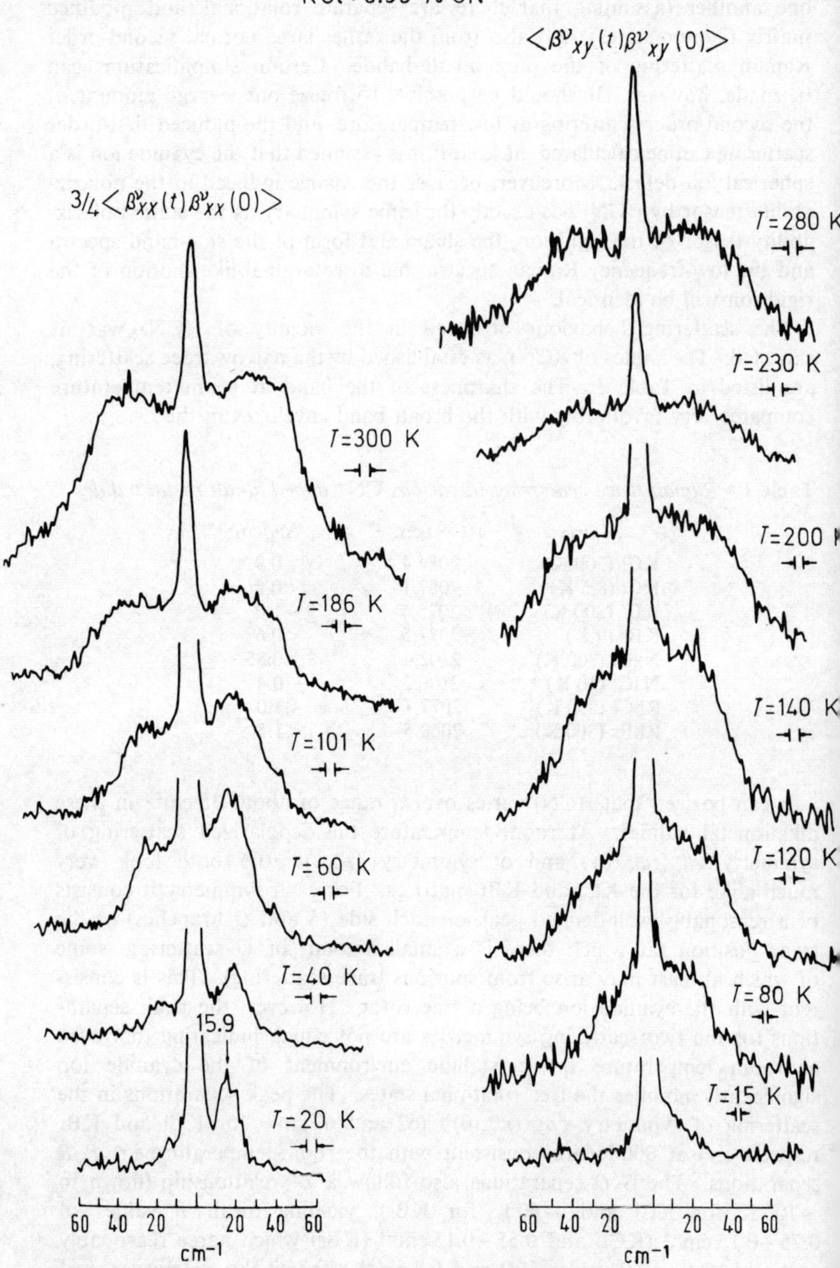

Figure 1 *Helium-temperature high-frequency data for* KCl:CN⁻
[Reproduced by permission from *Phys. Rev.* (*A*), 1970, **2**, 672]

ably well-defined sum satellite lines having e_g-like symmetry are observed in KCl [at 16 cm^{-1} from ν(CN) at 20 K; see Figure 1] and KBr (at 13 cm^{-1} and 12 K) and are assigned as $f_{1u}(J = 1) \rightarrow f_{1u}(J = 3)$. These lead to $\bar{K}$ values of 20 and 27 cm^{-1} and barrier heights in the $\langle 111 \rangle$ direction of 46 cm^{-1} (70 K) and 34 cm^{-1} (52 K) respectively. These characteristic temperatures agree quite well with the temperatures at which the *O–S* separations begin to deviate from $T^{\frac{1}{2}}$ dependence (100 and 60 K respectively).

The low-energy spectra of the doped KCl and KBr crystals are much less revealing. Only the second-order spectra of the matrix could be discerned at room temperature, and even at liquid-helium temperatures the only feature that could be attributed to matrix-rotational scattering is a structureless rise in intensity close to the exciting line.

The Raman study indicates, like the i.r. one, that the NaCl system is experimentally very different from those of KCl and KBr, and that the CN$^-$ ion is rather heavily trapped in the NaCl lattice in the sense that it is not free to rotate even at room temperature. However, from the Raman study

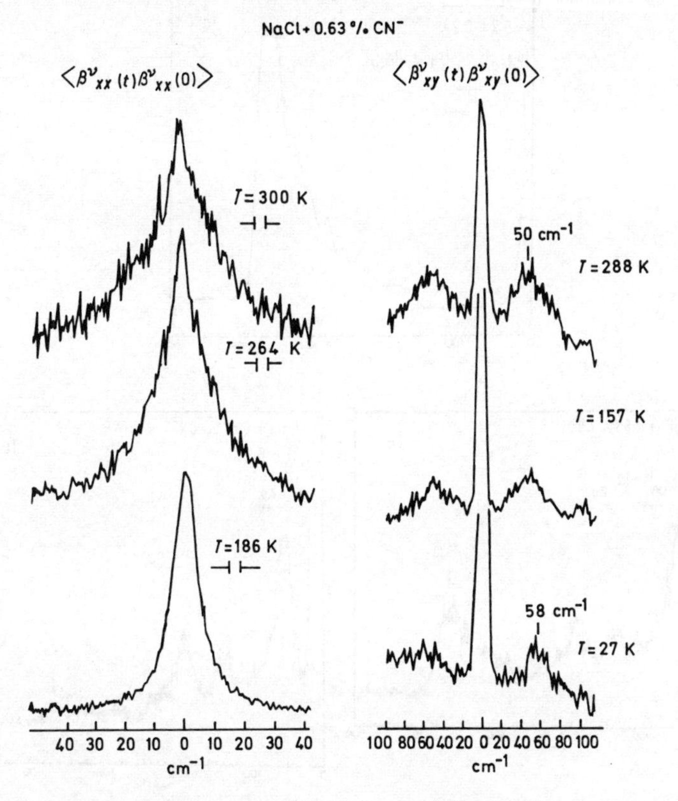

Figure 2 *Helium-temperature high-frequency data for* NaCl:CN$^-$
[Reproduced by permission from *Phys. Rev.* (*A*), 1970, **2**, 672]

relatively much more information is obtained and rather more precise conclusions are drawn than from thc i.r. investigation. First, side-band structure on ν(CN) is obtained and this structure is a single line, about 54 cm^{-1} from the trace position, throughout the temperature range, and of f_{2g}- rather than e_g-type symmetry (see Figure 2). Secondly, a line at 54 cm^{-1} is observed in the low-frequency spectrum where it seems to be of almost completely f_{2g} symmetry (see Figure 3). Thirdly, the scattering of $\langle\beta^{\nu}_{xx}(t)\beta^{\nu}_{xx}(0)\rangle$ symmetry reveals a Lorentzian line at the trace position whose linewidth has an exponential dependence on inverse temperature. These results can be understood in terms of a harmonic libration (*ca.* 54 cm^{-1}) about one of the six equivalent $\langle 100\rangle$ axes of the crystal and an occasional thermally activated 'hop' over a potential barrier (*ca.* 275 cm^{-1}) from one $\langle 100\rangle$ direction to another $\langle 100\rangle$ direction. The first motion gives rise to the f_{2g}-type scattering and the hopping process is consistent with the observation of

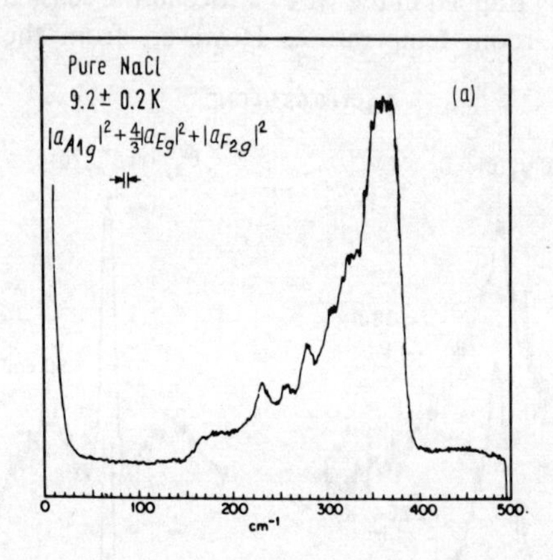

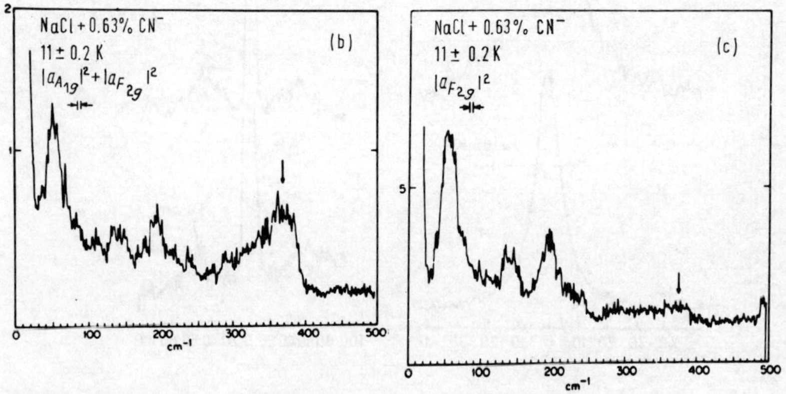

Figure 3 *Helium-temperature low-frequency data for* NaCl *and* NaCl:CN$^-$ [Reproduced by permission from *Phys. Rev.* (*A*), 1970, **2**, 672]

e_g-type scattering. The temperature dependence of the latter eliminates the possibility of quantum-mechanical tunnelling. The NaCl system seems to bear some relation to that proposed for KCN, though in the latter no preferred orientation direction has been established and cyanide-ion amplitudes cannot be reproduced by a dynamical model involving harmonic coupling.

3 The Cyano-radical

The free radical CN has been identified in the gas phase by electronic spectroscopy [18a] and in an argon matrix at 4 K by e.s.r. spectroscopy following vacuum-u.v. photolysis of HCN.[18b] It has subsequently been shown, from observation of the $B(^2\Sigma^+)$—$X(^2\Sigma^+)$ transition, that CN is produced in argon and nitrogen matrices at 14 K by vacuum-u.v. photolysis of several XCN species; the efficiency of the process is HCN > FCN > ClCN > BrCN $\gg$ C_2N_2. Examination of the i.r. spectra of the matrices indicates, not surprisingly, that several species are produced in these photolyses. Nevertheless, careful interpretation of the spectra, including those produced from $H^{13}CN$ (54%) and $HC^{15}N$ (95%), points to a ground-state vibrational fundamental of 2046.0 cm^{-1} for CN.[19] This value is certainly consistent with the value of 2080 ±5 cm^{-1} for ν(CN) in the cyanide ion, taking into account the diminution in bond-order involved in the loss of an electron from CN^-. However, in a recent study,[20] in which CN radicals (as established by visible and e.s.r. spectroscopy) were produced in an argon matrix at 4 K by photolysis of ICN, the 2046 cm^{-1} band was not observed. Photolysis yielded a strong, sharp, symmetric band (half-width of 1.5 cm^{-1}) at 2057 cm^{-1}, and gradual warming only produced irreversible changes. Further research is clearly needed here.

4 Cyanogen

Cyanogen, C_2N_2, a colourless gas at room temperature, liquefies at 252.1 K and solidifies at 245.4 K. All the early evidence (see *e.g.* refs. 21—23) pointed towards the molecule in the gas phase having a linear symmetrical NCCN structure. C_2N_2 is then expected, with $D_{\infty h}$ symmetry, to have five fundamental vibrations (see Figure 4). Three are non-degenerate stretching vibrations of which two (ν_1 and ν_2) are Raman-active (σ_g^+) and one, ν_3, is i.r.-active (σ_u^+). The two doubly degenerate bending vibrations (ν_4 and ν_5) are the π_g Raman-active and the π_u i.r.-active deformations respectively.

[18a] G. Herzberg, 'Spectra of Diatomic Molecules', Van Nostrand, New Jersey, 1950, pp. 31, 520; [b] E. L. Cochran, F. J. Adrian, and V. A. Bowers, *J. Chem. Phys.*, 1962, **36**, 1938.
[19] D. E. Milligan and M. E. Jacox, *J. Chem. Phys.*, 1967, **47**, 278.
[20] W. C. Easley and W. Welner, jun., *J. Chem. Phys.*, 1970, **52**, 197.
[21] H. E. Watson and K. L. Ramaswamy, *Proc. Roy. Soc.*, 1936, **A156**, 130.
[22] L. Pauling, H. Springall, and K. J. Palmer, *J. Amer. Chem. Soc.*, 1939, **61**, 927.
[23] A. Langseth and C. K. Møller, *Acta Chem. Scand.*, 1950, **4**, 725.

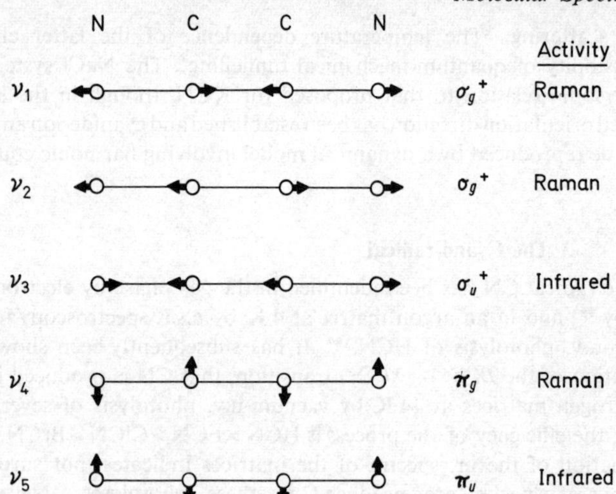

Figure 4 *Normal vibrations of linear symmetric* C_2N_2

(Since the unsymmetrical molecules $^{12}C^{13}C^{14}N_2$ and $^{12}C_2{}^{14}N^{15}N$ only have small isotope shifts, their fundamentals, with a prime to denote the isotopic substitution, are similarly numbered.)

Craine and Thompson [24] provided the first unambiguous proof of the linear symmetrical structure in their examination of the rotational structure of ν_3 in the i.r. spectrum of the gas. The liquid, with its narrow temperature range, has been little investigated. In the solid state at 178 K it has, however, been established by *X*-ray diffraction [25] that the molecule retains its centre of symmetry and effectively its linearity.

Most of the recent vibrational and vibrational–rotational spectroscopic interest [26–30] has centred on the i.r. spectrum, particularly of the gas phase. There are partial studies of the vibrational Raman spectrum of the gas published over forty years ago [31] and of the Raman spectrum of the solid.[27b] There appear to have been three investigations [4, 23, 32] of the Raman spectrum of the liquid, although the phase was not specified in the most recent of these studies.[23] However, in subsequent discussions of this work other workers, for example Cartwright *et al.*[33] and Schultz and

[24] G. D. Craine and H. W. Thompson, *Trans. Faraday Soc.*, 1953, **49**, 1273.
[25] A. S. Parkes and R. E. Hughes, *Acta Cryst.*, 1963, **16**, 734.
[26] F. D. Verderame, J. W. Negben, and E. R. Nixon, *J. Chem. Phys.*, 1963, **39**, 2274.
[27a] E. R. Nixon and F. D. Verderame, *J. Chem. Phys.*, 1964, **41**, 1684; [b] P. M. Richardson and E. R. Nixon, *ibid.*, 1968, **49**, 4276.
[28] A. G. Maki, *J. Chem. Phys.*, 1965, **43**, 3193.
[29] B. H. Thomas and W. J. Orville-Thomas, *J. Mol. Structure*, 1969, **3**, 191.
[30] A. Bersellini and C. Meyer, *Compt. rend.*, 1970, **270**, *B*, 1672.
[31] P. Daure and A. Kastler, *Compt. rend.*, 1931, **192**, 1721.
[32] A. W. Reitz and R. Sabathy, *Monatsh.*, 1938, **71**, 131.
[33] G. J. Cartwright, D. O'Hare, A. D. Walsh, and P. A. Warsop, *J. Mol. Spectroscopy*, 1971, **39**, 393; G. B. Fish, G. J. Cartwright, A. D. Walsh, and P. A. Warsop, *ibid.*, 1972, **41**, 20.

Eggers,[34] have assumed that the data referred to the liquid rather than to the gas. There appear to have been no i.r. or Raman matrix-isolation studies of cyanogen.

The Gaseous State.—There have been numerous i.r. studies of the gas phase since the pioneering work of Burmeister [2] and Rubens and von Wartenberg [3] published sixty years ago. Especially notable is the work of Maki,[28] who examined the vibration–rotation spectra of $^{12}C_2{}^{14}N_2$ and $^{12}C_2{}^{15}N_2$ (containing 97 atom % ^{15}N) with a resolution of 0.04—0.08 cm^{-1}.

Table 2 *Anharmonic fundamentals* (cm^{-1}) *of gaseous* $^{12}C_2{}^{14}N_2$, $^{12}C^{13}C^{14}N_2$, and $^{12}C_2{}^{15}N_2$

Vibration	$^{12}C_2{}^{14}N_2$	$^{12}C^{13}C^{14}N_2$	$^{12}C_2{}^{15}N_2$
ν_1, ν_1'	2329.94(15)[a, b, d]	2298(2)[f]	
ν_2	845.4[c, e, g]		
ν_3, ν_3'	2157.8271(5)[b, d]	2131(2)[f]	2121.922(3)[a, d]
ν_4, ν_4'	503.33[a, b, d, e, g]	495(2)[f]	
ν_5	233.15(15)[a, b, d]		

[a] Ref. 28. [b] Ref. 30. [c] Ref. 33. [d] ν_0 values. [e] The Raman vapour-phase values of ν_2 and ν_4 cited by Thomas and Orville-Thomas [29] refer, in fact, to the liquid.[31] [f] Ref. 34. [g] Unknown uncertainty.

The best values of the anharmonic fundamentals of $^{12}C_2{}^{14}N_2$, $^{12}C^{13}C^{14}N_2$, and $^{12}C_2{}^{15}N_2$ are listed in Table 2. The values of ν_1', ν_3', and ν_4' for $^{12}C^{13}C^{14}N_2$ were all obtained from the i.r. spectrum of cyanogen containing 26.3 atom % of this species and are not very precise. It is a reflection of the lack of interest shown in the Raman spectrum of gaseous cyanogen that the values of the three Raman-active fundamentals of $^{12}C_2{}^{14}N_2$ listed in Table 2 have not been obtained by Raman spectroscopy but indirectly from i.r. and electronic spectra. ν_1 and ν_4 were obtained from the i.r.-active combination bands $\nu_1+\nu_3$ and $\nu_3+\nu_4$ and the appropriate anharmonicity constants.[28, 30] It is interesting to note, nevertheless, that in 1931 Daure and Kastler [31] reported ν_1 of gaseous cyanogen at 2330 cm^{-1} in the Raman spectrum. ν_2 was obtained [33] from hot bands in the u.v. spectrum (the 300 and 220 nm systems); a ν_4 value derived in a similar way is 0.50 cm^{-1} smaller than that listed in Table 2. The ν_2 value should be compared with $\nu_0(010^0—000^0)+x_{23} = 842.98$ cm^{-1} derived [28] from the appropriate combination band. The value of ν_5 is also not based on direct observation but was obtained [28, 30] from the i.r.-active combination band $\nu_1+\nu_5$ and the appropriate anharmonicity constant. In fact, the i.r. spectrum of cyanogen in any phase does not appear to have been studied in the region of ν_5 since the measurements of Strong and Woo [35] on the gas forty years ago.

The vibrational fundamentals of cyanogen have been used as the basis

[34] J. W. Schultz and D. F. Eggers, jun., *J. Mol. Spectroscopy*, 1958, **2**, 113.
[35] J. Strong and S.-C. Woo, *Phys. Rev.*, 1932, **42**, 267.

Table 3 *Force-constant* N m^{-1}) *analysis of the vibrational fundamentals* (cm^{-1}) *of cyanogen*

k(CN)	= 1609.9	k(CN,CN)	= −164.4
k(CC)	= 791.6	k_δ	= 18.5
k(CC,CN)	= −19.1	$k_{\delta\delta}$	= −0.21

Fundamental	*Observed*	*Calculated*
ν_1	2328.5[a, c]	2328.5
ν_2	850.6[a, c]	850.6
ν_3	2159(2)[b, c]	2159.0
ν_4	507.2[a]	502.8
ν_5	240[a, c]	240.0
ν_1'	2298(2)[b, c]	2298.0
ν_2'	—	846.8
ν_3'	2131(2)[b]	2131.3
ν_4'	495(2)[b, c]	495.0
ν_5'	—	237.4

[a] Ref. 23; liquid-phase value. [b] Ref. 34. [c] Used in force-constant calculation.[34]

of several force-constant calculations.[23, 32, 34, 36, 37] Early calculations of this kind involved simple valence force fields. In a more recent analysis, Schultz and Eggers [34] were able to calculate values of all six force constants required for a complete valence force field from the five known fundamentals of $^{12}C_2{}^{14}N_2$ and ν_1', ν_3', and ν_4' of $^{12}C^{13}C^{14}N_2$. The results of this analysis are summarized in Table 3. They show that k(CC) is appreciably larger, k(CN) rather smaller, and k(CC,CN) very much smaller than previous calculations had suggested. Also, all three interaction constants are negative, k(CN,CN) having the greatest magnitude. The two deformation constants are less reliable than the stretching force constants because of the uncertainty attaching to the value of ν_5 used. Since the calculations of Schultz and Eggers, better values of the anharmonic fundamentals ν_1, ν_2, ν_3, ν_4, and ν_5 (especially ν_2, ν_4, and ν_5), all of which relate to the gas phase, have become available (see Table 2) and rather more is known about isotopically substituted cyanogen molecules. Improved force-constant calculations could thus be performed using revised anharmonic fundamentals but sufficient information is not yet available for the calculation of all the harmonic fundamentals.

There have been two i.r. studies of gaseous $^{12}C_2{}^{15}N_2$ (both samples containing about 97 atom % of ^{15}N), one at low resolution (about 1—2 cm^{-1}) by Nixon and Verderame [27a] and one at high resolution (0.04—0.08 cm^{-1}) by Maki.[28] Some of the available data are listed in Table 4, where they are compared with the corresponding and analogous $^{12}C_2{}^{14}N_2$ and $^{12}C_2{}^{14}N^{15}N$ data.

The high-resolution study by Maki included studies of $^{12}C_2{}^{14}N_2$ and

36 J. W. Linnett and H. W. Thompson, *J. Chem. Soc.*, 1937, 1399.
37 G. Herzberg, 'Infrared and Raman Spectra of Polyatomic Molecules', Van Nostrand, New Jersey, 1945, p. 180.

Table 4 *I.r. band positions* (cm^{-1}) *in gaseous* $^{12}C_2{}^{15}N_2$

(a) Low resolution [27a]

Assignment	$^{12}C_2{}^{15}N_2$	$^{12}C_2{}^{14}N_2$
$\nu_4+\nu_5{}^a$	729.9 s	732 s
$\nu_1+\nu_5$	2077.8 w	2092 w
$\nu_3{}^a$	2122.1 s	2157.6 s
$\nu_1+\nu_5$	2532.8 m	2563 s
$\nu_3+\nu_4$	2621.3 m	2663 s

(b) High resolution [28]

Assignment	$^{12}C_2{}^{15}N_2$	$^{12}C_2{}^{14}N_2$	$^{12}C_2{}^{14}N^{15}N$
$\nu_0(\nu_3)$	2124.922(3)	2157.849(5)	
$\nu_0(\nu_1+\nu_5)$	2536.021(3)	2561.339(2)	2549.63(5)
$\nu_0(\nu_3+\nu_4)$	2622.993(2)	2657.528(1)	2639.38(5)

[a] Observed as P–R doublets

$^{12}C_2{}^{14}N^{15}N$ and will now be considered in detail. Vibration–rotation band constants were determined for eight vibrational transitions of the $^{12}C_2{}^{14}N_2$ species and for three vibrational transitions of the $^{12}C_2{}^{15}N_2$ species. Only the ν_3 fundamental of $^{12}C_2{}^{15}N_2$ was observed and measured in the wave-number range covered in this investigation, but two combination bands, $\nu_1+\nu_5$ and $\nu_3+\nu_4$, were studied in detail for both species. In addition, three hot bands and two weaker combination bands, $\nu_1+\nu_3$ and $\nu_2+\nu_3$, were measured for the ^{14}N species. The intensity alternation in the P- and R-branch lines assisted in the unambiguous assignment of the J transitions and enabled Maki to re-analyse Craine and Thompson's data on ν_3 and re-adjust their B_0 value, from 0.1588 ± 0.0001 cm^{-1} to 0.1573 ± 0.00003 cm^{-1}. Spectra were obtained at room temperature and at 213 K. At the latter temperature the overall intensity was diminished due to the reduced vapour pressure of cyanogen, but the temperature-dependence of the Q branches provided further support for the assignments, in particular for the positions of the stronger of the many hot-band Q branches in $\nu_1+\nu_5$ and $\nu_3+\nu_4$. The occurrence of a Fermi resonance between $2\nu_4{}^0$ and ν_2 (the unperturbed levels are separated by about 150 cm^{-1}), which was first proposed by Langseth and Møller to explain the appreciable intensity of $2\nu_4{}^0$ in the Raman spectrum (of what has been taken to be the liquid), is also invoked by Maki, who discusses the effect of this resonance on the band constants. The rotational constants and band constants derived by Maki are summarized in Table 5.

The B_0 value of $^{12}C_2{}^{14}N_2$ listed in Table 5 [0.157124(12) cm^{-1}] is based on the corresponding experimental D_0 value, whereas the ^{15}N B_0 value is based on a D_0 value calculated (an experimental value could not be obtained) from the ^{14}N experimental D_0 value. Since the B_0 value is moderately sensitive to the value of D_0 used to fit the data, Maki also calculated D_0 for both isotopic species and used these to derive another set of B_0 values. However, the values listed in Table 5 were preferred, because there seemed to be

Table 5 *Rotational constants, band constants, and bond lengths of C_2N_2 from i.r. measurements*

Species	ν_3/cm^{-1}	B_0/cm^{-1}	$10^8 D_0$/cm^{-1}	$10^4 \alpha_3$/cm^{-1}	r_0(CC)/Å	r_0(CN)/Å
$^{12}C_2{}^{14}N_2$[a]	2157.849(5)	0.157124(12)	3.01(20)	−5.50(60)	1.389(30)	1.154(17)
$^{12}C_2{}^{15}N_2$[a]	2124.922(3)	0.14774(1)	2.61	4.94		
$^{12}C_2{}^{14}N_2$[b]	2157.8271(5)	0.157115(5)	3.0(3)	−5.355(6)	—	—

[a] Ref. 28. [b] Ref. 30.

Table 6 *Assignment of i.r. and Raman spectra* (cm^{-1}) *of liquid* C_2N_2

Assignment	*I.r.*[a]	*Raman*[b]
$\nu_2+\nu_3$	3000 vw	
$\nu_3+\nu_4$	2663 s	
$\nu_1+\nu_5$	2556 m	
ν_1		2328.5 vs
$\nu_1'(^{13}C)$		2298.4 w
ν_3	2158 vs	
$\nu_3'(^{13}C)$	2132 vw	
$\nu_1-\nu_5$	2089 vw	
$2\nu_4+\nu_5$	1262 vw	
$\nu_2+\nu_5$	1092 vw	
$2\nu_4$		1028.5 w
ν_2		850.6 m
$(\nu_2+\nu_5)-\nu_5$?		841 w
$\nu_4+\nu_5$	741 vs	750 vw
$\nu_2-\nu_5$	612 w	
$2\nu_4-\nu_5$		561.4 vw
$2\nu_4-\nu_4$		521.4 w
ν_4		507.2 s
$2\nu_5$		*ca.* 475 vw
ν_5		*ca.* 240 w

[a] Ref. 29. [b] Ref. 23.

systematic errors in the data which required the use of the experimental D_0 value. The correctness of this choice is demonstrated in a recent study by Bersellini and Meyer [30] of ν_3 and $\nu_3+\nu_5-\nu_5$ of $^{12}C_2{}^{14}N_2$ in the 2100—2200 cm^{-1} region with a resolution of 0.017 cm^{-1}. 119 vibration-rotation lines in the ν_3 band were measured and 79 lines in the $\nu_3+\nu_5-\nu_5$ band; other hot bands studied were $\nu_3+\nu_4-\nu_4$ and $\nu_3+2\nu_5-2\nu_5$. Values of B_0, D_0, and α_3 calculated from these results by Bersellini and Meyer are included in Table 5. Maki's preferred 'experimental' ^{14}N B_0 value is in better agreement with the more precise B_0 value of Bersellini and Meyer than is the alternative 'calculated' B_0 value. A value of B_0 [0.15752(15) cm^{-1}] obtained by Møller and Stoicheff [38] from a measurement of the pure rotational Raman spectrum of $^{12}C_2{}^{14}N_2$ is outside the error limits of the i.r. measurements. An explanation of this is afforded by the systematic overlapping of the rotational lines in the Raman spectrum from molecules in excited vibrational states, particularly the first excited state, which is only 233 cm^{-1} above the ground state. Since the molecules in these excited states will have smaller moments of inertia, the B_0 value given by the Raman measurements would be expected to be greater than the correct value. Recently, a preliminary report of an investigation of the pure rotational Raman spectra of $^{12}C_2{}^{14}N_2$ and $^{12}C_2{}^{15}N_2$, aimed at evaluating the C—C and C—N bond lengths with corrections for this complication, has been given in an article by Edwards.[39] At the moment, however, the best available cyanogen bond lengths derived by spectroscopic means are those obtained by Maki [28] (Bersellini and Meyer [30] did not study $^{12}C_2{}^{15}N_2$) and given in Table 5. These are in good agreement with the distances determined by electron diffraction [22, 23] and have about the same order of uncertainty ($\pm$0.02—0.03 Å). Because of insufficient data concerning $^{12}C_2{}^{15}N_2$ Maki was unable to estimate the B_e, and hence, the r_e values; however, it was observed that the B_e value for the ^{14}N species was some 0.00011 cm^{-1} *less* than the B_0 value, with an estimated uncertainty of $\pm$0.00007 cm^{-1}. If correct, this result is contrary to the usual situation where $B_e > B_0$, whence $r_e < r_0$. A final point worth noting is the discrepancy in the values of the anharmonicity constant x_{35} reported by Maki (-3.85 cm^{-1}) and by Bersellini and Meyer (-1.81 cm^{-1}).

The Liquid State.—There were two studies of the Raman spectrum of liquid cyanogen in the nineteen-thirties. As already explained, it has been inferred [33, 34] that the latest study [23] of the Raman spectrum of cyanogen was of the liquid phase, and the Raman lines observed and assignments made in this latest study are included in Table 6. The Raman lines at 521.4, 561.4, 841, and 2298.4 cm^{-1} were observed for the first time. The depolarization ratios of the fundamentals ν_1 and ν_4 were determined as 0.33 and 0.83

[38] C. K. Møller and B. P. Stoicheff, *Canad. J. Phys.*, 1954, **32,** 635.
[39] H. G. M. Edwards, in 'Essays in Structural Chemistry', ed. D. A. Long, A. J. Downs, and L. A. K. Staveley, Macmillan, London, 1971, p. 135.

respectively. (ν_2 was too weak for quantitative polarization measurements to be made.) It has been said [40] that ν_4 is more intense than ν_1 but this is obviously an error.[23, 32]

The i.r. spectrum of liquid cyanogen has recently been reported by Thomas and Orville-Thomas for the first time.[29] Twenty-seven bands were observed in the range 4000—250 cm^{-1} of which ten (see Table 6) were assigned. There is one possible coincidence between the Raman and i.r. spectrum but, in any event, the observation of the broad diffuse feature at *ca.* 240 cm^{-1} in the Raman spectrum and its assignment as ν_5 implies a relaxation in the liquid state of the selection rules which apparently operate in the gaseous and solid phases. Of the many bands not assigned in the i.r. spectrum, special note should be made of the very strong band at 791 cm^{-1} and the strong bands at 2339, 1713, and 766 cm^{-1}. Thomas and Orville-Thomas re-open the debate as to whether ν_5 has a value of ~750 cm^{-1} rather than the generall accepted ~230 cm^{-1}. In our view the weight of indirect evidence from gas-phase i.r. spectra points overwhelmingly to ν_5 at *ca.* 230 cm^{-1}, although the unassigned bands at 766 cm^{-1} and 791 cm^{-1} in the liquid-phase i.r. spectrum are a little disconcerting. Direct experimental evidence for the occurrence in the i.r. of the low-frequency band in the 200—250 cm^{-1} region for the gas, liquid, and solid states is certainly very desirable.

The Solid State.—Nixon and co-workers [26, 27a,b] have reported the i.r. spectra of $^{12}C_2{}^{14}N_2$ and $^{12}C_2{}^{15}N_2$ at 80 K in the region 4000—500 cm^{-1} (see Table 7), and the i.r. and Raman spectra of the lattice vibration region at 77 K in particular. A striking feature of ν_3 in the solid-state i r. spectrum is its weakness: its apparent integrated intensity is only about 5% of the integrated intensity reported [34] for ν_3 in the gas phase. In $^{12}C_2{}^{14}N_2$ and $^{12}C_2{}^{15}N_2$ ν_3 appears as two distinct peaks separated by 1.7 cm^{-1} in $^{12}C_2{}^{14}N_2$ and 1.8 cm^{-1} in $^{12}C_2{}^{15}N_2$ respectively. The lower-wavenumber peak is 2.4 and 2.1 times, respectively, more intense than the higher one. Initially, Nixon and co-workers, on the basis of a harmonic model with non-polarizable dipole–dipole coupling, interpreted the two peaks as b_{1u} (more intense) and b_{2u} factor-group components with a third and even less intense component (b_{3u}) lying either under b_{1u} or b_{2u} or elsewhere but undetected. (The site- and factor-group symmetries are C_i and D_{2h}.) Such a model, even including mutual polarization effects, does not, however, lead to a splitting as substantial as that observed. In order to obtain a calculated splitting of the right order and sense, Nixon and co-workers subsequently invoked a complex atom–atom potential function involving six constants. However, as the authors have themselves pointed out, the apparent inadequacy of the dipole–dipole model to account quantitatively for the observed bands casts some doubt upon the original assignment of the ν_3 components Some success has also been achieved in the interpretation of the frequencies and intensities of the lattice modes.

[40] Ref. 37, pp. 293, 294.

Table 7 *I.r. spectra* (cm^{-1}) *of solid* $^{12}C_2{}^{14}N_2$ *and* $^{12}C_2{}^{15}N_2$

Assignment	$^{12}C_2{}^{14}N_2$ [a]	$^{12}C_2{}^{14}N_2$ [b]	$^{12}C_2{}^{15}N_2$ [c]
$\nu_2+\nu_3$		3012.9 w	2959.6 w
$\nu_3+\nu_4$	2668 s	2673.4 m 2670.1 m	2638.0 m 2634.7 m
$(\nu_3+\nu_4)'(^{15}N)$			2651.4 vw
$(\nu_3+\nu_4)'(^{13}C)$		2635 vw	
$\nu_1+\nu_5$	2581 m 2569 m	2588.7 m 2573.8 m	2562.9 m 2548.6 m
ν_3	2163 s	2166.6 m 2164.9 m	2133.5 m 2131.7 m
$\nu_3'(^{15}N)$		2147.6 vw	2147.3 w
$\nu_3'(^{13}C)$		2138.7 vw	2105.4 vw
	1714 m		
	1387 vw		
	1367 w		
$2\nu_4+\nu_5$	1261 vw	1265.6 vw 1255.6 vw	
	1226 w		
	1184 vw		
$\nu_2+\nu_5$	1088 vw	1088.0 vw	1050.2 w
			800.7 w
	791 w		
$\nu_4+\nu_5$	754 m 744 m	755.5 s 745.6 vs	750.8 s 740.3 vs
$(\nu_4+\nu_5)'(^{13}C)$		735 w	737 sh

[a] Ref. 29. [b] Ref. 26. [c] Ref. 27[a].

Thomas and Orville-Thomas, who earlier reported the i.r. spectrum of liquid cyanogen, have recently reported [29] the spectrum of solid cyanogen at an unspecified temperature (see Table 7) This later work on solid cyanogen is less precise than, and published without reference to, the earlier work of Nixon *et al.*; moreover, whereas many features established in the earlier investigation are not present in the later one, six new features are observed in the later work but not assigned. Acceptance of the data and assignments of Nixon and co-workers leads to the deduction of the following *approximate* values of the unobserved fundamentals of solid $^{12}C_2N_2$: $\nu_1 = 2346$, $\nu_2 = 847$, $\nu_4 = 509$, and $\nu_5 = 246$ and 236 cm^{-1}.

5 Hydrogen Cyanide

Hydrogen cyanide boils at 298.8 K and freezes at 259.8 K. The vibrational and vibration–rotation i.r. spectra of the gaseous molecule have been intensively investigated at high resolution by several groups of workers.[41–56]

[41] K. N. Choi and E. F. Barker, *Phys. Rev.*, 1932, **42**, 780.
[42] D. H. Rank, R. P. Ruth, and K. L. Van der Sluis, *J. Opt. Soc. Amer.*, 1952, **42**, 693.

The vibration–rotation spectra, in particular, established many years ago that the molecule was linear in the gas phase. Recent high-resolution studies have centred on the obtaining of increasingly accurate data, especially of isotopically substituted species.[57–59] In addition, there have been a number of other recent spectroscopic studies. These include studies of the matrix-isolated species HCN, HNC, and $(HCN)_2$ at low temperatures,[19, 60, 61] solid-state intensities,[62, 63] and the Raman[64] and i.r.[65] spectra of the liquid and solid. The Raman spectrum of the gas has not been studied for many years—a further example of the relative paucity of the Raman studies of this class of compounds.

The Gaseous State.—Nakagawa and Morino[66] have recently made exhaustive calculations on the vibration and rotation parameters of HCN. They made a least-squares analysis of 124 observed values of vibration frequencies, ν (fundamentals, overtones, and combination bands), and rotational constants, B_v, for the six isotopic species $H^{12}C^{14}N$, $D^{12}C^{14}N$, $H^{13}C^{14}N$, $D^{13}C^{14}N$, $H^{12}C^{15}N$, and $D^{12}C^{15}N$. The observed fundamental frequencies for the six isotopic species used in these calculations are included in Table 8. Not all the ν (and ΔB) values were directly observed: some, *e.g.* those of 002^2, 102^2, and 004^2 of $H^{12}C^{14}N$, were obtained from hot bands and the appropriate ground-state transitions by application of the Ritz combination

---

[43] D. H. Rank, H. D. Rix, and R. E. Kagarise, *J. Chem. Phys.*, 1952, **20**, 1437.
[44] A. E. Douglas and D. Sharma, *J. Chem. Phys.*, 1953, **21**, 448.
[45] P. B. Checkland and H. W. Thompson, *Trans. Faraday Soc.*, 1955, **51**, 1.
[46] D. H. Rank, T. A. Wiggins, A. H. Guenther, and J. N. Shearer, *J. Opt. Soc. Amer.*, 1956, **46**, 953.
[47] H. C. Allen, jun., E. D. Tidwell, and E. K. Plyler, *J. Chem. Phys.*, 1956, **25**, 302.
[48] P. A. Staats, H. W. Morgan, and J. H. Goldstein, *J. Chem. Phys.*, 1956, **25**, 582.
[49] D. H. Rank, A. H. Guenther, J. N. Shearer, and T. A. Wiggins, *J. Opt. Soc. Amer.*, 1957, **47**, 148.
[50] D. H. Rank, G. Skorinko, D. P. Eastman, and T. A. Wiggins, *J. Opt. Soc. Amer.*, 1960 **50**, 421.
[51] W. W. Brim, J. M. Hoffmann, H. H. Nielsen, and K. N. Rao, *J. Opt. Soc. Amer.*, 1960, **50**, 1208.
[52] D. H. Rank, G. Skorinko, D. P. Eastman, and T. A. Wiggins, *J. Mol. Spectroscopy*, 1960, **4**, 518.
[53] D. H. Rank, D. P. Eastman, B. S. Rao, and T. A. Wiggins, *J. Opt. Soc. Amer.*, 1961, **51**, 929.
[54] A. G. Maki and L. R. Blaine, *J. Mol. Spectroscopy*, 1964, **12**, 45.
[55] A. G. Maki, E. K. Plyler, and R. Thibault, *J. Opt. Soc. Amer.*, 1964, **54**, 869.
[56] A. G. Maki and D. R. Lide, *J. Chem. Phys.*, 1967, **47**, 3206.
[57] A. G. Maki, W. B. Olson, and R. L. Sams, *J. Mol. Spectroscopy*, 1970, **36**, 433.
[58] B. D. Alpert, A. W. Mantz, and K. N. Rao, *J. Mol. Spectroscopy*, 1971, **39**, 159.
[59] P. K. L. Yin and N. Rao, *J. Mol. Spectroscopy*, 1972, **42**, 385.
[60] C. M. King and E. R. Nixon, *J. Chem. Phys.*, 1968, **48**, 1685.
[61] J. Pacansky and G. V. Calder, *J. Phys. Chem.*, 1972, **76**, 454.
[62] P. F. Krause and H. B. Friedrich, *J. Phys. Chem.*, 1972, **76**, 1140.
[63] M. Uyemura and S. Maeda, *Bull. Chem. Soc. Japan*, 1972, **45**, 1081.
[64] M. Pézolet and R. Savoie, *Canad. J. Chem.*, 1969, **47**, 3041.
[65] M. Pézolet and R. Savoie, *Canad. J. Spectroscopy*, 1972, **17**, 39.
[66] T. Nakagawa and Y. Morino, *Bull. Chem. Soc. Japan*, 1969, **42**, 2212.

Table 8 *Anharmonic fundamental vibrations* (cm^{-1}) *of hydrogen cyanide*[a]

Molecule	ν_1	ν_2	ν_3	*Ref.*
$H^{12}C^{14}N$		3311.473[b, e, f]	712.26[b, e, g]	50
			713.74[c, e, g]	50
		3311.476[b, e]		53
	2096.85(12)[b, f]			54
	2096.839(2)[b]			56
			711.982[b, e, f]	51, 52
			713.459(1)[c]	59
$D^{12}C^{14}N$	*1925.265(2)*[b, f]	*2630.303(2)*[b, f]	*569.039(11)*[b, f, h]	55
$T^{12}C^{14}N$	*1724(1)*[d]	*2460(1)*[d]	*513(1)*[d]	48
$H^{13}C^{14}N$	*2062.268*[b, e, h]	3293.506[b, e, f]	707.393[c, e, f]	50
		3293.515(2)[b]		55
			707.41(1)[c]	59
$D^{13}C^{14}N$	*1911.840(4)*[b, f]	*2590.073(4)*[b, f]		55
			[561.60][b, e]	66
$H^{12}C^{15}N$		3310.092(2)[b, f]		55
	2061.345(4)[b]	*3310.088(1)*[b]	*711.025(5)*[b]	58
$D^{12}C^{15}N$		2621.100(3)[b, f]		55
	1900.110(2)[b]	*2621.182(1)*[b]		58
			[568.06][b, e]	66
$H^{13}C^{15}N$		*3292.289(1)*[b]		58

[a] Best values in italic type; calculated values [66] in square brackets. [b] ν_0 value. [c] Unperturbed ν_0 value, *i.e.* corrected for *l*-type doubling. [d] ν_{obs} value and assumed to be not corrected for vacuum. [e] Uncertainty not stated. [f] Used in force-constant calculation of Nakagawa and Morino.[66] [g] Re-analysis of data of Choi and Barker.[41] [h] Not directly observed.

principle. Four of the B (000^0) values were determined by microwave spectroscopy (^{14}N species) [67] and two by i.r. spectroscopy (^{15}N species); [55] all the 51 ΔB values used appear to have been established by i.r. spectroscopy. Nakagawa and Morino obtained values of the harmonic fundamental frequencies, ω_i, the quadratic, cubic, and quartic force constants, the vibrational anharmonicity constants x_{ij}, the vibration–rotation constants α_i, and the Coriolis coupling constants, ζ_{ij}^{x}.

The harmonic fundamental frequencies, ω_i, and the vibrational anharmonicity constants, x_{ij}, of the six isotopic species are listed in Table 9. The quadratic force constants of hydrogen cyanide are given in Table 10. The values listed in Tables 9 and 10 were not considered in a recent discussion of the force constants of hydrogen cyanide.[68] An unfortunate inconsistency in these calculations was the use of values of ν_3 corrected for *l*-type doubling in $H^{13}C^{14}N$ but uncorrected in $H^{12}C^{14}N$ and $D^{12}C^{14}N$; the correction is of the order of 1.4 cm^{-1} in $H^{12}C^{14}N$ and will be of similar magnitude in $H^{13}C^{14}N$. It is probably not without significance that by far the largest difference between a calculated fundamental frequency and an observed frequency considered by Nakagawa and Morino was for ν_3 of $H^{13}C^{14}N$,

[67] J. W. Simmons, W. E. Anderson, and W. Gordy, *Phys. Rev.*, 1950, **77**, 77; 1952, **86**, 1055; C. A. Burrus and W. Gordy, *ibid.*, 1956, **101**, 599.
[68] Ref. 1 (*b*), p. 69.

Table 9 *Harmonic fundamental vibrations and vibrational anharmonicity constants* (cm^{-1}) *of hydrogen cyanide* [66]

	$H^{12}C^{14}N$[a]	$H^{13}C^{14}N$	$H^{12}C^{15}N$	$D^{12}C^{14}N$[a]	$D^{13}C^{14}N$	$D^{12}C^{15}N$
ω_1	2128.67 ±0.80	2094.16	2095.12	1952.12 ±0.63	1938.52	1925.83
ω_2	727.10 ±0.51	720.71	726.01	579.85 ±0.41	571.82	578.49
ω_3	3441.16 ±0.44	3422.04	3439.85	2703.34 ±0.72	2661.58	2794.30
x_{11}	−10.45 ±0.38	−10.27	−10.13	−6.84 ±0.29	−6.80	−6.69
x_{12}	−3.61 ±0.22	−4.13	−3.46	3.01 ±0.20	2.03	3.11
x_{13}	−14.61 ±0.22	−12.90	−14.08	−32.40 ±0.26	−30.29	−30.90
x_{22}	−2.44 ±0.17	−2.31	−2.46	−2.08 ±0.11	−1.91	−2.10
x_{23}	−18.98 ±0.21	−18.38	−18.91	−15.80 ±0.24	−14.91	−15.67
x_{33}	−51.71 ±0.21	−51.87	−51.88	−20.50 ±0.36	−20.74	−20.98
x_{ll}	5.35 ±0.09	5.24	5.40	3.29 ±0.64	3.13	3.33

[a] Standard error, $\pm\sigma$, calculated by taking into account the correlations among the force constants.

where the calculated value was found to be 1.05 cm^{-1} *smaller* than the observed value.

Table 8, as well as listing the observed fundamental frequencies used by Nakagawa and Morino, also includes other observed values mostly determined since their calculations were made. It can be seen that although there has been further progress in the establishment of accurate values of ν_1, ν_2, and ν_3, in no case has there been a gross change in a value used by Nakagawa and Morino in their force-constant calculation; this applies equally well to overtones and combination tones. The experimental values of ν_1 for $H^{12}C^{15}N$ and $D^{12}C^{15}N$ and ν_3 for $H^{12}C^{15}N$ made available by this later work are compared with the calculated values obtained by Nakagawa and Morino in Table 11. This Table also compares the calculated value of ν_1 of $H^{13}C^{14}N$ with an (indirectly determined) experimental value which, although available at the time the calculation was made, was not used by Nakagawa and Morino.

Table 10 *Quadratic force constants of hydrogen cyanide* [66]

Force constant	*Value*
k(CN)	1870.7(1.6)^a
k(HC)	624.4(0.3)^a
k(HC,CN)	−21.1(0.6)^a
k_δ	0.2598(4)^b

^{a} N m^{-1}. ^{b} 10^{-18} N m rad^{-2}.

An assessment of the accuracy of Nakawaga and Morino's calculated values of the fundamental vibrations is of considerable importance, particularly since it is these values rather than the experimental values that have been used [61] to assess the analogous experimental values of the matrix-isolated species and to conclude that the linear *x–y–z* vibrational model does not describe the vibrations of matrix-isolated HCN and DCN at all adequately (see p. 496).

It can be seen from Table 11 that the agreement between the observed and calculated fundamental values is very good for ν_1 of $D^{12}C^{15}N$, rather less so for ν_1 of $H^{13}C^{14}N$ and ν_3 of $H^{12}C^{15}N$, and completely unacceptable for ν_1 of $H^{12}C^{15}N$.

Referring now to Table 8, it should be noted that ν_1 for $H^{12}C^{15}N$ is just smaller than ν_1 for $H^{13}C^{14}N$ and ν_1 for $D^{12}C^{15}N$ is much smaller than ν_1 for $D^{13}C^{14}N$. Point-mass calculations (*i.e.* calculations assuming that hydrogen–carbon is a point mass in HCN and DCN) do not predict $\nu(^{12}C^{15}N) < \nu(^{13}C^{14}N)$ in either HCN or DCN. In fact, they predict identical frequencies for the two CN stretching vibrations in DCN. Such calculations neglect anharmonicity and coupling, and it would certainly be more appropriate to compare experimentally derived values* of the harmonic

* The calculated harmonic values of Nakagawa and Morino suggest that anharmonicity has little effect on this situation.

Table 11 *Experimental and calculated values of certain fundamental vibrations* (cm^{-1}) *of* $H^{13}C^{14}N$, $H^{12}C^{15}N$, *and* $D^{12}C^{15}N$

Species	*Fundamental*	*Experimental* ν_0	*Calculated* ν_0
$H^{13}C^{14}N$	ν_1	2062.268[a, b]	2063.05[c]
$H^{12}C^{15}N$	ν_1	2061.345(4)[d]	2064.35[c]
$H^{12}C^{15}N$	ν_3	711.025(5)[d]	711.41[c]
$D^{12}C^{15}N$	ν_1	1900.110(2)[d]	1900.12[c]

[a] Ref. 50. [b] Uncertainty not stated. [c] Ref. 66. [d] Ref. 58.

fundamentals. However, these values are not available for the ^{13}C and ^{15}N species; moreover there is some disagreement about these values for HCN and DCN. In fact, Suzuki *et al.*[69] found that, even when the best values of experimentally derived harmonic fundamentals, anharmonicity constants, and rotational constants of HCN and DCN are assumed, the harmonic fundamentals do not satisfy the theoretical isotope product-rule ratios. Satisfactory force constants cannot be calculated unless the dispersions and correlations of the experimental data are accounted for by taking the weight matrix to be the variance–covariance matrix obtained when the harmonic fundamentals and anharmonicity constants are determined by least squares. (For convenience these force constants will be designated as Suzuki Set A.) Another set of almost identical force constants (Suzuki Set B) could be more simply calculated by adjustment of the force constants to the observed ν and B_v values. The close agreement between Sets A and B, and the lack of experimental anharmonicity constants for the ^{13}C and ^{15}N species, is the essential justification for the method used by Nakagawa and Morino to obtain the values listed in Tables 9 and 10.

The principal improvement represented by the Nakagawa and Morino force constants over those of Suzuki *et al.* (Set A) is the greater precision of the former, though in the cubic and quartic constants there are also a few substantial changes. Both sets of results give a negative interaction constant k(HC,CN) [incidentally of a similar order of magnitude to k(CC,CN) of cyanogen], whereas the interaction displacement co-ordinates suggest a positive interaction.[68]

Before considering the application of the force constants of Nakagawa and Morino in the calculation of other molecular constants, it is worth noting that recent i.r. studies now allow a comparison between experimental ΔB values and 8 of the 22 calculated ΔB values of transitions for which there were no experimental values available at the time of the calculations (see Table 12). In absolute terms, it can be seen that the degree of agreement is very satisfactory for four of the transitions but rather less so for the remainder ($2\nu_3^2$ and $4\nu_3^0$ of $H^{12}C^{14}N$, ν_1 of $H^{12}C^{15}N$, and ν_3^1 of $D^{12}C^{15}N$).

Nakagawa and Morino[66] also calculated the vibration–rotation interaction constants, α_i, and the Coriolis coupling constants, ζ_{ij}^x, of the six

[69] I. Suzuki, M. A. Pariseau, and J. Overend, *J. Chem. Phys.*, 1966, **44**, 3561.

isotopic species. Their α values were uncorrected for γ terms and are thus merely the corresponding $-\Delta B_{calc}$ values.

In a subsequent paper, Nakagawa and Morino [70] calculated the v- and J-dependences of the centrifugal distortion constants, D_v, and the l-type

Table 12 *Experimental and calculated values of* ΔB/cm^{-1} *for certain transitions of* $H^{12}C^{14}N$, $H^{12}C^{15}N$, *and* $D^{12}C^{15}N$

Molecule	*Transition*	ΔB_{calc} [a]	B_{exp}	*Ref.*
$H^{12}C^{14}N$	200^0—000^0	−0.020135	−0.02019(16)	57
$H^{12}C^{14}N$	002^2—000^0	+0.007385	+0.006792[b]	57
$H^{12}C^{14}N$	004^0—000^0	+0.014770	+0.015636(12)	57
$H^{12}C^{14}N$	103^1—000^0	+0.001010	+0.000886(93)[c]	54, 57
$H^{12}C^{15}N$	100^0—000^0	−0.009655	−0.011471(76)	58
$H^{12}C^{15}N$	001^1—000^0	+0.003541	+0.003382(32)[d]	58
$D^{12}C^{15}N$	100^0—000^0	−0.006290	−0.006327(12)	58
$D^{12}C^{15}N$	001^1—000^0	+0.004236	+0.003733(148)[e]	58

[a] Ref. 66. [b] Calculated from $D(002^0)$ 1.485014 cm^{-1} (see section on $H^{12}C^{14}N$) and $B(000^0) = 1.47822162(1)$ cm^{-1}. Uncertainty probably ± 0.000060 cm^{-1}. [c] Calculated from $\Delta B(103^1—002^0) = —0.006712(36)$ cm^{-1} and $\Delta B(002^0—000^0) = +0.007598(57)$ cm^{-1}. [d] Calculated from ΔB values of 010^0—000^0, 010^0—001^1, 011^1—000^0, and 011^1—001^1; see section on $H^{12}C^{15}N$. [e] Calculated from $B(001^1) = 1.176920(112)$ cm^{-1} and $B(000^0) = 1.173187(36)$ cm^{-1}.

Table 13 *Calculated and experimental centrifugal distortion constants of hydrogen cyanide* (10^{-8} cm^{-1})

Constant	$H^{12}C^{14}N$	$H^{13}C^{14}N$	$H^{12}C^{15}N$	$D^{12}C^{14}N$	$D^{13}C^{14}N$	$D^{12}C^{15}N$
(calculated)[a]						
D_e [b]	284.77(20)[c]	271.39	268.91	186.06(12)	179.47	175.38
β_1 [b]	+0.72(40)	+0.69	+0.68	−0.11(24)	−0.05	−0.07
β_2 [b]	−2.76(09)	−2.58	−2.63	−0.90(06)	−0.90	−0.90
β_3 [b]	+7.11(12)	+6.31	+6.59	+7.01	+6.37	+6.52
D_0 [b]	290.87(09)	276.75	274.53	192.57	185.37	181.41
$H \times 10^4$ [b]	−2.61	−2.49		−0.85		
(experimental)						
D_0	291.29[d]	275(2)[e]	254(15)[e]	191.4[f]	183(13)[e]	130(20)[e]
$H \times 10^4$	5.73[d]					
(new experimental)						
D_0	291.0(2)[g] 291.00(4)[h]		275(5)[i]	192.9(1)[g]		182(3)[i]
$H \times 10^4$	3.3(0.8)[h]					
β_1 [j,k]	−0.9(1.0)			−1.4(2.0)		
β_2 [j,k]	−2.7(1.0)			−1.7(2.0)		
β_3 [j,k]	7.4(1.0)			10.8(2.0)		
D_e [j,k]	285.3			183.7		

[a] Ref. 70. [b] Constants defined by $D_v = D_e + \sum_i \beta_i (v_i + d_i/2) - H[J(J+1) - l^2]$. [c] Standard error $\pm\sigma$, estimated by taking into account the correlation among the force constants. [d] Ref. 53. [e] Ref. 55. [f] Ref. 67. [g] Ref. 71. [h] Ref. 57. [i] Ref. 58. [j] Ref. 72. [k] Constants in effect defined by $D_e = D_0 + k_1\beta_1 + k_2\beta_2 + k_3\beta_3$, where $D_0(H^{12}C^{14}N) = 290.9(1)$, $D_0(D^{12}C^{14}N) = 192.9(1)$, k_1 and $k_2 = -\frac{1}{2}$, and $k_3 = -1$.

[70] T. Nakagawa and Y. Morino, *J. Mol. Spectroscopy*, 1969, **31**, 208.

Table 14 *Calculated and experimental-type doubling constants*/10^{-3} cm^{-1} *of hydrogen cyanide*

Constant[a]	$H^{12}C^{14}N$	$H^{13}C^{14}N$	$H^{12}C^{15}N$	$D^{12}C^{14}N$	$D^{13}C^{14}N$	$D^{12}C^{15}N$
(calculated)[b]						
q_e	7.246(3)[c]	6.938	6.841	6.049(3)[c]	5.925	5.720
q_{v1}	−0.001(2)[c]	−0.008	+0.001	+0.151(3)[c]	+0.150	+0.125
q_{v2}	+0.070(2)[c]	+0.074	+0.068	−0.147(4)[c]	−0.145	−0.118
q_{v3}	+0.110(2)[c]	+0.094	+0.093	+0.075(1)[c]	+0.072	+0.071
$q(001^1)$	7.477(3)[c]	7.159	7.062	6.201(3)[c]	6.072	5.864
$q_J \times 10^5$	−7.956(30)[c]	−7.287	−7.260	−6.691(30)[c]	−6.326	−6.116
(observed)						
$q(001^1)$	7.487732[d]	7.1662[e]	7.0695[e]	6.21082[d]	6.0807[e]	5.8734[e]
$q_J(001^1)$	−8.866[d]	−8.14[e]	−8.07[e]	−7.362[d]	−6.94[e]	−6.74[e]

[a] Constants defined by $q_v = q_e + \sum_i q_{vi}(v_i + d_i/2) + q_J[J(J+1) - l^2]$. [b] Ref. 70. [c] Standard error, $\pm\sigma$, estimated by taking into account the correlations among the force constants. [d] Ref. 56. [e] Ref. 73.

doubling constants, q_v, of the six isotopic species using (*a*) their force constants and (*b*) the Set A force constants of Suzuki *et al.*[69] The two sets of calculated data were found to be nearly identical. The centrifugal distortion constants calculated from Nakagawa and Morino's force constants are listed in Table 13. This also contains experimental D_0 and H values, both those available at the time of Nakagawa and Morino's calculations and those subsequently determined. The v-dependences of the centrifugal distortion constants for hydrogen cyanide are seen to be as small as 0.3—3% of the D_e value; and the corrected D_0 values agree well with the observed i.r. values for $H^{12}C^{14}N$, $H^{13}C^{14}N$, and $D^{13}C^{14}N$ and the microwave value for $D^{12}C^{14}N$. Nakagawa and Morino suggested that the calculated values for $H^{12}C^{15}N$ and $D^{12}C^{15}N$ might be more accurate than the observed i.r. values. This cautious conclusion has been amply borne out by the recent work of Alpert, Mantz, and Rao,[58] who established new and much more precise experimental values which are in very close agreement with the calculated values. Alpert *et al.*[58] incidentally make no comment either on the previous experimental values or on those calculated by Nakagawa and Morino. The merit of the calculated values, and by implication the force constants, is further reinforced by the fact that the new experimental microwave values [71] for $H^{12}C^{14}N$ and $D^{12}C^{14}N$ and the combined microwave–i.r. value [57] for $H^{12}C^{14}N$ are significantly closer to the calculated values than the previous experimental values; moreover, most of the discrepancy between the calculated and recently observed β_i values derived from an analysis of i.r. ΔD data [72] can probably be accounted for by experimental error. On the other hand, the calculated value of the H constant, that is the J^6 term of the rotational energy, for $H^{12}C^{14}N$ is negative, whereas the experimental values are positive (see Table 13).

[71] F. DeLucia and W. Gordy, *Phys. Rev.*, 1969, **187**, 58.
[72] G. Winnewisser, A. G. Maki, and D. R. Johnson, *J. Mol. Spectroscopy*, 1971, **39**, 149.
[73] T. Törring, *Z. Physik*, 1961, **161**, 179.

Nakagawa and Morino's calculated *l*-type doubling constants are listed in Table 14, where they are compared with the experimental microwave values.[56, 73] It is worth noting that the *l*-type splitting is very large for hydrogen cyanide and is easily detectable in the i.r. spectrum. Nevertheless, microwave data are much more accurate. The calculated values of $q(001^1)$, however, agree well with the observed values; thus for $H^{12}C^{14}N$ the deviation, 0.0082×10^{-3} cm^{-1}, is only 3.5% of the correction term $q(001^1) - q_e = 0.2339 \times 10^{-3}$ cm^{-1}, and is about three times the standard error σ estimated by taking into account the correlation among the force constants. The calculations also suggest remarkable variations of v-dependence of q_v between HCN and DCN; thus, q_{v_1} is much larger for DCN than for HCN, and q_{v_2} is positive for HCN and negative for DCN. The calculated q_J values are smaller by about 10% than the experimental $q_J(001^1)$ values for all the six isotopic species; this may be due to the neglect of the higher-order v-dependence of the q_J constant in the calculation.

We now review briefly in turn each isotopic species for which vibration and/or vibration–rotation data have been reported.

$H^{12}C^{14}N$. For this species Nakagawa and Morino[66] were able to utilize 53 experimental values of ν and ΔB of 31 ground-state transitions in addition to the microwave $B(000^0)$ value. (The data for ten transitions were not directly observed but were derived *via* the Ritz combination principle, usually involving the 001^1 state.) Experimental ΔB values were unavailable for nine transitions. Since the publication of these calculations, there have been new i.r. and microwave studies.

A further stimulus to the study of the i.r. spectrum of gaseous hydrogen cyanide, and $H^{12}C^{14}N$ in particular, is that the vibrational and rotational constants may help to explain the HCN laser transitions,[74] many of which are as yet unassigned.* To this end Maki, Olson, and Sams[57] have

* It is now established[75] that the stimulated emission at 337 μm and the five other emission lines between 284 and 373 μm are associated with Coriolis perturbation between 101^1 and 004^0 near $J = 10$ of $H^{12}C^{14}N$. To explain the perturbation, Maki and Blaine[54] pointed out that whereas the band centre of $4\nu_3{}^0$ is 2.7 cm^{-1} below that of $\nu_1 + \nu_3{}^{1c}$ (101^1 is split by *l*-type doubling), the B value for $4\nu_3{}^0$ is greater, and eventually corresponding rotational levels must cross. This crossing occurs at $J = 10$ and Coriolis perturbation occurs because the two levels have the same J, the right symmetry properties, and virtually the same energy. A set of emission lines near 130 μm has been attributed[76] to transitions involving the 102^0, 102^2, and 005^1 states. The resonance responsible for the laser action is of the same type as that for the lower-frequency emissions but, because of the slight differences in anharmonicity, the crossing of levels occurs further up at $J = 26$. There is another group of emission lines between 200 and 223 μm. It is not surprising that the delicate balance of the vibrational and rotational parameters which ensures so many resonances for $H^{12}C^{14}N$ is destroyed by isotopic substitution. Indeed, it was the failure to observe emission at 337 μm from $H^{12}C^{15}N$ which indicated that resonance between vibrationally excited states was important in the laser mechanism. Likewise, with $D^{12}C^{14}N$ very few near coincidences occur; the only one significant for laser action[76] is that between 202^0 and 009^1, which cross near $J = 21$.

[74] G. W. Chantry, 'Submillimetre Spectroscopy', Academic Press, London, 1971.

[75] L. O. Hocker, A. Javan, D. R. Rao, L. Frenkel, and T. Sullivan, *Appl. Phys. Letters*, 1967, **10**, 147; L. O. Hocke and A. Javan, *Phys. Letters*, 1967, **A25**, 489; D. R. Lide and A. G. Maki, *Appl. Phys. Letters*, 1967, **11**, 62.

re-determined the rotational constants of the 000^0, 001^1, and 002^2 states and have made new i.r. measurements on the transitions 101^1—000^0, 004^0—000^0, 102^0—001^1, 102^2—001^1, 005^1—001^1, 200^0—000^0, 103^1—002^0, and 103^3—002^2. (The 004^0—000^0, 005^1—001^1, 103^1—002^0, and 103^3—002^2 transitions do not appear to have been measured before.)* The measurements of the first two transitions were combined with the precise frequency measurements of five of the six HCN laser transitions known to involve the 101^1 and 004^0 states,[75] and the measurements of the next three i.r. transitions with the less precise frequency measurements of the four HCN laser transitions involving the 102^0, 102^2, and 005^1 states.[76]

Previous microwave measurements[67, 71] of the frequencies of the three transitions $\nu(J = 1 \leftarrow 0)$, $\nu(J = 2 \leftarrow 1)$, and $\nu(J = 3 \leftarrow 2)$ and the new i.r. data on the 101^1—000^0 transition were combined with the i.r. data of Rank *et al.*[53] on the 010^0—000^0 and 020^0—000^0 transitions in a least-squares analysis to yield[57] a new value of $B(000^0)$; the value obtained is identical in frequency with, but apparently more precise than, the value obtained[71] from the fitting of eight line frequencies of the two transitions $J = 1 \leftarrow 0$ and $J = 2 \leftarrow 1$. Both values [44315975.7(1 and 4 respectively) kHz] are slightly smaller and at least two orders of magnitude more precise than the previous microwave value,[67] which was based on the $J = 1 \leftarrow 0$ transition only.† Best i.r. ΔB values were also given[57] for the transitions 001^1—000^0 and 002^2—000^0. (The latter appears to have been derived from the sum of the ΔB values of 002^0—000^0 and 002^2—002^0.) These data and the ΔB data for the newly measured transitions allow a comparison between calculated and subsequently measured experimental values for four of the nine transitions for which Nakagawa and Morino did not have experimental values (see Table 12).

Winnewisser, Maki, and Johnson[72] have recently determined microwave values of B for 100^0, 001^1, and 002^0. These new microwave data allow a check on the i.r. ΔB values of ν_1, $\nu_3{}^1$, and $2\nu_3{}^0$. In each case the i.r. value is numerically slightly smaller than the microwave value but the difference is not significant and the degree of agreement is highly satisfactory. (See Table 15.)

* Winnewisser, Maki, and Johnson[72] subsequently determined the rotational constants of the 002^0 and also 001^1 states by microwave spectroscopy (see p. 472).

† Attempts have been made to obtain the speed of light by comparison of the $B(000^0)$ value of $H^{12}C^{14}N$, measured in wavelength units by i.r. spectroscopy, with the microwave $B(000^0)$ value measured in frequency units (similarly with two of the sub-millimetre $H^{12}C^{14}N$ laser transitions). The microwave $B(000^0)$ value reported above is now about two orders of magnitude more accurate than the i.r. wavelength measurement, and is considerably more accurate than the presently accepted error in the value for the speed of light (1 in 3×10^6). A significant improvement in the i.r. wavelength measurement is needed to provide a better value for the speed of light.[71] It has been noted[74] that if it could be achieved to 1 part in 10^8 [*i.e.* the precision of the microwave value of $B(000^0)$ of $H^{12}C^{14}N$] it would be unnecessary to maintain separate standards of length and time.

[76] A. G. Maki, *Appl. Phys. Letters*, 1968, **12**, 122.

Table 15 *Rovibrational constants* (cm^{-1}) *for* $H^{12}C^{14}N$

Constant	ν_1	ν_2	$\nu_3{}^1$	$2\nu_3{}^0$	*Technique*
$\Delta B_{obs}{}^a$	−0.010081(4)		+0.003551(1)	+0.007607(2)	microwave
ΔB_{obs}	−0.010072(24)[b]	−0.010421(9)[c]	+0.003546(90)[d]	+0.007598(57)[e, f]	i.r.
			+0.003534(7)[g]		
$\Delta B_{calc} = -\alpha_i{}^h$	−0.010068(68)	−0.010418(43)	+0.003693(47)	+0.007385	
$\Delta B_{calc}{}^a$	−0.010071	−0.010423	+0.0035514	+0.0076091	
$-\alpha_i{}^{a,i}$	−0.009975(19)	−0.010446(9)	+0.0035616(25)		

[a] Ref. 72. [b] Ref. 56. [c] Ref. 53. [d] Ref. 57. [e] Ref. 54. [f] Ref. 51. [g] Ref. 59. [h] Ref. 66. [i] Corrected for higher-order terms.

Table 16 *Observed and calculated quadratic vibrational anharmonicity constants* (cm^{-1}) *of* $H^{12}C^{14}N$

x_{ij}	*Obs.*[a]	*Obs.*[b]	*Obs.*[c]	*Obs.*[d]	*Calc.*[e]	*Calc.*[f]
x_{11}	−9.0	−10.45	−7.0741		−11.16	−10.45(38)
x_{22}	−54.20	−52.20	−52.4901		−50.30	−51.71(21)
x_{33}	−2.47	−2.50	−2.6533		−2.39	−2.44(17)
x_{12}	−16.8	−14.43	−10.4434		−15.87	−14.61(22)
x_{13}	−2.15	−2.90	−2.5265		−3.24	−3.61(22)
x_{23}	−19.6	−19.19	−19.0055		−18.98	−18.98(21)
x_{ll}	3.63	3.63	5.160	5.252	5.06	5.35(09)

[a] Ref. 44. [b] Ref. 47. [c] Ref. 50. [d] Ref. 55. [e] Ref. 70; calculated from Set A force constants. [f] Ref. 66; uncertainties are the standard error, $\pm\sigma$, calculated by taking into account the correlations among the force constants.

Winnewisser *et al.*,[72] using essentially the same i.r. ΔB data as Nakagawa and Morino but also other i.r. data, evaluated the rotation–vibration constants in a least-squares analysis. The i.r. data related to seven ground-state transitions and 12 hot bands, six of which had been obtained in the recent work of Maki, Olson, and Sams.[57] The resulting calculated ΔB values for the three fundamental transitions and the corresponding α values (corrected for the effect of higher-order terms) are listed in Table 15, where they are compared with analogous calculated data obtained by Nakagawa and Morino and the relevant experimental data. It is clear that the least-squares analysis carried out by Winnewisser *et al.* represents a significant improvement over that of Nakagawa and Morino. In particular the α_i and γ_{ij} values allow the calculation of a B_e value [44512.36(23) MHz] from the $B(000^0)$ value [44315.9755(4) MHz].

In a very recent i.r. study ν_0 and ΔB values were reported for the 001^1—000^0 band and for the two difference bands 002^0—001^1, and 002^2—001^1. The ν_0 value of ν_3^1, which was corrected for *l*-type doubling, is the best value available for this fundamental; the ΔB value is much more precise than the recent value obtained by Maki, Olson, and Sams,[57] and in contrast appears slightly but significantly smaller than the most recent microwave value (see Table 15). Not surprisingly, the value of the *l*-type doubling constant $[q(001^1)]$ reported $[7.486(7) \times 10^{-3}\ \text{cm}^{-1}]$ [59] is much less precise than the corresponding microwave value (see Table 14). It should be noted that $\nu_0(002^2—001^1) - \nu_0(002^0—001^1)$ is reported to be $21.056(5)\ \text{cm}^{-1}$ whereas $\nu_0(002^2—000^0) - \nu_0(002^0—000^0)$ is $15.121(5)\ \text{cm}^{-1}$. On the other hand the ΔB values reported for the two difference bands are consistent with those calculable for these two transitions from known ΔB values of other transitions.

In the past forty years the anharmonicity constants of $H^{12}C^{14}N$ have been the subject of increasing investigation and no small controversy, and even now it cannot be said that the situation is completely free from doubt. In 1952 Douglas and Sharma [44] established that the available band positions could not be fitted uniquely even if one cubic constant, y_{222}, was included with the three harmonic fundamentals and the seven quadratic constants; they pointed out the need for inclusion of other cubic constants and consideration of the third-order resonance interaction between $(v_1, v_2, v_3{}^l)$ and $(v_1+3, v_2-2, v_3{}^l)$ which arises because $3\nu_1 \sim 2\nu_2$. Allen, Tidwell, and Plyler [47] then found that with four perturbation constants instead of y_{222} they were able to fit all the band positions below $12\,000\ \text{cm}^{-1}$ but not those above. (The perturbation constants affected most of the calculated band positions above $6800\ \text{cm}^{-1}$.) A little later Rank, Skorinko, Eastman, and Wiggins [50] successfully analysed 45 band positions below $18\,400\ \text{cm}^{-1}$ with one quartic and ten cubic anharmonicity constants in place of the four perturbation constants. Then in 1964 Maki and co-workers,[54, 55] in their analysis of *l*-type resonance in $H^{12}C^{14}N$, showed that the x_{ll} value of Rank *et al.* needed to be revised. Subsequently the force-constant calculations of Suzuki,

Pariseau, and Overend [69] and of Nakagawa and Morino [66] (which neglect third- and fourth-order terms and give a low weighting to states perturbed by Fermi resonance) have indicated that the value of x_{ll} of Rank *et al.*[50] is (numerically) too small. The relevant quadratic values are listed in Table 16.

The extent of our knowledge of the Raman spectrum of gaseous HCN is very small in comparison with that of the i.r. spectrum. In 1932 Kastler [77] measured ν_1 at 2089 cm^{-1} at 353 K. Twenty years later, Rank and co-workers [43] obtained, from a sample containing 60% $H^{13}C^{14}N$, photographs showing the doublet ν_1 ($H^{12}C^{14}N$) and ν_1 ($H^{13}C^{14}N$) but were unable to make a significant measurement of the isotopic shift in the vapour state. Douglas and Sharma [44] cite Stoicheff [78] as having photographed the Raman spectrum of HCN gas with a dispersion of 100 cm^{-1} per mm and observing ν_1 excited by both the 4358 and 4047 Å mercury lines. Douglas and Sharma [44] measured the spectra obtained by Stoicheff and found ν_1 to be 2095.5 cm^{-1}. In 1960 Rank *et al.*[50] referred to a measurement of ν_1 at very small dispersion (presumably either Stoicheff's or their own, but they do not give a value). They concluded that this was less accurate than the i.r. values obtained indirectly by themselves and by Allen *et al.*[47] directly.

$D^{12}C^{14}N$. For this species Nakagawa and Morino were able to make use of 34 experimental values of ν and ΔB of 18 ground-state transitions in addition to the microwave $B(000^0)$ value. (The data for half the transitions were not directly observed but were derived *via* the Ritz combination principle.) ΔB values were calculated for the two transitions (201^1—000^0 and 103^1—000^0) for which there were no experimental data (direct or indirect). A new and more precise microwave value of $B(000^0)$ was determined by DeLucia and Gordy [71] by fitting nine line frequencies of the $J = 1 \leftarrow 0$ and $J = 2 \leftarrow 1$ transitions; this value is very slightly larger than the previous value (which was based on the $J = 1 \leftarrow 0$ transition only), and it was suggested (along with the analogous HCN value) that this was the most accurate rotational constant determined for any molecule. Microwave B values for 100^0, 001^1, and 002^0 have been determined by Winnewisser *et al.* These new microwave data allow a check on the i.r. ΔB values of ν_1 and ν_3. The degree of agreement is highly satisfactory (see Table 17). Winnewisser *et al.*, using essentially the same i.r. ΔB data as Nakagawa and Morino (eight ground-state transitions and nine hot bands) plus microwave ΔB data for two hot-band transitions, evaluated the rotation–vibration constants from a least-squares analysis. The resulting calculated ΔB values for the three fundamental transitions and the corresponding α values (corrected for the effect of higher-order terms) are listed in Table 17, where they are compared with the analogous calculated data obtained by Nakagawa and Morino and with the relevant experimental data. It is clear that, just as with $H^{12}C^{14}N$, the least-squares analysis carried out by Winnewisser *et al.*

[77] A. Kastler, *Compt. rend.*, 1932, **194**, 858.
[78] B. P. Stoicheff, cited in ref. 44.

Table 17 *Rovibrational constants* (cm^{-1}) *for* $D^{12}C^{14}N$

Constant	ν_1	ν_2	ν_3	*Technique*	*Ref.*
ΔB_{obs}	−0.006550(4)		+0.004322(1)	microwave	72
ΔB_{obs}	−0.006553(24)	−0.010329(15)	+0.00434(15)	i.r.	55
$\Delta B_{calc} = -\alpha_i$	−0.006542(51)	−0.010329(15)	+0.004391(38)		66
ΔB_{calc}	−0.006558	−0.010350	+0.004323		72
$-\alpha_i$	−0.006005(114)	−0.010844(65)	+0.004233(10)	i.r.	72

Table 18 *Observed and calculated quadratic vibrational anharmonicity constants* (cm^{-1}) *of* $D^{12}C^{14}N$

x_{ij}	*Obs.*[a,c]	*Obs.*[b,d]	*Calc.*[a,e]	*Calc.*[a,]
x_{11}	−7.10	−7.05[a]	−7.03	−6.84
x_{12}	−32.88	−32.29[a]	−32.44	−32.40
x_{13}	+2.73	3.14[c]	2.68	3.01
x_{22}	−20.23	−20.16[a]	−20.56	−20.50
x_{23}	−15.71	−15.881[c]	−15.96	−15.80
x_{33}	−2.09	−2.085[a]	−2.08	−2.08
x_{ll}	3.21	3.22[a]	3.25	3.29

[a] Uncorrected for cubic terms. [b] Corrected for cubic terms. [c] Ref. 47. [d] Ref. 55; seven cubic and seven combinations of quadratic and cubic vibrational anharmonic constants are also given. [e] Ref. 69. [f] Ref. 70.

represents a significant improvement over that of Nakagawa and Morino. The B_e value (taking into account α and γ) is 36329.49(1.53) MHz for $B(000^0)$ = 36207.4627(2) MHz.

Although the derivation of the vibrational anharmonicity constants of $D^{12}C^{14}N$ has not reached the level of sophistication of that in $H^{12}C^{14}N$, there appear to be no significant disagreements between different groups of workers (see Table 18). Of the quadratic constants only x_{13} and x_{23} have been evaluated and corrected for cubic terms. There is a need for the measurement of more ternary or higher combinations or overtones in order to obtain all the vibrational constants explicitly.

$T^{12}C^{14}N$. The only spectroscopic investigation of this interesting species, which apart from water is one of the few inorganic compounds to have been tritiated at high specific activity, appears to be the i.r. study of Staats, Morgan, and Goldstein,[48] who prepared their sample by treating T_2O with $P(CN)_3$. Staats *et al.* reported that TCN in the vapour phase at room temperature and a pressure of 30 cmHg undergoes a spontaneous polymerization (believed to be a secondary radiation decomposition initiated by free radicals produced by the β-activity) which is almost complete (*ca.* 80%) after 12 h. Four bands were observed in the region 4000—400 cm^{-1}, three of which were assigned as fundamentals, *viz.* 1724 cm^{-1}(ν_1), 2460 cm^{-1}(ν_2), and 513 cm^{-1}(ν_3). If correct, ν_1 = 1724 cm^{-1} would represent the lowest C≡N stretch reported. It is generally accepted, however, that in DCN ν_1 and ν_2 are about 200 cm^{-1} higher and lower respectively than would be expected on a simple harmonic model with no coupling; by the same token, the unperturbed ν_2 in TCN is calculated to be about 40 cm^{-1} *lower* than the unperturbed ν_1. It seems likely, then, that there is very extensive coupling between ν_1 and ν_2 in TCN. The authors are of the opinion that this coupling could indeed be regarded as an example of Fermi resonance between two fundamentals. It is then not meaningful to regard the bands at 1724 and 2460 cm^{-1} as CN and CT stretches. It may be noted that, in view of the results obtained with HCN, matrix-isolation studies of TCN might be quite pertinent, particularly as dilution in an inert matrix and the associated low temperature should both inhibit secondary radiation decomposition; such studies might also allow the examination of the polymerization under controlled conditions.

$H^{13}C^{14}N$. In their calculations, Nakagawa and Morino were able to make use of twenty-four experimental values of ν_0 and ΔB of fourteen ground-state transitions of this species in addition to a value of $B(000^0)$. A new microwave value [72] of $B(000^0)$ [1.439999(2) cm^{-1}] is only slightly larger. No new transitions appear to have been examined by i.r. spectroscopy. Maki and co-workers [57] have re-investigated the 101^1—000^0 transition and have obtained a ΔB value [−0.006696(15) cm^{-1}] applicable to both *c* and *d* levels of the upper state. A recent re-examination of the 001^1—000^0 transition by Yin and Rao [59] has led to no change in the ΔB value. It is worth noting that ν_0 of ν_1 does not appear to have been obtained directly; the best value appears to be that of Rank, Skorinko, Eastman, and Wiggins [50] who used the $0Z0^0$—000^0 series and the 110^0—000^0 band.

The only Raman spectroscopic study of $H^{13}C^{14}N$ in the vapour phase appears to be that of Rank and his co-workers [43] previously mentioned under $H^{12}C^{14}N$.

$D^{13}C^{14}N$. The only experimental data on this species available to Nakagawa and Morino [66] were values of B of 000^0, ν_0 of 100^0—000^0 and 010^0—000^0 transitions, and ΔB of 010^0—000^0. Nakagawa and Morino calculated values of ν_0 and ΔB of the 001^1—000 transition, and ΔB of the 010^0—000^0 transition. There has been little improvement since then, with apparently no recent i.r. studies; the only new data are microwave values [72] of B for the 000^0 and 002^0 levels [1.187075(2) and 1.195290(8) cm^{-1} respectively]. The change in $B(000^0)$ compared with the earlier value used by Nakagawa and Morino is insignificant.

$H^{12}C^{15}N$. There are now considerably more experimental data on this species than were available in 1969 to Nakagawa and Morino,[66] who were only able to make use of an i.r. $B(000^0)$ value, ν_0, and ΔB for the 010^0—000^0 transition. Nakagawa and Morino calculated values of ν_0 and ΔB for the 100^0—000^0 and 001^1—000^0 transitions. Since then, Maki, Olson, and Sams [57] have measured ν_0 and ΔB (and ΔD) of the 101^1—000^0 transition and Winnewisser *et al.*[72] have determined a new microwave value for $B(000^0)$, slightly smaller and very much more precise than the previous i.r. value [55] used by Nakagawa and Morino. The most detailed study is that of Alpert *et al.*[58] who reported *inter alia* i.r. values for $B(000^0)$ and $D(000^0)$. The $B(000^0)$ value is rather smaller and less precise than the microwave value; it is also not compatible with the microwave value, in terms of the stated uncertainties. The $D(000^0)$ value is in very good agreement with that calculated by Nakagawa and Morino, in contrast to the previous experimental value. Alpert *et al.* also reported i.r. values of ν_0 and ΔB (and ΔD) for seven ground-state transitions (including 010^0—000^0 and 101^1—000^0) and one difference band and three hot bands (all involving 001^1 as the lower state) (see Table 19). As previously noted, Nakagawa and Morino's calculated ν_0 and ΔB of the 001^1—000^0 transition are in very good agreement with the indirectly determined experimental values; the situation is much less satisfactory for the 100^0—000^0 transition.

$D^{12}C^{15}N$. As with $H^{12}C^{15}N$ the only experimental data available to Nakagawa and Morino were the B value of the 000^0 level and ν_0 and ΔB of the 010^0—000^0 transition; values of ν_0 and ΔB of the other two fundamental transitions were calculated. New data (see Table 20) have been obtained in recent microwave [58] and i.r. [72] investigations. Just as with $H^{12}C^{15}N$, much improved values of $B(000^0)$ were determined; the new microwave value is much more precise, and this time is somewhat smaller than the i.r. value. Again, the new i.r. $D(000^0)$ value is in very good agreement with Nakagawa

Table 19 *Selected vibrational and vibrational–rotational data* (cm)$^{-1}$ *for* $H^{12}C^{15}N$[a]

Transition or State	*Calculated*[b]	*Observed*[b]	ν_0	ΔB	B	$D \times 10^8$	$q \times 10^5$	*Ref.*
000^0	f.c.					274.53		66, 70
000^0		m.w.			1.4352[illegible]9(2)			72
000^0		i.r.			1.43527[illegible](10)	252(15)		55
000^0		i.r.			1.4351[illegible]3(49)	275(5)		58
001^1		i.r.			1.4385[illegible]9(140)		705(14)	58
001^1		m.w.					706.95	73
100^0—000^0	f.c.		2064.35	−0.009655(68)				66, 70
100^0—000^0		i.r.	2061.345(4)	−0.011471(76)				58
010^0—000^0		i.r.	3310.088(1)	−0.010019(7)				58
001^1—000^0	f.c.		711.41	+0.003541(47)				66, 70
001^1—000^0	r.c.p.	i.r.	711.025(5)	+0.003382(32)[c]				58
010^0—001^1		i.r.	2599.066(5)[d]	−0.013394(33)[d]				58
011^1—000^0		i.r.	4001.845(3)[d]	−0.006422(7)[d]				58
011^1—001^1		i.r.	3290.815(1)[d]	−0.009808(17)[d]				58
101^1—000^0		i.r.	2772.223(3)	−0.00638(3)				57
101^1—000^0		i.r.	2772.230(8)[d]	−0.006427(24)[d]				58

[a] Other data reported in ref. 58 include ν_0 and ΔB values of $\nu_1+\nu_2$, $\nu_1+2\nu_3^0$, $2\nu_3^0$, $\nu_2+2\nu_3^0-\nu_3^1$ and $\nu_2+2\nu_3^2-\nu_3^1$. [b] f.c. = force constant; r.c.p. = Ritz combination principle; m.w. = microwave; i.r. = infrared. [c] Note that $\Delta B = B(001^1) - B(000^0) = +0.003326$ cm^{-1}. [d] Average of *c*- and *d*-level values, and sum of uncertainties.

Table 20 *Vibrational and vibrational–rotational data* (cm^{-1}) *for* $D^{12}C^{15}N$

Transition or State	*Calculated*[a]	*Observed*[a]	ν_0	ΔB	B	$D \times 10^8$	$q \times 10^5$	*Ref.*
000^0	f.c.					181.41		66, 70
000^0		i.r.			1.17300(14)	130(20)		55
000^0		i.r.			1.173187(36)	182(3)		58
000^0		m.w.			1.173140(2)			72
001^1		m.w.					587.34	73
001^1	pol.	i.r.			1.176920(112)	−377(112)	250(15)	58
002^0		m.w.						72
100^0—000^0	f.c.		1900.12	−0.006290				55
100^0—000^0		i.r.	1900.110(2)	−0.006327(12)				58
010^0—000^0		i.r.	2621.182(1)	−0.009984(6)				58
001^1—000^0	f.c.			+0.004236				55
001^1—000^0		i.r.		+0.003733(148)[b]				
011^1—001^1 }		i.r.	2605.602(3)	−0.009704(67)				58
011^1—001^1 }				−0.009104(27)				

[a] f.c. = force constant; pol. = polynomial analysis; i.r. = infrared; m.w. = microwave. [b] Calculated from $\Delta B = B(001^1) - B(000^0) = 1.176920(112) - 1.173187(36) cm^{-1}$.

and Morino's calculated value, in contrast to the previous experimental value. A microwave value of $B(002^0)$ and i.r. values of ν_0 and ΔB of one hot band and two fundamental transitions (including ν_2, again) have also been determined. In the i.r. investigation,[58] which was carried out on a sample of $H^{12}C^{15}N$, insufficient bands were observed to allow the determination of the B and D constants of the 001^1 state by application of the Ritz combination principle; a polynomial analysis, however, gave values of B and D in this level. The $B(001^1)$ value gives a ΔB for the transition 001^1—000^0 which does not agree very satisfactorily with Nakagawa and Morino's calculated value. The value of $q(001^1)$ cited in this investigation, unlike that of $H^{12}C^{15}N$, is in poor agreement with the microwave value, as well as being much less precise, but the authors make no comment. As noted previously, Nakagawa and Morino's calculated values for ν_0 and ΔB of the transition 100^0—000^0 are in very good agreement with the experimental values reported in this latest investigation.

$H^{13}C^{15}N$. Apparently, the only data available on this species are the recently determined [58] values of ν_0 and ΔB of the 010^0—000^0 transition [3292.289(1) and $-0.009243(5)$ cm^{-1} respectively] obtained from an analysis of bands observed in a sample of $H^{12}C^{15}N$.

Internuclear Distances. We conclude our account of gaseous monomeric hydrogen cyanide with a discussion of bond lengths. Of all the parameters associated with the vibrational and rotational spectroscopic investigations we have considered, internuclear separations are probably those of most interest to the chemist. Small changes in bond lengths from one molecule to another can, at least in principle, be correlated with other changes in the chemical bonds. A critical assessment of the bond distances that can be obtained for hydrogen cyanide from spectroscopic measurements is of particular interest, especially in view of the relatively large number of isotopic species for which rotational constants are available and the very high precision of those for $H^{12}C^{14}N$ and $D^{12}C^{14}N$. Such distances are expected to differ from those obtained from non-spectroscopic methods (*e.g.* electron diffraction), since different averaging processes over the atomic motions are involved. Two $B(000^0)$ values for different isotopic moieties of hydrogen cyanide allow the evaluation of r_0 values for the two distances r(CH) and r(CN). Thus, from $B(000^0)$ values for $H^{12}C^{14}N$ and $D^{12}C^{14}N$ [based on microwave ν_0 and i.r. $D(000^0)$ values], Gordy and co-workers [67] calculated $r_0(\mathrm{CH}) = 1.0637(40)$ and $r_0(\mathrm{CN}) = 1.1563(10)$ Å. There is a disturbingly large uncertainty in such calculations, thus possibly obscuring small but significant differences in the chemical bonds, because a $B(000^0)$ value refers to the ground state in which the molecule still retains the zero-point vibrational energy. For example, it has recently been claimed [74] that even microwave methods are unable to reveal the expected difference in r_0 value for CH and CD with certainty. The problem of zero-point vibrational energy can be removed by using B_e values; the resultant equilibrium distances r_e differ from the corresponding r_0 values owing to the effects of anharmonicity; also, r_e(CH) and r_e(CD) are theoretically identical. Less radical approaches are those put forward by Costain,[79] and by Laurie and Herschbach.[80] Costain has proposed the use of substitution distances (r_s) as a means of partially cancelling out the effects of zero-point vibration without actually removing it, the idea being that r_s distances are more consistent (have less scatter) regardless of which sets of isotopic species are used. Laurie and Herschbach suggested that an alternative approach might be to use bond-shrinkage parameters to take into account the fact that the average bond distances (for stretching-type vibrations) are smaller for heavier atoms because the zero-point vibrational amplitudes are smaller.

Winnewisser, Maki, and Johnson,[72] using the $B(000^0)$ values of DeLucia and Gordy [71] for $H^{12}C^{14}N$ and $D^{12}C^{14}N$ and their own B_v, α_i, and γ_{ij} values for the species $H^{12}C^{14}N$, $D^{12}C^{14}N$, $H^{13}C^{14}N$, $D^{13}C^{14}N$, $H^{12}C^{15}N$, and $D^{12}C^{15}N$, have evaluated various bond distances (see Table 21) and have discussed the values critically in the light of the above propositions. They found that γ terms may be ignored without significantly affecting the r_e values other than making them rather less precise. With H–D B_0 pairs a total spread of only 0.0008 Å is found for the r_0(CH) distances and 0.0002 Å for the r_0(CN) distances. [The $H^{12}C^{14}N$–$D^{12}C^{14}N$ pair gives 1.062407 and 1.156827 Å for the r_0(CH) and r_0(CN) distances. The total spread is much worse if the three H–H and the three D–D pairs are included.] Moreover, use of H–D pairs gives r_0(CH) distances which are consistently lower than the r_e distances by 0.003 Å

Table 21 *Comparison[a] of different types of interatomic distances[b]/Å in hydrogen cyanide*

Type	(CH)	(CN)	(HN)	*Footnote*
r_e	1.06549(24)	1.15321(5)	2.21870(20)	*c*
r_e	1.06573(80)	1.15317(25)	2.21876(60)	*d*
r_0	1.062376	1.156799		*e*
r_0	1.06825	1.15632		*f*
r_0	1.06557	1.15570		*g*
r_0	1.064(11)	1.1564(24)	2.2205(90)	*h*
r_s	1.06316(10)	1.15512(15)	2.21828(12)	*h*

[a] Ref. 72. [b] All calculations using $I = 505376/B$ (MHz) and unified mass units given by L. A. König, J. H. E. Mattauch, and A. H. Wapstra, *Nuclear Phys.*, 1962, **31**, 18. [c] Evaluated using $B_e(H^{12}C^{14}N)$ and $B_e(D^{12}C^{14}N)$, both calculated with α and γ; the uncertainty is the estimated error limit from the B_e values. [d] Calculated neglecting γ by using $B_e = \frac{5}{2}B(000^0) - \frac{1}{2}B(010^0) - \frac{1}{2}B(002^0)$ for $H^{12}C^{14}N$, $D^{12}C^{14}N$, and $D^{13}C^{14}N$; the uncertainty is the estimated error limit from the B_e values. [e] Averages of the nine H–D pairs; total spread = 0.0008 and 0.0002 Å for r_0(CH) and r_0(CN) respectively. [f] Averages of the three H–H pairs; total spread = 0.017 and 0.003 Å for r_0(CH) and r_0(CN) respectively. [g] Averages of the three D–D pairs; total spread = 0.014 and 0.005 Å for r_0(CH) and r_0(CN) respectively. [h] Averages of the values obtained for all possible pairs of B_0 values; the uncertainties are due to the dispersion of the calculated values and do not allow for measurement errors.

[79] C. C. Costain, *J. Chem. Phys.*, 1958, **29**, 864.
[80] V. W. Laurie and D. R. Herschbach, *J. Chem. Phys.*, 1962, **37**, 1687.

and r_0(CN) distances consistently higher by 0.003 Å. The C—D bond lengths calculated using the D–D pairs are consistently smaller by about 0.003 Å than the C—H bond lengths calculated using the H–H pairs, but there is very little to choose between the respective average C—N bond lengths. The $H^{12}C^{14}N$–$H^{12}C^{15}N$ pair gives the largest r_0(CH) distance whereas the $D^{13}C^{14}N$–$D^{12}C^{15}N$ pair gives the smallest r_0(CD) distance with experimental error contributing virtually nothing to the differences in these extremes. According to the Laurie–Herschbach bond-shortening principle, the C—N distance calculated for $H^{13}C^{14}N$–$H^{12}C^{15}N$ should be the same as that calculated for the pair $D^{13}C^{14}N$–$D^{12}C^{15}N$. This is not the case, and in fact these pairs give the maximum difference in C—N distance and this is of such a magnitude (nearly 0.001 Å) that it cannot be attributed to experimental error. On the other hand, the total spread in r_0 values can be significantly reduced by appropriate choice of bond-shrinkage parameter. However, the $H^{13}C^{14}N$–$H^{12}C^{15}N$ pair, for instance, illustrates the great sensitivity of this technique to the choice of what is essentially an empirical shrinkage parameter. Thus, the assumption that $r_0(^{12}C^{14}N) = r_0(^{13}C^{15}N)$ $+0.0001$ Å rather than $r_0(^{12}C^{14}N) = r_0(^{12}C^{15}N)$ results in diminishing the C—H distance by 0.026 Å and increasing the C—N distance by 0.005 Å. It would seem that substitution distances (see Table 21) provide a more effective way of dealing with the problem of zero-point vibrational energy in hydrogen cyanide than do bond-shrinkage parameters.

Non-monomeric Gas-phase Species. It is well established that the dominant species in the vapour phase of hydrogen cyanide is the monomer. There is, however, an accumulation of evidence that the vapour phase contains other species. For example, entropy studies [81] have given the fractions of molecules present as dimers and trimers at 298.8 K and 1 atm pressure as 0.0988 and 0.0268, and yielded values of -13.72 and -35.60 kJ mol^{-1} for the heats of formation of dimer and trimer respectively. Vapour pressure and heat of vaporization data have given a dimer : monomer ratio of 0.088 under the same conditions as above.[82] Dipole measurements [83] also yielded indirect evidence of association in the vapour phase.

The first vibrational evidence for polymeric species in the vapour phase was obtained by Hyde and Hornig [84] in 1952. It had been known for some time that the i.r. spectrum in the CN stretching region consists of three sharp, closely spaced peaks instead of the characteristic *P–R* structure of a parallel band in a linear molecule. In 1932, Choi and Barker [41] had assigned the three peaks to $3\nu_3{}^1$, $4\nu_3{}^0-\nu_3{}^1$, and $4\nu_3{}^2-\nu_3$, but it was subsequently realized that the Boltzmann factor for the last two transitions is only 0.06, whereas the three peaks are of approximately equal intensity. Hyde and Hornig studied the intensities of these peaks as a function of

[81] W. F. Giauque and R. A. Ruehrwein, *J. Amer. Chem. Soc.*, 1939, **61**, 2626.
[82] C. F. Curtiss and J. O. Hirschfelder, *J. Chem. Phys.*, 1942, **10**, 491.
[83] C. P. Smyth and K. B. McAlpine, *J. Amer. Chem. Soc.*, 1934, **56**, 1697.
[84] G. E. Hyde and D. F. Hornig, *J. Chem. Phys.*, 1952, **20**, 647.

pressure, and established that the intensity of the highest-wavenumber peak (2110 cm^{-1}) is directly proportional to the pressure, whereas the intensities of the other two peaks at 2096 and 2087 cm^{-1} depend on the square of the pressure. It was thus concluded that the absorption in this region is due to a dimer (2096 and 2087 cm^{-1}) in equilibrium with the monomer (2110 cm^{-1}). It was suggested that the dimer is only detectable in this region in HCN because a fortuitous cancellation of moments produced in the C—H and C—N bonds results in a very weak absorption for ν_1 of HCN monomer. This still leaves ν_1 with an unusual shape: a central peak at 2110 cm^{-1} with wings extending from the neighbourhood of 2025 cm^{-1} to 2150 cm^{-1}. It was inferred that this structure is almost certainly due to the superposition of the *Q* branch of $3\nu_3{}^1$ on the normal *P–R* structure of the parallel band ν_1 with a band centre near 2090 cm^{-1}, the unusual intensity of $3\nu_3{}^1$ arising from Coriolis interaction with ν_1.

Four years later, Dagg and Thompson [85] studied this same region of the i.r. spectrum but made no reference to the work of Hyde and Hornig. They too concluded that the unresolved band (at 2093 cm^{-1}), with an intensity roughly proportional to the square of the pressure and a pair of maxima as contour (separation about 12 cm^{-1}), is probably attributable to the *P* and *R* branches of a CN vibration of the dimer H—C≡N···H—C≡N, with assumed moment of inertia of 3.4×10^{-47} kg m^2.

Recently, a further study has been reported by Jones, Seel, and Sheppard.[86] These workers studied the absorption at a pressure of 129 Torr and temperature of 268 K in a 1 m path-length cell using a high-resolution single-beam instrument. They report the band at 2095 cm^{-1} with a *P–R* separation of 9.2(3) cm^{-1} and assign it to the CN stretching vibration of the dimer, probably that of the HCN molecule whose lone-pair is involved in hydrogen-bonding to the other molecule. They deduced a rotational constant, *B*, of 0.057(4) cm^{-1} from which they calculated the bond length (N···C) to be 3.34(20) Å, assuming unchanged dimensions for the constituent HCN molecules. Like Hyde and Hornig, they concluded that ν(CH) of the dimer was obscured by the rotational contour of the ν_2 monomer band. The displacement on hydrogen-bonding thus appears to be less than the amount of 160 cm^{-1} observed in the hydrogen-bonded complex $H_3N \cdots HCN$.[86]

The results of the above studies have been reported in some detail, not only because of their own intrinsic interest but also because of the implications they have for the monomer spectra and the structures of hydrogen cyanide in the liquid and solid states. Also, during the past ten or so years, Matthews and his co-workers, for example, have proposed in a number of publications [87, 88] that the hydrogen cyanide dimer has an important role

[85] I. R. Dagg and H. W. Thompson, *Trans. Faraday Soc.*, 1956, **53,** 455.
[86] W. J. Jones, R. M. Seel, and N. Sheppard, *Spectrochim. Acta*, 1969, **25A,** 385.
[87] R. M. Kliss and C. N. Matthews, *Proc. Nat. Acad. Sci. U.S.A.*, 1962, **48,** 1300.
[88] C. N. Matthews and R. E. Moser, *Proc. Nat. Acad. Sci. U.S.A.*, 1966, **56,** 1087; *Nature*, 1967, **215,** 1230; R. E. Moser and C. N. Matthews, *Experientia*, 1967, **24,** 658; R. E. Moser, A. R. Claggett, and C. N. Matthews, *Tetrahedron Letters*, 1968, 1599; *ibid.*, 1969, 1605.

in chemical evolution, acting as a key intermediate in the prebiotic synthesis of purines and proteins under simulated primitive earth conditions. The dimer is apparently the only small polymer of HCN which has never been isolated.[89] [The reviewers, at least, use the term 'isolated' in what might be called its classical orgânic sense: the dimer has been isolated at low temperature in a matrix[60] (see p. 495).] Seemingly, because the i.r. characterization of the dimer as HCN···HCN has gone unnoticed by the biologists, early inferences about its nature and its role in prebiotic synthesis were based on the assumption that the lowest energy form of the dimer is the isomer and 1,3-biradical aminocyanocarbene, $H_2N—\dot{C}=C=\dot{N}$, a highly reactive form which, it was argued, could easily dimerize to form the tetramer and higher hydrogen cyanide polymers. A theoretical calculation made in 1964 appeared to indicate that the linear conformation of the methylene dimer was indeed the lowest-energy isomer.[90a] Recently, however, it has been claimed on the basis of more extensive theoretical calculation[89] that the ground state is a singlet state of the *trans*-imino-dimer (1), but HCN···HCN

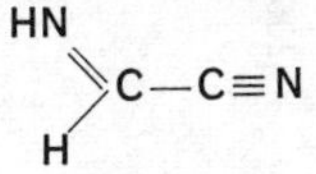

HN=C(H)—C≡N

(1)

was not included as a possibility,[90b] whereas it is this form of the dimer which by all accounts is present in the gas at room temperature and can be isolated[60] in a matrix at temperatures in the range 4.5—20.5 K.

The Liquid State.—Hydrogen cyanide has a liquid range of about 38 K. The very high dielectric constant of the liquid,[91] especially at low temperatures (107 at 25°C, 116 at 20 °C, and 194.4 at −13 °C) has been interpreted[92,93] in terms of a structure of hydrogen-bonded polymeric linear chains (heat of association = −19.2 kJ mol^{-1}) and quantitatively explained[94] in terms of a factor of 3 for the average degree of polymerization.

A Raman spectroscopic study of liquid HCN and DCN using Hg arc excitation has been reported recently by Pézolet and Savoie.[64] This makes no reference to the earlier work on HCN[43, 77, 95, 96] in which assignments were made for ν_1, $\nu_1(^{13}CN)$, and ν_2; indeed the band attributable to $\nu_1(^{13}CN)$ was not observed in the latest investigation. The various assignments are

[89] G. H. Loew, *J. Theor. Biol.*, 1971, **33**, 121.
[90a] J. Serre and F. Schneider, *J. Chim. phys.*, 1964, **61**, 1655; [b] A. Johansen, P. Kollman, and S. Rothenberg, *Theor. Chim. Acta*, 1972, **26**, 97, and references therein.
[91] K. Fredenhagen and J. Dahmlos, *Z. anorg. Chem.*, 1939, **179**, 77.
[92] G. E. Coates and J. E. Coates, *J. Chem. Soc.*, 1944, 77.
[93] R. H. Cole, *J. Amer. Chem. Soc.*, 1955, **77**, 2012.
[94] L. Pauling, 'The Nature of the Chemical Bond', Cornell University Press, Ithaca, New York, 1960, 3rd edn., p. 458.
[95] A. Dadieu and K. W. F. Kohlrausch, *Ber.*, 1930, **63**, 1657; A. Dadieu, *Naturwiss.*, 1930, **18**, 895; *Monatsh.*, 1931, **57**, 437.
[96] G. Herzberg, *J. Chem. Phys.*, 1940, **8**, 847.

Table 22 *Vibrational assignments for liquid* HCN *and* DCN

Compound	*Spectrum*	*Excitation*	*Temperature/K*	ν_1	$\nu_1(^{13}CN)$	ν_2	ν_3	*Ref.*
HCN	Raman	Hg	293	2098				72
	Raman	Hg	?	2094 (12)[a]	2062 (½)	3213 (2)		95, 96
	Raman	Hg	292	2096		3215	~794	64
	Raman	Hg	261	2097 [3][b]		3207 [95]	~798 [55]	64
	Raman	Ar^+	?				779	65
	i.r.		268	2098 [11][b]		3206 [108]	770 [66]	65
DCN	Raman	Hg	292	1909		2581	~615	64
	Raman	Hg	261	1908 [11][b]		2575 [50]	~622 [30]	64
	i.r.		268	1908 [17][b]		2578 [40]	615 [28]	65

[a] Numbers in parentheses refer to relative intensities. [b] Numbers in square brackets refer to bandwidths at half-intensities.

collected in Table 22. These data are consistent with a hydrogen-bonded liquid structure. Compared with the gas-phase values there are significant downward and upward shifts in the vibrations ν_2 and ν_3 respectively which formally involve hydrogen atoms, especially with HCN. There is only a small shift (negligible in HCN) in ν_1. Bandwidths in both liquid spectra are approximately proportional to the gas–liquid wavenumber shifts. The symmetrical band shapes observed provide no evidence for monomers and dimers being present in appreciable concentrations.

Very recently, Pézolet and Savoie have reported [65] the i.r. spectra of liquid HCN and DCN at 268 and 273 K respectively. As the authors themselves note, little is learned from these spectra that was not already known from the corresponding Raman spectra, though they resulted in the detection of an error of at least 20 cm^{-1} in the measurement of ν_3 (stated to be accurate to ± 1 cm^{-1} at the time) in the Raman spectrum of HCN. The ν_1 wavenumbers are almost identical in the i.r. and Raman spectra as are those of ν_2. In the i.r. spectra of HCN and DCN the ν_2 band of liquid HCN is only about ten times more intense than ν_1; in liquid DCN ν_2 is about 1.4 times more intense than ν_1. These results are consistent with the presence of polymeric units in the liquid phase. The corresponding gas-phase intensity ratios are $\sim$400 and $\sim$11.

The Solid State.—Until very recently there was little structural information available on crystalline HCN and very little on DCN. Smyth and McNeight [97] measured the dielectric constant of solid HCN and concluded from the very low values that there was no molecular rotation in the solid state; they could not accept the marked increase in dielectric constant near the melting point as being indicative of molecular rotation because there was no corresponding phase change. A reversible phase transition at *ca.* 170.5 K, well below the melting point, was established by heat-capacity studies a little later.[82] In what appears to be the only diffraction investigation of solid HCN, Dulmage and Lipscomb [98] deduced that the high-temperature form (phase I) is tetragonal with space group C_{4v}^{9} (*I4mm*) ($a = 4.63$, $c = 4.34$ Å, and $z = 2$; the primitive cell has $z = 1$), and that the transition to the orthorhombic low-temperature form (phase II) merely involves a contraction and expansion of the tetragonal a and b distances respectively [space group C_{2v}^{20} (*Imm*); $a = 4.13$, $b = 4.85$, $c = 4.34$ Å, $z = 2$; the primitive cell has $z = 1$]. From a vibrational point of view, this necessitates the site group, like the factor group, changing from C_{4v} to C_{2v}; a splitting of the doubly degenerate bending vibration ν_3 and the librational lattice mode ν_L is therefore expected in either approximation. It was inferred [98] that the crystal structure consisted of infinite linear chains parallel to the c axis. The hydrogen-bond distance (C—H···N) is then the difference between the length of the unit cell axis and the C≡N bond length,

[97] C. P. Smyth and S. A. McNeight, *J. Amer. Chem. Soc.*, 1936, **58**, 1723.
[98] W. J. Dulmage and W. N. Lipscomb, *Acta Cryst.*, 1951, **4**, 330.

Table 23 *Vibrational band positions* (cm^{-1}) *for solid* HCN

Spectrum:	*Raman (Hg excitation)*	*I.r. (abs.)*[a]	*I.r. (refl.)*[a]	*I.r. (abs.)*	*Raman (Hg excitation)*	*I.r. (abs.)*	*I.r. (refl.)*	*I.r. (abs.)*	*I.r. (abs.)*	*I.r. (abs.)*
*Temperature/*K	183	183	183	93	83	83	83	77	77	10
and phase[b]	I	I	I	II	II	II	II	II	II	II
ν_L ν_t[c]	161	162			173	176			173	176
ν_L ν_{ln}	245	239			255	249				
ν_3 ν_t	819	818	819	828	827	826	827	828		828
ν_3 ν_{ln}	840	836	841	838	849	847	848	840		
$2\nu_3$		1608		1632		1622				1626
$\nu_1(^{13}CN)$	2064	2064		2063	2064	2064				
ν_1 ν_t	2098	2098	2098	2097	2098	2098	2098	2099		2101
ν_1 ν_{ln}	2106		2107		2108		2108			
		3100								
ν_3 ν_t	3145	3144	3144	3132	3129	3130	3130	3130		3128
ν_3 ν_{ln}	3190		3188		3173		3172			
						3194				
		3240				3240				
Ref.	64	65	65	99	64	65	65	63	62	62

[a] abs. = absorption; refl. = reflection. [b] Transition between I and II occurs at *ca.* 170.5 K. [c] ν_t = transverse vibration; ν_{ln} = longitudinal vibration.

and is no doubt $\leqslant 3.2$ Å. (The *X*-ray diffraction analysis did not allow the determination of an accurate C≡N distance.) This probably provides the best example in the solid state of a C—H bond acting as a donor to a nitrogen atom.

The i.r. spectrum of phase II of HCN (see Table 23) was, in fact, studied as a thin film at 93 K, before the *X*-ray diffraction analysis.[99] A splitting of ν_3 (10 cm^{-1}) was deduced, and no anomalous features associated with ν_1 and ν_2 were apparently observed. A decrease in intensity of ν_3 relative to ν_1 and ν_2 of the order of 1000 times on going from the vapour phase to the solid was attributed to significant hydrogen-bond formation in the crystal. The upward frequency shift of ν_3 of about 120 cm^{-1} and the downward frequency shift in ν_1 of about 180 cm^{-1} were additional noteworthy features.

A recent study [64] of the Raman spectra of the high- and low-temperature forms of HCN and DCN (see Tables 23 and 24) leads, however, to the conclusion that the splitting of ν_3 observed in the i.r. spectrum of phase II of HCN) is not a site splitting but is related to the piezoelectric nature of the crystal. In particular the CN and CH (CD) stretching vibrations (ν_1 and ν_2) are complicated by high-wavenumber shoulders in both phases. These non-degenerate vibrations should not split even in the factor-group approximation because both phases have $z = 1$ for the primitive cells. Moreover, the bending mode (ν_3) yields an asymmetric peak even in the high-temperature phases, and the librational lattice mode appears substantially split in both phases. The possibilities of an incorrect crystal structure or positional and orientational disorder introduced through bad crystallization were considered and discarded. It was concluded that for all bands observed, the wavenumber of the band maximum corresponded to the tranverse vibration (ν_t) and the high-wavenumber cut-off of the anomalous shoulder to the longitudinal vibration (ν_{ln}). The unexpected occurrence of longitudinal modes of non-degenerate vibrations was thought to be due to the samples containing microcrystals of various sizes. The intensity ratios $I(\nu_2):I(\nu_1) = 7.2$ and $I(\nu_3):I(\nu_1) = 1.2$ for HCN calculated from the Raman frequencies according to the formula* of Haas and Hornig [100] were found to be in reasonable agreement with the experimental values

* The Haas–Hornig formula is

$$(\nu_{ln})^2 - (\nu_t)^2 \propto \left(\frac{\partial \mu}{\partial q}\right)^2 \propto I$$

Although this relation was derived for cubic crystals, with site symmetry not lower than T_d, it does seem that it can be used to estimate dipole derivatives from frequency values in crystals of lower than cubic symmetry, especially when there are only a small number of well separated fundamental bands in the spectrum, *e.g.* HCN. Also, as HCN has no centre of symmetry, the frequencies of ν_{ln} and ν_t can be obtained from the Raman spectrum and used to predict an i.r. intensity.

[99] R. E. Hoffman and D. F. Hornig, *J. Chem. Phys.*, 1949, **17,** 1163.
[100] C. Haas and D. F. Hornig, *J. Chem. Phys.*, 1957, **26,** 707.

Table 24 *Vibrational band positions* (cm^{-1}) *for solid* DCN

Spectrum:	*Raman (Hg excitation)*	*I.r. (abs.)*[a]	*I.r. (refl.)*[a]	*Raman (Hg excitation)*	*I.r. (abs.)*	*I.r. (refl.)*	*I.r. (abs.)*
Temperature/K	183	183	183	83	83	83	77
and phase[b]	I	I	I	II	II	II	II
ν_L ν_t[c]	149	153		163	165		*d*
ν_L ν_{ln}	235	226		240	236		*d*
ν_3 ν_t	643	642	643	648	648	648	649
ν_3 ν_{ln}	649	647	650	656	655	655	
$2\nu_3$		1270			1280		
$\nu_1(C^{15}N)$		1868			1865		
$\nu_1(^{13}CN)$		1882		1878	1878		
ν_1 ν_t	1890	1888	1888	1885	1885	1884	1885
ν_1 ν_{ln}	1908		1908	1906		1907	
$\nu_2(D^{13}C)$		2514			2509		
$\nu_2[D^{12}C(^{15}N)]$				2528	2531		
ν_2 ν_t	2553	2550	2552	2545	2545	2545	2545
ν_2 ν_{ln}	2565		2567	2558		2560	
Ref.	64	65	65	64	65	65	63

[a] abs. = absorption; refl. = reflection. [b] Transition between I and II occurs at *ca.* 170.5 K in HCN. [c] ν_t = transverse vibration; ν_{ln} = longitudinal vibration. [d] Ref. 62 reports ν_t and ν_{ln} of ν_L at 164 and 215 cm^{-1} respectively for a 70% DCN–30% HCN film.

of Hoffman and Hornig [99] (5.7 and 1.4 respectively). The corresponding calculated values for DCN are 1.0 and 0.17. This Raman spectroscopic investigation has prompted further and much more detailed work on the i.r. spectra of solid HCN and DCN. Three very recent publications include the study [65] of both phases of HCN and DCN using absorption and reflection techniques (see Tables 23 and 24) and intensity measurements of the internal modes [63] of HCN and DCN at 77 K and of the librational lattice mode [62] of pure HCN and mixed HCN–DCN films at 77 K and 10 K (see Table 25). These investigations provide confirmation of the interpretation of the Raman spectra and some insight into the changes in vibrational intensities and dipole moment in going from the gaseous to the crystalline state.

Pézolet and Savoie [65] investigated the lattice and internal modes of the high- and low-temperature forms of HCN and DCN in the i.r. Their estimated transverse and longitudinal vibrations are in excellent agreement with those derived from Raman spectra. Using these experimentally determined vibrational frequencies and the formula of Haas and Hornig,[100] they calculated values of $(\partial\mu/\partial q)$, the derivatives of the crystal dipole moment with respect to the appropriate internal co-ordinate. These calculated values were in quite good agreement with values deduced from intensity measurements.[62, 63] Also the ν_2/ν_1 intensity ratio calculated from these vibration-derived dipole derivatives is 6.5. This may be compared with the value of 7.2 obtained from the Raman frequency data alone and with the experimental values of 7.6 at 77 K [63] and 6.8 at 10 K [62] determined independently from i.r. intensity data. It is therefore interesting to note that the $(\partial\mu/\partial q)$ value for the librational mode in HCN is identical with that for ν_3, whereas in DCN $(\partial\mu/\partial q)$ for ν_L is twice as large as that of ν_3. It was suggested that this may indicate appreciable coupling between the librational and internal bending vibrations, which could explain *inter alia* why the isotopic shift of the lattice mode is appreciably less than expected. Other authors have noted, however, that the theoretical shift is calculated assuming a harmonic librational mode and no coupling to the translational acoustic modes.

Pézolet and Savoie [64, 65] observed, in their i.r. and Raman spectra of DCN, bands assignable to ν_1 and ν_2 of both $D^{13}C^{14}N$ and $D^{12}C^{15}N$, in addition to the transverse and longitudinal vibrations of ν_1 and ν_2 of $D^{12}C^{14}N$. (In the HCN spectra the only observed band attributed to an isotopically substituted species was ν_1 of $H^{13}C^{14}N$.) The assignments proposed are totally consistent with the corresponding gas-phase assignments; thus $\nu_1(^{12}C^{15}N)$ is less than $\nu_1(^{13}C^{14}N)$ and there is an appreciable (*ca.* 15 cm^{-1}) separation between ν_2 of $D^{12}C^{14}N$ and $D^{12}C^{15}N$.

Uyemura and Maeda [63] have reported integrated absorption coefficients and dipole-moment derivatives for ν_1, ν_2, and ν_3 of HCN and DCN at 77 K. No longitudinal vibrations were apparently observed apart from the well-established case of ν_3 in HCN. The relative intensities of ν_1, ν_2, and ν_3

Table 25 *Calculated and experimental intensity data of solid* HCN *and* DCN

	Calculated or experimental	ν_1	ν_2	ν_3	ν_L	*Temperature/*K	*Ref.*
(a) HCN							
$\frac{I(\nu_1)^a}{I(\nu_1)}$	Exp.	1.0	5.7	1.4		93	99
	Calc.	1.0	7.2	1.2		83—183	64
	Calc.	1.0	6.5			83—183	65
	Exp.	1.0	7.6	0.9		77	63
	Exp.	1.0	6.8	1.0	0.5	10	62
A^b	Exp.	5180	39 400	4620		77	63
	Exp.	6570	44 820	6440	3430	10	62
$\frac{\partial\mu^c}{\partial q}$	Calc.	77	206	70	70	83—183	65
	Exp.	81	236	57		77	63
	Exp.			75	70.4	10	62
(b) DCN							
$\frac{I(\nu_1)^a}{I(\nu_1)}$	Calc.	1.0	1.0	0.16		83—183	64
	Exp.	1.0	1.1	0.14		77	63
A^b	Exp.	9900	10 700	1410		77	63
$\frac{\partial\mu^c}{\partial q}$	Calc.	111	108	34	68	83—183	65
	Exp.	118	123	32		77	63

[a] Relative intensities of ν_2, ν_3, and ν_L with respect to ν_1. [b] Band intensity in cm^{-1} mol^{-3}; values cited by ref. 63 are corrected for the field effect.
[c] Electrostatic units.

predicted by Pézolet and Savoie are very well substantiated by Uyemura and Maeda's observed values; similarly there is quite good agreement between Pézolet and Savoie's calculated dipole-moment derivatives and Uyemura and Maeda's experimental values. Both investigations establish that the two stretching vibrations are very much intensified in going from the gas to the solid, whereas the intensity of the bending vibration is almost unchanged, despite its considerable shift in frequency. Uyemura and Maeda concluded that the general features of their intensity data do not suggest any remarkable change in the dipole moment due to the N···H hydrogen-bonding in the solid but point to a significant contribution of a charge-transfer structure.

At almost the same time as the publication of Uyemura and Maeda's work, Krause and Friedrich [62] reported i.r. absorption intensities of the librational modes of pure HCN and HCN–DCN mixtures at 77 K and 10 K. These workers argue that it is not too surprising that the expected factor-group splittings of these bands in the low-temperature phase are not observed since the orthorhombic *a* and *b* axes differ by only 0.7 Å. The total band intensity increased somewhat at 77 K but most of the increase is attributable to the high-wavenumber wing. The authors were unable to provide a quantitative explanation of this behaviour, and therefore only used the 10 K intensity data (*i.e.* that on pure HCN) in their analysis of the dielectric properties of crystalline hydrogen cyanide. Using a dipolar-coupling model they were able to infer a *significant reduction* in the intrinsic dipole moment of hydrogen cyanide in going from the gaseous to the crystalline state and to calculate values of the static dielectric constant and the longitudinal–transverse splitting of the librational modes in good agreement with the observed values.

Matrix-isolated Species.—There have been a number of i.r. studies of matrix-isolated hydrogen cyanide which have proved to be very revealing. Until the advent of matrix-isolation studies, all attempts to isolate hydrogen isocyanide or even to find evidence for it in equilibrium with hydrogen

Table 26 *Fundamental vibrations* (cm^{-1}) *of matrix-isolated* HNC

Species	*Matrix*	*Generating molecule*	ν_1	ν_2	ν_3	*Ref.*
$H^{14}N^{12}C$	Ar	CH_3N_3	2032	3583	535	101
$H^{14}N^{12}C$	N_2	HCN		3567	559	60
					538	
$H^{14}N^{12}C$	Ar	HCN	2029.2	3620	477	60
$D^{14}N^{12}C$	Ar	DCN	1940	2769	374	60
$H^{15}N^{12}C$	Ar	$HC^{15}N$	2003.6	3610	474.5	60
$H^{14}N^{13}C$	Ar	$N^{13}CN$	1986.5	3620	477	60

[101] D. E. Milligan and M. E. Jacox, *J. Chem. Phys.*, 1963, **39**, 712.

Table 27 *Vibrational assignments* (cm^{-1}) *of* HCN *and* DCN *in inert-gas matrices at* 20.5 K [60]

Mode	HCN						DCN			
	Ar		N_2		CO		Ar		N_2	
	M[a]	D[a, b]	M	D	M	D	M	D[b]	M	D
ν_1	3303.3	3301.3 3202.0	3287.6	3282.0 3204.9	3261.2	3244.4 3201.3	2631.3	2626.7 2573.9	2617.8	2616.0 2574.4
ν_2	2093.4	2114.4 2090.4	2097.3	2110.9 2092.9	2104	2112 2088	1922.7	1935.4 1900.0	1920.6	1927.0 1898.4
		797.3		{799.0 794.6}		{803.0 794.6}		622.0		{628.5 626.2}
ν_3	720.2		{745.6 736.0}		{761.3 739.0}		572.0		{594.0 588.4}	
		732.4		{757.6 749.6}		{778.6 756.1}		579.5		{603.0 598.0}
$2\nu_3$	1415	1555 1440	1455	1555 1480			1135	1230 1155	1155	1225 1175
ν_{lib}[c]				{144 137} 96						131 89

[a] M refers to the monomer and D to the dimer. [b] Extra bands at 3219.0 cm^{-1} and 791.8 cm^{-1} in HCN and at 2581.1 cm^{-1} and 618.0 cm^{-1} in DCN are thought to be due either to the dimer in a second site or environment, or to a dimer of a different structure. [c] Librational or bending mode about the hydrogen-bond.

cyanide either failed or were subsequently discredited, and hydrogen isocyanide was cited as a typical example of an inseparable tautomeric substance. In 1963 Milligan and Jacox [101] were able, however, to detect all three vibrational fundamentals of HNC (see Table 26) after extended photolysis of methyl azide isolated in an Ar matrix at 4 K. In 1967 the same authors [19] reported spectroscopic studies of photolysis products of $H^{12}C^{14}N$ and $D^{12}C^{14}N$ in both Ar and N_2 matrices at 14 K, and also of $H^{13}C^{14}N$ (54%) and $H^{12}C^{15}N$ (95%) in an Ar matrix at 14 K. They found that HNC is appreciably perturbed by the presence of N_2 but were able to obtain unperturbed values of the fundamental vibrations of HNC from the argon matrix experiments (see Table 26). It is interesting to note that for the three hydrogen species $\nu(^{14}N^{12}C) > \nu(^{15}N^{12}C) > \nu(^{14}N^{13}C)$. Revised thermodynamic properties of HNC were calculated from the unperturbed values.

A little later, King and Nixon [60] obtained the i.r. spectra of HCN and DCN in Ar, N_2, and CO matrices (M/S = 100—2000) at temperatures from 4.5 to 20.5 K. Bands assignable to the monomer (see Table 27) were characterized by an increasing optical density with increasing M/S. Of the many other bands observed, some had optical densities passing through a maximum at an M/S ratio in the neighbourhood of 500 and were attributed to the dimer (see Table 27). The appearance in Ar, N_2, and CO matrices of two peaks in the CN stretching region and of two in the CH stretching region and of just two bands (assignable to the dominant dimer) in the HCN bending region in Ar is consistent with a linear or nearly linear configuration of the dimer. The observation of four HCN bending modes for the dimer in both N_2 and CO is believed to be the result of a linear dimer in a site of symmetry lower than C_3 [the site symmetry of the N_2 molecule in the low-temperature (α) phase of the crystal] rather than to a bent dimer, because the corresponding ν_3 vibrations of the monomer occur as doublets. Such a linear dimer would have two degenerate low-wavenumber librational modes of the HCN (DCN) units and these are apparently observed in the far i.r. of the N_2 matrix, with at least one of these modes in HCN split by site effects. The hydrogen cyanide dimer is evidently hydrogen-bonded. The occurrence of dimer bands in the overtone region of ν_3 eliminates the possibility of the dimer having a cyclic structure with a centre of symmetry, and the fact that one of the CH (CD) vibrations of the dimer is almost coincident with that of the monomer whereas the other CH (CD) vibration is red-shifted by about 100 cm^{-1} is strong evidence for a hydrogen-bonded chain dimer. The far-i.r. bands could therefore be associated with a N···H(D) hydrogen-bond stretching vibration but the observed isotopic shifts are more in agreement with their assignment as torsional modes. A torsional vibration of 153 cm^{-1} for the HCN dimer, calculated according to the Lippincott–Schroeder one-dimensional model,[102] is in fair agreement with the observed band centred at about 141 cm^{-1}. It is clear that the matrix-

[102] E. R. Lippincott and R. Schroeder, *J. Chem. Phys.*, 1955, **23**, 1099; R Schroeder and E. R. Lippincott, *J. Phys. Chem.*, 1957, **61**, 921.

isolated hydrogen cyanide dimer has been characterized in considerably more detail than the vapour-phase dimer.

King and Nixon also obtained interesting information on the matrix-isolated monomer. From the lack of temperature dependence of monomer or dimer band intensities and comparison of certain monomer and dimer half-widths, they concluded that there is no evidence for rotation of the monomer at these temperatures. They also observed apparent multiple site effects manifested in ν_1 and ν_3 of the monomer [see Table 27, footnote (*b*)] in the Ar matrix.

By far the most detailed study of the monomer in an Ar matrix, however, is that very recently reported by Pacansky and Calder.[61] These workers made careful measurements (each band measured from 3 to 10 times) of $H^{12}C^{14}N$, $D^{12}C^{14}N$, $H^{13}C^{14}N$, $D^{13}C^{14}N$, $H^{12}C^{15}N$, and $D^{12}C^{15}N$ isolated in an Ar matrix (M/S = 333—667). All the fundamentals of the monomer were observed (see Table 28) except the very weak ν_1 vibration of the H species.

Table 28 *Vibrational fundamentals* (cm^{-1}) *of* $H^{12}C^{14}N$, $D^{12}C^{14}N$, $H^{13}C^{14}N$, $D^{13}C^{14}N$, $H^{12}C^{15}N$, *and* $D^{12}C^{15}N$ *in argon matrices*

Species	*Temperature/K*	ν_1	ν_2	ν_3	*Ref.*
$H^{12}C^{14}N$	20.5	2093.4	3303.3	720.2	60
$H^{12}C^{14}N$	8	—	3305.66(7)[a]	720.96(5)	61
$D^{12}C^{14}N$	20.5	1922.7	2631.3	572.0	60
$D^{12}C^{14}N$	8	1925.17(3)[a]	2626.43(5)	576.02(5)	61
$H^{13}C^{14}N$	8	—	3288.08(13)	714.94(3)	60
$D^{13}C^{14}N$	8	1911.91(1)	2585.85(3)	568.01(6)	60
$H^{12}C^{15}N$	8	—	3304.61(10)	719.74(4)	60
$D^{12}C^{15}N$	8	1900.16(1)	2616.99(2)	574.44(2)	60

[a] Figures in parentheses represent average deviation from the mean of 3—10 measurements.

The intensity of ν_1 increases markedly upon deuteriation. It is interesting to note that King and Nixon[60] had previously reported ν_1 of $H^{12}C^{14}N$ to be weak in an Ar matrix, as is the case in the gas phase, but were able to assign it at 2093.4 cm^{-1} by analogy with the N_2-matrix results where the identification of the fundamental is much less ambiguous.

Pacansky and Calder compared their matrix isotopic shifts [*e.g.* $\nu_2(H^{12}C^{14}N) - \nu_2(H^{13}C^{14}N)$] with the corresponding gas-phase shifts. For these latter values they chose to use Nakagawa and Morino's *calculated* values. The discrepancies between matrix and gas-phase shifts are small for isotopic combinations involving mass changes of the heavy atoms. However, for isotopic combinations involving H and D (*i.e.* ν_2 and ν_3) the discrepancies are about 2 cm^{-1}, values which far exceed the combined uncertainty of the gas and matrix data. This means that if the Redlich–Teller product rule is considered to be applicable to such matrix data, as is often assumed, then the calculated molecular geometry of the matrix species differs significantly from that of the gas-phase species. Strictly speaking the product rule should only be applied to harmonic data, but using anharmonic ν_3 values the average ratio of the C—N and C—H bond lengths is 1.00 in the matrix and 1.13 in the gas. No improvement in the matrix bond-length ratio is achieved by computing harmonic shifts using gas-phase anharmonicity constants. Pacansky and Calder, however, do not consider such a change in geometry plausible. They prefer to conclude that the discrepancy most likely reflects a failure of the linear *x–y–z* vibrational model to describe the vibrations of matrix-isolated hydrogen cyanide adequately, and point to the serious errors that may be involved in calculated molecular geometries of matrix-isolated species when light isotopic substitutions are involved. In view of the potential importance of these conclusions, the Reporters think it worthwhile emphasizing that the comparison of matrix and gas-phase data should preferably be made between sets of experimental data and ideally between sets of experimentally derived harmonic data. It should be noted that if the best experimental gas-phase values of ν_2 and ν_3 listed in Table 8 are used to calculate frequency differences of the type $\nu_2(H^aC^bN) - \nu_2(D^aC^bN)$ *etc.* for all possible H–D combinations, in no case is a discrepancy between matrix and calculated gas-phase data reduced. On the other hand, in the case of ν_3 only one of the three possible H–D calculated gas-phase frequency shifts can be checked against experimental data. In view of this and of the use of values of ν_3, some corrected and some uncorrected, for *l*-type doubling in Nakagawa and Morino's calculations and the appreciable discrepancy between observed and calculated values of ν_1 of $H^{12}C^{15}N$ (see Table 11), perhaps a little caution should be exercised over the apparent discrepancies of H–D matrix and gas-phase shifts for ν_3. Nevertheless, there is no doubt about the analogous discrepancies for ν_2. It can only be remarked in conclusion that, although there is at the moment no alternative to either using anharmonic shifts or obtaining harmonic matrix shifts by using gas-phase anharmonic constants, neither approximation is completely satisfactory.

An investigation of matrix-isolated hydrogen cyanide of a very different kind is the recent report of its reaction with hydrogen atoms; tentative evidence for the formation of $H_2C{=}NH$ was obtained.[103]

6 Cyanogen Fluoride

Cyanogen fluoride, FCN, first identified[104] in 1931 but not isolated[105] until 1964, is the least investigated of the four cyanogen halides. Monomeric cyanogen fluoride (b.p. 227 K, m.p. 191 K) is best prepared by the high-temperature pyrolysis of cyanuric fluoride, $(FCN)_3$, under reduced pressure and is purified by low-temperature fractional distillation in dry

[103] P. M. A. Sherwood and J. J. Turner, *J. Chem. Soc.* (*A*), 1971, 2474.
[104] V. E. Cosslett, *Z. anorg. Chem.*, 1931, **201**, 75.
[105] F. S. Fawcett and R. D. Lipscomb, *J. Amer. Chem. Soc.*, 1964, **86**, 2576.

glass equipment; it is stable indefinitely at solid carbon dioxide temperature in stainless steel cylinders. At room temperature, liquid cyanogen fluoride is converted rapidly into polymeric materials, including cyanuric fluoride and a high-melting, water-sensitive solid polymer, but in the gas phase at atmospheric pressure the monomer has been recovered partially after several weeks. The microwave spectra [106] of $F^{12}C^{14}N$, $F^{13}C^{14}N$, and $F^{12}C^{15}N$ (195 K; natural concentrations) establish the monomer to be linear, with a CN distance (r_s) of 1.159 Å, virtually identical to that of the other cyanogen halides, a CF distance (r_s) of 1.262 Å, the shortest yet found, and a nuclear quadrupole coupling constant for nitrogen-14, −2.67 MHz, the lowest yet measured for a cyanide. The only structural study on the liquid appears to be a ^{19}F n.m.r study [105] which showed a triplet peak centred at 3180 Hz to the high field of CF_3CO_2H with a 32—35 Hz splitting [attributed to $^{14}N(I = 1)$].

There have been no recent vibrational or vibrational–rotational spectroscopic studies of monomeric cyanogen fluoride, in contrast to the three other cyanogen halides. Also there appear to have been no Raman spectroscopic studies of any phase, no solid-state studies, and no published matrix-isolation study.

The Gaseous State.—The only vibrational and vibrational–rotational investigations of FCN have been of the gas phase (95% purity) in the i.r., at moderate resolution (0.4—0.5 cm^{-1}). It has been stated that high-resolution i.r. studies were to be reported [107] but this does not appear to have happened. It seems clear that the early measurements [105, 108, 109] of ν_1 were in serious error (as were predictions [110] of the values of ν_1, ν_2, and ν_3 based on force-constant extrapolations). The centre of ν_1 is at 2323 ±4 cm^{-1}; the large uncertainty is attributed to this band falling in a region of intense CO_2 absorption.[107] More accurate values of ν_2 [1076.52(5) cm^{-1}] and ν_3 [451.32(5) cm^{-1}] are known;[111] the latter value is uncorrected for *l*-type doubling. As expected, ν_1 and ν_2 have *PR*-doublet contours with no *Q* branch (separation 24 cm^{-1}) and ν_3 a strong *Q* branch and weak *P* and *R* maxima (separated by about 20 cm^{-1}). Rotational analysis of ν_2 and ν_3 in the i.r. has led to values of $B(000^0)$, $B(010^0)$, $B(001^1)$, and ΔB for the transitions ν_2 and ν_3 [109, 111] in agreement with the more precise microwave values [107] to within the stated errors (see Table 29). [It may be noted that several other microwave *B* values have been reported,[107] including $B(000^0)$ and $B(100^0)$ of $F^{13}CN$.] The perpendicular band due to the bending vibration ν_3 resembles that of N_2O very closely. The substantial population of the

[106] J. K. Tyler and J. Sheridan, *Trans. Faraday Soc.*, 1963, **59**, 2661.
[107] W. J. Lafferty and D. R. Lide, jun., *J. Mol. Spectroscopy*, 1967, **23**, 94.
[108] J. Sheridan, J. K. Tyler, E. E. Aynsley, R. E. Dodd, and R. Little, *Nature*, 1960, **185**, 96.
[109] R. E. Dodd and R. Little, *Spectrochim. Acta*, 1960, **16**, 1083.
[110] W. J. Orville-Thomas, *J. Chem. Phys.*, 1952, **20**, 920; N. W. Luft, *ibid.*, 1953, **21**, 1900.
[111] A. R. H. Cole, L. I. Isaacson, and R. C. Lord, *J. Mol. Spectroscopy*, 1967, **23**, 86.

Table 29 *Vibrational and vibrational–rotational data* (cm^{-1}) *for* FCN

(a) Wavenumbers

Mode	*Value*	*Ref.*
ν_1	2323(4)	107
ν_2	1076.52(5)[a]	111
ν_3	451.32(5)[b]	111
$\nu_1+\nu_2$	3370	111
$\nu_1+2\nu_3$	3190	111
$\nu_1+\nu_3$	2740	111
$2\nu_2$	2145	111

(b) Rotational Constants

$B(000^0)$	$B(010^0)$	$B(001^1)$	*Technique*	*Ref.*
0.35246(30)[c]	0.35070(30)	0.35330(30)	Infrared	111
0.3520502(7)	0.3506042(7)	0.3527370(7)[c]	Microwave	107
		0.3533924(7)[d]		

(c) $q(001^1)$

Calculated or observed	*Value*	*Ref.*
Calculated	0.0006511	112
Calculated	0.000656601	113
Observed	0.0006565945	112

[a] ν_0 value. [b] ν_0 value uncorrected for *l*-type doubling. [c] *c* level. [d] *d* level.

first excited state of ν_3 leads [111] to a clearly observable Σ—Π hot band at 441.36 cm^{-1}, which is probably 002^0—001^1, leading to a value of 892.68 cm^{-1} for $2\nu_3{}^0$. A band at about 892.5 ±1.0 cm^{-1} has been observed directly with a 10 cm cell at a pressure of about 500 mm Hg.[107, 111] Interference from upper-stage transitions of 001^1 is also present in the ν_2 band and precise analysis must await a higher-resolution study.

Here, as in other XCN examples, Fermi resonance exists between states with quantum numbers v_1, v_2, $v_3{}^l$ and v_1, v_2-1, v_3+2^l. It thus appears from microwave studies of vibrational satellite lines [107] that the unperturbed values of ν_2 and $2\nu_3{}^0$ are about 7 cm^{-1} higher and lower respectively than are directly observed in the i.r. Quite a good check on the values of these unperturbed vibrational fundamentals is provided by the fact that the *l*-type doubling constant $q(001^1)$, calculated by Lafferty,[112] agrees to better than 99% with the experimental microwave value.* This calculation neglected the off-diagonal force constant and involved taking $B(000^0)$ as an approximation to B_e. Much better agreement was apparently obtained in a

* It should be noted that in Lafferty's paper the ζ values are incorrectly printed. After correction, Lafferty's value for $\zeta_{12}{}^2$ is 0.04329 compared with Rüoff's value for the same constant of 0.05429.

[112] W. J. Lafferty, *J. Mol. Spectroscopy*, 1968, **25**, 359.

calculation by Rüoff[113] when an off-diagonal force constant was included (see Table 29).

Several features observed and assigned in the 2100—3400 cm^{-1} region of the i.r. spectrum include[111] the first overtone of ν_2 and a number of binary and ternary combination tones involving ν_1. The presence of impurities in the sample creates some uncertainty[111] in these assignments, which are listed along with other vibrational and vibrational–rotational data for FCN in Table 29.

The Matrix-isolated State.—It has been stated that ν_1 is at 2320 ±5 cm^{-1} in an argon matrix.[114] No further details of the vibrational properties of FCN in a matrix appear to have been made available, but upon vacuum-u.v. photolysis of argon- and nitrogen-matrix-isolated FCN, two i.r. absorptions appear which can be identified[19] with the stretching fundamentals of the species fluorine isocyanide, FNC:

	Ar	N_2
ν_1/cm^{-1}	2123	2132
ν_2/cm^{-1}	928	939

A careful search of the 250—900 cm^{-1} region yielded[19] no absorption which could be associated with the bending mode of FNC.

7 Cyanogen Chloride

Cyanogen chloride is a colourless gas at room temperature, liquefying at about 286 K. The liquid has a 20 K temperature range, melting at about 266 K. The gaseous state has been extensively investigated, particularly by microwave spectroscopy,[67, 106, 115–118] and the linear monomer is well characterized. An early Raman spectroscopic investigation[119] indicated that the liquid was composed of linear molecules, and a recent n.m.r. study[120] has established that the ^{14}N quadrupole coupling constant of ClCN in the liquid phase is the same, within experimental error, as the value in the solid. In 1956, Heiart and Carpenter[121] established the structure of solid ClCN at about 243 K, in a single-crystal *X*-ray diffraction analysis. The unit cell is orthorhombic with *Pmmn* (D_{2h}^{13}) space group and $z = 2$. The molecules are arranged in infinite linear chains parallel to the twofold *c*

113 A. Rüoff, *Spectrochim. Acta*, 1970, **26A**, 545.
114 D. E. Milligan and M. Jacox, personal communication cited in ref. 107.
115 C. H. Townes, A. N. Holden, J. Bardeen, and F. R. Merritt, *Phys. Rev.*, 1947, **71**, 664; C. H. Townes, A. N. Holden, and F. R. Merritt, *ibid.*, 1947, **72**, 513; 1948, **74**, 113; J. Bardeen and C. H. Townes, *ibid.*, 1947, **73**, 97.
116 A. G. Smith, H. Ring, W. V. Smith, and W. Gordy, *Phys. Rev.*, 1948, **74**, 370.
117 L. Yarmus, *Phys. Rev.*, 1957, **105**, 928.
118 J. J. Ewing, H. L. Tigelaar, and W. H. Flygare, *J. Chem. Phys.*, 1972, **56**, 1957.
119 W. West and M. Farnsworth, *J. Chem. Phys.*, 1933, **1**, 402.
120 K. T. Gillen and J. H. Noggle, *J. Chem. Phys.*, 1970, **52**, 4905.
121 R. B. Heiart and G. B. Carpenter, *Acta Cryst.*, 1956, **9**, 889.

Table 30 *Symmetry correlation for solid* ClCN

	Free molecule $C_{\infty v}$	*Site* C_{2v}	*Factor* D_{2h}	*Activity*
ν_1, ν_2	Σ^+	a_1	a_g	Raman
			b_{1u}	i.r.
ν_3	Π	b_1	b_{2g}	Raman
			b_{3u}	i.r.
		b_2	b_{3g}	Raman
			b_{2u}	i.r.

axis, with the repeat distance between adjacent molecules being the *c* axis cell side. This clearly identifies the existence of a significant intermolecular attraction along the chain. The sum of the gas-phase Cl—N length and the van der Waals radii of chlorine and nitrogen is 6.1 Å, compared with $c = 5.74$ Å. Most of this contraction in the solid state appears to be reflected in a shortened C—Cl bond length [1.57(2) Å compared with 1.630 Å in the gas phase]. The chlorine, carbon, and nitrogen atoms all occupy crystallographically equivalent sites (of symmetry C_{2v}) along the *c* axis and the molecules thus retain their linearity. There are two interesting implications of this crystal structure for the vibrational spectrum (see Table 30). First, in the site-group approximation the degeneracy of the bending vibration of the still linear molecules should be lifted. Secondly, as the primitive unit cell is centrosymmetric there should be no coincidences between i.r. and Raman spectra if the factor-group approximation is applicable.

Of the four cyanogen halides, cyanogen chloride has been most examined by vibrational and vibrational–rotational spectroscopy. Recent investigations have concentrated on obtaining data at higher resolution of the gaseous monomer with the object of estimating the Fermi resonance between the states $|\nu_1, \nu_2, \nu_3{}^l\rangle$ and $|\nu_1, \nu_2-1, \nu_3+2^l\rangle$, determining vibrational anharmonicity constants, and computing quadratic, cubic, and quartic force constants. The liquid, solid, and matrix-isolated states have also all received recent attention.

The Gaseous State.—ClCN has been the subject of a number of vibrational i.r. investigations at prism resolution, the most detailed of which is that of Freitag and Nixon.[122] These workers obtained band centres and assignments for 18 absorptions in the range 3200—300 cm^{-1}; they observed *P*- and/or *R*-branches for 13 of these bands. The band centres included values

[122] W. O. Freitag and E. R. Nixon, *J. Chem. Phys.*, 1956, **24**, 109.

Table 31 *Fundamental frequencies, anharmonic vibrational constants, and Fermi-resonance shifts for* ClCN[122]

Mode	*Wavenumber*/cm^{-1}	*Constant*	*Value*/cm^{-1}
ν_1	2219	x_{12}	−2.0
$\nu_1(^{13}C)$	2164	x_{13}	−11.0
$2\nu_3{}^0$	784	x_{33}	2.1
ν_2	714	x_{ll}	−0.1
ν_3	380		

(784 and 714 bracketed together)

Fermi-resonance shift in $(\nu_2, 2\nu_3{}^0) = 19.5$ cm^{-1}
Fermi-resonance shift in $(\nu_1+2\nu_3{}^0, \nu_1+\nu_2) = 29.5$ cm^{-1}

for the three fundamentals ν_1, ν_2, and ν_3, the isotopic band $\nu_1(^{13}CN)$, the first overtone $2\nu_3{}^0$ (all listed in Table 31), and also values for four summation bands [$\nu_1+2\nu_3{}^0$, $\nu_1+\nu_2$, $\nu_1+\nu_3$, and $\nu_1+\nu_3$ (^{13}C)], one binary and two ternary difference bands ($\nu_1-\nu_3$, $\nu_1+\nu_2-\nu_3$, and $\nu_1+2\nu_3{}^0-\nu_2$), and six hot bands, ($\nu_1+2\nu_3{}^2-\nu_3$, $\nu_1+2\nu_3{}^0-2\nu_3{}^2$, $\nu_1+\nu_3-\nu_3$, $\nu_1+2\nu_3{}^0-2\nu_3{}^0$, $\nu_1+\nu_3-2\nu_3{}^2$, $\nu_1+\nu_3-2\nu_3{}^0$, and $2\nu_3{}^0-\nu_3$). The difference and hot bands observed all involve ν_3, ν_2, or $2\nu_3{}^0$ as the lower level. These vibrations are, of course, all lower in wavenumber than the corresponding vibrations in FCN, where the available evidence suggests that there is less hot-band interference. Freitag and Nixon were able, utilizing nine of their band centres, to evaluate a number of approximate vibrational anharmonicity constants (x_{12}, x_{13}, x_{33}, and x_{ll}) and relationships (see Table 31). They were also able to derive approximate values for the Fermi-resonance shifts in the ν_3,$2\nu_3{}^0$ pair and in the $(\nu_1+2\nu_3{}^0)$,$(\nu_1+\nu_2)$ pair (see Table 31).

As indicated elsewhere, one of the objects of studying the high-resolution i.r. spectra of triatomic cyanides is to investigate Fermi resonance interactions. In order to treat a resonance, say of the $(\nu_1+2\nu_3{}^0)$,$(\nu_1+\nu_2)$ diad, it is necessary to know the α_1, α_2, and α_3 values. Values of α_2 and α_3 are usually available from microwave spectroscopy but it is necessary to evaluate α_1 from high-resolution i.r. spectroscopic data. With values of α_1, α_2, and α_3 together with $B(000^0)$ it is also possible to compute a value for B_e, and if all the data are available for the two isotopic species, such as ^{35}ClCN and ^{37}ClCN, r_e values can be calculated. The two chlorine isotopic species, useful as they thus are, do nevertheless cause complications in the i.r. work since the isotopic shifts are small and the *P* or *R* lines of one species often overlap the other.

There have been three significant high-resolution i.r. studies of ClCN of which the first was carried out by Lafferty, Lide, and Toth [123] in 1965. These workers obtained the i.r. spectrum of ClCN from 1710 cm^{-1} (0.10 cm^{-1} resolution) to 3300 cm^{-1} (*ca.* 0.03 cm^{-1} resolution) at pressures from 1 to 10 Torr, using an optical path of 16 m. In particular, ν_0 and ΔB values were obtained for ν_1 of $^{35}Cl^{12}C^{14}N$, $^{37}Cl^{12}C^{14}N$, and $^{35}Cl^{13}C^{14}N$ (see Table 32); ν_0, and sometimes ΔB and ΔD values, were also obtained for several com-

[123] W. J. Lafferty, D. R. Lide, and R. A. Toth, *J. Chem. Phys.*, 1965, **43**, 2063.

Table 32 ν_0, ΔB, *and* α *values* (cm^{-1}) *for* ClCN

Species	*Vibrational state*	ν_0	$10^5\Delta B$	$10^5\alpha$	*Footnote*
$^{35}Cl^{12}C^{14}N$	001^1	378.62	54.48	−54.47	*a*
	010^0	714.02	0.05	82.52	*a*
	100^0	2215.58	−107.70	107.20	*a*
	001^1	378.423(2)			*b*
	010^0	714.018(4)			*c*
	100^0	2215.582(2)	−107.7(2)		*d*
$^{37}Cl^{12}C^{14}N$	001^1	378.19	53.32		*a*
	010^0	708.32	−8.70		*a*
	100^0	2215.30	−105.90		*a*
	001^1	377.990(3)			*b*
	010^0	708.322(5)			*c*
	100^0	2215.301(3)	−105.9(5)		*d*
$^{35}Cl^{13}C^{14}N$	100^0	2163.58	−101.80		*a*
	100^0	2163.584(3)	−101.8(5)		*d*

[a] Ref. 126; used in force-constant analysis. [b] Ref. 125; corrected for *l*-type doubling and fitted to microwave rotational constants. [c] Ref. 124. [d] Ref. 123.

bination and hot bands of these species. From these data, estimates of the values of ν_2 and $\nu_3{}^1$ and the anharmonicity constants x_{12}, x_{13}, x_{33}, and x_{ll} were made. These data have been superseded in high-resolution investigations by Murchison and Overend,[124, 125] who have obtained a complete set of quadratic anharmonicity constants x_{ij} for ^{35}ClCN, corrected where appropriate for Fermi resonance (see Table 33). (There are insufficient data for an evaluation of a complete set for ^{37}ClCN, but Murchison and Overend have shown that x_{12} and x_{13} for this species are the same, within experimen-

Table 33 *Best harmonic frequencies,* ω_i, *and harmonic constants,* x_{ij}, *for* ^{35}ClCN *and harmonic frequencies for* ^{37}ClCN (cm^{-1})

	^{35}ClCN	^{37}ClCN[a]	*Ref.*
ω_1	2248.96	2248.67	125
ω_2	747.99	739.60	125
ω_3	381.91	381.46	125
x_{11}	−12.63	—	125
x_{22}	−3.19	—	125
x_{33}	0.56	—	125
x_{12}	−1.74	—	125
x_{13}	−7.24	—	125, 123
x_{23}	−2.46	—	125
x_{ll}	−0.11	—	125

[a] Calculated ω values using values of x_{11}, $4x_{33}+x_{23}$, $x_{33}+x_{ll}$, and x_{22} transferred from ^{35}ClCN.

[124] C. B. Murchison and J. Overend, *Spectrochim. Acta*, 1970, **26A**, 599.
[125] C. B. Murchison and J. Overend, *Spectrochim Acta*, 1971, **27A**, 2407.

Table 34 *Unperturbed vibrational energies, ν_0*, and Fermi-resonance matrix elements, W_{23}, for* $^{35}ClCN$ *and* $^{37}ClCN$ (cm^{-1}) [125]

Species	*Vibrational state*	ν_0*	W_{23}
$^{35}ClCN$	010^0	738.27	32.87
	002^0	758.58	
	011^1	1114.43	45.94
	003^1	1139.44	
	110^0	2952.11	33.88
	102^0	2959.68	
	020^0	1470.16	
	012^0	1491.93	
	004^0	1521.64	
$^{37}ClCN^a$	010^0	729.88	32.62
	002^0	757.68	
	110^0	2943.44	33.66
	102^0	2958.50	

[a] x-constants transferred from $^{35}ClCN$.

tal error, as for $^{35}ClCN$.) Lafferty *et al.* also calculated average values of α_1 for $^{35}Cl^{12}C^{14}N$ and $^{37}Cl^{12}C^{14}N$ (see Table 32) from values obtained from ν_1 (*i.e.* $-\Delta B$) and values obtained from $\nu_1+\nu_3{}^1$ using the microwave values for α_3 and $q(001^1)$, with the assumption that $q(101^1)$ is equal to $q(001^1)$, These average values of α_1 were used with the observed band origins and ΔB values to analyse the Fermi resonance between the levels 102^0 and 110^0. The more recent and detailed [124, 125] work of Murchison and Overend has, however, allowed the evaluation of more precise interaction constants, W_{23}, for this and other polyads of $^{35}ClCN$ and $^{37}ClCN$ (see Table 34). Later work has not modified these average values of α_1, and therefore the discussion of internuclear distances in ClCN given by Lafferty *et al.* is still of relevance. It is important to bear in mind, however, that the data available on ClCN to Lafferty *et al.* are more limited in extent and also less precise than those available to Winnewisser *et al.*[72] on HCN. Lafferty *et al.* found (see Table 35) that the Laurie–Herschbach average distances $\langle r \rangle$ in ClCN are probably a few thousandths of an ångström longer than the equilibrium distances (but no more than 0.01 Å different). The r_s structure is a good approximation to r_e (within, say, 0.005 Å) and r_0 is an excellent approximation to $\langle r \rangle$. Closer comparisons are limited by the experimental uncertainty in r_e and by the unknown effects of isotopic changes in $\langle r \rangle$. These isotopic changes limit the reliability of $\langle r \rangle$ in exactly the same way as they do the r_0 structures. These results indicate that the C—N bond is somewhat longer in ClCN than in HCN, as was noted [106] by Tyler and Sheridan. Whether r_e, r_s, or $\langle r \rangle$ distances are used in the comparison, this bond differs by about 0.004 Å in the two molecules.

Recent high-resolution work by Murchison and Overend [124, 125] has involved the measurement of ν_3, $2\nu_1$, $2\nu_2$, $\nu_2+2\nu_3{}^0$, and $4\nu_3{}^0$ bands of $^{35}Cl^{12}C^{14}N$

Table 35 *Bond parameters for* ClCN

(a) Internuclear distances/Å [123]

Isotopic pair	Cl—C r_e	Cl—C $\langle r \rangle$	Cl—C r_0	C—N r_e	C—N $\langle r \rangle$	C—N r_0	Cl⋯N r_e	Cl⋯N $\langle r \rangle$	Cl⋯N r_0
^{35}ClCN, ^{37}ClCN	1.629(6)	1.627(2)	1.627(1)	1.160(7)	1.166(2)	1.166(1)	2.789(2)	2.7941(7)	2.7934(3)
^{35}ClC^{15}N, ^{37}ClC^{15}N	—	1.627(3)	1.627(2)	—	1.167(8)	1.166(2)	—	2.7940(8)	2.7933(6)
^{35}ClC^{15}N, ^{35}ClC^{14}N	—	1.6349(9)	1.6340(5)	—	1.1570(5)	1.1574(3)	—	2.7918(7)	2.7913(4)
^{37}ClC^{15}N, ^{37}ClC^{14}N	—	1.6349(9)	1.6341(5)	—	1.1568(5)	1.1571(3)	—	2.7918(7)	2.7913(4)
		r_s(Cl—C) = 1.631			r_s(C—N) = 1.159			r_s(Cl⋯N) = 2.790	

(b) Force constants[a]

k(CN) = 1798.2(8.0) N m^{-1}
k(ClC) = 528.4(1.4) N m^{-1}
k(ClC,CN) = 39.5(4.5) N m^{-1}
k(ClCN) ≡ k_α = 0.3502(5) × 10^{-18} N m rad^{-2}

[a] Ref. 126; calculated assuming k_{11} = 2.674 mdyn Å^{-1}, k_{13} = 0.395 mdyn Å^{-1}, and k_{22} = 0.17[illegible]1 mdyn Å rad^{-2}.

and the Fermi doublet ν_2 and $2\nu_3$ of the ^{35}Cl and ^{37}Cl species in a natural sample. From the corresponding ν_0 and ΔB values, and also previous data, a complete set of ω_i and x_{ij} (see Table 33) corrected for the effects of Fermi resonance between the states $|\nu_1, \nu_2, \nu_3{}^l\rangle$ and $|\nu_1, \nu_2-1, \nu_3+2^l\rangle$ were evaluated for $^{35}Cl^{12}C^{14}N$. Also, Fermi-resonance operators (see Table 34) for the 010^0,002^0 and 011^1,003^1 diads of ^{35}ClCN and the 010^0,002^0 diad of ^{37}ClCN were estimated.

Murchison and Overend [126] have also attempted to adjust the force constants in the general quartic force field of ClCN to achieve a least-squares fit to 52 values of ν and ΔB of $^{35}Cl^{12}C^{14}N$, $^{37}Cl^{12}C^{14}N$, and $^{35}Cl^{13}C^{14}N$. Not too surprisingly, it was found that the general quartic force field contained too many (19) adjustable parameters. As a result, a Morse function was used for bond–bond stretching potentials and the interaction force constants were derived semi-empirically. The numerical values obtained for the quadratic force constants are listed in Table 35. More limited discussions of the force constants of ClCN have recently been given by Rüoff [113] and by Durig and Nagarajan.[127]

The Liquid State.—Apart from a brief report [128] of the far-i.r. absorption spectrum (20—180 cm^{-1}) of liquid ClCN at 268 and 300 K, which only showed the characteristic broad liquid band centred at 60—70 cm^{-1}, the spectroscopic investigation of the liquid state of ClCN has been by the Raman effect.[119, 128, 129] In the latest investigation [128] at 300 K using Toronto-arc and He–Ne-laser sources, low-wavenumber shoulders were observed for both ν_1 and ν_2 (see Table 36). The observed wavenumber splitting of ν_2 was of the right order (8 cm^{-1}) for the chlorine isotope effect, but in the ν_1 region the calculated splitting should be less than 1 cm^{-1}, whereas it was observed to be $\sim$5 cm^{-1}. This discrepancy led Pézolet and Savoie to attribute these shoulders to the presence of associated molecules in the liquid phase. This explanation is supported by the observation [128] of similar splittings in liquid BrCN which are much larger than calculated for the ^{79}Br and ^{81}Br isotopes. It is interesting to note that in liquid HCN and DCN,

Table 36 *Raman spectra of liquid* ClCN

Observed shift/cm^{-1}		*Assignment*
Ref. 119, *at* 291 K	*Ref.* 128, *at* 280 K	
397	398	ν_3
	720	ν_2 dimer?
729	728	ν_2
	2200	ν_1 dimer?
2201	2205	ν_1

[126] C. B. Murchison and J. Overend, *Spectrochim. Acta*, 1971, **27A,** 1801.
[127] J. R. Durig and G. Nagarajan, *Monatsh.*, 1970, **101,** 437.
[128] M. Pézolet and R. Savoie, *J. Chem. Phys.*, 1971, **54,** 5266.
[129] J. Wagner, *Z. phys. Chem.* (*Leipzig*), 1941, **B48,** 309; 1943, **A193,** 55.

Table 37 *I.r. and Raman spectra of solid* ClCN

I.r.			*Raman*[128]		*Assignment*
77 K[131]	93 K[122]	77 K[130]	203 K	78 K	
			59	65	T_x (b_{2g})
			59	65	T_z (a_g)
			77	90	T_y (b_{3g})
			84	97	R_y (b_{2g})
93 } (750 ± 90)[a]					R_y (b_{3u})
			120	130	R_x (b_{3g})
114					R_x (b_{2u})
			405	405	ν_3 (b_{2g})
			412	414	ν_3 (b_{3g})
	398	398 (652 ± 62) [±21.6][b]			ν_3 ($b_{2u}+b_{3u}$)
			728	728	ν_2 (^{37}Cl)
			736	736	ν_2 (a_g)
	734	734 (0 ± 20) [±0.0]			ν_2 (b_{1u})
			810	817 }	$2\nu_3$
	830.5	830 (~27)	833	839 }	$2\nu_3$
	2158		2155	2154	ν_1 (^{13}C)
	2182		2179	2179	ν_1 (^{15}N)
			2209	2210	ν_1 (a_g)
	2212	2209 (3636 ± 346) [±72.1]			ν_1 (b_{1u})

[a] Values in parentheses are absolute absorption intensities in darks. [b] Values in square brackets are $(\partial\mu/\partial q)$ in esu g$^{-\frac{1}{2}}$.

where there is an accumulation of evidence, both vibrational and otherwise, for association, such low-wavenumber shoulders were not observed; in fact, it was concluded from the symmetrical band shapes observed that dimers were not present in appreciable concentration.

The Solid State.—The band positions, band intensities, and assignments obtained in i.r. and Raman spectra of the solid are collected in Table 37. The Raman assignments, particularly in the lattice region, were considerably aided by analogy with those made for isostructural BrCN from the results of an oriented single-crystal study.[128] In so much as the Raman frequencies in the ν_1 and ν_2 stretching regions are almost identical to the i.r. frequencies, it would appear that the two stretching vibrations are not significantly perturbed by the environment. As previously noted, the unit cell of ClCN is centrosymmetric, and so in the factor-group approximation (see Table 30) the Raman-active CH and ClC stretching vibrations belong to symmetry species different from those of the corresponding i.r.-active vibrations. This lack of environmental perturbation is supported by the almost complete absence of temperature dependence of these vibrations (see Table 37). On the other hand, evidence of intermolecular coupling is possibly manifest in the intensities of ν_2. The observed intensity ratio $I[\nu_2(^{37}\text{ClC})] : I[\nu_2(^{35}\text{ClC})]$

is about 1:5, whereas it is predicted to be 1:3 from the natural abundances of the ^{37}Cl and ^{35}Cl isotopes. It does seem necessary to invoke the factor-group approximation (see Table 30) in the interpretation of the ν_3 bending region. Two well-separated peaks (b_{2g} and b_{3g} species) characterize this region of the Raman spectra of ClCN and BrCN; the two i.r.-active components (b_{2u} and b_{3u} species) have essentially identical frequencies some 7 cm^{-1} below the lower (b_{2g}) Raman component. The lattice region of ClCN is reasonably well assigned by analogy with the single-crystal BrCN Raman results. The absolute absorption intensities of the internal modes and the two i.r.-active librational modes of ClCN (and BrCN) have been recently measured.[130, 131] Even though the ClCN and BrCN crystal structures are isostructural, the observed internal mode intensities show quite different changes from the gas and crystal phases. The bending mode (ν_3) intensities do not change drastically upon condensation, and the changes that do occur are of the magnitude expected for the field effect. The stretching modes (ν_1 and ν_2) are affected much more by the change from gas to crystal, with the ν_1 intensities in the crystal being higher for ClCN (and BrCN and ICN—see p. 516). However, the intensity of ν_2 in ClCN is much lower in the crystal than in the gas phase, unlike BrCN and ICN (see p. 521). These varying intensity changes can be understood in terms of intermolecular coupling and necessitate the consideration of an additional resonance structure. In particular, the negligible changes in the intensities of ν_3 in going from the gas to the solid indicate that the major perturbation of the vibrational intensities in crystals of these molecules is parallel to the molecular chains. A dipolar-coupling model has been recently used [131] to compute the expected intensities of the two i.r.-active librational lattice modes and the frequencies of the Raman- and i.r.-active librations. This model is reasonably successful in explaining the optical spectra of ClCN (and BrCN), though it seems likely that the agreement between the calculated and observed frequencies for ClCN is fortuitous.

The Matrix-isolated State.—Murchison and Overend [132] have recently obtained the i.r. spectra of ClCN isolated in Ar and Ne matrices at temperatures in the range 4—20 K. They reported values of the fundamentals (see Table 38) and of a number of combination and overtone bands of the monomer. They were able to deduce values of the anharmonic constants x_{12} and x_{13}, which were close to the values they had found previously for ClCN in the gas phase. As there are no light-atom substitutions involved in ClCN it is perhaps not surprising that Murchison and Overend were able to account for the observed matrix shifts of the vibrational lines by adjusting the quadratic intramolecular force constants (*cf.* HCN). Their

[130] A. R. Bandy, H. B. Friedrich, and W. B. Person, *J. Chem. Phys.*, 1970, **53**, 674.
[131] H. B. Friedrich, *J. Chem. Phys.*, 1970, **52**, 3005.
[132] C. B. Murchison and J. Overend, *Spectrochim Acta*, 1971, **27A**, 1509.

Table 38 *Vibrational spectra of matrix-isolated* ClCN (*all values in* cm^{-1})

(a) Monomer [132]

Species	*Argon matrix*	*Neon matrix*	*Gas phase*	*Assignment*
$^{35}ClCN$	2208.71	2216.34	2215.58	ν_1
	717.80	719.64	714.02	ν_2
	386.95 } 384.62 }	384.53	378.62	$\nu_3{}^1$
$^{37}ClCN$	2208.47	2216.09	2215.30	ν_1
	711.38	713.44	708.32	ν_2
	386.58 } 384.20 }	—	378.20	$\nu_3{}^1$
$Cl^{13}C^{14}N$	2156.55[a]	2164.03[a]	2163.58	ν_1
$Cl^{12}C^{15}N$	—	2188.37[a]	—	ν_1

(b) Dimer [133] and monomer [133] (argon matrix)

Monomer	*Dimer*	*Assignment*
2208.6	2206.2, 2223.1	ν_1
791.6	800.1, 809.6	$2\nu_3$ (^{35}Cl)
788.7	796.9, 807.7	$2\nu_3$ (^{37}Cl)
717.8	722.6, 730.2	ν_2 (^{35}Cl)
711.4	715.6, 724.4	ν_2 (^{37}Cl)
384.4	388.6, 397.0	ν_3 }
386.8	390.8, 398.5	ν_3 }

[a] ^{35}Cl, ^{37}Cl isotopic species not resolved.

results indicate that the matrix cage does not perturb the anharmonic intramolecular force constants any more than the quadratic ones.

Very recently Freedman and Nixon [133] have obtained the i.r. spectra of ClCN (and BrCN) in Ar and Kr matrices at 20 K. They found good evidence that, for the monomer, ν_3 is split by the matrix environment (as was deduced by Murchison and Overend) and that the first overtones of these bending components form a Fermi-resonant triad with the ν_2 fundamental. The frequencies and intensities of the unperturbed levels and the Fermi-resonance coupling constants were calculated. (Murchison and Overend did not report the additional weak band at 772.0 cm^{-1} in the ν_2, $2\nu_3$ region, found by Freedman and Nixon in argon, but treated the Fermi resonance effect by taking the average of the ν_3 components.) In addition, Freedman and Nixon found extra absorption bands not reported by Murchison and Overend which they were able to attribute to a linear or nearly linear non-centrosymmetric dimer (see Table 38).

[133] T. B. Freedman and E. R. Nixon, *J. Chem. Phys.*, 1972, **56**, 698.

8 Cyanogen Bromide

Cyanogen bromide, BrCN, is a colourless solid which melts at about 225 K and boils at about 335 K. The linear symmetric structure of the monomeric molecule in the gas phase was established in an early electron-diffraction study.[134] Extensive microwave investigations have led to accurate molecular parameters.[67, 115, 123, 135] The liquid phase was the subject of an early Raman spectroscopic investigation;[119] no evidence of deviation from linearity was found. A single-crystal *X*-ray diffraction analysis [136] established BrCN to be isostructural with ClCN, and to have a slightly larger intermolecular contraction.

Recent i.r. spectroscopic interest has centred on high-resolution studies of the vibration–rotation spectrum, absolute absorption intensity measurements on the librational lattice mode, and an investigation of the matrix-isolated species. As with HCN and ClCN, the Raman spectra of the liquid and the solid, but not the gas, have received recent attention.

The Gaseous State.—The first i.r. spectroscopic investigation of gaseous BrCN was that of Freitag and Nixon,[122] who reported the vibrational spectrum in the range 3200—300 cm^{-1} in 1956. They were not able to observe any rotational fine structure with the resolution available to them. The positions of ν_1, ν_2, ν_3, ν_1(^{13}CN), and $2\nu_3{}^0$ are listed in Table 39. These and other band positions allowed evaluation of approximate values of the anharmonicity constants x_1, x_{12}, x_{13}, and x_{ll} (see Table 39).

Whereas the vibration spectrum of BrCN is similar in almost every detail to that of ClCN there are several factors which cause the vibration–rotation spectrum to be more complex. The two naturally occurring isotopes of bromine (^{79}Br and ^{81}Br) have nearly equal abundance, thus giving rise to two sets of spectra which nearly, but not quite, coincide. ν_3 occurs at a very low frequency and ν_2 at a relatively low frequency (see Table 39). As a result there are many quite prominent hot bands which cause considerable difficulty in following the main bands out to high *J*'s. Nevertheless there have been four significant high-resolution studies [137–140] of BrCN.

Table 39 *Vibrational frequencies and anharmonicity constants of* BrCN [122]

Mode	*Wavenumber*/cm^{-1}	*Constant*	*Value*/cm^{-1}
ν_1	2200	x_{11}	0.7
ν_2	575	x_{12}	−3.0
ν_3	342.5	x_{13}	−7.5
ν_1 (Br^{13}CN)	2147	x_{ll}	−0.4
$2\nu_3{}^0$	691		

[134] J. Y. Beach and A. Turkevich, *J. Amer. Chem. Soc.*, 1939, **61**, 299.
[135] S. J. Tetenbaum, *Phys. Rev.*, 1952, **86**, 440.
[136] S. Geller and A. L. Schawlow, *J. Chem. Phys.*, 1955, **23**, 779.
[137] A. G. Maki and C. T. Gott, *J. Chem. Phys.*, 1962, **36**, 2282.
[138] A. G. Maki, *J. Chem. Phys.*, 1963, **38**, 1261.
[139] M. Bellouard and C. Meyer, *Compt. rend.*, 1970, **270**, *B*, 1562.
[140] M. Bellouard, *Compt. rend.*, 1971, **273**, *B*, 1099.

Table 40 *Vibration–rotation analysis of* ^{79}BrCN *and* ^{81}BrCN

Molecule	*Transition*	ν_0	$\nu_0{}^{*a}$	B'	B'^{*}	ΔB $[B'-B(000^0)]$	$B(000^0)$	*Ref.*
^{79}BrCN	110^0—000^0	2771.541(2)	2784.92(1)	0.13643(3)	0.13627(3)	−0.00103(7)	0.13746(4)	140
		2771.548(20)	2784.93	0.136400	—	−0.001036	0.137435(5)	137
	101^1—000^0	2533.765(20)	—	0.137079	—	−0.000356		137
	102^0—000^0	2878.647(2)	2865.27(1)	0.13741(4)	0.1375[illegible](4)	−0.00005(8)		140
		2878.610(20)	2865.23	0.137398	—	−0.000048		137
^{81}BrCN	110^0—000^0	2769.898(2)	2783.12(1)	0.13564(2)	0.1354[illegible](3)	−0.00103(5)	0.13667(3)	140
		2769.905(20)	2783.13	0.135519	—	−0.001035	0.136654(5)	137
	101^1—000^0	2533.712(20)	—	0.136277	—	−0.000377		137
	102^0—000^0	2878.322(1)	2865.10(1)	0.13662(3)	0.136[illegible]8(3)	−0.00005(6)		140
		2878.298(20)	2865.07	0.136597	—	−0.000057		140

[a] Unperturbed by Fermi resonance.

Maki and Gott[137] measured the vibration–rotation structure of the $\nu_1+\nu_3^1$, $\nu_1+2\nu_3^0$, and $\nu_1+\nu_2$ bands for ^{79}BrCN and ^{81}BrCN (see Table 40). The positions of the Q branches of 21 other bands were also measured. Analysis of the vibration–rotation data of the three combination bands provided an average value of α_1 for both isotopic species (*cf.* ClCN) and individual values of $|W_{23}|$ and x_{ll}. The average α_1 value allowed the calculation of B_e values for the two isotopic species. Maki[138] was able to confirm the correctness of this average value a little later when he measured the vibration–rotation structure of ν_1. Rather surprisingly, however, the individual values obtained indicated that the heavier isotopic species has the larger α_1. Analysis of the hot band 101^1—001^1 in Maki's work[138] provided an average ν_0 value for the two isotopic species. This, when combined with the ν_0 for $\nu_1+\nu_3^1$, provided a better ν_0 value for ν_3 than was obtained by direct measurement at low resolution by either Freitag and Nixon or Maki and Gott (see Table 41). Maki was also able to give a revised value for x_{ll} and individual values for x_{33} and x_{13}. The values of x_{33}, x_{ll}, and α_3

Table 41 *Data* (cm^{-1}) *for* BrCN

(a) *Fundamental vibrations*

Species	ν_1	ν_2	ν_3			
		d, e	*a*	*b*	*c*	*d*
^{79}BrCN	2198.29(2)[i]	586.61	341.67(4)[i]			341.658(22)
	2198.3095(20)[j]			341.52 (both species)	342.5 (both species)	
^{81}BrCN	2198.29(2)[i]	584.84	341.61(4)[i]			341.632(22)
	2198.2783(20)[j]					

(b) *Vibrational anharmonic data and Fermi-resonance interaction constant*

	^{79}BrCN		^{81}BrCN	
	j	*k*	*j*	*k*
ν_2-x_{12}	586.59(1)	586.64(35)	584.84(1)	584.84(35)
x_{13}	−6.1925(40)	−6.19(4)	−6.198(4)	−6.19(4)
x_{33}	−0.85(8)	−0.86(10)	−0.85(8)	−0.86(10)
W_{23}	35.4(2)	−35.41	35.5(2)	−35.48
x_{ll}		1.14		1.14
$2x_{33}+x_{13}+x_{ll}$		−5.58		−5.60
$\nu_2+2x_{33}+x_{12}$		584.91		583.06

(c) *Rotational constants*

Species	$10^4\alpha_2$	$10^4\alpha_3$	$10^4\alpha_1$	B_e	10^4q[f, h]
^{79}BrCN	5.18[f]	−3.857[f]	6.786(3)[g]	0.137648(7)[j]	1.306
	5.19[f, h]	−3.851[f, h]	6.77[i]	0.137648[i]	
^{81}BrCN	5.16[f]	−3.831[f]	6.754(3)[g]	0.136867(7)[j]	1.291
	5.16[f, h]	−3.825[f, h]	6.77[i]	0.136869[i]	
	r_e(CN) = 1.157 Å[i]		r_e(CBr) = 1.790 Å[i]		

[a] From hot-band analysis of ref. 138. [b] Direct measurement; ref. 137. [c] Ref. 122. [d] From combination band; ref. 139. [e] Corrected for Fermi resonance. [f] Ref. 135. [g] Ref. 139. [h] Ref. 67. [i] Ref. 137. [j] Ref. 140. [k] Ref. 138.

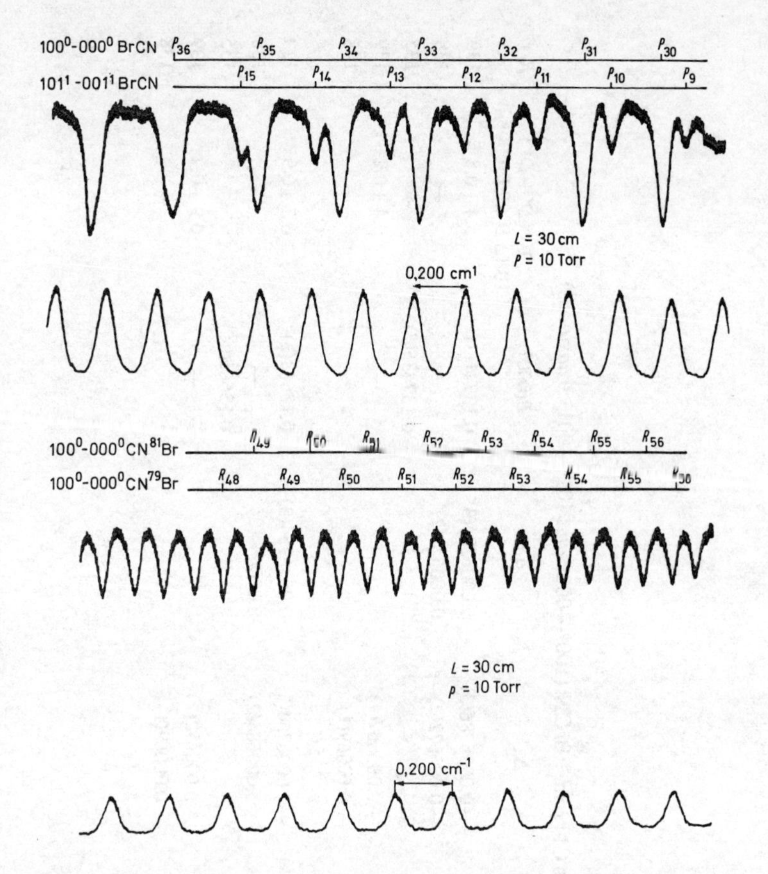

Figure 5 *Vibration–rotation i.r. spectrum of* ν_1 *and* $(\nu_1+\nu_3{}^1-\nu_3{}^1)$ *of* BrCN at 2200 cm^{-1} *with a resolution of* 0.017 cm^{-1}; (a) *part of the P branches*; (b) *part of the R branches*
(Reproduced by permission from *Compt. rend.*, 1970, **270**, *B*, 1562)

allowed the calculation of a value for W_{23} in satisfactory agreement with the observed value and, moreover, indicating the correct sign.

Recently, however, Bellouard and Meyer [139] have studied the same two transitions of both isotopic species as Maki did in his later work (10^00—00^00 and 10^11—00^11; see Figure 5), but with a resolution of 0.017 cm^{-1} compared with that of 0.09 cm^{-1} attained by Maki. The two sets of results are compared in Table 42. It can be seen that Bellouard and Meyer's more precise study indicates that α_1 (*i.e.* $-\Delta B$ of 10^00) is larger for the lighter isotopic species, in contrast to Maki's finding, and moreover that the separation between the two α_1 values is much smaller than suggested by Maki. In this context it is

Table 42 *Vibration–rotation analysis of* ^{79}BrCN *and* ^{81}BrCN (100^0—000^0 *and* 101^1—001^1 *transitions*)

Molecule	*Transition*	ν_0	ΔB	α_1	$B(000^0)$	$10^8(D'+D'')$ $10^8[\Delta D+2D(000^0)]$	*Ref.*
^{79}BrCN	100^0—000^0	2198.3095(20)	−0.0006786(3)	0.0006786(3)	0.13741(3)	5.4 ±0.8	139
		2198.29(2)	−0.000672(2)	0.000672(2)	—	—	138
					0.137435(5)	—	135
	101^1—001^1	2192.107(2)	−0.000667(1)		—	0.4 ±0.2	139
		2192.10(2)	−0.00069(1)		—	—	138
^{81}BrCN	100^0—000^0	2198.2783(20)	−0.0006754(3)	0.0006754(3)	0.13668(3)	6.7 ±0.9	139
		2198.29(2)	−0.000684(2)	0.000684(2)	—	—	138
					0.136654(5)		135
	101^1—001^1	2192.080(2)	−0.000672(2)		—	0.9 ±0.2	139
		2192.10(2)	−0.00069(1)		—	—	138

interesting to speculate whether similar higher-resolution studies on cyanogen might reveal that $B_e(^{12}C_2{}^{14}N_2)$ was *not* less than $B_0(^{12}C_2{}^{14}N_2)$ (see Section 4). Other noteworthy features of Bellouard and Meyer's investigation are the elucidation of the $[\Delta D+2D(000^0)]$ terms for both transitions and the derivation of values of $B(000^0)$ in excellent agreement with the microwave values, though an order of magnitude less precise. The improvement in measurement of the hot band, 101^1—001^1, results in the separation in the derived ν_0 values of the fundamental $\nu_3{}^1$ for the two isotopic species being no greater than the uncertainties of the individual values (see Table 41). Very recently Bellouard [140] has reported a high-resolution (0.025 cm^{-1}) study of the vibrations $\nu_1+\nu_2$ and $\nu_1+2\nu_3{}^0$ which are in Fermi resonance with each other; these bands were previously studied by Maki and Gott. The two sets of results together with Maki and Gott's data for the combination band $\nu_1+\nu_3{}^1$ are listed in Table 40. Again, it can be seen that the derived i.r. $B(000^0)$ values, particularly that for ^{81}BrCN, are in good agreement with those derived from the analysis of ν_1 and also with the more precise microwave values. This latest study results in very small revisions to the parameters ν_2-x_{12}, x_{13}, x_{33}, and W_{23} for the two isotopic species. These latest values together with the best values of ν_1, ν_2, $\nu_3{}^1$, α_1, α_2, α_3, and B_e for the two isotopic species are listed in Table 41. As the average of the revised α_1 values is the same as that originally proposed by Maki and Gott, there is little change in the B_e values of the two isotopic species, and none in r_e(BrC) and r_e(CN). These r_e values are incidentally almost identical to the r_s values obtained [106] entirely from microwave data.[141]

This account of gaseous BrCN may be concluded by noting that Rüoff,[113] using values of $\nu_0(100^0$—$000^0)$ and $\nu_0(010^0$—$000^0)$ obtained by Maki for both isotopic species, has calculated values of the force constants k(CN), k(BrC), and k(BrCN) and associated values of $D(000^0)$ and $q(001^1)$ in good agreement with the experimental microwave values (see Table 43).

Table 43 *Force constants, centrifugal distortion constants, and q constants of* ^{79}BrCN *and* ^{81}BrCN [113]

Force constant	*Value*/N m^{-1}
k(CN)	1750.5 ± 20
k(BrC)	417.4 ± 5
k(BrCN)	36.3 ± 16

Constant	*Value*/MHz	
	^{79}BrCN	^{81}BrCN
$D(000^0)$ calc.	0.8685	0.8592
$D(000^0)$ obs. (microwave)[a]	0.8844	0.8716
$q(001^1)$ calc.	3.9130	3.8679
$q(001^1)$ obs. (microwave)[a]	3.915	3.869

[a] Ref. 67.

[141] C. H. Townes and A. L. Schawlow, 'Microwave Spectroscopy', McGraw-Hill, New York, 1955.

The Liquid State.—There have been two Raman spectroscopic investigations [119, 128] of liquid BrCN. The results are listed in Table 44. The low-wavenumber shoulders of ν_1 and ν_3 observed in the latest investigation, which utilized Toronto-arc and He–Ne-laser sources, cannot be due to isotope effects as with ClCN. It is suggested that these shoulders arise from associated molecules.

Table 44 *Raman spectra of liquid* BrCN

Assignment	*Wavenumber*/cm^{-1} *Ref.* 119[a]	*Ref.* 128[b]
ν_1 BrCN	2186	2187
ν_1 $(BrCN)_2$?		2181
ν_2 BrCN	580	577
ν_2 $(BrCN)_2$?		554
ν_3	368	365

[a] Excitation: Hg 4358 Å. [b] Excitation: He–Ne laser and Hg 4358 Å.

The Solid State.—In view of the isostructural nature of solid BrCN and solid ClCN, it is not too surprising to find that the vibrational investigation of solid BrCN has followed the same pattern as with solid ClCN [122, 128–131] and led to similar results. The frequency and intensity results are listed in Table 45. Two points worth noting are that the assignments of the Raman spectrum were obtained from oriented single-crystal polarization data; and that whereas the intensity of ν_2 in ClCN is much lower in the crystal than in the gas phase, it is much higher for BrCN (and ICN—see p. 521).

The Matrix-isolated State.—We have already referred to a very recent investigation of ClCN in Ar and Kr matrices at 20 K by Freedman and Nixon.[133] These workers also investigated BrCN under the same conditions and obtained very similar results. Assignments of ν_1, ν_2 (^{79}BrC and ^{81}BrC), ν_3, and $2\nu_3$ for the monomer and non-centrosymmetric linear dimer are listed in Table 46.

9 Cyanogen Iodide

Cyanogen iodide, ICN, is a very poisonous solid crystallizing in long, white needles. The solid slowly volatilizes at ordinary temperatures (vapour pressure 0.9 Torr at 298 K) and melts at 419.7 K. It is decomposed at a moderate heat into cyanogen and iodine and is slightly soluble in water but very soluble in organic solvents such as ethanol or ether.

It is well established that the principal, if not exclusive, species in the gas phase is a linear monomer. In 1948 two sets of workers reported [115, 116] the microwave spectrum of ICN. The results were in excellent agreement and unequivocally established that ICN was linear in the gas phase with I—C = 1.995 Å and C—N = 1.159 Å. A more recent and detailed investi-

Table 45 *Frequencies and intensities of the fundamentals of* BrCN *in the solid state*

Assignment	*Raman*		*I.r.*					
	Ref. 128	*Ref.* 128	*Ref.* 130		*Ref.* 131		*Ref.* 122	
	310 K	78 K	77 K	*intensity*/dark	77 K	*intensity*/dark	93 K	$\Delta\nu^{\frac{1}{2}}$
$T_x\ (b_{2g})$	40	50						
$T_z\ (a_g)$	42	53						
$T_y\ (b_{3g})$	61	71						
$R_y\ (b_{2g})$	85	102						
$R_y\ (b_{3u})$	—	—			104			
$R_x\ (b_{3g})$	107	121				890 ±170		
$R_x\ (b_{2u})$	—	—			121			
$\nu_3\ (b_{2g})$	366	368					363.5 s	6
$\nu_3\ (b_{2u}+b_{3u})$	—	—	363	522 ±69				
$\nu_3\ (b_{3g})$	373	377						
$\nu_2\ (a_g)$	574	573					572.5 s	6
$\nu_2\ (b_{1u})$	—	—	573	675 ±41				
$2\nu_3$	745 760	755 770	740 762	0 ±20			740 m 762 m	
$\nu_1\ (^{13}C)$	2138	2138					2142 w	
$\nu_1\ (^{15}N)$	2159	2159					2163 vw	
$\nu_1\ (a_g)$	2190	2190					2194 vs	9
$\nu^1\ (b_{1u})$	—	—	2193	3514 ±414				

Table 46 *Vibrational spectra of matrix-isolated* BrCN [133]

Assignment	*Wavenumber*/cm^{-1} *Monomer*	*Dimer*
ν_1	2190.6	2189.3
		2211.0
$2\nu_3$	706.6	709.1
		732.3
	699.5	
ν_2 (^{79}Br)	575.0	569.8
		594.6
ν_2 (^{81}Br)	573.3	568.2
		593.2
ν_3	348.5	352.9
		362.4?
	350.9	354.7
		364.3

gation by Tyler and Sheridan [106] gave r_s(IC) = 1.994 Å, r_s(CN) = 1.159 Å, μ = 3.71 D, and a ^{14}N nuclear quadruple coupling constant of −3.8(1) MHz. In a very recent microwave investigation, the molecular Zeeman effect in ^{127}I^{12}C^{15}N has been analysed.[118]

The structural chemistry of ICN in the solid state is not so well established as in the vapour. In 1939 Ketelaar and Zwartsenberg [142] deduced, in a single-crystal *X*-ray diffraction analysis, that the structure is composed of linear chains of ICN molecules. Each atom occupies a crystallographically equivalent site (of symmetry C_{3v}). The chain lies parallel to the three-fold *c* axis of the trigonal unit cell of space-group C_{3v}^5 with $z = 3$. The primitive cell has $z = 1$. Although the accuracy of this structure determination is not high, it does lead to a reasonably satisfactory value for the repeat distance in the chain of 6.0 Å. This distance is certainly about 0.8 Å smaller than the sum of the I—C—N distances in the vapour phase and the I and N van der Waal's radii. It would thus seem that there is acceptable evidence of a fairly strong intermolecular interaction along the chain axis provided ICN···I is linear.[136, 143, 144] However, the presence of this intermolecular interaction has recently been contradicted by Pasternak and Sonnino,[145] though without explanation. Furthermore, these same authors dispute the suggestion of Townes and Dailey,[144] based on a nuclear quadrupole resonance study, that at room temperature molecules of ICN in the solid are joined by a bond with about 10% single covalent character. This explanation was put forward to explain the fact that the nuclear quadrupole coupling of iodine in ICN is 5% greater in the solid (at 293 K) than in the gas. Pasternak and Sonnino argue, however, that this increase in electric field gradient in iodine from the gaseous to the solid state can be attributed to intramolecular fac-

[142] J. A. A. Ketelaar and J. W. Zwartsenberg, *Rec. Trav. chim.*, 1939, **58**, 449.
[143] P. A. Casabella and P. J. Bray, *J. Chem. Phys.*, 1958, **28**, 1182.
[144] C. H. Townes and B. P. Dailey, *J. Chem. Phys.*, 1952, **20**, 35.
[145] M. Pasternak and T. Sonnino, *J. Chem. Phys.*, 1968, **48**, 2009.

tors if the nuclear quadrupole constants of BrCN are considered as well as those of ICN. They considered that this conclusion was supported by the absence of anisotropy in the recoilless fraction obtained in a Mössbauer ^{129}I study of ICN at 100 K. From this same investigation, they obtained an isomer shift consistent with no *sp*-hybridization at the iodine atom and a ^{127}I coupling constant at 100 K larger by 4.5% than that measured by nuclear quadrupole resonance spectroscopy.

Until recently there were only two publications of note on the vibrational spectrum of ICN, one on the Raman spectrum of ICN in MeOH [119] and the other on the i.r. spectrum of ICN at 93 K.[122] During the past five years, however, the i.r. spectrum of the vapour at moderate resolution, the Raman intensities and the depolarization ratios of the two stretching bands in solution, the absolute i.r. intensities of ν_1, ν_2, and ν_3 in the solid, and the far-i.r. and Raman spectra of the solid have all received attention. The vibrational properties of the liquid (and seemingly all its other structural properties) and of the matrix-isolated molecule and the Raman spectrum of the vapour all remain unexamined.

The Gaseous State.—The only report of a vapour-phase vibrational spectrum of ICN is that of Hemple and Nixon [146] in 1967. These workers examined the i.r. spectrum of the vapour in the range 4000—270 cm^{-1} at room temperature using a 10 m path-length cell; even then, the resultant absorption was found to be very weak. The three fundamentals (see Table 47a), the first overtone of ν_3, three combination bands, and nine hot bands (of which five involved 001^1 as the lower state) were assigned. These assignments, together with a microwave value for $|W_{23}|$, allowed the evaluation of a number of vibrational constants which are listed in Table 48, where they are compared with analogous data for ^{79}BrCN, ^{35}ClCN, and FCN. It is interesting to note that $|W_{23}|$ for ICN is significantly smaller than in the three other molecules. No separation of bands smaller than 4 cm^{-1} was observed.

The Solid State.—An early study of the i.r. spectrum of solid ICN as a thin film at 93 K in the range 3200—300 cm^{-1} was reported by Freitag and Nixon.[122] The three sharp internal fundamentals (see Table 47b), $\nu_1(^{13}CN)$, and $\nu_1(C^{15}N)$ were assigned. The bending overtone $2\nu_3$ was not observed, in

Table 47a *I.r. intensity data for* ICN

Fundamental	*Intensity*/dark [147]			$(\partial\mu/\partial q)$/esu g$^{-\frac{1}{2}}$	
	Ref. 147		*Ref.* 130	*Ref.* 147	*Ref.* 130
	$CHCl_3$	C_6H_6	*Solid*, 77 K		
ν_1	530		3023 ± 186	58	65.7
ν_2	530	880	3760 ± 420	61	73.3
ν_3	260	320	172 ± 20	<13	11.1

[146] S. Hemple and E. R. Nixon, *J. Chem. Phys.*, 1967, **47**, 4273.

Table 47b *I.r. spectra of* ICN (*wavenumbers*/cm^{-1})

Fundamental	*Vapour*	*Solid*			*Solution*								
	273 K	93 K	77 K	83 K	$CHCl_3$	C_6H_6	C_5H_5N	dioxan	DMF	Me_3N	CCl_4	$CHCl_3$	C_2Cl_4
	Ref. 146	122	130	148	147	147	147	147	147	147	29	29	29
$\nu_1(I^{12}C^{14}N)$	2188	2176s [6.5][a]	2176	2172.5	2168		2156	2159	2159		2168.1	2166.2	2168.2
$\nu_2(I^{12}C^{14}N)$	480.4*P* 491.3*R*	451.5s [10][a]	451		486	476	430	469	456	398			
$\nu_3(I^{12}C^{14}N)$	304.5	328.5s [6.5][a]	327		320	320	336	330		330			
$\nu_1(I^{12}C^{15}N)$		2142											
$\nu_1(I^{13}C^{14}N)$		2125											

[a] Values in square brackets are $\Delta\nu^{\frac{1}{2}}$.

Table 47c *Raman spectra of* ICN (*wavenumbers*/cm^{-1})

Fundamental	*Solid*		*Solution*			
	308 K	78 K	MeOH	C_6H_6	THF	CH_2Cl_2
	Ref. 148	148	119	149	149	149
$\nu_1(I^{12}C^{14}N)$	2169	2172 ν_t 2175 ν_{ln}	2158	2168 (0.33 ±0.03)[a]	2164 (0.33 ±0.04)[b]	2168 (0.32 ±0.025)
$\nu_2(I^{12}C^{14}N)$	456	449 ν_t 465 ν_{ln}	470	476 (0.31 ±0.025)	464 (0.33 ±0.04)	486 (0.31 ±0.025)
$\nu_3(I^{12}C^{14}N)$	330	329 ν_t <330 ν_{ln}	321			
$\nu_1(I^{12}C^{15}N)$	2139	2141				
$\nu_1(I^{13}C^{14}N)$	2122	2124				

[a] Figures in parentheses are depolarization ratios. [b] Estimated value.

Table 48 *Vibrational constants*[a] *of* ICN, BrCN, ClCN, *and* FCN

	ICN	^{79}BrCN	^{35}ClCN	FCN	^{35}ClCN
ω_1+2x_{11}	2207.5	—	2224.16	—	2223.70
ω_2+2x_{22}	499.2	—	—	—	741.61
ω_3	304.3	—	—	—	381.91
x_{33}	2.9	−0.86	−0.65	—	0.56
x_{ll}	1.8	1.15	0.95	—	−0.11
x_{23}	−6.6	—	—	—	−2.46
x_{12}	−8.5	[−12][b]	−2.7	—	−1.74
x_{13}	−14.0	−6.19	−7.23	—	−7.24
W_{23}	18	35.41	34.07	35.8	*c*
Ref.	146	137	123	111	125

[a] All values in cm^{-1}. [b] Approximate value calculated from ref. 122. [c] See Table 34.

contrast to solid ClCN and BrCN. Very similar values for ν_1, ν_2, and ν_3 at the same temperature have been reported recently by Bandy *et al.*[130] These workers also measured the absolute absorption intensities of these fundamentals. Unlike ClCN and BrCN, gas-phase intensities are not available for ICN. If, instead, the solid-state intensities are compared with those obtained in solution [147] (see Table 47a), it can be seen that the intensity of ν_3 is little affected upon condensation whereas those of the two stretching vibrations increase appreciably. Here, as with ClCN and BrCN, it was concluded that these intensity changes are not explainable by a change in the form of the internal-mode normal co-ordinate nor by the field effect but can be understood if an additional resonance structure is assumed to apply in the solid state. The form of the assumed structure seems to imply intermolecular interaction along the chain axis.

In the most recent vibrational study of solid ICN, Raman spectra at 308 and 78 K obtained using Hg-arc and He–Ne-laser sources, and the far-i.r. spectrum at 83 K were reported by Savoie and Pézolet (see Tables 47b and c).[148] As with solid HCN, and in contrast to solid ClCN and BrCN, the unit cell of ICN is non-centrosymmetric. Moreover, only in ICN is the degeneracy of the bending vibration expected to be retained in both site-group and factor-group approximations. As observed in the Raman spectrum, ν_1, ν_2, and ν_3 and the rotational lattice mode all exhibit high-frequency shoulders attributable to longitudinal vibrations; values of $(\partial\mu/\partial q)$ calculated from the vibrational frequencies using the formula of Haas and Hornig [100] are in satisfactory agreement with the experimental values of Bandy *et al.* Unlike the case of HCN, these high-frequency shoulders have not been satisfactorily observed in the i.r. spectrum of ICN at liquid-nitrogen temperature.

Solution Studies.—One of the first structural investigations of ICN was an early Raman spectroscopic study by West and Farnsworth [119] of the methanolic solution. They used a low-dispersion spectrograph and excited the

[147] W. B. Person, R. E. Humphrey, and A. I. Popov, *J. Amer. Chem. Soc.*, 1959, **81**, 273.
[148] R. Savoie and M. Pézolet, *Canad. J. Chem.*, 1971, **49**, 2459.

spectrum with the 4047 and 4358 Å lines of an Hg-arc source. These workers reported values for ν_1, ν_2, and ν_3 (see Table 47c), though the intensity of ν_3 was hardly above the noise level, and were able to satisfactorily conclude that the molecule was linear under these conditions. West and Farnsworth attempted to record the Raman spectrum of the liquid but found that there was sufficient decomposition at the melting point to colour the liquid dark red and so prevent the spectrum being obtained. Over 25 years later, i.r. wavenumber and intensity values for ν_1, ν_2, and ν_3 were reported in a variety of solvents.[147] It is clear from these results (see Tables 47a and b) that the ν_2 fundamental is very sensitive to 'solvent-complexing', decreasing in wavenumber and increasing in intensity. It is just this fundamental which shows a negligible shift in wavenumber in going from the gas phase to non-complexing [149] solvent ($CHCl_3$). On the other hand, ν_1 and ν_3, which only show relatively small wavenumber and intensity shifts on 'solvent-complexing', show relatively big wavenumber shifts, in *opposite* directions, on going from the gas phase to non-complexing solvent.

Quite recently, Bahnick and Person [149] studied the Raman spectrum of ICN in various solutions using a Cary 81 spectrometer and an Hg-arc source. They reported, *inter alia*, wavenumber and intensity measurements for the two stretching vibrations in CH_2Cl_2, C_6H_6, and THF (see Table 47c). Measurements were not made on ν_3 because of its weakness. Depolarization ratios were calculated according to the procedures developed by Koningstein and Bernstein.[150] On the basis of the observed band positions (see Table 47c), Bahnick and Person regarded ICN as 'free' in CH_2Cl_2 and 'complexed' in C_6H_6 and THF. The relative molar intensities of ν_1 and ν_2 led to formation constants of the C_6H_6 and THF complexes of $\sim$0.15 and 3.10 $\pm$1.1 l mol^{-1} respectively. In the case of low concentrations of THF, two overlapping bands are detectable in the 450—500 cm^{-1} region, one apparently due to 'free' ICN (486 cm^{-1}) and the other to 'complexed' ICN (464 cm^{-1}). The depolarization ratios are listed in Table 47c. In the case of C_6H_6 and THF they refer to the 'complexed' bands. The results confirm that, as expected, the stretching vibrations are polarized whether ICN is 'free' or complexed. However, because ν_3 was not studied, it is not possible to conclude whether ICN is linear or bent in 'free' or 'complexed' form. Finally, it may be noted that West and Farnsworth's values for ν_1 and ν_2 clearly indicate that ICN is 'complexed' in MeOH, in the light of Bahnick and Person's study, though the decrease in ν_1 (10 cm^{-1}) is too big relative to that in ν_2 (16 cm^{-1}) if the THF values are accepted as a guide (decreases of 4 cm^{-1} and 22 cm^{-1} respectively).

The latest vibrational study of ICN in solution is that of Thomas and Orville-Thomas,[29] who reported i.r. values of ν_1 in CCl_4, C_2Cl_4, and $CHCl_3$. Very little variation in ν_1 was observed for these solvents. This can be understood in terms of the proposition of Bahnick and Person that $CHCl_3$ and CH_2Cl_2 are non-complexing solvents for ICN.

[149] D. A. Bahnick and W. B. Person, *J. Chem. Phys.*, 1968, **48**, 5637.
[150] J. A. Koningstein and H. J. Bernstein, *Spectrochim. Acta.*, 1962, **18**, 1249.

9

Matrix Isolation

BY A. J. DOWNS AND S. C. PEAKE

1 Introduction

In its most commonly used form, matrix isolation entails the preparation of a rigid sample in which the molecular species of interest are at once isolated and effectively immobilized in an inert host lattice or matrix. The species which are thus preserved, typically at high dilution and at low temperatures, are denied both the opportunity for bimolecular collisions and the incentive for unimolecular decomposition. Hence, the lifespan even of many molecules which have no more than a transient existence under normal conditions can be extended almost indefinitely, thereby affording the leisure to use conventional methods of spectroscopic detection and characterization. Most widely exploited have been the techniques of i.r., u.v.–visible, and e.s.r. spectroscopy, but the spectroscopic armoury has lately been strengthened by successful recourse to the methods of Raman and Mössbauer spectroscopy. Matrix isolation is now acknowledged as one of the two primary methods of investigating molecular systems which are short-lived under conventional conditions. The alternative approach involves rapid spectroscopic observation of the system following its production in the gaseous or liquid phase, this being the essence of flash photolysis,[1] pulse radiolysis,[2] and related techniques. The spectra of 'high-temperature molecules' [3] in the matrix-isolated condition supplement substantially the information gleaned by *direct* measurements on high-temperature vapours involving, for example, vapour-pressure measurements,[4] mass spectrometry,[5] electron diffraction,[6] studies of molecular beams (*e.g.* under the

[1] A. B. Callear, *Endeavour*, 1967, **26,** 9; 'Fast Reactions and Primary Processes in Chemical Kinetics', ed. S. Claesson, Proceedings of the 5th Nobel Symposium, Almqvist and Wiksell, Stockholm, 1967; G. Porter, *Science*, 1968, **160,** 1299; J. W. Boag, *Photochem. and Photobiol.*, 1968, **8,** 565; M. R. Topp, 'Ultrafast Processes', ed. J. Jortner and P. M. Rentzepis, Plenum Press, New York, to be published.

[2] M. S. Matheson and L. M. Dorfman, 'Pulse Radiolysis', M.I.T. Research Monographs in Radiation Chemistry, M.I.T. Press, 1969.

[3] P. Goldfinger, *Adv. High Temp. Chem.*, 1967, **1,** 1.

[4] 'The Characterization of High-Temperature Vapors', ed. J. L. Margrave, Wiley, New York, 1967, pp. 19—192.

[5] 'Mass Spectrometry in Inorganic Chemistry', *Adv. Chem. Ser.*, 1968, no. 72; R. T. Grimley, 'The Characterization of High-Temperature Vapors', ed. J. L. Margrave, New York, 1967, p. 195; J. Berkowitz, *Adv. High Temp. Chem.*, 1971, **3,** 123.

[6] L. V. Vilkov, N. G. Rambidi, and V. P. Spiridonov, *J. Struct. Chem.*, 1967, **8,** 715;

action of an inhomogeneous electric field),[7] and microwave,[8] vibrational,[9] and electronic [10] spectroscopy.

Early reports [11] of luminescence following irradiation or electron-bombardment of solids at low temperatures imply that matrix isolation was discovered inadvertently long before the true nature and potential of the principle came to be appreciated. First to apply the technique were Lewis and Lipkin, who in 1942 reported [12] the visible spectra of large organic free radicals produced by photolysis and held in a glassy matrix at low temperatures. However, the matrix-isolation technique as we know it today was proposed independently by Norman and Porter [13] and by Whittle, Dows, and Pimentel [14] in 1954. The subsequent development of the technique owes its direction and impetus largely to the pioneering researches of Pimentel and his group. It was this group which was responsible for the first clear definition of the conditions and requirements that must be met for the successful isolation of molecules using nitrogen or a noble gas to form the rigid, inert matrix.[15] Hence came the first reported characterization, by i.r. spectroscopy, of a free radical (HCO) held in a matrix at low temperatures;[16] hence, too, came the first suggestion that 'high-temperature molecules' effusing from a Knudsen cell could profitably be trapped and characterized in a similar fashion.[15] Many of the early results are summarized in the book 'Formation and Trapping of Free Radicals' edited by Bass and Broida [17] and in review articles by Pimentel [18] and others.[19, 20] The use of the technique

---

P. A. Akishin, N. G. Rambidi, and V. P. Spiridonov, 'The Characterization of High-Temperature Vapors', ed. J. L. Margrave, Wiley, New York, 1967, p. 300.

[7] T. A. Milne and F. T. Greene, *Adv. High Temp. Chem.*, 1969, **2**, 107; A. Büchler, J. L. Stauffer, W. Klemperer, and L. Wharton, *J. Chem. Phys.*, 1963, **39**, 2299.

[8] A. H. Barrett and R. F. Curl, 'The Characterization of High-Temperature Vapors', ed. J. L. Margrave, Wiley, New York, 1967, p. 282; F. J. Lovas and D. R. Lide jun., *Adv. High Temp. Chem.*, 1971, **3**, 177.

[9] See, for example, J. M. Bassler and J. L. Margrave, 'The Characterization of High-Temperature Vapors', ed. J. L. Margrave, Wiley, New York, 1967, p. 264; I. R. Beattie, 'Essays in Structural Chemistry', ed. A. J. Downs, D. A. Long, and L. A. K. Staveley, Macmillan, London, 1971, p. 111; G. A. Ozin, *Progr. Inorg. Chem.*, 1971, **14**, 173.

[10] (*a*) F. T. Greene, 'The Characterization of High-Temperature Vapors', ed. J. L. Margrave, Wiley, New York, 1967, p. 244; (*b*) C. J. Cheetham and R. F. Barrow, *Adv. High Temp. Chem.*, 1967, **1**, 7; R. F. Barrow and C. Cousins, *ibid.*, 1971, **4**, 161; (*c*) R. F. Barrow, 'Essays in Structural Chemistry', ed. A. J. Downs, D. A. Long, and L. A. K. Staveley, Macmillan, London, 1971, p. 383; (*d*) D. M. Gruen, *Progr. Inorg. Chem.*, 1971, **14**, 119.

[11] J. Dewar, *Proc. Roy. Soc.*, 1901, **68**, 360; L. Vegard, *Nature*, 1924, **113**, 716.

[12] G. N. Lewis and D. Lipkin, *J. Amer. Chem. Soc.*, 1942, **64**, 2801.

[13] I. Norman and G. Porter, *Nature*, 1954, **174**, 508.

[14] E. Whittle D. A. Dows, and G. C. Pimentel, *J. Chem. Phys.*, 1954, **22**, 1943.

[15] E. D. Becker and G. C. Pimentel, *J. Chem. Phys.*, 1956, **25**, 224.

[16] G. E. Ewing, W. E. Thompson, and G. C. Pimentel, *J. Chem. Phys.*, 1960, **32**, 927.

[17] 'Formation and Trapping of Free Radicals', ed. A. M. Bass and H. P. Broida, Academic Press, New York, 1960.

[18] (*a*) G. C. Pimentel, *Pure Appl. Chem.*, 1962, **4**, 61; (*b*) G. C. Pimentel and S. W. Charles, *ibid.*, 1963, **7**, 111.

[19] M. E. Jacox and D. E. Milligan, *Appl. Optics.*, 1964, **3**, 873.

[20] F. J. Adrian, E. L. Cochran, and V. A. Bowers, 'Free Radicals in Inorganic Chemistry', *Adv. Chem. Ser.*, 1962, no. 36, p. 50.

was at first limited by the need for suitable cryogenic facilities, but with the increasing availability of liquid helium, and particularly with the commercial development of miniature cryostats, in which refrigeration is achieved by the Joule–Thomson expansion of compressed gases, matrix isolation has increased greatly in popularity and importance. In evidence of this, there has been a rich and ever growing harvest of research papers in recent years. The widening interest in the power and possibilities of the method is also made plain by the growing diversity of journals into which such papers find their way, although most continue to appear in journals oriented towards chemical physics or physical chemistry. Beyond the more immediate features of detecting and characterizing chemical entities which are short-lived under normal conditions (*e.g.* free radicals and other reactive intermediates and 'high-temperature molecules'), the technique has much intrinsic interest for the light it can shed on intermolecular interactions between the guest and its host lattice; the spectroscopist has a wealth of detail about the vibrational properties of natural and isotopically labelled molecules, whereby molecular structures and force fields have been successfully evaluated; the inorganic and organometallic chemist has also caught the first glimpse of molecules like $XeCl_2$, Br_2Cl_2, HOF, LiOF, LiO_2, BH_3, $Pd(CO)_4$, $Al(CO)_2$, and $Ni(CO)_3(N_2)$, some of which had previously been regarded as 'non-existent', and all of which pose intriguing problems in relation to current theories of structure and bonding. The extent of present interest in matrix-isolation studies may be judged by the appearance of a book about low-temperature spectroscopy [21] and by the abundance of review articles lately devoted to the topic; these include surveys dealing with the general scope of the method,[22–24] with more specific techniques as, for example, with the use of i.r.,[24–28] Raman,[29] optical,[24,30] and e.s.r.[31,32] spectroscopy to investigate matrix-trapped species, and with the study of free radicals [24,33–35] and high-temperature species [36,37] *via*

[21] B. Meyer, 'Low Temperature Spectroscopy', American Elsevier, New York, 1971.
[22] J. S. Ogden and J. J. Turner, *Chem. in Britain*, 1971, **7**, 186.
[23] J. W. Hastie, R. H. Hauge, and J. L. Margrave, 'Spectroscopy in Inorganic Chemistry', ed. C. N. R. Rao and J. R. Ferraro, Academic Press, 1970, vol. 1, p. 57.
[24] D. E. Milligan and M. E. Jacox, *Adv. High Temp. Chem.*, 1971, **4**, 1; also in 'MTP International Review of Science: Physical Chemistry Series One', Vol. 3, 'Spectroscopy', ed. D. A. Ramsay, Butterworths, London, 1972, p. 1.
[25] H. E. Hallam, 'Molecular Spectroscopy', ed. P. Hepple, Proceedings of the 4th Institute of Petroleum Hydrocarbon Research Conference, Institute of Petroleum, London, 1968, p. 329.
[26] A. J. Barnes and H. E. Hallam, *Quart. Rev.*, 1969, **23**, 392.
[27] H. E. Hallam, *Ann. Reports* (*A*), 1970, **67**, 117.
[28] T. S. Hermann and S. R. Harvey, *Appl. Spectroscopy*, 1969, **23**, 435; T. S. Hermann, S. R. Harvey, and C. N. Honts, *ibid.*, p. 451; T. S. Hermann, *ibid.*, pp. 461, 473.
[29] G. A. Ozin, *The Spex Speaker*, 1971, **XVI**, no. 4.
[30] B. Meyer, *Science*, 1970, **168**, 783.
[31] K. D. J. Root and M. T. Rogers, 'Spectroscopy in Inorganic Chemistry', ed. C. N. R. Rao and J. R. Ferraro, Academic Press, New York, 1971, vol. II, p. 115.
[32] P. H. Kasai, *Accounts Chem. Res.*, 1971, **4**, 329.
[33] B. Mile, *Angew. Chem. Internat. Edn.*, 1968, **7**, 507.
[34] D. E. Milligan and M. E. Jacox, 'Physical Chemistry, an Advanced Treatise', ed. D. Henderson, Academic Press, New York, 1970, vol. IV, p. 193.

matrix isolation. Elsewhere there are to be found reports containing brief sections alluding to matrix-isolation experiments.[10d, 38–40] Hallam[27] has also reviewed recent developments in spectroscopic studies of clathrate systems and of impurity centres in ionic crystals, which, through their preoccupation with trapped chemical species, have much ground in common with cryogenic matrix-isolation procedures.

The present Report aims to describe recent progress in the realm of matrix isolation. For reasons of space, it cannot be comprehensive, although an attempt has been made in Tables 1—3 (see Appendix) to touch on most of the research reported between January 1969 and July 1972; some of the more significant studies reported before 1969 are also listed here. Unless otherwise specified, references to individual compounds or groups of compounds cited in the subsequent discussion will be found in the appropriate Table. In view of the errors, uncertainties, and controversies that have attended the interpretation of certain experiments, the present report also seeks to take stock of the limitations and hazards, as well as the peculiar advantages, of the matrix-isolation technique. The studies reviewed have relied for detection primarily on the i.r., Raman, u.v.–visible or Mössbauer spectra of the trapped species. Somewhat arbitrarily, it has been necessary to exclude all but a few references to e.s.r. measurements and to the numerous investigations of optical as well as e.s.r. spectra of systems employing a glass or an ionic lattice as the host phase; cryogenic studies of this sort have been described in some detail in the new book by Meyer.[21]

2 Generation of Species

As indicated in Scheme 1, there exists a variety of methods for bringing the species of interest into a matrix environment. These fall into two broad categories:[22] (*i*) generation at ambient or high temperatures in the gas phase, followed by rapid quenching with excess inert gas, and (*ii*) generation *in situ*, *e.g.* by photolysis of matrix-isolated parent molecules. Fuller details of the relevant apparatus have been presented in references 21—26, 28, 32, 33, and 36 and in individual research papers.

Only the simplest provisions need be made for molecules which can be obtained at pressures greater than *ca.* 10^{-4} mmHg near or below room

---

35 L. Andrews, *Ann. Rev. Phys. Chem.*, 1971, **22**, 109.
36 (*a*) W. Weltner, jun., 'Condensation and Evaporation of Solids', ed. E. Rutner, P. Goldfinger, and J. P. Hirth, Gordon and Breach, New York, 1964, p. 243; (*b*) W. Weltner, jun., *Science*, 1967, **155**, 155; (*c*) W. Weltner, jun., *Adv. High Temp. Chem.*, 1969, **2**, 85.
37 J. W. Hastie, R. H. Hauge, and J. L. Margrave, *Ann. Rev. Phys. Chem.*, 1970, **21**, 475.
38 D. B. Powell, *Ann. Reports*, 1966, **63**, 112; D. A. Long, *Ann. Reports (A)*, 1968, **65**, 83.
39 J. Bridgwater, A. B. Callear, I. Fleming, J. S. Ogden, R. H. Ottewill, R. Pink, and S. G. Warren, *Chem. and Ind.*, 1969, 1381.
40 J. Overend, *Ann. Rev. Phys. Chem.*, 1970, **21**, 274.

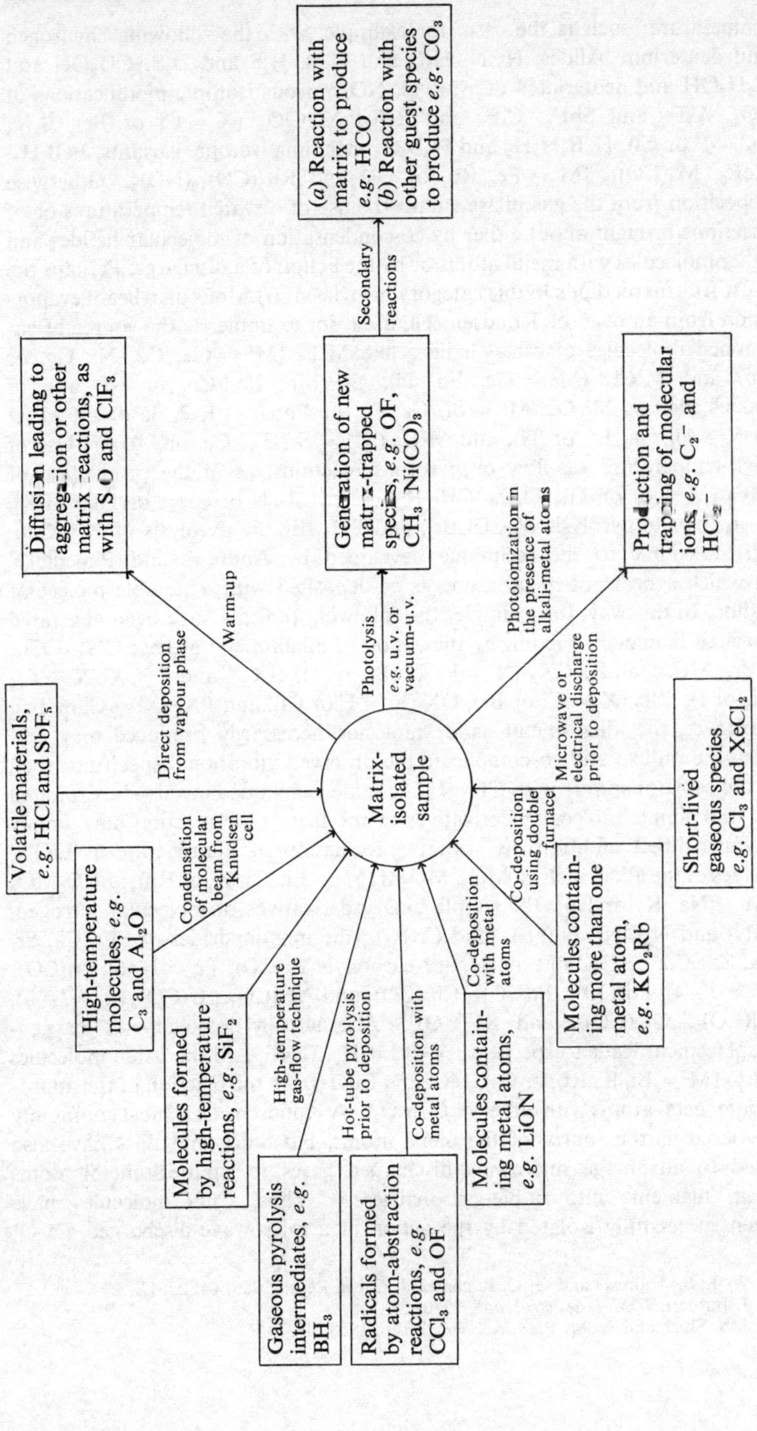

Volatile materials, *e.g.* HCl and SbF_5

Direct deposition from vapour phase

High-temperature molecules, *e.g.* C_3 and Al_2O

Condensation of molecular beam from Knudsen cell

Molecules formed by high-temperature reactions, *e.g.* SiF_2

High-temperature gas-flow technique

Gaseous pyrolysis intermediates, *e.g.* BH_3

Hot-tube pyrolysis prior to deposition

Radicals formed by atom-abstraction reactions, *e.g.* CCl_3 and OF

Co-deposition with metal atoms

Molecules containing metal atoms, *e.g.* LiON

Co-deposition with metal atoms

Molecules containing more than one metal atom, *e.g.* KO_2Rb

Co-deposition using double furnace

Short-lived gaseous species *e.g.* Cl_3 and $XeCl_2$

Microwave or electrical discharge prior to deposition

Matrix-isolated sample

Warm-up

Diffusion leading to aggregation or other matrix reactions, as with S O and ClF_3

Photolysis *e.g.* u.v. or vacuum-u.v.

Generation of new matrix-trapped species, *e.g.* OF, CH_3 $Ni(CO)_3$

Secondary reactions

(*a*) Reaction with matrix to produce *e.g.* HCO
(*b*) Reaction with other guest species to produce *e.g.* CO_3

Photoionization in the presence of alkali-metal atoms

Production and trapping of molecular ions, *e.g.* C_2^- and HCl_2^-

Scheme 1

temperature; such is the case, for example, with the following: hydrogen and deuterium halides, HCN, H_2O and D_2O, H_2S and D_2S, CH_3OH and C_2H_5OH and deuteriated derivatives, NO, various isotopic modifications of SO_2, AsF_5 and SbF_5, ClF_3 and BrF_3, $XOClO_3$ (X = Cl or Br), B_2X_4 (X = F or Cl), $H_3B_3N_3H_3$ and $H_3B_3O_3$ including isotopic variants, BeB_2H_8, XeF_6, $M_3(CO)_{12}$ (M = Fe, Ru, or Os) and $Ru_3(CO)_{10}(NO)_2$. Otherwise deposition from the gas phase entails the use of elevated temperatures or of reactions brought about either by co-condensation of molecular halides and other molecules with metal atoms or by the action of a discharge. Perhaps the most fruitful methods in this category have been: (*i*) Molecular-beam evaporation from an oven or Knudsen cell, used, for example, as the source of entrapped molecules of metal halides like M^1F_2 (M^1 = Ge, Co, Ni, Cu, or Zn) and M^2Cl_2 (M^2 = Ge, Sn, Pb, Cr, Mn, Fe, Co, or Ni) and of oxides such as $M^1{}_nO_n$ (M^1 = Si, Ge, Sn, or Pb; n = 1, 2, 3, or 4), $M^2{}_2O$ (M^2 = Al, Ga, In, or Tl), and M^3O_2 (M^3 = Se, Te, Ce, or U). (*ii*) Use of high-temperature gas-flow or pyrolysis reactions, as in the production of MF_2 (M = Si or Ti), Tl_xO_2, CH_3 [by the pyrolysis of CH_3I or $Hg(CH_3)_2$], CCl_3 (by the pyrolysis of CCl_3Br), and BH_3 (by the pyrolysis of H_3B,CO). (*iii*) Recourse to the technique developed by Andrews and Pimentel,[41] in which a beam of metal atoms is co-deposited with a suitable molecular halide. In this way, for example, the following radicals have been generated through halide-abstraction by the action of alkali-metal atoms: CH_3, CCl_3, CBr_3, HCX_2 and DCX_2 (X = F, Cl, Br, or I), H_2CX and D_2CX (X = Cl, Br, or I), CX_2 (X = Cl or Br), OX (X = F or Cl), and PX_2 (X = Cl or Br). However, the alkali-metal halide molecule necessarily produced may perturb the radical and so complicate the observed vibrational spectrum; such is the case apparently with CH_3.[42] The same technique may also be exploited to trap simple molecular derivatives of the metal; the reaction may simply involve direct addition, as with the formation of matrix-trapped LiON, oxides of the alkali metals, MO_2, MO_2M (M = Li, Na, K, or Rb), and O_2MO_2 (M = Na, K, or Rb), the simple binary derivatives of molecular nitrogen, LiN_2 and N_2LiN_2, $Ni(N_2)_x$, and $Cr(N_2)_x$, the monohydrides of Mg, Ca, Sr, Ba, Zn, Cd, or Hg, and the binary carbonyls $Ta(CO)_x$ (x = 1—6), $Ni(CO)_x$ (x = 1—4), $Pd(CO)_4$, $Pt(CO)_4$, $Cu_x(CO)_y$ and $Ag_x(CO)_y$, $U(CO)_x$ (x = 1—6), $Al(CO)_2$, $Ge_x(CO)_y$, and $Sn_x(CO)_y$. Alternatively, halogen- or oxygen-displacement leads to species such as $LiCH_3$, LiOF, and the oxide molecules M_2O (M = Li, K, Rb, or Cs), LiO, and CuO (from the reaction of the appropriate metal atoms with either N_2O or O_2). A Knudsen cell is most commonly favoured as the source of the metal atoms, but Shirk and Bass have also used to advantage microwave-discharged gases to 'sputter' metal atoms from filaments into noble-gas matrices.[43] Thus, CuO molecules have been successfully isolated by the action of a microwave-discharged Xe–O_2

[41] W. L. S. Andrews and G. C. Pimentel, *J. Chem. Phys.*, 1966, **44**, 2361.
[42] J. Burdett, *J. Mol. Spectroscopy*, 1970, **36**, 365.
[43] J. S. Shirk and A. M. Bass, *J. Chem. Phys.*, 1968, **49**, 5156.

mixture on a copper target. It has been suggested [35] that reaction occurs during the condensation of the gaseous species, diffusion taking place on the surface of the matrix and providing an opportunity for reaction with subsequent stabilization of the product within the matrix. In some cases the initial product may be implicated in secondary reactions, *e.g.*

$$CX_3 + Li \longrightarrow CX_2 + LiX \quad (X = Cl \text{ or } Br)$$

On this basis, the average interaction temperature approaches that of the cooled surface supporting the matrix, so that reaction is likely to occur only *via* pathways offering relatively low activation energies. However, the possibility of gas-phase reactions cannot always be excluded, and such reactions have certainly been exploited to isolate the molecules CaO and BaO_2.

Quite the most prolific method for producing free radicals and other species of interest has been photolysis *in situ* of the isolated parent molecule under the action of u.v. or even vacuum-u.v. radiation. Primary products isolated in this way include:

Parent molecule	*Product*	*Parent molecule*	*Product*
$H_2B_2O_3$	HBO	H_3CF	CF, HCF, H_2CF
CH_4	CH_3	H_3CCl, H_2CCl_2	CCl, CCl_2, H_2CCl
CF_3Br	CF_3	H_2CClF	CClF
HCX_3	CX_3 (X = Cl or Br)	Diazirine	CH_2
CF_2N_2	CF_2	XN_3	XN (X = H, F, Cl, Br, or CN)
SiH_4	Si_2, SiH, SiH_2, SiH_3, disilicon hydrides	HPX_2	PX_2 (X = H or F)
GeH_4	GeH_2, GeH_3, digermanium hydrides	OX_2	OX (X = H, F, or Cl)
H_3MN_3	HNM (M = Si or Ge)	ClF_3	ClF_2
H_3GeX	H_2GeX, HGeX (X = Cl or Br)	$Ni(CO)_4$	$Ni(CO)_3$
H_nGeCl_{4-n} (n = 0, 1, or 2)	$GeCl_3$, $GeCl_2$	$M(CO)_6$	$M(CO)_5$ (M = Cr, Mo, or W)

Photolysis of an isolated molecule may also initiate reaction either with an adjacent reactant in an inert matrix or, in certain cases, with the matrix itself. Representative products of such secondary reactions are

Source	*Product*	*Source*	*Product*
HI, HBr, or H_2S with CO	HCO	H_2O, CH_4, or HCl with NO_2	HONO
OF_2, NF_2, or *trans*-N_2F_2 with CO	FCO	HX (X = Cl, Br, I, or OH) with O_2	HO_2
HCl or Cl_2 with CO	ClCO	H_2O with F_2	HOF

Source	Product	Source	Product
H_2O with CO	HOCO	HX with O_3	HOX (X = Cl or Br)
N_3CN with CO	CCO	OF_2 or F_2 with O_2	O_2F, O_2F_2
CO_2 with O_3 or CO_2 alone	CO_3	$Ni(CO)_4$ with N_2	$Ni(CO)_3(N_2)$
HN_3 with F_2	HNF, HNF_2	$RMn(CO)_5$	$RCOMn(CO)_x$ (R = CH_3 or CF_3; $x < 5$)
HN_3 with O_2	HONO	$Fe(CO)_5$ with C_2H_4	$(C_2H_4)Fe(CO)_4$

Photoexcitation, *e.g.* by vacuum-u.v. irradiation in the presence of alkali-metal atoms, furnishes a simple and effective photoelectron source within a matrix;[24,32,44] an acceptor species may thus be converted into the corresponding molecular anion under conditions where it is effectively isolated and stabilized in an inert, non-ionic environment. For example, the molecular ions C_2^-, $HCCl_2^-$, $C_2(CN)_4^-$,[32] $B_2H_6^-$,[32] NO^-, NO_2^-, $N_2O_2^-$, SO_2^-, HCl_2^-, and HBr_2^- represent established or likely issues of this photoionization technique. In some instances, as with the formation of CO_2^-, CS_2^-, O_2^- and H_2S^-, no irradiation is necessary, whereas in others, as with the formation of CCl_3^+, $HCCl_2^+$, and $HCCl_2^-$ or the molecular halogen ions X_2^-,[45] ionization is realized through vacuum-u.v.- or γ-irradiation without the agency of metal atoms. Although numerous experiments have been performed using higher-energy sources of radiation for photolysis, *e.g.* γ-rays or electrons, these suffer, in general, by being less selective.[35]

The success of a photolysis experiment in which the precursor is isolated in a matrix at low temperature depends critically upon a number of factors. First, if the fission of a bond X—Y is to occur, not only must the frequency of the radiation ν_r coincide with that of an absorption band of the parent molecule, but the condition $h\nu_r \geqslant D(\text{X—Y})$ must also be met. Second, reactions of this type are often impeded by the surrounding 'cage' of the host lattice, which prevents the escape of the reaction products and may thus lead to regeneration of the parent molecule. To counteract the cage effect, the experimenter may take advantage of one or more of the following circumstances:

(*a*) The production of a non-reactive fragment, which remains trapped in an adjacent matrix site, *e.g.*

$$XN_3 \xrightarrow{h\nu} NX + N_2$$

In general, species can be generated and maintained in close proximity whenever the reverse reaction is attended by a significant kinetic barrier or by a

[44] D. E. Milligan and M. E. Jacox, *J. Chem. Phys.*, 1969, **51**, 1952; 1971, **54**, 3935; 1971, **55**, 3404.
[45] See, for example, D. M. Brown and F. S. Dainton, *Nature*, 1966, **209**, 195.

positive heat of reaction; thus, following the dissociation

$$M(CO)_6 \xrightarrow{h\nu} M(CO)_5 + CO \quad (M = Cr, Mo, or W)$$

the CO does not escape from the matrix cage, but the reverse reaction does not occur at 20 K, presumably because it presents a significant activation energy.

(*b*) Production of a fragment small enough to diffuse away from the photolysis site. Molecules of the type XCN (X = H, F, Cl, or Br) are all susceptible to photolysis in a matrix environment, but the concentration of the CN radical produced decreases rapidly in the series X = H > F > Cl > Br owing to the decrease in diffusion rates from hydrogen to bromine. The capacity of hydrogen and fluorine atoms to diffuse through matrices even at low temperatures is undoubtedly crucial to the successful isolation of numerous radicals produced by photolysis, *e.g.* MH_3 (M = C, Si, or Ge), PX_2 (X = H or F), and OF. A photolytic source of carbon atoms has also been found in N_3CN, by way of the initial photolysis product NCN. Secondary reactions of the diffusing atoms have also been characterized, *e.g.* CO with H, C, or F; O_2 with H or F; F_2, Cl_2, or HCl with C. Apart from the intrinsic interest of the product of such a reaction, it has thus been possible to assess the activation energy of the process. Further, when hydride molecules like CH_3F, $HSiF_3$, and $HGeCl_3$ are photolysed in a CO matrix, the amount of HCO formed testifies to the part played by H-atom detachment in the primary photodecomposition process.

(*c*) Stimulation of diffusion by warming the matrix. This may be engineered either during or after photolysis, for example, by the use of excess radiation to produce local heating, and may be assisted by the choice of matrices like N_2 or Ar which soften comparatively easily.[23, 26]

(*d*) Photolysis during deposition of the matrix to take advantage of the greater freedom attending surface reactions. This has proved effective, for instance, in the formation and isolation of HBO and the germanium hydride radicals GeH_3 and GeH_2.

Alternatively, cage effects may be obviated completely by the action of flash photolysis [46] or a microwave discharge [47] on gaseous molecules, followed by rapid quenching of the products. Species isolated in this way include CF_2, Cl_3, Br_3, Br_nCl_{4-n} (n = 1 or 2), $XeCl_2$, and possibly HX_2 (X = Cl, Br, or I). However, the advantage which accrues from the production of reactive species in the gas phase, rather than *in situ*, is more than offset by the difficulties posed by undesired reactions and by the heat released during condensation.[17]

[46] See, for example, L. J. Schoen and D. E. Mann, *J. Chem. Phys.*, 1964, **41**, 1514.

[47] See, for example, L. Y. Nelson and G. C. Pimentel, *J. Chem. Phys.*, 1967, **47**, 3671; *Inorg. Chem.*, 1968, **7**, 1695; D. H. Boal and G. A. Ozin, *J. Chem. Phys.*, 1971, **55**, 3598.

3 Cryogenic Considerations: Nature and Preparation of the Matrix

In early experiments the low temperatures needed for the deposition and maintenance of the matrices were attained by using liquid hydrogen (b.p. 20 K) or liquid helium (b.p. 4 K) in conventional double-Dewar cryostats.[17] A rotating cryostat has been successfully employed for e.s.r. measurements, notably of alkali-metal–halogenocarbon reactions and of ion-pair formation with electron-accepting species such as H_2S, CO_2, and CS_2;[33] the same principle has also been incorporated in an apparatus designed for i.r. studies.[48] However, one of the strongest influences in promoting the matrix-isolation technique has been the commercial development of miniature refrigerators which operate either in an open cycle by the Joule–Thomson expansion of high-pressure cylinder gases (as in the so-called 'Cryo-Tip') or in a closed cycle based on the Stirling or no-work extraction process (as in the so-called 'Cryogem' or 'Cryodyne'). Closed-cycle refrigerators suffer from a higher initial cost but gain over open-cycle units by their much reduced operating expenses. By variations of pressure or other operating conditions, it is possible with this sort of apparatus to vary the temperature over a maximum range of 4—300 K and to control it to ± 0.5 K or better. For the precise and reproducible study of temperature-dependent phenomena and of chemical reactions in matrices, this feature represents a major advance in design compared with liquid refrigerant cryostats. The majority of spectroscopic measurements have been made at temperatures in the range 4—20 K, but higher temperatures have commonly been required to allow softening of the matrix and so induce diffusion of the solute species. The increased availability and ease of operation of these cryogenic devices have now brought them into widespread use; technical details are to be found elsewhere.[25, 26, 28, 35, 49]

The matrix sample is deposited on a support at or near the temperature of the refrigerant; the nature of this support is dictated by the type of spectroscopic measurement to be made on the sample. For the most part, the choice has been restricted to the following materials:

Spectroscopic method	*Support*
U.v. absorption	Quartz or LiF (to 1050 Å) windows
U.v.–visible absorption	Quartz or CaF_2 windows
I.r.	CsI, CsBr, KBr, or NaCl windows
Far-i.r.	Crystalline Si or quartz windows
Raman	Metal surface (*e.g.* Cu, Ag, or Pt)
E.s.r.	Sapphire rod
Mössbauer	Be disc

[48] G. Mamantov, W. H. Fletcher, S. S. Cristy, C. T. Edwards, and R. E. Morton, *Rev. Sci. Instr.*, 1966, **37**, 836.

[49] D. White and D. E. Mann, *Rev. Sci. Instr.*, 1963, **34**, 1370; 'Principles of Operation of Cryodyne Helium Refrigerators', Cryogenic Technology Inc., Waltham, Mass., 1968.

Most studies have employed transmission windows which unfortunately possess relatively low thermal conductivities; for all their advantages, measurements of i.r. or u.v.–visible reflectance spectra from matrices held on cooled metal surfaces have been neglected with but few exceptions.[23, 50] To ensure rapid dissipation of the heat capacity of the sample during condensation, it is imperative that the spectroscopic surface should be in good thermal contact with the cooled metal block on which it is mounted; lead or indium spacers are generally used for this purpose.

The most obvious desiderata for the host lattice are rigidity, inertness, appropriate volatility, good thermal conductivity, and ability to accommodate the guest species. Possible host lattices include (*i*) the crystals of noble gases, N_2, CO, CO_2, CH_4, SF_6, and similar 'inert' gases, (*ii*) molecular crystals capable of forming inclusion compounds, *e.g.* β-quinol, and (*iii*) ionic crystals, *e.g.* alkali-metal halides. Each class provides a particular set of environmental conditions but (*ii*) and (*iii*) do not necessarily require cryogenic conditions. Since most studies to date have aimed at keeping solute–matrix interactions to a minimum, the host lattices most commonly favoured have been those of category (*i*). Descriptions of their relevant physical properties are available.[15, 17, 25, 26]

Deposition by rapid quenching from the gas phase appears to give rise to matrix samples in the form of collections of randomly oriented microcrystallites. Thus, whereas an ionic host lattice provides an ordered environment amenable to a fairly precise description in physical terms, but characterized by strong intermolecular forces, a matrix material like solid argon gives a highly imperfect solid deposit poorly defined by any physical model, but notable for the weakness of host–host and guest–host interactions. In such a deposit there are several possible sites for trapped species:[25, 26] substitutional sites formed by the displacement of a matrix molecule, interstitial sites, and dislocation sites. The dimensions of substitutional and interstitial sites have been tabulated for a number of matrix materials,[26] and relative isolation efficiencies have also been explored.[15] The choice of the matrix material must be determined, at least in part, by its spectroscopic properties and by the nature of the species to be isolated. The reactions of N_2 and CO with metal atoms or with metal carbonyl fragments give clear warning that the immunity to reaction even of comparatively inert molecules such as these cannot be taken for granted.

In the ideal experiment, it is desired to produce a sample with optimum isolation of the guest species within the host lattice, which is itself transparent to whatever radiation may be used for spectroscopic characterization. In practice, these requirements are apt to be mutually incompatible, and the following experimental factors have an important bearing on the achievement of the best possible compromise:

(*a*) The deposition rate. This affects both the effectiveness of the isolation

[50] J. W. Hastie, R. H. Hauge, and J. L. Margrave, *J. Inorg. Nuclear Chem.*, 1969, **31**, 281.

and the transparency of the matrix. Slow deposition rates (typically 1—5 mmol of matrix gas per hour) continue to be the rule for the matrix isolation of small molecules. In some experiments, notably those involving surface reactions, it has taken up to 48 hours to build up a suitable matrix. By contrast, in what has been termed the 'pseudo-matrix-isolation' technique, Rochkind [51] has shown that the whole process can be carried out in a matter of minutes by depositing the sample in a series of pulses. Reference to the infrared spectrum of the matrix so prepared is then advocated as a novel and effective method for the qualitative and quantitative analysis of multi-component gaseous mixtures. Illustrative of the efficiency of the method is the ease with which individual deuteriated ethylenes, *e.g.* [1,1-$^2H_2$]ethylene and *trans*-[1,2-$^2H_2$]ethylene can be identified in no more than μmol quantities. A computer-generated catalogue of frequencies and absorbances has been compiled to facilitate analysis of the results obtained under these conditions of 'pseudo-matrix-isolation'. The success of pulsed deposition is not easily explained, though local seeding and crystallizing effects are probably involved. Despite the inferior level of isolation accepted in the interests of rapid sample-preparation, the deposits obtained in this way are usually more transparent and so cause less scattering than those derived by conventional methods of protracted deposition. Accordingly, pulsed deposition has lately found favour with experimenters seeking to measure the Raman spectra of matrix-isolated molecules.

(*b*) The deposition temperature. The normal requirement is that the sample must be condensed fast enough to restrict diffusion of molecules during condensation and that it must be kept cold enough to inhibit subsequent diffusion of the trapped species. The problem of diffusion in solid matrices has been discussed by Pimentel in ref. 17. As a general, empirical rule, the deposition temperature should be well below half the melting point (T_m/K) of the matrix material.[25, 26] However, whereas optimum isolation requires the lowest temperatures attainable, the transparency of the matrix may thus be impaired. For example, with respect to i.r. absorption, condensation of argon at 20 K gives a more transparent deposit than condensation at 4 K, while, as condensed at 4 K, the noble gases become increasingly opaque in the sequence $Ar < Kr < Xe$. In fact, all but a few experiments have employed deposition temperatures in the range 4—20 K.

(*c*) The matrix ratio. The ratio of matrix to solute molecules (M/S) has a profound effect on the degree of isolation; values of M/S have ranged from *ca.* 25 to *ca.* 10^4, though most work has been done with $M/S \geqslant 100$. With many solutes, true isolation is achieved only at high M/S ratios, usually >1000; at lower ratios, molecular aggregates may be entrapped within the matrix. Such aggregates have been identified not only for molecules susceptible to polymerization, *e.g.* H_2O, HX (X = halogen or CN), CH_3OH,

[51] M. M. Rochkind, *Analyt. Chem.*, 1967, **39**, 567; 1968, **40**, 762; *Science*, 1968, **160**, 196; *Appl. Spectroscopy*, 1968, **22**, 313; *Spectrochim. Acta*, 1971, **27A**, 547.

LiF, SbF_5, and SiO, but also for systems like BF_3, CO, NO, SO_2, and SF_4, which are subject to comparatively weak intermolecular interactions. Thus, the i.r. spectrum of CO trapped in argon at 20 K is now believed, after some vicissitudes of interpretation, to denote the presence of $[CO]_n$ aggregates even at matrix ratios exceeding 1000:1. Spectroscopic features due to well-isolated monomeric species may be distinguished from those due to aggregates by investigating the effects (*i*) of systematic variations of matrix ratio and (*ii*) of diffusion initiated by controlled warming of the matrix.

There are at the present time few clearly established precepts governing the selection of the most suitable matrix material and the conditions appropriate to deposition of a given matrix. Most experiments lean heavily on the accumulated and largely unpublished experience of individual researchers, and practices vary significantly from one research group to another. Published results make it plain that due consideration has not always been given to the presence of impurities (notably O_2, H_2O, CO, and CO_2) or to the effects of aggregation. Failure to appreciate these complications has led to incorrect interpretation of more than one matrix-isolation experiment. Since the precise condition of a matrix-isolated molecule is subject to a variety of constraints, some of which are not fully understood, the wisest counsel must be to vary the nature of the matrix material and the conditions of deposition as widely and systematically as possible in order to eliminate or at least to identify possible complications.

4 Spectroscopic Properties of Matrix-trapped Systems

In the majority of matrix-isolation experiments information is sought about highly reactive species which are novel or comparatively unfamiliar. Thus, it is important to establish that the spectroscopic properties observed, *e.g.* vibrational frequencies, the energies of electronic transitions, *g*-values, and hyperfine coupling constants, are not significantly perturbed by the matrix environment. To identify any such perturbations, stable molecules have been characterized in the matrix-isolated condition, and the results compared with reliable gas-phase measurements. Certain of the spectral changes induced by a matrix environment resemble those which accompany the dissolution of molecules in liquid solvents. Thus, frequencies and absorption coefficients differ slightly from the corresponding gas-phase values; in addition, however, there are effects peculiar to matrices, which are revealed through multiplet spectral features. All of these matrix disturbances are liable to interfere with the identification and characterization of unusual molecular species; nevertheless, they all bear information about intermolecular forces within the matrix.

Frequency (or matrix) shifts.—The most habitual and easily identified matrix effect is the frequency shift of a spectroscopic feature relative to the value characteristic of the gas phase; this has been observed for rotational, vibrational, and electronic transitions. Vibrational frequencies are subject to matrix shifts which are usually less than 1% of the gas-phase values for non-polar species, though shifts as large as 10% have been reported for polar molecules. Although the frequencies reported for some electronic transitions may differ by as much as 4000 cm^{-1} from the gas-phase values, the frequency shifts are *proportionately* no greater than for vibrational transitions (see, for example, ref. 23). The g-values and hyperfine coupling constants, which define the e.s.r. spectra of trapped radicals, refer to transitions between spin states separated by no more than a few cm^{-1} and are relatively sensitive to changes of environment, though the most striking effects here arise from the rotation or orientation (*q.v.*) of the guest species.

Several attempts have been made to give a quantitative description of the matrix-induced perturbations of vibrational and rotational energy levels of a diatomic molecule,[23, 25, 26] which are a function of the polar properties of the guest and host molecules and the nature of the matrix cavity. The intermolecular potential energy can be expressed as the sum of dispersive, inductive, electrostatic, and repulsive contributions so that the matrix shift is given by

$$\Delta\nu = \nu_{\text{matrix}} - \nu_{\text{gas}} = \Delta\nu_{\text{dis}} + \Delta\nu_{\text{ind}} + \Delta\nu_{\text{elec}} + \Delta\nu_{\text{rep}}$$

Expressions for the individual terms have been derived on the basis of various idealized models, typically employing the semi-empirical Lennard-Jones potential to evaluate the interactions. As yet, however, there are insufficient data to provide a proper test of the predictive power of these expressions, though certain of them appear, on the limited evidence available, to give results of the correct order of magnitude. For the vibrational modes of polar molecules such as LiF and AlF, it has been found empirically that there is a red frequency shift; increasing the polarizability of the matrix or the charge separation of the vibrating bond tends to increase this shift. Vibrational frequencies for a range of such molecules isolated in neon and argon matrices have been represented, to a useful approximation (± 6 cm^{-1}), by the expression [23]

$$\nu_{\text{gas}} = \nu_{\text{Ne}} + (0.8 \pm 0.4)(\nu_{\text{Ne}} - \nu_{\text{Ar}})$$

Less polar polyatomic molecules exhibit both positive and negative shifts of frequency. Despite a large degree of scatter, the general tenor of the available data is that the shift becomes more positive as the vibrational frequency decreases.[18b] An explanation for this effect has been suggested in terms of the incompatibility of the guest molecule with its cage in the host lattice.[18b] In general, the guest is expected to accept a lowest-energy position that is a compromise between optimum distances at some points of contact with the host lattice and less than optimum distances at other points. There is thus a tendency for the host lattice to distort the molecule to fit the available cage, and the extent to which a given internal co-ordinate may be distorted depends upon the force constant associated with that co-ordinate. Co-ordinates characterized by low force constants are most likely to find themselves in a 'tight cage' as the matrix seeks out the most economical means of forcing the molecule to fit into the available matrix cage. These are the co-ordinates that participate in the low-frequency vibrations, the effect of the 'tight cage' thus being to induce a positive (blue) frequency shift; conversely, higher-frequency modes experience a 'loose cage' and show negative (red) shifts. By contrast, variations of vibrational amplitude are considered to make only a minor contribution. For all its persuasiveness, the cage model cannot always be reconciled with the matrix shifts observed for individual molecules.

Blue shifts appear to be more common for electronic spectra, though, as with vibrational transitions, there is a trend to lower frequencies as the polarizability of the matrix molecule increases; marked blue shifts have been reported for the Rydberg levels of some organic molecules. Furthermore, significant shifts to the blue or red appear commonly to be induced by variations of matrix temperature. There is also the intriguing possibility that matrix isolation may cause a reversal of electronic ground and low-lying excited states. The C_2 spectrum was mistakenly thought to provide an example of this, but at the present time there is no positive evidence for such a reversal of electronic levels. Matrix shifts in electronic spectra have been interpreted with the aid of models similar to those employed to describe the perturbation of vibrational levels,[52] but there is the additional complication that many electronic transitions are accompanied by a significant change in molecular geometry, and that work must be involved in accommodating this change within the matrix cage.[22]

Rotation.—The rotational motion of trapped molecules has been the subject of extensive research.[23, 25, 26] Where the matrix cavity is sufficiently large, it may be anticipated that the rotational energy levels of small guest species will not be unduly perturbed. However, at temperatures of 4—20 K only the lowest rotational levels will be appreciably populated. Thus, for a diatomic molecule, observations may typically be restricted to the $R(1)$ $(J = 2 \leftarrow J = 1)$, $R(0)$ $(J = 1 \leftarrow J = 0)$ and $P(1)$ $(J = 1 \rightarrow J = 0)$ transitions in the vibrational–rotational spectrum and to the $J = 1 \leftarrow J = 0$ and $J = 2 \leftarrow J = 1$ transitions in pure rotation. Depending upon the relative sizes of the guest species and the matrix site and their mutual interactions, it is evident that the behaviour of the guest may range from nearly free rotation in the matrix cage at one extreme to suppression of this motion at the other. In practice, for all but the smallest molecules, rotational motion is quenched by the matrix. This has the obvious and important advantage of simplifying and sharpening the features of the observed vibrational and

[52] M. McCarty, jun., and G. W. Robinson, *Mol. Phys.*, 1959, **2**, 415; G. W. Robinson, 'Methods of Experimental Physics', ed. D. Williams, Academic Press, New York, 1962, vol. 3, p. 155.

electronic spectra, but at the expense of accurate information about the moments of inertia of the trapped molecules. On the other hand, more-or-less convincing evidence of rotation has been advanced for a number of small molecules isolated in noble-gas matrices: these include HX, DX (X = F, Cl, Br, or I), H_2O and its deuteriated forms, NH_3,[53] NH_2, NO_2, SO_2, CH_4, CD_4, and CH_3. With SO_2 and NO_2 rotation is believed to be confined, in effect, to a single axis. Features due to rotational transitions are most clearly identified by the marked and reversible temperature-dependence of their intensities. Hence, it has been possible to distinguish between the effects of rotation and those of 'multiple trapping sites' (*q.v.*) and so resolve some of the earlier conflicts of interpretation, for example, concerning the i.r. spectrum of matrix-isolated HCl. Detailed studies of the hydrogen halides imply that rotation occurs in matrices which offer 'spherical' substitutional sites (noble gases, CH_4, CF_4, and SF_6) but not in matrices like N_2, CO, or CO_2 which are characterized by 'cylindrical' cavities. Two other effects which cause multiplet splitting of spectral features are associated with rotation, namely inversion and nuclear-spin conversion;[23, 26] these have been observed for H_2O, NH_3, and CH_4.

Attempts have been made to calculate the perturbation of the rotational levels of a diatomic molecule in an inert matrix. The models most widely favoured have involved either (*i*) rotation limited by a potential barrier, or (*ii*) perturbation of the rotational motion by the constrained translational motion of the solute molecule in the matrix lattice (embodied in the 'rotational–translational coupling' model).[26, 27] The hindered rotor model is successful in predicting the general trend of rotational spacings in noble-gas matrices and the splitting of the $J = 2$ level; it fails to predict that the spacings for DX should decrease less from gas to matrix than those for HX. The rotational–translational coupling model is more successful in accounting for the observed spectral effects, but is not consistent with the matrix shifts found for HF and DF, which, with other anomalous features, are better served by considerations of the anisotropies in the rotational potential function.[54] That the $J = 1$ level of HCl appears to be split by 0.8 cm^{-1} when the molecule is isolated in argon at 4 K may mean that the noble-gas atoms adopt an h.c.p. structure, instead of their normal f.c.c. structure, in the neighbourhood of the HCl impurity. Such a conclusion is consistent with the polymorphic changes reported for argon crystallized in the presence of impurities [55a] and with the zero-phonon lines observed for H_2-doped argon crystals.[55b] A recent study has extended the scope of theoretical treatments to take account of a tetrahedral guest molecule, the vibration–rotation states

[53] D. E. Milligan, R. M. Hexter, and K. Dressler, *J. Chem. Phys.*, 1961, **34**, 1009; H. P. Hopkins, jun., R. F. Curl, jun., and K. S. Pitzer, *J. Chem. Phys.*, 1968, **48**, 2959.

[54] M. G. Mason, W. G. Von Holle, and D. W. Robinson, *J. Chem. Phys.*, 1971, **54**, 3491.

[55] (*a*) L. Meyer, C. S. Barrett, and P. Haasen, *J. Chem. Phys.*, 1964, **40**, 2744; C. S. Barrett and L. Meyer, *ibid.*, 1965, **42**, 107; 1965, **43**, 3502; C. S. Barrett, L. Meyer, and J. Wasserman, *ibid.*, 1966, **44**, 998; (*b*) R. J. Kriegler and H. L. Welsh, *Canad. J. Phys.*, 1968, **46**, 1181.

having been determined for ν_3 and ν_4 of CD_4 entrapped in a noble-gas matrix to give results in good agreement with experimental findings.[56]

Analysis of the e.s.r. spectra of simple radicals has also testified to the incidence of free or hindered rotation in the matrix cage. For example, the species CN, CH_2, MH_3 (M = C, Si, Ge, or Sn), and PH_2 are all judged to undergo some form of rotation in noble-gas matrices at 4 K, and in neon, at this temperature, NO_2 is believed to occupy three distinct sites, which differ in the opportunities they afford for rotation of the guest molecule.

Molecular Orientation in Matrices. [23,36c].—The observation of free or hindered rotation in matrices suggests the possibility of preferred orientations for the trapped species in situations where rotation is debarred. E.s.r. investigations give grounds for believing that $Cu(NO_3)_2$,[36c] CuF_2,[36c] BO (in Ne but not in Ar at 4 K), VO, YO, MgF, ZnF, NF_2,[57] and CO_2^- (in CO_2 at 77 K) do assume preferred orientations with respect to the substrate on which condensation occurs; also, the i.r. spectra of matrix-trapped BF_3, AlF_3, and Li_2F_2 suggest that these species take up preferred orientations. It is found that AlF_3 tends to be aligned with its plane parallel to the trapping surface, but that VO favours a situation in which its internuclear axis is perpendicular to the surface. In neon matrices at 4 K, BO molecules lie in vertical planes perpendicular to the trapping surface; that they should also be preferentially distributed within these planes either parallel or perpendicular to the surface provides a rare commentary on the mode of crystallization. The effect has been exploited to analyse the anisotropic parts of the hyperfine interaction of the paramagnetic molecules NF_2[57] and VO. The precise mechanism of orientation effects is a matter for conjecture, but it is likely to vary with the nature of the guest and host species, depending, for example, on their size and shape. There is evidence that a certain amount of local diffusion or annealing is a prerequisite for the assumption of a preferred alignment; in keeping with this, it is significant that most cases where marked orientation is known to occur at 4 K have involved the least rigid matrix, *viz.* neon. Whether the guest is disposed in a random or a selective manner is an important issue that has been overlooked in all but a few matrix-isolation experiments. Thus, preferred orientation may result in discrimination against certain spectroscopic transitions, and interpretations based on an assumed random orientation may therefore be incorrect. Recent polarization measurements on the Raman scattering due to various matrix-isolated systems [29, 58] are indeed consistent with, but do not prove, random orientation of the trapped molecules. Described in idealized terms by the so-called 'cold gas' model,[21] this condition has not, unfortunately, been put to the test, being treated rather as axiomatic for the purposes of interpreta-

[56] K. Nishiyama, *J. Chem. Phys.*, 1972, **56**, 5096.
[57] P. H. Kasai and E. B. Whipple, *Mol. Phys.*, 1965, **9**, 497.
[58] (*a*) J. W. Nibler and D. A. Coe, *J, Chem. Phys.*, 1971, **55**, 5133; (*b*) H. Huber, G. A. Ozin, and A. Vander Voet, *Nature Phys. Sci.*, 1971, **232**, 166.

tion. Without a more scientific basis, the precise significance of the measured depolarization ratios is far from clear.

Matrix Site Effects.—Further complications can arise from a group of effects associated with the more intimate properties of the matrix sites enclosing the guest species. Thus, the observation of a multiplet where only a single spectroscopic feature is anticipated is commonly attributable to variations in the matrix environment experienced by the guest. These variations may reflect (*i*) different orientations of the guest within a particular matrix cage, (*ii*) population of more than one type of cage because of the opportunities offered by substitutional, interstitial, and dislocation sites within the host lattice, and (*iii*) occupation of cages which include guest, as well as host, molecules as nearest or next-nearest neighbours; this so-called 'aggregation site effect' differs from chemical aggregation in that guest–guest interactions are presumed to be weak, but the distinction is sometimes more apparent than real. The intermolecular forces between the guest and its environment will differ slightly according to changes in any one of these circumstances, and the resulting perturbations of the energy levels may generate discernible multiplet structure in the spectrum of the guest. Furthermore, the symmetry of the environmental interaction may remove the degeneracy of certain transitions and so provide yet another source of splitting. In general, matrix site effects cannot be predicted *a priori*, and have frequently been invoked to account for spectroscopic fine structure that finds no obvious explanation in other causes, *e.g.* chemical aggregation or rotation; more than once they have confused spectroscopic interpretation. In practice, careful studies, involving diffusion and variations of matrix ratio and of matrix material, are usually necessary to differentiate between molecular association, rotation, and matrix site effects, but detailed investigations of this sort have been evident in no more than a handful of experiments.

External and Lattice Vibrations.—With the introduction of an impurity into a host lattice, there is the prospect of inducing or modifying phonon bands of the solid, and of activating local or quasi-local vibrations of the impurity centre. Absorptions originating from fundamental modes may be expected to appear in the far-infrared and may thus confuse the interpretation of spectroscopic results in this region. However, there is also the possibility of coupling between the various motions, which may cause lattice vibrations or libration–translation motions of the impurity to subscribe to combination tones with internal vibrations of the impurity molecule. There have been numerous studies of these characteristics for small molecular ions isolated in ionic crystals.[27, 59a] For example, the i.r. spectra of potassium halide crystals doped with OCN^- exhibit combination tones of the vibrational fundamentals with external modes of the OCN^- ion and with lattice modes

[59] (*a*) D. A. Long, *Ann. Reports* (*A*), 1968, **65**, 89; (*b*) V. Schettino and I. C. Hisatsune, *J. Chem. Phys.*, 1970, **52**, 9; (*c*) W. F. Sherman and P. P. Smulovitch, *ibid.*, p. 5187.

of the host crystal.[59b] Similarly, when isolated in CsBr at 100 K, NH_4^+ exhibits weak combination bands which are believed to implicate external modes of the ion.[59c] With inert host lattices, such as those afforded by the noble gases, absorptions in the far-infrared have been attributed to libration–translation motions of the isolated hydrogen halide molecules, while the vibration of hydrogen atoms in the O_h interstitial sites of solid argon or krypton is judged to be responsible for the absorptions observed at 850—910 cm^{-1}.[60] In argon and nitrogen matrices at 8 K, H_2O_2 displays broad absorptions which are ascribed to sum-and-difference combinations of the internal torsional mode with translational and librational modes; there is also reason to believe that the torsional mode couples with the translational and/or librational motion in the nitrogen matrix. The lattice vibrations of argon crystals doped with H_2 appear to contribute to sum-and-difference tones with transitions of the H_2 molecule.[55b] Attention has also been focused on the fundamentals (or phonon bands) of argon and krypton crystals doped with various impurities.[61]

5 Spectroscopic Methods

At present, the three most important spectroscopic techniques employed in the study of matrix-trapped species are those of i.r., electronic, and e.s.r. spectroscopy, though measurements of the Raman and Mössbauer spectra of such species have lately been established as practical and fruitful propositions. Ideally, a combination of several techniques constitutes the most reliable approach to the identification of highly reactive species and to the exploration of intermolecular interactions within the matrix. However, this ideal has been realized in comparatively few instances. In general terms, the sensitivity of the techniques diminishes in the sequence e.s.r. > u.v.–visible > i.r. > Raman, and whereas the e.s.r. technique is highly selective and sensitive in singling out small concentrations of free radicals, i.r. and Raman methods suffer from the disadvantage of requiring substantial amounts of sample before a satisfactory spectrum can be secured. However, the methods of vibrational spectroscopy are probably the most versatile and informative, particularly in the identification of unfamiliar guest species; without some prior knowledge of the species, it is often difficult to interpret the u.v.–visible or e.s.r. spectrum. The precise concentration of the solute is not easily obtained. Where it is derived directly from the gas phase, its concentration is normally taken to be equivalent to the conditions of that phase, although condensation may be far from quantitative; where a radical is formed *in situ*, any attempt to improve upon order-of-magnitude estimates of concentration must normally involve its e.s.r. spectrum.

[60] V. E. Bondybey and G. C. Pimentel, *J. Chem. Phys.*, 1972, **56**, 3832.
[61] J. Obriot, P. Marteau, H. Vu, and B. Vodar, *Spectrochim. Acta.* 1970, **26A**, 2051; H. Vu, M. R. Atwood, and E. Staude, *Compt. rend.*, 1963, **257**, 1771; M. Jean-Louis, M. Bahreini, and H. Vu, *Compt. rend.*, 1969, **268**, *B*, 41, 390, 479.

I.R. Spectroscopy.—This continues to be the most popular method of detection and characterization, and has provided the bulk of chemically significant results about matrix-isolated systems. As already noted, the quenching of rotation for all but the simplest molecules has the effect of simplifying the i.r. spectrum and of producing sharp, line-like absorption bands corresponding, in effect, to pure vibrational transitions. There are several important consequences of such sharp bands: (*i*) the sensitivity is enhanced; (*ii*) vibrational frequencies can often be meaningfully determined to ± 0.1 cm^{-1}; (*iii*) the improved definition gives a much better chance of resolving nearly degenerate spectral features; in this respect, the effect of isotopic substitution is commonly displayed to striking advantage. Such characteristics compare most favourably with the diffuse bands observed in the vibrational spectra of high-temperature vapours, where complications arise from the wide spread of populated rotational levels and from the large amplitudes assumed by low-energy vibrations. Furthermore, population of excited vibrational levels at elevated temperatures is marked by the appearance of 'hot bands', which are an additional source of confusion. The low temperatures of matrix-isolation experiments provide for the thermal occupation of only ground vibrational states and so eliminate the complication of 'hot bands'. This simplification is particularly desirable for molecules having low-energy vibrational fundamentals. No evidence has yet been adduced to suggest significant departures from the selection rules appropriate to the gas phase, though the influence of site symmetry may lead to the splitting of absorptions due to degenerate vibrations of the trapped molecule. Selection rules have been widely enlisted for deductions about the geometry of such molecules (see Tables 1—3), despite the inherent weakness that failure to observe a spectroscopic transition is a negative and not a positive test of molecular symmetry.

Where comparisons are possible, a close correspondence is found between the vibrational frequencies of a typical matrix-isolated molecule and those of the same molecule in the gas phase. The frequency perturbations induced by the matrix cage (see the preceding section) are comparable in magnitude with the anharmonicity corrections, and relationships such as the isotope sum and product rules, which give a good account of gas-phase frequencies, are equally useful in characterizing matrix-isolated molecules. Little attention has been paid to the effect of the matrix cage on vibrational anharmonicities, but the close agreement between the anharmonic coefficients computed for ClCN in the gaseous and matrix-isolated condition certainly indicates that the cage does not perturb the anharmonic potential constants any more than the quadratic ones. Hence, with the simplicity and superior resolution of the spectra, matrix isolation may be useful in clarifying details in the vibrational analysis of stable molecules. This has proved to be the case, for example, with the molecules BeB_2H_8, $H_3B_3N_3H_3$, CH_3OH, C_2H_5OH, H_2S, CH_3SH, CH_3CO_2H, ClF_3 and BrF_3, $ClOClO_3$ and $BrOClO_3$, and $M_3(CO)_{12}$ (M = Fe, Ru, or Os), many of which have demonstrated the value of matrix isolation in resolving near-degenerate vibrational modes and in furnishing measurable isotopic shifts. There are clear signs that the method is gaining favour for the support it lends to proposed vibrational assignments and for the improved definition of molecular force fields made possible by the observations on isotopically distinct species. Furthermore, the possibility exists of exploiting the reduced entropy effects afforded by a low-temperature matrix to identify new isomeric forms of stable molecules. Such isomeric rearrangement is believed to occur, for example, when the following molecules are photolysed *in situ*: Cl_2O (to give ClClO), OClO (to give ClOO), XCN (to give XNC, where X = H or halogen), and diazomethane (to give diazirine and isodiazomethane). Matrix isolation has also been employed in attempts to resolve the perplexing behaviour of beryllium borohydride and xenon hexafluoride. The vapour of the borohydride is now believed to contain an equilibrium mixture of two distinct isomers, only one of which, probably with a C_{3v} configuration, is isolated *via* deposition in a nitrogen or argon matrix. By contrast, the i.r. spectrum of xenon hexafluoride depends on the thermal history of the sample, a peculiarity which has been attributed to the existence of three electronic isomers, interconversion of which is sluggish even at 100 °C.

To assist in the identification of unfamiliar species, it has been a common practice to draw on chemical intuition and spectroscopic analogy. Hence, the i.r. method has been used to good effect to monitor the course of reactions which occur during condensation or *in situ*; in principle, this approach may supplement the efforts recently made to synthesize macroscopic quantities of known or unusual materials by the co-condensation and subsequent diffusion of high-temperature species.[62] Qualitative spectroscopic evidence about chemical changes has been obtained, for example, following the co-condensation of metal atoms with molecular halides, CO, or N_2; likewise the photolysis of matrix-isolated metal carbonyls, either alone [*e.g.* $Ni(CO)_4$, $M(CO)_6$ (M = Cr, Mo, or W), and $RMn(CO)_5$ (R = CH_3 or CF_3)] or in admixture with other reagents [*e.g.* $Fe(CO)_5$ with C_2H_4 and $Ni(CO)_4$ with N_2], has been successfully monitored by variations in i.r. absorption; again, the same approach has been used to study the effects of bombarding frozen reagents such as HCN, N_2O_4, or an olefin with hydrogen atoms.[63] As a test of these qualitative deductions, two additional expedients have often proved helpful: (*i*) attempts are made to generate the unfamiliar species by different routes using two or more distinct chemical precursors; (*ii*) the effects of sample-warming are explored. The first measure is illustrated by the variety of routes used to produce and so characterize the free radical NCO:[64] *viz.* photolysis of HNCO and HN_3 in CO, HCN, and N_2O, and of mixtures of

[62] See, for example, J. B. Ezell, J. C. Thompson, J. L. Margrave, and P. L. Timms, *Adv. High Temp. Chem.*, 1967, **1**, 219; P. L. Timms, *Endeavour*, 1968, **27**, 133; *Chem. Comm.*, 1969, 1033; *J. Chem. Soc.* (*A*), 1970, 2526.

[63] P. M. A. Sherwood and J. J. Turner, *J. Chem. Soc.* (*A*), 1971, 2474; P. M. A. Sherwood, *ibid.*, p. 2478.

[64] D. E. Milligan and M. E. Jacox, *J. Chem. Phys.*, 1967, **47**, 5157.

N_3CN and NO in an argon matrix. Similarly, the co-condensation of alkali-metal atoms with a variety of halogenated derivatives of methane has served convincingly to identify the absorptions due to individual products: for example, those absorptions common to the reactions of the metal atoms with the precursors H_2CClF, H_2CCl_2, H_2CClBr, and H_2CClI are logically attributed to the H_2CCl free radical. Identification of the radicals CCl_nBr_{3-n} ($n = 0$—3) has been facilitated by studies of the species formed following halogen-abstraction from the precursors CCl_4, CCl_3Br, CCl_2Br_2, $CClBr_3$, and CBr_4. Sample-warming experiments may be helpful in identifying reactive molecules when diffusion provides the opportunity for simple bimolecular reactions such as

$$2CH_3 \longrightarrow C_2H_6$$

and

$$2CCl_2 \longrightarrow C_2Cl_4$$

which require little or no activation energy. Particularly conclusive are those experiments which have demonstrated the growth of absorption due to the stable oligomer at the expense of the bands tentatively identified with the monomer. Alternatively, new products may arise from polymerization or other reactions activated by diffusion; for example, the following oligomers have been characterized with varying degrees of confidence (Tables 1 and 2): C_n molecules, $[LiX]_n$ (X = F, Cl, Br, or I; $n = 2$ or 3), $[LiO_2]_2$, $[M^1O]_n$ (M^1 = Si, Ge, Sn, or Pb; $n = 2$, 3, or 4), $[M^2F_2]_2$ (M^2 = Ge or Sn), $[M^3O_2]_2$ (M^3 = S or Se), $[M^3F_4]_2$ (M^3 = S, Se, or Te), $[XF_3]_2$ (X = Cl or Br), and $[TCl_2]_2$ (T = Mn, Fe, Co, or Ni). Figure 1 illustrates clearly the behaviour on diffusion of $[SiO]_n$ species isolated from silicon oxide vapours; here the reaction

$$SiO + Si_2O_2 \longrightarrow Si_3O_3$$

evidently occurs when the matrix is warmed slightly, and its progress is readily followed by i.r. measurements. However, great care is necessary in the interpretation of many such experiments since molecules other than the species of interest (*e.g.* co-products or impurities) may become involved in matrix reactions.

The most decisive conclusions regarding the identity, vibrational properties, and structure of isolated species may be made on the basis of isotopic effects. Thus, the presence of a certain atom in a new species can be verified neatly by observing an isotopic shift. By using isotopic mixtures, the molecular formula of the species can frequently be determined with some confidence. Thus, introduction in a random manner of isotopes Y^1 and Y^2 into a molecule XY_n produces a multiplet structure of $n+1$ individual lines for each non-degenerate vibrational mode involving significant motion of the n equivalent Y atoms; a more complicated pattern results (*i*) if the atoms are not equivalent or (*ii*) if the vibration is a degenerate one for the isotopically pure species XY^1_n and XY^2_n, but the degeneracy is lifted for the mixed species

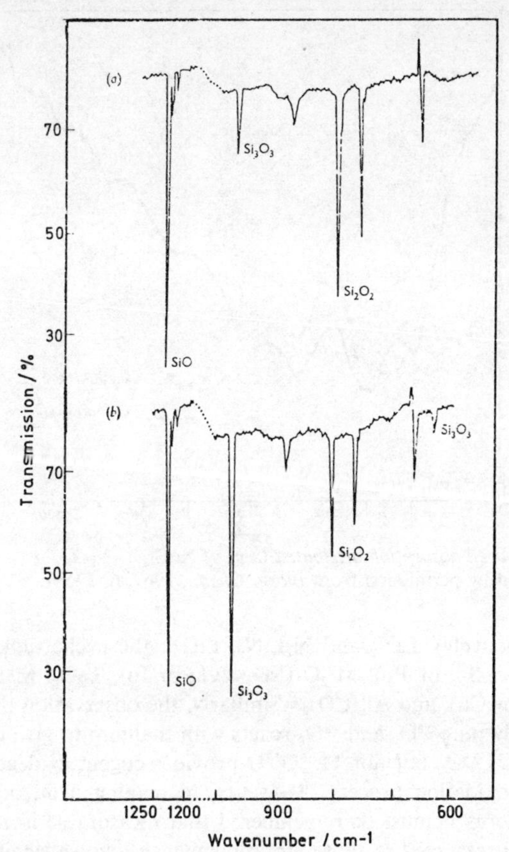

Figure 1 *Infrared spectra of silicon oxides in a nitrogen matrix:* (a) *at* 20 K *before diffusion;* (b) *after controlled warming to* 27 K
(Reproduced from *Chem. in Britain*, 1971, **7**, 186.)

$XY^1_{n-m}Y^2_m$. The method is well exemplified by the isolation of thallium oxide molecules using a roughly equimolar mixture of ^{16}O and ^{18}O. The resulting i.r. spectrum exhibits both doublet and triplet absorptions: the doublet is clearly due to a vibrational mode involving a single oxygen atom (ν_3 of Tl_2O), whereas the triplet must be caused by the vibration of two equivalent oxygen atoms [a Tl—O stretching mode of $(Tl_2O)_2$]. Failure to carry out a test of this sort has led, in one instance, to the erroneous conclusion that the Tl_2O molecule is responsible for all the principal bands observed. By way of further illustration, Figures 2 and 3 depict the isotopic patterns displayed by individual vibrational modes of matrix-isolated species thus identified as $XeCl_2$ and $Pd(CO)_4$, respectively. Isotopic effects have also been observed to striking advantage in the characterization, *inter alia*, of the following systems: HO_2 and LiO_2 (with non-equivalent and equivalent oxygen

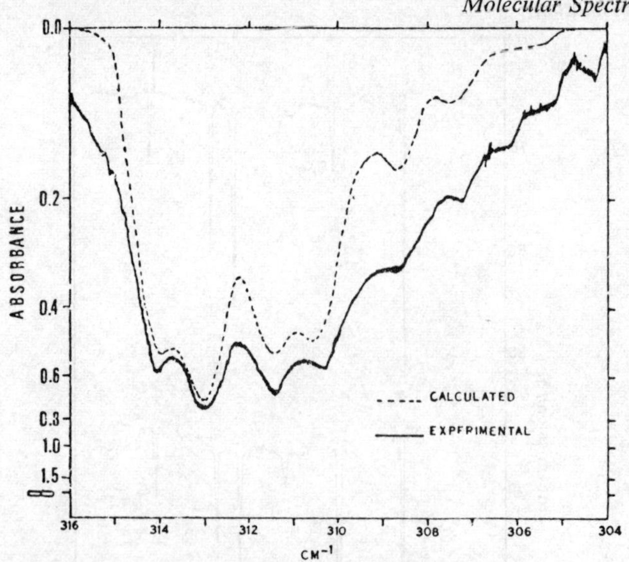

Figure 2 *Infrared absorption attributed to* ν_3 *of* $XeCl_2$
(Reproduced by permission from *Inorg. Chem.*, 1967, **6**, 1758)

atoms, respectively), LiN_2 and N_2LiN_2, LiOF, the cyclic molecules $M^1{}_2O_2$ (M^1 = Si, Ge, Sn, or Pb), $M^2{}_2O$ (M^2 = Ga or In), TeO_2, M^3F_2 (M^3 = Ti, Cr, Fe, Ni, or Cu), and $Al_x(CO)_2$. Similarly, the observation that a mixture of isotopically pure $^{16}O_2$ and $^{18}O_2$ reacts with thallium to give the molecules $Tl_x{}^{16}O_2$ and $Tl_x{}^{18}O_2$ but not $Tl_x{}^{16}O^{18}O$ provides cogent evidence for a non-dissociative oxidation process. However, in making deductions based on isotopic patterns it must be remembered that measurable isotopic splitting is necessarily restricted to those vibrations which involve significant motion of the isotopically modified atoms.

In many matrix-isolation studies, the measured vibrational frequencies of the isotopically distinct species have been used to analyse the molecular force field and to determine the relevant potential constants. Such normal-co-ordinate analysis admits, at least in principle, the possibility of determining certain of the molecular dimensions which contribute to the Wilson *G*-matrix. With the aid of the frequencies for two or more isotopically distinct molecules, bond angles have thus been deduced for a variety of simple molecules, *e.g.* XCO (X = F or Cl), CX_3 (X = H, F, or Cl), CFCl, CCl_2, CO_3, HO_2, LiON, M^1O_2 and $M^1{}_2O$ (M^1 = alkali metal), M^2F_2 (M^2 = Si, Ge, Mg, Ca, Sr, Ti, Cr, Fe, Ni, Cu, or Zn), SiX_3 (X = F or Cl), $M^3{}_2O$ (M^3 = Al, Ga, In, or Tl), YO_2 (Y = S, Se, or Te), and HOX (X = F, Cl, or Br). Plausible though this approach may appear, it is subject, none-the-less, to several limitations:

(*i*) Strictly one needs to know the anharmonicity corrections, neglect of which leads otherwise to a calculated bond angle which is either an

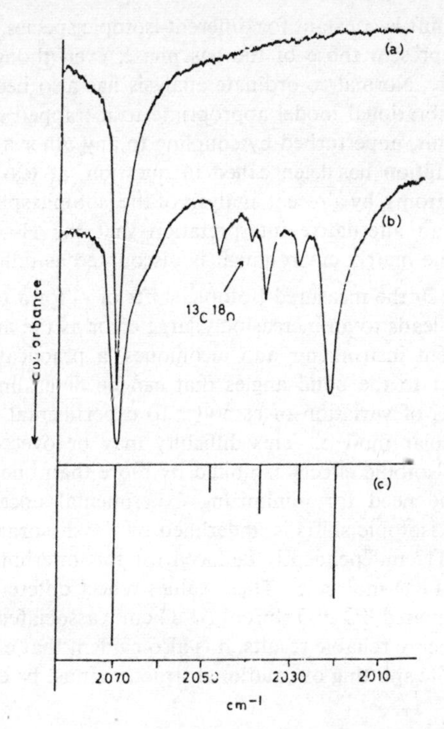

Figure 3 (a) *Infrared spectrum obtained from co-condensation of* Pd *atoms with* Ar–$C^{16}O$ *at* 27 K. (b) *Infrared spectrum obtained from co-condensation of* Pd *atoms with* Ar–$C^{16}O$–$C^{18}O$ *at* 27 K. (c) *Calculated spectrum for* $Pd(C^{16}O)_n(C^{18}O)_{4-n}$ *species*
(Reproduced by permission from *Inorg. Chem.*, 1972, **11**, 666)

upper or a lower limit to the true value. In the infrequent event that there are sufficient data to determine the anharmonicity terms, a precise bond angle may be calculated. This has been possible, for example, in the case of SO_2, for which a bond angle of 119° 37′ has been deduced, which is in extraordinarily close agreement with the value of 119° 19′ gained from the microwave spectrum of the gaseous molecule. For other species, the effect of anharmonicity is either neglected as being less than the experimental error or, better, estimated, for example, by reference to the known anharmonicity of the corresponding diatomic molecule. Where comparison with the bond angle of the gaseous molecule has been possible, as with SeO_2, SiF_2, and M_2O (M = Ga, In, or Tl), satisfactory agreement has generally been found, at least within the limits set by experimental uncertainty.

(*ii*) Beyond inducing small changes of vibrational frequency, the matrix is supposed not to perturb the isolated molecule. In particular, it is assumed, in accordance with predictions for the solvent shifts of diatomic species, that the matrix shift is constant for different isotopic species, so that the frequency ratios represent those of the gas phase, even though the absolute frequencies differ. Normal-co-ordinate analysis has also been based on the belief that the vibrational model appropriate to a trapped molecule is that of the isolated unit, unperturbed by coupling to any other agent. However, this latter assumption has been called in question, at least for molecules containing light atoms, by a recent analysis of the isotopic splitting of matrix-isolated HCN; an alternative interpretation that the H—C≡N skeleton is distorted by the matrix environment is discounted as unlikely.[65]

(*iii*) Uncertainty in the measured isotopic shifts of ν_3 for a triatomic species of the type XY_2 leads to an increasingly large error as the angle approaches 180°; with present instruments and techniques, a practical upper limit of about 160° is set to the bond angles that can be determined in this way. Similar problems of variation of response to experimental error are found for other molecular models. This difficulty may be overcome in part by considering the isotopic effects exhibited by more than one atomic species. Nevertheless, the need for minimizing experimental uncertainties in the measurement of isotopic shifts is underlined by the discordant values of *ca.* 90° and 131 ±11° independently deduced for the interbond angle of the matrix-isolated Tl_2O molecule. These values reflect differences of less than 2 cm^{-1} in an estimated ^{16}O–^{18}O shift of *ca.* 34 cm^{-1} associated with the fundamental ν_3. To secure reliable results, it is also evident that extraneous effects such as matrix site splitting or rotational structure must be excluded, if at all possible.

(*iv*) The response of isotopic shifts to variations of bond angle varies not only with the geometry but also with the relative masses of the vibrating atoms. For example, the ^{16}O–^{18}O isotopic shift for ν_3 of the MO_2 molecule varies by 13.0 cm^{-1} for M = S but by only 2.9 cm^{-1} for M = U, as the bond angle changes from 90 to 180°.

In a few instances, information about the dimensions of a trapped molecule has been gained from the relative intensities of selected i.r. absorptions. Thus, a value of *ca.* 105° has been deduced for the bond angle between the axes of the apical and basal CO groups of the $M(CO)_5$ molecules (M = Cr, Mo, or W), for which a square-pyramidal configuration is favoured; this has been based on an estimate of the i.r. intensity of the t_{1u} C—O stretching mode of $M(CO)_6$ relative to that of the corresponding *e* mode of $M(CO)_5$. For a symmetric-top molecule, use may be made of the expression $I_{sym}/I_{antisym} = \cos^2\theta/\sin^2\theta$ relating the i.r. intensities of the symmetric and antisymmetric stretching modes of the equivalent set of oscillators to the angle θ between the axis of any one oscillator and the top axis. An approximate analysis described for monomeric SbF_5, supposed to have C_{4v} symmetry, illustrates the method of attack, though it seems likely, in view of recent measurements,

[65] J. Pacansky and G. V. Calder, *J. Phys. Chem.*, 1972, **76**, 454; but see S. D. Gabelnick, *ibid.*, p. 2483.

that the conclusions rest on an erroneous reading of the observed spectrum (see Table 2).

The calculation of potential constants for matrix-isolated species provides an additional mechanism for testing the plausibility of vibrational assignments, for exploring quantitatively the interaction of normal vibrations of a given symmetry class, and, in some cases, for adjudicating on problems of stereochemistry (as with the CO_3 radical, for example). Furthermore, the potential constants themselves have often provided an acute commentary on the bonding of unfamiliar systems, examples of which are to be found in accounts relating to the following reactive species: C_n ($n = 4, 5, 6$, or 9), CNN, CCO, CH_2X, and CX_2 (X = F, Cl, or Br), $LiCH_3$, LiN_2, YO_2 (Y = H, alkali metal, F, or Cl), M^1F_2 (M^1 = first-row transition metal), M^2F_3 (M^2 = lanthanide), HOF, X_3 (X = Cl or Br), $XeCl_2$, and XeO_3F_2. Of particular note is the finding that the stretching force constant of an X_3 radical (X = Cl or Br) does not greatly exceed that of the X_3^- anion, but is markedly lower than that of the X_2 molecule. With due allowance for the effects of charge, this clearly implies that the electron removed in the process $X_3^- \rightarrow X_3$ originates from a non-bonding orbital, in agreement with the predictions of the three-centre MO model originally devised by Rundle and Pimentel.[66] Similarly, the stretching force constant of ClF_2 approaches shat of the ClF_2^- anion. However, the close correspondence of the i.r. tpectra attributed on the one hand to the HX_2 radical (X = Cl, Br, or I) and on the other to the corresponding HX_2^- anion has raised a bewildering problem of identification; hence, any conclusion about the bonding in HX_2 must be withheld until the ambiguities of the present position have been resolved. For molecules of the type YOO (Y = H, F, or Cl), the values of the O—O stretching force constants lend strong support to the bonding scheme of Spratley and Pimentel,[67] which invokes $(s\text{–}\pi^*)\sigma$– or $(p\text{–}\pi^*)\sigma$–interaction depending upon the nature of Y; in the event that Y is an alkali-metal atom, the system approximates closely to the formulation $Y^+O_2^-$ with virtually complete electron-transfer to an anti-bonding orbital of the O_2 molecule. Similarly, the potential constants of the species CH_2X and CX_2 (X = F, Cl, or Br) have been compared with those of related molecules in attempts to analyse the bonding contributions made by $(p \rightarrow p)\pi$– and $(p \rightarrow d)\pi$–interactions. Correlations have also been made between the potential constants and other bond parameters, *e.g.* electronegativity differences, dissociation energies, and internuclear distances, for series of molecules having a common structural unit. In this context, it is interesting to contrast the diversity exhibited by the M—F stretching force constants for the difluorides of the 3*d* transition metals with the comparatively narrow range and regular increase of the corresponding parameter for the lanthanide trifluoride molecules. Some investigators have taken advantage of one or

[66] G. C. Pimentel, *J. Chem. Phys.*, 1951, **19**, 446; R. J. Hach and R. E. Rundle, *J. Amer. Chem. Soc.*, 1951, **73**, 4321.
[67] R. D. Spratley and G. C. Pimentel, *J. Amer. Chem. Soc.*, 1966, **88**, 2394.

other of the empirical expressions devised to relate the length of a bond to its stretching force constant; thus, the logarithmic relation of Herschbach and Laurie[68] has been employed to calculate bond lengths in the molecules $M^1{}_2O$ (M^1 = Ga, In, or Tl), $M^2{}_2O_2$ (M^2 = Ge, Sn, or Pb), $M^2{}_3O_3$ (M^2 = Ge or Sn), $M^2{}_4O_4$ (M^2 = Sn or Pb), SbF_3, and TeF_4. However, it is important to appreciate that force constants are limited in practice by their dependence on the type of force-field assumed, and that there exists *a priori* no accurate means of relating them to the more familiar parameters of bond length and dissociation energy. Accordingly, it is not always easy to isolate the precise chemical significance of a change in the magnitude of a particular force constant.

The results of matrix-isolation experiments have frequently been applied to the calculation of thermodynamic functions appropriate to the trapped species, which may be of particular interest for the components of high-temperature vapours. With the knowledge gained about the geometry, vibrational frequencies and, in some cases, the electronic structure of the species, there may well be sufficient information to allow reliable statistical calculations of heat capacities, entropies, *etc.* The outcome of such calculations has been reported in numerous cases. However, for a complete definition of the thermodynamic properties, it is necessary to know the heats of formation, sublimation, *etc.*, parameters which have usually been deduced for high-temperature systems by mass-spectrometric estimates of the variation of partial pressure with temperature.[5] One weakness of this approach is that it does not distinguish between various molecular isomers, the existence of which is implied by certain matrix-isolation experiments involving, for example, Li_2F_2, LiO_2, and Li_2O_2; further studies may well show that these are not exceptional systems. It is of interest, therefore, that evidence has been presented in a few experiments to suggest that a molecular beam can be trapped in a matrix so that the relative concentrations of the guest molecules in the matrix are representative of those in the high-temperature vapour. In particular, Snelson has demonstrated that the intensity of an i.r. band due to a matrix-isolated species obeys the Beer–Lambert Law, despite the fact that the sticking probability of molecules striking the trapping surface is invariably less than unity.[69] Similar conclusions have been reached by Rochkind[51] using the less stringent conditions of isolation of pulsed deposition (*q.v.*), thus sustaining the potential of the method for quantitative analysis. Hence, by isolating the components of vapours maintained at various temperatures and measuring the absorption intensities, it is possible to deduce second-law heats of sublimation, vaporization, or reaction. The heat of sublimation of AlF_3 determined in this manner has been shown to agree well with reliable values obtained by other methods. In the case of Li_2F_2, moreover, it has been found that the heats of vaporization and energies of formation (from the monomer) differ significantly for the linear

[68] D. R. Herschbach and V. W. Laurie, *J. Chem. Phys.*, 1961, **35**, 458.
[69] A. Snelson, *J. Phys. Chem.*, 1969, **73**, 1919.

and cyclic forms. In a similar study, the thermodynamic characteristics of the reaction

$$2TiF_3(g) + Ti(s) \rightleftharpoons 3TiF_2(g)$$

have also been evaluated. It is curious, in view of the apparent success of such experiments, that more use should not have been made of intensity measurements as an index to thermodynamic properties.

Raman Spectroscopy.[29]—The advantages of being able to measure the Raman spectrum of a matrix-isolated system are easily recognized: thus, the additional knowledge about vibrational transitions complements that derived from i.r. measurements and may be expected to facilitate the identification of unfamiliar species, and to give added weight to deductions about their structural and vibrational characteristics. With a symmetrical molecule like S_2, CH_3, or $XeCl_2$, the Raman effect offers the opportunity, normally denied in i.r. absorption, of observing totally symmetric vibrational modes, which are commonly seen as the strongest and sharpest bands in the Raman spectrum. Hence, it may be possible to apply more rigorous tests to the operation of the selection rules, and so improve the reliability of any structural assignment proposed on this basis; with the additional frequencies, a complete vibrational assignment may also be realised, and the definition of the molecular force field thereby refined. There is, too, the rewarding prospect of being able to perform polarization measurements on the Raman scattering, though, without independent knowledge of how the trapped molecules are oriented within the matrix, the interpretation of the results is less easily accomplished (*q.v.*).

In practice, there are formidable experimental obstacles, even with the advent of the laser for excitation of the Raman spectra. The primary difficulty has been that of reconciling the inherent weakness of the Raman effect with the need to use high matrix : solute ratios to achieve efficient isolation. Other frustrating influences include (*i*) local heating of the matrix by the incident radiation, (*ii*) the inferior resolution imposed by the large spectral slit-widths which are normally required to achieve an acceptable signal-to-noise ratio, (*iii*) the light-scattering properties of the matrix, and (*iv*) fluorescence due to contaminants or to the matrix-isolated species itself. To what extent it has been possible to overcome the various effects is not yet clear, but that the method is practicable has been established beyond dispute by a minor flood-tide of publications. A review of some of the preliminary results has also been published,[29] and scientific jargon has been augmented if not enriched by the introduction of the ungainly portmanteau of 'matrix isolation laser Raman spectroscopy'. Many of the published reports are relatively coy about the practical details and about the reproducibility of the measurements; certain conclusions, it also appears, are born more in hope than in scientific conviction. The consensus of experimental practice involves the use of a highly polished metal surface as the substrate for the matrix, and excitation of the spectrum with near-grazing incidence of the focused laser beam has been recommended. Matrix : solute ratios as high as 10 000 : 1 are possible with solutes like CO_2 or CS_2, but, in general, ratios of 100 : 1—500 : 1 have been necessary to secure satisfactory results.[29, 58, 70] Some experiments have taken advantage of a time-averaging computer to improve the signal-to-noise ratio.[71]

Matrix-isolated molecules for which Raman spectra have been described include CH_4, COS, $CHCl_3$, Li_2, M^1O_2 (M^1 = Li, Na, or K), BeB_2H_8, GeF_2, M^2X_2 and M^2XY (M^2 = Ge, Sn, or Pb; X, Y = Cl or Br), M^3NO_3 and $[M^3NO_3]_x$ (M^3 = Li, K, Tl, or $\frac{1}{2}$Cu), PCl_2, S_2, SF_6, SeO_2 and $[SeO_2]_2$, Br_3, $XeCl_2$, XeO_3F_2, and PrF_3. Hence, for example, the linear structure of $XeCl_2$ is vindicated by the observation of a single, strong Raman line at 253 cm^{-1} which finds no counterpart in the i.r. spectrum. Likewise, reference to both the Raman and i.r. spectra indicates D_{3h} symmetry and allows a complete vibrational assignment for the newly identified molecule XeO_3F_2. A Raman line at 716 cm^{-1} has been attributed to the vibration of matrix-isolated S_2, a conclusion somewhat at variance with earlier spectroscopic claims regarding this molecule. Again, the failure to observe any features attributable to M—O stretching modes in the Raman spectra of entrapped MO_2 molecules (M = Li, Na, or K) lends support to the ion-pair formulation $M^+O_2^-$. The Raman spectrum of PrF_3 now suggests that the molecule is planar and not pyramidal as earlier infrared studies had indicated. However, anomalous features of both spectra may mean that the molecule enjoys an E' electronic ground-state.

Optical Spectroscopy.—The methods of u.v.–visible spectroscopy, which were used extensively in early matrix-isolation experiments, have been less popular in recent years, though such measurements continue to yield many noteworthy results. Apart from the experimental difficulties occasioned by the light-scattering properties of the matrix, particularly in the far-u.v., the optical spectrum is not so distinctive as the vibrational spectrum, and, of itself, is much less valuable as a means of identifying an unfamiliar matrix-trapped solute. Most experiments have therefore involved well-established molecules like CS_2, C_6H_6, and $Fe(C_5H_5)_2$ or have relied on the independent identification afforded, for example, by (*i*) the electronic spectrum of the gaseous species, or (*ii*) the vibrational spectrum of the matrix-trapped system. Thus, the observation of an intense absorption near 3290 Å corresponding to the electronic transition reported for gaseous NCN confirmed the formation of the radical by the photolysis of matrix-isolated N_3CN. Likewise, the u.v. absorption spectrum of matrices known to contain HCO confirm not only the identity of the ground-state, but also the origin of the complicated pattern of emission bands observed between 4100 and 2200 Å

[70] J. S. Shirk and H. H. Claassen, *J. Chem. Phys.*, 1971, **54,** 3237.
[71] D. A. Hatzenbuhler and L. Andrews, *J. Chem. Phys.*, 1972, **56,** 3398; R. R. Smardzewski and L. Andrews, *J. Chem. Phys.*, 1972, **57,** 1327.

in hydrocarbon flames. A protracted controversy concerning the electronic spectrum of C_2 in inert-gas matrices has only recently been resolved by cogent evidence that the u.v. bands previously ascribed to the Swan transition of C_2 actually arise from C_2^- (see Table 1).

Three main trends discernible in studies of the optical spectra of matrix-isolated species are as follows:

(*i*) Advantage is taken of the simplified pattern of sharpened absorption bands to assist the analysis of the spacings between both electronic and vibrational levels. In particular, the elimination of transitions originating from excited vibrational states greatly facilitates the interpretation of the electronic spectra exhibited by high-temperature molecules like ScO, TaO, and WO; hence, some of the excited states of these molecules have been brought to light for the first time. For example, a combination of gas-phase and matrix-isolation experiments has disclosed the existence of no less than 27 excited electronic states of TaO. It appears that a neon matrix usually provides the closest approach to the gaseous condition, giving rise to the smallest matrix shifts, though the finding that there is a difference of matrix shift for the transitions $A\,^2\Pi_{3/2} \leftarrow X\,^2\Sigma$ and $A\,^2\Pi_{1/2} \leftarrow X\,^2\Sigma$ of the oxides ScO, YO, and LaO clearly signifies that the matrix induces an appreciable change in the spin–orbit coupling constant. The gain in simplicity and definition brought about by matrix isolation is clearly evident, for example, in the results reported for MCl_2 (M = Cr, Mn, Fe, Co, or Ni), UX_4 (X = Cl or Br), $Fe(C_5H_5)_2$, and various alkali-metal halides. The charge-transfer spectra of the transition-metal dichloride molecules have been satisfactorily analysed in terms of ligand-field theory, while the numerous sharp absorption bands in the region 400—24 000 cm^{-1} exhibited by the uranium tetrahalides have been assigned to pure electronic transitions from the $^3H_4$ (Γ_5) ground-state to excited states of the f^2 configuration. Matrix isolation has also aided vibrational analysis in the u.v. spectra of some of the alkali-metal halides, and consideration has been given to the possible effects of matrix interactions on Franck–Condon factors and on the charge-transfer model of molecular absorption.

Since only the lowest electronic–vibrational energy state is appreciably populated at the low temperatures normally employed for matrix isolation, comparison of the matrix and gas-phase spectra affords the opportunity of establishing which of the gas-phase transitions originate from the ground state. Molecules whose ground states have thus been identified, with varying degrees of conviction, include BN, AlO, ScO, YO, LaO, ScF, ZrO, HfO, TaO, WO, and CuO. This approach is particularly valuable for simple molecules formed by transition metals, lanthanides and actinides, where it is commonly difficult to differentiate between the numerous electronic states. For most of the matrix-isolated monoxides MO, the absorption spectra have indicated the position of 0–0 transitions and given information about the vibrational frequencies of upper states, the results being underpinned, ideally, by reference to both $M^{16}O$ and $M^{18}O$.

(*ii*) The difficulty normally experienced in obtaining singlet–triplet energy separations has been partially overcome by studies of the fluorescence spectra of matrix-isolated molecules, since absorption in a singlet system may be followed either by triplet–triplet or by triplet–singlet fluorescence; the non-radiative decay necessary to sustain these processes is presumably facilitated by the matrix cage. Singlet–triplet transitions have thus been identified, it is believed, for the molecules MO and MS (M = Ge or Sn), GeSe, CeO_2, CS_2, and SeO_2. Other fluorescence studies have been concerned, *inter alia*, with C_2^-, CuO, and PrI_3. It has thus been demonstrated, for example, that the matrix-isolation technique can be used to 'tune' an absorption to bring it into coincidence with a suitable laser line and so excite fluorescence. Moreover, the many transitions within the $4f^2$ configuration observed in the fluorescence spectrum of PrI_3 offer unusually rich opportunities for studying the influence of matrix perturbation and for testing 'crystal-field' or MO models.

(*iii*) Measurements of u.v.–visible spectra have been used to monitor and, in some cases, to interpret the effects of photolysis on matrix-isolated systems. Illustrative of this principle are the studies carried out on the photolytically reversible process

$$M(CO)_6 \rightleftharpoons M(CO)_5 + CO \quad (M = Cr, Mo, \text{ or } W)$$

Irradiation of the supposedly square-pyramidal $M(CO)_5$ is thought to produce an excited state involving primarily the metal and $\pi^*(CO)$ for the unique apical CO group, which allows requisition of an additional CO molecule. Signs have also been found to suggest interaction between $Cr(CO)_5$ and a hydrocarbon host such as CH_4, and accordingly a new scheme has been advanced to account for the photochemistry of $Cr(CO)_6$. Although there are several instances where matrix isolation has been used to test photochemical mechanisms, it is questionable whether the conditions of such experiments allow a meaningful comparison with the behaviour in the gas phase or in liquid solution. Curiously, photolysis of the matrix-trapped monoxides ScO, YO, and LaO brings about the gradual disappearance of their characteristic absorption bands, a phenomenon attributed to diffusion and reaction of excited M—O molecules in the matrix.

Of the numerous reports concerning the optical spectra of matrix-trapped atoms, a selection is given in Table 2. The spectra have been interpreted on occasion in terms of the presence of specific aggregates, *e.g.* Ca_2 and Pb_2, but more than once the results leave doubts whether genuine isolation of the atoms has been accomplished.

E.S.R. Spectroscopy.—E.s.r. investigations of radicals trapped in inert, molecular solids at low temperatures continue to be greatly outnumbered by those employing ionic crystals, frozen solutions, or other relatively strongly interacting host phases. Nevertheless, some important findings have emerged from the e.s.r. spectra of radicals held in noble-gas matrices. A striking feature of the e.s.r. method is the sensitivity of its response to the free or partial rotation of the solute, to its orientation in the matrix, and to the number and nature of distinct matrix sites. Although this sensitivity creates difficulties in the characterization of unfamiliar radicals, it affords a unique advantage for exploring the more intimate aspects of the matrix-isolated condition. Thus, measurements concerned with the O_2 molecule as the solute in solid N_2 or CO imply that it occupies isolated sites essentially cylindrical in shape at low concentrations, but that different sites arise from clustering or defects at higher concentrations. Similarly, analysis of the e.s.r. spectra of copper, silver, or gold atoms in xenon matrices indicates that the metal atoms are substitutionally incorporated into the host lattice. Unfortunately, such experiments are commonly hampered by the polycrystalline nature of the samples. The effects of orientation and restricted rotation are well exemplified by the case of NO_2 in a neon matrix at 4 K (see pp. 538, 539). Again, the preferred orientation taken up by molecules like BO and VO in certain matrices (see p. 539) also has a conspicuous effect on the e.s.r. signal.

Despite its limitations as a technique for identifying new species, only the re.s.r. method has been successfully used to characterize the matrix-trapped CH adical, which has otherwise eluded numerous investigators for more than a, decade. Other radicals whose characterization rests significantly, if not exclusively, on their e.s.r. spectra include M^1H (M^1 = Mg, Ca, Sr, Ba, Zn, Cd, or Hg), M^1F (M^1 = Mg, Ca, Sr, or Ba), M^2H_3 (M^2 = C, Si, Ge, or Sn), CH_nF_{3-n} (n = 0, 1, or 2), PH_2, M^3O (M^3 = B, Al, Sc, Y, La, or V), HO_2 [72] and CN. The properties of the g-tensor and of the hyperfine interactions with magnetic nuclei deduced by analysis of the spectra have then served to elucidate one or more of the following parameters of the trapped radical: (*i*) the character of the electronic ground state, as with ScO, YO, and LaO (all $^2\Sigma$) and VO ($^4\Sigma$); (*ii*) the spin densities on the various atoms, which offer a quantitative test of MO descriptions or the possibility of evaluating the coefficients of LCAO wavefunctions, *e.g.* for the molecules ZnH, CdH and HgH; (*iii*) the bond angles of a polyatomic system like CH_2, CF_3, or SiH_3 (Tables 1 and 2).

Mössbauer Spectroscopy.—Recent investigations have confirmed the feasibility of measuring the Mössbauer spectra of atoms or molecules isolated in an inert, solid matrix. Published results have referred to the ^{57}Fe nucleus of iron atoms and $FeCl_2$, to the ^{119}Sn nucleus of tin atoms and SnO, and to the ^{129}I nucleus of I_2, ICl, and IBr; ^{83}Kr Mössbauer spectra have also been reported for the clathrate system formed by krypton and β-quinol (Table 2). A primary objective of these studies has been to gain well-defined calibration points for the scales relating isomer shift to electron density, and so to circumvent the complications inherent in theoretical treatments of bulk solids. The spectra of isolated iron and tin atoms in their essentially free

[72] F. J. Adrian, E. L. Cochran, and V. A. Bowers, *J. Chem. Phys.*, 1967, **47**, 5441.

states would appear to provide such calibration points. Well-defined quadrupole splitting has been observed, not only for the molecules Fe_2 and Sn_2, which are simultaneously trapped with the free atoms, but also for SnO, where it has been correlated with possible bonding schemes.

6 Caveats and Conclusions

The data of Tables 1—3 can leave little doubt about the importance of the matrix-isolation method and its capacity to yield significant results. However, the experimental riches should not disguise the comparative poverty of our understanding of the precise condition of matrix-trapped species. Until we have acquired more fundamental knowledge concerning matrix perturbations and their effect on conventional spectroscopic models for unperturbed systems (*cf.* the case of HCN),[65] empiricism is bound to play a part in the interpretation of the results. Furthermore, in matters of mechanism it is not realistic to suppose that processes shown to occur in a rigid, inert matrix will necessarily be favoured in the gas phase or in solution.[73] More specifically, the findings of various experiments have disclosed the following hazards:

(*a*) Ambiguities about molecular symmetry where arguments have been based on the vibrational selection rules which appear to operate (as with CCl_3 and PrF_3).
(*b*) Perturbation of the species of interest by a second guest molecule, which may be a co-product or an impurity (as with CH_3).[42]
(*c*) The effects of impurities like CO and H_2O, whose undisclosed presence has led, for example, to erroneous claims about the interaction of nickel atoms with CO_2.[74]
(*d*) Aggregation, the effects of which have clouded the interpretation of certain results, as with the molecules M_2O (M = Ga, In, or Tl) and SbF_5.
(*e*) Confusion between neutral and charged products of high-energy reactions [see, for example, the case histories of C_2/C_2^- and HX_2/HX_2^- (X = Cl, Br, or I)].
(*f*) The difficulty of unravelling the effects of rotation and aggregation, exemplified by the chequered history of CO.
(*g*) Contributions made by translational and librational motions of the solute and by lattice vibrations of the host phase (see, for example, the case of H_2O_2).

In the investigation of unfamiliar species, it is vital at least to recognize and, if possible, to eliminate the potential pitfalls. Wherever practicable, therefore, experiments should take account of the effects (*i*) of varying the synthetic route, matrix material, matrix ratio, and deposition rate, (*ii*) of sample purity,

[73] See for example J. F. Ogilvie, *Photochem. and Photobiol.*, 1969, **9**, 65; M. A. Graham, R. N. Perutz, M. Poliakoff, and J. J. Turner, *J. Organometallic Chem.*, 1972, **34**, C34.
[74] H. Huber, M. Moskovits, and G. A. Ozin, *Nature Phys. Sci.*, 1972, **236**, 127.

(*iii*) of controlled diffusion of the matrix, and (*iv*) of isotopic modifications (whether natural or contrived). It is also highly desirable to invoke more than one spectroscopic property of the trapped species. The meticulous practice that may be required to secure meaningful results is well illustrated by the reported isolation and characterization of the elusive molecule BH_3.

7 Appendix

Table 1 *Matrix-isolated carbon-containing species (excluding organometallic species)*

Species	*Method of preparation*	*Matrix*	*Means of characterization*	*Findings*	*Ref.*
C_2	*X*-Irradiation of HC≡CH	Ar, Kr, Xe; 4.2 K	Visible absorption and emission; u.v.–visible	Definitive assignment of bands to both C_2 and C_2^-; $A^3\Pi_g \rightarrow X\,^3\Pi_u$ Swan band of C_2 observed only in emission in certain matrices	75
C_2; C_3; C_4; C_5	Vaporization of graphite at 2300—2800 K	Ne, Ar, Kr, Xe, N_2, CO_2, SF_6; 4—20 K	I.r.; u.v.–visible; diffusion studies	Electronic states of C_2 and its aggregates investigated	76—78
C_3; C_4; C_5	Vaporization of graphite at 2300—2600 K; controlled diffusion of matrix	Ne, Ar, Xe; 4, 12, 20 K	U.v.–visible	$^1\Pi_u \leftarrow X\,^1\Sigma_g^+$ transition of C_3 studied; vibronic bands assigned; C_4 probably best represented as ·C≡C—C≡C· in agreement with theory	77, 79
C_4; C_5; C_6; C_9	Vaporization of graphite at 2900—3300 K; diffusion of matrix	Ar; 4 K	I.r.; u.v.–visible; ^{12}C, ^{13}C	Vibrational assignments for Σ_u^+ frequencies; stretching force constants computed	80
C_n	Careful diffusion of matrix containing C_2 to 12 K	Ne; 12 K	I.r.; u.v.–visible	New vibrational bands analysed in terms of linear-chain species	77
C_2^-	Photolysis of HC≡CH	Ar, Xe, N_2; 12 K	Laser-excited fluorescence	C_2^+ suggested as cationic product; ν_{0-0} for C_2^- in Ar = 19 193 cm^{-1}	81
	Photolysis of HC≡CH with photoelectron source, *e.g.* Cs atoms	Ar, 14 K	U.v.–visible	Band system at 5206 Å attributed to C_2^-; previously assigned to C_2 Swan transition	82
SiC_2; Si_2C; Si_2C_3	Pyrolysis of SiC_2	Ne, Ar; 4.2, 20 K	I.r.; u.v.–visible	SiC_2 identified; Si_2C and Si_2C_3 possibly also observed; vibrational assignments proposed	83
CH	Photolysis of CH_4	Ar; 4, 14 K	U.v.–visible	Three electronic transitions of CH observed	84
CH_2	Photolysis of diazirine (cyclic CH_2N_2)	Xe; 4.2 K	E.s.r.; ^1H, ^2H, ^{13}C	First convincing evidence of matrix-isolated CH_2; $\widehat{HCH}$ = 137.7° estimated	85
CH_3	Co-deposition of Li atoms with CH_3I	Ar; 15 K	I.r.; ^{12}C, ^{13}C, ^1H, ^2H, ^6Li, ^7Li	Only ν_2 and ν_4 observed	86
	Co-deposition of Li atoms with CH_3Br or CH_3I	Ar; 15 K	I.r.; ^{12}C, ^{13}C, ^1H, ^2H; diffusion studies	Data consistent with D_{3h} (planar) structure; deviation from planarity < 5°.	87
	Pyrolysis of CH_3I at 1625 K, or $(CH_3)_2Hg$ at 1575 K	Ne; 5 K	I.r.; ^1H, ^2H	ν_2, ν_3, and ν_4 observed; normal-co-ordinate analysis performed assuming a planar structure	88

	Photolysis of CH_4	Ar, N_2; 4, 19 K	I.r.; u.v.–visible; ^1H, ^2H	Electronic transitions and ν_2 observed for CH_3; evidence of possible rotation in Ar	89
	γ-Irradiation of CH_4	Kr; 4.2 K	E.s.r.	Free or restricted rotation occurs in matrix	90
CH_3, X^- (X = Cl, Br)	γ-Irradiation of CH_3X	CD_3CN; 88 K	E.s.r.	For CH_3,Br^- *ca.* 10% of the spin density resides on Br	91
CH_4	Single crystal of CH_4(5%)–Kr(95%) grown from liquid solution	Kr; ~13 K	Raman	Evidence of rotation of CH_4 in Kr matrix	92
	Direct deposition	N_2; 12 K	Raman; polarization measurements		93
	Direct deposition	Ar, Kr, Xe; 5—40 K	I.r.; ^1H, ^2H	Evidence of slightly hindered rotation; influence of size and shape of cavity considered	94
CD_4	Direct deposition	Ne, Kr; 0—10 K		Theoretical study of hindered rotational motion; good agreement with experimental findings	95

[75] R. P. Frosch, *J. Chem. Phys.*, 1971, **54**, 2660.
[76] W. Weltner, jun., P. N. Walsh, and C. L. Angell, *J. Chem. Phys.*, 1964, **40**, 1299.
[77] W. Weltner, jun. and D. McLeod, jun., *J. Chem. Phys.*, 1966, **45**, 3096.
[78] R. L. Barger and H. P. Broida, *J. Chem. Phys.*, 1965, **43**, 2371.
[79] W. Weltner, jun., P. N. Walsh, and C. L. Angell, *J. Chem. Phys.*, 1964, **40**, 1299.
[80] K. R. Thompson, R. L. DeKock, and W. Weltner, jun., *J. Amer. Chem. Soc.*, 1971, **93**, 4688.
[81] V. Bondybey and J. W. Nibler, *J. Chem. Phys.*, 1972, **56**, 4719.
[82] D. E. Milligan and M. E. Jacox, *J. Chem. Phys.*, 1969, **51**, 1952.
[83] W. Weltner, jun. and D. McLeod, jun., *J. Chem. Phys.*, 1964, **41**, 235.
[84] D. E. Milligan and M. E. Jacox, *J. Chem. Phys.*, 1967, **47**, 5146.
[85] R. A. Bernheim, H. W. Bernard, P. S. Wang, L. S. Wood, and P. S. Skell, *J. Chem. Phys.*, 1970, **53**, 1280; 1971, **54**, 3223; E. Wasserman, V. J. Kuck, R. S. Hutton, E. D. Anderson, and W. A. Yager, *J. Chem. Phys.*, 1971, **54**, 4120.
[86] W. L. S. Andrews and G. C. Pimentel, *J. Chem. Phys.*, 1966, **44**, 2527.
[87] L. Andrews and G. C. Pimentel, *J. Chem. Phys.*, 1967, **47**, 3637.
[88] A. Snelson, *J. Phys. Chem.*, 1970, **74**, 537.
[89] D. E. Milligan and M. E. Jacox, *J. Chem. Phys.*, 1967, **47**, 5146.
[90] R. L. Morehouse, J. J. Christiansen, and W. Gordy, *J. Chem. Phys.*, 1966, **45**, 1751.
[91] E. D. Sprague and F. Williams, *J. Chem. Phys.*, 1971, **54**, 5425.
[92] A. Cabana, A. Anderson, and R. Savoie, *J. Chem. Phys.*, 1965, **42**, 1122.
[93] J. W. Nibler and D. A. Coe, *J. Chem. Phys.*, 1971, **55**, 5133.
[94] A. Cabana, G. B. Savitsky, and D. F. Hornig, *J. Chem. Phys.*, 1963, **39**, 2942.
[95] K. Nishiyama, *J. Chem. Phys.*, 1972, **56**, 5096.

Table 1 (*continued*)

Species	*Method of preparation*	*Matrix*	*Means of characterization*	*Findings*	*Ref.*
CN	Photolysis of HCN	Ar, N_2; 14 K	I.r.; u.v.–visible; ^{12}C, ^{13}C, ^{14}N, ^{15}N	$B\,^2\Sigma^+ - X\,^2\Sigma^+$ ground-state transition and vibrational fundamental observed.	96
	Photolysis of BrCN or ICN	Ne, Ar, Kr; 4 K	E.s.r.; ^{13}C and ^{14}N hyperfine splitting	*g*- and *A*-values determined; evidence of hindered rotation	97
CNN	Photolysis of carbon atom source in N_2	Ar; 4.2 K	I.r.	Absorptions due to CNN and NCN radicals identified	98
	Photolysis of N_3CN	Ar, N_2; 14 K	I.r.; u.v.–visible; ^{12}C, ^{13}C, ^{14}N, ^{15}N	Vibrational assignments and potential constants determined assuming a linear structure C=N=N	99
	Co-deposition of C atoms with N_2	Ne; 4 K	U.v.–visible	Molecule identified by 4200 Å band system	77
NCN	Photolysis of N_3CN	Ar, N_2, CO, CO_2; 14, 20 K	I.r.; u.v.–visible; ^{12}C, ^{13}C, ^{14}N, ^{15}N	Vibrational assignments for linear species NCN; $^3\Pi_u - ^3\Sigma_g^-$ and $^3\Sigma_u^- - ^3\Sigma_g^-$ transitions observed; photodecomposition of NCN occurs at $\lambda < 2000$ Å to give C atoms	100
NCO	Photolysis of HNCO, CO/HN_3, HCN/N_2O, or NO/N_3CN	Ne, Ar, N_2, CO; 4, 14 K	I.r.; u.v.; ^{12}C, ^{13}C, ^{14}N, ^{15}N	Radical identified; electronic and vibrational transitions analysed	101
HCN	Direct deposition	Ar, N_2, CO; 4.5—20.5 K	I.r.; ^1H, ^2H	No evidence for rotation of monomer; dimer linear or nearly so: HCN·HCN	102
	Direct deposition	Ar; 8 K	I.r.; ^1H, ^2H, ^{12}C, ^{13}C, ^{14}N, ^{15}N	Anomalous isotopic shifts observed (see p. 548)	103
HNC	Photolysis of CH_3N_3	Ar; 4 K	I.r.; ^1H, ^2H, ^{12}C, ^{13}C, ^{14}N, ^{15}N	Vibrational analysis indicates a linear structure; produced *via* $H_2C{=}NH$	104
	Photolysis of HCN	Ar, N_2; 14 K	I.r.; u.v.–visible; ^1H, ^2H, ^{12}C, ^{13}C, ^{14}N, ^{15}N	Vibrational assignments for HNC, assumed to be linear; vibrational properties of HNC perturbed by N_2	96
HCNN	Photolysis of H_2CNN	Ar, Kr, N_2; 4 K	I.r.; ^1H, ^2H	Supplemented by gas-phase studies	105, 106
H_2CNN	Direct deposition	Ar, Kr, N_2; 4 K	I.r.; ^1H, ^2H	Different isomers detected; mechanism of isomerism explored	105, 107
H_2NCN	Direct deposition	Ar; 20 K	I.r.; ^1H, ^2H	NH_2 wagging frequencies observed and interpreted in terms of inversion transitions	108
HNCNH	Pyrolysis of H_2NCN	Ar; 20 K	I.r.; ^1H, ^2H	Positive identification realised	108

CH_2=NH	Reaction of HCN with H atoms	HCN; 77 K	I.r.	Tentative identification	109
	Photolysis of CH_3N_3	Ar; 4.2 K; CO_2; 48—50 K	I.r.; ^1H, ^2H	New absorption bands consistent with formation of CH_2=NH	110
CH_3NH_2	Direct deposition	Ar; 20 K	I.r.; ^1H, ^2H	Isotopic exchange during deposition	111
CO_2NH	Photolysis of HN_3 in CO_2 matrix	CO_2; 20—53 K	I.r.; u.v.–visible; ^1H, ^2H	Evidence of formation of $HNCO_2$ together with HNO; four possible structures discussed	112
CH_3NO_2	Direct deposition	N_2; ~4 K	I.r.; ^1H, ^2H	Absorptions assigned to monomer and dimer species	113
FNCN; F_2NCN	Photolysis of a mixture of F_2 with N_3CN	Ar; 14 K	I.r.; ^{12}C, ^{13}C, ^{14}N, ^{15}N	Vibrational assignments given	114
XNC (X = F, Cl, or Br)	Photolysis of XCN	Ar, N_2; 14 K	I.r.; u.v.; ^{12}C, ^{13}C, ^{14}N, ^{15}N	Stretching fundamentals tentatively identified	96
XCN (X = Cl or Br)	Direct deposition	Ar, Kr; 20 K	I.r.; ^{12}C, ^{13}C, ^{14}N, ^{15}N, ^{35}Cl, ^{37}Cl, ^{79}Br, ^{81}Br	Bending mode split by matrix environment; Fermi resonance effects studied; bands due to $(XCN)_2$ consistent with linear or nearly linear configuration	115

[96] D. E. Milligan and M. E. Jacox, *J. Chem. Phys.*, 1967, **47**, 278.
[97] W. C. Easley and W. Weltner, jun., *J. Chem. Phys.*, 1970, **52**, 197.
[98] N. G. Moll and W. E. Thompson, *J. Chem. Phys.*, 1966, **44**, 2684.
[99] D. E. Milligan and M. E. Jacox, *J. Chem. Phys.*, 1966, **44**, 2850.
[100] D. E. Milligan, M. E. Jacox, and A. M. Bass, *J. Chem. Phys.*, 1965, **43**, 3149.
[101] D. E. Milligan and M. E. Jacox, *J. Chem. Phys.*, 1967, **47**, 5157.
[102] C. M. King and E. R. Nixon, *J. Chem. Phys.*, 1968, **48**, 1685.
[103] J. Pacansky and G. V. Calder, *J. Phys. Chem.*, 1972, **76**, 454; but see S. D. Gabelnick, *ibid.*, p. 2483.
[104] D. E. Milligan and M. E. Jacox, *J. Chem. Phys.*, 1963, **39**, 712.
[105] J. F. Ogilvie, *Photochem. and Photobiol.*, 1969, **9**, 65.
[106] J. F. Ogilvie, *Canad. J. Chem.*, 1968, **46**, 2472.
[107] J. F. Ogilvie, *J. Mol. Structure*, 1969, **3**, 513.
[108] S. T. King and J. H. Strope, *J. Chem. Phys.*, 1971, **54**, 1289.
[109] P. M. A. Sherwood and J. J. Turner, *J. Chem. Soc. (A)*, 1971, 2474.
[110] D. E. Milligan, *J. Chem. Phys.*, 1961, **35**, 1491.
[111] J. R. Durig, S. F. Bush, and F. G. Baglin, *J. Chem. Phys.*, 1968, **49**, 2106.
[112] D. E. Milligan, M. E. Jacox, S. W. Charles, and G. C. Pimentel, *J. Chem. Phys.*, 1962, **37**, 2302.
[113] F. D. Verderame, J. A. Lannon, L. E. Harris, W. G. Thomas, and E. A. Lucia, *J. Chem. Phys.*, 1972, **56**, 2638.
[114] D. E. Milligan and M. E. Jacox, *J. Chem. Phys.*, 1968, **48**, 4811.
[115] T. B. Freedman and E. R. Nixon, *J. Chem. Phys.*, 1972, **56**, 698.

Table 1 (*continued*)

Species	*Method of preparation*	*Matrix*	*Means of characterization*	*Findings*	*Ref.*
ClCN	Direct deposition	Ne, Ar; 4—20 K	I.r.; ^{12}C, ^{13}C, ^{35}Cl, ^{37}Cl	Combinations and overtones used to deduce anharmonic coefficients close to the gas-phase values	116
CO	Direct deposition	Ar, CH_4, SF_6, neopentane; 20—40 K	I.r.	No rotation occurs in any of these matrices; aggregation occurs with matrix ratios less than 5000:1 in Ar	117
	Direct deposition	Ar; 20 K	I.r.; ^{16}O, ^{18}O, ^{12}C, ^{13}C	No rotation in Ar; additional bands observed due to aggregates	118
	Direct deposition	Kr, Xe; 20 K	I.r.	Hindered rotation suggested to occur; two-site model used to interpret the findings	119
CO–HX (X = Cl, Br or I)	Direct deposition	Ar; 20 K	I.r.	Three bands observed; attributed to HX···CO interactions	120
CO–DCl	Direct deposition	Ar; 20 K	I.r.	Bands assumed to arise from CO···DCl interactions	121
CO_2^-	Co-deposition of Na or K atoms with CO_2	CO_2; 77 K	E.s.r.; ^{13}C	$\widehat{OCO} = 134°$; evidence that electron transfer from M to CO_2 is not complete	122
CO_3	Photolysis of CO_2 or CO_2+H_2O	CO_2; 14 K	I.r.; u.v.–visible; ^1H, ^2H, ^{12}C, ^{13}C	C_{2v} Structure indicated, possible models being: (a) 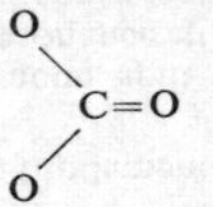 O / C=O \ O (two O atoms each bonded to the carbon of C=O) or (b) 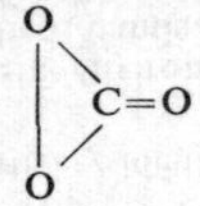 O—O ring with C=O (three-membered ring: O–O bonded, each O bonded to the carbon of C=O); normal-co-ordinate analysis suggests that (b) is more likely, with $\widehat{OCO} = \sim 65°$	123
	Photolysis of CO_2 or CO_2+O_3; radio-frequency discharge through CO_2	CO_2; 50—77 K	I.r.; ^{12}C, ^{13}C, ^{16}O, ^{18}O	Vibrational analysis favours C_{2v} model as in (b) above	124
	Photolysis of CO_2+O_3	CO_2, Ar; ~20 K	I.r.; ^{16}O, ^{18}O	Reaction: $CO_3+O_2 \rightarrow CO_2+O_3$ investigated using ^{18}O labelling	125
HCO	Photolysis of HY (Y = Cl, OH, or CH_3) with CO	Ar, CO; 14 K	I.r.; u.v.; ^{16}O, ^{18}O, ^1H, ^2H	Prominent u.v. absorptions between 2100 and 2600 Å; potential constants determined for excited and ground states	126

	Photolysis of HY (Y = I or SH)	CO; 14, 20 K	I.r.; ^{1}H, ^{2}H, ^{12}C, ^{13}C	Confirmation of the weakness of the C—H bond	127
HOCO	Photolysis of H_2O with CO	CO; 14 K	I.r.; ^{1}H, ^{2}H, ^{12}C, ^{13}C, ^{16}O, ^{18}O	*cis*- and *trans*-H—O—C=O identified; evidence of the reaction HOCO → H + CO_2 on photolysis	128
CH_3OH	Direct deposition	N_2, Ar; 20 K	I.r.; ^{1}H, ^{2}H, ^{16}O, ^{18}O	Evidence of cyclic dimer and trimer molecules; higher polymers with open-chain structures also observed; revised vibrational assignment for the monomer given; open-chain dimer and trimer also noted	129
FCO	Photolysis of fluorine atom source with CO; photolysis of F_2CO or HFCO	Ar, CO; 4, 14, 20 K	I.r.; u.v.; ^{12}C, ^{13}C, ^{16}O, ^{18}O	Assumed $\widehat{FCO} = 135°$	130
ClCO	Photolysis or Cl atom source with CO	Ar, CO; 14 K	I.r.; u.v.; ^{12}C, ^{13}C, ^{35}Cl, ^{37}Cl	Assumed $\widehat{ClCO} = 120$—$135°$; reaction C + CO → ClCO has essentially zero activation energy	131
CF	Photolysis of CH_3F	Ar, N_2; 14 K	I.r.; u.v.–visible; ^{1}H, ^{2}H, ^{12}C, ^{13}C	$k_{C-F} = 7.4$ mdyn Å^{-1}	132

116 C. B. Murchison and J. Overend, *Spectrochim Acta*, 1971, **27A**, 1509.
117 J. B. Davies and H. E. Hallam, *J. C. S. Faraday II*, 1972, **68**, 509.
118 G. E. Leroi, G. E. Ewing, and G. C. Pimentel, *J. Chem. Phys.*, 1964, **40**, 2298.
119 S. W. Charles and K. O. Lee, *Trans. Faraday Soc.*, 1965, **61**, 614.
120 A. J. Barnes, H. E. Hallam, and G. F. Scrimshaw, *Trans. Faraday Soc.*, 1969, **65**, 3172.
121 J. B. Davies and H. E. Hallam, *Trans. Faraday Soc.*, 1971, **67**, 3176.
122 J. E. Bennett, B. Mile, and A. Thomas, *Trans. Faraday Soc.*, 1965, **61**, 2357.
123 M. E. Jacox and D. E. Milligan, *J. Chem. Phys.*, 1971, **54**, 919.
124 N. G. Moll, D. R. Clutter, and W. E. Thompson, *J. Chem. Phys.*, 1966, **45**, 4469.
125 E. Weissberger, W. H. Breckenridge, and H. Taube, *J. Chem. Phys.*, 1967, **47**, 1764.
126 D. E. Milligan and M. E. Jacox, *J. Chem. Phys.*, 1969, **51**, 277.
127 D. E. Milligan and M. E. Jacox, *J. Chem. Phys.*, 1964, **41**, 3032.
128 D. E. Milligan and M. E. Jacox, *J. Chem. Phys.*, 1971, **54**, 927.
129 M. Van Thiel, E. D. Becker, and G. C. Pimentel, *J. Chem. Phys.*, 1957, **27**, 95; A. J. Barnes and H. E. Hallam, *Trans. Faraday Soc.*, 1970, **66**, 1920.
130 D. E. Milligan, M. E. Jacox, A. M. Bass, J. J. Comeford, and D. E. Mann, *J. Chem. Phys.*, 1965, **42**, 3187.
131 M. E. Jacox and D. E. Milligan, *J. Chem. Phys.*, 1965, **43**, 866.
132 M. E. Jacox and D. E. Milligan, *J. Chem. Phys.*, 1969, **50**, 3252.

Table 1 (*continued*)

Species	*Method of preparation*	*Matrix*	*Means of characterization*	*Findings*	*Ref.*
CF_2	Photolysis of F_2 with N_3CN	Ar; 14 K	I.r.; u.v.–visible; ^{12}C, ^{13}C	Vibrational assignments made	133
	Photolysis of CF_2N_2	Ar, N_2; 4, 14, 20 K	I.r.; u.v.–visible; ^{12}C, ^{13}C	$\widehat{FCF} = \sim 108°$	134
CF_2; CF_3	Pyrolysis of C_2F_4 or CF_3I	Ne; 4 K	I.r.	Radicals identified	135
CF_3	Photolysis of CF_3X, or a mixture of either CF_2N_2 or N_3CN with *trans*-N_2F_2	Ar, Kr, N_2; 14—20 K	I.r.; diffusion experiments	CF_3 identified by three of its fundamental modes; pyramidal geometry confirmed	136
	Photolysis of mixtures of F_2 with N_3CN	Ar; 14 K	I.r.; ^{12}C, ^{13}C	All four vibrational fundamentals identified; out-of-plane angle 13°	137
	γ-Irradiation of C_2F_6	Kr, Xe; 85 K	E.s.r.; ^{13}C	Pyramidal radical indicated with $\widehat{FCF} = 111.1°$ (out-of-plane angle = 17.8°)	138
HCF_2	γ-Irradiation of CH_2F_2	Kr, Xe; 85 K	E.s.r.	Radical has flattened pyramidal form; out-of-plane angle = 12.7°	138
	Co-deposition of Li atoms with HCF_2Br	Ar; 15 K	I.r.; ^1H, ^2H	Vibrational assignments made	139
H_2CF	Photolysis of CH_3F	Ar, N_2; 4 K	I.r.; u.v.–visible; ^1H, ^2H, ^{12}C, ^{13}C	Insufficient evidence for a definitive vibrational and structural assignment	140
	Co-deposition of Li atoms with CH_2FX (X = Cl or Br)	Ar; 15 K	I.r.; ^1H, ^2H	—	141
	γ-Irradiation of CH_3F	Kr, Xe; 85 K	E.s.r.; ^{13}C	Radical near to planar; < 5° out-of-plane angle	138
HCF	Photolysis of CH_3F or a mixture of HF with N_3CN	Ar, N_2; 14 K	I.r.; u.v.–visible; ^1H, ^2H, ^{12}C, ^{13}C	Assume bent H—C—F skeleton	140
FCCl	Photolysis of CH_2ClF	Ar; 14 K	I.r.; u.v.–visible; ^1H, ^2H, ^{35}Cl, ^{37}Cl	Bending fundamental estimated from the vibronic structure of an emission band system at 4000—4900 Å; $\widehat{ClCF} = \sim 105°$	142
CCl	Photolysis of CH_3Cl	Ar, N_2; 14 K	I.r.; u.v.; ^{35}Cl, ^{37}Cl	Ground-state confirmed to be $^2\Pi_{\frac{1}{2}}$	143
CCl_2	Photolysis of CH_2Cl_2	Ar, N_2; 14 K	I.r.; ^{35}Cl, ^{37}Cl	Revised force constants determined assuming $\widehat{ClCCl} = 100°$	143
	Photolysis of Cl_2 with N_3CN	Ar, N_2; 14 K	I.r.; u.v.–visible; ^{12}C, ^{13}C, ^{35}Cl, ^{37}Cl	$\widehat{ClCCl} = 90$—100°; indicated that: $CCl_2 + Cl_2 \rightarrow CCl_4$ with little or no activation energy at ~ 30 K	144

	Co-deposition of Li atoms with CCl_4	Ar; 15 K	I.r.; ^{12}C, ^{13}C, ^{35}Cl, ^{37}Cl	Isotope splitting suggests that $\widehat{ClCCl} = 100 \pm 9°$	145
CCl_3	Pyrolysis of CCl_3Br	Ar; ~4 K	I.r.	Results consistent with pyramidal C_{3v} geometry; no evidence of significant interaction with alkali-halide molecules	146
	Co-deposition of Li atoms with CCl_3X (X = Cl or Br)	Ar; 15 K	I.r.; ^{6}Li, ^{7}Li, ^{12}C, ^{13}C; diffusion studies	Only ν_3 observed; on diffusion: $2CCl_3 \rightarrow C_2Cl_6$	147
	Co-deposition of M atoms with CCl_3X (M = Li, Na, or K; X = Cl, Br, or I)	Ar; 15 K	I.r.; ^{6}Li, ^{7}Li, ^{12}C, ^{13}C, ^{35}Cl, ^{37}Cl; diffusion studies	Observation of ν_1 and ν_3; assume pyramidal geometry	148
	Co-deposition of Li atoms with CCl_4; photolysis of $CHCl_3$	Ar, N_2; 14, 20 K	I.r.; ^{1}H, ^{2}H, ^{12}C, ^{13}C, ^{35}Cl, ^{37}Cl	Earlier report of ν_1 in error (ref. 148); no band attributable to ν_1 observed, indicating that CCl_3 may be planar (*cf.* CH_3 and CF_3)	149
	Photolysis of $HCCl_3$	Ar; 14 K	I.r.; ^{1}H, ^{2}H, ^{12}C, ^{13}C	CCl_3 formed in high yield	150
CCl_3^+	Photolysis of $HCCl_3$	Ar; 14 K	I.r.; ^{12}C, ^{13}C, ^{35}Cl, ^{37}Cl	ν_1 not observed, therefore probably planar in accordance with Walsh's rules	150

133 D. E. Milligan and M. E. Jacox, *J. Chem. Phys.*, 1968, **48**, 2265.
134 D. E. Milligan, D. E. Mann, M. E. Jacox, and R. A. Mitsch, *J. Chem. Phys.*, 1964, **41**, 1199.
135 A. Snelson, *High Temp. Sci.*, 1970, **2**, 70.
136 D. E. Milligan, M. E. Jacox, and J. J. Comeford, *J. Chem. Phys.*, 1966, **44**, 4058.
137 D. E. Milligan and M. E. Jacox, *J. Chem. Phys.*, 1968, **48**, 2265.
138 R. W. Fessenden and R. H. Schuler, *J. Chem. Phys.*, 1965, **43**, 2704.
139 T. G. Carver and L. Andrews, *J. Chem. Phys.*, 1969, **50**, 5100.
140 M. E. Jacox and D. E. Milligan, *J. Chem. Phys.*, 1969, **50**, 3252.
141 J. I. Raymond and L. Andrews, *J. Phys. Chem.*, 1971, **75**, 3235.
142 C. E. Smith, D. E. Milligan, and M. E. Jacox, *J. Chem. Phys.*, 1971, **54**, 2780.
143 M. E. Jacox and D. E. Milligan, *J. Chem. Phys.*, 1970, **53**, 2688.
144 D. E. Milligan and M. E. Jacox, *J. Chem. Phys.*, 1967, **47**, 703.
145 L. Andrews, *J. Chem. Phys.*, 1968, **48**, 979; see also A. K. Maltsev, R. G. Mikaelyan, and O. M. Nefedov, *Bull. Acad. Sci., U.S.S.R.*, 1971, 188.
146 J. H. Current and J. K. Burdett, *J. Phys. Chem.*, 1969, **73**, 3504, 3505; A. K. Maltsev, O. M. Nefedov, R. H. Hauge, J. L. Margrave, and D. Seyferth, *J. Phys. Chem.*, 1971, **75**, 3984.
147 L. Andrews, *J. Phys. Chem.*, 1967, **71**, 2761.
148 L. Andrews, *J. Chem. Phys.*, 1968, **48**, 972.
149 E. E. Rogers, S. Abramowitz, M. E. Jacox, and D. E. Milligan, *J. Chem. Phys.*, 1970, **52**, 2198; see also R. Steudel, *Z. Naturforsch.*, 1971, **26b**, 475.
150 M. E. Jacox and D. E. Milligan, *J. Chem. Phys.*, 1971, **54**, 3935.

Table 1 (*continued*)

Species	*Method of preparation*	*Matrix*	*Means of characterization*	*Findings*	*Ref.*
CCl_4	Direct deposition	N_2; 12 K	Raman; polarization measurements	Demonstration of feasibility of Raman studies on matrix-trapped species	93
$CHCl_3$	Direct deposition	Ar; 4 K	Raman	Demonstration of feasibility of Raman studies on matrix-trapped species	151
$CHCl_2$	Co-deposition of M atoms with $HCCl_2X$ (M = Li or Na; X = Cl or Br)	Ar, Kr; 15 K	I.r.; ^1H, ^2H, ^6Li, ^7Li, ^{35}Cl, ^{37}Cl	Normal-co-ordinate calculations insensitive to bond angle; possibly some $(p\text{–}d)_\pi$ bonding contribution	152
	Photolysis of $HCCl_3$	Ar; 14 K	I.r.; ^1H, ^2H, ^{12}C, ^{13}C, ^{35}Cl, ^{37}Cl	Vibrational assignments and potential constants determined	150
$HCCl_2^+$	Photolysis of $HCCl_3$	Ar; 14 K	I.r.; ^1H, ^2H, ^{12}C, ^{13}C, ^{35}Cl, ^{37}Cl	Planar structure indicated by detailed analysis of isotopic data; consistent with MO treatment	150
$HCCl_2^-$	Photolysis of $HCCl_3$	Ar; 14 K	I.r.; ^1H, ^2H, ^{12}C, ^{13}C, ^{35}Cl, ^{37}Cl	Identity supported by enhanced intensity of bands when a photoelectron source is added; framework probably pyramidal	150
HCCl	Photolysis of CH_3Cl	Ar, N_2; 14 K	I.r.; u.v.–visible; ^1H, ^2H, ^{35}Cl, ^{37}Cl	Spectroscopic identity confirmed	143
H_2CCl	Photolysis of CH_3Cl	Ar, N_2; 14 K	I.r.; u.v.–visible; ^1H, ^2H, ^{35}Cl, ^{37}Cl	Planar skeleton implied; possibility of $(p\text{–}d)_\pi$ bonding in C—Cl unit explored	143
	Co-deposition of M atoms with $ClCH_2X$ (M = Li or Na; X = F, Cl, Br, or I)	Ar, N_2; 15 K	I.r.; ^1H, ^2H, ^6Li, ^7Li, ^{35}Cl, ^{37}Cl	Suggest planar skeleton; evidence of $(p\text{–}p)_\pi$ bonding	153
H_3CCl; H_2CCl_2 $XCCl_3$; $EtCH_2Cl$	Direct deposition	Ar; 20 K	I.r.; ^{35}Cl, ^{37}Cl	^{35}Cl–^{37}Cl isotopic splitting patterns studied in detail; normal-co-ordinate analysis; Fermi resonance effects studied	154
CCl_2Br; $CClBr_2$; CClBr	Co-deposition of Li atoms with CCl_3Br; pyrolysis reactions	Ar; ~4, 15 K	I.r.; ^{35}Cl, ^{37}Cl	Incomplete i.r. spectrum of CCl_2Br; no evidence of significant interaction with alkali-halide molecules	146, 147
CBr_2	Co-deposition of Li atoms with CBr_4	Ar; 15 K	I.r.; diffusion studies	ν_1 and ν_3 observed; normal-co-ordinate analysis suggests the presence of C—Br single bonds ($\widehat{BrCBr} = 100°$ assumed); on diffusion $2CBr_2 \rightarrow C_2Br_4$	155

CBr_3	Co-deposition of Li atoms with CBr_4	Ar; 15 K	I.r.	Observation of ν_1 and ν_3 claimed signifying a pyramidal framework	155
	Co-deposition of Li atoms with CBr_4; photolysis of $HCBr_3$	Ar, N_2; 14, 20 K	I.r.; ^1H, ^2H, ^{12}C, ^{13}C	Earlier report of ν_1 (ref. 155) in error; ν_1 not detected in i.r. absorption, hinting at planar geometry	149
$HCBr_2$	Co-deposition of M atoms with $HCBr_3$ (M = Li or Na)	Ar, Kr; 15 K	I.r.; ^1H, ^2H, ^6Li, ^7Li	Assume pyramidal geometry; $(p\text{–}d)_\pi$ bonding contribution considered	156
H_2CBr	Co-deposition of Li atoms with CH_2FBr	Ar; 15 K	I.r.; ^1H, ^2H, ^6Li, ^7Li	—	141
	Co-deposition of M atoms with CH_2BrX (M = Li, Na, or K; X = F, Cl, Br, or I)	Ar; 15 K	I.r.; ^1H, ^2H, ^6Li, ^7Li, ^{79}Br, ^{81}Br	Planar radical implied; $(p\text{–}p)_\pi$ and $(p\text{–}d)_\pi$ bonding suggested	157
CD_3I	Direct deposition	Ar; ~18 K	I.r.	Fundamentals and a number of overtones and combinations observed; anharmonic constants determined	158
CS_2; COS	Direct deposition	N_2; 10—12 K	Raman	Feasibility of Raman measurements demonstrated	93
CS_2	Direct deposition	Ar, N_2, CH_4; 20 K. Isopentane–cyclohexane glass; 77 K	U.v.–visible	$^1B_2 \leftarrow {}^1\Sigma_g^+$ transition observed; vibronic structure analysed	159
CS_2^-	Co-deposition of Na or K atoms with CS_2	CS_2; 77 K	E.s.r.; ^{13}C	g- and A-values determined; electron transfer from M to CS_2 not complete	160

[151] J. S. Shirk and H. H. Claassen, *J. Chem. Phys.*, 1971, **54**, 3237.
[152] T. G. Carver and L. Andrews, *J. Chem. Phys.*, 1969, **50**, 4235.
[153] L. Andrews and D. W. Smith, *J. Chem. Phys.*, 1970, **53**, 2956.
[154] S. T. King, *J. Chem. Phys.*, 1968, **49**, 1321.
[155] L. Andrews and T. G. Carver, *J. Chem. Phys.*, 1968, **49**, 896.
[156] T. G. Carver and L. Andrews, *J. Chem. Phys.*, 1969, **50**, 4223.
[157] D. W. Smith and L. Andrews, *J. Chem. Phys.*, 1971, **55**, 5295.
[158] D. R. Anderson and J. Overend, *Spectrochim. Acta*, 1972, **28A**, 1225.
[159] L. Bajema, M. Gouterman, and B. Meyer, *J. Phys. Chem.*, 1971, **75**, 2204.
[160] J. E. Bennett, B. Mile, and A. Thomas, *Trans. Faraday Soc.*, 1967, **63**, 262.

Table 1 (*continued*)

Species	*Method of preparation*	*Matrix*	*Means of characterization*	*Findings*	*Ref.*
C_2 species					
CCH	Photolysis of HC≡CH	Ne, Ar, N_2; 4, 14 K	I.r.; u.v.–visible; ^{1}H, ^{2}H, ^{12}C, ^{13}C	New i.r. absorption bands identified with CCH	161
CCO	Photolysis of C_3O_2	Ar; 4.2 K	I.r.	Intermediate in the formation of C atoms	98
	Photolysis of N_3CN with CO; photolysis of C_3O_2	Ar, CO, N_2; 14 K	I.r.; u.v.; ^{12}C, ^{13}C, ^{16}O, ^{18}O	Data consistent with a linear structure; thermodynamic functions computed; reaction $CCO + CO \rightarrow C_3O_2$ has little or zero activation energy	162
C_2H_5OH	Direct deposition	Ar; 20 K	I.r.; ^{1}H, ^{2}H	Bands attributable to *trans*- and *gauche*-conformers (ratio 2:1 in vapour at room temperature); open-chain dimer, trimer, and tetramer, and cyclic tetramer identified	163
C_2H_5SH; also CH_3SH	Direct deposition	Ar, N_2, CO; 20 K	I.r.	Bands attributable to both *trans*- and *gauche*-conformers (*trans* version more stable than *gauche*); open-chain dimer and cyclic tetramer species identified	164
CH_3CO_2H; CF_3CO_2H	Direct deposition	Ne, Ar, N_2; 4 K	I.r., ^{1}H, ^{2}H	Large anharmonic potential energy contribution evident from isotopic shifts	165, 166
$(CH_3CO_2H)_2$; $(CF_3CO_2H)_2$	Direct deposition	Ne, Ar; 4 K	I.r.; ^{1}H, ^{2}H	Frequencies assigned for cyclic dimer	166
	Direct deposition	Ar; 20 K	I.r.; ^{1}H, ^{2}H, ^{16}O, ^{18}O	Interpretation of results conflicts with assignments of ref. 166	167
$H_2NCH_2CO_2H$	Vaporization at 410—412 K	Ar; 20 K	I.r.; ^{1}H, ^{2}H, ^{14}N, ^{15}N	No evidence for zwitterion form in matrix	168
C_n systems					
$C_2H_5CH_2Cl$	See under H_3CCl	—	—	—	
Cyclopentane; benzene	Direct deposition	Ar; 10 K	I.r.	Evidence of intermolecular coupling in solids; demonstration of the use of the technique to analyse multiplets in the spectra of molecular crystals	169
Pyridine	Direct deposition	N_2; ~5—20 K	I.r.; ^{1}H, ^{2}H, ^{12}C, ^{13}C	Used to assist in the vibrational analysis of pyridine; evidence of pair-association	170, 171

Pyrazole; imidazole; dimethylphosphinic acid	Vaporization	Ar; 20 K	I.r.	Isolated monomeric molecules identified	172
Benzene	Direct deposition	HCl; 30 K	I.r.	Evidence that benzene is in a site of approximately D_{6h} symmetry but strongly perturbed	173
Aniline	Direct deposition	Ar; 6—32 K	U.v.–visible	$^1B_2 \leftarrow {}^1A_1$ band system observed; effects of inversion doubling in the ground-state investigated	174
Toluene; *m*- and *p*-xylene	Direct deposition	Kr; 20 K	U.v.; ^{1}H, ^{2}H	Three $\pi \rightarrow \pi^*$ transitions observed; vibronic structure analysed	175
1,3-Cyclohexadiene (a); *cis*-1,3,5-hexatriene (b); *trans*-1,3,5-hexatriene (c)	Photolysis of (a) to give (b), and further photolysis of (b) to give (c)	Ar; 20 K	I.r.; u.v.; ^{1}H, ^{2}H	Photochemical conversion studied spectroscopically; other products also detected	176
Other organic molecules	Various	Various	U.v.–visible	—	177

[161] D. E. Milligan, M. E. Jacox, and L. Abouaf-Marguin, *J. Chem. Phys.*, 1967, **46**, 4562.
[162] M. E. Jacox, D. E. Milligan, N. G. Moll, and W. E. Thompson, *J. Chem. Phys.*, 1965, **43**, 3734.
[163] A. J. Barnes and H. E. Hallam, *Trans. Faraday Soc.*, 1970, **66**, 1932.
[164] A. J. Barnes, H. E. Hallam, and J. D. R. Howells, *J. C. S. Faraday II*, 1972, **68**, 737.
[165] C. V. Berney, R. L. Redington, and K. C. Lin, *J. Chem. Phys.*, 1970, **53**, 1713.
[166] R. L. Redington and K. C. Lin, *J. Chem. Phys.*, 1971, **54**, 4111.
[167] Y. Grenie, J.-C. Cornut, and J.-C. Lassegues, *J. Chem. Phys.*, 1971, **55**, 5844.
[168] Y. Grenie, J.-C. Lassegues, and C. Garrigou-Lagrange, *J. Chem. Phys.*, 1970, **53**, 2980.
[169] A. LeRoy and E. Dayan, *Compt. rend.*, 1969, **268**, *B*, 48.
[170] G. Taddei, E. Castellucci, and F. D. Verderame, *J. Chem. Phys.*, 1970, **53**, 2407.
[171] E. Castellucci, G. Sbrana, and F. D. Verderame, *J. Chem. Phys.*, 1969, **51**, 3762.
[172] S. T. King, *J. Phys. Chem.*, 1970, **74**, 2133.
[173] K. Szczepaniak and W. B. Person, *Spectrochim. Acta*, 1972, **28A**, 15.
[174] J. C. D. Brand, V. T. Jones, B. J. Forrest, and R. J. Pirkle, *J. Mol. Spectroscopy*, 1971, **39**, 352.
[175] B. Katz, M. Brith, B. Sharf, and J. Jortner, *J. Chem. Phys.*, 1971, **54**, 3924.
[176] P. Datta, T. D. Goldfarb, and R. S. Boikess, *J. Amer. Chem. Soc.*, 1971, **93**, 5189.
[177] B. Meyer, 'Low Temperature Spectroscopy', American Elsevier, New York, 1971, p. 383.

Table 2 *Matrix-isolated inorganic species*

Species	*Method of preparation*	*Matrix*	*Means of identification and structural determination*	*Observations*	*Ref.*
H	Glow discharge through H_2	Ar, Kr; 4, 15 K	I.r.; ^1H, ^2H, ^{36}Ar, ^{40}Ar	Observed absorption bands assigned to H atoms in interstitial O_h sites. Ar isotopic splitting small	178
	Photolysis of glucose, sorbitol, inositol	Carbohydrate; 77 K	E.s.r.	—	179
H_2	Cooled solution in Ar	Ar; 82 K	I.r.	Induced vibrational fundamental with rotational structure observed	180
Group I/IA					
Li_2	Vaporization of Li solid	Ar; ~5 K	Raman; ^6Li, ^7Li	—	181
LiN_2; Li_2N_4	Co-deposition of Li atoms with N_2; diffusion to give dimer	N_2; 15 K	I.r.; ^6Li, ^7Li, ^{14}N, ^{15}N	Monomer: force-constant calculations suggest $Li^+ N_2^-$. Dimer: of the type $N_2Li_2N_2$	182
LiON	Co-deposition of Li atoms with NO	Ar; 4, 15 K	I.r.; ^6Li, ^7Li, ^{16}O, ^{18}O	$\widehat{LiON} = 100 \pm 10°$; NO bond is weaker than in HNO	183
Li_2O	Vaporization of Li_2O solid	Kr; 20 K	I.r.; ^6Li, ^7Li	Probably linear	184
LiO; Li_2O	Co-deposition of Li atoms with N_2O	N_2	I.r.; ^6Li, ^7Li, ^{16}O, ^{18}O	Linear geometry of Li_2O sustained	185
LiO_2Li	Co-deposition of Li atoms with O_2	Ar, Kr, O_2; 15 K	I.r.; ^6Li, ^7Li, ^{16}O, ^{18}O	Planar rhombus structure of peroxide ion between two Li^+ ions; $\widehat{OLiO} = 110°$ (gas phase value 116°)	186
LiLiOO	Co-deposition of Li_2 molecules with O_2	Ar, Kr, O_2; 15 K	I.r.	—	186
LiO_2	Co-deposition of Li atoms with O_2	Ar, Kr, O_2; 15 K	I.r.; ^6Li, ^7Li, ^{16}O, ^{18}O	Isosceles triangle structure of peroxide ion and Li^+ ion; $\widehat{OLiO} = 90 \pm 6°$. k_{O-O} very similar to that for free O_2^-	186
	Co-deposition of Li atoms with O_2	Ar, O_2; 4.2—15 K	Raman; ^{16}O, ^{18}O	No band arising from Li—O vibration observed; therefore ionic structure is reasonable	187
$(LiO_2)_2$	Co-deposition of Li atoms with O_2; diffusion to give dimer	Ar, Kr, O_2; 15 K	I.r.	—	186
	[illegible]	Ar; 15 K	I.r.; ^6Li, ^7Li, ^{16}O, ^{18}O	—	188

LiF; $(LiF)_2$; $(LiF)_3$	Vaporization of LiF solid	Ar, Kr, Xe; 5, 20 K	I.r.; ^{6}Li, ^{7}Li	Dimer: planar D_{2h} rhombus; B_{2u} and B_{3u} modes observed	189, 190
	Vaporization of LiF solid	Ar, Kr, Xe; 5, 20 K	I.r.; ^{6}Li, ^{7}Li	Observed B_{1u} mode of dimer	191
	Vaporization of LiF solid	Ar; 20 K	I.r.; ^{6}Li, ^{7}Li	Postulate Li—F—Li—F dimer structure in addition to cyclic rhombus	192
	Vaporization of LiF solid	Ne; ~5 K	I.r.; ^{6}Li, ^{7}Li	Trimer: possibly D_{3h} planar ring	193
LiX; $(LiX)_2$ (X = F, Cl, or Br)	Vaporization of LiX solid	Ar, Kr, Xe, N_2; 20 K	I.r.; ^{6}Li, ^{7}Li	Dimer: planar D_{2h} rhombus; B_{2u} and B_{3u} fundamentals observed	194
LiX; $(LiX)_2$ (X = Cl or Br)	Vaporization of LiX solid	Ar, Kr, Xe; 5, 20 K	I.r.; ^{6}Li, ^{7}Li	B_{1u} Fundamental of dimer observed	191
LiX (X = Br or I)	Vaporization of LiX solid	Ar, Kr, N_2; ~5 K	U.v.–visible	—	195
Na; Na_2	Vaporization of Na	Benzene; 55 K	U.v.–visible	Observed vibronic transitions assigned to Na–C_6H_6 complex with symmetry less than C_{6v}. Na_2 observed at low matrix-gas:Na ratios	196

[178] V. E. Bondybey and G. C. Pimentel, *J. Chem. Phys.*, 1972, **56**, 3832.
[179] A. Bos and A. S. Buchanan, *Nature Phys. Sci.*, 1971, **233**, 15.
[180] R. J. Kriegler and H. L. Welsh, *Canad. J. Phys.*, 1968, **46**, 1181.
[181] G. A. Ozin, *The Spex Speaker*, 1971, **XVI**, no. 4, 9.
[182] R. C. Spiker, jun., L. Andrews, and C. Trindle, *J. Amer. Chem. Soc.*, 1972, **94**, 2401.
[183] W. L. S. Andrews and G. C. Pimentel, *J. Chem. Phys.*, 1966, **44**, 2361.
[184] K. S. Seshadri, D. White, and D. E. Mann, *J. Chem. Phys.*, 1966, **45**, 4697.
[185] L. Andrews, *Ann. Rev. Phys. Chem.*, 1971, **22**, 109.
[186] L. Andrews, *J. Chem. Phys.*, 1969, **50**, 4288; *J. Amer. Chem. Soc.*, 1968, **90**, 7368.
[187] D. A. Hatzenbuhler and L. Andrews, *J. Chem. Phys.*, 1972, **56**, 3398; H. Hüber and G. A. Ozin, *J. Mol. Spectroscopy*, 1972, **41**, 595.
[188] L. Andrews and J. I. Raymond, *J. Chem. Phys.*, 1971, **55**, 3078.
[189] M. J. Linevsky, *J. Chem. Phys.*, 1963, **38**, 658.
[190] A. Snelson and K. S. Pitzer, *J. Phys. Chem.*, 1963, **67**, 882.
[191] M. Freiberg, A. Ron, and O. Schnepp, *J. Phys. Chem.*, 1968, **72**, 3526.
[192] S. Abramowitz, N. Acquista, and I. W. Levin, *J. Res. Nat. Bur. Stand.*, 1968, **72A**, 487.
[193] A. Snelson, *J. Chem. Phys.*, 1967, **46**, 3652.
[194] S. Schlick and O. Schnepp, *J. Chem. Phys.*, 1964, **41**, 463.
[195] M. Oppenheimer and R. S. Berry, *J. Chem. Phys.*, 1971, **54**, 5058.
[196] W. R. M. Graham and W. W. Duley, *J. Chem. Phys.*, 1971, **54**, 586.

Table 2 (*continued*)

Species	*Method of preparation*	*Matrix*	*Means of identification and structural determination*	*Observations*	*Ref.*
NaO_2Na	Co-deposition of Na atoms with O_2	Ar, O_2; 15 K	I.r.; ^{16}O, ^{18}O	Ionic rhombus structure, similar to LiO_2Li; $\widehat{ONaO} = 46—52°$	197
NaO_2	Co-deposition of Na atoms with O_2	Ar, O_2; 15 K	I.r.; Raman; ^{16}O, ^{18}O	Structurally analogous to LiO_2	197, 198
NaOH	Vaporization of NaOH solid at 1073 K	Ar; 20, 33 K	I.r.; ^{1}H, ^{2}H	Linear or almost linear; ν(OH) not observed	199
NaX (X = Cl, Br, or I)	Vaporization of NaX solid	Ar, Kr, N_2; 5 K	U.v.–visible	—	200
K_2O	Co-deposition of K atoms with N_2O	Details not available	I.r.; ^{16}O, ^{18}O	$\widehat{KOK} < 180°$	185
KO_2K	Co-deposition of K atoms with O_2	Ar; 15 K	I.r.; ^{16}O, ^{18}O; simultaneous deposition of two alkali metals	Planar D_{2h} rhombus; insufficient data for angle calculation	201
KO_2	Co-deposition of K atoms with O_2	Ar; 15 K	I.r.; ^{16}O, ^{18}O; Raman	Isosceles triangle structure, C_{2v}; $\widehat{OKO} = 66°$	198, 201
O_2KO_2	Co-deposition of K atoms with O_2	Ar; 15 K	I.r.; ^{16}O, ^{18}O; Raman	Probably D_{2d} symmetry	201
KI	Vaporization of KI solid	Ar, Kr, N_2; ~5 K	U.v.–visible	—	200
Rb_2O	Co-deposition of Rb atoms with N_2O	Details not available	I.r.; ^{16}O, ^{18}O	$\widehat{RbORb} < 180°$	185
RbO_2Rb	Co-deposition of Rb atoms with O_2	Ar; 15 K	I.r.; ^{16}O, ^{18}O; simultaneous deposition of two alkali metals	Planar D_{2h} rhombus; insufficient data for angle calculation	201
RbO_2	Co-deposition of Rb atoms with O_2	Ar; 15 K	I.r.; ^{16}O, ^{18}O	Isosceles triangle structure, C_{2v}; $\widehat{ORbO} = 56°$	201
O_2RbO_2	Co-deposition of Rb atoms with O_2	Ar; 15 K	I.r.; ^{16}O, ^{18}O	Probably D_{2d} symmetry	201
RbOH	Vaporization of RbOH solid at 773 K	Ar; 20—33 K	I.r.; ^{1}H, ^{2}H	Linear or almost linear; ν(OH) not observed	199
RbBr	Vaporization of RbBr solid	Ar, Kr, N_2; ~5 K	U.v.–visible	—	200

Cs_2O	Co-deposition of Cs atoms with N_2O	Details not available	I.r.; ^{16}O, ^{18}O	$\widehat{CsOCs}$ < 180°; $\widehat{MOM}$ decreases in the series Li > K > Rb > Cs	185
CsOH	Vaporization of CsOH solid at 773 K	Ar; 20—33 K	I.r.; ^{1}H, ^{2}H	Linear or almost linear; ν(OH) not observed	202
Group II/IIA					
BeF_2	Vaporization of BeF_2 at 875—1275 K	Ne, Ar, Kr; ~5 K	I.r.	Linear; ν_2 and ν_3 observed	203
$BeCl_2$	Vaporization of $BeCl_2$ at 725—1275 K	Ne, Ar, Kr; ~5 K	I.r.	Linear; only ν_3 observed	203
BeX_2 (X = Cl, Br, or I)	Vaporization of BeX_2	Ne, Ar; ~5 K	I.r.	Linear or presumed linear; ν_2 and ν_3 observed for $BeCl_2$ and $BeBr_2$; only ν_3 observed for BeI_2	204
MgH; CaH; SrH; BaH	Co-deposition of M atoms with H atoms	Ar; 4 K	E.s.r.; ^{1}H, ^{2}H, ^{25}Mg; visible	$^2\Sigma^+$ ground-state; *g*- and *A*-values determined; spin density predominantly on the metal atom	205
MgF; CaF; SrF; BaF	Vaporization of MF_2 with either B or Al at ~1475 K	Ne, Ar; 4 K	E.s.r.; ^{19}F, ^{87}Sr, ^{137}Ba; u.v.–visible	$^2\Sigma$ ground-state; *g*- and *A*-values determined; less than 4% spin density on F	206
MgF_2	Vaporization of MgF_2 solid	Ne, Ar, Kr; ~5 K	I.r.	Linear; ν_2 and ν_3 observed	203
	Vaporization of MgF_2 solid	Ar, Kr; 20 K	I.r.; ^{24}Mg, ^{25}Mg, ^{26}Mg	Bent; ν_1, ν_2, and ν_3 observed; $\widehat{FMgF}$ = 158 ± 2°	207, 208
Ca; Ca_2	Vaporization of Ca solid	Ar, Kr, Xe; 16—20 K	U.v.–visible	$^1P_1$ ← $^1S_0$ transition observed	209

197 L. Andrews, *J. Phys. Chem.*, 1969, **73**, 3922.
198 R. R. Smardzewski and L. Andrews, *J. Chem. Phys.*, 1972, **57**, 1327.
199 N. Acquista and S. Abramowitz, *J. Chem. Phys.*, 1969, **51**, 2911.
200 M. Oppenheimer and R. S. Berry, *J. Chem. Phys.*, 1971, **54**, 5058.
201 L. Andrews, *J. Chem. Phys.*, 1971, **54**, 4935.
202 N. Acquista, S. Abramowitz, and D. R. Lide, *J. Chem. Phys.*, 1968, **49**, 780.
203 A. Snelson, *J. Phys. Chem.*, 1966, **70**, 3208.
204 A. Snelson, *J. Phys. Chem.*, 1968, **72**, 250.
205 L. B. Knight, jun. and W. Weltner, jun., *J. Chem. Phys.*, 1971, **54**, 3875.
206 L. B. Knight, jun., W. C. Easley, W. Weltner, jun., and M. Wilson, *J. Chem. Phys.*, 1971, **54**, 322.
207 D. E. Mann, G. V. Calder, K. S. Seshadri, D. White, and M. J. Linevsky, *J. Chem. Phys.*, 1967, **46**, 1138.
208 V. Calder, D. E. Mann, K. S. Seshadri, M. Allavena, and D. White, *J. Chem. Phys.*, 1969, **51**, 2093.
209 J. E. Francis, jun. and S. E. Webber, *J. Chem. Phys.*, 1972, **56**, 5879.

Table 2 (*continued*)

Species	*Method of preparation*	*Matrix*	*Means of identification and structural determination*	*Observations*	*Ref.*
Ca	Vaporization of Ca solid	Kr, Xe; 20—40 K	U.v.–visible	$^1P \leftarrow {}^1S$ transition observed	210
CaO	Co-deposition of Ca atoms with O_2	Kr, Xe; 20 K	U.v.; ^{16}O, ^{18}O	—	210
CaF_2	Vaporization of CaF_2 at 1675—2175 K	Ar, Ne, Kr; ~5 K	I.r.; ^{40}Ca, ^{44}Ca	Bent; ν_1 and ν_3 observed	203
	Vaporization of CaF_2 at 1675—2175 K	Kr; 20 K	I.r.; ^{40}Ca, ^{44}Ca	Bent; ν_1, ν_2, and ν_3 observed; $\widehat{FCaF} = 140 \pm 5°$	208
SrF_2	Vaporization of SrF_2	Ne, Ar, Kr; ~5 K	I.r.	Bent; ν_1 and ν_3 observed	203
	Vaporization of SrF_2	Kr; 20 K	I.r.; ^{86}Sr, ^{88}Sr	Bent; ν_1, ν_2, and ν_3 observed; $\widehat{FSrF} = 108 \pm 3°$	208
BaO_2	Co-deposition of Ba atoms with O_2	Ar–O_2; 20 K	I.r.; ^{16}O, ^{18}O	Triangular structure, C_{2v}	211*a*
BaF_2	Vaporization of BaF_2 at 1475—2125 K	Ne, Ar, Kr; ~5 K	I.r.	Bent; ν_1 and ν_3 observed	203
	Vaporization of BaF_2 at 1475—2125 K	Kr; 20 K	I.r.	Bent; ν_1 and ν_3 observed, ν_2 calculated; $\widehat{FBaF} = \sim 100°$	208
MCl_2 (M = Ca, Sr, or Ba)	Vaporization of MCl_2	Ne, Ar, Kr, N_2; 5–20 K	I.r.; ^{35}Cl, ^{37}Cl	$\widehat{ClMCl}$ varies from 180 to 120°	211*b*
Group III/IIIB					
BH_3	Pyrolysis of H_3BCO	Ar, ~5 K	I.r.; ^1H, ^2H, ^{10}B, ^{11}B	In-plane and out-of-plane bending force constants calculated; $H_3B_3O_3$ and $H_2B_2O_3$ also observed as pyrolysis products	212
BN	Photolysis of H_3B,NH_3	Ne, Ar; 4 K	U.v.–visible	Ground state identified as $^3\Pi$; $A\,^3\Pi \leftarrow X\,^3\Pi$ transition observed	213
BO	Vaporization of MO (M = Sr or Ba) with B at 1800 K	Ne, Ar; 4 K	E.s.r.; ^{11}B	$^2\Sigma$ ground-state; *g*- and *A*-values determined; dipolar nuclear hyperfine coupling negative; highly orientated in Ne matrix	214
HBO	Photolysis of $H_2B_2O_3$	Ar; 5 K	I.r.; ^1H, ^2H, ^{10}B, ^{11}B	Linear	215

[illegible]	Vaporization of B_2O_3 with B at > 1500 K	Ar, Xe; 20 K	I.r.; ^{10}B, ^{11}B	Presumed linear	216
B_2O_3	Vaporization of B_2O_3 at 1400 K	Ar, Xe; 20 K	I.r.; ^{10}B, ^{11}B	Bent, C_{2v}	216
B_2X_4 (X = F or Cl)	Direct deposition	Ar; 20 K	I.r.; ^{10}B, ^{11}B; liquid-phase Raman spectrum also reported	Staggered V_d structure	217
BF_3; $(BF_3)_2$	Direct deposition	Ar, Kr; 20, 50 K	I.r.; ^{10}B, ^{11}B	Dimer may have D_{2h} symmetry	218
BeB_2H_8	Direct deposition	Ar, N_2; 12—20 K	I.r.; Raman; ^1H, ^2H; Raman depolarization measurements	Two equilibrium structures in vapour phase: (i) C_{3v}, (ii) D_{2d} or triangular	219
$H_3B_3N_3H_3$	Direct deposition	Ar, Xe, Ar–Xe; ~5 K	I.r.; ^1H, ^2H, ^{10}B, ^{11}B; gas-phase Raman spectrum reported	Complete vibrational assignment for D_{3h} symmetry	220
$H_3B_3O_3$	Electrical discharge through B_2H_6 and O_2	Ar, Xe, Ar–Xe; ~5 K	I.r.; ^1H, ^2H, ^{10}B, ^{11}B	Complete vibrational assignment for D_{3h} symmetry	220
MBO_2 (M = Li, Na, K, Rb, or Cs)	Vaporization of MBO_2	Ar, Kr; 20 K	I.r.; ^6Li, ^7Li, ^{10}B, ^{11}B	$\widehat{MOB} = \sim 90°$; $\widehat{OBO}$ assumed 180°	221
Al_2O	Vaporization of Al_2O_3 with Al at ~1500 K	Ne, Ar, Kr; ~5 K	I.r.; ^{16}O, ^{18}O	C_{2v} symmetry; $\widehat{AlOAl} = 145 \pm 5°$; ν_2 not observed, probably lies below 190 cm^{-1}	222, 223

210 L. Brewer and J. L.-F. Wang, *J. Chem. Phys.*, 1972, **56**, 4305.
211 (*a*) S. Abramowitz and N. Acquista, *J. Res. Nat. Bur. Stand.*, 1971, **75A**, 23; (*b*) J. W. Hastie, R. H. Hauge, and J. L. Margrave, *High Temp. Sci.*, 1971, **3**, 56, 257.
212 A. Kaldor and R. F. Porter, *J. Amer. Chem. Soc.*, 1971, **93**, 2140.
213 O. A. Mosher and R. P. Frosch, *J. Chem. Phys.*, 1970, **52**, 5781.
214 L. B. Knight, jun., W. C. Easley, and W. Weltner, jun., *J. Chem. Phys.*, 1971, **54**, 1610.
215 E. R. Lory and R. F. Porter, *J. Amer. Chem. Soc.*, 1971, **93**, 6301.
216 W. Weltner, jun. and J. R. W. Warn, *J. Chem. Phys.*, 1962, **37**, 292.
217 L. A. Nimon, K. S. Seshadri, R. C. Taylor, and D. White, *J. Chem. Phys.*, 1970, **53**, 2416.
218 J. M. Bassler, P. L. Timms, and J. L. Margrave, *J. Chem. Phys.*, 1966, **45**, 2704.
219 J. W. Nibler, *J. Amer. Chem. Soc.*, 1972, **94**, 3349.
220 A. Kaldor and R. F. Porter, *Inorg. Chem.*, 1971, **10**, 775.
221 K. S. Seshadri, L. A. Nimon, and D. White, *J. Mol. Spectroscopy*, 1969, **30**, 128.
222 M. J. Linevsky, D. White, and D. E. Mann, *J. Chem. Phys.*, 1964, **41**, 542.
223 A. Snelson, *J. Phys. Chem.*, 1970, **74**, 2574.

Table 2 (*continued*)

Species	*Method of preparation*	*Matrix*	*Means of identification and structural determination*	*Observations*	*Ref.*
AlO	Vaporization of Al_2O_3 at 2500—2700 K	Ne, Ar, Kr; 4 K	I.r.; u.v.–visible; e.s.r.	$g_{\parallel} = 2.0015$, $A_{\parallel}(^{27}Al) = 872$ MHz; $g_{\perp} = 2.0004$, $A_{\perp}(^{27}Al) = 713$ MHz	224
AlF; AlF_3; Al_2F_6	Vaporization of AlF_3 at 1150—1875 K	Ne, Ar, Kr; 4.2 K	I.r.	—	225
$AlCl_3$; Al_2Cl_6	Vaporization of Al_2Cl_6 at 1175 K	Ar; 20 K	I.r.; results compared with gas-phase i.r. and Raman spectra of $AlCl_3$	Data taken to imply that $AlCl_3$ monomer is pyramidal, with $\widehat{ClAlCl} = \sim 112°$	226
$NaAlF_4$; $LiAlF_4$	Vaporization of MF with AlF_3 at 1025—1225 K	Ne; ~5 K	I.r.; ^{6}Li, ^{7}Li	Structures not conclusively established; C_{2v} symmetry assumed for both	227
Ga_2O	Vaporization of Ga_2O_3, or reaction of Ga with O_2	N_2; 15 K	I.r.; ^{16}O, ^{18}O	Only ν_3 observed; $\widehat{GaOGa} = 143 \pm 5°$	228
	Vaporization of Ga_2O_3	Ar; 10 K	I.r.	Bands assigned to ν_1, ν_2, and ν_3	229
	Vaporization of Ga_2O_3	N_2; 15 K	I.r.; ^{16}O, ^{18}O	Assignments for ν_1 and ν_2 of monomer in ref. 229 reassigned to polymer	230
Ga_2S	Details not available	Details not available	I.r.; ^{32}S, ^{34}S	C_{2v} symmetry; $\widehat{GaSGa} = 112 \pm 8°$	230
In_2O	Vaporization of In_2O_3 or reaction of In with O_2	N_2; 15 K	I.r.; ^{16}O, ^{18}O	Only ν_3 observed; $\widehat{InOIn} = 135 \pm 7°$	228
	Vaporization of In_2O_3	Ar; 10 K	I.r.	Bands assigned to ν_1, ν_2, and ν_3	229
	Vaporization of In_2O_3	N_2; 15 K	I.r.; ^{16}O, ^{18}O	Bands assigned to ν_1 and ν_2 of monomer by ref. 229 re-assigned to polymer	230
Tl_2O	Vaporization of Tl_2O_3 or reaction of Tl with O_2	N_2; 15 K	I.r.; ^{16}O, ^{18}O	Only ν_3 observed; $\widehat{TlOTl} = 131 \pm 11°$	228
	Vaporization of Tl_2O_3	Ar, Kr, N_2; 10 K	I.r.; ^{16}O, ^{18}O	Only ν_3 observed; $\widehat{TlOTl} = \sim 90°$; strong Tl—Tl bond proposed: Tl—Tl = 2.63 Å (estimated)	231
	Vaporization of Tl_2O_3	Ar; 10 K	I.r.	Bands assigned to ν_1, ν_2, and ν_3	229
	Vaporization of Tl_2O_3	N_2; 15 K	I.r.; ^{16}O, ^{18}O	Bands assigned to ν_1 and ν_2 of monomer by ref. 229 re-assigned to polymer	230
Tl_xO_2	Reaction of Tl with O_2	Details not available	I.r.; ^{16}O, ^{18}O, using $^{16}O_2$–$^{18}O_2$ mixture containing no $^{16}O^{18}O$	No isotope scrambling found to occur; sole products are $Tl_x\,^{18}O_2$ and $Tl_x\,^{16}O_2$; oxidation is therefore non-dissociative (*cf.* reactions with Ge, Sn, Ga, or In)	232

TlX (X = F, Cl, Br, or I); Tl_2X_2 (X = F or Cl)	Vaporization of TlX	Ar, Kr; ~5 K	I.r.; ^{35}Cl, ^{37}Cl	Linear X—Tl—X—Tl structure proposed for dimer	233
Group IV/IVB (*for carbon-containing species see Table* 1)					
Si_2	Photolysis of SiH_4	Ar; 4—14 K	U.v.–visible	—	234
SiH	Photolysis of SiH_4	Ar; 4—14 K	I.r.; u.v.–visible; ^1H, ^2H	—	234
SiH_2	Photolysis of SiH_4	Ar; 4—14 K	I.r.; u.v.–visible; ^1H, ^2H	Bent, C_{2v}; $\widehat{HSiH}$ = 92°	234
SiH_3	Photolysis of SiH_4	Ar; 4—14 K	I.r.; u.v.–visible; ^1H, ^2H	Pyramidal, C_{3v}	234
	γ-Irradiation of SiH_4	Kr; 4.2 K	E.s.r.	Pyramidal; bond angles = 110.6°	235
SiH_4	Direct deposition	Ne, Ar, Kr, Xe, N_2, CO, CH_4	I.r.; ^1H, ^2H	No evidence to suggest rotation in matrix sites	236
HNSi	Photolysis of SiH_3N_3	Ar; 4 K	I.r.; ^1H, ^2H	Assumed to be linear	237, 238
SiO	Vaporization of SiO_2 at 2000 K, or mixture of Si with SiO_2 at 1600 K	N_2, Ar; 15—20 K	I.r.; ^{16}O, ^{18}O	—	239
Si_2O_2	Vaporization of SiO_2 at 2000 K, or mixture of Si with SiO_2 at 1600 K	N_2, Ar; 15—20 K	I.r.; ^{16}O, ^{18}O	Cyclic D_{2h} rhombus implied; $\widehat{OSiO}$ = ~87°	239, 240

[224] L. B. Knight, jun. and W. Weltner, jun., *J. Chem. Phys.*, 1971, **55**, 5066.
[225] A. Snelson, *J. Phys. Chem.*, 1967, **71**, 3202.
[226] M. L. Lesiecki and J. S. Shirk, *J. Chem. Phys.*, 1972, **56**, 4171.
[227] S. J. Cyvin, B. N. Cyvin, and A. Snelson, *J. Phys. Chem.*, 1971, **75**, 2609.
[228] A. J. Hinchcliffe and J. S. Ogden, *Chem. Comm.*, 1969, 1053.
[229] D. M. Makowiecki, D. A. Lynch, jun., and K. D. Carlson, *J. Phys. Chem.*, 1971, **75**, 1963.
[230] A. J. Hinchcliffe and J. S. Ogden, *J. Phys. Chem.*, 1971, **75**, 3908.
[231] J. M. Brom, jun., T. Devore, and H. F. Franzen, *J. Chem. Phys.*, 1971, **54**, 2742.
[232] J. S. Ogden, A. J. Hinchcliffe, and J. S. Anderson, *Nature*, 1970, **226**, 940.
[233] J. M. Brom, jun. and H. F. Franzen, *J. Chem. Phys.*, 1971, **54**, 2874.
[234] D. E. Milligan and M. E. Jacox, *J. Chem. Phys.*, 1970, **52**, 2594.
[235] R. L. Morehouse, J. J. Christiansen, and W. Gordy, *J. Chem. Phys.*, 1966, **45**, 1751.
[236] R. E. Wilde, T. K. K. Srinivasan, R. W. Harral, and S. G. Sankar, *J. Chem. Phys.*, 1971, **55**, 5681.
[237] J. F. Ogilvie and S. Cradock, *Chem. Comm.*, 1966, 364.
[238] J. F. Ogilvie and M. J. Newlands, *Trans. Faraday Soc.*, 1969, **65**, 2602.
[239] J. S. Anderson and J. S. Ogden, *J. Chem. Phys.*, 1969, **51**, 4189.
[240] J. S. Anderson, J. S. Ogden, and M. J. Ricks, *Chem. Comm.*, 1968, 1585.

Table 2 (*continued*)

Species	*Method of preparation*	*Matrix*	*Means of identification and structural determination*	*Observations*	*Ref.*
Si_3O_3	As above, with diffusion of matrix	N_2, Ar; 27—30 K	I.r.; ^{16}O, ^{18}O	Planar D_{3h} ring; $\widehat{OSiO} = \sim 100°$	239
$(SiO)_n$ ($n = 1, 2, 3$, or 5)	Vaporization of SiO_2	Ne, Ar; 5—15 K	I.r.; ^{28}Si, ^{29}Si, ^{30}Si	Conclusions of refs. 239 and 240 verified for SiO, Si_2O_2, and Si_3O_3; $(SiO)_5$ also identified.	241
SiF_2	Photolysis of H_2SiF_2	Ne, Ar; 4—14 K	I.r.; u.v.–visible; ^{28}Si, ^{29}Si, ^{30}Si	—	242
	Reaction of SiF_4 with Si at 1150 K	Ne, Ar; ~5 K	I.r.; ^{28}Si, ^{29}Si, ^{30}Si	$\widehat{FSiF} = 97.5 \pm 1°$ (Ne); $102.0 \pm 2°$ (Ar) (gas phase value 100.9°)	243
SiF_3	Photolysis of $HSiF_3$	Ar, N_2, CO; 14 K	I.r.; ^{28}Si, ^{29}Si, ^{30}Si	Pyramidal; semi-vertical angle = $71 \pm 2°$ (close to that for sp^3 hybridisation)	244
	No details available	—	Raman	No details available	245
$SiCl_2$	Photolysis of H_2SiCl_2	Ar; 14 K	I.r.; ^1H, ^2H, ^{28}Si, ^{29}Si, ^{30}Si, ^{35}Cl, ^{37}Cl	$\widehat{ClSiCl}$ not calculated accurately, but in the range 90—120°	246
$SiCl_3$	Photolysis of $HSiCl_3$	Ar, N_2; 14 K	I.r.; ^{28}Si, ^{29}Si, ^{30}Si, ^{35}Cl, ^{37}Cl	Pyramidal; semi-vertical angle = $72 \pm 5°$	247
GeH_2	Photolysis of GeH_4	Ar; 4—25 K	I.r.; ^1H, ^2H	Bent; bond angle not calculated	248
GeH_3	γ-Irradiation of GeH_4	Kr; 4.2 K	E.s.r.	Pyramidal	235
	Photolysis of GeH_4	Ar; 4—25 K	I.r.; ^1H, ^2H	Pyramidal; flatter than SiH_3	248
HNGe	Photolysis of GeH_3N_3	Ar; 4 K	I.r.; ^1H, ^2H	Presumed linear	238
GeH_2Cl	Photolysis of GeH_3Cl	Ar, CO; 4—25 K	I.r.; ^1H, ^2H, ^{35}Cl, ^{37}Cl	Pyramidal radical with approximately sp^3 hybridization bond angles; summary given of data for halogenomethyl, halogenosilyl, and halogenogermyl radicals	249
GeHBr; GeH_2Br	Photolysis of GeH_3Br	Ar; 8—24 K	I.r.; ^1H, ^2H	GeHBr: bent; GeH_2Br: pyramidal	250
GeO	Vaporization of GeO at 1100 K	N_2, Ar; 20 K	I.r.; ^{16}O, ^{18}O, ^{70}Ge, ^{72}Ge, ^{73}Ge, ^{74}Ge, ^{76}Ge	—	251
	Vaporization of GeO_2 at 780 K	Ar, Kr, Xe, O_2, SF_6, CH_4; 20 K	U.v.–visible	$A\,^1\Pi$–$X\,^1\Sigma$ system observed; bands assigned to new $a\,^3\Pi$–$X\,^1\Sigma$ system	252
Ge_2O_2	Vaporization of GeO at 1100 K	N_2, Ar; 20 K	I.r.; ^{16}O, ^{18}O	D_{2h} rhombus; $\widehat{OGeO} = 83°$	251

Ge_3O_3	Vaporization of GeO at 1100 K	N_2, Ar; 20 K	I.r.; ^{16}O, ^{18}O	D_{3h} planar six-membered ring; $\widehat{OGeO} = 100°$	251
Ge_4O_4	Vaporization of GeO at 1100 K	N_2, Ar; 36 K	I.r.; ^{16}O, ^{18}O	Possibly puckered eight-membered ring with C_{4v} symmetry	251
GeS	Vaporization of GeS	Ar, Kr, Xe, O_2, SF_6, CH_4; 20 K	U.v.–visible	$A\,^1\Pi–X\,^1\Sigma$ system and new $a\,^3\Pi–X\,^1\Sigma$ system observed	252
GeSe	Vaporization of GeSe	SF_6, Kr; 20 K	U.v.–visible	$^1\Sigma \leftarrow X^1\Sigma$ and $^3\Pi \leftarrow X^1\Sigma$ systems observed	253
GeF_2	Vaporization of GeF_2 at temperatures up to 420 K	Ne, Ar; 4.2 K	I.r.; ^{70}Ge, ^{72}Ge, ^{73}Ge, ^{74}Ge, ^{76}Ge	C_{2v} symmetry; $\widehat{FGeF} = 94 \pm 4°$; comparison made with CF_2 and SiF_2	254
	Vaporization of GeF_4 with Ge at 575 K	N_2; ~5 K	Raman; polarization measurements	ν_1 identified by polarization measurements	255
$(GeF_2)_2$	As above, with diffusion of matrix	Details not available	I.r.	Bands due to vibrations of terminal and bridging Ge—F bonds observed	256
	As above	N_2; ~5 K	Raman; polarization measurements	Probably has the C_{2h} *trans*-configuration: 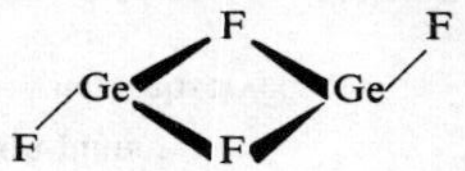	255

241 J. W. Hastie, R. H. Hauge, and J. L. Margrave, *Inorg. Chim. Acta*, 1969, **3**, 601.
242 D. E. Milligan and M. E. Jacox, *J. Chem. Phys.*, 1968, **49**, 4269.
243 J. W. Hastie, R. H. Hauge, and J. L. Margrave, *J. Amer. Chem. Soc.*, 1969, **91**, 2536.
244 D. E. Milligan, M. E. Jacox, and W. A. Guillory, *J. Chem. Phys.*, 1968, **49**, 5330
245 G. A. Ozin, *The Spex Speaker*, 1971, **XVI**, no. 4, 5.
246 D. E. Milligan and M. E. Jacox, *J. Chem. Phys.*, 1968, **49**, 1938.
247 M. E. Jacox and D. E. Milligan, *J. Chem. Phys.*, 1968, **49**, 3130.
248 G. R. Smith and W. A. Guillory, *J. Chem. Phys.*, 1972, **56**, 1423.
249 R. J. Isabel and W. A. Guillory, *J. Chem. Phys.*, 1971, **55**, 1197.
250 R. J. Isabel and W. A. Guillory, *J. Chem. Phys.*, 1972, **57**, 1116.
251 J. S. Ogden and M. J. Ricks, *J. Chem. Phys.*, 1970, **52**, 352.
252 B. Meyer, Y. Jones, J. J. Smith, and K. Spitzer, *J. Mol. Spectroscopy*, 1971, **37**, 100.
253 B. Meyer, 'Low Temperature Spectroscopy', American Elsevier, New York, 1971, p. 331.
254 J. W. Hastie, R. H. Hauge, and J. L. Margrave, *J. Phys. Chem.*, 1968, **72**, 4492.
255 G. A. Ozin, *The Spex Speaker*, 1971, **XVI**, no. 4, 7.
256 J. W. Hastie, R. H. Hauge, and J. L. Margrave, in 'Spectroscopy in Inorganic Chemistry', ed. C. N. R. Rao and J. R. Ferraro, Academic Press, 1970, vol. 1, p. 57.

Table 2 (*continued*)

Species	*Method of preparation*	*Matrix*	*Means of identification and structural determination*	*Observations*	*Ref.*
$GeCl_2$	Vaporization of $GeCl_2$	Ar; 15 K	I.r.; ^{35}Cl, ^{37}Cl	Bent; Ge—Cl force constant lower than for $GeCl_4$	257
	Photolysis of H_2GeCl_2	Ar, N_2, CO; ~4 K	I.r.	—	258
	Vaporization of $GeCl_2$	N_2, Ar; ~5 K	Raman; polarization measurements	C_{2v}; previous i.r. assignments confirmed	259
$GeCl_3$	Photolysis of $HGeCl_3$	Ar, N_2, CO; ~4 K	I.r.	Pyramidal C_{3v}; semi-vertical angle ~73° (slightly wider than the sp^3 hybridisation angle)	258
Sn	Vaporization of Sn	Ar; 4.2 K	U.v.–visible	—	260
^{119}Sn; $^{119}Sn_2$	Vaporization of Sn 84% enriched in ^{119}Sn at 1345 K	Ar, Kr, Xe; Be disc at 4.2 K	^{119}Sn Mössbauer	Isomer shifts (mm s^{-1} rel. $BaSnO_3$ at 300 K): Sn, +3.21; Sn_2, +3.05. $\delta R/R$ calculated; quadrupole splitting observed for Sn_2	261
SnH_3	γ-Irradiation of SnH_4	Kr; 4.2 K	E.s.r.	Probably non-planar	235
SnO	Vaporization of SnO	Ar, Kr, Xe; 20 K	U.v.–visible	$D^1\Pi$—$X^1\Sigma$ system observed	262
^{119}SnO	Vaporization of SnO_2 at 1200—1300 K	Ar, N_2; Be disc at 4.2 K	^{119}Sn Mössbauer	Isomer shifts: 3.02 (Ar) and 3.04 (N_2); large quadrupole splitting observed and related to possible bonding schemes	263
SnO	Vaporization of SnO_2; vaporization of SnO_2 with Sn; reaction of Sn with O_2	N_2, Ar; 20 K	I.r.; ^{16}O, ^{18}O, ^{116}Sn, ^{118}Sn, ^{120}Sn	—	264
Sn_2O_2	As above	N_2, Ar; 20 K	I.r.; ^{16}O, ^{18}O	D_{2h} rhombus; $\widehat{OSnO}$ = ~80°; $\widehat{SnOSn}$ = ~100°	264
Sn_3O_3	As above	N_2, Ar; 20 K	I.r.; ^{16}O, ^{18}O	D_{3h} planar six-membered ring; $\widehat{OSnO}$ = ~100°; $\widehat{SnOSn}$ = ~140°	264
Sn_4O_4	As above	N_2, Ar; 20 K	I.r.; ^{16}O, ^{18}O	D_{4h} or T_d unit indicated; $\widehat{OSnO}$ = ~81.5°; $\widehat{SnOSn}$ = ~98° assuming T_d geometry	264
SnO_2	Co-deposition of Sn atoms with O_2	Kr; 16 K	I.r.; ^{16}O, ^{18}O	Isostructural with CO_2; force constants calculated	265

SnS	Vaporization of SnS	Ar, Kr, Xe; 20 K	U.v.–visible	$D\ ^1\Pi - X\ ^1\Sigma$ system observed	262
SnF_2; $(SnF_2)_2$	Details not available	Details not available	I.r.; ^{116}Sn, ^{118}Sn, ^{120}Sn	Monomer: $\widehat{FSnF} = 92 \pm 3°$. Dimer: vibrational features due to terminal and bridging Sn—F bonds observed	256
$KSnF_3$	Details not available	Details not available	I.r.	C_{3v} unit having K—F—Sn bridges	256
$SnCl_2$	Vaporization of $SnCl_2$	Ar; 15 K	I.r.; ^{35}Cl, ^{37}Cl	Bent molecule	257
	Vaporization of $SnCl_2$	Ar, N_2; 4 K	Raman	Bent molecule	259, 266
SnClBr	Vaporization of $SnCl_2$ with $SnBr_2$	Ar, N_2; 4 K	Raman	Bent molecule	259
$SnBr_2$	Vaporization of $SnBr_2$	Ar, N_2; 4 K	Raman	Bent molecule	259, 267
Pb; Pb_2	Vaporization of Pb at ~870—1070 K; diffusion of matrix for Pb_2	Ar, Kr, Xe, SF_6; 20 K	U.v.–visible	$A \leftarrow X$ system observed for Pb_2	268
PbO	Vaporization of PbO at ~1040 K	Ar, N_2; 20 K	I.r.; ^{16}O, ^{18}O	—	269
Pb_2O_2	Vaporization of PbO at ~1040 K	Ar, N_2; 20 K	I.r.; ^{16}O, ^{18}O	D_{2h} rhombus; $\widehat{OPbO} = \sim 79°$	269
Pb_4O_4	As above, with diffusion of matrix	Ar, N_2; 20 K	I.r.; ^{16}O, ^{18}O	Data consistent with structure having T_d symmetry; $\widehat{OPbO} = \sim 81°$	269

257 L. Andrews and D. L. Frederick, *J. Amer. Chem. Soc.*, 1970, **92,** 775.
258 W. A. Guillory and C. E. Smith, *J. Chem. Phys.*, 1970, **53,** 1661.
259 G. A. Ozin and A. Vander Voet, *J. Chem. Phys.*, 1972, **56,** 4768.
260 D. M. Mann and H. P. Broida, *J. Chem. Phys.*, 1971, **55,** 84.
261 H. Micklitz and P. H. Barrett, *Phys. Rev.* (*B*), 1972, **5,** 1704.
262 J. J. Smith and B. Meyer, *J. Mol. Spectroscopy*, 1968, **27,** 304.
263 A. Bos, A. T. Howe, B. W. Dale, and L. W. Becker, *Chem. Comm.*, 1972, 730.
264 J. S. Ogden and M. J. Ricks, *J. Chem. Phys.*, 1970, **53,** 896.
265 J. S. Anderson, A. Bos, and J. S. Ogden, *Chem. Comm.*, 1971, 1381.
266 H. Huber, G. A. Ozin, and A. Vander Voet, *J. Mol. Spectroscopy*, 1971, **40,** 421.
267 H. Huber, G. A. Ozin, and A. Vander Voet, *Nature Phys. Sci.*, 1971, **232,** 166.
268 L. Brewer and C.-A. Chang, *J. Chem. Phys.*, 1972, **56,** 1728.

Table 2 (*continued*)

Species	*Method of preparation*	*Matrix*	*Means of identification and structural determination*	*Observations*	*Ref.*
$PbCl_2$	Vaporization of $PbCl_2$	Ne, Ar, Kr, N_2; 5–20 K	I.r.; ^{35}Cl, ^{37}Cl	Bent molecule; $\widehat{ClPbCl} = 96 \pm 3°$	211*b*, 257
	Vaporization of $PbCl_2$	Ar, N_2; 4 K	Raman; ^{35}Cl, ^{37}Cl; polarization measurements	Bent molecule	259, 267
$CsPbCl_3$	Details not available	Details not available	I.r.	Data suggest unit with C_{3v} symmetry having Cs—Cl—Pb bridges	256
PbClBr	Vaporization of $PbCl_2$ with $PbBr_2$	Ar, N_2; 4 K	Raman; polarization measurements	Bent molecule	259
$PbBr_2$	Vaporization of $PbBr_2$	Ar, N_2; 4 K	Raman; polarization measurements	Bent molecule; ν_2 not observed	259
Group V/VB					
NH	Photolysis of HN_3	Ar, N_2, CO; 4—20 K	I.r.; ^1H, ^2H	—	270
	Photolysis of NH_3	Ar, N_2, CO; 14 K	I.r.; ^1H, ^2H, ^{14}N, ^{15}N	—	271
	Photolysis of HN_3	Ar, Xe, N_2; 20 K	I.r.	Bands tentatively assigned to NH and N_3	272, 273
N_2H_2	Photolysis of HN_3	N_2; 20 K	I.r.; ^1H, ^2H, ^{14}N, ^{15}N	*trans*-N_2H_2 (planar C_{2h}) and *cis*-N_2H_2 both observed	273
NH_2	Photolysis of NH_3	Ar, CO, N_2; 14 K	I.r.; ^1H, ^2H, ^{14}N, ^{15}N	Rotation may occur in Ar matrix	271
N_2O	Direct deposition	N_2; 15 K	I.r.; ^{14}N, ^{15}N, ^{16}O, ^{18}O	Matrix cage appears to perturb force field by a few percent	274
NO; $(NO)_2$	Direct deposition	Ar, N_2, O_2, CO_2, H_2, N_2O; ~4 K	I.r.	Dimer exists in *cis*- and *trans*-configurations; *cis*-configuration is more stable: N—N with O= on each N (O=N—N=O, cis)	275
	Direct deposition	N_2; 15 K	I.r.; ^{14}N, ^{15}N, ^{16}O, ^{18}O	Dimer exists in three forms (all O—N—N—O type): (*i*) stable *cis*-form; (*ii*) less stable *trans*-form; (*iii*) unstable *cis*-form (note: MO calculations (ref. 277) suggest that the planar	276

				trans-structure has the lowest energy, 3.5 kcal mol^{-1} less than that of the optimum *cis*-form)	
N_2O_3	Direct deposition	N_2; 15 K	I.r.	Exists in two forms: (*i*) normal stable ON—NO_2 molecule; (*ii*) ONONO molecule	275
NO_2	Direct deposition	Kr; 20 K	I.r.	Data incomplete; evidence of one-dimensional rotation of NO_2 molecule	277
NO_2; $(NO_2)_2$	Direct deposition	N_2; 15 K	I.r.	Dimer exists in three forms: (*i*) stable planar molecule; (*ii*) possibly twisted version of above (V_d); (*iii*) O—N—O—NO_2 molecule	275
NO_2	Direct deposition	Ne; 4 K	E.s.r.	Exists in three different sites in Ne matrix	278
NO_2^-; M_x^+ NO_2^-	Photolysis or electron bombardment of NO_2 in matrix; co-deposition of alkali-metal atoms with NO_2	Ar; 4, 14 K	I.r.; ^{14}N, ^{15}N, ^{16}O, ^{18}O	—	279
	Co-deposition of alkali-metal atoms with NO_2	Ar; 4, 14 K	I.r.	M_x^+ NO_2^- ion-pairs as well as isolated NO_2^- observed	280
M_x^+ NO^-	Co-deposition of alkali-metal atoms with NO	Ar; 4, 14 K	I.r.	M_x^+ NO^- ion-pairs identified	280
$N_2O_2^-$	Co-deposition of Na atoms with NO + N_2O mixture	Ar; 4, 14 K	I.r.	New absorption attributed to planar $O_2N{=}N^-$ anion	280
NF	Photolysis of mixture of HN_3 with F_2	Ar; 14 K	I.r.; ^{14}N, ^{15}N	—	281, 282

[269] J. S. Ogden and M. J. Ricks, *J. Chem. Phys.*, 1972, **56**, 1658.
[270] D. E. Milligan and M. E. Jacox, *J. Chem. Phys.*, 1964, **41**, 2838.
[271] D. E. Milligan and M. E. Jacox, *J. Chem. Phys.*, 1965, **43**, 4487.
[272] E. D. Becker, G. C. Pimentel, and M. Van Thiel, *J. Chem. Phys.*, 1957, **26**, 145.
[273] K. Rosengren and G. C. Pimentel, *J. Chem. Phys.*, 1965, **43**, 507.
[274] D. F. Smith, jun., J. Overend, R. C. Spiker, and L. Andrews, *Spectrochim. Acta*, 1972, **28A**, 87.
[275] W. G. Fateley, H. A. Bent, and B. Crawford jun., *J. Chem. Phys.*, 1959, **31**, 204.
[276] W. A. Guillory and C. E. Hunter, *J. Chem. Phys.*, 1969, **50**, 3516.
[277] J. E. Williams and J. N. Murrell, *J. Amer. Chem. Soc.*, 1971, **93**, 7149.
[278] G. H. Myers, W. C. Easley, and B. A. Zilles, *J. Chem. Phys.*, 1970, **53**, 1181.
[279] D. E. Milligan, M. E. Jacox, and W. A. Guillory, *J. Chem. Phys.*, 1970, **52**, 3864.
[280] D. E. Milligan and M. E. Jacox, *J. Chem. Phys.*, 1971, **55**, 3404.
[281] M. E. Jacox and D. E. Milligan, *J. Chem. Phys.*, 1967, **46**, 184.

Table 2 (*continued*)

Species	*Method of preparation*	*Matrix*	*Means of identification and structural determination*	*Observations*	*Ref.*
NF_2	Pyrolysis of N_2F_4	N_2; 20 K	I.r.	—	283
NX (X = Cl or Br)	Photolysis of XN_3	Ar, N_2; 4, 20 K	I.r.	—	282
HNO	Photolysis of a mixture of HN_3 with CO_2	CO_2; 20, 53 K	I.r.; ^1H, ^2H	—	284
HONO	Photolysis of a mixture of HN_3 with O_2	N_2; 20 K	I.r.; ^1H, ^2H, ^{16}O, ^{18}O	*cis*- and *trans*-HONO identified; u.v. light causes *cis* → *trans* isomerization; near-i.r. light causes *trans* → *cis* conversion; no *rationale* given	285
	Photolysis of a mixture of HN_3 with O_2	N_2; 20 K	I.r.	*cis–trans* isomerization induced by i.r. radiation in the ranges 3200—3650 cm^{-1} for HONO and 3500—4100 cm^{-1} for DONO; mechanism proposed	286
	Photolysis of a mixture of NO_2 with H_2O, CH_4, or HCl	Ar; 4—14 K	I.r.; ^{14}N, ^{15}N, ^{16}O, ^{18}O	Vibrational assignments of previous work (refs. 285 and 286) confirmed; normal-co-ordinate analysis performed	287
HNF; HNF_2	Photolysis of a mixture of F_2 with HN_3	Ar; 14 K	I.r.; u.v.–visible; ^1H, ^2H, ^{14}N, ^{15}N	—	281
MNO_3; $(MNO_3)_x$ (M = Li, Na, K, Rb, Tl, or ½Cu)	Vaporization of the appropriate metal nitrate	Ar, CCl_4, CO_2; ~20 K	I.r.; Raman for M = K, ½Cu	Monomer and dimer species observed; vibrational properties of the NO_3 unit related to the polarization effects of different cations	288
$(MNO_3)_x$ (M = Li, K, or Tl)	As above	CO_2; ~93 K	I.r.; Raman	MNO_3 aggregates identified	289
$(SiH_3)_3N$	Direct deposition	Ar; 20 K	I.r.	Suggest C_{3v} symmetry for Si_3N skeleton (gas-phase spectra were assigned on the basis of D_{3h} symmetry)	290
P; PH_2	Photolysis of PH_3	Kr; 4 K	E.s.r.	—	291
PF_2	Photolysis of PF_2H or P_2F_4; pyrolysis of P_2F_4	Ar; 4 K	I.r.	ν_1 and ν_3 observed; $\widehat{FPF}$ = ~99° (assumed); absorptions attributed to $F_3P{=}PF$ in photolysis of P_2F_4	292

PCl_2	Co-deposition of Li atoms with PCl_3	Ar; 15 K	I.r.; ^{35}Cl, ^{37}Cl	ν_1 and ν_3 observed; consistent with $\widehat{ClPCl} = \sim 120°$	293
	As above	Ar; ~5 K	Raman	Reversal of assignment for ν_1 and ν_3 given in ref. 293	294
PBr_2	Co-deposition of Li atoms with PBr_3	Ar; 15 K	I.r.	ν_1 and ν_3 observed	293
As	Photolysis of AsH_3	Kr; 4 K	E.s.r.	—	291
AsF_5	Direct deposition	Ne, Ar; ~5 K	I.r.	Consistent with gas-phase data	295
SbF_3	Vaporization of SbF_3	N_2; 20 K	I.r.	C_{3v} monomer	296
SbF_5	Direct deposition	Ne, Ar; ~5 K	I.r.	C_{4v} symmetry suggested for SbF_5 unit	295
	Direct deposition	N_2, Ar; 15 K	I.r.; diffusion studies	Certain bands assigned to monomer in ref. 295 shown to be due to aggregate; in agreement with gas-phase Raman spectrum in predicting D_{3h} symmetry for isolated SbF_5 unit	297
Group VI/VIB					
O_2	Direct deposition	N_2, CO, Ar, CD_4; 4.2 K	E.s.r.	Information gained about trapping sites	298

[282] D. E. Milligan and M. E. Jacox, *J. Chem. Phys.*, 1964, **40**, 2461.
[283] M. D. Harmony and R. J. Myers, *J. Chem. Phys.*, 1962, **37**, 636.
[284] D. E. Milligan, M. E. Jacox, S. W. Charles, and G. C. Pimentel, *J. Chem. Phys.*, 1962, **37**, 2302.
[285] J. D. Baldeschwieler and G. C. Pimentel, *J. Chem. Phys.*, 1960, **33**, 1008.
[286] R. T. Hall and G. C. Pimentel, *J. Chem. Phys.*, 1963, **38**, 1889.
[287] W. A. Guillory and C. E. Hunter, *J. Chem. Phys.*, 1971, **54**, 598.
[288] D. Smith, D. W. James, and J. P. Devlin, *J. Chem. Phys.*, 1971, **54**, 4437.
[289] G. Pollard, N. Smyrl, and J. P. Devlin, *J. Phys. Chem.*, 1972, **76**, 1826.
[290] T. D. Goldfarb and B. N. Khare, *J. Chem. Phys.*, 1967, **46**, 3379.
[291] R. L. Morehouse, J. J. Christiansen, and W. Gordy, *J. Chem. Phys.*, 1966, **45**, 1747.
[292] J. K. Burdett, L. Hodges, V. Dunning, and J. H. Current, *J. Phys. Chem.*, 1970, **74**, 4053.
[293] L. Andrews and D. L. Frederick, *J. Phys. Chem.*, 1969, **73**, 2774.
[294] G. A. Ozin, *The Spex Speaker*, 1971, **XVI**, no. 4, 9.
[295] A. L. K. Aljibury and R. L. Redington, *J. Chem. Phys.*, 1970, **52**, 453.
[296] C. J. Adams and A. J. Downs, *J. Chem. Soc.* (*A*), 1971, 1534.
[297] A. J. Downs and P. J. Tyrrell, unpublished work.
[298] G. M. Graham, J. S. M. Harvey, and H. Kiefte, *J. Chem. Phys.*, 1970, **52**, 2235; R. Simoneau, J. S. M. Harvey, and G. M. Graham, *ibid.*, 1971, **54**, 4819.

Table 2 (*continued*)

Species	*Method of preparation*	*Matrix*	*Means of identification and structural determination*	*Observations*	*Ref.*
O_2^-	Co-deposition of alkali metal atoms with O_2	H_2O, alcohols; 77 K	E.s.r.	$g_{\parallel}$ varies with matrix	299
O_3	Direct deposition	Ar, Kr, Xe; 16, 20 K	I.r.; Raman; ^{16}O, ^{18}O	$\widehat{OOO} = 116.3 \pm 4°$ calculated	300
OH	Photolysis of H_2O	Ar; 4.2, 20.4 K	I.r.; ^1H, ^2H, ^{16}O, ^{18}O	No evidence to suggest rotation in Ar	301
H_2O; $(H_2O)_2$	Direct deposition	Various	I.r.; ^1H, ^2H, ^{16}O, ^{18}O	Vibrational, rotational–vibrational, and pure rotational transitions observed; aggregation studies; 'open' structure favoured for $(H_2O)_2$	302*a*—*i*
HO_2	Photolysis of a mixture of HI with O_2	Ar; 4 K	I.r.; ^1H, ^2H, ^{16}O, ^{18}O	—	303
	Photolysis of a mixture of O_2 with either HCl or H_2O	Ar; 14 K	I.r.; ^1H, ^2H, ^{16}O, ^{18}O; u.v.–visible	$\widehat{HOO} = 105 \pm 5°$; electronic structure probably approaches that of O_2^-	304
H_2O_2	Direct deposition	Ar, N_2; ~8 K	I.r.; ^1H, ^2H	Hindered internal rotation occurs in both matrices; broad torsional bands observed	305
S_2	Microwave discharge through solid sulphur; vaporization of sulphur	Ar, N_2; 20 K	I.r.	Absorption at 668 cm^{-1} thought to be due to S_2 perturbed by matrix environment	306
	As above	Ar, Kr, Xe, N_2; 4, 20 K	I.r.; u.v.–visible	$^3\Sigma_u^- \leftarrow {}^3\Sigma_g^-$ absorption system observed; i.r. absorptions depend on nature of matrix	307
	As above	Ar; ~7 K	Raman	Band observed at 716 cm^{-1}; casts doubt on the i.r. observations of refs. 306 and 307	308
SH	Photolysis of H_2S	Ar; 20.4 K	I.r.; u.v.–visible; ^1H, ^2H	Comparison made with OH work of ref. 301; $A\,^2\Sigma^+ - X\,^2\Pi$ transition observed in u.v.	309
H_2S; $(H_2S)_2$	Direct deposition	N_2; 20 K	I.r.; ^1H, ^2H	Monomer: ν_2 not observed; dimer: suggest open-chain structure with one hydrogen bond (*cf.* $(H_2O)_2$): H–S(–H)–H····S(–H)–H	310

H_2S	Direct deposition	Kr, Ar; 8 K	I.r.; ^{1}H, ^{2}H	Observe all three vibrational fundamentals of the monomer	316
	Direct deposition	Ar, Kr, N_2, CO; 20 K	I.r.	Assignments for the three fundamentals differ from those of ref. 316; monomer, dimer, possibly trimer, and polymer detected	317
H_2S^-	Co-deposition of Na or K atoms with H_2S	H_2S; 77 K	E.s.r.; ^{1}H, ^{2}H, ^{33}S	$g_{\parallel} = 2.0023 \pm 0.0005$ $g_{\perp} = 2.0164 \pm 0.001$	311
S_2O	Microwave discharge through S and SO_2	Solid film; 77 K	I.r.	—	312
	As above	Ar, Kr, Xe, N_2; 20 K	U.v.–visible	—	313
SO_2	Direct deposition	Kr; 20 K	I.r.; ^{16}O, ^{18}O, ^{32}S, ^{34}S	$\widehat{OSO} = 119°37'$ in good agreement with the results of microwave measurements in the gas phase; some evidence of rotational effects	314
SO_2; $(SO_2)_2$	Direct deposition	Ne, Ar; 4, 30 K	I.r.; ^{32}S, ^{34}S; diffusion studies	Dimer observed on diffusion; no definite evidence for SOO isomer on photolysis of matrix	315

[299] J. E. Bennett, B. Mile, and A. Thomas, *Trans. Faraday Soc.*, 1968, **64**, 3200.
[300] L. Brewer and J. L.-F. Wang, *J. Chem. Phys.*, 1972, **56**, 759; L. Andrews and R. C. Spiker jun., *J. Phys. Chem.*, 1972, **76**, 3208.
[301] N. Acquista, L. J. Schoen, and D. R. Lide jun., *J. Chem. Phys.*, 1968, **48**, 1534.
[302] (*a*) K. B. Harvey and H. F. Shurvell, *J. Mol. Spectroscopy*, 1968, **25**, 120; (*b*) A. J. Tursi and E. R. Nixon, *J. Chem. Phys.*, 1970, **52**, 1521; (*c*) M. Van Thiel, E. D. Becker, and G. C. Pimentel, *ibid.*, 1957, **27**, 486; (*d*) E. Catalano and D. E. Milligan, *ibid.*, 1959, **30**, 45; (*e*) J. A. Glasel, *ibid.*, 1960, **33**, 252; (*f*) R. L. Redington and D. E. Milligan, *ibid.*, 1962, **37**, 2162; (*g*) R. L. Redington and D. E. Milligan, *ibid.*, 1963, **39**, 1276; (*h*) H. P. Hopkins, jun., R. F. Curl, jun., and K. S. Pitzer, *ibid.*, 1968, **48**, 2959; (*i*) D. W. Robinson, *ibid.*, 1963, **39**, 3430.
[303] D. E. Milligan and M. E. Jacox, *J. Chem. Phys.*, 1963, **38**, 2627.
[304] M. E. Jacox and D. E. Milligan, *J. Mol. Spectroscopy*, 1972, **42**, 495.
[305] J. A. Lannon, F. D. Verderame, and R. W. Anderson, jun., *J. Chem. Phys.*, 1971, **54**, 2212.
[306] B. Meyer, *J. Chem. Phys.*, 1962, **37**, 1577.
[307] L. Brewer, G. D. Brabson, and B. Meyer, *J. Chem. Phys.*, 1965, **42**, 1385.
[308] R. E. Barletta, H. H. Claassen, and R. L. McBeth, *J. Chem. Phys.*, 1971, **55**, 5409.
[309] N. Acquista and L. J. Schoen, *J. Chem. Phys.*, 1970, **53**, 1290.
[310] J. Pacansky and V. Calder, *J. Chem. Phys.*, 1970, **53**, 4519.
[311] J. E. Bennett, B. Mile, and A. Thomas, *Chem. Comm.*, 1966, 182.
[312] U. Blukis and R. J. Myers, *J. Phys. Chem.*, 1965, **69**, 1154.
[313] L. F. Phillips, J. J. Smith, and B. Meyer, *J. Mol. Spectroscopy*, 1969, **29**, 230.
[314] J.M. Allavena, R. Rysnik, D. White, V. Calder, and D. E. Mann, *J. Chem. Phys.*, 1969, **50**, 3399.
[315] W. Hastie, R. Hauge, and J. L. Margrave, *Inorg. J.Nuclear Chem.*, 1969, **31**, 281.

Table 2 (*continued*)

Species	*Method of preparation*	*Matrix*	*Means of identification and structural determination*	*Observations*	*Ref.*
SO_2^-	Co-deposition of alkali-metal atoms with SO_2	Ar; 4, 14 K	I.r.; ^{16}O, ^{18}O, ^{32}S, ^{34}S	$\widehat{OSO} = 110 \pm 5°$; frequencies dependent on nature of metal atom; S—O stretching force constant significantly lower than that of SO_2	318
SF_4	Direct deposition	Ar; 4 K	I.r.	Monomer and dimer observed; structure proposed for dimer:	319
	—	—	—	Data from ref. 319 re-analysed; monomer bands re-assigned; new dimer structure proposed:	320
SF_6	Direct deposition	Ar; 4 K	Raman	Test of the sensitivity of the technique	321
SeO_2; $(SeO_2)_2$	Vaporization of SeO_2	Ne, Ar; 4.2, 30 K	I.r.; diffusion studies; ^{74}Se, ^{76}Se, ^{77}Se, ^{78}Se, ^{80}Se, ^{82}Se	$\widehat{OSeO} = 110 \pm 2°$ (gas-phase microwave spectra give 113° 30′); dimer present in vapour and formed on diffusion	322
	Vaporization of SeO_2	CO_2; 4.2 K	Raman	*trans*-C_{2h} structure proposed for dimer:	323, 324
	Vaporization of SeO_2	Ar; 20 K	I.r.	Value reported for ν_2; thermodynamic parameters given	325

	Vaporization of SeO_2	Ar, Kr, Xe, N_2, CO, CH_4, SF_6, C_4H_8; 20 K	U.v.–visible	$S_3 \leftarrow S_0$, $S_1 \leftarrow S_0$, $T_1 \leftarrow S_0$(?), $T_1 \rightarrow S_0$, and $S_3 \rightarrow S_0$ transitions observed	326
SeF_4; $(SeF_4)_2$	Direct deposition	Ar, Kr, N_2; 15 K	I.r.; ^{76}Se, ^{77}Se, ^{78}Se, ^{80}Se, ^{82}Se	C_{2v} monomer isolated; dimer involves bridging via axial F atoms (*cf.* SF_4)	327
TeO; $(TeO)_2$	Vaporization of TeO_2 at 970 K	Ne, Ar; 5, 15 K	I.r.; ^{122}Te, ^{124}Te, ^{126}Te, ^{128}Te, ^{130}Te	Dimer possibly cyclic D_{2h} with $\widehat{OTeO} = \sim 100°$	328
TeO_2; $(TeO_2)_2$	Vaporization of TeO_2 at 970 K	Ne, Ar; 5, 15 K	I.r.; ^{122}Te, ^{124}Te, ^{126}Te, ^{128}Te, ^{130}Te	Monomer: $\widehat{OTeO} = 110 \pm 2°$	328
TeF_4; $(TeF_4)_2$	Direct deposition of TeF_4	Ar, Kr, Xe, N_2; 15 K	I.r.; ^{126}Te, ^{128}Te, ^{130}Te	C_{2v} monomer isolated; dimer involves bridging via axial F atoms	327
Group VII/VIIB					
HF	Direct deposition	Ne, Ar, Kr, Xe; ~4 K	Far-i.r.; ^1H, ^2H	$J = 1 \leftarrow J = O$ pure rotational spectra observed	329
	—	—	—	Shifts of rotational bands of HF in inert gas matrices calculated	330
	Direct deposition	Ne, Ar, Kr, Xe; 5 K	Mid- and far-i.r.; ^1H, ^2H	Vibration–rotation and pure rotation spectra observed	331

[316] A. J. Tursi and E. R. Nixon, *J. Chem. Phys.*, 1970, **53**, 518.
[317] A. J. Barnes and J. D. R. Howells, *J. C. S. Faraday II*, 1972, **68**, 729.
[318] D. E. Milligan and M. E. Jacox, *J. Chem. Phys.*, 1971, **55**, 1003.
[319] R. L. Redington and C. V. Berney, *J. Chem. Phys.*, 1965, **43**, 2020.
[320] R. A. Frey, R. L. Redington, and A. L. K. Aljibury, *J. Chem. Phys.*, 1971, **54**, 344.
[321] J. S. Shirk and H. H. Claassen, *J. Chem. Phys.*, 1971, **54**, 3237.
[322] J. W. Hastie, R. Hauge, and J. L. Margrave, *J. Inorg. Nuclear Chem.*, 1969, **31**, 281.
[323] D. Boal, G. Briggs, H. Hüber, G. A. Ozin, E. A. Robinson, and A. Vander Voet, *Chem. Comm.*, 1971, 686.
[324] D. Boal, G. Briggs, H. Hüber, G. A. Ozin, E. A. Robinson, and A. Vander Voet, *Nature Phys. Sci.*, 1971, **231**, 174.
[325] S. N. Cesaro, M. Spoliti, A. J. Hinchcliffe, and J. S. Ogden, *J. Chem. Phys.*, 1971, **55**, 5834.
[326] E. M. Voigt, B. Meyer, A. Morelle, and J. J. Smith, *J. Mol. Spectroscopy*, 1970, **34**, 179.
[327] C. J. Adams and A. J. Downs, *Spectrochim. Acta*, 1972, **28A**, 1841.
[328] D. W. Muenow, J. W. Hastie, R. Hauge, R. Bautista, and J. L. Margrave, *Trans. Faraday Soc.* 1969, **65**, 3210.
[329] D. W. Robinson and W. G. Von Holle, *J. Chem. Phys.*, 1966, **44**, 410.
[330] G. K. Pandey and S. Chandra, *J. Chem. Phys.*, 1966, **45**, 4369.
[331] M. G. Mason, W. G. Von Holle, and D. W. Robinson, *J. Chem. Phys.*, 1971, **54**, 3491.

Table 2 (*continued*)

Species	*Method of preparation*	*Matrix*	*Means of identification and structural determination*	*Observations*	*Ref.*
OF	Photolysis of OF_2	N_2, Ar; 4, 16 K	I.r.; Raman; ^{16}O, ^{18}O	On diffusion OF reacts to give O_3 and OF_2	332
	Photolysis of mixtures of OF_2 with N_2O or CO_2, or F_2 with N_2O	N_2, Ar; 4 K	I.r.	Obtained an increased yield of OF on addition of N_2O or CO_2 to OF_2 prior to photolysis; may be due to removal of F atoms, preventing the reaction: $OF+F \rightarrow OF_2$	333
	Co-deposition of Li, Na, K or Mg atoms with OF_2	Ar; 15 K	I.r.; ^{6}Li, ^{7}Li, ^{16}O, ^{18}O	Bands due to OF and weakly bonded MF···OF observed	334
O_2F_2	Direct deposition	Ar, N_2, CO_2; 20 K	I.r.; Raman spectrum of solid also reported	Results consistent with C_2 symmetry, and confirm the strength of the O—O bond.	335
O_2F	Photolysis of a mixture of O_2 with F_2	Ar, N_2, O_2; 20 K	I.r.; ^{16}O, ^{18}O	Bent FOO structure; assumed $\widehat{FOO} = 109°$; high stretching force constant for O—O bond.	336
	Vapour-phase decomposition of O_2F_2	Ar, N_2, CO_2; 20 K	I.r.	—	335
OF_2	Photolysis of a mixture of O_2 with F_2	Ar, N_2, O_2; 4.2, 20 K	I.r.; ^{16}O, ^{18}O	—	337
HOF; HOF···HF	Photolysis of a mixture of H_2O with F_2	N_2; 14, 20 K	I.r.; ^{1}H, ^{2}H, ^{16}O, ^{18}O; diffusion studies	Assumed $\widehat{HOF} = 104°$ [note: gas-phase microwave study (ref. 339) gives $\widehat{HOF} = 97.2 \pm 0.6°$]; observed band shifts due to HOF···HF species	338
	Direct deposition	N_2; 8 K	I.r.	Confirmed results of ref. 338	340
Cl_3	Microwave discharge through Cl_2 with Kr	Kr; 20 K	I.r.; ^{35}Cl, ^{37}Cl	Linear, slightly asymmetric unit ($C_{\infty v}$); stretching force constant similar to that for Cl_3^-	341
HCl	Direct deposition	Various matrices	I.r.; ^{1}H, ^{2}H, ^{35}Cl, ^{37}Cl	Arguments proposed for and against rotation of HCl in different matrices	342*a*—*f*
	Direct deposition	CH_4, CF_4, N_2, CO, CO_2, C_2H_4, SF_6; 20 K	I.r.	Rotation of HCl monomer occurs in noble-gas matrices and in CH_4, CF_4, and SF_6, but not in N_2, CO, CO_2, or C_2H_4	343
DCl	Direct deposition	Ar, Kr, N_2, SF_6, CO_2, CO; 20 K	I.r.	Rotation of DCl monomer occurs in Ar, Kr, and SF_6, but not in N_2, CO_2, or CO	344

$(HCl)_2$	Direct deposition and diffusion	Ne, Ar, Kr, Xe; 7, 22 K	I.r.	Cyclic C_{2h} structure most likely	345
$ClHCl/ClHCl^-$	Glow discharge through Cl_2, HCl and Ar	Ar; 14 K	I.r.; ^1H, ^2H, ^{35}Cl, ^{37}Cl	Results interpreted in terms of linear ($D_{\infty h}$) ClHCl radical	346
	Photolysis of $(HCl)_2$ in matrix alone or with K or Cs atoms	Ar; 14 K	I.r.; u.v.–visible; ^1H, ^2H, ^{35}Cl, ^{37}Cl	Suggest that linear symmetric anion $ClHCl^-$ is responsible for absorptions assigned to ClHCl by ref. 346; effect of alkali-metal photoelectron source is to increase in intensity these absorptions	347
ClClO	Photolysis of Cl_2O	N_2, Ar; 20 K	I.r.; ^{16}O, ^{18}O, ^{35}Cl, ^{37}Cl	Isotopic shifts imply a structure analogous to that of ClNO; ν_1 and ν_2 observed	348
Cl_2O	Direct deposition	Ar, N_2; 20 K	I.r.; ^{16}O, ^{18}O, ^{35}Cl, ^{37}Cl	Reassignment of stretching fundamentals; first report of ν_2; Cl—O stretching force constant identified with prototype Cl—O single bond	349

[332] A. Arkell, R. R. Reinhard, and L. P. Larson, *J. Amer. Chem. Soc.*, 1965, **87**, 1016; L. Andrews, *J. Chem. Phys.*, 1972, **57**, 51.
[333] A. Arkell, *J. Phys. Chem.*, 1969, **73**, 3877.
[334] L. Andrews and J. I. Raymond, *J. Chem. Phys.*, 1971, **55**, 3078.
[335] D. J. Gardiner, N. J. Lawrence, and J. J. Turner, *J. Chem. Soc.(A)*, 1971, 400.
[336] R. D. Spratley, J. J. Turner, and G. C. Pimentel, *J. Chem. Phys.*, 1966, **44**, 2063.
[337] A. Arkell, *J. Amer. Chem. Soc.*, 1965, **87**, 4057.
[338] P. N. Noble and G. C. Pimentel, *Spectrochim. Acta*, 1968, **24A**, 797.
[339] H. Kim, E. F. Pearson, and E. H. Appelman, *J. Chem. Phys.*, 1972, **56**, 1.
[340] J. A. Goleb, H. H. Claassen, M. H. Studier, and E. H. Appelman, *Spectrochim. Acta*, 1972, **28A**, 65.
[341] L. Y. Nelson and G. C. Pimentel, *J. Chem. Phys.*, 1967, **47**, 3671.
[342] (*a*) K. B. Harvey and H. F. Shurvell, *Canad. J. Chem.*, 1967, **45**, 2689; *Chem. Comm.*, 1967, 490; *Canad. Spectroscopy*, 1969, **14**, 32, 42, 84, 91; (*b*) G. C. Pimentel and S. W. Charles, *Pure Appl. Chem.*, 1963, **7**, 111; (*c*) M. T. Bowers and W. H. Flygare, *J. Chem. Phys.*, 1966, **44**, 1389; (*d*) B. Katz and A. Ron, *Chem. Phys. Letters*, 1970, **7**, 357; (*e*) T. E. Whyte, jun., Ph.D. thesis, Howard University, 1965; (*f*) L. F. Keyser and G. W. Robinson, *J. Chem. Phys.*, 1966, **44**, 3225.
[343] A. J. Barnes, H. E. Hallam, and G. F. Scrimshaw, *Trans. Faraday Soc.*, 1969, **65**, 3150, 3159, 3172.
[344] J. B. Davies and H. E. Hallam, *Trans. Faraday Soc.*, 1971, **67**, 3176.
[345] B. Katz, A. Ron, and O. Schnepp, *J. Chem. Phys.*, 1967, **47**, 5303.
[346] P. N. Noble and G. C. Pimentel, *J. Chem. Phys.*, 1968, **49**, 3165.
[347] D. E. Milligan and M. E. Jacox, *J. Chem. Phys.*, 1970, **53**, 2034.
[348] M. M. Rochkind and G. C. Pimentel, *J. Chem. Phys.*, 1967, **46**, 4481.
[349] M. M. Rochkind and G. C. Pimentel, *J. Chem. Phys.*, 1965, **42**, 1361.

Table 2 (*continued*)

Species	*Method of preparation*	*Matrix*	*Means of identification and structural determination*	*Observations*	*Ref.*
ClO; $(ClO)_2$	Photolysis of Cl_2O	Ar, N_2; 20 K	I.r.	Monomer: reassignment of vibrational fundamental suggested; dimer: consists of weakly bonded pairs of ClO units	348, 350
ClO	Co-deposition of Cl_2O with alkali-metal atoms	Ar; 15 K	I.r.; ^{16}O, ^{18}O, ^{35}Cl, ^{37}Cl	Earlier results (ref. 348) for ClO confirmed; suggest $(p\text{-}p)\pi$-bonding for ClO unit	351
ClOO	Photolysis of OClO, or a mixture of Cl_2 and O_2	Ar, O_2; 4 K	I.r.; ^{16}O, ^{18}O, ^{35}Cl, ^{37}Cl	Assumed $\widehat{ClOO} = 110°$; k_{O-O} force constant large and k_{O-Cl} small	352
$ClOClO_3$	Direct deposition	Ar; ~4 K	I.r.; ^{35}Cl, ^{37}Cl	Results consistent with C_s symmetry	353
ClF_2	Photolysis of ClF_3 or mixtures of F_2 with ClF	Ar, N_2; 16 K	I.r.; ^{35}Cl, ^{37}Cl	Bent FClF molecule with $\widehat{FClF} = 136 \pm 15°$; normal-co-ordinate analysis performed (21-electron molecule, *cf.* Cl_3)	354
ClF_3; $(ClF_3)_2$	Direct deposition	Ne, Ar, N_2; ~5 K	I.r.; ^{35}Cl, ^{37}Cl	Dimer thought to involve bridging *via* axial F atoms: may provide a mechanism for exchange of F atoms in liquid phase	320
ClF_5	Direct deposition	Ar, N_2; 4 K	I.r.; ^{35}Cl, ^{37}Cl	Evidence of aggregation	355
HOCl	Photolysis of a mixture of O_3 with HCl	Ar; 4 K	I.r.; ^{1}H, ^{2}H, ^{16}O, ^{18}O, ^{35}Cl, ^{37}Cl	Assumed $\widehat{HOCl} = 113°$	356
Br_3	Microwave discharge through Br_2 with a noble gas	Ar, Kr, Xe; 4.2 K	Raman	Linear centrosymmetric ($D_{\infty h}$)	357, 358
Br_4/Br_5	As above, with thermal diffusion of matrix	Ar; 35 K	Raman	New features compatible with either Br_4 or Br_5	357
BrHBr/$BrHBr^-$	Glow discharge through a mixture of Br_2 with HBr	Ar; 20 K	I.r.; ^{1}H, ^{2}H	Linear centrosymmetric ($D_{\infty h}$); review of evidence against formulation as $BrHBr^-$ anion	359
	Photolysis of HBr in matrix with or without alkali-metal atoms	Ar; 14 K	I.r.; ^{1}H, ^{2}H	I.r. spectra identical to those reported by ref. 359, but assigned to $BrHBr^-$; arguments presented against formulation as BrHBr	360

BrF_3; $(BrF_3)_2$	Direct deposition	Ne, Ar, N_2; ~5 K	I.r.	Monomer: C_{2v} unit; dimer: conclusions analogous to those reached for ClF_3	320
BrF_5	Direct deposition	Ar; ~5 K	I.r.	Vibrational frequencies compared with those for ClF_3 and BrF_3	320
$BrBrCl_2$ and related species	Microwave discharge through Cl_2, Br_2 and noble gas	Ar, Kr, Xe; 20 K	I.r.; ^{35}Cl, ^{37}Cl	Evidence for the presence of several 'T'-shaped molecules (analogous to ClF_3) possessing either Cl—Br—Cl or Cl—Br—Br linear units; $BrBrCl_2$ most definitely identified species	361
HOBr	Photolysis of O_3 and HBr mixture	Ar; 4 K	I.r.; ^1H, ^2H, ^{16}O, ^{18}O	Assumed $\widehat{HOBr} = 110°$	356
$BrOClO_3$	Direct deposition	Ar; ~4 K	I.r.; ^{35}Cl, ^{37}Cl	Results consistent with C_s symmetry	353
$^{129}I_2$	Direct deposition or freezing of solutions	Ar; 22 K; C_6H_6, C_6H_{12}, CCl_4; 88 K	^{129}I Mössbauer	Isomer shift ($mm\,s^{-1}$ rel. ZnTe source): +0.94 (Ar, hexane, CCl_4), +0.76 (C_6H_6); quadrupole coupling constants calculated	362
IHI	Glow discharge through H_2, I_2 and Ar	Ar; 16 K	I.r.; ^1H, ^2H	Absorptions attributed to linear symmetric IHI radical; summary of force field data for all known XHX and XHX^- species given	363

[350] W. G. Alcock and G. C. Pimentel, *J. Chem. Phys.*, 1968, **48**, 2373.
[351] L. Andrews and J. I. Raymond, *J. Chem. Phys.*, 1971, **55**, 3087.
[352] A. Arkell and I. Schwager, *J. Amer. Chem. Soc.*, 1967, **89**, 5999.
[353] K. O. Christe, C. J. Schack, and E. C. Curtis, *Inorg. Chem.*, 1971, **10**, 1589.
[354] G. Mamantov, E. J. Vasini, M. C. Moulton, D. G. Vickroy, and T. Maekawa, *J. Chem. Phys.*, 1971, **54**, 3419.
[355] K. O. Christe, *Spectrochim. Acta*, 1971, **27A**, 631.
[356] I. Schwager and A. Arkell, *J. Amer. Chem. Soc.*, 1967, **89**, 6006.
[357] G. A. Ozin, *The Spex Speaker*, 1971, **XVI**, no. 4, 6.
[358] D. H. Boal and G. A. Ozin, *J. Chem. Phys.*, 1971, **55**, 3598.
[359] V. Bondybey, G. C. Pimentel, and P. N. Noble, *J. Chem. Phys.*, 1971, **55**, 540.
[360] D. E. Milligan and M. E. Jacox, *J. Chem. Phys.*, 1971, **55**, 2550.
[361] L. Y. Nelson and G. C. Pimentel, *Inorg. Chem.*, 1968, **7**, 1695.
[362] S. Bukshpan, C. Goldstein, and T. Sonnino, *J. Chem. Phys.*, 1968, **49**, 5477.
[363] P. N. Noble, *J. Chem. Phys.*, 1972, **56**, 2088.

Table 2 (*continued*)

Species	*Method of preparation*	*Matrix*	*Means of identification and structural determination*	*Observations*	*Ref.*
^{129}ICl; ^{129}IBr	Direct deposition	Ar; Al foil at 20.4 K	^{129}I Mössbauer	Isomer shifts (mm s^{-1} rel. ZnTe): ICl, +1.20; IBr, +0.98; quadrupole coupling constants calculated	364
Group VIII					
^{83}Kr	Direct deposition	Solid Kr and β-quinol; 32—300 K	^{83}Kr Mössbauer	—	365
KrF_2	Photolysis of Kr and F_2 mixture with Ar	Ar–Kr; 20 K	I.r.	First evidence of linear centrosymmetric ($D_{\infty h}$) KrF_2 molecule; no evidence of Ar—F compounds under these conditions	366, 367
XeF_2; XeF_4	Photolysis of Xe and F_2 mixture with Ar	Ar–Xe; 20 K	I.r.	ν_1 and ν_3 of XeF_2 observed; ν_2 and ν_6 of XeF_4 observed	367
XeF_6	Direct deposition	Ar, N_2; 5—57 K	I.r.	Results suggest the existence of three electronic isomers of XeF_6 in equilibrium; populations vary with temperature	368
$XeCl_2$	Microwave discharge through Xe and Cl_2 mixture	Xe; 20 K	I.r.; ^{35}Cl, ^{37}Cl, ^{128}Xe, ^{129}Xe, ^{130}Xe, ^{131}Xe, ^{132}Xe	Only ν_3 (Σ_u^+) of $XeCl_2$ observed; isotopic splitting consistent with linear centrosymmetric molecule; $k_{Xe-Cl} = \sim 1.3$ mdyn Å^{-1}	369
	As above	Xe; 4.2 K	Raman	Only ν_1 (Σ_g^+) observed in support of linear geometry for $XeCl_2$	370, 371
XeO_3F_2	Direct deposition	Ne, Ar; 4 K	Raman; i.r.	D_{3h} symmetry indicated for the XeO_3F_2 molecule	372
Transition elements					
MCl_2 (M = Sc, Ti, V, Cr, Mn, Fe, or Ni)	Various Knudsen-cell techniques	Ne, Ar, N_2; 5–20 K	I.r.; ^{35}Cl, ^{37}Cl, metal isotopes	All molecules judged to be linear	211*b*
ScO	Vaporization of Sc_2O_3 at 2600 K	Ne, Ar; 4, 20 K	I.r.; u.v.–visible; e.s.r.	Ground state defined as $^2\Sigma$; $g = 2.00$; $A = 2010$ MHz; $A^2\Pi \leftarrow X^2\Sigma$ and $B^2\Sigma \leftarrow X^2\Sigma$ transitions observed	373

ScF	Vaporization of ScF_3 with Sc at 1700 K; vapour superheated to 2900 K	Ne; 4 K	I.r.; visible (absorption and emission)	$^1\Sigma^+$ identified as ground state; $E\,^1\Pi \leftarrow X\,^1\Sigma$, $C\,^1\Sigma \leftarrow X\,^1\Sigma$, and $B\,^1\Pi \leftarrow X\,^1\Sigma$ transitions observed	374
ScF_2; ScF_3	Vaporization of ScF_3 +Sc (1600 K) or ScF_3 (1900 K)	Ne, Ar; 4–15 K	I.r.; u.v.	Only ν_3 of ScF_2 observed; $\widehat{FScF} = 135°$ (estimated)	374, 375
ScF_3	Vaporization of ScF_3 at 1575—1775 K	Ne, Ar, N_2; 5—15 K	I.r. and far-i.r.	Possibly planar; ν_2, ν_3, and ν_4 observed	376
YO	Vaporization of Y_2O_3 at 2680 K	Ne, Ar; 4, 20 K	I.r.; u.v.–visible; e.s.r.	Ground state defined as $^2\Sigma$; $g = 2.003$; $A = 803$ MHz; $A\,^2\Pi \leftarrow X^2\Sigma$ and $B\,^2\Sigma \leftarrow X\,^2\Sigma$ transitions observed	373
YF_3	Vaporization of YF_3 at 1575—1775 K	Ne, Ar, N_2; 5—15 K	I.r. and far-i.r.	ν_3 and possibly ν_1 observed; non-planar geometry inferred	376
TiO	Vaporization of TiO_2 at 2200—2400 K; reaction of O_2 with heated Ti	Ne, Kr, Xe; 4 K	I.r.; u.v.–visible; ^{16}O, ^{18}O	—	377, 378
TiO_2	As above	Ne; 4 K	I.r.; visible emission; ^{16}O, ^{18}O, ^{46}Ti, ^{47}Ti, ^{48}Ti, ^{49}Ti, ^{50}Ti	ν_1 and ν_3 observed; $\widehat{OTiO} = 110 \pm 15°$	377

[364] C. Goldstein and T. Barnoi, *Chem. Phys. Letters*, 1971, **10**, 136.
[365] Y. Hazoni, P. Hillman, M. Pasternak, and S. Ruby, *Phys. Letters*, 1962, **2**, 337; M. Pasternak, A. Simopoulos, S. Bukshpan, and T. Sonnino, *Phys. Letters*, 1966, **22**, 52.
[366] J. J. Turner and G. C. Pimentel, *Science*, 1963, **140**, 974.
[367] J. J. Turner and G. C. Pimentel, 'Noble-Gas Compounds', ed. H. H. Hyman, University of Chicago Press, Chicago, 1963, p. 101.
[368] H. H. Claassen, G. L. Goodman, and H. Kim, *J. Chem. Phys.*, 1972, **56**, 5042.
[369] L. Y. Nelson and G. C. Pimentel, *Inorg. Chem.*, 1967, **6**, 1758.
[370] G. A. Ozin, *The Spex Speaker*, 1971, **XVI**, no. 4, 5.
[371] D. Boal and G. A. Ozin, *Spectroscopy Letters*, 1971, **4**, 43.
[372] H. H. Claassen and J. L. Huston, *J. Chem. Phys.*, 1971, **55**, 1505.
[373] W. Weltner, jun., D. McLeod, jun., and P. H. Kasai, *J. Chem. Phys.*, 1967, **46**, 3172.
[374] D. McLeod, jun. and W. Weltner, jun., *J. Phys., Chem.*, 1966, **70**, 3293.
[375] J. W. Hastie, R. Hauge, and J. L. Margrave, *Chem. Comm.*, 1969, 1452.
[376] R. H. Hauge, J. W. Hastie, and J. L. Margrave, *J. Less-Common Metals*, 1971, **23**, 359.
[377] N. S. McIntyre, K. R. Thompson, and W. Weltner, jun., *J. Phys. Chem.*, 1971, **75**, 3243.
[378] W. Weltner, jun. and D. McLeod, jun., *J. Phys. Chem.*, 1965, **69**, 3488.

Table 2 (*continued*)

Species	*Method of preparation*	*Matrix*	*Means of identification and structural determination*	*Observations*	*Ref.*
TiF_2	Vaporization of TiF_3 with Ti at ~1275 K	Ne, Ar; 5—15 K	I.r.; ^{46}Ti, ^{47}Ti, ^{48}Ti, ^{49}Ti, ^{50}Ti	ν_1 and ν_2 observed for TiF_2; $\widehat{FTiF} = 130 \pm 5°$; equilibrium constants for $2\ TiF_3(g) + Ti(s) \rightarrow 3\ TiF_2(g)$ calculated from i.r. absorption intensities	379
TiF_3	Vaporization of TiF_3 at ~925 K	Ne, Ar; 5—15 K	I.r.	No definite conclusions reached concerning geometry; ν_2 not observed	379
ZrO	Vaporization of ZrO_2 at 2600 K	Ne, Ar; 4, 20 K	I.r.; u.v.–visible in absorption or emission; ^{16}O, ^{18}O	$^1\Sigma^+$ ground state indicated	378
HfO	Vaporization of HfO_2 at 2700 K	Ne, Ar; 4, 20 K	I.r.; u.v.–visible in absorption or emission; ^{16}O, ^{18}O	Ground state probably $^1\Sigma^+$; spectra compared with those of ZrO	378
VO	Vaporization of V_2O_5 with V at ~2075 K	Ar; 4 K	E.s.r.	Electronic ground state $^4\Sigma$; $g_\parallel = 2.0023 \pm 0.001$; $A_\parallel = 714.7 \pm 0.5$ MHz; $g_\perp = 1.9804 \pm 0.001$; $A_\perp = 837.1 \pm 0.5$ MHz	380
VF_2	Not specified	Ne, Ar; 5—15 K	I.r.	ν_2 and ν_3 observed; $\widehat{FVF} = 150°$ (estimated)	375
NbF_5	Deposition of NbF_5 vapour heated to 675 K	Ar; 20 K	I.r.	Results interpreted on the basis of C_{4v} symmetry for monomeric NbF_5	381
$NbCl_5$; $(NbCl_5)_2$	Vaporization of $NbCl_5$	N_2, C_6H_{12}; 5, 77 K	I.r. and far-i.r.	Vibrational assignments for $NbCl_5$ (D_{3h}) and $(NbCl_5)_2$ (D_{2h}) given; thermodynamic functions calculated for $NbCl_5$	382
Ta	Vaporization of Ta metal at ~2675 K	Ar; 4.2 K	U.v.–visible	Transitions correspond well with those observed in the gas phase; matrix shifts 963—1451 cm^{-1}	383
TaO	Vaporization of Ta_2O_5 at 2270 K; reaction of O_2 with Ta at 2270 K	Ne, Ar; 4, 20 K	I.r.; u.v.–visible; ^{16}O, ^{18}O	$^2\Delta_r$ ground state indicated for TaO	384
TaO_2	As above	Ne, Ar; 4, 20 K	I.r.; u.v.–visible; ^{16}O, ^{18}O	Evidence for bent molecule; two electronic transitions observed together with ν_1 and ν_3 for the ground state	384

Cr	Vaporization of Cr metal	Ar; 4.2 K	U.v.	Matrix shifts from gas-phase transitions evaluated	260
$Cr(N_2)_x$	Co-deposition of Cr atoms with N_2	N_2, N_2–Ar; 17—26 K	I.r.; ^{14}N, ^{15}N	—	385
CrF_2	Not specified	Ne, Ar; 5—15 K	I.r.; ^{50}Cr, ^{52}Cr, ^{53}Cr, ^{54}Cr	ν_2 and ν_3 observed; $\widehat{FCrF} = 180 \pm 8°$	375
$CrCl_2$	Vaporization of $CrCl_2$	Ar; 14 K	I.r.; u.v.–visible; ^{35}Cl, ^{37}Cl, ^{50}Cr, ^{52}Cr, ^{53}Cr, ^{54}Cr	May be linear or bent with $\widehat{ClCrCl} = 120—150°$ but see ref. 211*b*	386
WO; WO_2	Vaporization of $WO_{2\cdot96}$ at 1600 K; reaction of O_2 with heated W at 1900—2950 K	Ne, Ar; 4, 20 K	I.r.; u.v.–visible; ^{16}O, ^{18}O	$^3\Sigma^-$ ground state suggested for WO; two electronic transitions observed for WO_2 together with ν_1 and ν_3 for the ground state	387
WO_3; W_2O_6; W_3O_8; W_3O_9; W_4O_{12}	As above	Ne, Ar; 4, 20 K	I.r.; u.v.–visible; ^{16}O, ^{18}O; variation of conditions of vaporization	Spectroscopic features tentatively identified with the species listed	387
Mn	Vaporization of Mn metal	Ar; 4.2 K	U.v.	Observed transitions correlated with gas-phase transitions; matrix shifts evaluated	260
MnF_2	Not specified	Ne, Ar; 5—15 K	I.r.	ν_2 and ν_3 observed; $\widehat{FMnF} = 180°$ (estimated)	375
$MnCl_2$	Vaporization of $MnCl_2$; reaction of Cl_2 with heated Mn	Ar; 4, 14 K	I.r.; u.v.–visible; ^{35}Cl, ^{37}Cl	Only ν_3 observed; results consistent with a linear structure	386
	Vaporization of $MnCl_2$	Ar; 4 K	I.r.; far-i.r.; ^{35}Cl, ^{37}Cl	ν_2 and ν_3 observed; data compatible with a linear structure; see also ref. 211*b*	388

379 J. W. Hastie, R. H. Hauge, and J. L. Margrave, *J. Chem. Phys.*, 1969, **51**, 2648.
380 P. H. Kasai, *J. Chem. Phys.*, 1968, **49**, 4979.
381 N. Acquista and S. Abramowitz, *J. Chem. Phys.*, 1972, **56**, 5221.
382 R. D. Werder, R. A. Frey, and H. H. Günthard, *J. Chem. Phys.*, 1967, **47**, 4159.
383 W. R. M. Graham and W. Weltner, jun., *J. Chem. Phys.*, 1972, **56**, 4400.
384 W. Weltner, jun. and D. McLeod, jun., *J. Chem. Phys.*, 1965, **42**, 882.
385 J. K. Burdett and J. J. Turner, *Chem. Comm.*, 1971, 885.
386 M. E. Jacox and D. E. Milligan, *J. Chem. Phys.*, 1969, **51**, 4143.
387 W. Weltner, jun. and D. McLeod, jun., *J. Mol. Spectroscopy*, 1965, **17**, 276.
388 K. R. Thompson and K. D. Carlson, *J. Chem. Phys.*, 1968, **49**, 4379.

Table 2 (*continued*)

Species	*Method of preparation*	*Matrix*	*Means of identification and structural determination*	*Observations*	*Ref.*
Mn_2Cl_4	Vaporization of $MnCl_2$ with matrix diffusion	Ar; 4 K	I.r.; far-i.r.	D_{2h} rhombus structure suggested: Cl—Mn(μ-Cl)₂Mn—Cl	388
Fe	Vaporization of Fe metal	Ar, Kr, Xe; 4.2, 20 K	U.v.	Matrix shifts from gas-phase transitions evaluated	260
^{57}Fe; $^{57}Fe_2$	Vaporization of Fe metal 90% enriched with ^{57}Fe	Ar, Kr, Xe; Be disc at 1.45—20.5 K	^{57}Fe Mössbauer	Fe isomer shift (mm s^{-1} rel. Fe metal at 300 K): Fe, -0.75 ± 0.03; Fe_2, -0.14 ± 0.02	389, 390
^{57}Fe; ^{57}FeO; $^{57}Fe_2O_3$	Vaporization of Fe from Al_2O_3 crucible	Ne; 4 K	Optical absorption; ^{57}Fe Mössbauer	Clusters of Fe_2O_3 and FeO identified in the matrix	391
FeF_2	Not specified	Ne, Ar; 5—15 K	I.r.; Fe isotopes	ν_2 and ν_3 observed; $\widehat{FFeF} = 180\pm8°$	375
$^{57}FeCl_2$	Vaporization of $FeCl_2$ 90% enriched in $^{57}FeCl_2$	Ar, Xe; 4.2 K	^{57}Fe Mössbauer	Isomer shift for $FeCl_2$ (mm s^{-1} rel. Fe metal at 300 K), 0.88 ± 0.05; quadrupole splitting given; site effects noted in Xe	392
$FeCl_2$	Vaporization of $FeCl_2$; reaction of Cl_2 with Fe at 1085 K	Ar; 4, 14 K	I.r.; u.v.–visible; ^{35}Cl, ^{37}Cl, ^{54}Fe, ^{56}Fe, ^{57}Fe	Only ν_3 observed; linear molecule indicated; see also ref. 211*b*	386, 393
	Vaporization of $FeCl_2$	Ar; 4 K	I.r.; far-i.r.; ^{35}Cl, ^{37}Cl, Fe isotopes	ν_2 and ν_3 observed; linear geometry sustained	388
	Vaporization of $FeCl_2$	Ar, N_2, C_6H_{12}; 5 K	I.r.; far-i.r.; ^{54}Fe, ^{56}Fe, ^{35}Cl, ^{37}Cl	Only ν_3 observed; linear molecule; thermodynamic functions calculated	394
Fe_2Cl_4	Vaporization of $FeCl_2$ with diffusion of matrix	Ar; 4 K	I.r.; far-i.r.	D_{2h} structure suggested, analogous to that of Mn_2Cl_4	388
	Vaporization of $FeCl_2$ with diffusion of matrix	Ar, N_2, C_6H_{12}; 5 K	I.r.; far-i.r.	Assignment of i.r.-active modes on basis of D_{2h} symmetry	394
$FeCl_3$	Deposition of Fe_2Cl_6 vapour heated to 705—785 K	Ar, N_2, C_6H_{12}; 5 K	I.r.; far-i.r.	Assumed D_{3h}; report ν_2, ν_3, and ν_4; thermodynamic functions calculated	394

Fe_2Cl_6	Deposition of Fe_2Cl_6 vapour (440 K)	Ar, N_2, C_6H_{12}; 5 K	I.r.; far-i.r.	D_{2h} symmetry; thermodynamic functions calculated	394
Co	Vaporization of Co metal	Ar; 4.2 K	U.v.	Observed transitions correlated with gas-phase transitions; matrix shifts evaluated	260
CoF_2	Vaporization of CoF_2 at 1290 K	Ne, Ar; ~5—15 K	I.r.	ν_2 and ν_3 observed; $\widehat{FCoF} = 170°$ (estimated)	395
$CoCl_2$	Vaporization of $CoCl_2$ at ~825 K	Ar; 4, 14 K	I.r.; u.v.–visible; ^{35}Cl, ^{37}Cl	Only ν_3 observed; results consistent with a linear structure	386, 393
	Vaporization of $CoCl_2$ at ~825 K	Ar; 4 K	I.r.; far-i.r.; ^{35}Cl, ^{37}Cl	ν_2 and ν_3 observed; results support linear structure	388
Co_2Cl_4	As above, with diffusion of matrix	Ar; 4 K	I.r.; far-i.r.	D_{2h} structure indicated, analogous to that of Mn_2Cl_4	388
Ni	Vaporization of Ni metal	Ar; 4.2 K	U.v.	Observed transitions correlated with gas-phase transitions; matrix shifts evaluated	260
$Ni(N_2)_x$	Co-deposition of Ni atoms with N_2	N_2, N_2—Ar; 17—26 K	I.r.; ^{14}N, ^{15}N	Probably $x = 1$ or 2	385
$Ni_x(N_2)_y$	Co-deposition of Ni atoms with N_2	N_2, N_2—Ar; 17—26 K	I.r.; ^{14}N, ^{15}N	$x > 1$	385
$(NiF_2)_2$	Details not available	—	I.r.	D_{2h} structure analogous to that of Mn_2Cl_4	256
NiF_2	Vaporization of NiF_2 at 1075—1175 K	Ar; 14 K	I.r.; u.v.–visible; ^{58}Ni, ^{60}Ni, ^{62}Ni, ^{64}Ni	Only ν_3 observed; linear structure	396
	Vaporization of NiF_2 at 1075—1175 K	Ne, Ar; 5—15 K	I.r.; ^{58}Ni, ^{60}Ni, ^{62}Ni, ^{64}Ni	ν_2, ν_3, and possibly ν_1 observed; bent structure indicated with $\widehat{FNiF} = 165 \pm 8°$	395

[389] P. H. Barrett and T. K. McNab, *Phys. Rev. Letters*, 1970, **25**, 1601.
[390] T. K. McNab, H. Micklitz, and P. H. Barrett, *Phys. Rev.(B)*, 1971, **4**, 3787.
[391] W. Keune and E. Lüscher, *Ann. Univ. Saraviensis, Sci.*, 1970, **8**, 92.
[392] T. K. McNab, D. H. W. Carstens, D. M. Gruen, and R. L. McBeth, *Chem. Phys. Letters*, 1972, **13**, 600.
[393] C. W. DeKock and D. M. Gruen, *J. Chem. Phys.*, 1968, **49**, 4521.
[394] R. A. Frey, R. D. Werder, and H. H. Günthard, *J. Mol. Spectroscopy*, 1970, **35**, 260.
[395] J. W. Hastie, R. H. Hauge, and J. L. Margrave, *High Temp. Sci.*, 1969, **1**, 76.
[396] D. E. Milligan, M. E. Jacox, and J. D. McKinley, *J. Chem. Phys.*, 1965, **42**, 902.

Table 2 (*continued*)

Species	*Method of preparation*	*Matrix*	*Means of identification and structural determination*	*Observations*	*Ref.*
$NiCl_2$	Vaporization of $NiCl_2$	Ar; 14 K	I.r.; u.v.–visible; ^{35}Cl, ^{37}Cl, ^{58}Ni, ^{60}Ni	Only ν_3 observed; linear geometry indicated	396
	As above, and reaction of Cl_2 with heated Ni	Ar; 4, 14 K	I.r.; u.v.; ^{35}Cl, ^{37}Cl, ^{58}Ni, ^{60}Ni, ^{61}Ni, ^{62}Ni, ^{64}Ni	Data and conclusions of ref. 396 sustained	386, 393
	Vaporization of $NiCl_2$	Ar; 4 K	I.r.; far-i.r.; ^{35}Cl, ^{37}Cl, ^{58}Ni, ^{60}Ni	ν_2 and ν_3 observed; results support linear structure; see also ref. 211*b*	388
Ni_2Cl_4	As above, with diffusion of matrix	Ar; 4 K	I.r.; far-i.r.	D_{2h} structure indicated, analogous to that of Mn_2Cl_4	388
$NiX_2(N_2)$; $NiX_2(CO)$ (X = F or Cl)	Co-deposition of NiX_2 vapour with N_2 or CO	Ar; temp. not available	I.r.; ^{12}C, ^{13}C	Little change in XNiX bond angle on formation of either species	397
$NiBr_2$	Vaporization of $NiBr_2$	Ar; 4 K	I.r.; far-i.r.; ^{58}Ni, ^{60}Ni, ^{79}Br, ^{81}Br	ν_2 and ν_3 observed; linear structure indicated	388
Pd	Vaporization of Pd metal	Ar; 4.2 K	U.v.	Observed transitions correlated with gas-phase transitions; matrix shifts evaluated	260
Cu	Vaporization of Cu metal	Ar, Kr, Xe, SF_6; 4.2, 20 K	U.v.	As above	260, 398
	Vaporization of Cu metal	Ne, Ar, Kr, Xe; ~4 K	E.s.r.	*g*- and *A*-values measured and matrix effects examined	399
CuO	Photolysis of Cu atoms in O_2 matrix; sputtering of Cu metal, with discharge through O_2–Xe	Ar, Kr, Xe; 15 K	U.v.–visible absorption; laser-excited fluorescence	Ground state identified as $^2\Pi$; vibrational frequencies of ground and excited states determined	400
CuF_2	Vaporization of CuF_2	Ne, Ar; 5—15 K	I.r.; ^{63}Cu, ^{65}Cu	ν_2 and ν_3 observed; bent molecule indicated with $\widehat{FCuF} = 165 \pm 8°$	395
Ag; Au	Vaporization of metal	Ar, Kr, Xe, SF_6; 20 K	U.v.	Observed transitions correlated with gas-phase transitions; matrix shifts evaluated	398
	Vaporization of metal	Ne, Ar, Kr, Xe; ~4 K	E.s.r.	*g*- and *A*-values measured and matrix effects examined	399
ZnH; CdH; HgH	Co-deposition of M atoms with H atoms	Ar; 4 K	E.s.r.; ^1H, ^2H, ^{111}Cd, ^{113}Cd, ^{199}Hg, ^{201}Hg	$^2\Sigma^+$ ground-state; *g*- and *A*-values determined; coefficients of LCAO wave-functions derived	401

ZnF_2	Vaporization of ZnF_2	Ne, Ar; 5, 15 K	I.r.; ^{64}Zn, ^{66}Zn, ^{67}Zn, ^{68}Zn, ^{70}Zn	Only ν_3 observed; bent molecule suggested with $\widehat{FZnF} = 165 \pm 8°$	395
	Vaporization of ZnF_2	Ar, Kr, Xe; 20 K	I.r.; far-i.r.; ^{64}Zn, ^{66}Zn, ^{68}Zn	ν_2 and ν_3 identified; consistent with $D_{\infty h}$ symmetry	402
ZnX_2 (X = Cl, Br, or I)	Vaporization of ZnX_2	Ar, Kr, Xe; 20 K	I.r.; far-i.r.; ^{35}Cl, ^{37}Cl, ^{64}Zn, ^{66}Zn, ^{68}Zn	ν_2 and ν_3 identified in each case; consistent with $D_{\infty h}$ symmetry	402
CdX_2 (X = F, Cl, Br, or I)	Vaporization of CdX_2	Kr; 20 K	I.r.; far-i.r.; ^{35}Cl, ^{37}Cl, Cd isotopes	ν_2 and ν_3 observed in each case; consistent with $D_{\infty h}$ symmetry; thermodynamic functions calculated; evidence of Cd_2X_4 molecules	403
HgX_2 (X = F, Cl, Br, or I)	Vaporization of HgX_2	Kr; 20 K	I.r.; far-i.r.; ^{35}Cl, ^{37}Cl, Hg isotopes	ν_2 and ν_3 observed in each case; consistent with $D_{\infty h}$ symmetry; thermodynamic functions calculated; evidence of Hg_2X_4	403
Lanthanides and Actinides					
MO (M = Ce, Pr, Nd, Sm, Gd, Tb, Dy, Ho, Er, Tm, or Lu)	Reaction of O_2 with Ce; vaporization of Pr_6O_{11}, Tb_4O_7, and M_2O_3 where M = Nd, Sm, Gd, Tb, Dy, Ho, Er, Tm, or Lu	Ne, Ar; 4 K	I.r.; u.v.–visible–near-i.r.; ^{16}O, ^{18}O for CeO	Vibrational frequencies of ground-state molecules all lie in the range 808—832 cm^{-1} (*cf.* transition-metal monoxides); electronic spectra compared with gas-phase spectra where available	404
LaO	Vaporization of La_2O_3	Ne, Ar; 4 and 20 K	I.r.; u.v.–visible; e.s.r.	Ground state defined as $^2\Sigma$; $A\,^2\Pi \leftarrow X\,^2\Sigma$ and $B\,^2\Sigma \leftarrow X\,^2\Sigma$ transitions observed; *g*- and *A*-values quoted	373
MO_2 (M = Ce, Pr, or Tb); $(CeO_2)_2$	Vaporization of CeO_2, Ce_2O_3, Pr_6O_{11}, or Tb_4O_7	Ne, Ar; 4 K	I.r.; u.v.–visible–near-i.r.; ^{16}O, ^{18}O for CeO_2	ν_1 and ν_3 observed in support of bent CeO_2 and TbO_2 molecules; only ν_3 observed for PrO_2, which may thus be linear or nearly linear; absorptions identified with $(CeO_2)_2$: probably D_{2h}	404

[397] C. W. DeKock and D. A. VanLeirsburg, *J. Amer. Chem. Soc.*, 1972, **94**, 3235.
[398] L. Brewer and B. King, *J. Chem. Phys.*, 1970, **53**, 3981; L. Brewer, C.-A. Chang, and B. King, *Inorg. Chem.*, 1970, **9**, 814.
[399] P. H. Kasai and D. McLeod, jun., *J. Chem. Phys.*, 1971, **55**, 1566.
[400] J. S. Shirk and A. M. Bass, *J. Chem. Phys.*, 1970, **52**, 1894.
[401] L. B. Knight, jun. and W. Weltner, jun., *J. Chem. Phys.*, 1971, **55**, 2061.
[402] A. Loewenschuss, A. Ron, and O. Schnepp, *J. Chem. Phys.*, 1968, **49**, 272.
[403] A. Loewenschuss, A. Ron, and O. Schnepp, *J. Chem. Phys.*, 1969, **50**, 2502.
[404] R. L. DeKock and W. Weltner, jun., *J. Phys. Chem.*, 1971, **75**, 514.

Table 2 (*continued*)

Species	*Method of preparation*	*Matrix*	*Means of identification and structural determination*	*Observations*	*Ref.*
MF_3 (M = La, Ce, Pr, Nd, Sm, or Eu)	Vaporization of MF_3 at 1475—1600 K	Ar, Kr, N_2; 21 K	I.r.; far-i.r.	For LaF_3, CeF_3, SmF_3, and EuF_3 only the antisymmetric stretching and bending modes observed, indicating a planar configuration (D_{3h}); no satisfactory interpretation of the results for NdF_3; force constants k_{M-F} in the range 2.12—2.67 mdyn Å^{-1}	405
PrF_3	Vaporization of PrF_3 at 1475—1600 K	Ne, Ar, N_2; 5—15 K	Raman	Earlier conclusions concerning the pyramidal structure of PrF_3 (ref. 405) shown to be erroneous; ν_1, ν_3 and ν_4 observed in Raman spectrum	40
MF_3 (M = La, Ce, Nd, Eu, Gd, Tb, Ho, Yb, or Lu)	Vaporization of MF_3 at 1575—1775 K	Ne, Ar, N_2; 5—15 K	I.r.; far-i.r.	ν_1, ν_2, ν_3, and ν_4 observed for each molecule suggesting that all molecules are non-planar (*cf.* ref. 405); k_{M-F} increases regularly across lanthanide series	376
PrI_3	Vaporization of PrI_3 at 976—1006 K	Ar, Xe; ~10 K	U.v.–visible fluorescence	More than 40 transitions measured and assigned to transitions within the $4f^2$ configuration; assumed D_{3h} symmetry	407
UO	Vaporization of UO_2 with or without U	Ar, Kr; 4, 20 K	I.r.; ^{16}O, ^{18}O	$U^{16}O$ and $U^{18}O$ identified giving a harmonic force constant of 5.32 mdyn Å^{-1}	408
UO_2	Vaporization of UO_2 with U	Ar; 20 K	I.r.; ^{16}O, ^{18}O	Only ν_3 observed; $^{16}O/^{18}O$ isotopic shifts insensitive to $\widehat{OUO}$ because of the large mass of U	408
	Vaporization of UO_2 with or without U	Ar, Kr; 4, 20 K	I.r.; far-i.r.; ^{16}O, ^{18}O	ν_2 and ν_3 observed; data consistent with $\widehat{OUO} = 105°$	408, 409
UCl_4; UBr_4	Vaporization of UX_4 at 675—725 K	N_2; 4 K	I.r.; u.v.–visible	More than 30 narrow absorption bands observed for each molecule in the 400—24 000 cm^{-1} region; assigned to electronic transitions from the $^3H_4$ (Γ_5) ground state to excited states of the f^2 configuration	410
UCl_2; EuF_2; $EuCl_2$	Various methods	Ne, Ar, Kr, N_2; 5—20 K	I.r.; ^{35}Cl, ^{37}Cl	$\widehat{XMX} = 100—135°$	211*b*

Table 3 *Matrix-isolated organometallic species*

Species	*Method of preparation*	*Matrix*	*Means of identification and structural determination*	*Observations*	*Ref.*
$LiCH_3$	Co-deposition of Li atoms with CH_3Br or CH_3I	Ar; 15 K	I.r.; ^1H, ^2H, ^6Li, ^7Li, ^{12}C, ^{13}C	Fundamental frequencies assigned for $LiCH_3$ and $LiCD_3$; normal-co-ordinate analysis	411
$NaCH_3$; KCH_3	Co-deposition of Na or K atoms with CH_3I	N_2, Ar; 15—20 K	I.r.; ^1H, ^2H	Vibrational assignments made for discrete MCH_3 and MCD_3 molecules; normal-co-ordinate analysis	412
CH_3,MX	Co-deposition of M atoms with CH_3X (M = Li, Na, or K; X = Br or I)	Ar; 15 K	I.r.; ^1H, ^2H	CH_3 radical perturbed by MX molecule; interaction may be due to a new molecular type CH_3MX assumed to have a linear C—M—X unit	413
$K_x(CO)_3$	Co-deposition of K atoms with CO	CO; 15 K	I.r.; ^{16}O, ^{18}O; diffusion studies	Data attributable to the isolation of the aggregate $K_x(CO)_3$ having C_{3v} symmetry; no analogous derivative of Na formed under similar conditions	414
$Al_x(CO)_2$	Co-deposition of Al atoms with CO	Kr; 20 K	I.r.; ^{16}O, ^{18}O	Bands previously assigned to Al_2O now attributed to $Al(CO)_2$ having a bent C—Al—C skeleton	415
$Me_2Si{=}CH_2$	Flash photolysis of Me_2Si◇ at 875 K	C_2H_4; 77 K	I.r.; diffusion to give starting material	Band observed at 1407 cm^{-1} assigned to stretching mode of Si=C unit	416

[405] R. D. Wesley and C. W. DeKock, *J. Chem. Phys.*, 1971, **55**, 3866.
[406] M. Lesiecki, J. W. Nibler, and C. W. De Kock, *J. Chem. Phys.*, 1972, **57**, 1352.
[407] J. R. Clifton, D. M. Gruen, and A. Ron, *J. Mol. Spectroscopy*, 1971, **39**, 202.
[408] S. Abramowitz, N. Acquista, and K. R. Thompson, *J. Phys. Chem.*, 1971, **75**, 2283; H. J. Leary jun., T. A. Rooney, E. D. Cater, and H. B. Friedrich, *High Temp. Sci.*, 1971, **3**, 433.
[409] S. Abramowitz and N. Acquista, *J. Phys. Chem.*, 1972, **76**, 648.
[410] J. R. Clifton, D. M. Gruen, and A. Ron, *J. Chem. Phys.*, 1969, **51**, 224.
[411] L. Andrews, *J. Chem. Phys.*, 1967, **47**, 4834.
[412] K. J. Burczyk and A. J. Downs, unpublished work.
[413] L. Y. Tan and G. C. Pimentel, *J. Chem. Phys.*, 1968, **48**, 5202.
[414] M. A. Hooper and A. J. Downs, unpublished work.
[415] A. J. Hinchcliffe, J. S. Ogden, and D. D. Oswald, *Chem. Comm.*, 1972, 338.
[416] T. J. Barton and C. L. McIntosh, *Chem. Comm.*, 1972, 861.

Table 3 (*continued*)

Species	*Method of preparation*	*Matrix*	*Means of identification and structural determination*	*Observations*	*Ref.*
$Ge_x(CO)_y$; $Sn_x(CO)_y$	Co-deposition of Ge or Sn atoms with CO	Kr; 20 K	I.r.; ^{16}O, ^{18}O	New absorptions near 2000 cm^{-1} identified with the formation of carbonyl derivatives of Ge and Sn, including possibly GeCO and SnCO (*cf.* CCO)	417
Transition Elements and Actinides					
$Ta(CO)_x$ ($x = 1$—6)	Co-deposition of Ta atoms with CO	Ar; 4.2 K	I.r.; u.v.–visible; ^{16}O, ^{18}O; diffusion studies	Tentative assignments to C—O stretching modes made for all the species $Ta(CO)_x$ ($x = 1$—6); frequencies increase with increasing co-ordination number	418
$Cr(CO)_x$ ($x = 1$—4)	Prolonged photolysis of $Cr(CO)_6$ in dilute matrices; co-deposition of Cr atoms with CO	Ar; 20 K	I.r.	Species believed to adopt the most symmetrical structure in each case	419
$Cr(CO)_5$; $Cr(CO)_6$	Photolysis of $Cr(CO)_6$ in inert matrices	Ar, N_2; 20 K	I.r.; u.v.-visible; ^{16}O, ^{18}O	$Cr(CO)_5$ square pyramidal (C_{4v}); apical–basal angle = $\sim 105°$; $Cr(CO)_6$ regenerated by irradiation at appropriate frequency	420
$Cr(CO)_5R$ (R = hydrocarbon)	Photolysis of $Cr(CO)_6$ in hydrocarbon matrices	CH_4; 20 K	U.v.–visible	Probably C_{4v} symmetry; new scheme for photochemistry of $Cr(CO)_6$ proposed	419
$Mo(CO)_x$ ($x = 1$—4)	Prolonged photolysis of $Mo(CO)_6$ in dilute inert matrices; co-deposition of Mo atoms with CO	Ar; 20 K	I.r.	Species believed to adopt the most symmetrical structure in each case	419
$Mo(CO)_5$; $Mo(CO)_6$	Photolysis of $Mo(CO)_6$ in inert matrices	Ar, N_2; 20 K	I.r.; u.v.–visible; ^{16}O, ^{18}O	$Mo(CO)_5$ square pyramidal (C_{4v}); apical–basal bond angle = $\sim 105°$; photolytic act reversible	420
$W(CO)_x$ ($x = 1$—4)	Prolonged photolysis of $W(CO)_6$ in dilute inert matrices; co-deposition of W atoms with CO	Ar; 20 K	I.r.	Species believed to adopt the most symmetrical structures in each case	419

$W(CO)_5$; $W(CO)_6$	Photolysis of $W(CO)_6$ in inert matrices	Ar; 20 K	I.r.; u.v.–visible	$W(CO)_5$ square pyramidal (C_{4v}); photolytic act reversible; mechanism discussed	421
	Photolysis of $W(CO)_6$ in inert matrices	Ar, N_2; 20 K	I.r.; u.v.–visible; ^{16}O, ^{18}O	Apical–basal bond angle for $W(CO)_5 = \sim 105°$	420
$HMn(CO)_5$; $HMn(CO)_4$	Photolysis of $HMn(CO)_5$ in an inert matrix	Ar; 15 K	I.r.	$HMn(CO)_4$ probably takes the form of a trigonal bipyramid (C_{3v}) [cf. $HCo(CO)_4$]; photolytic act reversible	422
$RMn(CO)_5$; $RCOMn(CO)_x$ ($R = CH_3$ or CF_3; x < 5)	Photolysis of $RMn(CO)_5$	Ar; 17 K	I.r.; ^1H, ^2H	Evidence of carbonyl insertion	423
$Fe(C_5H_5)_2$	Vaporization at 575 K	Ar, Kr, Xe, N_2, CH_4; 20 K	U.v.–visible; absorption and phosphorescence	Five electronic transitions observed, three exhibiting vibrational structure; simpler than room-temperature spectrum; phosphorescence observed in the 5000 Å region	424
$Fe_2(CO)_9$; $Fe_2(CO)_8$	Photolysis of $Fe_2(CO)_9$	Ar, N_2; 20 K	I.r.	Formation of bridged and non-bridged forms of $Fe_2(CO)_8$ suggested; evidence of N_2-containing carbonyl species	425
$Fe_3(CO)_{12}$	Direct deposition	Ar, N_2; 20 K	I.r.	Isolated molecule believed to have the doubly-bridged C_{2v} structure; spectrum of matrix completely different from that of solution at room temperature	426, 427
$Fe(CO)_4(C_2H_4)$	Photolysis of $Fe(CO)_5$ and C_2H_4 mixture	Ar; 17 K	I.r.	C_2H_2 reacts under similar conditions to give a complex of $CH_2{=}CH{-}C{\equiv}CH$	428

[417] A. Bos, *Chem. Comm.*, 1972, 26.
[418] R. L. DeKock, *Inorg. Chem.*, 1971, **10**, 1205.
[419] M. A. Graham, R. N. Perutz, M. Poliakoff, and J. J. Turner, *J. Organometallic Chem.*, 1972, **34**, C34.
[420] M. A. Graham, M. Poliakoff, and J. J. Turner, *J. Chem. Soc.(A)*, 1971, 2939.
[421] M. A. Graham, A. J. Rest, and J. J. Turner, *J. Organometallic Chem.*, 1970, **24**, C54.
[422] A. J. Rest and J. J. Turner, *Chem. Comm.*, 1969, 375.
[423] J. F. Ogilvie, *Chem. Comm.*, 1970, 323.
[424] J. J. Smith and B. Meyer, *J. Chem. Phys.*, 1968, **48**, 5436.
[425] M. Poliakoff and J. J. Turner, *J. Chem. Soc.(A)*, 1971, 2403.
[426] M. Poliakoff and J. J. Turner, *Chem. Comm.*, 1970, 1008.
[427] M. Poliakoff and J. J. Turner, *J. Chem. Soc.(A)*, 1971, 654.
[428] M. J. Newlands and J. F. Ogilvie, *Canad. J. Chem.*, 1971, **49**, 343.

Table 3 (*continued*)

Species	*Method of preparation*	*Matrix*	*Means of identification and structural determination*	*Observations*	*Ref.*
$Ru_3(CO)_{12}$	Direct deposition	Ar, N_2; 20 K	I.r.	Evidence that the molecule is somewhat distorted from an idealised D_{3h} structure; degree of distortion varies with the matrix	427
$Ru_3(CO)_{10}(NO)_2$	Direct deposition	Ar, N_2; 20 K	I.r.	Spectrum of matrix resembles solution spectrum; doubly bridged C_{2v} structure implied	426, 427
$Os_3(CO)_{12}$	Direct deposition	Ar, N_2; 20 K	I.r.	Conclusions analogous to those for $Ru_3(CO)_{12}$	426, 427
CO adsorbed on Ni	Co-deposition of Ni particles with CO	Ar; 44 K	I r.	Broad bands attributed to chemisorbed CO	429
NiCO	Co-deposition of Ni atoms with CO	Ar; 4.2 K	I.r.; ^{16}O, ^{18}O	$C_{\infty v}$ linear structure indicated	418
$Ni(CO)_2$	Co-deposition of Ni atoms with CO	Ar; 4.2 K	I.r.; ^{16}O, ^{18}O	Structure probably linear $D_{\infty h}$	418
$Ni(CO)_3$	As above, with diffusion of matrix	Ar; 4.2 K	I.r.; ^{16}O, ^{18}O	Structure probably planar D_{3h}	418
	Photolysis of $Ni(CO)_4$	Ar, Kr, Xe–Kr; 15 K	I.r.; ^{16}O, ^{18}O	C_{3v} symmetry proposed; $Ni(CO)_4$ regenerated on annealing the matrix	430
$Ni(CO)_4$	Co-deposition of Ni atoms with CO	Ar; 4.2 K	I.r.; u.v.–visible; ^{16}O, ^{18}O	C—O stretching frequencies for $Ni(CO)_x$ increase as x increases	418
$Ni(CO)_3(N_2)$	Photolysis of $Ni(CO)_4$ in N_2 matrix	N_2; 20 K	I.r.; ^{14}N, ^{15}N	C_{3v} structure with terminal N≡N unit believed to be more likely than C_s structure (with π-donor N_2 unit)	431
$Pd(CO)_4$	Co-deposition of Pd atoms with CO	Ar; ~20 K	I.r.; ^{16}O, ^{18}O	New band near 2000 cm^{-1} attributed to the f_2 mode of $Pd(CO)_4$; supported by matrix isotopic splitting pattern	432
$Pd(CO)_x$ ($x = 1$—4)	Co-deposition of Pd atoms with CO	CO, Ar; 12—14 K	I.r.	New features ascribed to the C—O stretching modes of $Pd(CO)_x$ molecules	433
$Pt(CO)_x$ ($x = 1$—4)	Co-deposition of Pt atoms with CO	CO, Ar; 12—14 K	I.r.	New features ascribed to the C—O stretching modes of $Pt(CO)_x$ molecules	433

$Cu_x(CO)_y$; $Ag_x(CO)_y$	Co-deposition of Cu or Ag atoms with CO	CO; 20 K	I.r.; ^{16}O, ^{18}O	Bands due to Ag and Cu carbonyls observed in the range 1900—2000 cm^{-1}	434
$HgCH_3$	Pyrolysis of $Hg(CH_3)_2$ at ~1600 K	Ne; ~4 K	I.r.; ^{1}H, ^{2}H	All vibrational fundamentals except ν_4 observed	435
$U(CO)_x$ ($x = 1$—6)	Co-deposition of U atoms with CO	Ar; 4 K	I.r.; ^{12}C, ^{13}C; diffusion studies	Assignments tentatively made for all the species $U(CO)_x$ ($x = 1$—6); spectra resemble those of $Ta(CO)_x$	436

429 G. Blyholder, M. Tanaka, and J. D. Richardson, *Chem. Comm.*, 1971, 499.
430 A. J. Rest and J. J. Turner, *Chem. Comm.*, 1969, 1026.
431 A. J. Rest, *J. Organometallic Chem.*, 1972, **40**, C76.
432 J. H. Darling and J. S. Ogden, *Inorg. Chem.*, 1972, **11**, 666.
433 H. Huber, P. Kündig, M. Moskovits, and G. A. Ozin, *Nature Phys. Sci.*, 1972, **235**, 98.
434 J. S. Ogden, *Chem. Comm.*, 1971, 978.
435 A. Snelson, *J. Phys. Chem.*, 1970, **74**, 537.
436 J. L. Slater, R. K. Sheline, K. C. Lin, and W. Weltner, jun., *J. Chem. Phys.*, 1971, **55**, 5129.

Author Index